The Changing Earth: Exploring Geology and Evolution

With Geology of Texas

6th Edition

James S. Monroe / Reed Wicander

CENGAGE
Learning™

Australia • Brazil • Japan • Korea • Mexico • Singapore • Spain • United Kingdom • United States

CENGAGE
Learning™

The Changing Earth: Exploring Geology and Evolution: With Geology of Texas, 6th Edition

Executive Editors:
Maureen Staudt
Michael Stranz

Senior Project Development Manager:
Linda deStefano

Marketing Specialist:
Courtney Sheldon

Senior Production/Manufacturing Manager:
Donna M. Brown

PreMedia Manager:
Joel Brennecke

Sr. Rights Acquisition Account Manager:
Todd Osborne

Cover Image:
Getty Images*

*Unless otherwise noted, all cover images used by Custom Solutions, a part of Cengage Learning, have been supplied courtesy of Getty Images with the exception of the Earthview cover image, which has been supplied by the National Aeronautics and Space Administration (NASA).

The Changing Earth: Exploring Geology and Evolution, 6th Edition
James S. Monroe / Reed Wicander

© 2012 Cengage Learning. All rights reserved.

The Geology of Texas: University of North Texas
C. Reid Ferring

© 2009 Cengage Learning. All rights reserved.

For product information and technology assistance, contact us at
Cengage Learning Customer & Sales Support, 1-800-354-9706

For permission to use material from this text or product,
submit all requests online at **cengage.com/permissions**
Further permissions questions can be emailed to
permissionrequest@cengage.com

This book contains select works from existing Cengage Learning resources and was produced by Cengage Learning Custom Solutions for collegiate use. As such, those adopting and/or contributing to this work are responsible for editorial content accuracy, continuity and completeness.

Compilation © 2011 Cengage Learning
ISBN-13: 978-1-133-15357-3

ISBN-10: 1-133-15357-7

Cengage Learning
5191 Natorp Boulevard
Mason, Ohio 45040
USA
Cengage Learning is a leading provider of customized learning solutions with office locations around the globe, including Singapore, the United Kingdom, Australia, Mexico, Brazil, and Japan. Locate your local office at:
international.cengage.com/region.

Cengage Learning products are represented in Canada by Nelson Education, Ltd.
For your lifelong learning solutions, visit **www.cengage.com/custom.**
Visit our corporate website at **www.cengage.com.**

Printed in the United States of America

BRIEF CONTENTS

The Changing Earth: Exploring Geology and Evolution: With Geology of Texas
Sixth Edition
Contents-Index

The Geology of Texas: University of North Texas

Brief Contents

1. Understanding Earth: A Dynamic and Evolving Planet   2

2. Plate Tectonics: A Unifying Theory   26

3. Minerals—The Building Blocks of Rocks   58

4. Igneous Rocks and Plutons   84

5. Volcanoes and Volcanism   108

6. Weathering, Erosion, and Soil   134

7. Sediment and Sedimentary Rocks   156

8. Metamorphism and Metamorphic Rocks   180

9. Earthquakes and Earth's Interior   198

10. Deformation, Mountain Building, and Earth's Crust   232

11. Mass Wasting   258

12. Running Water—Streams and Rivers   284

13. Groundwater   312

14. Glaciers and Glaciation   338

15. The Work of Wind and Deserts   364

16. Oceans, Shorelines, and Shoreline Processes   386

17. Geologic Time: Concepts and Principles   414

18. Organic Evolution—The Theory and Its Supporting Evidence   448

19. Precambrian Earth and Life History   474

20. Paleozoic Earth History   502

21. Paleozoic Life History   540

22. Mesozoic Earth and Life History   574

23. Cenozoic Earth and Life History   612

24. Geology in Perspective   654

Contents

1 Understanding Earth: A Dynamic and Evolving Planet 2

1.1 Introduction   4

1.2 What Is Geology?   4

1.3 Geology and the Formulation of Theories   4

1.4 How Does Geology Relate to the Human Experience?   6

1.5 How Does Geology Affect Our Everyday Lives?   6

1.6 Global Geologic and Environmental Issues Facing Humankind   7

1.7 Origin of the Universe and Solar System, and Earth's Place in Them   9

   Origin of the Universe—Did It Begin with a Big Bang?   10
   Our Solar System—Its Origin and Evolution   11
   Earth—Its Place in Our Solar System   12

1.8 Why Earth Is a Dynamic and Evolving Planet   12

GEO-INSIGHT: The Terrestrial and Jovian Planets   14

   Plate Tectonic Theory   16

1.9 The Rock Cycle   17

   How Are the Rock Cycle and Plate Tectonics Related?   20

1.10 Organic Evolution and the History of Life   21

1.11 Geologic Time and Uniformitarianism   21

1.12 How Does the Study of Geology Benefit Us?   22

GEO-RECAP   23

2 Plate Tectonics: A Unifying Theory 26

2.1 Introduction   28

2.2 Early Ideas About Continental Drift   28

   Alfred Wegener and the Continental Drift Hypothesis   29

2.3 What Is the Evidence for Continental Drift?   29

   Continental Fit   29
   Similarity of Rock Sequences and Mountain Ranges   29
   Glacial Evidence   30
   Fossil Evidence   31

2.4 Features of the Seafloor   32

   The Continental Shelf, Slope, and Rise   32
   Abyssal Plains, Oceanic Ridges, Submarine Hydrothermal Vents, and Oceanic Trenches   33

   Seamounts, Guyots, and Aseismic Ridges   34

2.5 Earth's Magentic Field   35

2.6 Paleomagnetism and Polar Wandering   36

2.7 Magentic Reversals and Seafloor Spreading   36

   Deep-Sea Drilling and the Confirmation of Seafloor Spreading   39

2.8 Plate Tectonics: A Unifying Theory   39

2.9 The Three Types of Plate Boundaries   40

   Divergent Boundaries   40

GEO-INSIGHT: Tectonics of the Terrestrial Planets   42

   Convergent Boundaries   44
   Transform Boundaries   47

2.10 Hot Spots and Mantle Plumes   48

2.11 Plate Movement and Motion   48

2.12 The Driving Mechanism of Plate Tectonics   50

2.13 Plate Tectonics and the Distribution of Natural Resources   52

   Petroleum   52
   Mineral Deposits   52

2.14 Plate Tectonics and the Distribution of Life   53

GEO-RECAP   55

3 Minerals—The Building Blocks of Rocks 58

3.1 Introduction   60

3.2 Matter—What Is It?   60

   Atoms and Elements   61
   Bonding and Compounds   62

3.3 Explore the World of Minerals   64

   Naturally Occurring Inorganic Substances   64
   Mineral Crystals   64
   Chemical Composition of Minerals   66
   Physical Properties of Minerals   66

3.4 Mineral Groups Recognized by Geologists   66

   Silicate Minerals   67
   Carbonate Minerals   69
   Other Mineral Groups   70

3.5 Mineral Identification   70

   Luster and Color   70
   Crystal Form   71

GEO-INSIGHT: The Precious Metals   72

Cleavage and Fracture   74
Hardness   74
Specific Gravity (Density)   74
Other Useful Mineral Properties   75

3.6 Rock-Forming Minerals   75

GEO-FOCUS: Gems and Gemstones   76

3.7 How Do Minerals Form?   78

3.8 Natural Resources and Reserves   78

GEO-RECAP   80

4 **Igneous Rocks and Plutons   84**

4.1 Introduction   86

4.2 The Properties and Behavior of Magma
      and Lava   86
Composition of Magma   87
How Hot Are Magma and Lava?   87
Viscosity—Resistance to Flow   88

4.3 How Does Magma Originate and Change?   88
Bowen's Reaction Series   89
The Origin of Magma at Spreading Ridges   90
Subduction Zones and the Origin of Magma   91
Hot Spots and the Origin of Magma   91
Compositional Changes in Magma   92

4.4 Igneous Rocks—Their Characteristics and
      Classification   93
Igneous Rock Textures   93
Composition of Igneous Rocks   95
Classifying Igneous Rocks   95

4.5 Intrusive Igneous Bodies—Plutons   98
Dikes, Sills, and Laccoliths   99

GEO-FOCUS: Granite—Common, Attractive, and Useful   100
Volcanic Pipes and Necks   102
Batholiths and Stocks   102

4.6 The Origin of Batholiths   103

GEO-RECAP   105

5 **Volcanoes and Volcanism   108**

5.1 Introduction   110

5.2 Volcanoes and Volcanism   110
Volcanic Gases   110
Lava Flows   112
Pyroclastic Materials   113

5.3 Types of Volcanoes   114
Shield Volcanoes   116
Cinder Cones   118

Composite Volcanoes (Stratovolcanoes)   118
Lava Domes   119
Supervolcano Eruptions   121

GEO-FOCUS: Do Volcanic Gases Cause Ozone Depletion?   122

5.4 Other Volcanic Landforms   122
Fissure Eruptions and Basalt Plateaus   122
Pyroclastic Sheet Deposits   124

5.5 Volcano Belts   125

5.6 North America's Active Volcanoes   126
Alaska's Volcanoes   126
The Cascade Range   126

5.7 Plate Tectonics, Volcanoes, and Plutons   126
Igneous Activity at Divergent Plate Boundaries   127
Igneous Activity at Convergent Plate Boundaries   128
Intraplate Volcanism   128

5.8 Volcanic Hazards, Volcano Monitoring, and Forecasting
      Eruptions   128
*How Large Is an Eruption, and How Long
      Do Eruptions Last?   128*
Is It Possible to Forecast Eruptions?   129

GEO-RECAP   131

6 **Weathering, Erosion, and Soil   134**

6.1 Introduction   136

6.2 Alteration of Minerals and Rocks   137

6.3 Mechanical Weathering—Disaggregation of Earth
      Materials   137
Frost Action   137
Pressure Release   138
Thermal Expansion and Contraction   139
Growth of Salt Crystals   139
Activities of Organisms   139

6.4 Chemical Weathering—Decomposition of Earth
      Materials   139
Solution   140
Oxidation   141
Hydrolysis   141

6.5 Soil and Its Origin   142
The Soil Profile   143

GEO-FOCUS: Industrialization and Acid Rain   144
Factors That Control Soil Formation   145

6.6 Expansive Soils and Soil Degradation   148
Expansive Soils   148
Soil Degradation   149

GEO-INSIGHT: Arches National Park, Utah   150

6.7 Weathering and Resources   153

GEO-RECAP   154

7 Sediment and Sedimentary Rocks 156

7.1 Introduction   158

7.2 Sediment Sources, Transport, and Deposition   158

7.3 How Does Sediment Become Sedimentary Rock?   160

7.4 Types of Sedimentary Rocks   161
Detrital Sedimentary Rocks   161
Chemical and Biochemical Sedimentary Rocks   162

7.5 Sedimentary Facies   165

7.6 Reading the Story in Sedimentary Rocks   167
Sedimentary Structures   167
Fossils—Remains and Traces of Ancient Life   169

GEO-INSIGHT: Fossilization   170
Determining the Environment of Deposition   172

7.7 Important Resources in Sedimentary Rocks   172
Petroleum and Natural Gas   173

GEO-FOCUS: Concretions and Geodes   174
Uranium   176
Banded Iron Formation   176

GEO-RECAP   177

8 Metamorphism and Metamorphic Rocks 180

8.1 Introduction   182

GEO-FOCUS: Asbestos: Good or Bad?   183

8.2 The Agents of Metamorphism   184
Heat   184
Pressure   184
Fluid Activity   185

8.3 The Three Types of Metamorphism   185
Contact Metamorphism   185
Dynamic Metamorphism   187
Regional Metamorphism   188
Index Minerals and Metamorphic Grade   188

8.4 How Are Metamorphic Rocks Classified?   189
Foliated Metamorphic Rocks   189
Nonfoliated Metamorphic Rocks   192

8.5 Metamorphic Zones and Facies   193

8.6 Plate Tectonics and Metamorphism   194

8.7 Metamorphism and Natural Resources   195

GEO-RECAP   196

9 Earthquakes and Earth's Interior 198

9.1 Introduction   200

9.2 Elastic Rebound Theory   201

9.3 Seismology   202
The Focus and Epicenter of an Earthquake   202

9.4 Where Do Earthquakes Occur, and How Often?   203

9.5 Seismic Waves   205
Body Waves   206
Surface Waves   206

9.6 Locating an Earthquake   207

9.7 Measuring the Strength of an Earthquake   208
Intensity   208
Magnitude   211

9.8 What Are the Destructive Effects of Earthquakes?   212
Ground Shaking   212
Fire   213
Tsunami: Killer Waves   213

GEO-INSIGHT: The San Andreas Fault   214
Ground Failure   217

9.9 Earthquake Prediction   217
Earthquake Precursors   217
Earthquake Prediction Programs   219

9.10 Earthquake Control   219

GEO-FOCUS: Paleoseismology   222

9.11 What Is Earth's Interior Like?   223

9.12 The Core   224
Density and Composition of the Core   224

9.13 Earth's Mantle   225
The Mantle's Structure, Density, and Composition   225

9.14 Seismic Tomography   226

9.15 Earth's Internal Heat   227

9.16 Earth's Crust   227

GEO-RECAP   228

10 Deformation, Mountain Building, and Earth's Crust 232

10.1 Introduction   234

10.2 Rock Deformation—How Does It Occur?   235
Stress and Strain   235
Types of Strain   235

10.3 Strike and Dip—The Orientation of Deformed Rock
Layers   236

10.4 Deformation and Geologic Structures   237
Folded Rock Layers   237
Joints   241
Faults   242

10.5 Deformation and the Origin of Mountains   245
Mountain Building   245

GEO-FOCUS: Engineering and Geology   246
Plate Tectonics and Mountain Building   246
Terranes and the Origin of Mountains   252

10.6 Earth's Crust   252
Principle of Isostasy   252
Isostatic Rebound   254

GEO-RECAP   255

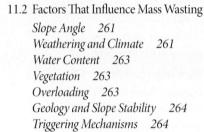

11 Mass Wasting   258

11.1 Introduction   260

11.2 Factors That Influence Mass Wasting   260
Slope Angle   261
Weathering and Climate   261
Water Content   263
Vegetation   263
Overloading   263
Geology and Slope Stability   264
Triggering Mechanisms   264

11.3 Types of Mass Wasting   265
Falls   265
Slides   265

GEO-FOCUS: Southern California Landslides   268
Flows   271
Complex Movements   277

11.4 Recognizing and Minimizing the Effects of Mass
Wasting   277

GEO-RECAP   281

12 Running Water—Streams and Rivers   284

12.1 Introduction   286

12.2 Water on Earth   287
The Hydrologic Cycle   287

GEO-FOCUS: Dams, Reservoirs, and Hydroelectric Power
Plants   288
Fluid Flow   290

12.3 Running Water   290
Sheet Flow and Channel Flow   290
Gradient, Velocity, and Discharge   290

12.4 Running Water, Erosion, and Sediment Transport   292

12.5 Deposition by Running Water   293
The Deposits of Braided and Meandering Channels   293
Floodplain Deposits   296
Deltas   296
Alluvial Fans   298

12.6 Can Floods Be Predicted and Controlled?   298

GEO-INSIGHT: Floods and Flooding   300

12.7 Drainage Systems   302
The Significance of Base Level   304
What Is a Graded Stream?   305

12.8 The Evolution of Valleys   305
Incised Meanders   306
Superposed Streams   308

GEO-RECAP   309

13 Groundwater   312

13.1 Introduction   314

13.2 Groundwater and the Hydrologic
Cycle   314

13.3 Porosity and Permeability   314

13.4 The Water Table   316

13.5 Groundwater Movement   316

13.6 Springs, Water Wells, and Artesian Systems   317
Springs   317
Water Wells   317
Artesian Systems   319

13.7 Groundwater Erosion and Deposition   320
Sinkholes and Karst Topography   320
Caves and Cave Deposits   322

13.8 Modification of the Groundwater System and Its Effects   323
Lowering the Water Table   323
Saltwater Incursion   325
Subsidence   325
Groundwater Contamination   326

GEO-FOCUS: Arsenic and Old Lace   328
Groundwater Quality   331

13.9 Hydrothermal Activity   331
Hot Springs   331
Geysers   332
Geothermal Energy   334

GEO-RECAP   335

14 Glaciers and Glaciation   338

14.1 Introduction   340

14.2 The Kinds of Glaciers   340
    Valley Glaciers   340
    Continental Glaciers   341

GEO-INSIGHT: The Little Ice Age   342

14.3 Glaciers—Moving Bodies of Ice on Land   344
    Glaciers—Part of the Hydrologic Cycle   344
    How Do Glaciers Originate and Move?   344
    Distribution of Glaciers   345

14.4 The Glacial Budget   345
    How Fast Do Glaciers Move?   347

GEO-FOCUS: Glaciers and Global Warming   348

14.5 Erosion and Sediment Transport by Glaciers   350
    Erosion by Valley Glaciers   351
    Continental Glaciers and Erosional Landforms   354

14.6 Glacial Deposits   355
    Landforms Composed of Till   355
    Landforms Composed of Stratified Drift   358
    Deposits in Glacial Lakes   359

14.7 What Causes Ice Ages?   359
    The Milankovitch Theory   359
    Short-Term Climatic Events   360

GEO-RECAP   361

15 The Work of Wind and Deserts   364

15.1 Introduction   366

15.2 Sediment Transport by Wind   366
    Bed Load   366
    Suspended Load   367

15.3 Wind Erosion   367
    Abrasion   367
    Deflation   368

15.4 Wind Deposits   368
    The Formation and Migration of Dunes   369
    Dune Types   369
    Loess   372

15.5 Air-Pressure Belts and Global Wind Patterns   372

15.6 The Distribution of Deserts   374

15.7 Characteristics of Deserts   376
    Temperature, Precipitation, and Vegetation   376
    Weathering and Soils   376
    Mass Wasting, Streams, and Groundwater   377
    Wind   377

GEO-FOCUS: Windmills and Wind Power   378

15.8 Desert Landforms   378

GEO-RECAP   383

16 Oceans, Shorelines, and Shoreline Processes   386

16.1 Introduction   388

16.2 Exploring the Oceans   389

16.3 Seawater and Oceanic Circulation   390
    Seawater—Its Composition and Color   390
    Oceanic Circulation   390
    Sediments and Sedimentary Rocks on the Seafloor   391

16.4 Shorelines and Shoreline Processes   392
    Tides, Waves, and Nearshore Currents   393

16.5 Erosion and Deposition Along Shorelines   398
    Erosion   398
    Deposition   400
    The Nearshore Sediment Budget   403

16.6 Types of Coasts   403
    Depositional and Erosional Coasts   404
    Submergent and Emergent Coasts   404

16.7 The Perils of Living Along a Shoreline   405
    Storm Waves and Coastal Flooding   405
    Coastal Management as Sea Level Rises   407

16.8 Resources from the Oceans   409

GEO-FOCUS: The *Deepwater Horizon* Oil Spill   410

GEO-RECAP   412

17 Geologic Time: Concepts and Principles   414

17.1 Introduction   416

17.2 How Geologic Time Is Measured   416

17.3 Early Concepts of Geologic Time and Earth's Age   416

17.4 James Hutton and the Recognition of Geologic Time   417

17.5 Relative Dating Methods   418
    Fundamental Principles of Relative Dating   418
    Unconformities   420
    Applying the Principles of Relative Dating   427

17.6 Correlating Rock Units   428
    Subsurface Correlation   431

17.7 Absolute Dating Methods   432
    Atoms, Elements, and Isotopes   432
    Radioactive Decay and Half-Lives   432
    Sources of Uncertainty   434
    Long-Lived Radioactive Isotope Pairs   435
    Fission-Track Dating   436

Radiocarbon and Tree-Ring Dating Methods   436

17.8 Development of the Geologic Time Scale   437

17.9 Stratigraphy and Stratigraphic Terminology   437

17.10 Geologic Time and Climate Change   440

GEO-FOCUS: Denver's Weather—280 Million Years Ago!   442

GEO-RECAP   445

18 Organic Evolution—The Theory and Its Supporting Evidence   448

18.1 Introduction   450

18.2 Organic Evolution: What Does It Mean?   450

Jean-Baptiste de Lamarck and His Ideas on Evolution   450
The Contributions of Charles Robert Darwin and Alfred Russel Wallace   451

GEO-FOCUS: The Tragic Lysenko Affair   453

Natural Selection—What Is Its Significance?   453

GEO-INSIGHT: Artificial Selection, Natural Selection, Fossils, and Evolution   454

18.3 Mendel and the Birth of Genetics   456

Mendel's Experiments   456
Genes and Chromosomes   457

18.4 The Modern View of Organic Evolution   458

What Brings About Variation?   458
Speciation and the Rate of Evolution   458
Divergent, Convergent, and Parallel Evolution   459
Microevolution and Macroevolution   461
Cladistics and Cladograms   461
Mosaic Evolution and Evolutionary Trends   462
Extinctions   463

18.5 Evidence Supporting Evolutionary Theory   463

Biological Evidence Supporting Evolution   465
Fossils: What Do We Learn from Them?   467
The Evidence—A Summary   468

GEO-RECAP   471

19 Precambrian Earth and Life History   474

19.1 Introduction   476

19.2 What Happened During the Hadean?   476

19.3 The Foundations of Continents—Shields, Platforms, and Cratons   478

19.4 Archean Earth History   480

Archean Rocks   480
Archean Plate Tectonics and the Origin of Cratons   480

19.5 Proterozoic Earth History   482

Paleoproterozoic History of Laurentia   484
Paleo- and Mesoproterozoic Igneous Activity   486
Mesoproterozoic Orogeny and Rifting   486
Meso- and Neoproterozoic Sedimentation   487
Proterozoic Supercontinents   488
Proterozoic Rocks   488

19.6 Origin and Evolution of the Atmosphere and Hydrosphere   489

The Atmosphere   490
The Hydrosphere   491

19.7 Life—Its Origin and Early History   491

The Origin of Life   492
Archean Organisms   493
Life of the Proterozoic   494

19.8 Resources in Precambrian Rocks   497

Archean Resources   497

GEO-FOCUS: The Sudbury Meteorite Impact and Its Aftermath   498

Proterozoic Resources   499

GEO-RECAP   499

20 Paleozoic Earth History   502

20.1 Introduction   504

20.2 Continental Architecture: Cratons and Mobile Belts   504

20.3 Paleozoic Paleogeography   505

Early-Middle Paleozoic Global History   506
Late Paleozoic Global History   509

20.4 Paleozoic Evolution of North America   509

20.5 The Sauk Sequence   511

20.6 The Tippecanoe Sequence   514

Tippecanoe Reefs and Evaporites   515

GEO-INSIGHT: The Grand Canyon—A Geologist's Paradise   518

The End of the Tippecanoe Sequence   520

20.7 The Kaskaskia Sequence   520

Reef Development in Western Canada   520
Black Shales   521
The Late Kaskaskia—A Return to Extensive Carbonate Deposition   522

20.8 The Absaroka Sequence   523

What Are Cyclothems, and Why Are They Important?   524
Cratonic Uplift—The Ancestral Rockies   526
The Late Absaroka—More Evaporite Deposits and Reefs   526

20.9 History of the Paleozoic Mobile Belts   526

Appalachian Mobile Belt   526
Cordilleran Mobile Belt   532
Ouachita Mobile Belt   532

20.10 What Role Did Microplates and Terranes Play in the Formation of Pangaea?   533

20.11 Paleozoic Mineral Resources   534

GEO-RECAP   535

21 Paleozoic Life History 540

21.1 Introduction   542

21.2 What Was the Cambrian Explosion?   542

21.3 The Emergence of a Shelly Fauna   543

21.4 Paleozoic Invertebrate Marine Life   544

The Present Marine Ecosystem   544
Cambrian Marine Community   546
The Burgess Shale Biota   547

GEO-INSIGHT: Trilobites—Paleozoic Arthropods   548

Ordovician Marine Community   550
Silurian and Devonian Marine Communities   551
Carboniferous and Permian Marine Communities   552
The Permian Marine Invertebrate Mass Extinction   553

21.5 Vertebrate Evolution   554

21.6 Fish   555

21.7 Amphibians—Vertebrates Invade the Land   560

21.8 Evolution of the Reptiles—The Land Is Conquered   562

21.9 Plant Evolution   564

Silurian and Devonian Floras   565
Late Carboniferous and Permian Floras   566

GEO-RECAP   569

22 Mesozoic Earth and Life History 574

22.1 Introduction   576

22.2 The Breakup of Pangaea   576

The Effects of the Breakup of Pangaea on Global Climates and Ocean Circulation Patterns   578

22.3 Mesozoic History of North America   580

Continental Interior   581
Eastern Coastal Region   581
Gulf Coastal Region   583
Western Region   583

22.4 What Role Did Accretion of Terranes Play in the Growth of Western North America?   589

22.5 Mesozoic Mineral Resources   590

22.6 Life of the Mesozoic Era   591

Marine Invertebrates and Phytoplankton   592
Plants—Primary Producers on Land   593
The Diversification of Reptiles   593

GEO-INSIGHT: Dinosaurs   598

The Origin and Evolution of Birds   601
Evolution of Mammals   602

22.7 Mesozoic Climates and Paleogeography   604

GEO-FOCUS: Mary Anning's Contributions to Paleontology   605

22.8 Mass Extinctions—A Crisis in the History of Life   606

GEO-RECAP   607

23 Cenozoic Earth and Life History 612

23.1 Introduction   614

23.2 Cenozoic Plate Tectonics   615

23.3 Cenozoic Orogenic Belts   615

The Alpine–Himalayan Orogenic Belt   615
The Circum-Pacific Orogenic Belt   615

23.4 Paleogene and Neogene Evolution of North America   616

The North American Cordillera   617
Cordilleran Igneous Activity   618
Basin and Range Province   620
Colorado Plateau and Rio Grande Rift   621

GEO-INSIGHT: Cascade Range Volcanoes   622

Pacific Coast   624
The Continental Interior   625
Cenozoic History of the Appalachian Mountains   626
North America's Southern and Eastern Continental Margins   627
The Gulf Coastal Plain   627
The Atlantic Continental Margin   628

23.5 The Pleistocene and Holocene Epochs   628

Pleistocene and Holocene Tectonism and Volcanism   628
Pleistocene Glaciation   628

23.6 Cenozoic Mineral Resources   631

23.7 Paleogene and Neogene Life History   632

Marine Invertebrates and Phytoplankton   632
Cenozoic Vegetation and Climate   632
Paleogene and Neogene Birds   634
Diversification of Mammals   635
Cenozoic Mammals   636

GEO-FOCUS: The Messel Pit Fossil Site in Germany   638

23.8 Pleistocene Faunas 639
   Mammals and Birds 639
   Primate Evolution 641
   Hominids 642
   Pleistocene Extinctions 649
GEO-RECAP 650

24 Geology in Perspective 654

24.1 Introduction 656

24.2 Geology and the Environment 656

24.3 A Final Word 659

Appendices

A: English-Metric Conversion Chart 660

B: Topographic Maps 661
   Contours 661
   Map Scales 661
   Map Locations 662
   Where to Obtain Topographic Maps 663

C: Mineral Identification 664

Answers

Multiple-Choice Questions 666

Selected Short Answer Questions 667

Glossary 671

Index 687

Preface

Earth is a dynamic planet that has changed continuously during its 4.6 billion years of existence. The size, shape, and geographic distribution of the continents and ocean basins have changed through time, as have the atmosphere and biota. As scientists and concerned citizens, we have become increasingly aware of how fragile our planet is and, more importantly, how interdependent all of its various systems and subsystems are.

We have also learned that we cannot continually pollute our environment and that our natural resources are limited and, in most cases, nonrenewable. Furthermore, we are coming to realize how central geology is to our everyday lives. For example, in 2010 a 7.0 magnitude earthquake struck Haiti, killing more than 200,000 people, and in 2011 a 9.0 magnitude earthquake struck Japan, killing more than 13,000 people and generating a tsunami that wreaked destruction along the coast of northeastern Japan, as well as damaging three nuclear power plants and causing radioactive leakage in one of them. A major oil spill in the Gulf of Mexico in 2010 resulted in much ecological damage along the shorelines of the Gulf Coast of the United States, as well as to the biota in the Gulf of Mexico. Numerous volcanic eruptions also occurred in 2010, one of which caused disruption to air traffic over the Atlantic Ocean. And finally, major flooding resulted in extensive damage to Brisbane, Australia in January, 2011.

All of these events point out how much geology affects our lives, as well as the global economy. For these and other reasons, geology is one of the most important college or university courses that a student can take.

The Changing Earth: Exploring Geology and Evolution, Sixth Edition, is designed for an introductory course in geology that can serve both majors and nonmajors in geology and the earth sciences. One of the problems with any introductory science course is that students are overwhelmed by the amount of material that must be learned. Furthermore, most of the material does not seem to be linked by any unifying theme and does not always appear to be relevant to their lives. This book, however, is written to address that problem in that it shows, in its easy-to-read style, that geology is an exciting and ever-changing science, and one in which new discoveries and insights are continually being made.

The goals of this book are to provide students with a basic understanding of geology and its processes and, most importantly, an understanding of how geology relates to the human experience—that is, how geology affects not only individuals, but society in general. It is also our intent to present the geologic and biologic history of Earth, not as a set of encyclopedic facts to memorize, but rather as a continuum of interrelated events that reflect the underlying geologic and biologic principles and processes that have shaped our planet and life upon it.

Instead of emphasizing individual, and seemingly unrelated, events, we seek to understand the underlying causes of why things happened the way they did and how all of Earth's systems and subsystems are interrelated. Using this approach, students will gain a better understanding of how everything fits together, and why geology is such an important course.

With these goals in mind, we introduce the major themes of the book in the first chapter to provide students with an overview of the subject and to enable them to see how the various systems, subsystems, and cycles of Earth are interrelated. We then cover the unifying theme of geology—plate tectonics—in the second chapter. Plate tectonic theory is central to the study of geology because it links many aspects of geology together. It is a theme that is woven throughout this edition.

The economic and environmental aspects of geology are discussed throughout the book rather than treating these topics in separate chapters. In this way, students can see, through topical and interesting examples, how geology impacts our lives. Climate change is an especially relevant and important topic that is currently in the news and being discussed and debated by scientists, politicians, and citizens alike. Because of its importance, we introduce the topic somewhat in detail in the first chapter, and integrate it throughout the book as it relates to the various topics covered. Geology is unique in that it can provide the perspective of geologic time in this important debate as to what, and the degree of, impact humans have had on climate change.

New and Retained Features in the Sixth Edition

Just as Earth is dynamic and evolving, so too is *The Changing Earth: Exploring Geology and Evolution.* The sixth edition has undergone considerable rewriting and updating, resulting in a volume that is easy to read and contains a high level of current information. Drawing on the comments and suggestions of reviewers, we have retained those features that were both relevant and popular in the fifth edition, and have incorporated numerous new features into this edition.

- Chapter content has been updated and rewritten to help clarify concepts and emphasize underlying processes. Much of the material, especially on seafloor spreading, in Chapter 9 (The Seafloor), has been incorporated into Chapter 2 (Plate Tectonics: A Unifying Theory) to give students more background for understanding the processes involved with plate movement. The remaining material from Chapter 9 has been moved to Chapter 16 (Oceans, Shorelines, and Shoreline Processes). In addition, the previous Chapter 6 (Weathering, Soil, and Sedimentary Rocks) has been divided into two chapters; Chapter 6 (Weathering, Erosion, and Soils) and Chapter 7 (Sediment and Sedimentary Rocks). This was done at the suggestion of reviewers to, in part, increase content for a better understanding of these important topics. Current events, of which there were numerous ones, especially during 2010 and 2011, have been incorporated throughout the text.

- Two new features develop critical thinking skills while enhancing visual learning and observational skill development.

  - *Critical Thinking Questions* have been added to many of the figures. Selected figures within each chapter are now accompanied by a critical thinking question to encourage active student learning, guide student observational skill development, and deepen their understanding of geologic processes.

  - At the end of each chapter is a new feature called *Geo-Images*. In this section, a photo or graphic is presented that challenges students to describe the geologic process being depicted and to support their conclusions with logical reasoning. These photos or graphics were chosen to challenge students to develop strong observational and critical thinking skills.

- *Connection Link* boxes help students see the big picture. Throughout each chapter *Connection Link* boxes help students see how Earth systems are interrelated and connected. These *Connection Link* boxes are identified by a small box next to the pertinent concepts in the text, and they refer the reader to other locations in the book where more information can be found and connections made with other related important topics.

- Expanded Learning Tools. New *Have You Ever Wondered?* questions follow the chapter *Outline* at the beginning of each chapter. These intriguing questions are designed to catch student interest about what is covered in the chapter and motivate them to find the answers by reading the chapter.

The following features have been retained and improved in the sixth edition.

- Many new, bold, and dramatic photos open each chapter, a number of which are geologic events that happened as recently as 2010 and 2011. These add relevancy to the text and emphasize the theme of how geology relates to humans.

- Many of the popular *Geo-Focus* and *Geo-Insight* features either contain new topics or have been updated.

- The successful *What Would You Do?* boxes in each chapter continue to encourage students to think critically about what they're learning by asking open-ended questions related to the chapter material. Many of these boxes either are new or have been rewritten to reflect current topics in the news.

- The format of 10 multiple-choice questions and 10 short-answer questions has been retained in the *Review Questions* section at the end of each chapter. Answers to all of the multiple-choice questions, as well as several of the short-answer questions in each chapter, are provided at the back of the book.

It is our strong belief that the rewriting and updating done in the text, as well as the new features introduced, significantly improve the sixth edition of *The Changing Earth: Exploring Geology and Evolution*. We think that these changes and the new enhancements make this textbook easier to read and comprehend, as well as a more effective teaching tool.

Text Organization

Plate tectonic theory is the unifying theme of geology and also for this book. This theory has revolutionized geology because it provides a global perspective of Earth and allows geologists to treat many seemingly unrelated geologic phenomena as part of a total planetary system.

A second, and equally important, theme is that Earth is a complex, dynamic planet that has changed continuously since its origins some 4.6 billion years ago. We can better understand this complexity by using a systems approach to the study of Earth and emphasizing this concept throughout the book.

We have therefore organized *The Changing Earth: Exploring Geology and Evolution,* Sixth Edition, into the following informal categories:

- Chapter 1 is an introduction to geology and Earth systems, geology's relevance to the human experience, and the debate about climate change and humans' possible role and effect, as well as the origin of the solar system and Earth's place in it.

- Chapter 2 deals with plate tectonics in detail and sets the stage for its integration throughout the rest of the book. Pertinent material from the previous edition's Chapter 9 (The Seafloor) has been moved to this chapter to better clarify and illustrate the concept of seafloor spreading and how it relates to plate movement.

- Chapters 3–8 examine Earth's materials (minerals, and igneous, sedimentary, and metamorphic rocks) and the geologic processes accounting for them, including the role of plate tectonics in their origin and distribution.

- Chapters 9–10 deal with the related topics of Earth's interior, earthquakes, and deformation and mountain building.

- Chapters 11–16 cover Earth's surface processes and features such as mass wasting, running water, groundwater, glaciers and glaciation, the work of wind and deserts, and shorelines and shoreline processes.

- Chapter 17 discusses geologic time, and Chapter 18 explores fossils and evolution.

- Chapters 19–23 constitute our chronological treatment of the geologic and biologic history of Earth.

- Chapter 24 summarizes and synthesizes the concepts, themes, and major topics covered in this book.

Of particular assistance to students are the end-of-chapter summary tables found in Chapters 20–22. These tables are designed to give an overall perspective of the geologic and biologic events that occurred during the particular time interval covered in that chapter and to show how the events are interrelated.

We have found that presenting the material in the order discussed above works well for most students. We are also pleased to follow the suggestions of the reviewers who have used the book by redistributing the material in Chapter 9 and dividing the former Chapter 6 into two separate chapters. We know, however, that many professors prefer an entirely different order of topics, depending on the emphasis in their course. We have therefore written this book so that instructors can present the chapters in any order that suits the needs of a particular course.

Chapter Organization

All chapters have the same organizational format as follows:

- Each chapter opens with a dramatic photograph, many of which are new, followed by an *Outline* of the topics covered and a series of questions under the title *Have You Ever Wondered?* that are designed to pique student interest in the topics covered in that chapter.

- An *Introduction* follows that is intended to stimulate interest in the chapter and show how the chapter material fits into the larger geologic perspective.

- The text is written in a clear, informal style, making it easy for students to comprehend.

- Within each chapter are several *Connection Link* boxes that refer the reader to other locations in the book where more information can be found, and connections are made to important concepts and topics.

- Numerous *Critical Thinking Questions* are found associated with selected figures in each chapter. These are designed to encourage active student learning and deepen understanding of geologic processes.

- Many newly rendered color diagrams and photographs complement the text and provide a visual representation of the concepts and information presented.

- Each chapter contains at least one *Geo-Focus* and/or *Geo-Insight* feature that presents a brief discussion or visual representation of an interesting aspect of geology or geologic research. Several of these features are new to this edition and emphasize the climate change debate.

- At least one *What Would You Do?* box per chapter encourage students to engage in critical thinking by solving hypothetical problems or issues that are related to the chapter material.

- Topics related to environmental issues, such as climate change, and economic geology, are discussed throughout the text. Integrating economic geology and environmental issues with the chapter material helps students relate the importance and relevance of geology to their lives.

- The end-of-chapter *Geo-Recap* begins with a new *Geo-Image* feature that challenges students to develop strong observational and critical thinking skills based on chapter material. This is followed by a concise review of important concepts and ideas presented in the chapter.

- The *Important Terms*, which are printed in boldface type in the chapter text, are listed at the end of each chapter for easy review, along with the page numbers on which they are first defined.

- The *Review Questions* are another important feature of this book and include multiple-choice questions with answers as well as short-answer questions, some of which have the answers provided at the end of the book. Many new questions have been added to each chapter of the sixth edition.

- A full *Glossary* of important terms appears at the end of the text.

Ancillary Materials

For Instructors

We are pleased to offer a full suite of text and multimedia products to accompany *The Changing Earth: Exploring Geology and Evolution*, Sixth Edition.

New to this edition, the *Geology CourseMate* features a rich array of study tools and learning resources for your students. This text-specific companion website includes quizzing, flashcards, and other web-based activities that will help students explore the concepts presented in the text.

PowerLecture with JoinIn™ *on Turning Point*® A complete all in one reference for instructors, the PowerLecture DVD contains PowerPoint slides of images from the text, stepped art based on figures from the text, zoomable art figures from the text, Active Figures that interactively demonstrate concepts, and lectures that outline the main points of each chapter. In addition to providing you with fantastic course presentation material, the PowerLecture DVD also contains electronic files of the Test Bank and Instructor's Manual as well as JoinIn, the easiest Audience Response System to use, featuring instant classroom assessment and learning. ISBN-10: 0840068042 | ISBN-13: 9780840068040

Online Instructor's Manual with Test Bank This guide contains resources designed to streamline your course preparation. The Instructor's Manual includes Chapter Outlines, Learning Objectives, Chapter Summaries, Enrichment Topics, Common Misconceptions, Lecture Suggestions, "Consider This" questions, Important Terms, and Weblinks/Videos/Software suggestions. The Test Bank includes Short Answer, Fill in the Blank, Multiple Choice, and True/False questions. Available on our password protected instructor companion website. ISBN-10: 084006666X, ISBN-13: 9780840066664

ExamView® Create, deliver, and customize tests and study guides (both print and online) in minutes with this easy-to-use assessment and tutorial system. ExamView offers both a Quick Test Wizard and an Online Test Wizard that guide you step by step through the process of creating tests, whereas its "what you see is what you get" interface allows you to see the test you are creating on the screen exactly as it will print or display online. You can build tests of up to 250 questions using up to 12 question types. Using ExamView's complete word processing capabilities, you can enter an unlimited number of new questions or edit existing questions. ISBN-10: 0840068050, ISBN-13: 9780840068057

Global Geoscience Watch This online resource center provides access to a rich array of media resources to help you keep up with current events and highlight the concepts taught in class—as well as show the human impact on our planet. Features include:

- Articles
- Case studies
- Podcasts
- Videos
- World Map (searchable by topic and country)
- Citation tools
- Sharing options (email, Twitter, Facebook, etc.)
- Topic browsing and advanced searching
- And more!

For Students

eCompanion An eCompanion is now available to accompany the eBook for *The Changing Earth,* Sixth Edition. Carry this light-weight manual to class and use it to help synthesize your understanding of key concepts from the text. Features include chapter objectives and summaries, key terms, an interactive "Concept Check" section, review questions,

and space for note-taking. A comprehensive study tool, the eCompanion assists in exam preparation, allows you to follow along in class without the printed book or computer, and reinforces the concepts presented in the text. ISBN-10: 0840062133, ISBN-13: 9780840069047

Geology CourseMate *The Changing Earth* now features Geology CourseMate, which helps you make the grade.

Geology CourseMate includes:

- an interactive eBook, with highlighting, note taking, and search capabilities
- interactive learning tools including:
  - Quizzes
  - Flashcards
  - Video Exercises
  - Animations
  - and more!

Go to login.cengagebrain.com to access these resources. ISBN-10: 1111675686, ISBN-13: 9781111675684

Virtual Field Trips in Geology The Virtual Field Trips in Geology are concept-based modules that teach you geology by using famous locations throughout the United States. Grand Canyon, Arches, and Hawaii Volcanoes National Parks are included, as well as many others. Designed to be used as homework assignments or lab work, the modules use a rich array of multimedia to demonstrate concepts. High definition videos, images, animations, quizzes, and Google Earth activities work together in Virtual Field Trips to bring the concepts to life.

Global Geoscience Watch Updated several times a day, the Global Geoscience Watch is a focused portal into GREENR—our Global Reference on the Environment, Energy, and Natural Resources—an ideal one-stop site for current events and research projects for all things geoscience! Broken into the four key course areas (Geography, Geology, Meteorology, and Oceanography), you can easily find the most relevant information for the course you are taking.

You will have access to the latest information from trusted academic journals, news outlets, and magazines. You also will receive access to statistics, primary sources, case studies, podcasts, and much more! ISBN-10: 1111429065, ISBN-13: 9781111429065

Acknowledgments

As the authors, we are, of course, responsible for the organization, style, and accuracy of the text, and any mistakes, omissions, or errors are our responsibility. The finished product is the culmination of many years of work during which we received numerous comments and advice from many geologists who reviewed all or parts of the text for the first four editions. They are: Kenneth Beem, Montgomery College; David Berry, California State Polytechnic

University, Pomona; Wesley A. Brown, Stephen F. Austin State University; Patricia J. Bush, Dèlgado Community College; Paul J. Bybee, Utah Valley State College; Deborah Caskey, El Paso Community College; Renee M. Clary, University of Louisiana at Lafayette; Michael Conway, Arizona Western College; David Cordero, Lower Columbia College; William C. Cornell, University of Texas at El Paso; Kathleen Devaney, El Paso Community College; Richard Diecchio, George Mason University; Robert Ewing, Portland Community College; David J. Fitzgerald, St. Mary's University; Yongli Gao, East Tennessee State University; Dann M. Halverson, University of Southwestern Louisiana; Kristi Higginbotham, San Jacinto College; Ray Kenny, New Mexico Highlands University; Gary L. Kinsland, University of Louisiana at Lafayette; Jorg Maletz, University at Buffalo-SUNY; Kevin McCartney, University of Maine-Presque Isle; Bob Mims, Richland College; Roger Steinberg, Del Mar College; Michelle Stoklosa, Boise State University; Glenn B. Stracher, East Georgia College; Azam M. Tabrizi, Tide Water Community College; Thomas J. Weiland, Georgia Southwestern State University; Monte D. Wilson, Boise State University; and Guy Worthey, St. Ambrose University.

We especially wish to express our sincere appreciation to the reviewers of the fifth edition who made numerous helpful and useful comments that led to many improvements and new features seen in this sixth edition. They are: Brian Campbell, Southwestern Oklahoma State University; Michael O'Donnell, Lord Fairfax Community College; Ken Griffin, Tarrant County College; Clair Ossian, Tarrant County College; Susan Grandy, St. Clair County Community College; Gary Kinsland, University of Louisiana at Lafayette; James MacDonald, Florida Gulf Coast University; Joseph Sarnecki, Macomb Community College; and Lynn Milwood, Mountain View College.

We also wish to thank Kathy Benison, R. V. Dietrich (Professor Emeritus), David J. Matty, Jane M. Matty, Wayne E. Moore (Professor Emeritus), and Sven Morgan of the Geology Department, and Bruce M. C. Pape (Emeritus) of the Geography Department of Central Michigan University, as well as Eric Johnson (Hartwick College, New York), and Stephen D. Stahl (St. Bonaventure, New York) for providing us with photographs and answering our questions concerning various topics. We are also grateful for the generosity of the various agencies and individuals from many countries that provided photographs.

Special thanks must go to Yolanda Cossio, Publisher for the Earth Sciences at Cengage Learning, who initiated this sixth edition, and to our Development Editor, Jake Warde, who not only kept us on task, but who superbly edited and managed the content for this edition, as well as providing a fresh perspective on this edition, and excellent suggestions throughout the continuing gestation of this new edition. We are indebted to Annie Beck for her attention to detail and consistency throughout the book production process. We would also like to thank John Walker and Lisa Buckley for the fresh design. We thank Parvinder Sethi for his help in locating appropriate photographs. We would also like to recognize Hal Humphrey, Cengage Learning Production Project Manager; Alexandria Brady, Media Editor, for developing the media program; Jack Cooney, Marketing Manager; Darlene Macanan Marketing Communications Manager; and Kristina Chiapella, Assistant Editor, who managed the accompanying ancillary package. We also extend thanks to the artists at Lachina Publishing Services, who are responsible for updating much of the art program.

As always, our families were very patient and encouraging when much of our spare time and energy were devoted to this book. We again thank them for their continued support and understanding.

James S. Monroe
Reed Wicander

The Changing Earth

Satellite-based image of Earth. North America is clearly visible in the center of this view, as well as Central America and the northern part of South America. The present locations of continents and ocean basins are the result of plate movements. The interaction of plates through time has affected the physical and biological history of Earth.

CHAPTER 1

Understanding Earth: A Dynamic and Evolving Planet

OUTLINE

1.1 Introduction

1.2 What Is Geology?

1.3 Geology and the Formulation of Theories

1.4 How Does Geology Relate to the Human Experience?

1.5 How Does Geology Affect Our Everyday Lives?

1.6 Global Geologic and Environmental Issues Facing Humankind

1.7 Origin of the Universe and Solar System, and Earth's Place in Them

1.8 Why Earth Is a Dynamic and Evolving Planet

GEO-INSIGHT: The Terrestrial and Jovian Planets

1.9 The Rock Cycle

1.10 Organic Evolution and the History of Life

1.11 Geologic Time and Uniformitarianism

1.12 How Does the Study of Geology Benefit Us?

GEO-RECAP

HAVE YOU EVER WONDERED?

- How all of Earth's different components are interconnected and how the interactions among them are what make Earth a dynamic and ever-changing planet?

- Why a theory is not an unsubstantiated wild guess at explaining various natural phenomena?

- How geology relates to the human experience and affects our everyday lives?

- How such environmental issues as overpopulation, climate change, and rising sea level, to name a few, directly affect and impact you?

- How the universe, solar system, and the planets came into being and evolved to what we now see when we look beyond our own planet Earth?

- What Earth is composed of and how its history has changed through time?

- Why you should study geology and the benefits that you will derive from a better understanding of this most important science?

1.1 Introduction

A major benefit of the space age has been the ability to look back from space and view our planet in its entirety. Every astronaut has remarked in one way or another on how Earth stands out as an inviting oasis in the otherwise black void of space (see the chapter opening photo). We are able to see not only the beauty of our planet but also its fragility. We can also decipher Earth's long and frequently turbulent history by reading the clues preserved in the geologic record.

A major theme of this book is that Earth is a complex, dynamic planet that has changed continuously since its origin some 4.6 billion years ago. These changes and the present-day features we observe result from the interactions among Earth's internal and external systems, subsystems, and cycles. By viewing Earth as a whole—that is, thinking of it as a system—we not only see how its various components are interconnected, but we can also better appreciate its complex and dynamic nature.

The system concept makes it easier for us to study a complex subject such as Earth because it divides the whole into smaller components that we can easily understand, without losing sight of how the components fit together as a whole. In the same way, you can think of this book as a large, panoramic landscape painting. Each chapter fills in the details of the landscape, thereby enhancing the overall enjoyment and understanding of the entire painting.

A **system** is a combination of related parts that interact in an organized manner. An automobile is a good example of a system. Its various components or subsystems, such as the engine, transmission, steering, and brakes, are all interconnected in such a way that a change in any one of them affects the others.

We can examine Earth in the same way we view an automobile—that is, as a system of interconnected components that interact and affect one another in many ways. The principal subsystems of Earth are the *atmosphere*, *biosphere*, *hydrosphere*, *lithosphere*, *mantle*, and *core* (▮ Figure 1.1). The complex interactions among these subsystems result in a dynamically changing planet in which matter and energy are continuously recycled into different forms.

We must also not forget that humans are part of the Earth system, and our activities can produce changes with potentially wide-ranging consequences. When people discuss and debate such environmental issues as pollution and global warming, it is important to remember that these are not isolated issues, but are part of the larger Earth system. Furthermore, remember that Earth goes through longer time cycles than humans are used to. Although they may have disastrous short-term effects on the human species, global warming and cooling are also part of a longer-term cycle that has resulted in many glacial advances and retreats during the past 2.6 million years.

As you study the various topics covered in this book, keep in mind the themes discussed in this chapter and how, like the parts of a system, they are interrelated. By relating each chapter's topic to its place in the entire Earth system, you will gain a greater appreciation of why geology is so integral to our lives.

1.2 What Is Geology?

Geology, from the Greek *geo* and *logos*, is defined as the study of Earth, but now must also include the study of the planets and moons in our solar system. The discipline of geology is generally divided into two broad areas—physical geology and historical geology. *Physical geology* is the study of Earth materials, such as minerals and rocks, as well as the processes operating within Earth and on its surface. *Historical geology* examines the origin and evolution of Earth, its continents, oceans, atmosphere, and life.

Although the discipline of geology is very broad and subdivided into numerous fields or specialties, nearly every aspect of geology has some economic or environmental relevance. For example, many geologists are involved in exploration for mineral and energy resources, using their specialized knowledge to locate the natural resources on which our industrialized society is based. As the demand for these nonrenewable resources increases, geologists apply the basic principles of geology in increasingly sophisticated ways by focusing their attention on areas that have the greatest chance for economic success.

Other geologists, however, use their expertise to help solve environmental problems. Finding adequate sources of groundwater for the ever-burgeoning needs of communities and industries is becoming increasingly important, as is the monitoring of surface and underground water pollution and its cleanup. Geologic engineers help find safe locations for dams, waste-disposal sites, and power plants, as well as designing earthquake-resistant buildings.

Geologists are also engaged in making short- and long-range predictions about earthquakes and volcanic eruptions and the potential destruction that may result. In fact, geologists are now more involved than ever in working with various governmental agencies and civil defense planners to ensure that timely warnings are given to potentially affected regions when natural disasters such as tsunami occur and that contingency plans are in place.

1.3 Geology and the Formulation of Theories

The term **theory** has various meanings. In colloquial usage, it means a speculative or conjectural view of something—hence, the widespread belief that scientific theories are little more than unsubstantiated guesses. In scientific usage, however, a theory is a coherent explanation for one natural phenomenon or several related phenomena supported by a large body of objective evidence. From a theory, scientists derive predictive statements that can be tested by observations and/or experiments so that their validity can be assessed. The law of universal gravitation is an example of a theory that describes the attraction between masses (an apple and Earth in the popularized account of Newton and his discovery).

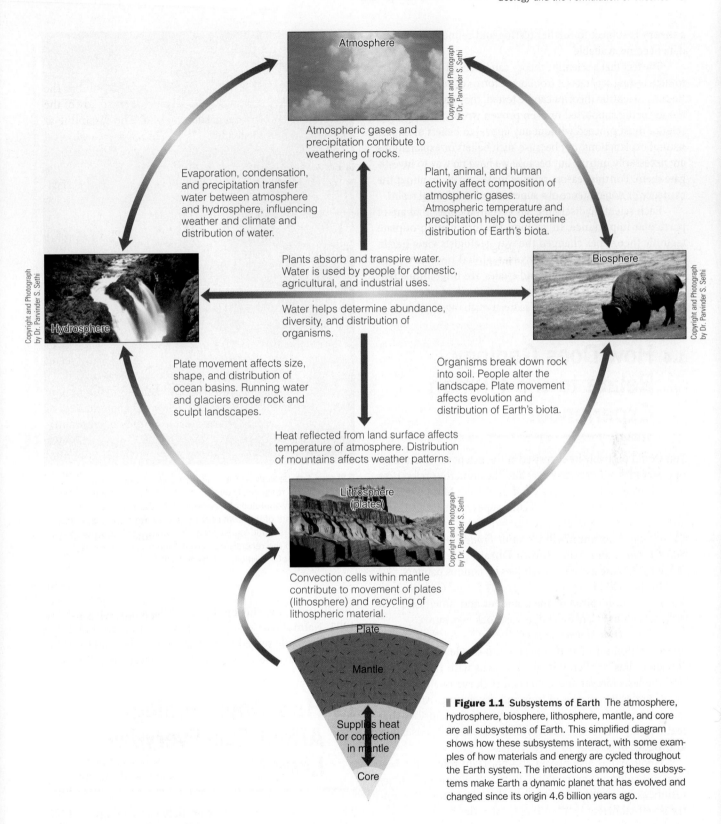

Atmospheric gases and precipitation contribute to weathering of rocks.

Evaporation, condensation, and precipitation transfer water between atmosphere and hydrosphere, influencing weather and climate and distribution of water.

Plant, animal, and human activity affect composition of atmospheric gases. Atmospheric temperature and precipitation help to determine distribution of Earth's biota.

Plants absorb and transpire water. Water is used by people for domestic, agricultural, and industrial uses.

Water helps determine abundance, diversity, and distribution of organisms.

Plate movement affects size, shape, and distribution of ocean basins. Running water and glaciers erode rock and sculpt landscapes.

Organisms break down rock into soil. People alter the landscape. Plate movement affects evolution and distribution of Earth's biota.

Heat reflected from land surface affects temperature of atmosphere. Distribution of mountains affects weather patterns.

Convection cells within mantle contribute to movement of plates (lithosphere) and recycling of lithospheric material.

Plate
Mantle
Supplies heat for convection in mantle
Core

Figure 1.1 Subsystems of Earth The atmosphere, hydrosphere, biosphere, lithosphere, mantle, and core are all subsystems of Earth. This simplified diagram shows how these subsystems interact, with some examples of how materials and energy are cycled throughout the Earth system. The interactions among these subsystems make Earth a dynamic planet that has evolved and changed since its origin 4.6 billion years ago.

Theories are formulated through the process known as the **scientific method**. This method is an orderly, logical approach that involves gathering and analyzing facts or data about the problem under consideration. Tentative explanations, or **hypotheses**, are then formulated to explain the observed phenomena. Next, the hypotheses are tested to see whether what was predicted actually occurs in a given situation. Finally, if one of the hypotheses is found, after repeated tests, to explain the phenomena, then the hypothesis is proposed as a theory. Remember, however, that in science, even a theory is subject to further testing and refinement as new data become available.

The fact that a scientific theory can be tested and is subject to such testing separates it from other forms of human inquiry. Because scientific theories can be tested, they have the potential for being supported or even proven wrong. Accordingly, science must proceed without any appeal to beliefs or supernatural explanations, not because such beliefs or explanations are necessarily untrue, but because we have no way to investigate them. For this reason, science makes no claim about the existence or nonexistence of a supernatural or spiritual realm.

Each scientific discipline has certain theories that are of particular importance. In geology, the formulation of plate tectonic theory has changed the way geologists view Earth. Geologists now view Earth from a global perspective in which all of its subsystems and cycles are interconnected, and Earth history is seen to be a continuum of interrelated events that are part of a global pattern of change.

1.4 How Does Geology Relate to the Human Experience?

You would probably be surprised at the extent to which geology pervades our everyday lives and the numerous references to geology in the arts, music, and literature. Many sketches and paintings depict rocks and landscapes realistically. Leonardo da Vinci's *Virgin of the Rocks* and *Virgin and Child with Saint Anne*, Giovanni Bellini's *Saint Francis in Ecstasy* and *Saint Jerome*, and Asher Brown Durand's *Kindred Spirits* (▌Figure 1.2) are just a few examples by famous painters.

In the field of music, Ferde Grofé's *Grand Canyon Suite* was no doubt inspired by the grandeur and timelessness of Arizona's Grand Canyon and its vast rock exposures.

Jules Verne's *Journey to the Center of the Earth* describes an expedition into Earth's interior. On one level, the poem "Ozymandias" by Percy B. Shelley deals with the fact that nothing lasts forever and even solid rock eventually disintegrates under the ravages of time and weathering. And, there is even a series of mystery books by Sarah Andrews that features the fictional geologist Em Hansen, who uses her knowledge of geology to solve crimes.

Geology has also played an important role in the history and culture of humankind. Because mineral and energy resources are not equally distributed, and because no country is self-sufficient in all of them, countries have, throughout history, fought wars to secure such natural resources as oil and gas and valuable minerals like gold, silver, and diamonds.

The configuration of Earth's surface, or its topography, which is shaped by geologic agents, has played a critical role in military tactics. For example, Napoleon included two geologists in his expeditionary forces when he invaded Egypt in 1798, and the Russians used geologists as advisors in selecting

Kindred Spirits, 1849 by Asher B. Durand. Oil on Canvas, 44 x 36 36 inches. Courtesy Crystal Bridges Museum of American Art, Bentonville, Arkansas. Photography by the Metropolitan Museum of Art.

▌**Figure 1.2** Geology and Art *Kindred Spirits* by Asher Brown Durand (1849) realistically depicts the layered rocks along gorges in the Catskill Mountains of New York State. Durand was one of numerous artists of the 19th-century Hudson River School, which was known for realistic landscapes. This painting shows Durand conversing with the recently deceased Thomas Cole, the original founding force of the Hudson River School.

fortification sites during the Russo-Japanese war of 1904–1905. Natural barriers such as mountain ranges and rivers have frequently served as political boundaries, and the shifting of river channels has sparked numerous border disputes.

1.5 How Does Geology Affect Our Everyday Lives?

The most obvious connection between geology and our everyday lives is when natural disasters strike. Destructive events such as volcanic eruptions, earthquakes, landslides, tsunami, floods, and droughts make headlines and affect many people in obvious ways. Although we cannot prevent most of these natural disasters from happening, the more we learn about what causes them, the better we will be able to predict and mitigate the severity of their impact.

Less apparent, but equally significant, are the connections between geology and economic, social, and political issues. Although most readers of this book will not become professional geologists, everyone should have a basic understanding of the geologic processes that ultimately affect all of us. For example, you may become involved in geologic decisions as a member of a planning board or as a property owner with mineral rights. As government plays a greater role in environmental issues and regulations, members of Congress have increased the number of staff devoted to studying issues related to the environment and geology.

If issues such as nonrenewable energy resources, waste disposal, and pollution seem too far removed or too complex to be fully appreciated, consider for a moment just how dependent we are on geology in our daily routines (▌ Figure 1.3).

Much of the electricity for our appliances comes from the burning of coal, oil, natural gas, or uranium consumed in nuclear-generating plants. Geologists locate the coal, petroleum (oil and natural gas), and uranium. The copper or other metal wires through which electricity travels are manufactured from materials found as the result of mineral exploration. The concrete foundation (concrete is a mixture of clay, sand or gravel, and limestone), drywall (made largely from the mineral gypsum), and windows (the mineral quartz is the principal ingredient in the manufacture of glass) of the buildings we live and work in owe their very existence to geologic resources.

When we go to work, the car or public transportation we use is powered and lubricated by some type of petroleum by-product and is constructed of metal alloys and plastics. And the roads or rails we ride over come from geologic materials, such as gravel, asphalt, concrete, or steel. All of these items are the result of processing geologic resources.

As individuals and societies, we enjoy a standard of living that is obviously directly dependent on the consumption of geologic materials. We therefore need to be aware of how our use and misuse of geologic resources may affect the environment, and develop policies that not only encourage management of our natural resources, but also allow for continuing economic development among all the world's nations.

1.6 Global Geologic and Environmental Issues Facing Humankind

Most scientists would argue that overpopulation is the greatest environmental problem facing the world today. The world's population reached 7 billion in 2011, and projections indicate that this number will grow to 9 billion people by 2045. Although this may not seem to be a geologic problem, remember that these people must be fed, housed, and clothed, and all with a minimal impact on the environment. Much of this population growth will be in areas that are already at risk from such hazards as earthquakes, tsunami, volcanic eruptions, and floods. Adequate water supplies must be found and protected from pollution. Additional energy resources will be needed to help fuel the economies of nations with ever-increasing populations. New techniques must be developed to reduce the use of our dwindling nonrenewable resource base and to increase our recycling efforts so that we can decrease our dependence on new sources of these materials.

5,454 kg Clays
245 kg Zinc
274,438 l Petroleum
499,500 kg Stone, sand, and gravel
14,301 kg Salt
166,040 m³ Natural gas
18,531 kg Cement
+19,720 kg Other minerals and metals
6,539 kg Iron ore
350 kg Lead
1,818 kg Bauxite (Aluminum)
244,336 kg Coal
6,818 kg Phosphate rock
43 g Gold
419 kg Copper

▌ **Figure 1.3 Lifetime Mineral Usage** According to the Mineral Information Institute in Golden, Colorado, the average American born in 2009 has a life expectancy of 77.7 years and will need 1,305,000 kg of minerals, metals, and fuels to sustain his or her standard of living over a lifetime. That is an average of 16,959 kg of mineral and energy resources per year for every man, woman, and child in the United States.

The problems of overpopulation and how it affects the global ecosystem vary from country to country. For many poor and nonindustrialized countries, the problem is too many people and not enough food. For the more developed and industrialized countries, it is too many people rapidly depleting both the nonrenewable and renewable natural resource base. And in the most industrially developed countries, it is people producing more pollutants than the environment can safely recycle on a human time scale. The common thread tying these varied situations together is an environmental imbalance created by a human population exceeding Earth's short-term carrying capacity.

Other global issues in the news all the time are the greenhouse effect, global warming, and climate change. These topics affect not only all of us but also the planet we live on. But just how will global warming and the resultant climate change personally affect you? Should you really be concerned about them? After all, you have exams to worry about, graduation, finding a job, let alone the everyday issues we all must deal with, not to mention your personal life. Yet, part of the college experience is examining and debating the "big picture" and issues facing society today. So what about global warming and you?

The relationship between the greenhouse effect and global warming is an excellent example of how Earth's various subsystems are interrelated. As a by-product of respiration and the burning of organic material, carbon dioxide is a component of the global ecosystem and is constantly being recycled as part of the carbon cycle. The concern in recent years over the increase in atmospheric carbon dioxide levels is related to its role in the greenhouse effect.

The recycling of carbon dioxide between Earth's crust and atmosphere is an important climate regulator because carbon dioxide and other gases, such as methane, nitrous oxide, chlorofluorocarbons, and water vapor, allow sunlight to pass through them, but trap the heat reflected back from Earth's surface. This retention of heat is called the greenhouse effect. It results in an increase in the temperature of Earth's surface and, more importantly, its atmosphere, thus producing global warming (❚ Figure 1.4). The issue is not whether we have a greenhouse effect, because we do, but rather the degree to which human activity, such as the burning of fossil fuels, is increasing the greenhouse effect and thus contributing to global warming.

Because of the increase in human-produced greenhouse gases during the past 200 years, many scientists are concerned that a global warming trend has already begun and will result in severe global climatic shifts. Presently, most climate researchers use a range of scenarios for greenhouse

What Would You Do ?

An important environmental issue facing the world today is global warming. How can this problem be approached from a global systems perspective? What are the possible consequences of global warming, and can we really do anything about it? Are there ways to tell whether global warming occurred in the geologic past?

❚ **Figure 1.4** The Greenhouse Effect and Global Warming

a Short-wavelength radiation from the Sun that is not reflected back into space penetrates the atmosphere and warms Earth's surface.

b Earth's surface radiates heat in the form of long-wavelength radiation back into the atmosphere, where some of it escapes into space. The rest is absorbed by greenhouse gases and water vapor and reradiated back toward Earth.

c Increased concentrations of greenhouse gases trap more heat near Earth's surface, causing a general increase in surface and atmospheric temperatures, which leads to global warming.

gas emissions when predicting future warming rates. State-of-the-art climate model simulations published in the *2007 Fourth Intergovernmental Panel on Climate Change* show a predicted increase in global average temperature from 2000 to 2100 of 1–3°C under the best conditions, to a 2.5–6.5°C rise under business as usual conditions. These predicted increases in temperatures are based on various scenarios that explore different global development pathways.

Regardless which scenario is followed, the global temperature change will be uneven, with the greatest warming occurring in the higher latitudes of the Northern Hemisphere. As a consequence of this warming, rainfall patterns will shift dramatically. This realignment will have a major effect on the largest grain-producing areas of the world, such as the American Midwest. Drier and hotter conditions will intensify the severity and frequency of droughts, leading to more crop failures and higher food prices (▌ Figure 1.5). With such shifts in climate, Earth's deserts may expand, with a resulting decrease in valuable crop and grazing land.

As climates change, diseases such as malaria are easily spreading into areas of warmer, wetter climates. Disease-carrying mosquitoes are expanding their range as climate changes allow them to survive in formerly inhospitable regions.

▌ **Figure 1.5** Withered Corn Crop as a Result of Drought Conditions

Higher temperature will also affect regional water supplies, creating potential water crises in the western United States within the next 20 years, as well as other areas such as Peru and western China. Just as many regions will experience longer and hotter summers, other areas will suffer from intense and increased rainfall, which will result in severe flooding and landslides.

Continued global warming will result in a rise in mean sea level, as icecaps and glaciers melt and contribute their water to the world's oceans. It is predicted that at the current rate of glacial melting, sea level will rise 21 cm by around 2050, thus increasing the number of people at risk from flooding in coastal areas by approximately 20 million!

We would be remiss, however, if we did not point out that many other scientists are not convinced that the global warming trend is the direct result of increased human activity related to industrialization. They indicate that although the level of greenhouse gases has increased, we are still uncertain about their rate of generation and rate of removal, and whether the rise in global temperatures during the past century resulted from normal climatic variations through time or from human activity. Furthermore, they conclude that even if a general global warming occurs during the next hundred years, it is not certain that the dire predictions made by proponents of global warming will come true.

Earth, as we know, is a remarkably complex system, with many feedback mechanisms and interconnections throughout its various subsystems and cycles. It is very difficult to predict all of the consequences that global warming would have for atmospheric and oceanic circulation patterns and its ultimate effect on Earth's biota.

However, it is important to remember that everyone is vulnerable to weather-related disasters; large-scale changes brought about by climate change will impact people in poor countries more than those in the more industrialized countries. However, whether these climate changes are part of a natural global cycle taking place over thousands or hundreds of thousands of years—that is, on a geological time scale—or are driven, in part, by human activities, is immaterial. The bottom line is that we already are, or eventually will be, affected in some way, be it economic or social, by the climate changes that are taking place.

1.7 Origin of the Universe and Solar System, and Earth's Place in Them

How did the universe begin? What has been its history? What is its eventual fate, or is it infinite? These are just some of the basic questions people have asked and wondered about since they first looked into the nighttime sky and saw the vastness of the universe beyond Earth.

Origin of the Universe—Did It Begin with a Big Bang?

Most scientists think that the universe originated about 14 billion years ago in what is popularly called the **Big Bang**. The Big Bang is a model for the evolution of the universe in which a dense, hot state was followed by expansion, cooling, and a less dense state.

According to modern *cosmology* (the study of the origin, evolution, and nature of the universe), the universe has no edge and therefore no center. Thus, when the universe began, all matter and energy were compressed into an infinitely small high-temperature and high-density state in which both time and space were set at zero. Therefore, there is no "before the Big Bang," only what occurred after it. As demonstrated by Einstein's Theory of Relativity, space and time are unalterably linked to form a space–time continuum—that is, without space, there can be no time.

How do we know that the Big Bang took place approximately 14 billion years ago? Why couldn't the universe have always existed as we know it today? Two fundamental phenomena indicate that the Big Bang occurred: (1) the universe is expanding, and (2) it is permeated by background radiation.

When astronomers look beyond our own solar system, they observe that everywhere in the universe galaxies are moving away from each other at tremendous speeds. Edwin Hubble first recognized this phenomenon in 1929. By measuring the optical spectra of distant galaxies, Hubble noted that the velocity at which a galaxy moves away from Earth increases proportionally to its distance from Earth. He observed that the spectral lines (wavelengths of light) of the galaxies are shifted toward the red end of the spectrum; that is, the lines are shifted toward longer wavelengths. Galaxies receding from each other at tremendous speeds would produce such a redshift. This is an example of the *Doppler effect*, which is a change in the frequency of a sound, light, or other wave caused by movement of the wave's source relative to the observer (▌ Figure 1.6).

One way to envision how velocity increases with increasing distance is by reference to the popular analogy of a rising loaf of raisin bread in which the raisins are uniformly distributed throughout the loaf (▌ Figure 1.7). As the dough rises, the raisins are uniformly pushed away from each other at velocities directly proportional to the distance between any two raisins. The farther away a given raisin is to begin with, the farther it must move to maintain the regular spacing during the expansion, and hence the greater its velocity must be. In the same way that raisins move apart in a rising loaf of bread, galaxies are receding from each other at a rate proportional to the distance between them, which is exactly what astronomers see when they observe the universe. By measuring this expansion rate, astronomers can calculate how long ago the galaxies were all together at a single point, which turns out to be about 14 billion years, the currently accepted age of the universe.

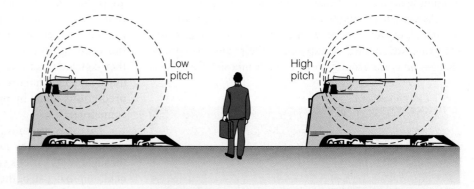

▌ **Figure 1.6** The Doppler Effect The sound waves of an approaching whistle are slightly compressed so that the individual hears a shorter-wavelength, higher-pitched sound. As the whistle passes and recedes from the individual, the sound waves are slightly spread out, and a longer-wavelength, lower-pitched sound is heard.

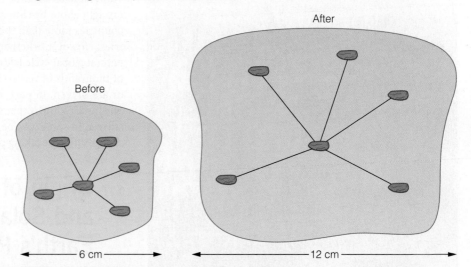

▌ **Figure 1.7** The Expanding Universe The motion of raisins in a rising loaf of raisin bread illustrates the relationship that exists between distance and speed and is analogous to an expanding universe. In this diagram, adjacent raisins are located 2 cm apart before the loaf rises. After one hour, any raisin is now 4 cm away from its nearest neighbor and 8 cm away from the next raisin over, and so on. Therefore, from the perspective of any raisin, its nearest neighbor has moved away from it at a speed of 2 cm per hour, and the next raisin over has moved away from it at a speed of 4 cm per hour. In the same way that raisins move apart in a rising loaf of bread, galaxies are receding from each other at a rate proportional to the distance between them.

Arno Penzias and Robert Wilson of Bell Telephone Laboratories made the second important observation that provided evidence of the Big Bang in 1965. They discovered that there is a pervasive background radiation of 2.7 Kelvin (K) above absolute zero (absolute zero equals −273°C; 2.7 K = −270.3°C) everywhere in the universe. This background radiation is thought to be the fading afterglow of the Big Bang.

Currently, cosmologists cannot say what it was like at time zero of the Big Bang because they do not understand the physics of matter and energy under such extreme conditions. However, it is thought that during the first second following the Big Bang, the four basic forces—(1) *gravity* (the attraction of one body toward another), (2) *electromagnetic force* (combines electricity and magnetism into one force and binds atoms into molecules), (3) *strong nuclear force* (binds protons and neutrons together), and (4) *weak nuclear force* (responsible for the breakdown of an atom's nucleus, producing radioactive decay)—separated and the universe experienced enormous expansion.

As the universe continued expanding and cooling, stars and galaxies began to form and the chemical makeup of the universe changed. Initially, the universe was 100% hydrogen and helium, whereas today it is 98% hydrogen and helium and 2% all other elements by weight.

How did such a change in the universe's composition occur? Throughout their life cycle, stars undergo many nuclear reactions in which lighter elements are converted into heavier elements by nuclear fusion. When a star dies, often explosively, the heavier elements that were formed in its core are returned to interstellar space and are available for inclusion in new stars. In this way, the composition of the universe is gradually enhanced by heavier elements.

Our Solar System—Its Origin and Evolution

Our solar system, which is part of the Milky Way galaxy, consists of the Sun, eight planets, one dwarf planet (Pluto), 101 known moons or satellites (although this number keeps changing with the discovery of new moons and satellites surrounding the Jovian planets), a tremendous number of asteroids—most of which orbit the Sun in a zone between Mars and Jupiter—and millions of comets and meteorites, as well as interplanetary dust and gases (■ Figure 1.8). Any theory formulated to explain the origin and evolution of our solar system must, therefore, take into account its various features and characteristics.

Many scientific theories for the origin of the solar system have been proposed, modified, and discarded since the French scientist and philosopher René Descartes first proposed, in 1644, that the solar system formed from a gigantic whirlpool within a universal fluid. Today, the **solar nebula theory** for the origin of our solar system involves the condensation and collapse of interstellar material in a spiral arm of the Milky Way galaxy (■ Figure 1.9).

The collapse of this cloud of gases and small grains into a counterclockwise-rotating disk concentrated about 90% of the material in the central part of the disk and formed an embryonic Sun, around which swirled a rotating cloud of material called a *solar nebula*. Within this solar nebula were localized eddies in which gases and solid particles condensed. During the condensation process, gaseous, liquid, and solid particles began to accrete into ever-larger masses called *planetesimals*, which collided and grew in size and mass until they eventually became planets.

The composition and evolutionary history of the planets are a consequence, in part, of their distance from the

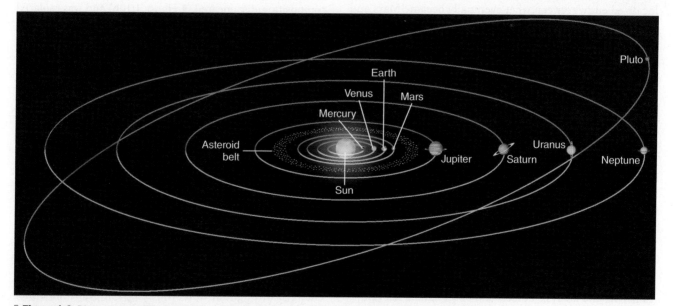

■ **Figure 1.8 Diagrammatic Representation of the Solar System** This representation of the solar system shows the planets and their orbits around the Sun. On August 24, 2006, the International Astronomical Union downgraded Pluto from planet to a dwarf planet. A dwarf planet has the same characteristics as a planet, except that it does not clear the neighborhood around its orbit. Pluto orbits among the icy debris of the Kuiper Belt, and therefore does not meet the criteria for a true planet.

a A huge rotating cloud of gas contracts and flattens

b to form a disk of gas and dust with the Sun forming in the center

c and eddies gathering up material to form planets.

▌ **Figure 1.9 Solar Nebula Theory** According to the currently accepted theory for the origin of our solar system, the planets and the Sun formed from a rotating cloud of gas.

Sun (see the Geo-Insight on pages 14 and 15). The **terrestrial planets**—Mercury, Venus, Earth, and Mars—so named because they are similar to *terra*, Latin for "earth," are all small and composed of rock and metallic elements that condensed at the high temperatures of the inner nebula. The **Jovian planets**—Jupiter, Saturn, Uranus, and Neptune—so named because they resemble Jupiter (the Roman god was also called *Jove*), all have small rocky cores compared to their overall size, and are composed mostly of hydrogen, helium, ammonia, and methane, which condense at low temperatures.

While the planets were accreting, material that had been pulled into the center of the nebula also condensed, collapsed, and was heated to several million degrees by gravitational compression. The result was the birth of a star, our Sun.

During the early accretionary phase of the solar system's history, collisions between various bodies were common, as indicated by the craters on many planets and moons. Asteroids probably formed as planetesimals in a localized eddy between what eventually became Mars and Jupiter in much the same way that other planetesimals formed the terrestrial planets. The tremendous gravitational field of Jupiter, however, prevented this material from ever accreting into a planet. Comets, which are interplanetary bodies composed of loosely bound rocky and icy materials, are thought to have condensed near the orbits of Uranus and Neptune.

The solar nebula theory for the formation of the solar system thus accounts for most of the characteristics of the planets and their moons, the differences in composition between the terrestrial and Jovian planets, and the presence of the asteroid belt. Based on the available data, the solar nebula theory best explains the features of the solar system and provides a logical explanation for its evolutionary history.

Recall from our earlier discussion that theories are formulated on the basis of the scientific method, that is, the gathering and analysis of facts and observations relevant to the problem, in this case the formation of the solar system. A hypothesis is then proposed and tested to account for the observed phenomenon, and if the hypothesis explains the phenomenon, then a theory is proposed. This is exactly how

the solar nebula theory came to explain the formation of the solar system. However, new observations or data may show that the theory is incomplete or doesn't explain some newly discovered aspect of the solar system, and, thus it will be modified. This is how science works.

Earth—Its Place in Our Solar System

Some 4.6 billion years ago, various planetesimals in our solar system gathered enough material together to form Earth and the other planets. Scientists think that this early Earth was probably cool, of generally uniform composition and density throughout, and composed mostly of silicates, compounds consisting of silicon and oxygen, iron and magnesium oxides, and smaller amounts of all the other chemical elements (▌ Figure 1.10a). Subsequently, when the combination of meteorite impacts, gravitational compression, and heat from radioactive decay increased the temperature of Earth enough to melt iron and nickel, this homogeneous composition disappeared (▌ Figure 1.10b) and was replaced by a series of concentric layers of differing composition and density, resulting in a differentiated planet (▌ Figure 1.10c).

This differentiation into a layered planet is probably the most significant event in Earth's history. Not only did it lead to the formation of a crust and eventually continents, but it also was probably responsible for the emission of gases from the interior that led to the formation of the oceans and atmosphere.

1.8 Why Earth Is a Dynamic and Evolving Planet

Earth is a dynamic planet that has continuously changed during its 4.6-billion-year existence. The size, shape, and geographic distribution of continents and ocean basins have changed throughout time; the composition of the atmosphere has evolved; and life-forms existing today differ from those that lived during the past. Mountains and hills have been worn away by erosion, and the forces of wind, water, and ice have sculpted a diversity of landscapes. Volcanic

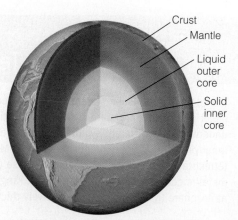

a Early Earth probably had a uniform composition and density throughout.

b The temperature of early Earth reached the melting point of iron and nickel, which, being denser than silicate minerals, settled to Earth's center. At the same time, the lighter silicates flowed upward to form the mantle and the crust.

c In this way, a differentiated Earth formed, consisting of a dense iron–nickel core, an iron-rich silicate mantle, and a silicate crust with continents and ocean basis.

Figure 1.10 Homogeneous Accretion Theory for the Formation of a Differentiated Earth

eruptions and earthquakes reveal an active interior, and folded and fractured rocks are testimony to the tremendous power of Earth's internal forces.

Earth consists of three concentric layers: the core, the mantle, and the crust (**Figure 1.11**). This orderly division results from density differences between the layers as a function of variations in composition, temperature, and pressure.

The **core** has a calculated density of 10–13 grams per cubic centimeter (g/cm^3) and occupies about 16% of Earth's total volume. Seismic (earthquake) data indicate that the core consists of a small, solid inner region and a larger, apparently liquid, outer portion. Both are thought to consist mostly of iron and a small amount of nickel.

The **mantle** surrounds the core and comprises about 83% of Earth's volume. It is less dense than the core (3.3–5.7 g/cm^3) and is thought to be composed mostly of *peridotite*, a dark, dense igneous rock containing abundant iron and magnesium. The mantle can be divided into three distinct zones based on physical characteristics. The lower mantle is solid and forms most of the volume of Earth's interior. The **asthenosphere** surrounds the lower mantle. It has the same composition as the lower mantle, but behaves plastically and flows slowly. Partial melting within the asthenosphere generates *magma* (molten material), some of which rises to the surface because it is less dense than the rock from which it was derived. The upper mantle surrounds the asthenosphere. The solid upper mantle and the overlying crust constitute the **lithosphere**, which is broken into numerous individual pieces called **plates** that move over the asthenosphere, partially as a result of underlying *convection cells* (**Figure 1.12**). Interactions of these plates are responsible for such phenomena as earthquakes, volcanic eruptions, and the formation of mountain ranges and ocean basins.

The **crust**, Earth's outermost layer, consists of two types. *Continental crust* is thick (20–90 km), has an average density of 2.7 g/cm^3, and contains considerable silicon and aluminum. *Oceanic crust* is thin (5–10 km), denser than continental crust (3.0 g/cm^3), and composed of the dark igneous rocks *basalt* and *gabbro*.

> **CONNECTION LINK**
>
> You can learn more about Earth's core, mantle, and crust, how they were discovered, and how we know their composition and physical properties in Chapter 9, pages 224–227.

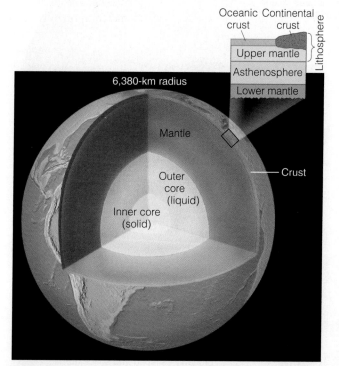

Figure 1.11 Cross Section of Earth Illustrating the Core, Mantle, and Crust The enlarged portion shows the relationship between the lithosphere (composed of the continental crust, oceanic crust, and solid upper mantle) and the underlying asthenosphere and lower mantle.

GEO-INSIGHT   The Terrestrial and Jovian Planets

The planets of our solar system can be divided into two major groups that are quite different, indicating that the two underwent very different evolutionary histories. The four inner planets—Mercury, Venus, Earth, and Mars—are the terrestrial planets; they are small and dense (composed of a metallic core and silicate mantle-crust), ranging from no atmosphere (Mercury) to an oppressively thick one (Venus).

The outer four planets (Pluto is now considered a dwarf planet)—Jupiter, Saturn, Uranus, and Neptune—are the Jovian planets; they are large, ringed, low-density planets with liquid cores surrounded by thick atmospheres.

► **1.** The relative sizes of the planets and the Sun. (Distances between planets are not to scale.)

100,000 km

► **2. Venus** is surrounded by an oppressively thick atmosphere that completely obscures its surface. However, radar images from orbiting spacecraft reveal a wide variety of terrains, including volcanic features, folded mountain ranges, and a complex network of faults.

▲ **3.** The **Moon** is one-fourth the diameter of Earth, has a low density relative to the terrestrial planets, and is extremely dry. Its surface is divided into low-lying dark-colored plains and light-colored highlands that are heavily cratered, attesting to a period of massive meteorite bombardment in our solar system more than 4 billion years ago. The hypothesis that best accounts for the origin of the Moon has a giant planetesimal, the size of Mars or larger, crashing into Earth 4.6 to 4.4 billion years ago, causing ejection of a large quantity of hot material that cooled and formed the Moon.

◄ **4. Mercury** has a heavily cratered surface that has changed very little since its early history. Because Mercury is so small, its gravitational attraction is insufficient to retain atmospheric gases; any atmosphere that it may have held when it formed probably escaped into space quickly.

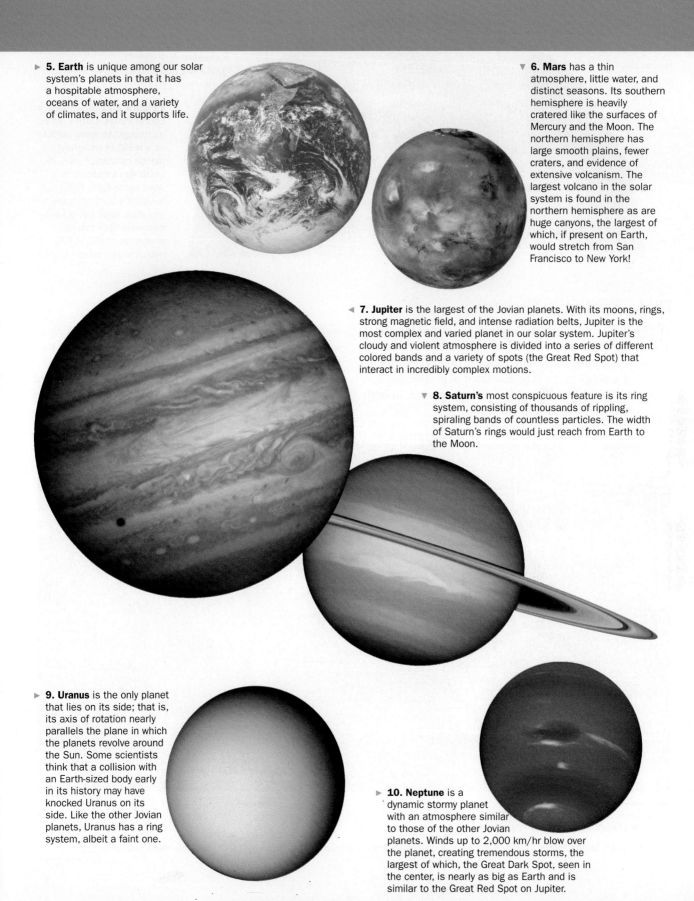

▶ **5. Earth** is unique among our solar system's planets in that it has a hospitable atmosphere, oceans of water, and a variety of climates, and it supports life.

▼ **6. Mars** has a thin atmosphere, little water, and distinct seasons. Its southern hemisphere is heavily cratered like the surfaces of Mercury and the Moon. The northern hemisphere has large smooth plains, fewer craters, and evidence of extensive volcanism. The largest volcano in the solar system is found in the northern hemisphere as are huge canyons, the largest of which, if present on Earth, would stretch from San Francisco to New York!

◀ **7. Jupiter** is the largest of the Jovian planets. With its moons, rings, strong magnetic field, and intense radiation belts, Jupiter is the most complex and varied planet in our solar system. Jupiter's cloudy and violent atmosphere is divided into a series of different colored bands and a variety of spots (the Great Red Spot) that interact in incredibly complex motions.

▼ **8. Saturn's** most conspicuous feature is its ring system, consisting of thousands of rippling, spiraling bands of countless particles. The width of Saturn's rings would just reach from Earth to the Moon.

▶ **9. Uranus** is the only planet that lies on its side; that is, its axis of rotation nearly parallels the plane in which the planets revolve around the Sun. Some scientists think that a collision with an Earth-sized body early in its history may have knocked Uranus on its side. Like the other Jovian planets, Uranus has a ring system, albeit a faint one.

▶ **10. Neptune** is a dynamic stormy planet with an atmosphere similar to those of the other Jovian planets. Winds up to 2,000 km/hr blow over the planet, creating tremendous storms, the largest of which, the Great Dark Spot, seen in the center, is nearly as big as Earth and is similar to the Great Red Spot on Jupiter.

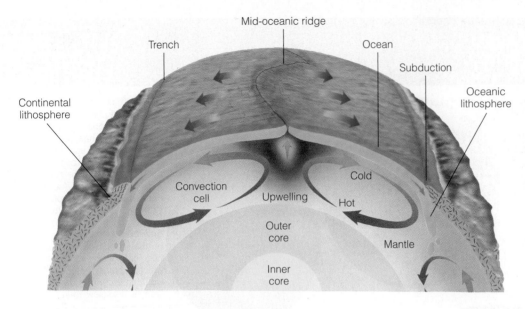

Figure 1.12 Movement of Earth's Plates Earth's plates are thought to move partially as a result of underlying mantle convection cells, in which warm material from deep within Earth rises toward the surface, cools, and then, upon losing heat, descends back into the interior, as shown in this diagrammatic cross section.

Plate Tectonic Theory

The recognition that the lithosphere is divided into rigid plates that move over the asthenosphere (∥ Figure 1.13) forms the foundation of **plate tectonic theory**, a unifying theory of geology that will be discussed in greater detail in Chapter 2. Zones of volcanic activity, earthquakes, or both mark most plate boundaries. Along these boundaries, plates separate (diverge), collide (converge), or slide sideways past each other (∥ Figure 1.14).

The acceptance of plate tectonic theory is recognized as a major milestone in the geologic sciences, comparable

Ridge axis
Divergent boundary

Transform fault
Transform boundary

Subduction zone
Convergent boundary

Zones of extension within continents

Uncertain plate boundary

∥ **Figure 1.13** Earth's Plates Earth's lithosphere is divided into rigid plates of various sizes that move over the asthenosphere.

Upwelling

Asthenosphere

Lithosphere

Upwelling

▌ Figure 1.14 **Relationship Between Lithosphere, Asthenosphere, and Plate Boundaries** An idealized block diagram illustrating the relationship between the lithosphere and the underlying asthenosphere and the three principal types of plate boundaries: divergent, convergent, and transform.

to the revolution that Darwin's theory of evolution caused in biology. Plate tectonics has provided a framework for interpreting the composition, structure, and internal processes of Earth on a global scale. The theory has led to the realization that the continents and ocean basins are part of a lithosphere–atmosphere–hydrosphere system that evolved together with Earth's interior.

A revolutionary concept when it was proposed in the 1960s, plate tectonic theory has had far-reaching consequences in all fields of geology because it provides the basis for relating many seemingly unrelated phenomena, such as the formation and occurrence of Earth's natural resources, as well as the distribution and evolution of the world's biota.

Furthermore, the impact of plate tectonic theory has been particularly notable in the interpretation of Earth's history. For example, the Appalachian Mountains in eastern North America and the mountain ranges of Greenland, Scotland, Norway, and Sweden are not the result of unrelated mountain-building episodes, but, rather, are part of a larger mountain-building event that involved the closing of an ancient Atlantic Ocean and the formation of the supercontinent Pangaea approximately 251 million years ago.

> **CONNECTION LINK**
>
> You can find more information on Earth's history in Chapters 19 through 23.

1.9 The Rock Cycle

A **rock** is an aggregate of **minerals**, which are naturally occurring, inorganic, crystalline solids that have definite physical and chemical properties. Minerals are composed of elements such as oxygen, silicon, and aluminum, and elements are made up of atoms, the smallest particles of matter that retain the characteristics of an element. More than 3,500 minerals have been identified and described, but only about a dozen make up the bulk of the rocks in Earth's crust.

Geologists recognize three major groups of rocks—*igneous*, *sedimentary*, and *metamorphic*—each of which is characterized by its mode of formation. The **rock cycle** is a pictorial representation of events leading to the origin, destruction and/or changes, and reformation of rocks as a consequence of Earth's internal and surface processes (▌ Figure 1.15). Furthermore, it shows that the three major rock groups are interrelated; that is, any rock type can be derived from the others.

Igneous rocks result when magma or lava crystallizes, or volcanic ejecta, such as ash, accumulate and consolidate. As magma cools, minerals crystallize, and the resulting rock is characterized by interlocking mineral grains. Magma that cools slowly beneath the surface produces *intrusive igneous rocks* (▌ Figure 1.16a); magma that cools at the surface produces *extrusive igneous rocks* (▌ Figure 1.16b).

▌**Figure 1.15** **The Rock Cycle** This cycle shows the interrelationships between Earth's internal and external processes and how the three major rock groups are related. An ideal cycle includes the events on the outer margin of the cycle, but interruptions, indicated by internal arrows, are common.

Rocks exposed at Earth's surface are broken into particles and dissolved by various weathering processes. The particles and dissolved materials may be transported by wind, water, or ice and eventually deposited as *sediment*. This sediment may then be compacted or cemented (lithified) into sedimentary rock.

Sedimentary rocks form in one of three ways: (1) consolidation of mineral or rock fragments, (2) precipitation of mineral matter from solution, or (3) compaction of plant or animal remains (▌ Figure 1.16c, d). Because sedimentary rocks form at or near Earth's surface, geologists can infer the environment in which they were deposited, the transporting agent, and perhaps even something about the source from which the sediments were derived (see Chapter 7). Accordingly, sedimentary rocks are especially useful for interpreting Earth history.

Metamorphic rocks result from the alteration of other rocks, usually beneath the surface, by heat, pressure, and the chemical activity of fluids. For example, marble—a rock preferred by many sculptors and builders—is a metamorphic rock produced when the agents of metamorphism are applied to the sedimentary rocks limestone or dolostone. Metamorphic rocks are either *foliated* (▌ Figure 1.16e) or *nonfoliated* (▌ Figure 1.16f). Foliation, the parallel alignment of minerals due to pressure, gives the rock a layered or banded appearance.

a **Granite,** an intrusive igneous rock.

b **Basalt,** an extrusive igneous rock.

c **Conglomerate,** a sedimentary rock formed by the consolidation of rounded rock fragments.

d **Limestone,** a sedimentary rock formed by the extraction of mineral matter from seawater by organisms or by the inorganic precipitation of the mineral calcite from seawater.

e **Gneiss,** a foliated metamorphic rock.

f **Quartzite,** a nonfoliated metamorphic rock.

▌**Figure 1.16** Hand Specimens of Common Igneous, Sedimentary, and Metamorphic Rocks

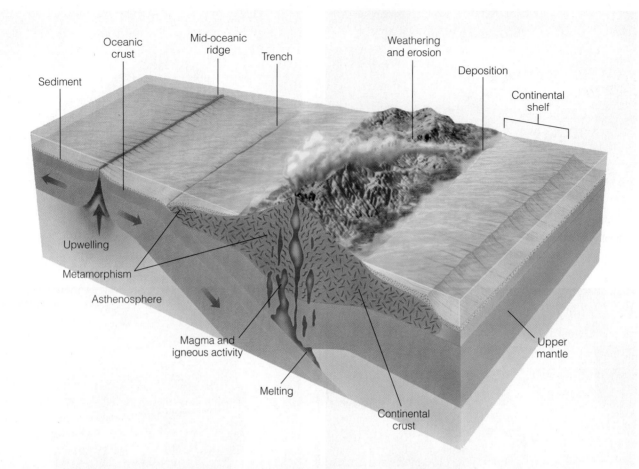

Figure 1.17 Plate Tectonics and the Rock Cycle Plate movement provides the driving mechanism that recycles Earth materials. This block diagram shows how the three major rock groups—igneous, sedimentary, and metamorphic—are recycled through both the continental and oceanic regions. Subducting plates are partially melted to produce magma, which rises and either crystallizes beneath Earth's surface as intrusive igneous rock, or spills out on the surface, solidifying as extrusive igneous rock. Rocks exposed at the surface are weathered and eroded to produce sediments that are transported and eventually lithified into sedimentary rocks. Metamorphic rocks result from pressure generated along converging plates or adjacent to rising magma.

How Are the Rock Cycle and Plate Tectonics Related?

Interactions between plates determine, to some extent, which of the three rock groups will form (Figure 1.17). For example, when plates converge, heat and pressure generated along the plate boundary may lead to igneous activity and metamorphism within the descending oceanic plate, thus producing various igneous and metamorphic rocks.

Some of the sediments and sedimentary rocks on the descending plate are melted, whereas other sediments and sedimentary rocks along the boundary of the nondescending plate are metamorphosed by the heat and pressure generated along the converging plate boundary. Later, the mountain range or chain of volcanic islands formed along the convergent plate boundary will be weathered and eroded, and the new sediments will be transported to the ocean, where they will be deposited, to begin yet another cycle.

The interrelationship between the rock cycle and plate tectonics is just one of many examples of how Earth's various subsystems and cycles are interrelated. Heating within Earth's interior results in convection cells that power the movement of plates and magma, which forms intrusive and extrusive igneous rocks. Movement along plate boundaries may result in volcanic activity, earthquakes, and, in some cases, mountain building. The interaction between the atmosphere, hydrosphere, and biosphere contributes to the weathering of rocks exposed on Earth's surface. Plates descending back into Earth's interior are subjected to increasing heat and pressure, which may lead to metamorphism, as well as the generation of magma and yet another recycling of materials.

1.10 Organic Evolution and the History of Life

Plate tectonic theory provides us with a model for understanding the internal workings of Earth and their effect on Earth's surface. The theory of **organic evolution** (whose central theses are that all present-day organisms are related and that they have descended with modifications from organisms that lived in the past) provides the conceptual framework for understanding the history of life. Together, the theories of plate tectonics and organic evolution have changed the way we view our planet, and we should not be surprised at the intimate association between them. Although the relationship between plate tectonic processes and the evolution of life is incredibly complex, paleontological data provide indisputable evidence of the influence of plate movement on the distribution of organisms.

> **CONNECTION LINK**
>
> You can learn more about Charles Darwin, the theory of evolution, and its supporting evidence in Chapter 18.

The publication in 1859 of Darwin's *On the Origin of Species by Means of Natural Selection* revolutionized biology and marked the beginning of modern evolutionary biology. With its publication, most naturalists recognized that evolution provided a unifying theory that explained an otherwise encyclopedic collection of biologic facts.

When Darwin proposed his theory of organic evolution, he cited a wealth of supporting evidence, including the way organisms are classified, embryology, comparative anatomy, the geographic distribution of organisms, and, to a limited extent, the fossil record. Furthermore, Darwin proposed that *natural selection*, which results in the survival to reproductive age of those organisms best adapted to their environment, is the mechanism that best accounts for evolution.

Perhaps the most compelling evidence in favor of evolution can be found in the fossil record. Just as the geologic record allows geologists to interpret physical events and conditions in the geologic past, **fossils**, which are the remains or traces of once-living organisms, not only provide evidence that evolution has occurred, but also demonstrate that Earth has a history extending beyond that recorded by humans. The succession of fossils in the rock record provides geologists with a

> **CONNECTION LINK**
>
> You can learn how the various principles of relative dating and the discovery of radioactivity played an important role in the construction of the geologic time scale, and its importance in placing climatic changes in a geologic time perspective in Chapter 17.

means for dating rocks and allowed for a relative geologic time scale to be constructed in the 1800s.

1.11 Geologic Time and Uniformitarianism

An appreciation of the immensity of geologic time is central to understanding the evolution of Earth and its biota. Indeed, time is one of the main aspects that sets geology apart from the other sciences, except astronomy. Most people have difficulty comprehending geologic time, because they tend to think in terms of the human perspective—seconds, hours, days, and years. Ancient history is what occurred hundreds or even thousands of years ago. When geologists talk of ancient geologic history, however, they are referring to events that happened hundreds of millions or even billions of years ago.

It is equally important to remember that Earth goes through cycles of much longer duration than the human perspective of time. Although they may have disastrous effects on the human species, global warming and cooling are part of a larger cycle that has resulted in numerous glacial advances and retreats during the past 2.6 million years. Because of their perspective on time and the interrelationships of Earth's various systems and cycles, geologists are in a unique position to make valuable contributions to many of the current environmental debates such as global warming and sea-level changes.

The **geologic time scale** subdivides geologic time into a hierarchy of increasingly shorter time intervals; each time subdivision has a specific name. The geologic time scale resulted from the work of many 19th-century geologists who pieced together information from numerous rock exposures and constructed a chronology based on changes in Earth's biota through time. Subsequently, with the discovery of radioactivity in 1895 and the development of various radiometric dating techniques, geologists have been able to assign numerical ages (also known as absolute ages) in years to the subdivisions of the geologic time scale (▮ Figure 1.18).

One of the cornerstones of geology is the **principle of uniformitarianism**, which is based on the premise that present-day processes have operated throughout geologic time. Therefore, to understand and interpret geologic events from evidence preserved in rocks, we must first understand present-day processes and their results. In fact, uniformitarianism fits completely with the system approach that we are following for the study of Earth.

Uniformitarianism is a powerful principle that allows us to use present-day processes as the basis for interpreting the past and for predicting potential future events. We should keep in mind, however, that uniformitarianism does not exclude sudden or catastrophic events such as volcanic eruptions, earthquakes, tsunami, landslides, or floods.

What uniformitarianism means is that even though the rates and intensities of geologic processes have varied during

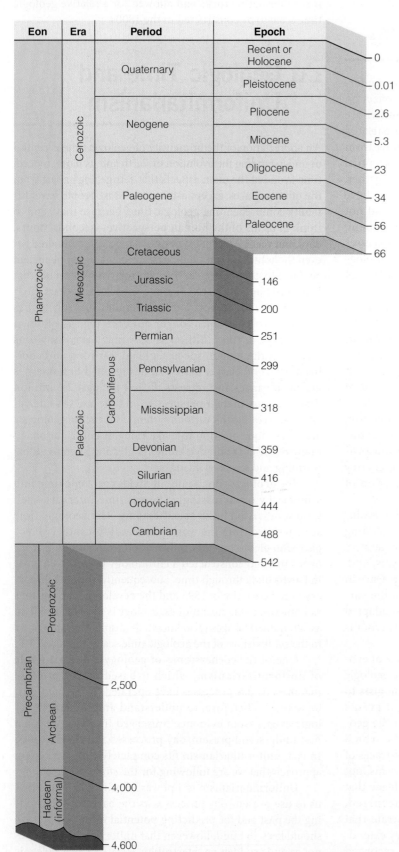

Eon	Era	Period	Epoch	
Phanerozoic	Cenozoic	Quaternary	Recent or Holocene	0
			Pleistocene	0.01
		Neogene	Pliocene	2.6
			Miocene	5.3
		Paleogene	Oligocene	23
			Eocene	34
			Paleocene	56
	Mesozoic	Cretaceous		66
		Jurassic		146
		Triassic		200
	Paleozoic	Permian		251
		Carboniferous / Pennsylvanian		299
		Carboniferous / Mississippian		318
		Devonian		359
		Silurian		416
		Ordovician		444
		Cambrian		488
				542
Precambrian	Proterozoic			
	Archean			2,500
	Hadean (informal)			4,000
				4,600

▌ **Figure 1.18 The Geologic Time Scale** The numbers to the right of the columns are the ages in millions of years before the present. Dates are from the 2009 International Stratigraphic Chart, © 2009 by the International Commission on Stratigraphy.

the past, the physical and chemical laws of nature have remained the same. Although Earth is in a dynamic state of change and has been ever since it formed, the processes that shaped it during the past are the same ones operating today.

1.12 How Does the Study of Geology Benefit Us?

The most meaningful lesson to learn from the study of geology is that Earth is an extremely complex planet in which interactions are taking place between its various subsystems and have been for the past 4.6 billion years. If we want to ensure the survival of the human species, we must understand how the various subsystems work and interact with each other and, more importantly, how our actions affect the delicate balance between these systems. We can do this, in part, by studying what has happened in the past, particularly on the global scale, and by using that information to determine how our actions might affect the delicate balance between Earth's various subsystems in the future.

The study of geology goes beyond learning numerous facts about Earth. In fact, we don't just study geology—we *live* it. Geology is an integral part of our lives. Our standard of living depends directly on our consumption of natural resources, which formed millions and billions of years ago. However, the way we consume natural resources and interact with the environment, as individuals and as a society, also determines our ability to pass on this standard of living to the next generation.

As you study the various topics covered in this book, keep in mind the themes discussed in this chapter and how, like the parts of a system, they are interrelated and responsible for the 4.6-billion-year history of Earth. View each chapter's topic in the context of how it fits in the whole Earth system, and remember that Earth's history is a continuum and the result of interaction among its various subsystems. By relating each chapter's topic to its place in the Earth system, you will gain a greater appreciation of why geology is so integral to our lives.

GEO-RECAP

» Geo-Image

One of the major environmental issues facing human-kind today is global warming and what effect, if any, human activity, such as the burning of fossil fuels, has had on rising global average temperatures. Proponents on both sides of the argument point to increasing mean global temperatures change from 1880 to 2008

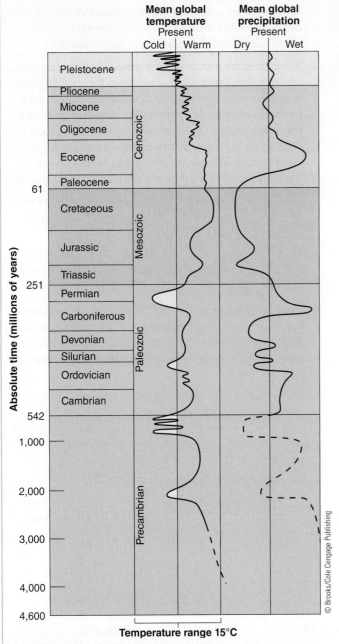

Figure 2 Mean Global Temperature and Precipitation Changes Throughout Earth History There have been several episodes of global cooling throughout Earth's history, notably at the end of the Precambrian (Proterozoic Eon), the end of the Ordovician Period, the Late Carboniferous to Permian periods, and a series of glacial

Figure 1 Mean Global Temperature Changes from 1880 to 2008 The zero line represents the average from 1951 to 1980. The plus and minus values represent deviations (in C°) from the average for the annual mean (in black) and 5-year mean (in red). Note the dramatic increase in temperature from approximately 1970 to 2008.

(▮ Figure 1), which is a historical time interval. On the other hand, mean global temperature and precipitation have changed on a geologic time scale (▮ Figure 2). In both cases, scientists use a variety of methods to determine historical and ancient climates (▮ Figure 3).

Discuss why an accurate geologic time scale is particularly important for geologists in examining changes in global temperatures during the past, and how an understanding of geologic time is crucial to the current debate on global warming and its consequences.

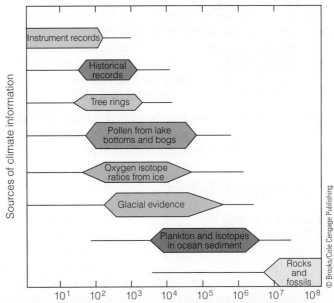

Figure 3 Sources of Climate Data Some of the methods used by scientists to determine historical and ancient climates. Each method has its own useful time range and accuracy within that time range.

Chapter Summary

- Earth can be viewed as a system of interconnected components that interact with and affect one another. The principal subsystems of Earth are the atmosphere, hydrosphere, biosphere, lithosphere, mantle, and core.

- Earth is considered a continually changing and dynamic planet because of the interactions among its various subsystems and cycles.

- Geology, the study of Earth, is divided into (1) physical geology, which is the study of Earth materials and the processes that operate both within Earth and on its surface; and (2) historical geology, which examines the origin and evolution of Earth, its continents, oceans, atmosphere, and life.

- The scientific method is an orderly, logical approach that involves gathering and analyzing facts about a particular phenomenon, formulating hypotheses to explain the phenomenon, testing the hypotheses, and finally proposing a theory. A theory is a testable explanation for some natural phenomenon that has a large body of supporting evidence.

- Geology not only is part of the human experience, examples of which can be found in art, music, and literature, but it also affects our daily lives as individuals, societies, and nation-states. A basic understanding of geology is critical in dealing with the many environmental problems and issues facing humankind.

- The universe began, in what is popularly called the Big Bang, approximately 14 billion years ago. Astronomers have deduced this age by observing that celestial objects are moving away from each other in an ever-expanding universe. Furthermore, the universe has a pervasive background radiation of 2.7 K above absolute zero (2.7 K = −270.3°C), which is thought to be the faint afterglow of the Big Bang.

- About 4.6 billion years ago, our solar system formed from a rotating cloud of interstellar matter. As this cloud condensed, it eventually collapsed under the influence of gravity and flattened into a counterclockwise-rotating disk. Within this rotating disk, the Sun, planets, and moons formed from the turbulent eddies of nebular gases and solids.

- Earth formed from a swirling eddy of nebular material 4.6 billion years ago, accreting as a solid body and soon thereafter differentiating into a layered planet.

- Earth's outermost layer is the crust, which is divided into continental and oceanic portions. The crust and underlying solid upper mantle, together known as the lithosphere, overlie the asthenosphere, a zone that behaves plastically and flows slowly. The asthenosphere is underlain by the solid lower mantle. Earth's core consists of an outer liquid portion and an inner solid portion.

- The lithosphere is divided into a series of plates that diverge, converge, and slide sideways past one another.

- Plate tectonic theory provides a unifying explanation for many geologic features and events. The interaction between plates is responsible for volcanic eruptions, earthquakes, the formation of mountain ranges and ocean basins, and the recycling of rock materials.

- The three major rock groups are igneous, sedimentary, and metamorphic. Igneous rocks result from the crystallization of magma or the consolidation of volcanic ejecta. Sedimentary rocks are typically formed by the consolidation of rock fragments, precipitation of mineral matter from solution, or compaction of plant or animal remains. Metamorphic rocks result from the alteration of other rocks, usually beneath Earth's surface, by heat, pressure, and chemically active fluids.

- The rock cycle illustrates the interactions between Earth's internal and external processes and how the three rock groups are interrelated.

- The central thesis of the theory of organic evolution is that all living organisms evolved (descended with modifications) from organisms that existed in the past.

- Time sets geology apart from the other sciences except astronomy, and an appreciation of the immensity of geologic time is central to understanding Earth's evolution. The geologic time scale is the calendar geologists use to date past events.

- The principle of uniformitarianism is basic to the interpretation of Earth history. This principle holds that the laws of nature have been constant through time and that the same processes operating today have operated in the past, although not necessarily at the same rates.

- Geology is an integral part of our lives. Our standard of living depends directly on our consumption of natural resources, resources that formed millions and billions of years ago.

Important Terms

asthenosphere (p. 13)
Big Bang (p. 10)
core (p. 13)
crust (p. 13)
fossil (p. 21)
geologic time scale (p. 21)
geology (p. 4)
hypothesis (p. 5)
igneous rock (p. 17)

Jovian planets (p. 12)
lithosphere (p. 13)
mantle (p. 13)
metamorphic rock (p. 18)
mineral (p. 17)
organic evolution (p. 21)
plate (p. 13)
plate tectonic theory (p. 16)
principle of uniformitarianism (p. 21)

rock (p. 17)
rock cycle (p. 17)
scientific method (p. 5)
sedimentary rock (p. 18)
solar nebula theory (p. 11)
system (p. 4)
terrestrial planets (p. 12)
theory (p. 4)

Review Questions

1. Rocks that result from the crystallization of magma beneath Earth's surface are
   a. ____ volcanic.
   b. ____ metamorphic.
   c. ____ igneous.
   d. ____ sedimentary.
   e. ____ none of these.

2. A combination of related parts interacting in an organized fashion is
   a. ____ a cycle.
   b. ____ a theory.
   c. ____ uniformitarianism.
   d. ____ a hypothesis.
   e. ____ a system.

3. The premise that present-day processes have operated throughout geologic time is the principle of
   a. ____ organic evolution.
   b. ____ plate tectonics.
   c. ____ uniformitarianism.
   d. ____ geologic time.
   e. ____ scientific deduction.

4. The change in frequency of a sound wave caused by movement of its source relative to an observer is known as the
   a. ____ Curie point.
   b. ____ Hubble shift.
   c. ____ Doppler effect.
   d. ____ Quasar force.
   e. ____ none of these.

5. The study of Earth materials, as well as the processes operating within Earth and on its surface is
   a. ____ geomorphology.
   b. ____ cosmology.
   c. ____ historical geology.
   d. ____ seismology.
   e. ____ physical geology.

6. Which of the following statements about a scientific theory is *not* true?
   a. ____ it is an explanation for some natural phenomenon.
   b. ____ predictive statements can be derived from it.
   c. ____ it is a conjecture or guess.
   d. ____ it has a large body of supporting evidence.
   e. ____ it is testable.

7. That all living organisms are the descendents of different life-forms that existed in the past is the central claim of
   a. ____ the principle of fossil succession.
   b. ____ plate tectonics.
   c. ____ the principle of uniformitarianism.
   d. ____ organic evolution.
   e. ____ none of these.

8. What two observations lead scientists to conclude that the Big Bang occurred approximately 14 billion years ago?
   a. ____ a steady-state universe and 2.7 K background radiation.
   b. ____ a steady-state universe and opaque background radiation.
   c. ____ an expanding universe and opaque background radiation.
   d. ____ an expanding universe and 2.7 K background radiation.
   e. ____ a shrinking universe and opaque background radiation.

9. The movement of plates is thought to result from
   a. ____ density differences between the inner and outer core.
   b. ____ rotation of the mantle around the core.
   c. ____ gravitational forces.
   d. ____ the Coriolis effect.
   e. ____ convection cells.

10. Plates are composed of
    a. ____ the crust and upper mantle.
    b. ____ the asthenosphere and upper mantle.
    c. ____ the crust and asthenosphere.
    d. ____ continental and oceanic crust only.
    e. ____ the core and mantle.

11. Why is viewing Earth as a system a good way to study it? Are humans a part of the Earth system? If so, what role, if any, do we play in Earth's evolution?

12. How does the solar nebula theory account for the formation of our solar system, its features, and evolutionary history?

13. Explain how the principle of uniformitarianism allows for catastrophic events.

14. Why is it important that everyone have a basic understanding of geology, even if they aren't going to become geologists?

15. What is the Big Bang? What evidence do we have that the universe began approximately 14 billion years ago?

16. Using plate movement as the driving mechanism of the rock cycle, explain how the three rock groups are related and how each rock group can be converted into a different rock group.

17. Why is plate tectonic theory a unifying theory of geology, and how does it fit into a systems approach to the study of Earth?

18. In what ways does geology affect our everyday lives at the individual, local, societal, and nation-state levels?

19. Why is Earth considered a dynamic planet?

20. Why do most scientists think overpopulation is the greatest environmental problem facing the world today? Do you agree? If not, what do you think is the greatest threat to our existence as a species?

People walking through the debris of houses in Iwaki, Fukushima Prefecture, northeastern Japan, that were destroyed by a 9.0-magnitude earthquake that struck on March 11, 2011. This earthquake and tsunami resulted from the interaction of two plates coming together along an oceanic-oceanic convergent plate boundary. The movement of plates is responsible for earthquakes, volcanic eruptions, and tsunami, particularly in the Pacific Ocean basin in what is aptly named the "Ring of Fire."

CHAPTER 2

Plate Tectonics: A Unifying Theory

OUTLINE

2.1 Introduction

2.2 Early Ideas About Continental Drift

2.3 What Is the Evidence for Continental Drift?

2.4 Features of the Seafloor

2.5 Earth's Magnetic Field

2.6 Paleomagnetism and Polar Wandering

2.7 Magnetic Reversals and Seafloor Spreading

2.8 Plate Tectonics: A Unifying Theory

2.9 The Three Types of Plate Boundaries

GEO-INSIGHT: Tectonics of the Terrestrial Planets

2.10 Hot Spots and Mantle Plumes

2.11 Plate Movement and Motion

2.12 The Driving Mechanism of Plate Tectonics

2.13 Plate Tectonics and the Distribution of Natural Resources

2.14 Plate Tectonics and the Distribution of Life

GEO-RECAP

HAVE YOU EVER WONDERED?

- If Earth looked different in the past than it does today in terms of the position of continents and ocean basins?

- Why such natural disasters as earthquakes, volcanic eruptions, and tsunami always seem to occur repeatedly in the same areas of the world?

- If there is a unifying theory in geology, similar to the theory of evolution in biology?

- Why there is so much oil in the Middle East?

- Why there is a seemingly uneven distribution of natural resources throughout the world?

- Why the present distribution of plants and animals is not random?

Kyodo/Landov

2.1 Introduction

Imagine it is the day after Christmas, December 26, 2004, and you are vacationing on a beautiful beach in Thailand. You look up from the book you're reading to see the sea suddenly retreat from the shoreline, exposing a vast expanse of seafloor that had moments before been underwater and teeming with exotic and colorful fish. It is hard to believe that within minutes of this unusual event, a powerful tsunami will sweep over your resort and everything in its path for several kilometers inland. Within hours, the coasts of Indonesia, Sri Lanka, India, Thailand, Somalia, Myanmar, Malaysia, and the Maldives will be inundated by the deadliest tsunami in history. More than 220,000 people will die, and the region will incur billions of dollars in damage.

Now go forward to October 25, 2010, when Mount Merapi in Central Java, Indonesia, began a series of violent eruptions that lasted into late November. Huge quantities of ash, lava, and ash flows were emitted during these eruptions. Although warnings were issued, at least 350 people were killed by the eruptions, and more than 350,000 were evacuated from areas around the volcano.

Almost five months later, on March 11, 2011, a catastrophic earthquake struck Japan, killing more than 13,000 people, and leaving thousands injured and homeless. This devastating earthquake also generated a tsunami that sent walls of water up to 37 m high in some locations, crashing into the northeast shores of the island, and causing further damage and casualities, as well as forcing tens of thousands of people to flee their homes. This magnitude 9.0 earthquake makes it the world's fourth largest since 1900, and the largest in Japan in the last 130 years.

What do these three recent tragic events have in common? They are part of the dynamic interactions involving Earth's plates. When two plates come together, one plate is pushed or pulled under the other plate, triggering large earthquakes such as the one that shook Pakistan in 2005, Indonesia in 2006, Haiti in 2010, and New Zealand and Japan in 2011. If conditions are right, earthquakes can produce a tsunami such as the one in Indonesia in 2004, and Japan in 2011.

As the descending plate moves downward and is assimilated into Earth's interior, magma is generated. Being less dense than the surrounding material, the magma rises toward the surface, where it may erupt as a volcano. It therefore should not be surprising that the distribution of volcanoes and earthquakes closely follows plate boundaries.

If you're like most people, you probably have only a vague notion of what plate tectonic theory is. Yet plate tectonics affects all of us. Volcanic eruptions, earthquakes, and tsunami are the result of interactions between plates. Global weather patterns and oceanic currents are caused, in part, by the configuration of the continents and ocean basins. The formation and distribution of many natural resources are related to plate movement and, thus, have an impact on the economic well-being and political decisions of nations. It is therefore important to understand this unifying theory,

not only because it affects us as individuals and as citizens of nation-states, but also because it ties together many aspects of the geology you will be studying.

2.2 Early Ideas About Continental Drift

The idea that Earth's past geography was different from today's is not new. The earliest maps showing the east coast of South America and the west coast of Africa probably provided people with the first evidence that continents may have once been joined, then broken apart and moved to their present positions. As far back as 1620, Sir Francis Bacon commented on the similarity of the shorelines of western Africa and eastern South America. However, he did not make the connection that the Old and New Worlds might once have been joined.

During the late 19th century, the Austrian geologist Edward Suess noted the similarities between the Late Paleozoic plant fossils of India, Australia, South Africa, and South America, as well as evidence of glaciation in the rock sequences of these continents. The plant fossils comprise a unique flora that occurs in the coal layers just above the glacial deposits of these southern continents. This flora is very different from the contemporaneous coal swamp flora of the northern continents and is collectively known as the *Glossopteris* **flora** after its most conspicuous genus (❙ Figure 2.1).

Suess also proposed the name *Gondwanaland* (or **Gondwana** as we will use here) for a supercontinent composed of the aforementioned southern continents. Abundant fossils of the *Glossopteris* flora are found in coal beds in Gondwana, a province in India. Suess thought these southern continents

❙ **Figure 2.1 Fossil *Glossopteris* Leaves** Plant fossils, such as these *Glossopteris* leaves from the Upper Permian Dunedoo Formation in Australia, are found on all five of the Gondwana continents. Their presence on continents with widely varying climates today is evidence that the continents were at one time connected. The distribution of the plants at that time was in the same climatic latitudinal belt.

were at one time connected by land bridges over which plants and animals migrated. Thus, in his view, the similarities of fossils on these continents were due to the appearance and disappearance of the connecting land bridges.

Alfred Wegener and the Continental Drift Hypothesis

Alfred Wegener, a German meteorologist (■ Figure 2.2), is generally credited with developing the hypothesis of **continental drift**. In his monumental book, *The Origin of Continents and Oceans* (first published in 1915), Wegener proposed that all landmasses were originally united in a single supercontinent that he named **Pangaea**, from the Greek meaning "all land." Wegener portrayed his grand concept of continental movement in a series of maps showing the breakup of Pangaea and the movement of the various continents to their present-day locations. Wegener amassed a tremendous amount of geologic, paleontologic, and climatologic evidence in support of continental drift; however, initial reaction of scientists to his then-heretical ideas can best be described as mixed.

Nevertheless, the South African geologist Alexander du Toit further developed Wegener's arguments and gathered more geologic and paleontologic evidence in support of continental drift. In 1937, du Toit published *Our Wandering Continents*, in which he contrasted the glacial deposits of Gondwana with coal deposits of the same age found in the continents of the Northern Hemisphere. To resolve this apparent climatologic paradox, du Toit moved the Gondwana continents to the South Pole and brought the northern continents together such that the coal deposits were located at the equator. He named this northern landmass **Laurasia**. It consisted of present-day North America, Greenland, Europe, and Asia (except for India).

2.3 What Is the Evidence for Continental Drift?

What then was the evidence Wegener, du Toit, and others used to support the hypothesis of continental drift? It includes the fit of the shorelines of continents, the appearance of the same rock sequences and mountain ranges of the same age on continents now widely separated, the matching of glacial deposits and paleoclimatic zones, and the similarities of many extinct plant and animal groups whose fossil remains are found today on widely separated continents. Wegener and his supporters argued that this vast amount of evidence from a variety of sources surely indicated that the continents must have been close together in the past.

Continental Fit

Wegener, like some before him, was impressed by the close resemblance between the coastlines of continents on opposite sides of the Atlantic Ocean, particularly South America

■ **Figure 2.2** Alfred Wegener Alfred Wegener, a German meteorologist, proposed the continental drift hypothesis in 1912 based on a tremendous amount of geologic, paleontologic, and climatologic evidence. He is shown here waiting out the Arctic winter in an expedition hut in Greenland.

Bildarchiv Preussischer Kulturbesitz/Art Resource, NY

and Africa. He cited these similarities as partial evidence that the continents were at one time joined together as a supercontinent that subsequently split apart. As his critics pointed out, though, the configuration of coastlines results from erosional and depositional processes and therefore is continuously being modified. So, even if the continents had separated during the Mesozoic Era, as Wegener proposed, it is not likely that the coastlines would fit exactly.

A more realistic approach is to fit the continents together along the continental slope where erosion would be minimal. In 1965, Sir Edward Bullard, an English geophysicist, and two associates showed that the best fit between the continents occurs at a depth of about 2,000 m (■ Figure 2.3). Since then, other reconstructions using the latest ocean basin data have confirmed the close fit between continents when they are reassembled to form Pangaea.

Similarity of Rock Sequences and Mountain Ranges

If the continents were at one time joined, then the rocks and mountain ranges of the same age in adjoining locations on the opposite continents should closely match. Such is the case for the Gondwana continents (■ Figure 2.4). Marine, nonmarine, and glacial rock sequences of the Pennsylvanian to Jurassic periods are almost identical on all five Gondwana continents, strongly indicating that they were joined at one time.

Furthermore, the trends of several major mountain ranges also support the hypothesis of continental drift. These mountain ranges seemingly end at the coastline of one continent only to apparently continue on another continent across the ocean. The folded Appalachian Mountains of North America, for example, trend northeastward through the eastern United States and Canada and terminate abruptly at the Newfoundland coastline. Mountain ranges of the same age and deformational style are found in eastern Greenland, Ireland, Great Britain, and Norway. So, even though the

Figure 2.3 Continental Fit When continents are placed together based on their outlines, the best fit isn't along their present-day coastlines but, rather, along the continental slope at a depth of about 2,000 m.

Appalachian Mountains and their equivalent-age mountain ranges in Great Britain are currently separated by the Atlantic Ocean, they form an essentially continuous mountain range when the continents are positioned next to each other as they were during the Paleozoic Era.

Glacial Evidence

During the Late Paleozoic Era, massive glaciers covered large continental areas of the Southern Hemisphere. Evidence for this glaciation includes layers of till (sediments deposited by glaciers) and striations (scratch marks) in the bedrock beneath the till (▊ Figure 2.5). Fossils and sedimentary rocks of the same age from the Northern Hemisphere, however, give no indication of glaciation. Fossil plants found in coals indicate that the Northern Hemisphere had a tropical climate during the time that the Southern Hemisphere was glaciated.

All of the Gondwana continents except Antarctica are currently located near the equator in subtropical to tropical climates. Mapping of glacial striations in bedrock in Australia, India, and South America indicates that the glaciers moved from the areas of the present-day oceans onto land. This would be highly unlikely because large continental glaciers (such as occurred on the Gondwana continents during the Late Paleozoic Era) flow outward from their central area of accumulation toward the sea.

Crystalline basement rocks	Coal beds	Basalt lava flows	Glacial deposits
Sandstone	Jurassic (Jʀ)	Triassic (Tʀ)	Permian (P)
Pennsylvanian (UC)	Carboniferous (C) (Mississippian and Pennsylvanian)	Devonian (D)

Figure 2.4 Similarity of Rock Sequences on the Gondwana Continents Sequences of marine, nonmarine, and glacial rocks of Pennsylvanian (UC) to Jurassic (Jʀ) age are nearly the same on all five Gondwana continents (South America, Africa, India, Australia, and Antarctica). These continents are widely separated today and have different environments and climates ranging from tropical to polar. Thus, the rocks forming on each continent are very different. When the continents were all joined in the past, however, the environments of adjacent continents were similar, and the rocks forming in those areas were similar. The range indicated by G in each column is the age range (Carboniferous-Permian) of the *Glossopteris* flora.

Figure 2.5 Glacial Evidence Indicating Continental Drift

a When the Gondwana continents are placed together so that South Africa is located at the South Pole, the glacial movements indicated by striations (red arrows) found on rock outcrops on each continent make sense. In this situation, the glacier (white area) is located in a polar climate and has moved radially outward from its thick central area toward its periphery.

b Glacial striations (scratch marks) on an outcrop of Permian-age bedrock exposed at Hallet's Cove, Australia, indicate the general direction of glacial movement more than 200 million years ago. As a glacier moves over a continent's surface, it grinds and scratches the underlying rock. The glacial striations that are preserved on a rock's surface thus provide evidence of the direction (red arrows) the glacier moved at that time.

If the continents did not move during the past, one would have to explain how glaciers moved from the oceans onto land and how large-scale continental glaciers formed near the equator. But if the continents are reassembled as a single landmass with South Africa located at the South Pole, the direction of movement of Late Paleozoic continental glaciers makes sense (Figure 2.5a). Furthermore, this geographic arrangement places the northern continents nearer the tropics, which is consistent with the fossil and climatologic evidence from Laurasia.

Fossil Evidence

Some of the most compelling evidence for continental drift comes from the fossil record (❚ Figure 2.6). Fossils of the *Glossopteris* flora are found in equivalent Pennsylvanian- and Permian-age coal deposits on all five Gondwana continents. The *Glossopteris* flora is characterized by the seed fern *Glossopteris* (Figure 2.1) as well as by many other distinctive and easily identifiable plants. Pollen and spores of plants can be dispersed over great distances by wind; however, *Glossopteris*-type plants produced seeds that are too large to have been carried by winds. Even if the seeds had floated across the ocean, they probably would not have remained viable for any length of time in saltwater.

The present-day climates of South America, Africa, India, Australia, and Antarctica range from tropical to polar and are much too diverse to support the type of plants in the *Glossopteris* flora. Wegener therefore reasoned that these continents must once have been joined so that these widely separated localities were all in the same latitudinal climatic belt (Figure 2.6).

The fossil remains of animals also provide strong evidence for continental drift. One of the best examples is *Mesosaurus*, a freshwater reptile whose fossils are found in Permian-age rocks in certain regions of Brazil and South Africa and nowhere else in the world (Figure 2.6). Because the physiologies of freshwater and marine animals are completely different, it is hard to imagine how a freshwater reptile could have swum across the Atlantic Ocean and found a freshwater environment nearly identical to its former habitat. Moreover, if *Mesosaurus* could have swum across the ocean, its fossil remains should be widely dispersed. It is more logical to assume that *Mesosaurus* lived in lakes in what were once adjacent areas of South America and Africa when it was united into a single continent. Discoveries of fossils from other land-dwelling animals from various other Gondwana continents further solidify the argument that these landmasses were in proximity during the Early and Middle Mesozoic Era.

Notwithstanding all the empirical evidence presented by Wegener and later by du Toit and others, most geologists simply refused to entertain the idea that continents might have moved in the past. The geologists were not necessarily being obstinate about accepting new ideas; rather, they found the evidence for continental drift inadequate and unconvincing. In part, this was because no one could provide a suitable mechanism to explain how continents could move over Earth's surface.

Interest in continental drift waned until new evidence from oceanographic research and studies of Earth's magnetic field showed that the present-day ocean basins were not as old as the continents, but were geologically young features that resulted from the breakup of Pangaea.

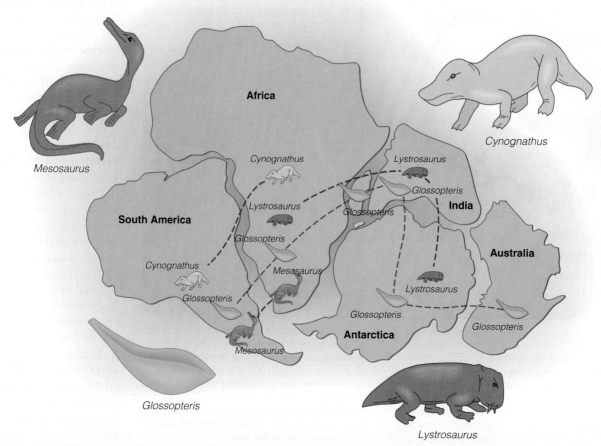

▌**Figure 2.6** **Fossil Evidence Supporting Continental Drift** Some of the plants and animals whose fossils are found today on the widely separated continents of South America, Africa, India, Australia, and Antarctica. During the Late Paleozoic Era, these continents were joined to form Gondwana, the southern landmass of Pangaea. Plants of the *Glossopteris* flora are found on all five continents, which today have widely different climates; however, during the Pennsylvanian and Permian periods, they were all located in the same general climatic belt. *Mesosaurus* is a freshwater reptile whose fossils are found only in similar nonmarine Permian-age rocks in Brazil and South Africa. *Cynognathus* and *Lystrosaurus* are land reptiles that lived during the Early Triassic Period. Fossils of *Cynognathus* are found in South America and Africa, whereas fossils of *Lystrosaurus* have been recovered from Africa, India, and Antarctica. It is hard to imagine how a freshwater reptile and land-dwelling reptiles could have swum across the wide oceans that presently separate these continents. It is more logical to assume that the continents were once connected.

2.4 Features of the Seafloor

At this point, it is useful to discuss some of the various features of Earth's seafloor. Many of the topographic features found on the seafloor and along the continental margins are the manifestations of Earth's internal processes and activity taking place along plate margins. Thus, it is important to know how these features relate to plate tectonic theory.

Most people think of continents as land areas outlined by the oceans; however, the true geologic margin of a continent—where granitic continental crust changes to basalt and gabbro oceanic crust—is below sea level. A **continental margin** is made up of a gently sloping continental shelf, a more steeply inclined continental slope, and, in some cases, a deeper, gently sloping continental rise (▌Figure 2.7). Seaward

of the continental margin lies the deep ocean basin. Thus, the continental margins extend to increasingly greater depths until they merge with the deep seafloor. Continental crust changes to oceanic crust somewhere beneath the continental rise, so part of the continental slope and the continental rise actually rest on oceanic crust.

The Continental Shelf, Slope, and Rise

As one proceeds seaward from the shoreline across the continental margin, the first area encountered is the gently sloping **continental shelf** lying between the shore and the more steeply dipping continental slope (Figure 2.7). The width of the continental shelf varies considerably, ranging from a few

▌Figure 2.7 **Features of Continental Margins** A generalized profile showing features of the continental margins. The vertical dimensions of the features in this profile are greatly exaggerated, because the vertical and horizontal scales differ.

tens of meters to more than 1,000 km; the shelf terminates where the inclination of the seafloor increases abruptly from 1 degree or less to several degrees.

The seaward margin of the continental shelf is marked by the *shelf–slope break* (at an average depth of 135 m), where the more steeply inclined **continental slope** begins (Figure 2.7). In most areas around the margins of the Atlantic, the continental slope merges with a more gently sloping **continental rise**. This rise is absent around the margins of the Pacific, where continental slopes descend directly into an oceanic trench (Figure 2.7).

Abyssal Plains, Oceanic Ridges, Submarine Hydrothermal Vents, and Oceanic Trenches

Beyond the continental rises are **abyssal plains**, flat surfaces covering vast areas of the seafloor. In some areas, they are interrupted by peaks rising more than 1 km, but abyssal plains are nevertheless the flattest, most featureless areas on Earth (Figure 2.7). Their flatness is a result of sediment deposition covering the rugged topography of the seafloor.

A renewed interest in oceanographic research led to extensive mapping of the ocean basins during the 1960s. Such mapping revealed an **oceanic ridge** system more than 65,000 km long, constituting the most extensive mountain range in the world (Figure 2.7). This system runs from the Arctic Ocean through the middle of the Atlantic and curves around South Africa, where the Indian Ridge continues into the Indian Ocean; the Atlantic–Pacific Ridge extends eastward, and a branch of it, the East Pacific Rise, trends northeast until it reaches the Gulf of California. Perhaps the best-known part of the ridge system is the Mid-Atlantic Ridge, which divides the Atlantic Ocean basin into two nearly equal parts (see Figure 1.13).

Oceanic ridges are composed almost entirely of the igneous rocks basalt and gabbro and possess features produced by tensional forces. Thus, they are the sites where new oceanic crust is generated and plates move away from each other along divergent plate boundaries.

First seen on the seafloor in 1979, **submarine hydrothermal vents** are found at or near spreading ridges. Here, cold seawater seeps through oceanic crust, is heated by the hot rocks at depth, and then rises and discharges into the seawater as plumes of hot water with temperatures as high as 400°C. Many of the plumes are black because dissolved minerals give them the appearance of black smoke—hence the name *black smoker* (▌ Figure 2.8).

Submarine hydrothermal vents are interesting from the biologic, geologic, and economic points of view. Near the vents live communities of organisms, such as bacteria, crabs, mussels, starfish, and tube worms, many of which have never been seen before. No sunlight is available, so these organisms depend on bacteria that oxidize sulfur compounds for their ultimate source of nutrients. The vents are also interesting because of their economic potential. The heated seawater reacts with oceanic crust, transforming it into a metal-rich solution that discharges into seawater and cools, precipitating iron, copper, and zinc sulfides and other minerals (Figure 2.8a). A chimney-like vent forms that eventually collapses and forms a mound of sediments rich in the elements just mentioned (Figure 2.8b).

Oceanic trenches are long, steep-sided depressions on the seafloor near convergent plate boundaries and constitute no more than 2% of the seafloor (Figure 2.7). However, it is here that oceanic lithosphere is consumed by subduction; that is, oceanic lithosphere plunges into Earth's interior along convergent plate boundaries (see Figure 1.14). It should also be noted that the greatest oceanic depths are found in trenches; the Challenger Deep of the Marianas Trench in the Pacific is more than 11,000 m deep!

❚ Figure 2.8 Submarine Hydrothermal Vents

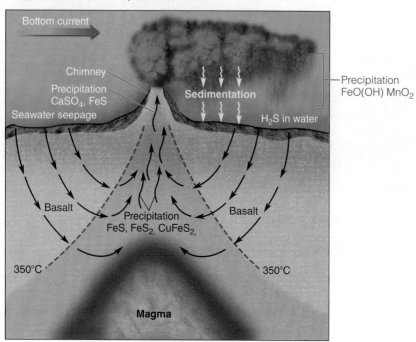

a Cross section showing the origin of a submarine hydrothermal vent called a black smoker.

b This black smoker on the East Pacific Rise is at a depth of 2,800 m. The plume of "black smoke" is heated seawater with dissolved minerals.

Seamounts, Guyots, and Aseismic Ridges

Except for the abyssal plains, the seafloor is not a flat, featureless expanse. In fact, a large number of volcanic hills, seamounts, and guyots rise above the seafloor in all ocean basins and are particularly abundant in the Pacific. All are of volcanic origin and differ mostly in size. *Seamounts* rise more than 1 km above the seafloor, and if flat-topped, they are called *guyots* (❚ Figure 2.9). Guyots are volcanoes that originally extended above sea level. However, as the plate upon which they were located continued to move, they were carried away from a spreading ridge, and as the oceanic crust cooled, it descended to greater depths. Thus, what was once an island slowly sank beneath the sea, and as it did, wave erosion produced the typical flat-topped appearance (Figure 2.9).

Other common features in the ocean basins are long, narrow ridges and broad, plateau-like features rising as much as 2 to 3 km above the surrounding seafloor. These *aseismic ridges* are so called because they lack seismic (earthquake) activity (see Hot Spots and Mantle Plumes on page 48). A few of these ridges are probably small fragments separated from continents during rifting and are referred to as *microcontinents*. Avalonia is a good example of a Paleozoic microcontinent (see Figure 20.2b).

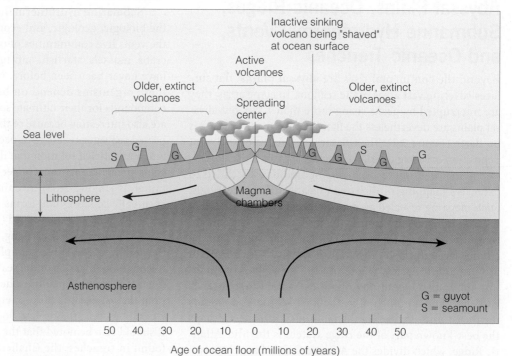

❚ Figure 2.9 The Origin of Seamounts and Guyots As the plate on which a volcano rests moves into greater water depths, the submerged volcanic island is called a *seamount*. Those that are flat-topped are called *guyots*.

2.5 Earth's Magentic Field

Having just discussed those features of the seafloor that are related to plate movement, we now turn our attention to the phenomenon of magnetism and its role in the formulation of plate tectonic theory.

Magnetism is a physical phenomenon resulting from the spin of electrons in some solids—particularly those of iron—and moving electricity. A **magnetic field** is an area in which magnetic substances such as iron are affected by lines of magnetic force emanating from a magnet (❙ Figure 2.10). The magnetic field shown in Figure 2.10 is *dipolar*, meaning that it possesses two unlike magnetic poles referred to as the north and south poles.

Although Earth's interior is too hot for a permanent magnet to exist, it is thought that thermal and compositional convection within the liquid metallic outer core, coupled with Earth's rotation, produce complex electrical currents (known as a *self-exciting dynamo*) that, in turn, generate the magnetic field. A useful analogy is to think of Earth as a giant dipole magnet in which the magnetic poles are in proximity to the geographic poles (❙ Figure 2.11). This arrangement means that the strength of the magnetic field is not constant, but varies.

Notice in Figure 2.11 that the lines of magnetic force around Earth parallel its surface only near the equator, just as the iron filings do around a bar magnet (Figure 2.10). As the lines of force approach the poles, they are oriented at increasingly larger angles with respect to the surface, and the strength of the magnetic field increases; it is strongest at the poles and weakest at the equator.

Another important aspect of the magnetic field is that the magnetic poles, where the lines of force leave and enter Earth, do not coincide with the geographic (rotational) poles. Currently, an 11.5-degree angle exists between the two (Figure 2.11). Studies of Earth's magnetic field show that the locations of the magnetic poles vary slightly over time, but that they still correspond closely, on average, with the locations of the geographic poles.

❙ **Figure 2.10** **Magnetic Field** Iron filings align along the lines of magnetic force radiating from a bar magnet.

❙ **Figure 2.11** Earth's Magnetic Field

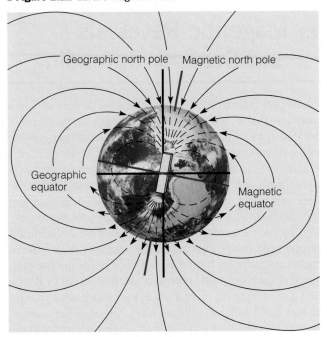

a Earth's magnetic field has lines of force like those of a bar magnet.

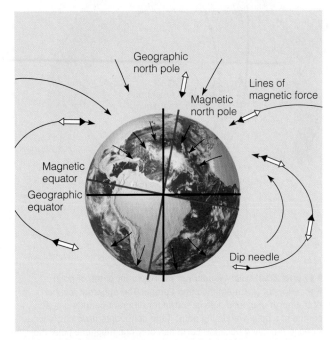

b The strength of the magnetic field changes from the magnetic equator to the magnetic poles. This change in strength causes a dip needle (a magnetic needle that is balanced on the tip of a support so that it can freely move vertically) to be parallel to Earth's surface only at the magnetic equator, where the strength of the magnetic north and south poles are equally balanced. Its inclination or dip with respect to Earth's surface increases as it moves toward the magnetic poles, until it is at 90 degrees or perpendicular to Earth's surface at the magnetic poles.

2.6 Paleomagnetism and Polar Wandering

Interest in continental drift revived during the 1950s as a result of evidence from paleomagnetic studies, a relatively new discipline at the time. **Paleomagnetism** is the remanent magnetism in ancient rocks recording the direction and intensity of Earth's magnetic poles at the time of the rock's formation.

When magma cools, the magnetic iron-bearing minerals align themselves with Earth's magnetic field, recording both its direction and strength. The temperature at which iron-bearing minerals gain their magnetization is called the **Curie point**. As long as the rock is not subsequently heated above the Curie point, it will preserve that remanent magnetism. Thus, an ancient lava flow provides a record of the orientation and strength of Earth's magnetic field at the time the lava flow cooled.

As paleomagnetic research progressed during the 1950s, some unexpected results emerged. When geologists measured the paleomagnetism of geologically recent rocks from different continents, they found that it was generally consistent with Earth's current magnetic field.

The paleomagnetism of ancient rocks, though, showed different orientations. For example, paleomagnetic studies of Silurian lava flows in North America indicated that the north magnetic pole was located in the western Pacific Ocean at that time, whereas the paleomagnetic evidence from Permian lava flows pointed to yet another location in Asia. When plotted on a map, the paleomagnetic readings of numerous lava flows from all ages in North America trace the apparent movement of the magnetic pole (called *polar wandering*) through time (❚ Figure 2.12).

Furthermore, analysis of lava flows from all continents indicated that each continent seemingly had its own series of magnetic poles! How could this be? Does it really mean there were different north magnetic poles for each continent? That would be highly unlikely and difficult to reconcile with the theory accounting for Earth's magnetic field.

The best explanation for such data is that the magnetic poles have remained near their present locations at the geographic north and south poles and the continents have moved. When the continental margins are fit together so that the paleomagnetic data point to only one magnetic pole, we find, just as Wegener did, that the rock sequences and glacial deposits match and that the fossil evidence is consistent with the reconstructed paleogeography.

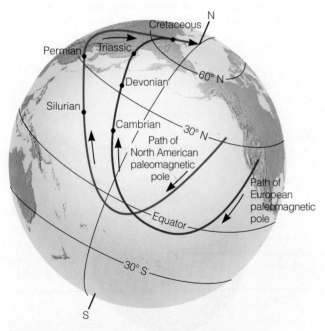

❚ **Figure 2.12 Polar Wandering** The apparent paths of polar wandering for North America and Europe. The apparent location of the north magnetic pole is shown for different periods on each continent's polar wandering path. If the continents have not moved through time, and because Earth has only one magnetic pole, the paleomagnetic readings for the same time in the past, taken on different continents, should all point to the same location. However, the north magnetic pole has different locations for the same time in the past when measured on different continents, indicating multiple north magnetic poles. The logical explanation for this dilemma is that the magnetic north pole has remained at the same approximate geographic location during the past, and the continents have moved.

2.7 Magentic Reversals and Seafloor Spreading

Geologists refer to Earth's present magnetic field as being normal—that is, with the north and south magnetic poles located approximately at the north and south geographic poles. At various times in the geologic past, however, Earth's magnetic field has completely reversed; that is, the magnetic north and south poles reverse positions, so that the magnetic north pole becomes the magnetic south pole, and the magnetic south pole becomes the magnetic north pole. During such a reversal, the magnetic field weakens until it temporarily disappears. When the magnetic field returns, the magnetic poles have reversed their position. The existence of such **magnetic reversals** was discovered by dating and determining the orientation of the remanent magnetism in lava flows on land (❚ Figure 2.13). Although the cause of magnetic reversals is still uncertain, their occurrence in the geologic record is well documented.

As a result of oceanographic research conducted during the 1950s, Harry Hess of Princeton University proposed, in a 1962 landmark paper, the theory of **seafloor spreading** to account for continental movement. He suggested that continents do not move through oceanic crust

South magnetic pole (normal position)

North magnetic pole (normal position)

a

North magnetic pole (reversed)

South magnetic pole (reversed)

b

South magnetic pole (normal)

North magnetic pole (normal)

c

North magnetic pole (reversed)

South magnetic pole (reversed)

d

Figure 2.13 Magnetic Reversals During the time period shown (a–d), volcanic eruptions produced a succession of overlapping lava flows. At the time of these volcanic eruptions, Earth's magnetic field completely reversed—that is, the magnetic north pole moved to the geographic south pole, and the magnetic south pole moved to the geographic north pole. Thus, the end of the needle of a magnetic compass that today would point to the North Pole would point to the South Pole if the magnetic field should again reverse. We know that Earth's magnetic field has reversed numerous times in the past because when lava flows cool below the Curie point, magnetic minerals within the flow orient themselves parallel to the magnetic field at the time. They thus record whether the magnetic field was normal or reversed at that time. The white arrows in this diagram show the direction of the north magnetic pole for each individual lava flow, thus confirming that Earth's magnetic field has reversed during the past.

objection of the opponents of continental drift—namely, how could continents move through oceanic crust? In fact, the continents moved with the oceanic crust as part of a lithospheric system.

As a mechanism to drive this system, Hess revived the idea (first proposed in the late 1920s by the British geologist Arthur Holmes) of a heat transfer system—or **thermal convection cells**—within the mantle to move the plates. According to Hess, hot magma rises from the mantle, intrudes along fractures defining oceanic ridges, and thus forms new crust. Cold crust is subducted back into the mantle at oceanic trenches, where it is heated and recycled, thus completing a thermal convection cell (see Figure 1.12).

How could Hess's hypothesis be confirmed? Magnetic surveys of the oceanic crust revealed a pattern of striped **magnetic anomalies** (deviations from the average strength of Earth's present-day magnetic field) in the rocks that are both parallel to and symmetric around the oceanic ridges (Figure 2.14). A positive magnetic anomaly results when Earth's magnetic field at the time of oceanic crust formation along an oceanic ridge summit was the same as today, thus yielding a stronger than normal (positive) magnetic signal. A negative magnetic anomaly results when Earth's magnetic field at the time of oceanic crust formation was reversed, therefore yielding a weaker than normal (negative) magnetic signal.

Thus, as new oceanic crust forms at oceanic ridge summits and records Earth's magnetic field at the time, the previously formed crust moves laterally away from the ridge. These magnetic stripes therefore represent times of normal and reversed polarity at oceanic ridges (where upwelling magma forms new oceanic crust) and conclusively confirm Hess's theory of seafloor spreading.

One of the consequences of the seafloor spreading theory is its confirmation that ocean basins are geologically young features whose openings and closings are partially responsible for continental movement (Figure 2.15). Radiometric dating

as do ships plowing through sea ice, but, rather, that the continents and oceanic crust move together as a single unit. Thus, the theory of seafloor spreading answered a major

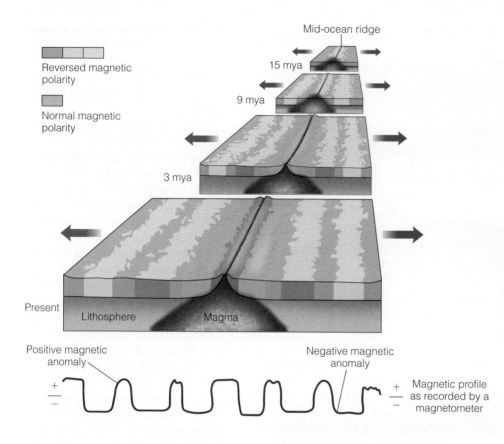

Figure 2.14 **Magnetic Anomalies and Seafloor Spreading** The sequence of magnetic anomalies preserved within the oceanic crust is both parallel to and symmetric around oceanic ridges. Basaltic lava intruding into an oceanic ridge today and spreading laterally away from the ridge records Earth's current magnetic field or polarity (considered by convention to be normal). Basaltic intrusions 3, 9, and 15 million years ago record Earth's reversed magnetic field at those times. This schematic diagram shows how the solidified basalt moves away from the oceanic ridge (or spreading ridge), carrying with it the magnetic anomalies that are preserved in the oceanic crust. Magnetic anomalies are magnetic readings that are either higher (positive magnetic anomalies) or lower (negative magnetic anomalies) than Earth's current magnetic field strength. The magnetic anomalies are recorded by a magnetometer, which measures the strength of the magnetic field.

Figure 2.15 **Age of the World's Ocean Basins** The age of the world's ocean basins has been determined from magnetic anomalies preserved in oceanic crust. The red colors adjacent to the oceanic ridges are the youngest oceanic crust. Moving laterally away from the ridges, the red colors grade to yellow at 48 million years ago, to green at 68 million years ago, and to dark blue some 155 million years ago. The darkest blue color is adjacent to the continental margins and is just somewhat less than 180 million years old.

>> **Critical Thinking Question**
How does the age of the oceanic crust confirm the seafloor spreading theory?

reveals that the oldest oceanic crust is somewhat younger than 180 million years old, whereas the oldest continental crust is 3.96 billion years old. Although geologists do not universally accept the idea of thermal convection cells as the sole driving mechanism for plate movement, most accept that plates are created at oceanic ridges and destroyed at deep-sea trenches (Figure 2.7), regardless of the driving mechanism involved.

Deep-Sea Drilling and the Confirmation of Seafloor Spreading

For many geologists, the paleomagnetic data amassed in support of continental drift and seafloor spreading were convincing. Moreover, results obtained from the Deep-Sea Drilling Project later confirmed the interpretations made from earlier paleomagnetic studies. Cores of deep-sea sediments and seismic profiles obtained by the *Glomar Challenger* and other research vessels have provided much of the data that support the seafloor spreading theory.

According to the theory of seafloor spreading, oceanic crust continuously forms at mid-oceanic ridges, moves away from these ridges by seafloor spreading, and is consumed at subduction zones. If this is the case, then oceanic crust should be youngest at the ridges and become progressively older with increasing distance away from them. Furthermore, the age of the oceanic crust should be symmetrically distributed about the ridges. As we have just noted, paleomagnetic data confirm these statements. In addition, fossils from sediments overlying the oceanic crust and radiometric dating of rocks found on oceanic islands both substantiate this predicted age distribution.

Sediments in the open ocean accumulate, on average, at a rate of less than 0.3 cm in 1,000 years. If the ocean basins were as old as the continents, we would expect deep-sea sediments to be several kilometers thick. However, data from numerous drill holes indicate that deep-sea sediments are, at most, only a few hundred meters thick and are thin or absent at oceanic ridges.

Their near-absence at the ridges should come as no surprise because these are the areas where new crust is continuously produced by volcanism and seafloor spreading. Accordingly, sediments have had little time to accumulate at or very close to spreading ridges. However, their thickness increases with distance away from the ridges because of the longer amount of time sediment has had to accumulate. Hence, the farther away one is from a ridge, the thicker the sediments.

2.8 Plate Tectonics: A Unifying Theory

Plate tectonic theory is based on a simple model of Earth. The rigid lithosphere, composed of both oceanic and continental crust, as well as the underlying upper mantle, consists of numerous variable-sized pieces called *plates* (❚ Figure 2.16). There are seven major plates (Eurasian, Indian-Australian, Antarctic, North American, South American, Pacific, and African) and numerous smaller ones ranging from only a few tens to several hundreds of kilometers in width. Plates also vary in thickness; those composed of upper mantle and continental crust are as much as 250 km thick, whereas those of upper mantle and oceanic crust are up to 100 km thick.

The lithosphere overlies the hotter and weaker semiplastic asthenosphere. It is thought that movement resulting from some type of heat-transfer system within the asthenosphere causes the overlying plates to move. As plates move over the asthenosphere, they separate, mostly at oceanic

❚ **Figure 2.16** Earth's Plates A world map showing Earth's plates, their boundaries, their relative motion, average rates of movement in centimeters per year, and hot spots.

ridges; in other areas, such as at oceanic trenches, they collide and are subducted back into the mantle.

An easy way to visualize plate movement is to think of a conveyer belt moving luggage from an airplane's cargo hold to a baggage cart. The conveyer belt represents convection currents within the mantle, and the luggage represents Earth's lithospheric plates. The luggage is moved along by the conveyer belt until it is dumped into the baggage cart in the same way that plates are moved by convection cells until they are subducted into Earth's interior. Although this analogy allows you to visualize how the mechanism of plate movement takes place, remember that this analogy is limited. The major limitation is that, unlike the luggage, plates consist of continental and oceanic lithosphere, which have different densities, and only oceanic lithosphere is subducted into Earth's interior.

Most geologists accept plate tectonic theory because the evidence for it is overwhelming and because it ties together many seemingly unrelated geologic features and events and shows how they are interrelated. Consequently, geologists now view geologic processes from the global perspective of plate tectonic theory in which plate interaction along plate margins is responsible for such phenomena as mountain building, earthquakes, and volcanism. Furthermore, because all of the inner planets have had a similar origin and early history, geologists are interested in determining whether plate tectonics is unique to Earth or whether it operates in the same way on other planets (see the Geo-Insight on pages 42 and 43).

2.9 The Three Types of Plate Boundaries

Because it appears that plate tectonics has operated since at least the Proterozoic Eon, it is important that we understand how plates move and interact with each other and how ancient plate boundaries are recognized. After all, the movement of plates has profoundly affected the geologic and biologic history of this planet.

What Would You Do ?

You've been selected to be part of the first astronaut team to go to Mars. While your two fellow crew members descend to the Martian surface, you'll be staying in the command module and circling the Red Planet. As part of the geologic investigation of Mars, one of the crew members will be mapping the geology around the landing site and deciphering the geologic history of the area. Your job will be to observe and photograph the planet's surface and try to determine whether Mars had an active plate tectonic regime in the past, and whether there is current plate movement. What features would you look for, and what evidence might reveal current or previous plate activity?

Geologists recognize three major types of plate boundaries: *divergent*, *convergent*, and *transform*. Along these boundaries, new plates are formed, are consumed, or slide laterally past one another. Interaction of plates at their boundaries accounts for most of Earth's volcanic eruptions and earthquakes, as well as the formation and evolution of its mountain systems.

Divergent Boundaries

Divergent plate boundaries, or *spreading ridges*, occur where plates are separating and new oceanic lithosphere is forming. Divergent boundaries are places where the crust is extended, thinned, and fractured as magma, derived from the partial melting of the mantle, rises to the surface. The magma is almost entirely basaltic and intrudes into vertical fractures to form dikes and pillow lava flows (see Figure 5.6). As successive injections of magma cool and solidify, they form new oceanic crust and record the intensity and orientation of Earth's magnetic field (Figure 2.14).

Divergent boundaries most commonly occur along the crests of oceanic ridges—for example, the Mid-Atlantic Ridge. Oceanic ridges are thus characterized by rugged topography with high relief resulting from displacement of rocks along large fractures, shallow-depth earthquakes, high heat flow, and basaltic flows or pillow lavas.

Divergent boundaries are also present under continents during the early stages of continental breakup. When magma wells up beneath a continent, the crust is initially elevated, stretched, and thinned, producing fractures, faults, rift valleys, and volcanic activity (Figure 2.17a). As magma intrudes into faults and fractures, it solidifies or flows out onto the surface as lava flows, the latter often covering the rift valley floor (Figure 2.17b). The East African Rift Valley is an excellent example of continental breakup at this stage (Figure 2.18).

As spreading proceeds, some rift valleys continue to lengthen and deepen until the continental crust eventually breaks and a narrow linear sea is formed, separating two continental blocks (Figure 2.17c). The Red Sea, separating the Arabian Peninsula from Africa (Figure 2.18), and the Gulf of California, which separates Baja California from mainland Mexico, are good examples of this more advanced stage of rifting.

As a newly created narrow sea continues to enlarge, it may eventually become an expansive ocean basin such as the Atlantic Ocean basin is today, separating North and South America from Europe and Africa by thousands of kilometers (Figure 2.17d). The Mid-Atlantic Ridge is the boundary between these diverging plates; the American plates are moving westward, and the Eurasian and African plates are moving eastward.

It should be noted that extending outward from the eastern coasts of North and South America, as well as the western coasts of Europe and Africa, are broad continental shelves, continental slopes, and rises (Figure 2.7, right side). These features are referred to as *passive continental margins*, and although they are found within a plate, rather than at a plate boundary, they result from the breakup of a continent and the subsequent movement away from a divergent plate boundary (Figure 2.17).

Volcanic activity

Magma

Continental crust

a Rising magma beneath a continent pushes the crust up, producing numerous fractures, faults, rift valleys, and volcanic activity.

Rift valley

b As the crust is stretched and thinned, rift valleys develop and lava flows onto the valley floors, such as seen today in the East African Rift Valley.

Coastal mountain range

Narrow fault-bounded sea

c Continued spreading further separates the continent until it splits apart and a narrow seaway develops. The Red Sea, which separates the Arabian Peninsula from Africa, is a good example of this stage of development.

Continental "seaboard" (coastal mountains gone)

Wide ocean

d As spreading continues, an oceanic ridge system forms, and an ocean basin develops and grows. The Mid-Atlantic Ridge illustrates this stage in a divergent plate boundary's history.

▌ **Figure 2.17** History of a Divergent Plate Boundary

The four inner, or terrestrial, planets—Mercury, Venus, Earth, and Mars—all had a similar early history involving accretion, differentiation into a metallic core and silicate mantle and crust, and formation of an early atmosphere by outgassing. Their early history was also marked by widespread volcanism and meteorite impacts, both of which helped modify their surfaces. Whereas the other three terrestrial planets as well as some of the Jovian moons display internal activity, Earth appears to be unique in that its surface is broken into a series of plates.

◄ **1.** A color-enhanced photomosaic of Mercury shows its heavily cratered surface, which has changed very little since its early history.

Images of **Mercury** sent back by *Mariner 10* show a heavily cratered surface with the largest impact basins filled with what appear to be lava flows similar to the lava plains on Earth's Moon. The lava plains are not deformed, however, indicating that there has been little or no tectonic activity.

Courtesy of Victor Royer

JPL/ NASA

Another feature of Mercury's surface is a large number of scarps, a feature usually associated with earthquake activity. Yet, some scientists think that these scarps formed when Mercury cooled and contracted.

▲ **2.** Seven scarps (indicated by arrows) can clearly be seen in this image. These scarps might have formed when Mercury cooled and contracted early in its history.

Of all the planets, **Venus** is the most similar in size and mass to Earth, but it differs in most other respects. Whereas Earth is dominated by plate tectonics, volcanism seems to have been the dominant force in the evolution of the Venusian surface. Even though no active volcanism has been observed on Venus, the various-sized volcanic features and what appear to be folded mountains indicate a once-active planetary interior. All these structures appear to be the products of rising convection currents of magma pushing up under the crust and then sinking back into the Venusian interior.

NASA/JPL/USGS

▲ **3.** A color-enhanced photomosaic of Venus based on radar images beamed back to Earth by the *Magellan* spacecraft. This image shows impact craters and volcanic features characteristic of the planet.

JPL/ NASA

▼ **4.** Arrows point to a 600-km segment of Venus's 6,800-km long Baltis Vallis, the longest known lava flow channel in our solar system.

▲ **5.** Venus's Aine Corona, about 200 km in diameter, is ringed by concentric faults, suggesting that it was pushed up by rising magma. A network of fractures is visible in the upper right of this image as well as a recent lava flow at the center of the corona, several volcanic domes in the lower portion of the image, and a large volcanic pancake dome in the upper left of the image.

NASA/JPL

◄ **6.** Volcano Sapas Mons contains two lava-filled calderas and is flanked by lava flows, attesting to the volcanic activity that was once common on Venus.

Mars, the Red Planet, has numerous features that indicate an extensive early period of volcanism. These include Olympus Mons, the solar system's largest volcano, lava flows, and uplifted regions thought to have resulted from mantle convection. In addition to volcanic features, Mars displays abundant evidence of tensional tectonics, including numerous faults and large fault-produced valley structures. Whereas Mars was tectonically active during the past, no evidence indicates that plate tectonics comparable to that on Earth has ever occurred there.

NASA/JPL

NASA/USGS

▲ **7.** A vertical view of Olympus Mons, a shield volcano and the largest volcano in our solar system. The edge of the Olympus Mons caldera is marked by a cliff several kilometers high rather than a moat as in Mauna Loa, Earth's largest shield volcano.

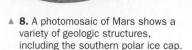

NASA/USGS

▲ **8.** A photomosaic of Mars shows a variety of geologic structures, including the southern polar ice cap.

NASA/JPL/University of Arizona

Although not a terrestrial planet, **Io,** the innermost of Jupiter's Galilean moons, must be mentioned. Images from the *Voyager* and *Galileo* spacecrafts show that Io has no impact craters. In fact, more than a hundred active volcanoes are visible on the moon's surface, and the sulfurous gas and ash erupted by these volcanoes bury any newly formed meteorite impact craters. Because of its proximity to Jupiter, the heat source of Io is probably tidal heating, in which the resulting friction is enough to at least partially melt Io's interior and drive its volcanoes.

NASA/JPL/University of Arizona

NASA/JPL/University of Arizona

NASA/JPL/University of Arizona

50 km

► **9.** Volcanic features of Io, the innermost moon of Jupiter. As shown in these digitally enhanced color images, Io is a very volcanically active moon.

Figure 2.18 East African Rift Valley and the Red Sea—Present-Day Examples of Divergent Plate Boundaries The East African Rift Valley and the Red Sea represent different stages in the history of a divergent plate boundary. The East African Rift Valley is being formed by the separation of eastern Africa from the rest of the continent along a divergent plate boundary. The Red Sea represents a more advanced stage of rifting, in which two continental blocks (Africa and the Arabian Peninsula) are separated by a narrow sea.

An Example of Ancient Rifting What features in the geologic record can geologists use to recognize ancient rifting? Associated with regions of continental rifting are faults, dikes (vertical intrusive igneous bodies), sills (horizontal intrusive igneous bodies), lava flows, and thick sedimentary sequences within rift valleys, all features that are preserved in the geologic record. The Triassic fault basins of the eastern United States are a good example of ancient continental rifting (see Figure 22.7). These fault basins mark the zone of rifting that

occurred when North America split apart from Africa. The basins contain thousands of meters of continental sediment and are riddled with dikes and sills (see Chapter 22).

Pillow lavas, in association with deep-sea sediment, are also evidence of ancient rifting. The presence of pillow lavas marks the formation of a spreading ridge in a narrow linear sea (Figures 2.17c and 2.18). A narrow linear sea forms when the continental crust in the rift valley finally breaks apart, and the area is flooded with seawater. Magma, intruding into the sea along this newly formed spreading ridge, solidifies as pillow lavas, which are preserved in the geologic record, along with the sediment being deposited on them.

Convergent Boundaries

Whereas new crust forms at divergent plate boundaries, older crust must be destroyed and recycled in order for the entire surface area of Earth to remain the same. Otherwise, we would have an expanding Earth. Such plate destruction occurs at **convergent plate boundaries** (Figure 2.19), where two plates collide and the leading edge of one plate is subducted beneath the margin of the other plate and eventually incorporated into the asthenosphere. A dipping plane of earthquake foci, called a *Benioff zone*, defines subduction zones (see Figure 9.5). Most of these planes dip from oceanic trenches beneath adjacent island arcs or continents, marking the surface of slippage between the converging plates.

Deformation, volcanism, mountain building, metamorphism, earthquake activity, and deposits of valuable minerals characterize convergent boundaries. Three types of convergent plate boundaries are recognized: *oceanic–oceanic, oceanic–continental,* and *continental–continental.*

> **CONNECTION LINK** ‹
>
> More information about the relationship between plate tectonics and mountain building can be found on pages 246–252 in Chapter 10.

Oceanic–Oceanic Boundaries When two oceanic plates converge, one is subducted beneath the other along an **oceanic–oceanic plate boundary** (Figure 2.19a). The subducting plate bends downward to form the outer wall of an oceanic trench. A subduction complex, composed of wedge-shaped slices of highly folded and faulted marine sediments and oceanic lithosphere scraped off the descending plate, forms along the inner wall of the oceanic trench. As the subducting plate descends into the mantle, it is heated and partially melted, generating magma commonly of andesitic composition (see Chapter 4). This magma is less dense than the surrounding mantle rocks and rises to the surface of the nonsubducted plate to form a curved chain of volcanic islands called a volcanic island arc (any plane intersecting a sphere makes an arc). This arc is nearly parallel to the oceanic trench and is separated from it by a distance of up to several hundred kilometers—the distance depends on the angle of dip of the subducting plate (Figure 2.19a).

a Oceanic–oceanic plate boundary. An oceanic trench forms where one oceanic plate is subducted beneath another. On the nonsubducted plate, a volcanic island arc forms from the rising magma generated from the subducting plate. The Japanese Islands are a volcanic island arc resulting from the subduction of one oceanic plate beneath another oceanic plate.

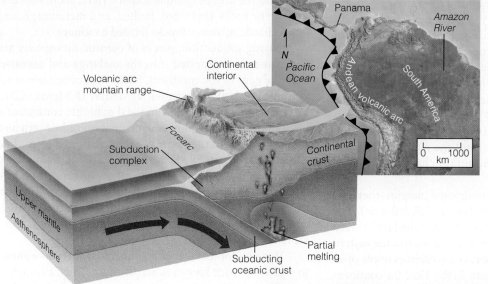

b Oceanic–continental plate boundary. When an oceanic plate is subducted beneath a continental plate, an andesitic volcanic mountain range is formed on the continental plate as a result of rising magma. The Andes Mountains in Peru are one of the best examples of continuing mountain building at an oceanic–continental plate boundary.

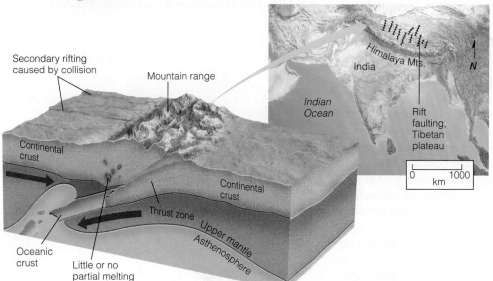

c Continental–continental plate boundary. When two continental plates converge, neither is subducted because of their great thickness and low and equal densities. As the two continental plates collide, a mountain range is formed in the interior of a new and larger continent. The Himalayas in central Asia resulted from the collision between India and Asia approximately 40 to 50 million years ago.

| Figure 2.19 Three Types of Convergent Plate Boundaries

In those areas where the rate of subduction is faster than the forward movement of the overriding plate, the lithosphere on the landward side of the volcanic island arc may be subjected to tensional stress and stretched and thinned, resulting in the formation of a *back-arc basin*. This back-arc basin may grow by spreading if magma breaks through the thin crust and forms new oceanic crust (Figure 2.19a). A good example of a back-arc basin associated with an oceanic–oceanic plate boundary is the Sea of Japan between the Asian continent and the islands of Japan.

Most present-day active volcanic island arcs are in the Pacific Ocean basin and include the Aleutian Islands, the Kermadec–Tonga arc, and the Japanese (Figure 2.19a) and Philippine Islands. The Scotia and Antillean (Caribbean) island arcs are in the Atlantic Ocean basin.

Oceanic–Continental Boundaries

When an oceanic and a continental plate converge, the denser oceanic plate is subducted under the continental plate along an **oceanic-continental plate boundary** (Figure 2.19b). Just as at oceanic–oceanic plate boundaries, the descending oceanic plate forms the outer wall of an oceanic trench.

The magma generated by subduction rises beneath the continent and either crystallizes as large intrusive bodies, called *plutons*, before reaching the surface or erupts at the surface to produce a chain of andesitic volcanoes, also called a *volcanic arc*. An excellent example of an oceanic–continental plate boundary is the Pacific Coast of South America where the oceanic Nazca plate is currently being subducted under South America (Figure 2.19b; see also Chapter 10). The Peru–Chile Trench marks the site of subduction, and the Andes Mountains are the resulting volcanic mountain chain on the non-subducting plate.

Just as there are passive continental margins, there are also *active continental margins* (Figure 2.7, left side). The aforementioned oceanic–continental plate boundary between the west coast of the South American plate and the eastern side of the oceanic Nazca plate is an excellent example of an active continental margin (Figure 2.19b). Here the continental shelf is narrow, and the continental slope descends directly into the Peru–Chile Trench, so sediment is dumped into the trench and no continental rise develops, such as along a passive continental margin (Figure 2.7, right side).

Continental–Continental Boundaries

Two continents approaching each other are initially separated by an ocean floor that is being subducted under one continent. The edge of that continent displays the features characteristic of oceanic–continental convergence along an active continental plate margin. As the ocean floor continues to be subducted, the two continents come closer together until they eventually collide. Because continental lithosphere, which consists of continental crust and the upper mantle, is less dense than oceanic lithosphere (oceanic crust and upper mantle), it cannot sink into the asthenosphere. Although one continent may partially slide under the other, it cannot be pulled or pushed down into a subduction zone (Figure 2.19c).

When two continents collide, they are welded together along a zone marking the former site of subduction. At this **continental–continental plate boundary**, an interior mountain belt is formed consisting of deformed sediments and sedimentary rocks, igneous intrusions, metamorphic rocks,

and fragments of oceanic crust. In addition, the entire region is subjected to numerous earthquakes. The Himalayas in central Asia, the world's youngest and highest mountain system, resulted from the collision between India and Asia that began 40 to 50 million years ago and is still continuing (Figure 2.19c; see Chapter 10).

Recognizing Ancient Convergent Plate Boundaries

How can former subduction zones be recognized in the geologic record? Igneous rocks provide one such clue. The magma erupted at the surface, forming island arc volcanoes and continental volcanoes, is of andesitic composition. Another clue is the zone of intensely deformed rocks between the deep-sea trench where subduction is taking place and the area of igneous activity. Here, sediments and submarine rocks are folded, faulted, and metamorphosed into a chaotic mixture of rocks termed a *mélange*.

During subduction, pieces of oceanic lithosphere are sometimes incorporated into the mélange and accreted onto the edge of the continent. Such slices of oceanic crust and upper mantle are called *ophiolites* (▌ Figure 2.20). Detailed studies reveal that an ideal ophiolite consists of a layer of deep-sea sediments deposited on pillow lava and sheet lava flows of the upper oceanic crust. A sheeted dike complex consisting of vertical basaltic dikes, followed by massive gabbro (a dark intrusive igneous rock), and layered gabbro, form the rest of the oceanic crust. Beneath the gabbro is peridotite (a dark intrusive igneous rock composed of the mineral olivine), which probably represents the upper mantle.

This peridotite is sometimes altered by metamorphism to a greenish rock known as serpentinite. Thus, a complete ophiolite consists of deep-sea sedimentary rocks underlain

Deep-sea sediments

Pillow lava and sheet lava

Sheeted dikes

Massive gabbro

Layered gabbro

Upper mantle peridotite

▌ **Figure 2.20 Ophiolites** Ophiolites are sequences of rock on land consisting of deep-sea sediments, oceanic crust, and upper mantle. Ophiolites are one feature used to recognize ancient convergent plate boundaries.

a Most transform faults connect two oceanic ridge segments.

Labels: Trench, Fracture zone, Transform fault, Fracture zone, Magma, Mid-ocean ridge, Oceanic crust

b A transform fault can connect a ridge and a trench.

Labels: "Subducted" oceanic ridge segment, Transform fault

c A transform fault can also link two trenches.

Labels: Fracture zone, Transform fault

❚ Figure 2.21 **Transform Plate Boundaries** Horizontal movement between plates occurs along transform faults. Extensions of transform faults on the seafloor form fracture zones. Note that relative motion between the plates only occurs between the two ridges.

Appalachians, Alps, Himalayas, and Andes. The combination of such features is significant evidence that these mountain ranges resulted from deformation along convergent plate boundaries.

Transform Boundaries

The third type of plate boundary is a **transform plate boundary**. These mostly occur along fractures in the seafloor, known as *transform faults*, where plates slide laterally past one another roughly parallel to the direction of plate movement. Although lithosphere is neither created nor destroyed along a transform boundary, the movement between plates results in a zone of intensely shattered rock and numerous shallow-depth earthquakes.

Transform faults "transform" or change one type of motion between plates into another type of motion. Most commonly, transform faults connect two oceanic ridge segments; however, they can also connect ridges to trenches and trenches to trenches (❚ Figure 2.21). Although the majority of transform faults are in oceanic crust and are marked by distinct fracture zones, they may also extend into continents.

One of the best-known transform faults is the San Andreas fault in California. It separates the Pacific plate from the North American plate and connects spreading ridges in the Gulf of California with the Juan de Fuca and Pacific plates off the coast of northern California (❚ Figure 2.22).

> **CONNECTION LINK** ◀
>
> Learn more about the San Andreas fault in the Geo-Insight feature on pages 214 and 215 in Chapter 9.

Many of the earthquakes that affect California are the result of movement along this fault (see Chapter 9).

Unfortunately, transform faults generally do not leave any characteristic or diagnostic features except for the obvious displace-ment of the rocks with which they are associated. This displacement is usually large, on the order of tens to hundreds of kilometers. Such large displacements in ancient rocks can sometimes be related to transform fault systems.

by rocks of the oceanic crust and upper mantle. The presence of ophiolites in an outcrop or drilling core is a key indicator of plate convergence along a subduction zone.

Elongated belts of folded and faulted marine sedimentary rocks, andesites, and ophiolites are found in the

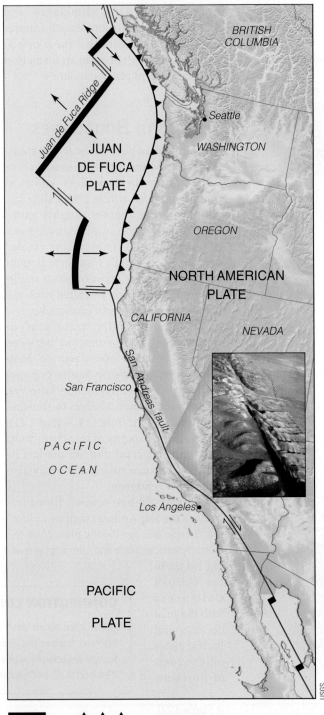

Oceanic ridge | Zone of subduction | Transform faults

Figure 2.22 The San Andreas Fault—A Transform Plate Boundary The San Andreas fault is a transform fault separating the Pacific plate from the North American plate. It connects the spreading ridges in the Gulf of California with the Juan de Fuca and Pacific plates off the coast of northern California. Movement along the San Andreas fault has caused numerous earthquakes. The insert photograph shows a segment of the San Andreas fault as it cuts through the Carrizo Plain, California.

» Critical Thinking Question
If the movement along the San Andreas fault, which separates the Pacific plate from the North American plate, averages 5.5 cm per year, how long will it take before Los Angeles is opposite San Francisco?

2.10 Hot Spots and Mantle Plumes

Before leaving the topic of plate boundaries, we should mention an intraplate feature found beneath both oceanic and continental plates. A **hot spot** (Figure 2.16) is the location on Earth's surface where a stationary column of magma, originating deep within the mantle (*mantle plume*), has slowly risen to the surface and formed a volcano. Because the mantle plumes apparently remain stationary (although some evidence suggests that they might not) within the mantle while the plates move over them, the resulting hot spots leave a trail of extinct and progressively older volcanoes (aseismic ridges) that record the movement of the plate.

One of the best examples of aseismic ridges and hot spots is the Emperor Seamount–Hawaiian Island chain (▌Figure 2.23). This chain of islands and seamounts extends from the island of Hawaii to the Aleutian Trench off Alaska, a distance of some 6,000 km, and consists of more than 80 volcanic structures.

Currently, the only active volcanoes in this island chain are on the island of Hawaii and the Loihi Seamount. The rest of the islands are extinct volcanic structures that become progressively older toward the north and northwest. This means that the Emperor Seamount–Hawaiian Island chain records the direction that the Pacific plate traveled as it moved over an apparently stationary mantle plume. In this case, the Pacific plate first moved in a north-northwesterly direction and then, as indicated by the sharp bend in the chain, changed to a west-northwesterly direction approximately 43 million years ago. The reason that the Pacific plate changed directions is not known, but the shift might be related to the collision of India with the Asian continent at about the same time (see Figure 10.21).

Mantle plumes and hot spots help geologists explain some of the geologic activity occurring within plates as opposed to activity occurring at or near plate boundaries. In addition, if mantle plumes are essentially fixed with respect to earth's rotational axis, they can be used to determine not only the direction of plate movement but also the rate of movement.

2.11 Plate Movement and Motion

How fast and in what direction are Earth's plates moving? Do they all move at the same rate? Rates of plate movement can be calculated in several ways. The least accurate method is to determine the age of the sediments immediately above any portion of the oceanic crust and then divide the distance

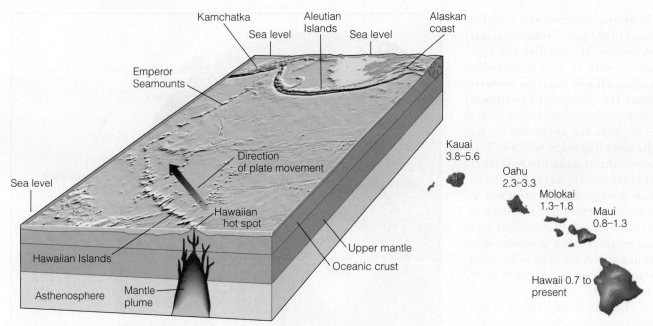

Figure 2.23 Hot Spots A hot spot is the location where a stationary mantle plume has risen to the surface and formed a volcano. The Emperor Seamount–Hawaiian Island chain formed as a result of the Pacific plate moving over a mantle plume, and the line of volcanic islands in this chain traces the direction of plate movement. The island of Hawaii and the Loihi Seamount are the only current hot spots of this island chain. The numbers indicate the age of the islands in millions of years.

Figure 2.24 Reconstructing Plate Positions Using Magnetic Anomalies

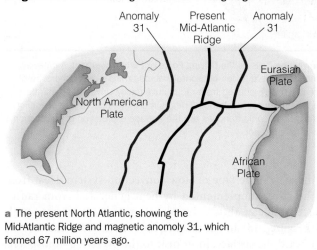

a The present North Atlantic, showing the Mid-Atlantic Ridge and magnetic anomoly 31, which formed 67 million years ago.

b The Atlantic 67 million years ago. Anomaly 31 marks the plate boundary 67 million years ago. By moving the anomalies back together, along with the plates they are on, we can reconstruct the former positions of the continents.

from the spreading ridge by that age. Such calculations give an average rate of movement.

A more accurate method of determining both the average rate of movement and relative motion is by dating the magnetic anomalies in the crust of the seafloor. The distance from an oceanic ridge axis to any magnetic anomaly indicates the width of new seafloor that formed during that time interval. For example, if the distance between the present-day Mid-Atlantic Ridge and anomaly 31 is 2,010 km, and anomaly 31 formed 67 million years ago (Figure 2.24), then the average rate of movement during the past 67 million years has been 3 cm per year (2,010 km, which equals 201 million cm divided by 67 million years; 201,000,000 cm/67,000,000 years = 3 cm/year). Thus, for a given interval of time, the wider the strip of seafloor, the faster the plate has moved. In this way, not only can the present average rate of movement and relative motion be determined (Figure 2.16), but the average rate of movement in the past can also be calculated by dividing the distance between anomalies by the amount of time elapsed between anomalies.

Geologists use magnetic anomalies not only to calculate the average rate of plate movement but also to determine plate positions at various times in the past. Because magnetic anomalies are parallel and symmetric with respect to spreading ridges, all one must do to determine the position of continents when particular anomalies formed is to move the anomalies back to the spreading ridge, which will also move the continents with them (Figure 2.24). Because subduction destroys oceanic crust and the magnetic record that it carries, we have an excellent record of plate movements since the breakup of Pangaea, but not as good an understanding of plate movement before that time.

The average rate of movement, as well as the relative motion between any two plates, can also be determined by satellite-laser ranging techniques. Laser beams from a station on one plate are bounced off a satellite (in geosynchronous orbit) and returned to a station on a different plate. As the plates move away from each other, the laser beam takes more time to go from the sending station to the stationary satellite and back to the receiving station. This difference in elapsed time is used to calculate the rate of movement and the relative motion between plates.

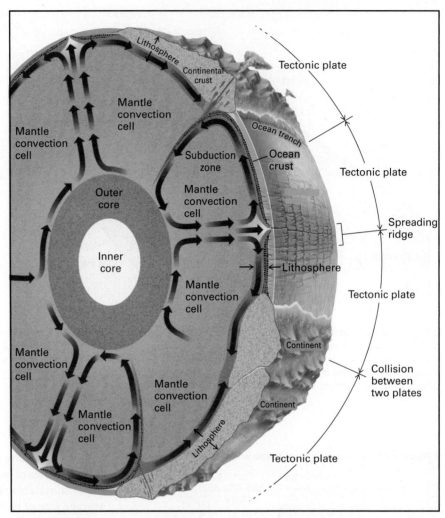

▌Figure 2.25 Thermal Convection Cells as the Driving Force of Plate Movement A cutaway view of Earth shows that the lithosphere glides horizontally across the asthenosphere. Heat from the core, supplemented by heat produced from radioactive decay, drives huge mantle convection cells that move the lithosphere. Spreading ridges mark the location of ascending limbs of the convection cells, and oceanic trenches are the surface expression of subduction, where plates descend into Earth's interior.

2.12 The Driving Mechanism of Plate Tectonics

A major obstacle to the acceptance of the continental drift hypothesis was the lack of a driving mechanism to explain continental movement. When it was shown that continents and ocean floors moved together, not separately, and that new crust is formed at spreading ridges by rising magma, most geologists accepted some type of convective heat system (convection cells) as the basic process responsible for plate motion. However, the question of what exactly drives the plates remains.

Most of the heat from Earth's interior results from the decay of radioactive elements, such as uranium (see Chapter 17), in the core and lower mantle. The most efficient way for this heat to escape Earth's interior is through some type of slow convection system. Heat from the core, supplemented by heat generated from radioactive decay, thus drives large mantle convection cells (▌Figure 2.25). In this manner, hot rock from the interior rises toward the surface, loses heat to the overlying lithosphere, becomes denser as it cools, and then sinks back into the interior where it is heated, and the process repeats itself. This type of convective heat system is analogous to a pot of stew cooking on a stove (▌Figure 2.26).

In this mantle convection cell model, spreading ridges mark the ascending limbs of adjacent convection cells, and trenches are present where convection cells descend back into Earth's interior. The convection cells therefore determine the location of spreading ridges and trenches, with the lithosphere lying above the thermal convection cells. Thus, each plate corresponds to a single convection cell that moves as a result of the convective movement of the cell itself (Figure 2.25).

Although most geologists agree that Earth's internal heat plays an important role in plate movement, two other processes, referred to as "slab-pull" and "ridge-push," might also help facilitate the movement of plates (▌ Figure 2.27). Both mechanisms are gravity driven, but still depend on thermal differences within Earth.

In slab-pull, the subducting cold slab of lithosphere, being denser than the surrounding warmer asthenosphere, pulls the rest of the plate along as it descends into the asthenosphere (Figure 2.27). As the lithosphere moves downward, there is a corresponding upward flow back into the spreading ridge.

Operating in conjunction with slab-pull is the ridge-push mechanism. As a result of rising magma, the oceanic ridges are higher than the surrounding oceanic crust. It is thought that gravity pushes the oceanic lithosphere away from the higher spreading ridges and toward the trenches, where it is subducted back into Earth's interior (Figure 2.27).

CONNECTION LINK ◀

More can be learned about Earth's interior on pages 223–227 in Chapter 9.

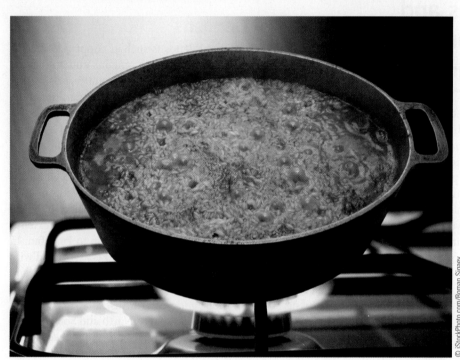

▌**Figure 2.26** Convection in a Pot of Stew Heat from the stove is applied to the base of the stew pot causing the stew to heat up. As heat rises through the stew, pieces of the stew are carried to the surface, where the heat is dissipated, the pieces of stew cool, and then sink back to the bottom of the pot. The bubbling seen at the surface of the stew is the result of convection cells churning the stew. In the same manner, heat from the decay of radioactive elements produces convection cells within Earth's interior.

© iStockPhoto.com/Roman Sigaev

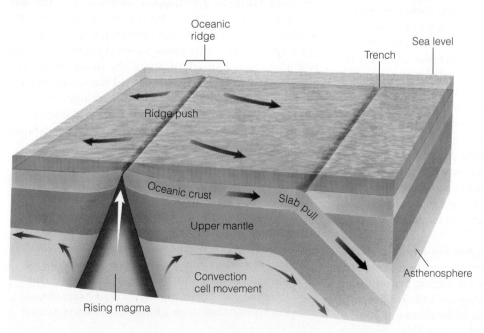

▌**Figure 2.27** Plate Movement Resulting from Gravity-Driven Mechanisms Plate movement is also thought to occur, at least partially, from gravity-driven "slab-pull" or "ridge-push" mechanisms. In slab-pull, the edge of the subducting plate descends into Earth's interior, and the rest of the plate is pulled downward. In ridge-push, rising magma pushes the oceanic ridges higher than the rest of the oceanic crust. Gravity thus pushes the oceanic lithosphere away from the ridges and toward the trenches.

Currently, geologists are fairly certain that some type of convective system is involved in plate movement, but the extent to which other mechanisms, such as slab-pull and ridge-push, are involved is still unresolved. However, the fact that plates have moved in the past and are still moving today has been proven beyond a doubt. And although a comprehensive theory of plate movement has not yet been developed, more and more of the pieces are falling into place as geologists learn more about Earth's interior.

2.13 Plate Tectonics and the Distribution of Natural Resources

In addition to being responsible for the major features of Earth's crust and influencing the distribution and evolution of the world's biota, plate movement affects the formation and distribution of some natural resources. Consequently, geologists are using plate tectonic theory in their search for petroleum and mineral deposits and in explaining the occurrence of these natural resources. It is becoming increasingly clear that if we are to keep up with the continuing demands of a global industrialized society, the application of plate tectonic theory to the origin and distribution of natural resources is essential.

Petroleum

Although significant concentrations of petroleum occur in many areas of the world, more than 50% of all proven reserves are in the Persian Gulf region. It should therefore not be surprising that many of the conflicts in the Middle East have shared as their underlying cause the desire to control these vast amounts of petroleum. Most people, however, are not aware of *why* there is so much oil in this region of the world. The answer lies in the paleogeography and plate movement of this region during the Mesozoic and Cenozoic eras.

During the Mesozoic Era, and particularly the Cretaceous Period when most of the petroleum formed, the Persian Gulf area was a broad marine shelf extending eastward from Africa. This passive continental margin lay near the equator where countless microorganisms lived in the surface waters. The remains of these organisms accumulated with the bottom sediments and were buried, beginning the complex process of petroleum generation and the formation of source beds in which petroleum forms.

As a result of rifting in the Red Sea and the Gulf of Aden during the Cenozoic Era, the Arabian plate is moving northeast away from Africa and subducting beneath Iran (Figure 2.18). During the early stages of collision between Arabia and Iran, as the sediments of the passive continental margin were initially subducted, heating broke down the organic molecules and led to the formation of petroleum.

The continued subduction and collision with Iran folded the rocks, creating traps for petroleum to accumulate, so much so that the vast area south of the collision zone is now a major oil-producing region. Elsewhere in the world, plate tectonics is also responsible for concentrations of petroleum.

CONNECTION LINK ◀

To learn more about petroleum and natural gas, go to pages 173 and 176 in Chapter 7.

Mineral Deposits

Many metallic mineral deposits such as copper, gold, lead, silver, tin, and zinc are related to igneous and associated hydrothermal (hot water) activity. So it is not surprising that a close relationship exists between plate boundaries and the occurrence of these valuable deposits.

The magma generated by partial melting of a subducting plate rises toward the surface, and as it cools, it precipitates and concentrates various metallic ores. Many of the world's major metallic ore deposits are associated with convergent plate boundaries, including those in the Andes of South America and the Coast Ranges and Rockies of North America, Japan, the Philippines, Russia, and a zone extending from the eastern Mediterranean region to Pakistan. In addition, the majority of the world's gold is associated with deposits located at ancient convergent plate boundaries in such areas as Canada, Alaska, California, Venezuela, Brazil, Russia, southern India, and western Australia.

The copper deposits of western North and South America are an excellent example of the relationship between convergent plate boundaries and the distribution, concentration, and exploitation of valuable metallic ores (❙ Figure 2.28a). The world's largest copper deposits are found along this belt. The majority of the copper deposits in the Andes and the southwestern United States were formed less than 60 million years ago when oceanic plates were subducted under the North and South American plates. The rising magma and associated hydrothermal fluids carried minute amounts of copper, which were originally widely disseminated but eventually became concentrated in the cracks and fractures of the surrounding andesites. These low-grade copper deposits

What Would You Do ?

You are part of a mining exploration team that is exploring a promising and remote area of central Asia. You know that former convergent and divergent plate boundaries frequently are sites of ore deposits. What evidence would you look for to determine whether the area you're exploring might be an ancient convergent or divergent plate boundary? Is there anything you can do before visiting the area that might help you to determine the geology of the area?

Figure 2.28 Copper Deposits and Convergent Plate Boundaries

a Valuable copper deposits are located along the west coasts of North and South America in association with convergent plate boundaries. The rising magma and associated hydrothermal activity resulting from subduction carried small amounts of copper, which became trapped and concentrated in the surrounding rocks through time.

b Bingham Copper Mine, near Salt Lake City, Utah, is a huge open-pit copper mine with reserves estimated at 1.7 billion tons. More than 400,000 tons of rock are removed for processing each day. Note the small specks toward the middle of the photograph that are the 12-foot high dump trucks!

contain from 0.2 to 2% copper and are extracted from large open-pit mines (⦚ Figure 2.28b).

Divergent plate boundaries also yield valuable ore resources. The island of Cyprus in the Mediterranean is rich in copper and has been supplying all or part of the world's needs for the past 3,000 years. The concentration of copper on Cyprus formed as a result of precipitation adjacent to hydrothermal vents along a divergent plate boundary (Figure 2.8). This deposit was brought to the surface when the copper-rich seafloor collided with the European plate, warping the seafloor and forming Cyprus.

2.14 Plate Tectonics and the Distribution of Life

Plate tectonic theory is as revolutionary and far-reaching in its implications for geology as the theory of evolution was for biology when it was proposed. Interestingly, it was the fossil evidence that convinced Wegener, Suess, and du Toit, as well as many other geologists, of the correctness of continental drift. Together, the theories of plate tectonics and evolution have changed the way we view our planet, and we should not be surprised at the intimate association between them. Although the relationship between plate tectonic processes and the evolution of life is incredibly

complex, paleontological data provide convincing evidence of the influence of plate movement on the distribution of organisms.

The present distribution of plants and animals is not random, but is controlled mostly by climate and geographic barriers. The world's biota occupy *biotic provinces*, which are regions characterized by a distinctive assemblage of plants and animals. Organisms within a province have similar ecological requirements, and the boundaries separating provinces are therefore natural ecological breaks. Climatic or geographic barriers are the most common province boundaries, and these are mostly controlled by plate movement.

The complex interaction between wind and ocean currents has a strong influence on the world's climates. Wind and ocean currents are thus strongly influenced by the number, distribution, topography, and orientation of continents. For example, the southern Andes Mountains act as an effective barrier to moist, easterly blowing Pacific winds, resulting in a desert east of the southern Andes that is virtually uninhabitable.

The distribution of continents and ocean basins not only influences wind and ocean currents but also affects provinciality by creating physical barriers to, or pathways for, the migration of organisms. Intraplate volcanoes, island arcs, mid-oceanic ridges, mountain ranges, and subduction zones all result from the interaction of plates, and their orientation and distribution strongly influence the number of provinces and hence total global diversity. Thus, provinciality and diversity will be highest where numerous small continents are spread across many zones of latitude.

When a geographic barrier separates a once-uniform fauna, species may undergo divergence. If conditions on opposite sides of the barrier are sufficiently different, then species must adapt to the new conditions, migrate, or become extinct. Adaptation to the new environment by various species may involve enough change that new species eventually evolve.

The marine invertebrates found on opposite sides of the Isthmus of Panama provide an excellent example of divergence caused by the formation of a geographic barrier. Prior to the rise of this land connection between North and South America, a homogeneous population of bottom-dwelling invertebrates inhabited the shallow seas of the area. After the rise of the Isthmus of Panama by subduction of the Pacific plate approximately 5 million years ago, the original population was divided. In response to the changing environment, new species evolved on opposite sides of the isthmus.

The formation of the Isthmus of Panama also influenced the evolution of the North and South American mammalian faunas (❙ Figure 2.29). During most of the Cenozoic Era, South America was an island continent, and its mammalian fauna evolved in isolation from the rest of the world's faunas. When North and South America were connected by the Isthmus of Panama, most of the indigenous South American mammals were replaced by migrants from North America. Surprisingly, only a few South American mammal groups migrated northward.

> **CONNECTION LINK**

You can learn more about divergent, convergent, and parallel evolution on pages 459–461 in Chapter 18.

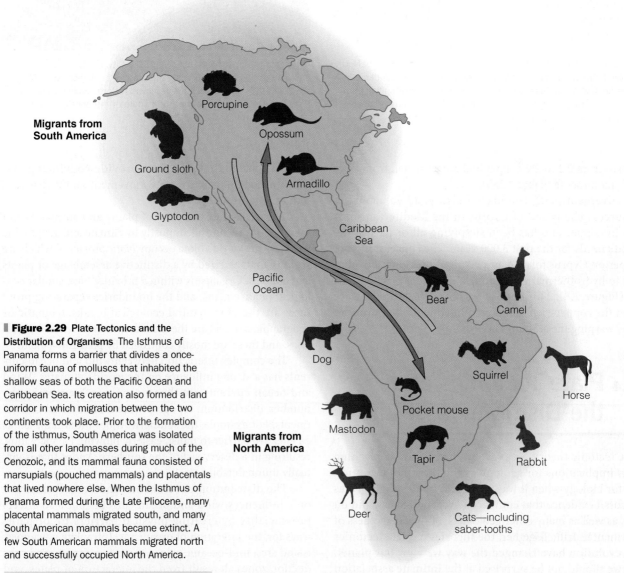

❙ **Figure 2.29** Plate Tectonics and the Distribution of Organisms The Isthmus of Panama forms a barrier that divides a once-uniform fauna of molluscs that inhabited the shallow seas of both the Pacific Ocean and Caribbean Sea. Its creation also formed a land corridor in which migration between the two continents took place. Prior to the formation of the isthmus, South America was isolated from all other landmasses during much of the Cenozoic, and its mammal fauna consisted of marsupials (pouched mammals) and placentals that lived nowhere else. When the Isthmus of Panama formed during the Late Pliocene, many placental mammals migrated south, and many South American mammals became extinct. A few South American mammals migrated north and successfully occupied North America.

>> **Critical Thinking Question**
Why is the mammalian fauna of Australia so different from elsewhere?

GEO-RECAP

» Geo-Image

Figure 1 Using the ages (the numbers represent ages in millions of years) for each of the Hawaiian Islands, as well as the scale given, calculate the average rate of movement per year for the Pacific plate since each island formed. Is the average rate of movement the same for each island? Would you expect it to be? Explain why it may not be and why there are different ages for some of the islands.

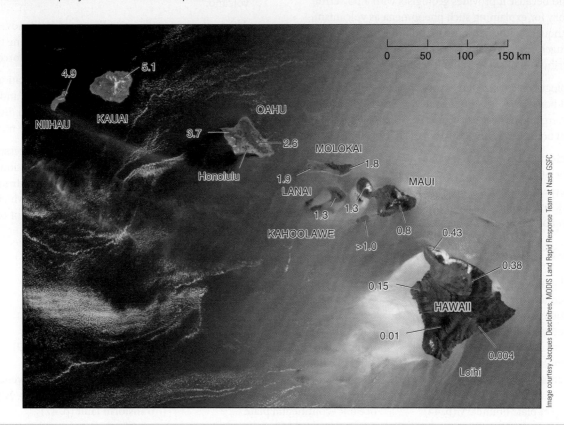

Image courtesy Jacques Descloitres, MODIS Land Rapid Response Team at Nasa GSFC

Chapter Summary

- The concept of continental movement is not new. The earliest maps showing the similarity between the east coast of South America and the west coast of Africa provided the first evidence that continents may once have been united and subsequently separated from each other.

- Alfred Wegener is generally credited with developing the hypothesis of continental drift. He provided abundant geologic and paleontologic evidence to show that the continents were once united in one supercontinent, which he named Pangaea. Unfortunately, Wegener could not explain how the continents moved, and most geologists ignored his ideas.

- Various features of the continental margins and the seafloor are a reflection of plate movement. Continental margins are active or passive, depending on their relationship to plate boundaries. Oceanic trenches are long, steep-sided depressions on the seafloor near convergent plate boundaries where oceanic lithosphere is consumed by subduction. Submarine hydrothermal vents are found at or near spreading ridges and associated with divergent plate boundaries.

- The hypothesis of continental drift was revived during the 1950s when paleomagnetic studies of rocks indicated the presence of multiple magnetic north poles instead of just one as there is today. This paradox was resolved by constructing a map in which the continents could be moved into different positions such that the paleomagnetic data would then be consistent with a single magnetic north pole.

- Seafloor spreading was confirmed by the discovery of magnetic anomalies in the oceanic crust that were both parallel to and symmetric around oceanic ridges, indicating that new oceanic crust must have formed as the seafloor was spreading. The pattern of oceanic magnetic anomalies matched the pattern of magnetic reversals already known from continental lava flows and showed that Earth's magnetic field has reversed itself numerous times during the past.

- Plate tectonic theory became widely accepted by the 1970s because the evidence overwhelmingly supports it and because it provides geologists with a powerful theory for explaining such phenomena as volcanism, earthquake activity, mountain building, global climatic changes, the distribution of the world's biota, and the distribution of some mineral resources.

- Geologists recognize three types of plate boundaries: divergent boundaries, where plates move away from each other; convergent boundaries, where two plates collide; and transform boundaries, where two plates slide past each other.

- Ancient plate boundaries can be recognized by their associated rock assemblages and geologic structures. For divergent boundaries, these may include rift valleys with thick sedimentary sequences and numerous dikes and sills. For convergent boundaries, ophiolites and andesitic rocks are two characteristic features. Transform faults generally do not leave any characteristic or diagnostic features in the geologic record.

- The average rate of movement and relative motion of the plates can be calculated in several ways. The results of these different methods all agree and indicate that the plates move at different average velocities.

- The absolute motion of plates can be determined by the movement of plates over mantle plumes. A mantle plume is an apparently stationary column of magma that rises to the surface where it becomes a hot spot and forms a volcano.

- Although a comprehensive theory of plate movement has yet to be developed, geologists think that some type of convective heat system is the major driving force.

- A close relationship exists between the formation of petroleum as well as some mineral deposits, and plate boundaries. Furthermore, the formation and distribution of some natural resources are related to plate movement.

- The relationship between plate tectonic processes and the evolution of life is complex. The distribution of plants and animals is not random, but is controlled mostly by climate and geographic barriers, which are influenced, to a great extent, by the movement of plates.

Important Terms

abyssal plain (p. 33)
continental–continental plate boundary (p. 46)
continental drift (p. 29)
continental margin (p. 32)
continental rise (p. 33)
continental shelf (p. 32)
continental slope (p. 33)
convergent plate boundary (p. 44)
Curie point (p. 36)
divergent plate boundary (p. 40)

Glossopteris flora (p. 28)
Gondwana (p. 28)
hot spot (p. 48)
Laurasia (p. 29)
magnetic anomaly (p. 37)
magnetic field (p. 35)
magnetic reversal (p. 36)
magnetism (p. 35)
oceanic–continental plate boundary (p. 46)
oceanic–oceanic plate boundary (p. 44)

oceanic ridge (p. 33)
oceanic trench (p. 33)
paleomagnetism (p. 36)
Pangaea (p. 29)
plate tectonic theory (p. 39)
seafloor spreading (p. 36)
submarine hydrothermal vent (p. 33)
thermal convection cell (p. 37)
transform fault (p. 47)
transform plate boundary (p. 47)

Review Questions

1. The man credited with developing the continental drift hypothesis is
   a. ____ Wilson.
   b. ____ Wegener.
   c. ____ Hess.
   d. ____ du Toit.
   e. ____ Vine.

2. The areas on the seafloor where oceanic lithosphere is subducted are called
   a. ____ submarine canyons.
   b. ____ oceanic trenches.
   c. ____ black smokers.
   d. ____ guyots.
   e. ____ seamounts.

3. The southern part of Pangaea, consisting of South America, Africa, India, Australia, and Antarctica, is called
   a. ____ Laurasia.
   b. ____ Gondwana.
   c. ____ Panthalassa.
   d. ____ Laurentia.
   e. ____ Pacifica.

4. Hot spots and aseismic ridges can be used to determine the
   a. ____ location of divergent plate boundaries.
   b. ____ absolute motion of plates.
   c. ____ location of magnetic anomalies in oceanic crust.
   d. ____ relative motion of plates.
   e. ____ location of convergent plate boundaries.

5. Along what type of boundary does subduction occur?
   a. ____ divergent.
   b. ____ transform.
   c. ____ convergent.
   d. ____ answers a and b.
   e. ____ answers a and c.

6. The most common biotic province boundaries are
   a. ____ geographic barriers.
   b. ____ biologic barriers.
   c. ____ climatic barriers.
   d. ____ answers a and b.
   e. ____ answers a and c.

7. The driving mechanism of plate movement is thought to be
   a. ____ isostasy.
   b. ____ Earth's rotation.
   c. ____ thermal convection cells.
   d. ____ magnetism.
   e. ____ polar wandering.

8. Magnetic surveys of the ocean basins indicate that
   a. ____ the oceanic crust is youngest adjacent to mid-oceanic ridges.
   b. ____ the oceanic crust is oldest adjacent to mid-oceanic ridges.
   c. ____ the oceanic crust is youngest adjacent to the continents.
   d. ____ the oceanic crust is the same age everywhere.
   e. ____ answers b and c.

9. Which one of the following is characteristic of a passive continental margin?
   a. ____ broad continental shelf.
   b. ____ volcanic activity.
   c. ____ narrow abyssal plain.
   d. ____ submarine hydrothermal vents.
   e. ____ aseismic ridges.

10. The Andes Mountains are a good example of what type of plate boundary?
    a. ____ continental–continental.
    b. ____ oceanic–oceanic.
    c. ____ oceanic–continental.
    d. ____ divergent.
    e. ____ transform.

11. Explain why global diversity increases with an increase in the number of biotic provinces. How does plate movement affect the number of biotic provinces?

12. What evidence convinced Wegener and others that continents must have moved in the past and at one time formed a supercontinent?

13. Why was the continental drift hypothesis proposed by Wegener rejected by so many geologists for so long?

14. How did the theory of seafloor spreading, proposed by Harry Hess in 1962, overcome the objections of those opposed to continental drift?

15. Why is some type of thermal convection system thought to be the major force driving plate movement? How have slab-pull and ridge-push, both mainly gravity driven, modified a purely thermal convection model for plate movement?

16. The most distant part of a 30-million-year-old aseismic ridge is 1,000 km from an oceanic ridge. How fast, on average, did the plate with this ridge move in centimeters per year?

17. In addition to the volcanic eruptions and earthquakes associated with convergent and divergent plate boundaries, why are these boundaries associated with the formation and accumulation of various metallic ore deposits?

18. Based on your knowledge of biology and the distribution of organisms throughout the world, how do you think plate tectonics has affected this distribution both on land and in the oceans?

19. Estimate the age of the seafloor crust and the age and thickness of the oldest sediment off the East Coast of the United States (e.g., Virginia). In so doing, refer to Figure 2.15 for the ages and to the deep-sea sediment accumulation rate stated in this chapter.

20. Plate tectonic theory builds on the continental drift hypothesis and the theory of seafloor spreading. As such, it is a unifying theory of geology. Explain why it is a unifying theory.

This object is a geode, which forms when minerals grow along the margins of a cavity, partially filling it. The outermost layers are thin, sometimes discontinuous layers of agate, which are composed of tiny crystals. Next inward in this geode toward the cavity's center are quartz crystals that point inward. Many geodes are found in volcanic rocks, but they also occur in other types of rocks.

CHAPTER 3

Minerals—The Building Blocks of Rocks

OUTLINE

3.1 Introduction

3.2 Matter—What Is It?

3.3 Explore the World of Minerals

3.4 Mineral Groups Recognized by Geologists

3.5 Mineral Identification

GEO-INSIGHT: The Precious Metals

3.6 Rock-Forming Minerals

GEO-FOCUS: Gems and Gemstones

3.7 How Do Minerals Form?

3.8 Natural Resources and Reserves

GEO-RECAP

HAVE YOU EVER WONDERED?

- What minerals are and how they form?
- How to distinguish minerals from other naturally occurring substances?
- Why so few minerals are abundant and yet there are so many of them?
- What criteria geologists use to identify and classify minerals?
- Which minerals are valued as gemstones?
- What kinds of minerals are important to industrialized societies?

James Monroe

3.1 Introduction

Liquid water and water vapor are not minerals, but ice is because it meets the criteria geologists use to define the term **mineral:** a naturally occurring, inorganic, crystalline solid, with a narrowly defined chemical composition and characteristic physical properties. Of course, ice is naturally occurring and inorganic, that is, it is not composed of the complex carbon-based molecules found in organisms. Chemically, it is composed of hydrogen and oxygen (H_2O), and it has distinguishing properties such as hardness, color, and density. The only part of the definition of a mineral that may be unfamiliar is *crystalline solid,* which means that the atoms in a mineral are arranged in a specific three-dimensional framework, as opposed to the atoms in glass, which have no such orderly arrangement.

We cannot overstate the importance of minerals in many human endeavors. Ores of iron, copper, tin, and aluminum, as well as minerals and rocks used to make glass, wallboard, fertilizers, and animal feed supplements, are essential to our economic well-being. The United States and Canada owe much of their economic success and standard of living to abundant mineral and energy resources. Nevertheless, both nations must import all of some essential commodities such as aluminum ore and manganese. Indeed, the distribution of economic minerals, rocks, and energy resources has important implications in foreign relations and accounts for economic ties among nations. Even some rather common minerals are valuable: Quartz is used for sandpaper and glass, whereas muscovite is used in eye liner and glitter, and it accounts for the lustrous sheen of paint used on automobiles and appliances.

One reason to study minerals is their economic importance, but for students it is important to know that minerals are the essential constituents of rocks (Figure 3.1). Granite, for instance, is made up of specific percentages of quartz, potassium feldspars, and plagioclase feldspars, whereas most sandstone is made up primarily of quartz. Many minerals are attractive and eagerly sought out by collectors and for museum displays, and some are known as gemstones—that is, minerals and rocks that are used for decorative purposes, especially jewelry (see Geo-Focus on pages 76 and 77).

From our discussion so far, we have a formal definition of the term *mineral,* and we know that rocks are made up of minerals. Now let's delve deeper into what minerals are made of by considering matter, atoms, bonding, and elements.

3.2 Matter—What Is It?

Anything that has mass and occupies space is matter, and accordingly includes water, all organisms, and the atmosphere, as well as minerals and rocks. Physicists recognize four states of matter; *plasma* (composed of ionized gases as in the Sun and stars), *liquids, gases,* and *solids.* Liquids and gases are important in our understanding of several surface processes such as running water and wind, but here our main concern is solids, because, by definition, minerals are solids. That is, they are rigid substances that resist changes in shape or volume.

Figure 3.1 Minerals and Rocks

a Quartz is a common mineral that is found in many rocks. It is also the most common mineral in sand.

© blickwinkel/Alamy

b This metamorphic rock lies on the shoreline of Lake Superior at Marquette, Michigan. It is made up of several minerals as are most rocks.

James S. Monroe

》 Critical Thinking Question
You must explain to an interested audience the distinction between minerals and rocks. How would you do so, and can you think of any analogies that might clarify the points you make?

Atoms and Elements

Matter is made up of chemical **elements,** which, in turn, are composed of **atoms,** the smallest units of matter that retain the characteristics of a particular element (❚ Figure 3.2). That is, elements cannot be changed into different substances except through radioactive decay (discussed in Chapter 17). Thus, an element is made up of atoms, all of which have the same properties. Scientists have discovered 92 naturally occurring elements, some of which are listed in Figure 3.2. All naturally occurring elements have a name and a symbol—for example, oxygen (O), aluminum (Al), and potassium (K) (❚ Figure 3.3).

			Distribution of Electrons			
		Atomic	First	Second	Third	Fourth
Element	Symbol	Number	Shell	Shell	Shell	Shell
Hydrogen	H	1	1	—	—	—
Helium	He	2	2	—	—	—
Carbon	C	6	2	4	—	—
Oxygen	O	8	2	6	—	—
Neon	Ne	10	2	8	—	—
Sodium	Na	11	2	8	1	—
Magnesium	Mg	12	2	8	2	—
Aluminum	Al	13	2	8	3	—
Silicon	Si	14	2	8	4	—
Phosphorus	P	15	2	8	5	—
Sulfur	S	16	2	8	6	—
Chlorine	Cl	17	2	8	7	—
Potassium	K	19	2	8	8	1
Calcium	Ca	20	2	8	8	2
Iron	Fe	26	2	8	14	2

HYDROGEN
1 p⁺, 1 e⁻

HELIUM
2 p⁺, 2 e⁻

OXYGEN
8 p⁺, 8 e⁻

NEON
10 p⁺, 10 e⁻

SILICON
14 p⁺, 14 e⁻

IRON
26 p⁺, 26 e⁻

❚ **Figure 3.2 Shell Models for Common Atoms** The shell model for several atoms and their electron configurations. A blue circle represents the nucleus of each atom, but remember that atomic nuclei are made up of protons and neutrons, as shown in Figure 3.4.

Note: Atomic masses are 1993 IUPAC values (up to four decimal places). Numbers in parentheses are atomic masses or mass numbers of the most stable isotope of an element.

❚ **Figure 3.3 The Periodic Table of Elements** Only about a dozen elements are common in minerals and rocks, but many uncommon ones are important sources of natural resources. For example, lead (Pb) is not found in many minerals, but it is present in the mineral galena, the main ore of lead. Silicon (Si) and oxygen (O), in contrast, are important elements in most of the minerals in Earth's crust.

At the center of an atom is a **nucleus** made up of one or more particles known as **protons,** which have a positive electrical charge, and **neutrons,** which are electrically neutral. The nucleus is only about 1/100,000 of the diameter of an atom, yet it contains virtually all of the atom's mass. **Electrons,** particles with a negative electrical charge, orbit rapidly around the nucleus at specific distances in one or more **electron shells.** The electrons control how an atom interacts with other atoms, but the nucleus determines how many electrons an atom has, because the positively charged protons attract and hold the negatively charged electrons in their orbits.

The number of protons in its nucleus determines an atom's identity and its **atomic number.** Hydrogen (H), for instance, has one proton in its nucleus and thus has an atomic number of 1. The nuclei of helium (He) atoms possess 2 protons, whereas those of carbon (C) have 6, and uranium (U) 92, so their atomic numbers are 2, 6, and 92, respectively. Atoms also have an **atomic mass number,** which is the sum of protons and neutrons in the nucleus (electrons contribute negligible mass to atoms). However, atoms of the same chemical element might have different atomic mass numbers because the number of neutrons can vary. All carbon (C) atoms have six protons—otherwise they would not be carbon—but the number of neutrons can be 6, 7, or 8. Thus, we recognize three types of carbon, or what are known as *isotopes* (❙ Figure 3.4), each with a different atomic mass number.

The isotopes of carbon, or those of any other element, behave the same chemically; carbon 12 and carbon 14 are both present in carbon dioxide (CO_2), for example. However, some isotopes are radioactive, meaning that they spontaneously decay or change to other elements. Carbon 14 is radioactive, whereas both carbon 12 and carbon 13 are not. Radioactive isotopes are important for determining the absolute ages of rocks (see Chapter 17).

> **CONNECTION LINK**
>
> You can learn more about the role of radioactive isotopes in absolute dating on pages 432 and 433 in Chapter 17.

Bonding and Compounds

Interactions among electrons around atoms can result in two or more atoms joining together, a process known as **bonding.** If atoms of two or more elements bond, the resulting substance is a **compound.** Gaseous oxygen consists of only oxygen atoms and is thus an element, whereas the mineral quartz, consisting of silicon and oxygen atoms, is a compound. Most minerals are compounds, although gold, platinum, and several others are important exceptions.

To understand bonding, it is necessary to delve deeper into the structure of atoms. Recall that negatively charged electrons orbit the nuclei of atoms in electron shells. With the exception of hydrogen, which has only one proton and one electron, the innermost electron shell of an atom contains only two electrons. The other shells contain various numbers of electrons, but the outermost shell never has more than eight (Figure 3.2), and it is these that are usually involved in chemical bonding.

Two types of chemical bonds, *ionic* and *covalent,* are particularly important in minerals, and many minerals contain both types of bonds. Two other types of chemical bonds, *metallic* and *van der Waals,* are much less common, but are very important in determining the properties of some useful minerals.

Ionic Bonding   Notice in Figure 3.2 that most atoms have fewer than eight electrons in their outermost electron shell. However, some elements, including neon and argon, have complete outer shells with eight electrons; because of this electron configuration, these elements, known as the *noble gases,* do not react readily with other elements to form compounds. Interactions among atoms tend to produce electron configurations similar to those of the noble gases. That is, atoms interact so that their outermost electron shell is filled with eight electrons, unless the first shell (with two electrons) is also the outermost electron shell, as in helium.

One way that the noble gas configuration is attained is by the transfer of one or more electrons from one atom to another. Common salt is composed of the elements sodium (Na) and chlorine (Cl); each element is poisonous, but when combined chemically they form the compound sodium chloride (NaCl), the mineral halite. Notice in ❙ Figure 3.5a that sodium has 11 protons and 11 electrons; thus the positive electrical charges of the protons are exactly balanced by the negative charges of the electrons, and the atom is electrically neutral. Likewise, chlorine with 17 protons and 17 electrons is electrically neutral (Figure 3.5a). However, neither sodium nor chlorine has eight electrons in its outermost

^{12}C (Carbon 12)

^{13}C (Carbon 13)

^{14}C (Carbon 14)

❙ **Figure 3.4 Carbon Isotopes**
Schematic representation of the isotopes of carbon. Carbon has an atomic number of 6 and an atomic mass number of 12, 13, or 14, depending on the number of neutrons in its nucleus.

electron shell; sodium has only one, whereas chlorine has seven. To attain a stable configuration, sodium loses the electron in its outermost electron shell, leaving its next shell with eight electrons as the outermost one (Figure 3.5a). Sodium now has one fewer electron (negative charge) than it has protons (positive charge), so it is an electrically charged **ion** and is symbolized Na^+.

The electron lost by sodium is transferred to the outermost electron shell of chlorine, which had seven electrons to begin with. The addition of one more electron gives chlorine an outermost electron shell of eight electrons, the configuration of a noble gas. But its total number of electrons is now 18, which exceeds by 1 the number of protons. Accordingly, chlorine also becomes an ion, but it is negatively charged (Cl^-). An **ionic bond** forms between sodium and chlorine because of the attractive force between the positively charged sodium ion and the negatively charged chlorine ion (Figure 3.5a).

In ionic compounds, such as sodium chloride (the mineral halite), the ions are arranged in a three-dimensional framework that results in overall electrical neutrality. In halite, sodium ions are bonded to chlorine ions on all sides, and chlorine ions are surrounded by sodium ions (▍ Figure 3.5b).

Covalent Bonding

Covalent bonds form between atoms when their electron shells overlap and they share electrons. For example, atoms of the same element, such as carbon, cannot

bond by transferring electrons from one atom to another. Carbon (C), which forms the minerals graphite and diamond, has four electrons in its outermost electron shell (▍ Figure 3.6a). If these four electrons were transferred to another carbon atom, the atom receiving the electrons would have the noble gas configuration of eight electrons in its outermost electron shell, but the atom contributing the electrons would not.

In such situations, adjacent atoms share electrons by overlapping their electron shells. A carbon atom in diamond shares all four of its outermost electrons with a neighbor to produce a stable noble gas configuration (Figure 3.6a). Diamond and graphite, both crystalline forms of carbon, have covalent bonds, but in diamond the atoms form a three-dimensional framework whereas in graphite the carbon atoms form sheets that are weakly bonded together (▍ Figures 3.6b and c).

Covalent bonds are not restricted to substances composed of atoms of a single kind. Among the most common minerals, the silicates (discussed later in this chapter), the element silicon forms partly covalent and partly ionic bonds with oxygen.

Metallic and van der Waals Bonds

Metallic bonding involves an extreme type of electron sharing. The electrons of the outermost electron shell of metals such as gold, silver, and copper readily move about from one atom to another. This electron mobility accounts for the metallic luster of

▍ **Figure 3.5** Ionic Bonding and the Origin of the Mineral Halite (NaCl)

a Transfer of the electron in the outermost shell of sodium to the outermost shell of chlorine. After electron transfer, the sodium and chlorine atoms are positively and negatively charged ions, respectively.

b This diagram shows the relative sizes of the sodium and chlorine atoms and their locations in a crystal of halite.

c Tiny crystals of halite.

❚ Figure 3.6 Covalent Bonds

a The orbits in the outermost electron shell overlap, so electrons are shared in diamond.

b Covalent bonding of carbon atoms in diamond forms a three-dimensional framework.

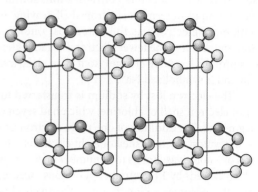

c Covalent bonds in graphite form strong sheets, but the van der Waals bonds between sheets are weak.

metals (their appearance in reflected light), their good electrical and thermal conductivity, and their ability to be easily reshaped. Only a few minerals possess metallic bonds, but those that do are very useful; copper, for example, is used for electrical wiring because of its high electrical conductivity.

Some electrically neutral atoms and molecules* have no electrons available for ionic, covalent, or metallic bonding. They nevertheless have a weak attractive force between them, called a *van der Waals* or *residual bond,* when in proximity. The carbon atoms in the mineral graphite are covalently bonded to form sheets, but the sheets are weakly held together by van der Waals bonds (Figure 3.6c). This type of bonding makes graphite useful for pencil lead—when a pencil point is moved across a piece of paper, small pieces of graphite flake off along the planes held together by van der Waals bonds and adhere to the paper.

3.3 Explore the World of Minerals

We defined a *mineral* as a naturally occurring, inorganic crystalline solid with a narrowly defined chemical composition and characteristic physical properties. In the following sections, we will examine each part of the formal definition of the term *mineral*.

Naturally Occurring Inorganic Substances

The criterion *naturally occurring* excludes from minerals all manufactured substances such as synthetic diamonds and rubies. This criterion is particularly important to those who buy and sell gemstones, most of which are minerals, because

some human-made substances are very difficult to distinguish from natural gem minerals.

Some geologists think the term *inorganic* in the mineral definition is unnecessary, but it does remind us that animal matter and vegetable matter are not minerals. Nevertheless, some organisms, including corals, clams, and a number of other animals and some plants, construct their shells of calcium carbonate ($CaCO_3$), which is either the mineral aragonite or calcite, or their shells are made of silicon dioxide (SiO_2) as in quartz.

Mineral Crystals

By definition, minerals are **crystalline solids,** in which the constituent atoms are arranged in a regular, three-dimensional framework (Figure 3.5b). Under ideal conditions, such as in a cavity, these crystalline solids can grow and form perfect **crystals** that possess planar surfaces (crystal faces), sharp corners, and straight edges (❚ Figure 3.7). In other words, the regular geometric shape of a well-formed mineral crystal is the exterior manifestation of an ordered internal atomic arrangement. Not all rigid substances are crystalline solids; natural and manufactured glass lack the ordered arrangement of atoms and is said to be *amorphous,* meaning "without form." Minerals are crystalline solids; however, crystalline solids do not always yield well-formed crystals. The reason is that when crystals form, they may grow in proximity and form an interlocking mosaic in which individual crystals are not apparent or easily discerned (❚ Figure 3.8).

So how do we know that the mineral in Figure 3.8b is actually crystalline? X-ray beams and light transmitted through mineral crystals or crystalline solids behave in a predictable manner, which provides compelling evidence for an internal orderly structure. Another way we can determine that minerals with no obvious crystals are actually crystalline is by their *cleavage,* the property of breaking or splitting repeatedly along smooth, closely spaced planes. Not all minerals have cleavage planes, but many do, and such regularity certainly indicates that splitting is controlled by internal structure.

---

*A molecule is the smallest unit of a substance that has the properties of that substance. A water molecule (H_2O), for example, possesses two hydrogen atoms and one oxygen atom.

Figure 3.7 A Variety of Mineral Crystal Shapes

a Cubic crystals are typical of the minerals halite and galena.

b Pyritohedron crystals such as those of pyrite have 12 sides.

c Diamond has octahedral, or eight-sided, crystals.

d A prism terminated by a pyramid is found in quartz.

As early as 1669, the Danish scientist Nicholas Steno determined that the angles of intersection of equivalent crystal faces on different specimens of quartz are identical. Since then, this *constancy of interfacial angles* has been demonstrated for many other minerals, regardless of their size,

shape, age, or geographic occurrence (Figure 3.8c). Steno postulated that mineral crystals are made up of very small, identical building blocks, and that the arrangement of these building blocks determines the external form of mineral crystals, a proposal that has since been verified.

Figure 3.8 Quartz and the Constancy of Interfacial Angles

a Well-shaped crystal of smoky quartz.

b Specimen of rose quartz in which no obvious crystals can be discerned.

c Side views and cross sections of quartz crystals showing the constancy of interfacial angles. A well-shaped crystal (left), a larger well-shaped crystal (middle), and a poorly shaped crystal (right). The angles formed between equivalent crystal faces on different specimens of the same mineral are the same regardless of size, shape, age, or geographic occurrence of the specimens.

Chemical Composition of Minerals

The chemical composition of a mineral is shown by a formula, which is a shorthand way of indicating the numbers of atoms of different elements. The mineral quartz consists of one silicon (Si) atom for every two oxygen (O) atoms and thus has the formula SiO_2; the subscript number indicates the number of atoms. Orthoclase is composed of one potassium, one aluminum, three silicon, and eight oxygen atoms, so its formula is $KAlSi_3O_8$. Some minerals known as *native elements* consist of a single element and include silver (Ag), platinum (Pt), gold (Au), and graphite and diamond, both of which are composed of carbon (C).

For many minerals, the chemical composition does not vary. Quartz is made up only of silicon and oxygen $(SiO)_2$, and halite contains only sodium and chlorine (NaCl). However, some minerals have a range of compositions because one element can substitute for another if the atoms of two or more elements are nearly the same size and the same charge. Thus, the definition of the term *mineral* has the phrase *narrowly defined chemical composition*. Iron and magnesium meet these criteria and can substitute for one another (∎ Figure 3.9). The chemical formula for olivine is $(Mg,Fe)_2SiO_4$, meaning that in addition to silicon and oxygen, it may contain magnesium, iron, or both. A number of other minerals also have ranges of composition, so they are actually mineral groups with several members.

Physical Properties of Minerals

The last criterion in our definition of a mineral, *characteristic physical properties*, refers to such properties as hardness, color, and crystal form. These properties are controlled by composition and structure. We will have more to say about the physical properties of minerals later in this chapter.

3.4 Mineral Groups Recognized by Geologists

Geologists have identified and described more than 3,500 minerals, but only a few—perhaps two dozen—are common. One might think that an extremely large number of minerals could form from 92 naturally occurring elements; however, the bulk of Earth's crust is made up of only eight chemical elements, and even among these, silicon and oxygen are by far the most common. In fact, most common minerals in Earth's crust consist of silicon, oxygen, and one or more of the elements in ∎ Figure 3.10.

Geologists recognize mineral classes or groups, each with members that share the same negatively charged ion or ion group (Table 3.1). We have mentioned that ions are atoms that have either a positive or negative electrical

Negatively charged ions		Positively charged ions			
2^-	1^-	1^+	2^+	3^+	4^+
1.40 Oxygen	1.36 Fluorine	0.99 Sodium	1.00 Calcium	0.39 Aluminum	0.26 Silicon
1.84 Sulfur	1.81 Chlorine	1.37 Potassium	0.63 Iron^{2+}	0.49 Iron^{3+}	0.15 Carbon
			0.72 Magnesium		

∎ **Figure 3.9 Sizes and Charges of Ions** Electrical charges and relative sizes of ions common in minerals. The numbers within the ions are the radii shown in Ångstrom units. [1 Ångstrom = 10^{-8} cm]

charge resulting from the loss or gain of electrons in their outermost shell. In addition to ions, some minerals contain tightly bonded, complex groups of different atoms known as *radicals* that act as single units. A good example is the carbonate radical, consisting of a carbon atom bonded to three oxygen atoms and thus having the formula CO_3 and a -2 electrical charge. Other common radicals and their charges are sulfate (SO_4, -2), hydroxyl (OH, -1), and silicate (SiO_4, -4) (▌Figure 3.11).

Silicate Minerals

Because silicon and oxygen are the two most abundant elements in Earth's crust, it is not surprising that many minerals contain these elements. A combination of silicon and oxygen is known as **silica,** and minerals that contain silica are **silicates.** Quartz (SiO_2) is pure silica because it is composed entirely of silicon and oxygen. But most silicates have one or more additional elements, as in orthoclase ($KAlSi_3O_8$)

▌**Figure 3.10** Common Elements in Earth's Crust

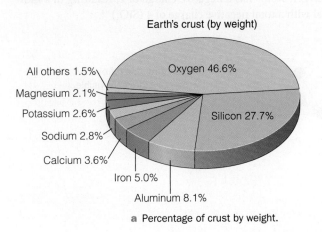

a Percentage of crust by weight.

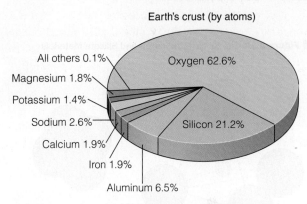

b Percentage of crust by atoms.

Source: From G. T. Miller, 1996. *Living in the Environment: Principles, Concepts, and Solutions.* Wadsworth Publishing. Figure 8.3.

TABLE 3.1	Mineral Groups Recognized by Geologists		
Mineral Group	**Negatively Charged Ion or Radical**	**Examples**	**Composition**
Carbonate	$(CO_3)^{-2}$	Calcite	$CaCO_3$
		Dolomite	$CaMg(CO_3)_2$
Halide	Cl^{-1}, F^{-1}	Halite	$NaCl$
		Fluorite	CaF_2
Hydroxide	$(OH)^{-1}$	Brucite	$Mg(OH)_2$
Native element	—	Gold	Au
		Silver	Ag^*
		Diamond	C
Phosphate	$(PO_4)^{-3}$	Apatite	$Ca_5(PO_4)_3(F,Cl)$
Oxide	O^{-2}	Hematite	Fe_2O_3
		Magnetite	Fe_3O_4
Silicate	$(SiO_4)^{-4}$	Quartz	SiO_2
		Potassium feldspar	$KAlSi_3O_8$
		Olivine	$(Mg,Fe)_2SiO_4$
Sulfate	$(SO_4)^{-2}$	Anhydrite	$CaSO_4$
		Gypsum	$CaSO_4 \cdot 2H_2O$
Sulfide	S^{-2}	Galena	PbS
		Pyrite	FeS_2
		Argentite	Ag_2S^*

*Note that silver is found as both a native element and a sulfide mineral.

Carbonate   Hydroxyl   Sulfate   Silica
CO_3 (−2)   OH (−1) SO_4 (−2) SiO_4 (−4)

▌Figure 3.11 **Radicals** Many minerals contain radicals, which are complex groups of atoms tightly bonded together. The silica and carbonate radicals are particularly common in many minerals, such as quartz (SiO_2) and calcite ($CaCO_3$).

and olivine [$(Mg,Fe)_2SiO_4$]. Silicate minerals include about one-third of all known minerals, but their abundance is even more impressive when one considers that they make up perhaps 95% of Earth's crust.

The basic building block of all silicate minerals is the **silica tetrahedron,** consisting of one silicon atom and four oxygen atoms (▌ Figure 3.12a). These atoms are arranged so that the four oxygen atoms surround a silicon atom, which occupies the space between the oxygen atoms, thus forming a four-faced pyramidal structure (▌ Figure 3.12b). The silicon atom has a positive charge of 4, and each of the four oxygen atoms has a negative charge of 2, resulting in a radical with a total negative charge of 4 $(SiO_4)^{-4}$.

▌Figure 3.12 The Silica Tetrahedron and Silicate Materials

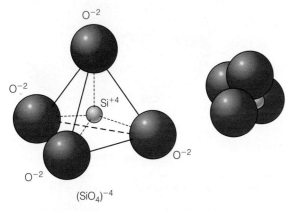

O^{-2}

O^{-2} Si^{+4}

O^{-2}

O^{-2}

$(SiO_4)^{-4}$

a Expanded view of the silica tetrahedron (left) and how it actually exists with its oxygen atoms touching.

b View of the silica tetrahedron from above. Only the oxygen atoms are visible.

		Formula of negatively charged ion group		Example
c	Isolated tetrahedra	$(SiO_4)^{-4}$	No oxygen atoms shared	Olivine
d	Continuous chains of tetrahedra	$(SiO_3)^{-2}$ Single chain	Each tetrahedra shares two oxygen atoms with adjacent tetrahedra	Pyroxene group (augite)
		$(Si_4O_{11})^{-6}$ Double chain	Single chains linked by sharing oxygen atoms	Amphibole group (hornblende)
e	Continuous sheets	$(Si_4O_{10})^{-4}$	Three oxygen atoms shared with adjacent tetrahedra	Micas (muscovite)
f	Three-dimensional networks	$(SiO_2)^0$ SiO_2	All four oxygen atoms in tetrahedra shared	Quartz Potassium feldspars Plagioclase feldspars

c–f Structures of the common silicate minerals shown by various arrangements of the silica tetrahedra.

▌Figure 3.13 Common Rock-Forming Silicate Minerals

a The ferromagnesian silicates.

b The nonferromagnesian silicates.

Because the silica tetrahedron has a negative charge, it does not exist in nature as an isolated ion group; rather, it combines with positively charged ions or shares its oxygen atoms with other silica tetrahedra. In the simplest silicate minerals, the silica tetrahedra exist as single units bonded to positively charged ions. In minerals that contain isolated tetrahedra, the silicon-to-oxygen ratio is 1:4, and the negative charge of the silica ion is balanced by positive ions (▌Figure 3.12c). Olivine [$(Mg,Fe)_2SiO_4$], for example, has either two magnesium (Mg^{+2}) ions, two iron (Fe^{+2}) ions, or one of each to offset the -4 charge of the silica ion.

Silica tetrahedra may also join together to form chains of indefinite length (▌Figure 3.12d). Single chains, as in the pyroxene minerals, form when each tetrahedron shares two of its oxygens with an adjacent tetrahedron, resulting in a silicon-to-oxygen ratio of 1:3. Enstatite, a pyroxene-group mineral, reflects this ratio in its chemical formula $MgSiO_3$. Individual chains, however, possess a net -2 electrical charge, so they are balanced by positive ions, such as Mg^{+2}, that link parallel chains together (Figure 3.12d).

The amphibole group of minerals is characterized by a double-chain structure in which alternate tetrahedra in two parallel rows are cross-linked (Figure 3.12d). The formation of double chains results in a silicon-to-oxygen ratio of 4:11, so each double chain possesses a -6 electrical charge. Mg^{+2}, Fe^{+2}, and Al^{+3} are usually involved in linking the double chains together.

In sheet-structure silicates, three oxygens of each tetrahedron are shared by adjacent tetrahedra (▌Figure 3.12e). Such structures result in continuous sheets of silica tetrahedra with silicon-to-oxygen ratios of 2:5. Continuous sheets also possess a negative electrical charge satisfied by positive ions located between the sheets. This particular structure accounts for the characteristic sheet structure of the *micas*, such as biotite and muscovite, and the *clay minerals*.

Three-dimensional networks of silica tetrahedra form when all four oxygens of the silica tetrahedra are shared by adjacent tetrahedra (▌Figure 3.12f). Such sharing of oxygen atoms results in a silicon-to-oxygen ratio of 1:2, which is electrically neutral. Quartz is a common framework silicate.

Geologists define two subgroups of silicates: ferromagnesian and nonferromagnesian silicates. The **ferromagnesian silicates** are those that contain iron (Fe), magnesium (Mg), or both. These minerals are commonly dark and more dense than nonferromagnesian silicates. Some of the common ferromagnesian silicate minerals are olivine, the pyroxenes, the amphiboles, and biotite (▌Figure 3.13a).

The **nonferromagnesian silicates** lack iron and magnesium, are generally light colored, and are less dense than ferromagnesian silicates (▌Figure 3.13b). The most common minerals in Earth's crust are nonferromagnesian silicates known as *feldspars*. Feldspar is a general name, however, and includes two distinct groups, each of which consists of several species. The *potassium feldspars* are represented by microcline and orthoclase ($KAlSi_3O_8$). The second group of feldspars, the *plagioclase feldspars*, range from calcium-rich ($CaAl_2Si_2O_8$) to sodium-rich ($NaAlSi_3O_8$) varieties.

Quartz (SiO_2) is another common nonferromagnesian silicate. It is a framework silicate that can usually be recognized by its glassy appearance and hardness. Another fairly common nonferromagnesian silicate is muscovite, which is a mica (Figure 3.13b).

Carbonate Minerals

Carbonate minerals, are those that contain the negatively charged carbonate radical (CO_3)$^{-2}$ and include calcium carbonate ($CaCO_3$) as the minerals *aragonite* or *calcite* (▌Figure 3.14a). Aragonite is unstable and commonly changes to calcite, the main constituent of the sedimentary rock *limestone*.

Several other carbonate minerals are known, but only one of these need concern us: *Dolomite* [CaMg(CO$_3$)$_2$] forms by the chemical alteration of calcite by the addition of magnesium. Sedimentary rock composed of the mineral dolomite is *dolostone* (see Chapter 7).

Other Mineral Groups

Even though minerals from the other groups in Table 3.1 are less common, many are found in rocks in small quantities and others are important resources. In the oxides, an element combines with oxygen as in hematite (Fe$_2$O$_3$) and magnetite (Fe$_3$O$_4$). Rocks with high concentrations of these minerals in the Lake Superior region of Canada and the United States are sources of iron ores for the manufacture of steel. The related hydroxides form mostly by the chemical alteration of other minerals.

We have noted that the *native elements* are minerals composed of a single element, such as diamond and graphite (C) and the precious metals gold (Au), silver (Ag), and platinum (Pt). Some elements, such as silver and copper, are found both as native elements and as compounds and are thus also included in other mineral groups; argentite (Ag$_2$S), a silver sulfide, is an example.

Several minerals and rocks that contain the phosphate radical (PO$_4$)$^{-3}$ are important sources of phosphorus for fertilizers. The sulfides, such as galena (PbS), the ore of lead, have a positively charged ion combined with sulfur (S^{-2}) (∎ Figure 3.14b), whereas the sulfates have an element combined with the complex radical (SO$_4$)$^{-2}$, as in gypsum (CaSO$_4$·2H$_2$O) (∎ Figure 3.14c). The halides contain the halogen elements, fluorine (F^{-1}) and chlorine (Cl^{-1}); examples are halite (NaCl) (∎ Figure 3.14d) and fluorite (CaF$_2$).

3.5 Mineral Identification

Many physical properties of minerals are remarkably constant for a given mineral species, but some, especially color, may vary. Although professional geologists use sophisticated techniques to study and identify minerals, you can identify most common minerals by using the physical properties described next.

∎ **Figure 3.14** Representative Specimens from Four Mineral Groups

a Calcite (CaCO$_3$) is the most common carbonate mineral.

b The sulfide mineral galena (PbS) is the ore of lead.

c Gypsum (CaSO$_4$·2H$_2$O) is a common sulfate mineral.

d Halite (NaCl) is a good example of a halide mineral.

》 Critical Thinking Question
Which of these minerals has a metallic luster and nonmetallic luster?

Luster and Color

Luster (not to be confused with *color*) is the quality and intensity of light reflected from a mineral's surface. Geologists define two types of luster: *metallic,* having the appearance of a metal, and *nonmetallic.* Notice that of the four minerals shown in Figure 3.14, only galena has a metallic luster. Among the several types of nonmetallic luster are glassy or vitreous (as in quartz), dull or earthy, waxy, greasy, and brilliant (as in diamond).

Beginning geology students are distressed by the fact that the color of some minerals varies, making the most obvious physical property of little use for mineral identification. In any case, we can make some helpful generalizations about color. Ferromagnesian silicates are typically black, brown, or dark green, although olivine is olive green (Figure 3.13a). Nonferromagnesian silicates, on the other hand, vary in color, but are rarely very dark. White, cream, colorless, and shades of pink and pale green are typical (Figure 3.14b).

Another helpful generalization is that the color of minerals with a metallic luster is more consistent than is the color of nonmetallic minerals. For example, galena is always lead-gray (Figure 3.13b) and pyrite is invariably brassy yellow. In contrast, quartz, a nonmetallic mineral, may be colorless, smoky brown to almost black, rose, yellow-brown, milky white, blue, or violet to purple (Figure 3.8a, b).

Crystal Form

As we noted, many mineral specimens do not show the perfect crystal form typical of that mineral species. Nevertheless, some minerals do commonly occur as crystals. For example, 12-sided crystals of garnet are common, as are 6- and 12-sided crystals of pyrite. Minerals that grow in cavities or are precipitated from circulating hot water (hydrothermal solutions) in cracks and crevices in rocks also commonly occur as crystals and, under ideal conditions, some minerals grow to incredible sizes (▌ Figure 3.15).

Crystal form is a useful characteristic for mineral identification, but a number of minerals have the same crystal form. Pyrite (FeS_2), galena (PbS), and halite ($NaCl$) all occur as cubic crystals, but they can be easily identified by other properties such as color, luster, hardness, and density (▌ Figure 3.16).

▌ **Figure 3.16** Cubic Crystals of Pyrite, Fluorite, and Halite

a Pyrite (FeS_2).

b Fluorite (CaF_2).

c Halite ($NaCl$).

▌ **Figure 3.15** Giant Gypsum Crystals In April 2000, giant mineral crystals were found in a silver and lead mine in Mexico. A cavity there is lined with hundreds of gypsum ($CaSO_4 \cdot 2H_2O$) crystals more than 1 m long with what one author called "crystal moonbeams," which are gypsum crystals 15.2 m long and 1.2 m in diameter.

>> **Critical Thinking Question**
Refer to the rock cycle in Figure 1.15 and explain at what points in the cycle you would expect minerals to form.

>> **Critical Thinking Question**
These three minerals all occur as cubic crystals, so how can you tell them apart?

GEO-INSIGHT   The Precious Metals

The discovery of gold by James Marshall at Sutter's Mill near Coloma in 1848 sparked the California gold rush (1849–1853) during which $200 million in gold was recovered.

▲ **1.** Specimen of gold from Grass Valley, California. Gold is too heavy and too soft for tools and weapons, so it has been prized for jewelry and as a symbol of wealth, but it is also used in glass making, electrical circuitry, gold plating, the chemical industry, and dentistry.

Gold rush belt

PACIFIC OCEAN

NEVADA
Carson City
Lake Tahoe
Placerville
Sacramento
San Francisco
Stockton
Sierra Nevada
Yosemite National Park
CALIFORNIA
Colorado River

▼ **2.** A miner pans for gold (foreground) by swirling water, sand, and gravel in a broad, shallow pan. The heavier gold sinks to the bottom. At the far left, a miner washes sediment in a cradle. As in panning, the cradle separates the heavier gold from other materials.

▲ **3.** Gold miners on the American River near Sacramento, California. Most of the gold came from placer deposits in which running water separated and concentrated minerals and rock fragments by their density.

4. Hydraulic mining in California in which strong jets of water washed gold-bearing sand and gravel into sluices. In this image taken in 1905 at Junction City, California, water is directed through a monitor onto a hillside. Hydraulic mining was efficient from the mining point of view, but caused considerable environmental damage.

© Bettmann/CORBIS

5. Reports in 1876 of gold in the Black Hills of South Dakota resulted in a flood of miners that led to hostilities with the Sioux Indians and the annihilation of Lt. Col. George Armstrong Custer and 260 of his men at the Battle of the Little Big Horn in Montana. This view shows the headworks (upper right) of the Homestake Mine at Lead, South Dakota, in 1900. The headworks is the cluster of buildings near the opening to a mine.

John C.H. Grabill/CORBIS

© Ken Lucas/Visuals Unlimited

6. Like gold, silver is found as a native element as in this specimen, but it also occurs as a compound in the sulfide mineral argentite (Ag_2S). Silver is used in North America for silver halide film, jewelry, flatware, surgical instruments, and backing for mirrors.

7. ▶ 8. This image shows the headworks of the Yellowjacket Mine at Gold Hill, Nevada, and the inset shows silver-bearing quartz (white) in volcanic rock. This largest silver discovery in North America, called the Comstock Lode, was responsible for bringing Nevada into the Union in 1864 during the Civil War, even though it had too few people to qualify for statehood. The Comstock Lode was mined for silver and gold from 1859 until 1898.

James S. Monroe

Sue. Monroe

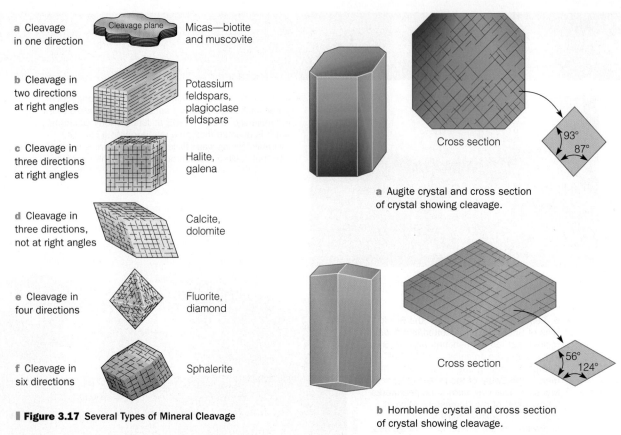

Figure 3.17 Several Types of Mineral Cleavage

a Augite crystal and cross section
of crystal showing cleavage.

b Hornblende crystal and cross section
of crystal showing cleavage.

Figure 3.18 Cleavages in Augite and Hornblende

Cleavage and Fracture

Not all minerals possess **cleavage,** but those that do break, or split, along smooth planes of weakness determined by the strength of their chemical bonds. Cleavage is characterized in terms of quality (perfect, good, poor), direction, and angles of intersection of cleavage planes. Biotite, a common ferromagnesian silicate, has perfect cleavage in one direction (◗ Figure 3.17a). Biotite is a sheet silicate with the sheets of silica tetrahedra weakly bonded to one another by iron and magnesium ions.

Feldspars possess two directions of cleavage that intersect at right angles (◗ Figure 3.17b), and the mineral halite has three directions of cleavage, all of which intersect at right angles (◗ Figure 3.17c). Calcite also possesses three directions of cleavage, but none of the intersection angles is a right angle, so cleavage fragments of calcite are rhombohedrons (◗ Figure 3.17d). Minerals with four directions of cleavage include fluorite and diamond (◗ Figure 3.17e). Ironically, diamond, the hardest mineral, can be cleaved easily. A few minerals, such as sphalerite, an ore of zinc, have six directions of cleavage (◗ Figure 3.17f).

Cleavage is an important diagnostic property of minerals, and recognizing it is essential in distinguishing between some minerals. The pyroxene mineral augite and the amphibole mineral hornblende, for example, look much alike: Both are dark green to black, have the same hardness, and possess two directions of cleavage. But the cleavage planes of augite intersect at about 90 degrees, whereas the cleavage planes of hornblende intersect at angles of 56 degrees and 124 degrees (◗ Figure 3.18).

In contrast to cleavage, *fracture* is mineral breakage along irregular surfaces. Any mineral will fracture if enough force is applied, but the fracture surfaces are uneven or conchoidal (curved) rather than smooth.

Hardness

An Austrian geologist, Friedrich Mohs (1773–1839), devised a relative hardness scale for 10 minerals. He arbitrarily assigned a hardness value of 10 to diamond, the hardest mineral known, and lesser values to the other minerals. Relative hardness is easily determined by the use of Mohs hardness scale (Table 3.2). Quartz will scratch fluorite but cannot be scratched by fluorite, gypsum can be scratched by a fingernail, and so on. So **hardness** is defined as a mineral's resistance to abrasion and is controlled mostly by internal structure. For example, both graphite and diamond are composed of carbon, but the former has a hardness of 1 to 2, whereas the latter has a hardness of 10.

Specific Gravity (Density)

Specific gravity and density are two separate concepts, but here we will use them more or less as synonyms. A mineral's **specific gravity** is the ratio of its weight to the weight of an equal volume of pure water at 4°C. Thus, a mineral with a specific gravity of 3.0 is three times as heavy as water. **Density,** in contrast, is a mineral's mass (weight) per unit of volume expressed in grams per

TABLE 3.2	Mohs Hardness Scale	
Hardness	Mineral	Hardness of Some Common Objects
10	Diamond	
9	Corundum	
8	Topaz	
7	Quartz	
		Steel file (6½)
6	Orthoclase	
		Glass (5½–6)
5	Apatite	
4	Fluorite	
3	Calcite	Copper penny (3)
		Fingernail (2½)
2	Gypsum	
1	Talc	

TABLE 3.3	Important Rock-Forming Minerals
Mineral	Primary Occurrence
Ferromagnesian silicates	
Olivine	Igneous and metamorphic rocks
Pyroxene group Augite most common	Igneous and metamorphic rocks
Amphibole group Hornblende most common	Igneous and metamorphic rocks
Biotite	All rock types
Nonferromagnesian silicates	
Quartz	All rock types
Potassium feldspar group	
Orthoclase, microcline	All rock types
Plagioclase feldspar group	All rock types
Muscovite	All rock types
Clay mineral group	Soils, sedimentary rocks, and some metamorphic rocks
Carbonates	
Calcite	Sedimentary rocks
Dolomite	Sedimentary rocks
Sulfates	
Anhydrite	Sedimentary rocks
Gypsum	Sedimentary rocks
Halides	
Halite	Sedimentary rocks

cubic centimeter. So the specific gravity of galena (Figure 3.14b) is 7.58 and its density is 7.58 g/cm³. In most instances, we will refer to a mineral's density, and in some of the following chapters, we will mention the density of various rocks.

Structure and composition control a mineral's specific gravity and density. Because ferromagnesian silicates contain iron, magnesium, or both, they tend to be denser than nonferromagnesian silicates. In general, the metallic minerals, such as galena and hematite, are denser than nonmetals. Pure gold with a density of 19.3 g/cm³ is about 2.5 times as dense as lead. Diamond and graphite, both of which are composed of carbon (C), illustrate how structure controls specific gravity or density. The specific gravity of diamond is 3.5, whereas that of graphite varies from 2.09 to 2.33.

Other Useful Mineral Properties

Talc has a distinctive soapy feel, graphite writes on paper, halite tastes salty, and magnetite is magnetic. Calcite possesses the property of *double refraction,* meaning that an object when viewed through a transparent piece of calcite will have a double image. Some sheet silicates are plastic and, when bent into a new shape, will retain that shape; others are flexible and, if bent, will return to their original position when the forces that bent them are removed.

A simple chemical test to identify the minerals calcite and dolomite involves applying a drop of dilute hydrochloric acid to the mineral specimen. If the mineral is calcite, it will react vigorously with the acid and release carbon dioxide, which causes the acid to bubble or effervesce. Dolomite, in contrast, will not react with hydrochloric acid unless it is powdered.

3.6 Rock-Forming Minerals

The term **rock** refers to a solid aggregate of one or more minerals, but it also refers to masses of mineral-like matter, as in the natural glass obsidian (see Chapter 4), and masses of solid

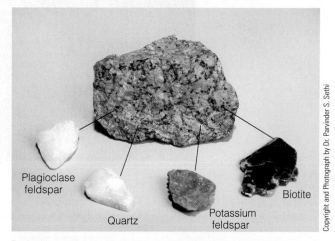

▌Figure 3.19 Minerals in Granite The igneous rock granite (see Chapter 4) is made up of mostly three minerals—quartz, potassium feldspar, and plagioclase feldspar—but it may also contain small amounts of biotite, muscovite, and hornblende.

organic matter, as in coal (see Chapter 7). And even though some rocks may contain many minerals, only a few, designated **rock-forming minerals,** are sufficiently common for rock identification and classification (Table 3.3 and ▌ Figure 3.19). Others, known as *accessory minerals,* are present in such small quantities that they can be disregarded.

GEO-FOCUS   Gems and Gemstones

Most gemstones are minerals—more rarely, rocks—that are cut, polished, or both for jewelry. Some gemstones, as well as fossils, have served as symbols of wealth and religious expression, been used as talismans, or been worn or ingested for their presumed mystical or curative powers. Their brilliance, durability, and scarcity are important qualities for gemstones, but relating them to birth month or some kind of lore gives them even more appeal.

Gemstones are actually the raw materials for gems. That is, gems have been fashioned in some way from gemstones by shaping, polishing, or cutting, the most notable exception being pearls which are ready to use when found. Many minerals and rocks are attractive but fail to qualify as gemstones. For instance, cut and polished fluorite is alluring, but it is fairly common and too soft to be durable. Diamond, in contrast, is a gemstone, although not all diamonds

are of gem quality; most are used in industrial applications such as diamond-tipped drill bits or saw blades.

Only about two dozen minerals and rocks are used as gemstones, termed either *precious* or *semiprecious*. Diamond, ruby and sapphire (red and blue varieties of the mineral corundum), and emerald (❚ Figure 1) are precious gemstones, but some people also include precious opal in this category. Some of the semiprecious gemstones are peridot

❚ **Figure 1** Emerald is a green variety of the mineral beryl, a beryllium aluminum silicate $(Be_2Al_2(SiO_3)_6$. This specimen on display at the Natural History Museum in Vienna, Austria, measures about 17 cm high.

Sue Monroe

What Would You Do ❓

The distinction between minerals and rocks is not easy for beginning students to understand. As a teacher, you know that minerals are made up of chemical elements and that rocks consist of one or more minerals, but despite your best efforts to clearly define them, your students commonly mistake one for the other. Can you think of analogies that might help students understand the difference between minerals and rocks?

Given that silicate minerals are by far the most common ones in Earth's crust, it follows that most rocks are composed of these minerals. Indeed, feldspar minerals (plagioclase feldspars and potassium feldspars) and quartz make up more than 60% of Earth's crust. So, even though there are hundreds of silicates, only a few are particularly common in rocks.

(olivine), garnet, topaz, tourmaline, turquoise, and amber, which is not a true mineral but rather hardened tree resin. Pearls are also semiprecious gemstones even though they are produced by organisms. A mollusk such as a clam or oyster forms pearls by depositing successive layers of tiny mineral crystals around some irritant such as a sand grain.

Transparent gemstones are most often cut to yield small, polished plane surfaces called facets that enhance the quality of reflected light. Gem cutters maximize the brilliance of diamond by faceting the gem so that it reflects as much light as possible (Figure 2). Opaque and translucent gemstones are rarely faceted. Usually they are shaped and polished into dome-shaped stones known as *cabochons,* or they are simply polished by tumbling.

Most people are surprised to learn that diamond, the most sought-after gemstone, and graphite consist of the same chemical element, carbon. Other than chemical composition, though, they have little in common; gem-quality diamond is most often colorless and the hardest substance known, whereas graphite is gray, not very attractive, and so soft you can scratch it with your fingernails. The atomic structure is quite different in these two varieties of carbon. In diamond, the atoms bond to form a complex three-dimensional framework, but in graphite the carbon atoms bond to form sheets that are held together by weak bonds (Figures 3.6b and c).

"Cutting" a diamond is actually done by cleaving the mineral along its internal planes of weakness, or its cleavage planes. If perfectly cleaved, diamond yields a "stone" shaped like two pyramids placed base to base. Large diamonds are preshaped by cleaving them into smaller pieces that are then further shaped by sawing and grinding with diamond dust. A diamond's value is determined by evaluating its color (colorless ones are most desirable), clarity (lack of internal flaws such as microfractures), and carat, a measure of a precious stone's weight (1 carat = 200 milligrams).

Many people own small gemstones, but many of the truly remarkable ones are in museum displays or in collections of crown jewels. The British Crown Jewels consist of a variety of objects, including the crown worn by Queen Elizabeth II at her coronation in 1953 and set with 2,868 diamonds, 17 sapphires, 11 emeralds, 5 rubies, and 273 pearls.

Copyright and Photograph by Dr. Parvinder S. Sethi

Figure 2 The pendant in this necklace is the Victoria Transvaal diamond n the Smithsonian Institution in Washington, D.C. The pendant is 67.89 carats and was retrieved from a stone found in South Africa in 1951. A total of 106 other diamonds make up the necklace.

CONNECTION LINK

To learn more about carbonate, sulfate, and halite in sedimentary rocks, see pages 162–165 in Chapter 7.

The most common nonsilicate rock-forming minerals are the carbonates calcite ($CaCO_3$) and dolomite [$CaMg(CO_3)_2$], the main constituents of the sedimentary rocks limestone and dolostone, respectively. Among the sulfates and halides, gypsum ($CaSO_4 \cdot 2H_2O$) in rock gypsum and halite (NaCl) in rock salt are common enough to qualify as rock-forming minerals. Even though these minerals and their corresponding rocks might be common in some areas, their overall abundance is limited compared to the silicate and carbonate rock-forming minerals.

3.7 How Do Minerals Form?

Thus far, we have discussed the composition, structure, and physical properties of minerals, but have not addressed how they originate. One phenomenon that accounts for the origin of minerals is the cooling of molten rock material known as *magma* (magma that reaches the surface is called *lava*).

As magma or lava cools, minerals crystallize and grow, thereby determining the mineral composition of various igneous rocks such as basalt (dominated by ferromagnesian silicates) and granite (dominated by nonferromagnesian silicates) (see Chapter 4). Hot-water solutions derived from magma commonly invade cracks and crevasses in adjacent rocks, and from these solutions, several minerals crystallize, some of economic importance. Minerals also originate when water in hot springs cools (see Chapter 13), and when hot, mineral-rich water discharges onto the seafloor at hot springs known as hydrothermal vents (see Figure 2.8).

> **CONNECTION LINK**
>
> To learn more about magma and lava, go to pages 86–88 in Chapter 4.

Dissolved materials in seawater, more rarely in lake water, combine to form minerals such as halite (NaCl), gypsum ($CaSO_4 \cdot 2H_2O$), and several others when the water evaporates. Aragonite and/or calcite, both varieties of calcium carbonate ($CaCO_3$), might also form from evaporating water, but most originate when organisms such as clams, oysters, corals, and floating microorganisms use this compound to construct their shells. A few plants and animals use silicon dioxide (SiO_2) for their skeletons, which accumulate as mineral matter on the seafloor when the organisms die (see Chapter 16).

Some clay minerals form when chemical processes compositionally and structurally alter other minerals (see Chapter 6), and others originate when rocks are changed during metamorphism (see Chapter 8). In fact, the agents that cause metamorphism—heat, pressure, and chemically active fluids—are responsible for the origin of many minerals. A few minerals even originate when gases such as hydrogen sulfide (H_2S) and sulfur dioxide (SO_2) react at volcanic vents to produce sulfur.

3.8 Natural Resources and Reserves

Geologists at the U.S. Geological Survey define a **resource** as a naturally occurring concentration of solid, liquid, or gaseous material in or on Earth's crust in such form and amount that economic extraction of a commodity is currently or potentially feasible. Many natural resources are concentrations of minerals, rocks, or both, but liquid petroleum, and natural gas are also included. In fact, we refer to *metallic resources* (copper, tin, iron ore, etc.), *nonmetallic resources* (sand and gravel, crushed stone, sulfur, salt, etc.), and *energy resources* (petroleum, natural gas, coal, and uranium). All are resources, but it is important to make a distinction between a resource, the total amount of a commodity whether discovered or not—and a **reserve**, which is only that part of the resource base that is known and can be economically recovered.

In principal the distinction between a resource and a reserve is simple, but in practice it depends on several factors, not all of which remain constant. For instance, a resource in a remote region might not be mined because transport expenses are too high, whereas what might be deemed a resource rather than a reserve might be mined in a developing nation where labor costs are low. The value of gold and diamonds makes it profitable to mine them just about anywhere, whereas deposits of sand and gravel for construction must be close to their market areas. Changes in technology are important, too. For example, iron ore currently mined in the Great Lakes region is not as rich as that mined in decades past, but a technique for separating the iron from its host rock and shaping it into pellets has made mining of low-grade deposits profitable.

TABLE 3.4	The 10 Leading States in U.S. Nonfuel Mineral Production for 2009	
State	**$**	**Most Important Commodities in Order of Value**
1. Nevada	5,510,000,000	Gold, copper, sand and gravel, lime, stone
2. Arizona	5,420,000,000	Copper, molybdenum, sand and gravel, cement, stone
3. Utah	4,000,000,000	Copper, molybdenum, gold, magnesium metal, potash
4. California	3,600,000,000	Sand and gravel, cement, boron minerals, stone, gold
5. Texas	2,900,000,000	Cement, stone, sand and gravel, salt
6. Alaska	2,480,000,000	Zinc, gold, lead, silver, sand and gravel
7. Florida	2,170,000,000	Phosphate rock, stone, cement, sand and gravel
8. Colorado	1,960,000,000	Molybdenum, sand and gravel, gold, cement, stone
9. Wyoming	1,940,000,000	Soda ash, clays, helium, sand and gravel, cement
10. Missouri	1,800,000,000	Stone, cement, lead, lime, sand and gravel
All States	57,100,000,000*	

*The total nonfuel mineral production for 2009 was 22% less than it was for 2008.

In addition to resources such as petroleum, gold, and ores of iron, copper, and aluminum, some common minerals and rocks are important. Pure quartz sand is used to make glass, sandpaper, and optical instruments; clay minerals are needed to make ceramics and paper; feldspars are used for porcelain, ceramics, enamel, and glass; and phosphate-bearing rocks are used for fertilizers. Muscovite is an ingredient in wallboard joint compound, lipstick, glitter, and eye shadow, as well as the lustrous paints on appliances and automobiles. In 2009, the extraction of nonfuel mineral resources in the United States totaled more than $57 billion (Table 3.4).

Access to natural resources is essential for industrialization and the high standard of living enjoyed in the developed countries. However, most resources are *nonrenewable,* meaning that there is a limited supply and they cannot be replenished by natural processes as fast as they are depleted. Accordingly, as the amount of a resource diminishes, suitable substitutes, if available, must be found, or we continue to exploit the resource and simply pay more and more as further depletion takes place. For some essential resources, the United States is totally dependent on imports, some from politically unstable parts of the world. For example, no cobalt was mined in the United States during 2009, so all that was needed for use in gas-turbine aircraft engines, magnets, and corrosion- and wear-resistant alloys had to be imported. Manganese is an essential element for making steel, and all that is used in the United States is imported.

The United States also imports all of the aluminum ore it uses, as well as all or some of many other resources (❙ Figure 3.20). For example, as of June 2010, 57.5% of all

Commodity	Percentage imported 0 25 50 75 100	Major Import Sources	Uses
Bauxite		Jamaica, Brazil, Guinea, Australia	Ore of aluminum
Niobium		Brazil, Canada, Germany, Estonia	Carbon steel, superalloys
Graphite		China, Mexico, Canada, Brazil	Brake linings, lubricants
Manganese		South Africa, Gabon, China, Australia	Steel production, dry cell batteries
Vanadium		Czech Republic, Rep. of Korea, Canada, Austria	Steel alloys
Platinum		South Africa, Germany, United Kingdom, Canada	Catalytic converters, jewelry
Tin		Peru, China, Bolivia, Indonesia	Tin cans and containers
Cobalt		Norway, Russia, China, Canada	Superalloys
Tungsten		China, Germany, Canada, Bolivia	Carbide parts for cutting tools
Chromium		South Africa, Kazakhstan, Russia	Stainless and heat-resistant steel
Silver		Mexico, Canada, Peru, Chile	Silver halide film, jewelry
Zinc		Canada, Peru, Mexico, Ireland	Galvanized metal, zinc alloys
Gold		Canada, Dominican Republic, Peru, Brazil	Jewelry and arts, electrical industry
Nickel		Canada, Russia, Australia, Norway	Stainless steel, electroplating
Copper		Chile, Canada, Peru, Mexico	Copper and copper alloys, wiring
Lead		Canada, China, Mexico, Australia	Lead for batteries, protective coatings
Iron ore		Canada, European Union, China, Mexico	Steel, cast iron

Sources: USGS Minerals Information: http://minerals.usgs.gov/minerals/
USGS Mineral Commodity Summaries 2011: http://minerals.usgs.gov/minerals/pubs/mcs/

❙ **Figure 3.20** **Mineral Commodities** The dependence of the United States on imports of various mineral commodities is apparent from this chart. The lengths of the green bars correspond to the amounts of the resources imported by the United States in 2011.

≫ Critical Thinking Question
Why do you think that the United States and Canada import bauxite (the ore of aluminum) given that both countries have vast exposures of igneous rocks and clay-rich rock from which we can extract aluminum? What factors(s) may change the status of these rocks so that we do extract aluminum from them?

> **CONNECTION LINK**

To find out more about petroleum and uranium, see pages 173 and 176 in Chapter 7.

petroleum needed is imported. Canada, in contrast, is more self-reliant, meeting most of its domestic mineral and energy needs. Nevertheless, it must import aluminum ore, chromium, manganese, and phosphate. Canada produces more crude oil and natural gas than it uses, and it is among the world's leaders in the production and export of uranium.

To ensure continued supplies of mineral and energy resources, scientists, government agencies, and executives in business and industry continually assess the status of resources in view of changing economic and political conditions and changes in science and technology. The U.S. Geological Survey keeps detailed statistical records of mine

What Would You Do ?

Some reputable businesspeople tell you of opportunities to invest in natural resources. Two ventures look promising: a gold mine and a sand and gravel pit. Given that gold sells for more than $1,500 per ounce (as of April, 2011), whereas sand and gravel are worth a few dollars per ton, would it be more prudent to invest in the gold mine? Explain not only how market price would influence your decision but also what other factors you might need to consider.

production, imports, and exports, and regularly publishes reports on the status of numerous resources in its *Mineral Commodity Summaries*. Similar reports also appear regularly in the *Canadian Minerals Yearbook*.

GEO-RECAP

» Geo-Image

Figure 1 If diamond is perfectly cleaved, it would yield geometric figures like the ones shown here, but this mineral is fluorite. From the image alone you should be able to tell that these specimens are not diamond. How? Look up other properties of these two minerals that you can use to differentiate them.

Sue Monroe

Chapter Summary

- Matter is composed of chemical elements, each of which consists of atoms. Protons and neutrons are present in an atom's nucleus, and electrons orbit around the nucleus in electron shells.

- The number of protons in an atom's nucleus determines its atomic number. The atomic mass number is the number of protons plus neutrons in the nucleus.

- Bonding results when atoms join with other atoms; different elements bond to form compounds. With few exceptions, minerals are compounds.

- Ionic and covalent bonds are most common in minerals, but metallic and van der Waals bonds are found in some.

- Minerals are crystalline solids, which means that they possess an ordered internal arrangement of atoms.

- Mineral composition is indicated by a chemical formula, such as SiO_2 for quartz.

- Some minerals have a range of compositions because some elements substitute for one another if their atoms are about the same size and have the same electrical charge.

- More than 3,500 minerals are known, and most of them are silicates. The two types of silicates are ferromagnesian and nonferromagnesian.

- In addition to silicates, geologists recognize carbonates, native elements, hydroxides, oxides, phosphates, halides, sulfates, and sulfides.

- Structure and composition control the physical properties of minerals, such as luster, crystal form, hardness, color, cleavage, fracture, and specific gravity.

- Several processes account for the origin of minerals, including cooling magma, weathering, evaporation of seawater, metamorphism, and organisms using dissolved substances in seawater to build their shells.

- A few minerals, designated rock-forming minerals, are common enough in rocks to be essential in their identification and classification. Most rock-forming minerals are silicates, but some carbonates are also common.

- Many resources are concentrations of minerals or rocks of economic importance. They are further characterized as metallic resources, nonmetallic resources, and energy resources.

- Reserves are that part of the resource base that can be extracted profitably. Distinguishing a resource from a reserve depends on market price, labor costs, geographic location, and developments in science and technology.

- The United States must import many resources to maintain its industrial capacity. Canada is more self-reliant, but it too must import some commodities

Important Terms

atom (p. 61)
atomic mass number (p. 62)
atomic number (p. 62)
bonding (p. 62)
carbonate mineral (p. 69)
cleavage (p. 74)
compound (p. 62)
covalent bond (p. 63)
crystal (p. 64)
crystalline solid (p. 64)
density (p. 74)
electron (p. 62)
electron shell (p. 62)
element (p. 61)
ferromagnesian silicate (p. 69)
hardness (p. 74)
ion (p. 63)
ionic bond (p. 63)
luster (p. 70)
mineral (p. 60)
neutron (p. 62)
nonferromagnesian silicate (p. 69)
nucleus (p. 62)
proton (p. 62)
reserve (p. 78)
resource (p. 78)
rock (p. 75)
rock-forming mineral (p. 75)
silica (p. 67)
silica tetrahedron (p. 68)
silicate (p. 67)
specific gravity (p. 74)

Review Questions

1. In covalent bonding
   a. ____electrons are shared by adjacent atoms.
   b. ____the noble gases form compounds.
   c. ____sodium and chlorine atoms attract one another.
   d. ____chemical bonds form between carbonate minerals and silicate minerals.
   e. ____atoms of one atom are transferred to another atom.

2. An atom with 8 protons and 14 neutrons in its nucleus has an atomic mass number of
   a. ____14.
   b. ____6.
   c. ____8.
   d. ____22.
   e. ____112.

3. A ferromagnesian silicate is one that contains
   a. ____calcium and manganese.
   b. ____sodium and chlorine.
   c. ____silica and potassium.
   d. ____iron and magnesium.
   e. ____pheromones and endorphins.

4. Which one of the following is a carbonate mineral?
   a. ____quartz.
   b. ____olivine.
   c. ____muscovite.
   d. ____halite.
   e. ____calcite.

5. A few minerals are composed of only one chemical element. They are known as
   a. ____native elements.
   b. ____phosphates.
   c. ____calcium carbonates.
   d. ____iron hydroxides.
   e. ____sulfides.

6. The two most abundant elements in Earth's crust are
   a. ____iron and calcium.
   b. ____potassium and sodium.
   c. ____silicon and oxygen.
   d. ____aluminum and magnesium.
   e. ____carbon and phosphorous.

7. Which one of the following is part of the definition of the term *mineral*?
   a. ____economically important.
   b. ____crystalline solid.
   c. ____atomic mass number.
   d. ____natural resource.
   e. ____ionic bond.

8. The chemical formula of the mineral olivine is $(Fe,Mg)_2SiO_4$, which means that
   a. ____iron and magnesium can substitute for one another.
   b. ____silicon and oxygen may or may not be present.
   c. ____olivine is a nonferromagnesian silicate.
   d. ____nonferromagnesian silicates are denser than ferromagnesian silicates.
   e. ____olivine is common in sedimentary rocks.

9. The quality and intensity of light reflected from a mineral is called its
   a. ____specific gravity.
   b. ____hardness.
   c. ____atomic number.
   d. ____density.
   e. ____luster.

10. The carbonate minerals are those with the carbonate radical (CO_3) as in
    a. ____basalt.
    b. ____galena.
    c. ____granite.
    d. ____calcite.
    e. ____olivine.

11. Give three examples of how minerals form.

12. Why are some minerals designated rock-forming minerals, whereas others are called accessory minerals? Name two rock-forming silicate minerals.

13. What is the silica tetrahedron, and how may they bond to form silicate minerals?

14. An igneous rock is made up mostly of ferromagnesian silicates and another one consists predominantly of nonferromagnesian silicates. Which one do you think will be densest and darkest? Explain.

15. How does mineral cleavage differ from fracture? Why do some minerals lack cleavage?

16. What accounts for the fact that some minerals such as plagioclase feldspars have a range of compositions rather than one specific composition?

17. Explain ionic bonding and give an example of how it takes place.

18. Although about 3,500 minerals have been described and named, only a few are very common. Why and which ones?

19. How do resources and reserves differ, and why must the United States import all of some resources?

20. What are atoms, elements, and compounds? Are all minerals compounds?

This exposure of the Harney Peak Granite, called The Needles, is in Custer State Park in the Black Hills of South Dakota. It is part of a large body of granite that cooled and crystallized from magma about 12 km below Earth's surface 1.7 billion years ago. The granite is now well exposed following uplift and deep erosion. The 18-m-high images of Presidents Washington, Jefferson, T. Roosevelt, and Lincoln are carved in the Harney Park Granite at Mount Rushmore National Memorial, which is also in the Black Hills.

CHAPTER 4

Sue Monroe

Igneous Rocks and Plutons

OUTLINE

4.1   Introduction

4.2   The Properties and Behavior of Magma and Lava

4.3   How Does Magma Originate and Change?

4.4   Igneous Rocks—Their Characteristics and Classification

4.5   Intrusive Igneous Bodies—Plutons

GEO-FOCUS: Granite—Common, Attractive, and Useful

4.6   The Origin of Batholiths

GEO-RECAP

HAVE YOU EVER WONDERED?

- Why geologists make a distinction between magma and lava?
- How hot lava is and how fast it flows?
- How and why igneous rocks differ from other types of rocks?
- Whether plate tectonics plays a role in the origin of magma and lava?
- How the vast bodies of granite in many parts of North America originated?
- What natural resources are found in or near igneous rocks?

4.1 Introduction

Before we begin a new topic, let's review some terms. You now know that the term *rock* applies to aggregates of minerals, mineral-like matter as in obsidian, and coal, which is altered organic matter. You also know from Chapter 1 that the three groups of rocks are igneous, sedimentary, and metamorphic (see Figure 1.16). Here we are concerned only with **igneous rocks**—that is, rocks made up of minerals that crystallized from cooling molten rock matter, and rocks composed of particulate matter ejected from volcanoes during explosive eruptions. We use the term **magma** for any mass of molten rock below Earth's surface, whereas **lava** is simply magma that has reached the surface. So, magma and lava are the same material, although the gas content of lava is less because of decreased pressure.

Most people have given little thought to magma, but are nevertheless familiar with volcanoes, which are simply hills or mountains where magma reaches the surface. In any case, volcanism is an observable phenomenon, but most magma never reaches the surface. In fact, it cools and crystallizes underground, a phenomenon called *intrusive igneous activity,* thus forming several types of igneous rock bodies known as *plutons,* named for Pluto, the Roman god of the underworld.

> CONNECTION LINK

You can find more information about volcanoes on pages 114–121 in Chapter 5.

Granite, composed of potassium feldspars, plagioclase feldspars, and quartz, is common in plutons, especially the larger ones such as those in Idaho, Montana, British Columbia, and South Dakota (see the chapter opening photo). The central part of the Sierra Nevada of California and Nevada is made up of a huge composite body of granite and related rocks (▌ Figure 4.1). In all these examples, magma was intruded into Earth's crust where it crystallized to form huge composite bodies of granite that were later exposed at the surface following uplift and erosion.

Even though we discuss intrusive igneous activity and the origin of plutons in this chapter and volcanism in Chapter 5, both processes are related. In fact, the same kinds of magmas are involved, but magma varies in its mobility, so only some of it reaches the surface. Furthermore, plutons lie beneath areas of volcanism and are the source of the lava flows and particulate matter erupted at volcanoes or along fissures.

An important reason to study igneous rocks and plutons is that igneous rocks make up parts of the continents and all of the oceanic crust, which forms continuously at divergent plate boundaries. As a matter of fact, most plutons and volcanoes are found at or near divergent and convergent plate boundaries, so the presence of igneous rocks in the geologic record is one criterion for recognizing ancient plate boundaries. Another reason to study these topics is that when plutons form, especially the larger ones, fluids emanating from them follow cracks and crevasses in adjacent rocks where important mineral resources such as copper may form.

CONNECTION LINK ◀

You can find more information about the origin of oceanic crust on page 252 in Chapter 10.

4.2 The Properties and Behavior of Magma and Lava

Magma is less dense than the rock from which it formed, so it tends to rise toward the surface, but much of it cools and crystallizes far beneath the surface, thereby forming plutons. Some magma does reach the surface as **lava flows,** which also cool and crystallize. As we mentioned in Chapter 3, cooling magma and lava is one way in which minerals form, thereby determining the composition of igneous rocks. Magma may also reach the surface during explosive eruptions when it is forcefully ejected into the atmosphere as

▌ **Figure 4.1 Granite in the Sierra Nevada** View of the Sierra Nevada taken west of Lone Pine, California. The rocks in this view are part of the Sierra Nevada batholith, a huge mass of granite and related rocks made up of many intrusive bodies. The high peak toward the right is Mount Whitney, which at 4,421 m is the highest peak in the continental United States.

particles called **pyroclastic materials** (from the Greek *pyro,* "fire," and *klastos,* "broken").

All igneous rocks ultimately derive from magma, but two separate processes account for them. **Plutonic rocks,** or **intrusive igneous rocks,** are those resulting from the cooling and crystallization of magma below Earth's surface—that is, from magma intruded into the crust. In contrast, **volcanic rocks,** or **extrusive igneous rocks,** form when lava flows cool or when pyroclastic materials become consolidated, both of which are extruded onto the surface.

Composition of Magma

By far the most abundant minerals in Earth's crust are silicates such as quartz, feldspars, and several ferromagnesian silicates, all made up of silicon and oxygen, and other elements shown in Figure 3.10. As a result, melting of the crust yields mostly silica-rich magmas that also contain considerable aluminum, calcium, sodium, iron, magnesium, and potassium, and several other elements in lesser quantities. Another source of magma is Earth's upper mantle, which is composed of rocks that contain mostly ferromagnesian silicates. Thus, magma from this source contains comparatively less silicon and oxygen (silica) and more iron and magnesium.

In most magmas, the primary constituent is silica, which varies enough to distinguish magmas called felsic, intermediate, mafic, and ultramafic* (Table 4.1). **Felsic magma,** with more than 65% silica, is silica rich and contains considerable sodium, potassium, and aluminum, but little calcium, iron, and magnesium. As you would expect, **intermediate magma** (53 to 65% silica) has a composition between felsic and mafic magma (Table 4.1). **Mafic magma** has 45 to 52% silica, but it contains proportionately more calcium, iron, and magnesium than does felsic magma. Any magma with less than 45% silica is termed **ultramafic,** which cools to form rocks made up mostly of ferromagnesian silicates.

TABLE 4.1	The Most Common Types of Magmas and Their Characteristics		
Type of Magma	Silica Content (%)	Sodium, Potassium, and Aluminum	Calcium, Iron, and Magnesium
Ultramafic	<45		Increase
Mafic	45–52		
Intermediate	53–65		
Felsic	>65	Increase	

Most lava temperatures are taken at volcanoes that show little or no explosive activity, so our best information comes from mafic lava flows such as those in Hawaii (❙ Figure 4.2). In contrast, eruptions of felsic lava are not as common, and the volcanoes that these flows issue from tend to be explosive and thus cannot be approached safely. Nevertheless, the temperatures of some bulbous masses of felsic lava in lava domes have been measured at a distance with an optical pyrometer. The surfaces of these lava domes are as hot as 900°C, but their interiors must surely be even hotter.

When Mount St. Helens erupted in 1980 in Washington State, it ejected felsic magma as particulate matter in pyroclastic flows. Two weeks later, these flows had temperatures between 300° and 420°C, and a steam explosion took place more than a year later when water encountered some of the still-hot pyroclastic materials.

Lava and magma retain heat so well because rock conducts heat so poorly. Accordingly, the interiors of thick lava flows and pyroclastic flow deposits may remain hot for months or years. In 1959, lava filled a crater to 85 m deep in Hawaii, but when the lava was drilled in 1988, it still had not completely solidified near its base. Plutons, depending on their size and depth, may not cool completely for thousands to millions of years.

How Hot Are Magma and Lava?

Everyone knows that lava is very hot, but how hot is hot? Erupting lava has a temperature in the range of 700° to 1,200°C, although a temperature of 1,350°C was recorded above a lava lake in Hawaii where volcanic gases reacted with the atmosphere. Magma must be even hotter than lava but no direct measurements of magma temperatures have ever been made.

*Lava from some volcanoes in Africa cools to form carbonitite, an igneous rock with at least 50% carbonate minerals, mostly calcite and dolomite.

❙ **Figure 4.2** A Geologist Takes the Temperature of an Active Lava Flow in Hawaii

Viscosity—Resistance to Flow

Viscosity, or resistance to flow, is a property of all liquids. Water's viscosity is very low so it is highly fluid and flows readily. For other liquids, viscosity is so high that they flow much more slowly. Good examples are cold motor oil and syrup, both of which are quite viscous and thus flow only with difficulty. But when these liquids are heated, their viscosity is much lower and they flow more easily; that is, they become more fluid with increasing temperature. Accordingly, you might suspect that temperature controls the viscosity of magma and lava, and this inference is partly correct. We can generalize and say that the hotter the magma or lava, the more readily it moves, but we must qualify this statement by noting that temperature is not the only control of viscosity.

Silica content also strongly controls magma and lava viscosity. With increasing silica content, numerous networks of silica tetrahedra form and retard flow because for flow to take place, the strong bonds of the networks must be ruptured. Mafic magma and lava with 45 to 52% silica have fewer silica tetrahedra networks and, as a result, are more mobile than felsic magma and lava flows (❚ Figure 4.3). One mafic flow in 1783 in Iceland flowed nearly 80 km, and geologists traced ancient flows in Washington State for more than 500 km. Felsic magma, because of its higher viscosity, does not reach the surface as often as mafic magma. And when felsic lava flows do occur, they tend to be slow moving and thick, and move only short distances. A thick, pasty lava flow that erupted in 1915 from Lassen Peak in California flowed only about 300 m before it ceased moving.

Temperature and silica content are important controls on the viscosity of magma and lava, but other factors include gases, mostly CO_2, as well as the presence of mineral crystals and friction from the surface over which lava flows. Lava with a high content of dissolved gases flows more readily than one with a lesser amount of gases, whereas lava that has many crystals or that flows over a rough surface tends to be more viscous.

4.3 How Does Magma Originate and Change?

Most people are unaware of how and where magma originates, how it rises, and how it might change. Indeed, many believe the misconception that lava comes from a continuous layer of molten rock beneath the crust or that it comes from Earth's molten outer core.

First, let us address how magma originates. We know that the atoms in a solid are in constant motion and that, when a solid is heated, the energy of motion exceeds the binding forces and the solid melts. We are all familiar with this phenomenon, and we are also aware that not all solids melt at the same temperature. Once magma forms, it tends to rise because as rocks become hotter, they expand and their density decreases.

Magma may come from 100 to 300 km deep, but most forms at much shallower depths in the upper mantle

❚ **Figure 4.3** **Viscosity of Magma and Lava** Temperature is an important control on viscosity, but so is composition. Mafic lava tends to be fluid, whereas felsic lava is much more viscous.

》 Critical Thinking Question
Which one of these volcanoes would you expect to erupt explosively? Explain.

a A mafic lava flow in 1984 on Mauna Loa Volcano in Hawaii. These flows move rapidly and form thin layers.

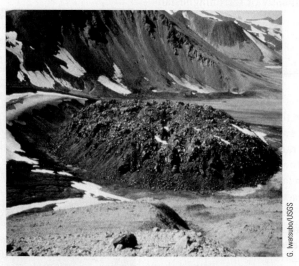

b The Novarupta lava dome in Katmai National Park in Alaska. The lava is felsic and viscous, so it was extruded as a bulbous mass. This image was taken in 1987.

or lower crust and accumulates in reservoirs known as **magma chambers.** Beneath spreading ridges, where the crust is thin, magma chambers exist at a depth of only a few kilometers, but along convergent plate boundaries, they are commonly a few tens of kilometers deep. The volume of a magma chamber ranges from a few to many hundreds of cubic kilometers of molten rock within the otherwise solid lithosphere. Some simply cools and crystallizes within Earth's crust, thus accounting for the origin of plutons, whereas some rises to the surface and is erupted as lava flows or pyroclastic materials.

What Would You Do ?

You are a high school teacher interested in developing experiments to show your students that (1) composition and temperature affect the viscosity of a lava flow and (2) when magma or lava cools, some minerals crystallize before others. Describe the experiments you might devise to illustrate these points.

Bowen's Reaction Series

During the early 1900s, N. L. Bowen knew that some minerals occurred together in igneous rocks but others did not. Based on his observations and laboratory experiments with cooling molten material, Bowen determined a predictable sequence in which specific minerals crystallized, rather than all crystallizing simultaneously. Furthermore, he proposed how intermediate and felsic magma could be derived from mafic magma. **Bowen's reaction series,** as it is now called, actually consists of two branches: a *discontinuous branch* and a *continuous branch* (❙ Figure 4.4). As magma cools, minerals crystallize along both branches simultaneously, but for convenience, we will discuss them separately.

In the discontinuous branch, which contains only ferromagnesian silicates, one mineral changes to another over specific temperature ranges (Figure 4.4). As the temperature decreases, a temperature range is reached in which a given mineral begins to crystallize. A previously formed mineral reacts with the remaining liquid magma (the melt) so that it forms the next mineral in the sequence. For instance, olivine [$(Mg,Fe)_2SiO_4$] is the first ferromagnesian silicate to crystallize. As the magma continues to cool, it reaches the temperature range at which pyroxene

is stable; a reaction occurs between the olivine and the remaining melt, and pyroxene forms.

With continued cooling, a reaction takes place between pyroxene and the melt, and the pyroxene structure is rearranged to form amphibole. Further cooling causes a reaction between the amphibole and the melt, and its structure is rearranged so that the sheet structure of biotite mica forms. Although the reactions just described tend to convert one mineral to the next in the series, the reactions are not always complete. Olivine, for example, might have a rim of pyroxene, indicating an incomplete reaction. If magma cools rapidly enough, the early-formed minerals do not have time to react with the melt, and thus all the ferromagnesian silicates in the discontinuous branch can be in one rock. In any case, by the time biotite has crystallized, essentially all the magnesium and iron present in the original magma have been used up.

Plagioclase feldspars, which are nonferromagnesian silicates, are the only minerals in the continuous branch of Bowen's reaction series (Figure 4.4). Calcium-rich plagioclase crystallizes first. As the magma continues to cool, calcium-rich plagioclase reacts with the melt, and plagioclase containing proportionately more sodium crystallizes until all of the calcium and sodium are used up. In many

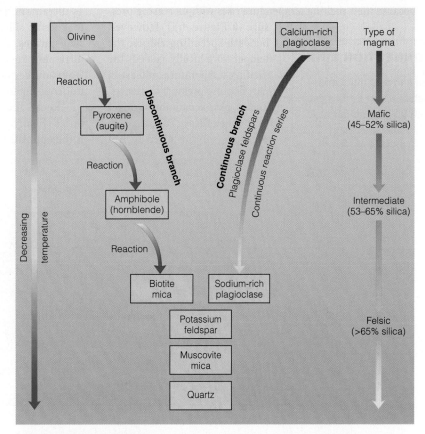

❙ **Figure 4.4** Bowen's Reaction Series Bowen's reaction series consists of a discontinuous branch along which a succession of ferromagnesian silicates crystallize as the magma's temperature decreases, and a continuous branch along which plagioclase feldspars with increasing amounts of sodium crystallize. Notice also that the composition of the initial mafic magma changes as crystallization takes place along the two branches.

cases, cooling is too rapid for a complete transformation from calcium-rich to sodium-rich plagioclase to take place. Plagioclase forming under these conditions is *zoned,* meaning that it has a calcium-rich core surrounded by zones progressively richer in sodium.

As minerals crystallize simultaneously along the two branches of Bowen's reaction series, iron and magnesium are depleted because they are used in ferromagnesian silicates, whereas calcium and sodium are used up in plagioclase feldspars. At this point, any leftover magma is enriched in potassium, aluminum, and silicon, which combine to form orthoclase ($KAlSi_3O_8$), a potassium feldspar, and if water pressure is high, the sheet silicate muscovite forms. Any remaining magma is enriched in silicon and oxygen (silica) and forms the mineral quartz (SiO_2). The crystallization of orthoclase and quartz is not a true reaction series because they form independently rather than by a reaction of orthoclase with the melt.

The Origin of Magma at Spreading Ridges

One fundamental observation regarding the origin of magma is that Earth's temperature increases with depth. Known as the *geothermal gradient,* this temperature increase averages about 25°C/km. Accordingly, rocks at depth are hot but remain solid because their melting temperature rises with increasing pressure (❙ Figure 4.5). However, beneath spreading ridges, the temperature locally exceeds the melting temperature, at least in part because pressure decreases. That is, plate separation at ridges probably causes a decrease in pressure on the already hot

> **CONNECTION LINK**

You can find more information about Earth's geothermal gradient on page 227 in Chapter 9.

❙ **Figure 4.5** The Effects of Pressure and Water on Melting

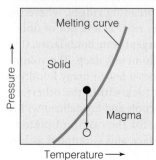

a As pressure decreases, even when temperature remains constant, melting takes place. The black circle represents rock at high temperature. The same rock (open circle) melts at lower pressure.

b If water is present, the melting curve shifts to the left because water provides an additional agent to break chemical bonds. Accordingly, rocks melt at a lower temperature (green melting curve) if water is present.

rocks at depth, thus initiating melting (❙ Figure 4.6). In addition, the presence of water decreases the melting temperature beneath spreading ridges because water aids thermal energy in breaking the chemical bonds in minerals.

Magma formed beneath spreading ridges is invariably mafic (45–52% silica). However, the upper mantle rocks from which this magma is derived are characterized as ultramafic (<45% silica), consisting mostly of ferromagnesian silicates and lesser amounts of nonferromagnesian silicates. To explain how mafic magma originates from ultramafic rock, geologists propose that the magma forms from source rock that only partially melts because not all minerals melt at the same temperature.

Recall the sequence of minerals in Bowen's reaction series (Figure 4.4). The order in which these minerals melt is the opposite of their order of crystallization. Accordingly,

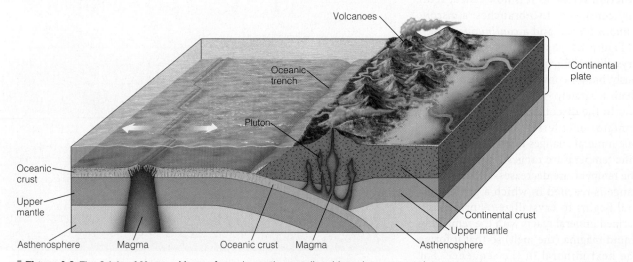

❙ **Figure 4.6** The Origin of Magma Magma forms beneath spreading ridges, because as plates separate, pressure is reduced on the hot rocks and partial melting of the upper mantle begins. Invariably, the magma formed is mafic. Magma also forms at subduction zones where water from the subducted plate aids partial melting of the upper mantle. This magma is also mafic, but as it rises, melting of the lower crust makes it more felsic.

rocks made up of quartz, potassium feldspar, and sodium-rich plagioclase begin melting at lower temperatures than those composed of ferromagnesian silicates and the calcic varieties of plagioclase. So when ultramafic rock starts to melt, the minerals richest in silica melt first, followed by those containing less silica. Therefore, if melting is not complete, mafic magma containing proportionately more silica than the source rock results.

Subduction Zones and the Origin of Magma

Another fundamental observation regarding magma is that, where an oceanic plate is subducted beneath either a continental plate or another oceanic plate, a belt of volcanoes and plutons is found near the leading edge of the overriding plate (Figure 4.6). It would seem, then, that subduction and the origin of magma must be related in some way, and indeed they are. Furthermore, magma at these convergent plate boundaries is mostly intermediate (53–65% silica) or felsic (>65% silica).

Once again, geologists invoke the phenomenon of partial melting to explain the origin and composition of magma at subduction zones. As a subducted plate descends toward the asthenosphere, it eventually reaches the depth where the temperature is high enough to initiate partial melting. In addition, the oceanic crust descends to a depth at which dewatering of hydrous minerals takes place, and as the water rises into the overlying mantle, it enhances melting and magma forms (Figure 4.5).

Recall that partial melting of ultramafic rock at spreading ridges yields mafic magma. Similarly, partial melting of mafic rocks of the oceanic crust yields intermediate (53–65% silica) and felsic (>65% silica) magmas, both of which are richer in silica than the source rock. Moreover, some of the silica-rich

sediments and sedimentary rocks of continental margins are probably carried downward with the subducted plate and contribute their silica to the magma. Also, mafic magma rising through the lower continental crust must be contaminated with silica-rich materials, which changes its composition.

Hot Spots and the Origin of Magma

Most volcanism occurs at divergent and convergent plate boundaries, but there are some chains of volcanic outpourings in the ocean basins and on continents that are not near either of these boundaries. The Emperor Seamount–Hawaiian Islands, for instance, form a chain of volcanic islands 6,000 km long, and the volcanic rocks become progressively older toward the northwest (see Figure 2.23). In 1963, Canadian geologist J. Tuzo Wilson proposed that the Hawaiian Islands and other areas showing similar trends lay above a **hot spot** over which a plate moves, thereby yielding a succession of volcanoes (▮ Figure 4.7a).

Many geologists now think that hot-spot volcanism results from a rising **mantle plume,** a cylindrical plume of hot mantle rock that rises from perhaps near the core–mantle boundary. As it rises toward the surface, the pressure decreases on the hot rock and melting begins, thus yielding magma. Hot-spot volcanism may also account for vast flat-lying areas of overlapping lava flows or what geologists call *flood basalts* (▮ Figure 4.7b). Figure 2.16 shows the locations of many hot spots, but of particular interest to us is the Yellowstone hot spot in Wyoming. We will have more to say about the Yellowstone region in Chapter 24.

Not all geologists agree with the mantle plume theory. They cite evidence based on studies of earthquake waves that do not seem to be consistent with the theory. Nevertheless, mantle plumes and hot-spot volcanism are the most widely accepted explanation for chains of volcanoes in the oceans, as well as linear associations of volcanic rocks on land.

▮ **Figure 4.7** Mantle Plume and Hot Spot

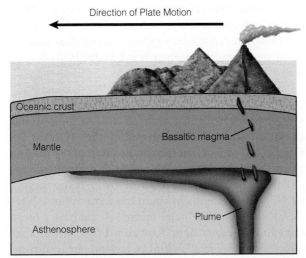

a A mantle plume beneath oceanic crust with a hot spot. Rising magma forms a series of volcanoes that become younger in the direction of plate movement.

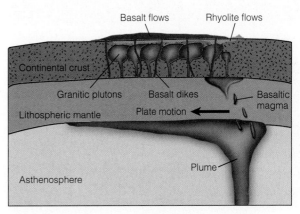

b A mantle plume with an overlying hot spot yields flood basalts, and some of the continental crust melts to form felsic magma.

Compositional Changes in Magma

Obviously the composition of any magma depends on what was melted in the first place. Should mafic rock melt completely, the resulting mama would also be mafic. However, as we discussed previously, partial melting yields magma that differs from its parent rock. In any case, once magma forms, it may change by **crystal settling,** which involves gravitational settling of minerals as they crystallize (▮ Figure 4.8). A good example is olivine, the first ferromagnesian silicate in the discontinuous branch of Bowen's reaction series, which has a density greater than the remaining melt and tends to sink. As a result, the remaining melt becomes richer in silica, sodium, and potassium, because much of the iron and magnesium were removed as olivine and perhaps pyroxene

a Early formed ferromagnesian silicates such as olivine crystallize and because of their density settle to the bottom of the magma chamber.

b Ferromagnesian silicates continue to form and settle.

c The remaining melt becomes richer in silicon, sodium, and potassium because much of the iron and magnesium originally present is now in the ferromagensian minerals that settled.

▮ **Figure 4.8 Crystal Settling Is a Process That Changes the Composition of Magma**

minerals crystallized. In other words, the remaining melt becomes more felsic.

In some thick, sheet-like plutons called *sills,* the first-formed ferromagnesian silicates are in fact concentrated in their lower parts, thus making their upper parts less mafic. But even so, very little felsic magma forms by this process. Calculations show, that to yield a given volume of granite (a felsic igneous rock), about 10 times as much mafic magma would have to be present initially for crystal settling to yield the volume of granite in question. If this were so, mafic igneous rocks should be much more common than felsic ones, but just the opposite is true, so it seems that something other than crystal settling must account for the large volume of felsic magma.

Once again, geologists refer to partial melting to solve this apparent dilemma. Remember that partial melting of oceanic crust and silica-rich sediments of continental margins yields magma richer in silica (more felsic) than the source rock. In addition, magma rising through continental crust changes in composition by **assimilation** as it reacts with preexisting rock, called **country rock,** with which it comes in contact (▮ Figure 4.9a). Country rock may be heated to 1,300°C and partially or completely melt, provided that its melting temperature is lower than that of the magma. Because assimilated rocks rarely have the same composition as the magma, the composition of the magma changes.

The fact that assimilation takes place is clearly indicated by *inclusions,* which are incompletely melted pieces of country rock that are fairly common in many igneous rocks (▮ Figure 4.9b). As magma rises it forces its way into cracks and crevasses and wedges loose pieces of country rock. Assimilation certainly takes place, but its effect on the bulk composition of magma must be slight. The reason is that the heat for melting must come from the magma which has the effect of cooling the magma, so only a limited amount of country rock can be assimilated.

Neither crystal settling nor assimilation can produce a significant amount of felsic magma from a mafic one. But both processes, if operating concurrently, can bring about greater changes than either process acting alone. Some geologists think that this is one way that intermediate magma forms where oceanic lithosphere is subducted beneath continental lithosphere.

A single volcano can erupt lavas of different composition, indicating that magmas of differing composition are present. It seems likely that some of these magmas would come into contact and mix with one another. If this is the case, we would expect that the composition of the magma resulting from **magma mixing** would be a modified version of the parent magmas. Suppose rising mafic magma mixes with felsic magma of about the same volume (Figure 4.9a). The resulting "new" magma would have a more intermediate composition. To account for the fact that the eruptions of Eyjafjallajökull in Iceland changed from mafic to more felsic from March to April 2010, geologists propose either that magma mixing occurred or that the later eruptions tapped into a different magma chamber.

4.4 Igneous Rocks—Their Characteristics and Classification

We have already defined *plutonic* or *intrusive igneous rocks* and *volcanic* or *extrusive igneous rocks.* Here we will have considerably more to say about the texture, composition, and classification of these rocks, which constitute one of the three major rock groups depicted in the rock cycle (see Figure 1.15).

Igneous Rock Textures

The term *texture* refers to the size, shape, and arrangement of the minerals that make up igneous rocks. Size is the most important because mineral crystal size is related to the cooling history of magma or lava and usually indicates whether an igneous rock is volcanic or plutonic. The atoms in magma and lava are in constant motion, but when cooling begins, some atoms bond to form small nuclei. As other atoms in the liquid chemically bond to these nuclei, they do so in an orderly geometric arrangement and the nuclei grow into crystalline *mineral grains,* the individual particles that make up igneous rocks.

If cooling takes place rapidly as in lava flows, the rate at which mineral nuclei form exceeds the rate of growth and an aggregate of many small mineral grains forms. The result is a fine-grained or **aphanitic texture,** in which individual minerals are too small to be seen without magnification (❙ Figure 4.10a,b). If cooling is slow, the rate of growth exceeds the rate of nuclei formation, and large mineral grains form, thus yielding a coarse-grained or **phaneritic texture,** in which minerals are clearly visible (❙ Figure 4.10c,d). Aphanitic textures usually indicate an extrusive origin, whereas rocks with phaneritic textures are usually intrusive. However, shallow plutons might have an aphanitic texture, and the rocks that form in the interiors of thick lava flows might be phaneritic.

Another common texture in igneous rocks is one termed **porphyritic,** in which minerals of markedly different size are present in the same rock. The larger minerals are *phenocrysts* and the smaller ones collectively make up the *groundmass,* which is simply the grains between phenocrysts (❙ Figure 4.10e,f). The only requirement for a porphyritic texture is that the phenocrysts be considerably larger than the minerals in the groundmass, which may be either aphanitic or phaneritic. Igneous rocks with porphyritic textures are designated *porphyry,* as in basalt porphyry. These rocks have more complex cooling histories than those with aphanitic or phaneritic textures and might involve, for example, magma partly cooling beneath the surface followed by its eruption and rapid cooling at the surface.

Lava may cool so rapidly that its constituent atoms do not have time to become arranged in the ordered, three-dimensional frameworks of minerals. As a consequence, *natural glass,* such as *obsidian,* forms (❙ Figure 4.10g). Even though obsidian with its glassy texture is not composed of minerals, geologists nevertheless classify it as an igneous rock.

❙ **Figure 4.9** Assimilation and Magma Mixing

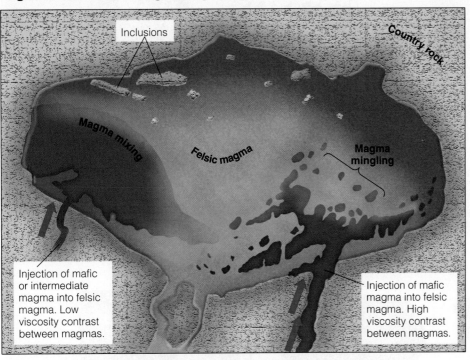

Inclusions

Country rock

Magma mixing

Felsic magma

Magma mingling

Injection of mafic or intermediate magma into felsic magma. Low viscosity contrast between magmas.

Injection of mafic magma into felsic magma. High viscosity contrast between magmas.

a Fragments of rock dislodged by rising magma may melt and become incorporated into the magma, a process called assimilation, or they may remain as inclusions.

Sue Monroe

b Dark inclusions in granitic rocks in the Sierra Nevada in California.

Some magmas contain large amounts of water vapor and other gases. These gases may be trapped in cooling lava where they form numerous small holes or cavities known as **vesicles;** rocks with many vesicles are termed *vesicular,* as in vesicular basalt (∎ Figure 4.10h).

A **pyroclastic** or **fragmental texture** characterizes igneous rocks formed by explosive volcanic activity (∎ Figure 4.10i). For example, ash discharged high into the atmosphere eventually settles to the surface where it accumulates; if consolidated, it forms pyroclastic igneous rock.

∎ **Figure 4.10** **The Various Textures of Igneous Rocks** Texture is one criterion used to classify igneous rocks.
a,b Rapid cooling as in lava flows results in many small minerals and an aphanitic (fine-grained) texture.
c,d Slower cooling in plutons yields a phaneritic texture.
e,f These porphyritic textures indicate a complex cooling history.
g Obsidian has a glassy texture because magma cooled too quickly for mineral crystals to form.
h Gases expand in lava to yield a vesicular texture.
i Microscopic view of a rock with a fragmental texture. The colorless, angular objects are pieces of volcanic glass that measure up to 2 mm.

》Critical Thinking Question
How can you determine which of these rocks formed in a pluton, a lava flow, and an accumulation of volcanic ash?

Composition of Igneous Rocks

Most igneous rocks, like the magma from which they originate, are mafic (45–52% silica), intermediate (53–65% silica), or felsic (>65% silica). A few are *ultramafic* (<45% silica), but these are probably derived from mafic magma by a process we discuss later. The parent magma plays an important role in determining the mineral composition of igneous rocks, yet it is possible for the same magma to yield a variety of igneous rocks because its composition can change as a result of the sequence in which minerals crystallize, or by crystal settling, assimilation, and magma mixing (Figures 4.4, 4.8, and 4.9).

Classifying Igneous Rocks

Geologists use texture and composition to classify igneous rocks, although a few are classified mostly by texture only. Notice in ▮ Figure 4.11 that all rocks except peridotite are in pairs; the members of a pair have the same composition but different textures. Basalt and gabbro, andesite and diorite, and rhyolite and granite are compositional (mineralogical) pairs, but basalt, andesite, and rhyolite are aphanitic and most commonly volcanic, whereas gabbro, diorite, and granite are phaneritic and mostly plutonic. The volcanic and plutonic members of each pair can usually be distinguished by texture, but remember that rocks in some shallow plutons may be aphanitic and rocks that formed in thick lava flows may be phaneritic. In other words, all these rocks exist in a textural continuum.

The igneous rocks in Figure 4.11 are also differentiated by composition—that is, by their mineral content. Reading across

▮ Figure 4.12 Peridotite This specimen of the ultramafic rock peridotite is made up mostly of olivine. Notice in Figure 4.11 that peridotite is the only phaneritic rock that does not have an aphanitic counterpart. Peridotite is rare at Earth's surface, but is very likely the rock that makes up the mantle.

the chart from rhyolite to andesite to basalt, for example, we see that the proportions of nonferromagnesian and ferromagnesian silicates change. The differences in composition, however, are gradual along a compositional continuum. In other words, there are rocks with compositions that correspond to the lines between granite and diorite, basalt and andesite, and so on. For example, Lassen Peak in California, which erupted from 1914 through 1917, is made up mostly of *dacite*, a rock with a composition between andesite and rhyolite.

Ultramafic Rocks Ultramafic rocks (<45% silica) are composed mostly of ferromagnesian silicates. *Peridotite* contains olivine, lesser amounts of pyroxene, and usually a little plagioclase feldspar (Figure 4.11 and ▮ Figure 4.12), and *pyroxenite* is composed predominately of pyroxene. Because these minerals are dark, the rocks are black or green. Peridotite is a likely candidate for the rock that makes up the upper mantle (see Chapter 9). Ultramafic rocks in Earth's crust probably originate by concentration of the early-formed ferromagnesian minerals that separated from mafic magmas.

Ultramafic lava flows, called *komatiites*, are known in rocks older than 2.5 billion years, but are rare or absent in younger ones. The reason is that to erupt, ultramafic lava must have a near-surface temperature of about 1,600°C; the surface temperatures of present-day mafic lava flows are rarely more than 1,200°C. During early Earth history, though, more radioactive decay heated the mantle to as much as 300°C hotter than now, and ultramafic lavas could erupt onto the surface. Because the amount of heat has decreased over time, Earth has cooled, and eruptions of ultramafic lava flows ceased.

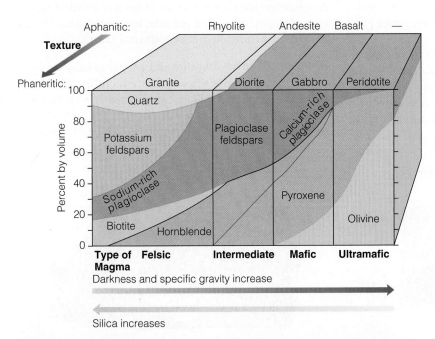

▮ Figure 4.11 Classification of Igneous Rocks This diagram shows the percentages of minerals, as well as the textures of common igneous rocks. For example, an aphanitic (fine-grained) rock of mostly calcium-rich plagioclase and pyroxene is basalt.

Figure 4.13 Mafic Igneous Rocks—Basalt and Gabbro

a Basalt is aphanitic.

b Gabbro is phaneritic. Notice the light reflected from the crystal faces.

Basalt-Gabbro *Basalt* and *gabbro* are the aphanitic and phaneritic rocks that crystallize from mafic magma (45–52% silica) (Figure 4.13). Both have the same composition—mostly calcium-rich plagioclase and pyroxene, with smaller amounts of olivine and amphibole (Figure 4.11). Because they contain a large proportion of ferromagnesian silicates, basalt and gabbro are dark; those that are porphyritic contain calcium plagioclase or olivine phenocrysts.

Extensive basalt lava flows cover vast areas in Washington, Oregon, Idaho, and northern California (see Chapters 5 and 23). Oceanic islands such as Iceland, the Galápagos, the Azores, and the Hawaiian Islands are composed mostly of basalt, and basalt makes up the upper part of the oceanic crust.

Gabbro is much less common than basalt, at least in the continental crust or where it can be easily observed. Small intrusive bodies of gabbro are present in the continental crust, but intermediate to felsic intrusive rocks are much more common. However, the lower part of the oceanic crust is composed of gabbro.

Andesite-Diorite Intermediate-composition magma (53–65% silica) crystallizes to form *andesite* and *diorite,* which are compositionally equivalent fine- and coarse-grained igneous rocks (Figure 4.14). Andesite and diorite are composed predominately of plagioclase feldspar, with the typical ferromagnesian component being amphibole or biotite (Figure 4.11). Andesite is medium to dark gray, but diorite has a salt-and-pepper appearance because of its white to light gray plagioclase and dark ferromagnesian silicates (Figure 4.14).

Andesite is a common extrusive igneous rock formed from lava erupted in volcanic chains at convergent plate boundaries. The volcanoes of the Andes Mountains of South America and the Cascade Range in western North America are composed, in part, of andesite. Intrusive bodies of diorite are fairly common in the continental crust.

Rhyolite-Granite *Rhyolite* and *granite* crystallize from felsic magma (>65% silica) and are therefore silica-rich rocks (Figure 4.15). They consist mostly of potassium feldspar, sodium-rich plagioclase, and quartz, with perhaps some biotite and, rarely, amphibole (Figure 4.11). Because

Figure 4.14 Intermediate Igneous Rocks—Andesite and Diorite

a This specimen of andesite has hornblende phenocrysts that are so numerous the rock may be classified as an andesite hornblende porphyry.

b Diorite has a salt-and-pepper appearance because it contains light-colored nonferromagnesian silicates and dark-colored ferromagnesian silicates.

a Rhyolite.

Sue Monroe

Figure 4.15 Felsic Igneous Rocks—Rhyolite and Granite These rocks are typically light colored because they contain mostly nonferromagnesian silicate minerals. The dark spots in the granite specimen are biotite mica. The white and pinkish minerals are feldspars, whereas the glassy-appearing minerals are quartz.

>> **Critical Thinking Question**
Why do you think these two rocks are so light colored whereas those in Figure 4.13 are much darker?

b Granite.

nonferromagnesian silicates predominate, rhyolite and granite are typically light colored. Rhyolite is fine grained, although most often it contains phenocrysts of potassium feldspar or quartz, and granite is coarse grained. Granite porphyry is also fairly common.

Rhyolite lava flows are much less common than andesite and basalt flows. Recall that one control of magma viscosity is silica content. Thus, if felsic magma rises to the surface, it begins to cool, the pressure on it decreases, and gases are released explosively, usually yielding rhyolitic pyroclastic materials. The rhyolitic lava flows that do occur are thick and highly viscous and move only short distances.

Granite is a coarsely crystalline igneous rock with a composition corresponding to that of the field shown in Figure 4.11. Strictly speaking, not all rocks in this field are granites. For example, a rock with a composition close to the line separating granite and diorite is called *granodiorite*. To avoid the confusion that might result from introducing more rock names, we will follow the practice of referring to rocks to the left of the granite-diorite line in Figure 4.11 as *granitic*.

Granitic rocks are by far the most common plutonic rocks, although they are restricted to the continents (see Geo-Focus on page 100). Most granitic rocks were intruded at or near convergent plate margins during mountain-building episodes. When these mountainous regions are uplifted and eroded, the vast bodies of granitic rocks forming their cores are exposed. The granitic rocks of the Sierra Nevada of California form a composite body measuring approximately 640 km long and 110 km wide, and the granitic rocks of the Coast Ranges of British Columbia, Canada, are even more voluminous.

Pegmatite The term *pegmatite* refers to a particular texture rather than a specific composition, but most pegmatites are

composed mostly of quartz, potassium feldspar, and sodium-rich plagioclase, thus corresponding closely to granite. A few pegmatites are mafic or intermediate in composition and are appropriately called *gabbro* and *diorite pegmatites*. The most remarkable feature of pegmatites is the size of their minerals, which measure at least 1 cm across, and in some pegmatites they measure tens of centimeters or meters (Figure 4.16). Many pegmatites are adjacent to large granite plutons and are composed of minerals that formed from the water-rich magma that remained after most of the granite crystallized.

When felsic magma cools and forms granite, the remaining water-rich magma has properties that differ from the magma from which it separated. It has a lower density and viscosity and invades cracks in the nearby rocks where minerals crystallize. This water-rich magma also contains elements that rarely enter into the common minerals that form granite. Pegmatites that are essentially very coarsely crystalline granite are simple pegmatites, whereas those with minerals containing elements such as lithium, beryllium, cesium, boron, and several others are complex pegmatites. Some complex pegmatites contain 300 mineral species, a few of which are important economically. In addition, several gem minerals such as emerald and aquamarine, both of which are varieties of the silicate mineral beryl, and tourmaline are found in some pegmatites.

The formation and growth of mineral-crystal nuclei in pegmatites are similar to those processes in other magmas, but with one critical difference: The water-rich magma from which pegmatites crystallize inhibits the formation of nuclei. However, some nuclei do form, and because the appropriate ions in the liquid can move easily and attach themselves to a growing crystal, individual minerals may grow very large.

a This pegmatite, the light-colored rock, is exposed in the Black Hills of South Dakota.

Figure 4.16 Pegmatite

b Close-up view of a specimen from a pegmatite with minerals measuring 8 to 10 cm across.

What Would You Do ❓

As the only member of your community with any geology background you are considered the local expert on minerals and rocks. Suppose one of your friends brings you a rock specimen with the following features—composition: mostly potassium feldspar and plagioclase feldspar with about 10% quartz and minor amounts of biotite; texture: minerals average 3 mm across, but several potassium-feldspars are up to 5 cm. Give the specimen a rock name and tell your friend as much as you can about the rock's history.

Other Igneous Rocks Geologists classify the igneous rocks in Figure 4.11 by texture and composition, but a few others are identified primarily by their textures (▌Figure 4.17). Much of the fragmental material erupted by volcanoes is *ash,* a designation for pyroclastic materials measuring less than 2.0 mm, most of which consists of pieces of minerals or shards of volcanic glass (Figure 4.10i). The consolidation of ash forms the pyroclastic rock *tuff* (▌Figure 4.18a). Most tuff

is silica rich and light colored and is appropriately called *rhyolite tuff.* Some ash flows are so hot that as they come to rest, the ash particles fuse together and form a *welded tuff.* Consolidated deposits of larger pyroclastic materials, such as cinders, blocks, and bombs, are *volcanic breccia* (Figure 4.17).

Both *obsidian* and *pumice* are varieties of volcanic glass (▌Figure 4.18b, c). Obsidian may be black, dark gray, red, or brown, depending on the presence of iron. Obsidian breaks with the conchoidal (smoothly curved) fracture typical of glass. Analyses of many samples indicate that most obsidian has a high silica content and is compositionally similar to rhyolite.

Pumice is a variety of volcanic glass containing numerous vesicles that develop when gas escapes through lava and forms a froth (Figure 4.18c). If pumice falls into water, it can be carried great distances because it is so porous and not very permeable so it floats. Another vesicular rock is *scoria.* It is more crystalline and denser than pumice, but it has more vesicles than solid rock (▌Figure 4.18d).

> **CONNECTION LINK** ‹
>
> To learn more about porosity and permeability see pages 314 and 315 in Chapter 13.

Composition	Felsic ⟷ Mafic	
Texture — Vesicular	Pumice	Scoria
Glassy	Obsidian	
Pyroclastic or fragmental	⟷ Volcanic Breccia ⟷	
	Tuff/welded tuff	

▌**Figure 4.17** **Texture Classification** Classification of igneous rocks for which texture is the main consideration. Composition is shown, but it is not essential for naming these rocks.

4.5 Intrusive Igneous Bodies—Plutons

Unlike volcanism and the origin of volcanic rocks, we can study intrusive igneous bodies, collectively called **plutons,** only indirectly because intrusive rocks form when magma cools and crystallizes within Earth's crust. We can observe these rock bodies only when uplift and deep erosion have taken place, thereby exposing them at the surface. Furthermore, geologists cannot duplicate the conditions under which intrusive rocks form except in small laboratory experiments.

❚ **Figure 4.18** Examples of Igneous Rocks Classified Primarily by Their Texture

>> Critical Thinking Question
Why do pumice and scoria have so many vesicles but there are none in obsidian?

a Tuff is composed of pyroclastic materials such as those in Figure 4.10i.

Courtesy of David J. Matty

b The natural glass obsidian.

Sue Monroe

c Pumice is glassy and extremely vesicular.

Sue Monroe

d Scoria is also vesicular, but it is darker, heavier, and more crystalline than pumice.

Sue Monroe

Geologists recognize several types of plutons based on their geometry (three-dimensional shape) and relationships to the country rocks. In terms of their geometry, plutons are tabular, cylindrical, or irregular (massive). Furthermore, they may be **concordant,** meaning they have boundaries that parallel the layering in the country rock, or **discordant,** with boundaries that cut across the country rock's layering (❚ Figure 4.19).

Dikes, Sills, and Laccoliths

Dikes and **sills** are tabular or sheetlike igneous bodies that differ only in that dikes are discordant and sills are concordant (❚ Figure 4.20). Dikes are quite common and range from a few centimeters to more than 100 m thick. Invariably, they are intruded into preexisting fractures or where fluid pressure is great enough for them to form their own fractures as they move upward into county rock. Erosion of the Hawaiian volcanoes exposes dikes in fracture zones, and the Columbia River basalts in Washington State issued from long fissures in which magma solidified to form dikes. Dikes underlie the Eldgja fissure and Laki fissure in Iceland, both of which have been the sites of voluminous eruptions in 934 and 1783–1784, respectively.

Sills are tabular just as dikes are, but they are concordant (Figure 4.20). Many sills are a meter or less thick, although some are much thicker. A well-known sill in the United States is the Palisades sill that forms the Palisades along the west side of the Hudson River in New York and New Jersey (see Chapter 22). This 300-m-thick sill is exposed for 60 km along the river. Most sills were intruded into sedimentary rocks, but eroded volcanoes also reveal that sills are injected into piles of volcanic rocks. In fact, some of the inflation of a volcano that precedes an eruption may be caused by the injection of sills. In contrast to dikes, which follow zones of weakness, sills are intruded between layers in country rock when the fluid pressure is great enough for the magma to actually lift the overlying rocks.

Under some circumstances, a sill inflates and causes the overlying rocks to bow upward, forming an igneous body called a **laccolith.** A laccolith has a flat floor and is domed in its central part, giving it a mushroom-like geometry (Figure 4.19). Like sills, laccoliths are rather shallow intrusions that lift the overlying rocks. Well-known laccoliths in the United States are in the Henry Mountains of southeastern Utah, and several buttes in Montana are eroded laccoliths.

GEO-FOCUS   Granite—Common, Attractive, and Useful

You already know that granite and related rocks make up huge plutons called batholiths and that granite is found in many areas in North America. Those of you from the Northeast are probably aware of New Hampshire's rock outcrop known as the Old Man of the Mountain, which was a jagged profile of a face before it collapsed in 2003. Students living near Atlanta, Georgia, have likely visited Stone Mountain (see Figure 6.3b), a huge granite monolith in which are carved the images of Robert E. Lee, Jefferson Davis, and Stonewall Jackson. Mount Rushmore National Memorial in South Dakota has the faces of Presidents Washington, Jefferson, T. Roosevelt, and Lincoln carved from the Harney Peak Granite (see the chapter opening photo). A large mass of pink granite is found in the Enchanted Rock State Natural Area in Texas, and similar granite exposures are present in several other western states (Figure 4.1) and several areas in Canada.

Today, no one doubts that granite and related rocks formed when magma cooled beneath Earth's surface. This has not always been the case, though. In 1787, a German professor of mineralogy, Abraham Gottlob Werner (1749–1817), proposed that all rocks, including basalt and granite, had precipitated in an orderly sequence from a primeval, worldwide ocean. In fact, according to Werner, granite and metamorphic rocks were the oldest ones on Earth. His concept, called Neptunism, was popular for many years, but eventually scientists recognized the igneous origin of basalt and, based on their observations, concluded that granite formed from magma as we have outlined.

Granite and similar rocks are very attractive, especially when sawed and polished. These rocks are used for tombstones, mantelpieces, countertops, facing stones on buildings, monuments, pedestals for statues, and statuary itself. A 23-m-high, red granite obelisk weighing 250 metric tons (▍ Figure 1) was carved from a single piece of granite at Aswan, Egypt, about 3,300 years ago and transported nearly 200 km to its present location at Luxor, Egypt. Granite from Aswan was also used for parts of the interiors of the pyramids at Giza near Cairo, Egypt; the pyramids themselves, however, are made up mostly of limestone.

Plymouth Rock at Plymouth, Massachusetts, has great symbolic value, but otherwise is a rather ordinary rock (▍ Figure 2). Actually it is a rather small stone from the 600-million-year-old Dedham Granodiorite, which was transported by glaciers to its present location; as its name implies, granodiorite has a composition between granite and diorite (Figure 4.11). Legend has it that Plymouth Rock is the landing place where the Pilgrims first set foot in the New World in 1620. In fact, the Pilgrims first landed at Provincetown on Cape Cod and later went to the Plymouth area. But even then, they probably landed north of Plymouth rather than at Plymouth Rock. Incidentally, Plymouth Rock is only one-third to one-half its original size; much of it has been lost to breakage and souvenir hunters.

The term *granite* has a specific meaning in geology, but in commercial applications it is a generic term for a variety of rocks, including gabbro and basalt, as in black granite. And although true granite is used for many purposes, many granite-like substances are manufactured. This cultured granite, as it is called, is made by chemically bonding materials and molding them into various shapes, which cannot be done with natural stone.

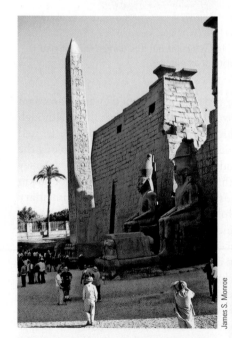

▍ **Figure 1** This 23-m-high granite obelisk is at Luxor Temple in the City of Luxor, Egypt. It was carved at Aswan and transported along the Nile River to its location in Luxor about 3,300 years ago. The source of the granite at Aswan has an unfinished obelisk that was abandoned when it cracked.

James S. Monroe

▍ **Figure 2** Plymouth Rock is the first piece of land on which the Pilgrims supposedly set foot when they arrived in Massachusetts in 1620. The rock first became a patriotic icon during the Revolutionary War.

Mark E. Gibson/CORBIS

▌Figure 4.19 Block Diagram Showing Various Plutons Some plutons cut across the layering in country rock and are discordant, whereas others parallel the layering and are concordant.

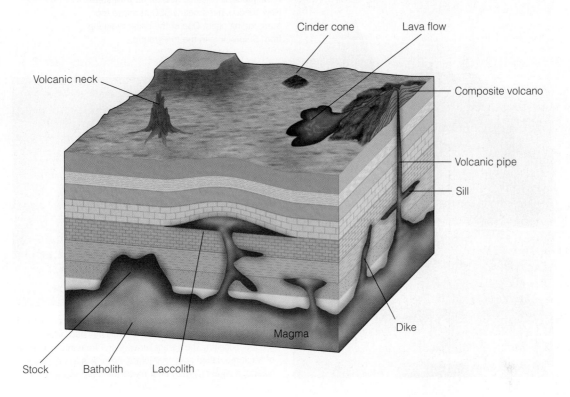

▌Figure 4.20 Dike and Sill The dark materials in this image are igneous rocks, whereas the light layers are sedimentary. Notice that the sill parallels the layering, so it is concordant. The dike, though, clearly cuts across the layering and is discordant. Sills and dikes have sheetlike geometry, but in this view we can see them in only two dimensions.

▌Figure 4.21 Volcanic Necks

a Shiprock, a volcanic neck in northwestern New Mexico, rises nearly 550 m above the surrounding plain. One of the dikes radiating from Shiprock is in the foreground.

b Devil's Tower in northeastern Wyoming rises approximately 260 m above its base. It may be a volcanic neck or an eroded laccolith. The vertical lines result from intersections of fractures called columnar joints. According to Cheyenne legend, a gigantic grizzly bear made the deep scratches.

Volcanic Pipes and Necks

Volcanoes have a cylindrical conduit known as a **volcanic pipe** that connects to an underlying magma chamber. Magma rises through this structure; however, when a volcano ceases to erupt, its slopes are attacked by weathering and erosion, but the magma in the pipe is commonly more resistant to erosion and is left as a remnant called a **volcanic neck** (▌ Figure 4.21). Several volcanic necks are found in the southwestern United States, especially in Arizona and New Mexico, and others are recognized elsewhere.

Batholiths and Stocks

A **batholith,** the largest of all plutons, must have at least 100 km² of surface area, and most are far larger. A **stock,** in contrast, is similar but smaller (▌ Figure 4.22). Some stocks are simply parts of large plutons that once exposed by erosion become batholiths (Figure 4.19). Both batholiths and stocks are mostly discordant, although locally they may be concordant, and batholiths, especially, consist of multiple intrusions. In other words, a batholith is a large composite body produced by repeated, voluminous

intrusions of magma in the same region. The coastal batholith of Peru, for instance, was emplaced during a period of 60 to 70 million years and is made up of as many as 800 individual plutons.

The igneous rocks that make up batholiths are mostly granitic, although diorite may also be present. Batholiths and stocks are emplaced mostly near convergent plate boundaries during episodes of mountain building. One example is the Sierra Nevada batholith of California (Figure 4.1), which formed over millions of years. Other large batholiths in North America include the Idaho batholith, the Boulder batholith in Montana, and the Coast Range batholith in British Columbia, Canada.

Mineral resources are found in rocks of batholiths and stocks and in the adjacent country rocks. The copper deposits at Butte, Montana, are in rocks near the margins of the granitic rocks of the Boulder batholith. Near Salt Lake City, Utah, copper is mined from the mineralized rocks of the Bingham stock, a composite pluton composed of granite and granite porphyry. Granitic rocks also are the primary source of gold, which forms from mineral-rich solutions moving through cracks and fractures of the igneous body.

James S. Monroe

∥ Figure 4.22 Castle Crags Granitic rocks of a small stock at Castle Crags State Park in California.

4.6 The Origin of Batholiths

Geologists realized long ago that the origin of batholiths posed a space problem. What happened to the rock that was once in the space now occupied by a batholith? One solution to the space problem is that these large igneous bodies melted their way into the crust. In other words, they simply assimilated the country rock as they moved upward (Figure 4.9a). The presence of inclusions, especially near the tops of some plutons, indicates that assimilation does occur. Nevertheless, as we noted, assimilation is a limited process because magma cools as country rock is assimilated. Calculations indicate that far too little heat is available in magma to assimilate the huge quantities of country rock necessary to make room for a batholith.

Geologists now generally agree that batholiths were emplaced by *forceful injection* as magma moved upward (∥ Figure 4.23). Recall that granite is derived from viscous felsic magma and therefore rises slowly. It appears that the magma deforms and shoulders aside the country rock, and as it rises farther, some of the country rock fills the space beneath the magma.

Some batholiths do indeed show evidence of having been emplaced forcefully by shouldering aside and deforming the country rock. This mechanism probably occurs in the deeper parts of the crust where temperature and pressure are high and the country rocks are easily deformed in the manner described. At shallower depths, the crust is more rigid and tends to deform by fracturing. In this environment, batholiths may move upward by **stoping,** a process in which rising magma detaches and engulfs pieces of country rock (∥ Figure 4.24). According to this concept, magma moves up along fractures and the planes separating layers of country rock. Eventually, pieces of country rock detach and settle into the magma. No new room is created during stoping; the magma simply fills the space formerly occupied by country rock (Figure 4.24).

❚ Figure 4.23 Emplacement of a Hypothetical Batholith As the magma rises, it shoulders aside and deforms the country rock.

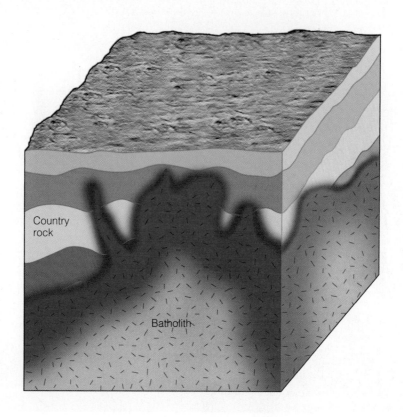

Country rock

Batholith

❚ Figure 4.24 Emplacement of a Batholith by Stoping

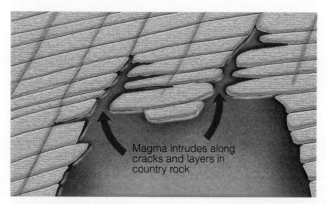

Magma intrudes along cracks and layers in country rock

a Stoping takes place when magma rises into the crust by detaching and engulfing pieces of country rock.

b Some of the detached blocks may be assimilated, and some may remain as inclusions (Figure 4.9).

» Geo-Image

Figure 1 The image below shows an imaginary landscape made up of sandstone and igneous rocks. Many features discussed in this chapter appear here (dike, stock, etc.). Identify as many as you can. Rock types include basalt (B), diorite (D), gabbro (GB), granite (GR), pegmatite (PG), and sandstone (SS). After making your identifications, decipher as much of the geologic history as you can. That is, what event took place first, second, and so on, and how do you know? Do not be frustrated if you cannot determine all the relationships among the rocks. Geologists face this dilemma all the time!

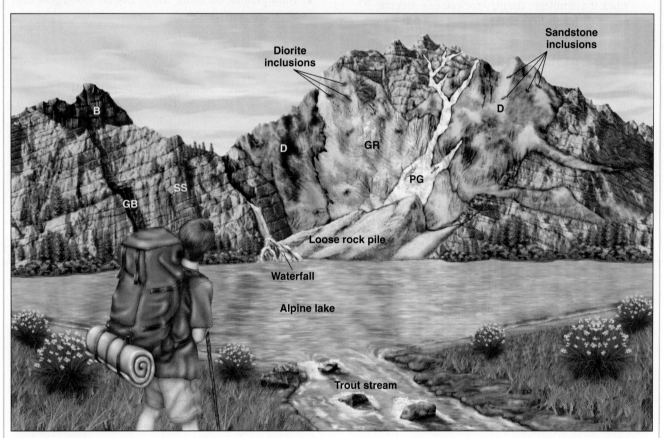

Chapter Summary

- Magma is the term for molten rock below Earth's surface, whereas the same material at the surface is lava.

- Silica content distinguishes among ultramafic (<45% silica), mafic (45–52% silica), intermediate (53–65% silica), and felsic (>65% silica) magmas.

- Magma and lava viscosity depends mostly on temperature and composition: The higher the temperature, the lower the viscosity; the more silica, the greater the viscosity.

- Minerals crystallize from magma and lava when small crystal nuclei form and grow.

- Rapid cooling accounts for the aphanitic textures of volcanic rocks, whereas comparatively slow cooling yields the phaneritic textures of plutonic rocks. Igneous rocks with markedly different-sized minerals are porphyritic.

- Igneous rock composition is determined mostly by the composition of the parent magma, but magma composition can change so that the same magma may yield more than one kind of igneous rock.

- According to Bowen's reaction series, cooling mafic magma yields a sequence of minerals, each of which is stable within specific temperature ranges. Only ferromagnesian silicates are found in the discontinuous branch of Bowen's reaction series. The continuous branch yields only plagioclase feldspars that become increasingly enriched with sodium as cooling occurs.

- A chemical change in magma may take place as early ferromagnesian silicates form and, because of their density, settle in the magma.

- Compositional changes also take place in magma when it assimilates country rock or one magma mixes with another.

- Geologists recognize two broad categories of igneous rocks: volcanic or extrusive and plutonic or intrusive.

- Texture and composition are the criteria used to classify igneous rocks, although a few are defined only by texture.

- Crystallization from water-rich magma results in very large minerals that form rocks known as pegmatite. Most pegmatite has an overall composition similar to granite.

- Intrusive igneous bodies known as plutons vary in their geometry and their relationship to country rock: Some are concordant, whereas others are discordant.

- The largest plutons, known as batholiths, consist of multiple intrusions of magma during long periods of time.

- Most plutons, including batholiths, are found at or near divergent and convergent plate boundaries.

Important Terms

aphanitic texture (p. 93)
assimilation (p. 92)
batholith (p. 102)
Bowen's reaction series (p. 89)
concordant pluton (p. 99)
country rock (p. 92)
crystal settling (p. 92)
dike (p. 99)
discordant pluton (p. 99)
felsic magma (p. 87)
hot spot (p. 91)
igneous rock (p. 86)

intermediate magma (p. 87)
laccolith (p. 99)
lava (p. 86)
lava flow (p. 86)
mafic magma (p. 87)
magma (p. 86)
magma chamber (p. 89)
magma mixing (p. 92)
mantle plume (p. 91)
phaneritic texture (p. 93)
pluton (p. 98)
plutonic (intrusive igneous) rock (p. 87)

porphyritic texture (p. 93)
pyroclastic (fragmental) texture (p. 94)
pyroclastic materials (p. 87)
sill (p. 99)
stock (p. 102)
stoping (p. 103)
ultramafic magma (p. 87)
vesicle (p. 94)
viscosity (p. 88)
volcanic neck (p. 102)
volcanic pipe (p. 102)
volcanic (extrusive igneous) rock (p. 87)

Review Questions

1. An igneous rock such as basalt with minerals too small to see without magnification has a _____ texture.
   a. ____porphyritic.
   b. ____aphanitic.
   c. ____vesicular.
   d. ____glassy.
   e. ____phaneritic.

2. An igneous rock made up of pyroclastic materials called ash is
   a. ____tuff.
   b. ____rhyolite.
   c. ____diorite.
   d. ____obsidian.
   e. ____peridotite.

3. The viscosity of magma or lava
   a. ____is a measure of its silica content.
   b. ____is determined mostly by its elevation.
   c. ____decreases as the magma or lava cools.
   d. ____causes it to cool more quickly.
   e. ____is its resistance to flow.

4. Lava characterized as mafic
   a. ____has 45–52% silica.
   b. ____crystallizes to form granite.
   c. ____makes up most of the continental crust.
   d. ____flows more slowly than felsic lava.
   e. ____cools to form plutons.

5. Which one of the following is a concordant, tabular pluton?
   a. ____stock.
   b. ____dike.
   c. ____batholith.
   d. ____sill.
   e. ____volcanic neck.

6. Which pair of the following igneous rocks have the same mineral composition?
   a. ____gabbro-granite.
   b. ____basalt-andesite.
   c. ____rhyolite-granite.
   d. ____peridotite-tuff.
   e. ____obsidian-volcanic breccia.

7. Andesite is found in abundance in
   a. _____ the Hawaiian Islands.
   b. _____ the Andes Mountains and Cascade Range.
   c. _____ batholiths and stocks.
   d. _____ the lower oceanic crust.
   e. _____ Earth's core.

8. The phenomenon by which pieces of country rock are detached and engulfed by rising magma is
   a. _____ viscosity.
   b. _____ stoping.
   c. _____ Bowen's principle.
   d. _____ crystal settling.
   e. _____ endosymbiosis.

9. One process that brings about chemical changes in magma is
   a. _____ recapitulation.
   b. _____ crystal inversion.
   c. _____ magma mixing.
   d. _____ ultramafic accumulation.
   e. _____ discordant separation.

10. Which one of the following statements about batholiths is correct?
    a. _____ they are composed mostly of granitic rocks.
    b. _____ they formed from a single huge mass of magma.
    c. _____ they are mostly concordant.
    d. _____ they form at divergent plate boundaries.
    e. _____ they consist primarily of basalt and gabbro.

11. Refer to Bowen's reaction series and explain why you would not expect to find quartz phenocrysts in basalt.

12. Mafic magma at divergent plate boundaries continuously forms oceanic crust, but the lower part of the crust is gabbro whereas the upper part is basalt. Why?

13. How does pegmatite form, and why are its minerals so large?

14. How do assimilation and stoping contribute to the emplacement of batholiths in continental crust?

15. What are aphanitic and phaneritic textures in igneous rock, and what do they indicate about the cooling history of these rocks?

16. What is a volcanic neck, and how does one form?

17. Explain why and what kinds of magmas form at divergent and convergent plate boundaries.

18. Two aphanitic rocks have the following compositions: Specimen 1—5% biotite, 20% sodium-rich plagioclase, 65% potassium feldspar, and 10% quartz; specimen 2—10% olivine, 55% pyroxene, 5% hornblende, and 30% calcium-rich plagioclase. Use Figure 4.11 and classify these rocks. Which one would you expect to be the darkest and densest? How do you know?

19. What accounts for the volcanoes in the Hawaiian Islands, and why are they older toward the northwest?

20. Explain how it is possible to derive mafic magma from ultramafic source rock.

Eyjafjallajökull in Iceland erupting a huge cloud of volcanic ash and steam on April 17, 2010. The eruption, although not very large, took place beneath a glacier that added to the cloud over the volcano. Air traffic over the North Atlantic was disrupted for several days.

bt3/ZUMA Press/Newscom

Volcanoes and Volcanism

OUTLINE

5.1   Introduction

5.2   Volcanoes and Volcanism

5.3   Types of Volcanoes

GEO-FOCUS: Do Volcanic Gases Cause Ozone Depletion?

5.4   Other Volcanic Landforms

5.5   Volcano Belts

5.6   North America's Active Volcanoes

5.7   Plate Tectonics, Volcanoes, and Plutons

5.8   Volcanic Hazards, Volcano Monitoring, and Forecasting Eruptions

GEO-RECAP

HAVE YOU EVER WONDERED?

- What besides lava is emitted from erupting volcanoes?
- How and why volcanoes differ from one another?
- What the most dangerous types of volcanoes are?
- Whether volcanoes are really as dangerous as depicted in movies and TV shows?
- Where the active volcanoes are in the United States?
- Whether there is a relationship between plate boundaries and volcanoes?
- How large volcanic eruptions are and whether they can be forecast with any certainty?

5.1 Introduction

About 550 volcanoes on Earth are active, that is, they have erupted during historic time, but only a dozen or so are erupting at any one time. Most eruptions are minor and go unreported in the popular press unless they occur in or near populated areas or have tragic consequences. A good recent example is the April 2010 eruptions of Eyjafjallajökull in Iceland (see the chapter opening photo). The eruptions were not particularly large but they took place beneath 200 m of glacial ice resulting in steam explosions that injected huge quantities of volcanic ash into the atmosphere, disrupting air travel over the North Atlantic for several days.

In addition to Earth's active volcanoes, there are many others that are *dormant,* meaning they have not erupted in historic time but may do so again in the future. In 1991, Mount Pinatubo in the Philippines erupted for the first time in 600 years, and similarly, Mount Sinabung in Indonesia lay dormant for 400 years and then began erupting in August 2010. Thousands of volcanoes have not erupted during historic time and show no signs of doing so again so they are *extinct* or *inactive* volcanoes. The Sutter Buttes in northern California last erupted about 1.5 million years ago and now show no indications of any activity.

No other geologic phenomenon has captured the public imagination more than erupting volcanoes, especially fiery streams of lava and particulate matter blasted into the atmosphere in huge clouds. What better subject for a disaster movie, many of which show lava flows as a great danger to humans. In 2002, lava flows in Goma, Zaire (Democratic Republic of the Congo), caused gasoline storage tanks to explode killing 147 people. And of course lava flows may destroy buildings and cover productive land, but overall they are the least dangerous manifestation of volcanic eruptions. Explosive eruptions, in contrast, are quite dangerous, especially if they occur near populated areas.

A well known volcanic catastrophe is the AD 79 eruption of Mount Vesuvius that destroyed the thriving Roman communities of Pompeii, Herculaneum, and Stabiae in what is now Italy (▌Figure 5.1). Pompeii was gradually buried in nearly 3 m of pyroclastic materials that covered all but the tallest buildings. About 2,000 bodies have been discovered in the city, but certainly far more perished. Surges of incandescent glowing avalanches swept through Herculaneum, quickly burying the town to a depth of about 20 m. Since AD 79, Mount Vesuvius has erupted 80 times, most violently in 1631 and 1906. It last erupted in 1944.

Erupting volcanoes do cause fatalities and considerable damage, but when considered in the context of Earth history, volcanism is actually a constructive process.

> **CONNECTION LINK**
>
> You can find more information about the origin of Earth's atmosphere and surface waters on pages 489–491 in Chapter 19.

Emissions from volcanoes during Earth's early history played an important part in the origin of the atmosphere and surface waters, and oceanic crust is continuously produced by volcanism at spreading ridges. Oceanic islands such as the Hawaiian Islands, Iceland, the Azores, and many others owe their existence to volcanic eruptions, and weathering of lava, pyroclastic materials, and volcanic mudflows in areas such as Indonesia converts them to productive soils.

An excellent reason to study volcanism is that it illustrates the complex interactions among Earth's systems. The emission of gases and pyroclastic materials has an immediate and profound impact on the atmosphere, hydrosphere, and biosphere, at least in the vicinity of an eruption. And in some cases the effects are worldwide, as they were following the eruptions of Tambora in 1815, Krakatau in 1883, and Pinatubo in 1991.

5.2 Volcanoes and Volcanism

A **volcano** is a hill or mountain that forms around a vent where lava, pyroclastic materials, and gases erupt. Thus, it is a landform—a feature on Earth's surface. The term **volcanism** refers to all processes related to the rise and discharge of magma and gases at the surface or into the atmosphere. Thus, volcanism accounts for the origin of all volcanic (extrusive igneous) rocks, such as basalt, tuff, and obsidian, as well as volcanoes, but not all volcanism results in the origin of volcanoes. Some eruptions take place along fissures and build up basalt plateaus (discussed later in this chapter).

All of the other terrestrial planets (Mercury, Venus, and Mars) and Earth's Moon were volcanically active during their early histories, but now the only one with possible active volcanoes is Venus (see the Geo-Insight, Chapter 2). As recently as April 2010, Suzanne Smerkar of the Jet Propulsion Laboratory in Pasadena, California, claimed that three active volcanoes are present on Venus. In addition, volcanoes are known on two other bodies in the solar system. Triton, one of Neptune's moons, probably has active volcanoes, and Jupiter's moon Io is by far the most volcanically active body in the solar system. Many of its hundred or so volcanoes are erupting at any given time.

Volcanic Gases

Volcanic gases from present-day volcanoes are 50–80% water vapor, with lesser amounts of carbon dioxide, nitrogen, sulfur gases, especially sulfur dioxide and hydrogen sulfide, and very small amounts of carbon monoxide, hydrogen, and chlorine. Indeed, some people claim that chlorine gas from volcanoes is responsible for ozone depletion in the atmosphere (see Geo-Focus on pages 122 and 123). In areas

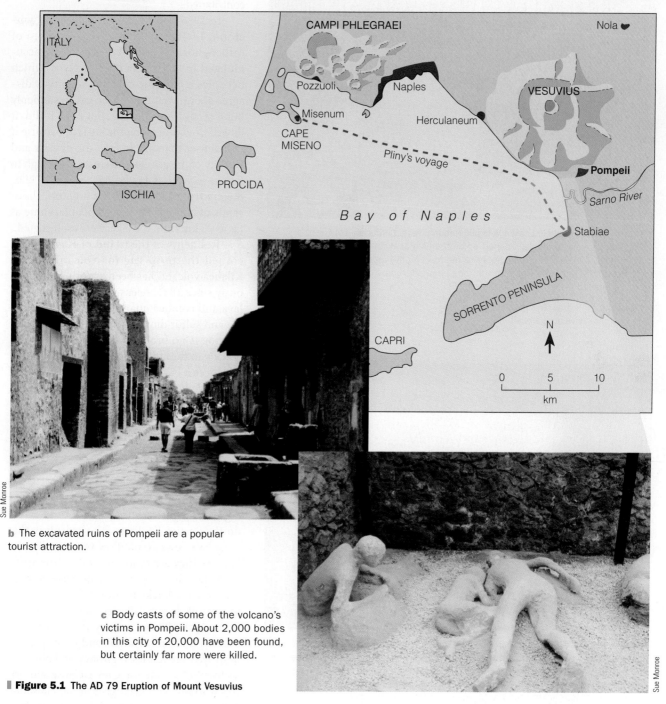

a The Mount Vesuvius region on the shore of the Bay of Naples in what is now Italy. Vesuvius erupted in AD 79 and destroyed the cities of Pompeii, Herculaneum, and Stabiae. The eruption was recorded by Pliny the Younger whose uncle, Pliny the Elder, died trying to investigate the event. Pliny the Elder's route is shown by a dashed line.

b The excavated ruins of Pompeii are a popular tourist attraction.

c Body casts of some of the volcano's victims in Pompeii. About 2,000 bodies in this city of 20,000 have been found, but certainly far more were killed.

▌ Figure 5.1 The AD 79 Eruption of Mount Vesuvius

of recent volcanism, emission of gases continues, and one cannot help noticing the rotten-egg odor of hydrogen sulfide gas (▌ Figure 5.2).

As magma rises toward the surface, the pressure is reduced and the contained gases begin to expand. In highly viscous, felsic magma, expansion is inhibited and gas pressure increases. Eventually, the pressure may become great enough to cause an explosion and produce pyroclastic materials. In contrast, low-viscosity mafic magma allows gases to expand and escape easily, so mafic magma generally erupts rather quietly.

Most volcanic gases quickly dissipate in the atmosphere and pose little danger to humans, but on several occasions they have had far-reaching climatic effects or have caused

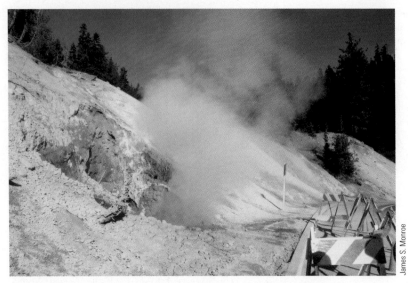

Figure 5.2 Fumerole in Lassen Volcanic National Park, California This vent discharging gases (fumerole) opened up on the hillside a few years ago and has progressively moved closer and closer to the road on the right. The light-colored rock around the fumerole was lava that has been altered by the gases and water, a process called weathering (see Chapter 6).

》 Critical Thinking Question
Are volcanic gases dangerous to humans?

Figure 5.3 Gases Billowing from Halema 'uma'u Crater on Kilauea Volcano in Hawaii Volcanic gases are mostly water vapor, but carbon dioxide, sulfur dioxide, and several others are emitted, too. Sulfur dioxide produces a haze and the unpleasant odor of sulfur, or what people in Hawaii call *vog*.

of Tambora in 1815 in Indonesia, although an eruption of Mayon volcano in the Philippines during the previous year also probably contributed.

In 1986, in the African nation of Cameroon, 1,746 people died when a cloud of carbon dioxide engulfed them. The gas accumulated in the waters of Lake Nyos, which occupies a volcanic caldera. Scientists disagree about what caused the gas to suddenly burst forth from the lake, but once it did, it flowed downhill along the surface because it was denser than air. In fact, the density and velocity of the gas cloud were great enough to flatten vegetation, including trees, a few kilometers from the lake. Unfortunately, thousands of animals and many people, some as far as 23 km from the lake, were asphyxiated.

Residents of the island of Hawaii have coined the term *vog* for volcanic smog. Kilauea volcano has been erupting continuously since 1983, releasing small amounts of lava and copious quantities of carbon dioxide and sulfur dioxide (Figure 5.3). Carbon dioxide is no problem because it dissipates quickly in the atmosphere, but sulfur dioxide produces a haze and the unpleasant odor of sulfur. Vog probably poses little or no health risk for tourists, but a long-term threat exists for residents of the west side of the island where vog is most common.

Lava Flows

Movies and TV shows depict lava flows as a great danger to humans, but only rarely do they cause fatalities. The reason is that most lava flows do not flow very rapidly, and because they are fluid they follow low areas. So once a flow begins moving, determining the path it will take is fairly easy, and anyone in areas likely to be affected can be evacuated. From April 1990 to January 1991, lava flows covered Kalapana, Hawaii, and in the process destroyed 180 homes, highways, and points of archaeological interest. Civil Defense authorities working with geologists at the Hawaiian Volcano Observatory made important decisions regarding evacuations and road closures, and as a result there were no injuries or fatalities.

Even low-viscosity flows, such as those in Hawaii, do not move very rapidly. They can move much more rapidly, though, when their margins cool to form a channel, and especially when insulated on all sides as in a **lava tube,** where geologists have recorded speeds of more than 50 km/hr. A lava tube forms within a flow when its margins and upper surface solidify, thereby forming a

fatalities. In 1783, sulfur gases erupting from Laki fissure in Iceland were responsible for the Blue Haze famine, crop failures, and the deaths of 24% of Iceland's population. In 1816, a persistent "dry fog" caused unusually cold spring and summer weather in Europe and eastern North America. In North America 1816 was called "The Year Without a Summer" or "Eighteen Hundred and Froze to Death." Killing frosts during the summer resulted in crop failures and food shortages. This particularly cold summer was attributed to the huge eruption

Figure 5.4 Lava Tubes Lava tubes consisting of hollow spaces beneath the surfaces of lava flows are common in many areas.

J. B. Judd/USGS

a A lava tube in Hawaii after the lava has drained out.

Hawaii Volcanoes National Park, USGS

b An active lava tube in Hawaii. Part of the tube's roof has collapsed, forming a skylight.

conduit through which lava can move rapidly and for great distances. When an eruption ceases, the tube drains, leaving an empty tunnel-like structure (▐ Figure 5.4a). Part of the roof of a lava tube may collapse to form a *sky-light* through which an active flow can be observed (▐ Figure 5.4b), or an inactive lava tube can be accessed.

Geologists in Hawaii characterized basalt lava flows as pahoehoe or aa, although these terms are now used elsewhere, too. **Pahoehoe** (pronounced *pay-hoy-hoy*) has a smooth ropy surface much like taffy (▐ Figure 5.5a). The surface of an **aa** (pronounced *ah-ah*) flow consists of rough, jagged, angular blocks and fragments (▐ Figure 5.5b). Pahoehoe flows are hotter and thinner than aa flows; indeed, aa flows are viscous enough to break into blocks and move forward as a wall of rubble. One further point: A pahoehoe flow may change along its length to aa as its viscosity increases, partly because it cools, but aa does not change to pahoehoe.

Much of the igneous rock in the upper oceanic crust is a distinctive type made up of bulbous masses of basalt that resemble pillows, hence the name **pillow lava.** Geologists realized long ago that pillow lava forms when lava is rapidly chilled underwater, but its formation was not observed until 1971. Divers near Hawaii saw pillows form when a blob of lava broke through the crust of an underwater lava flow and cooled very quickly, forming a pillow- or loaf-shaped structure with a glassy exterior. The remaining fluid lava inside then broke though the crust of an already-formed pillow and formed another pillow, and so on (▐ Figure 5.6).

Mafic lava flows, and some intermediate ones, as well as some rocks in dikes, sills, and volcanic necks, show a pattern of columns bounded by fractures, or what geologists call **columnar joints.** For columnar joints to form, a lava flow must cease moving, and cool and contract thereby setting up forces that cause fractures, known as *joints,* to open. On a lava flow's surface the fractures are commonly polygonal (often six-sided) cracks that extend downward, thus

outlining columns with their long axes perpendicular to the cooling surface (▐ Figure 5.7).

Pyroclastic Materials

In addition to lava flows, erupting volcanoes eject pyroclastic materials, especially **volcanic ash,** a designation for pyroclastic particles that measure less than 2.0 mm. In some cases, ash is ejected into the atmosphere and settles to the surface as an *ash fall.* In 1947, ash that erupted from Mount Hekla in Iceland fell 3,800 km away on Helsinki, Finland. In contrast to an ash fall, an *ash flow* is a cloud of ash and gas that flows along or close to the land surface. Ash flows can move faster than 100 km/hr, and some cover vast areas.

In populated areas adjacent to volcanoes, ash falls and ash flows pose serious problems, and volcanic ash in the atmosphere is a hazard to aviation. The most recent incident took place in April 2010 when an eruption in Iceland disrupted air traffic across the North Atlantic. Since 1980, approximately 80 aircraft have been damaged when they encountered clouds of volcanic ash. The most serious incident took place in 1989 when ash from Redoubt volcano in Alaska caused all four jet engines to fail on KLM Flight 867. The plane, carrying 231 passengers, nearly crashed when it fell more than 3 km before the crew could restart the engines. The plane landed safely in Anchorage, Alaska, but it required $80 million in repairs.

In addition to volcanic ash, volcanoes erupt *lapilli,* consisting of pyroclastic materials that measure from 2 to 64 mm, and *blocks* and *bombs,* both larger than 64 mm. Bombs have a twisted, streamlined shape, which indicates that they were erupted as globs of magma that cooled and solidified during their flight through the air. Blocks, in contrast, are angular pieces of rock ripped from a volcanic conduit or pieces of a solidified crust of a lava flow. Because of their size, lapilli, bombs, and blocks are confined to the immediate area of an eruption.

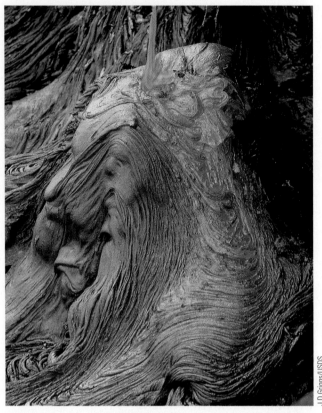

J.D. Griggs/USDS

a An excellent example of the taffylike appearance of pahoehoe.

Figure 5.5 **Pahoehoe and Aa Lava Flows** Pahoehoe and aa were named for lava flows in Hawaii, but the same kinds of flows are found in many other areas.

Robert Tilling/USGS

b An aa lava flow advances over an older pahoehoe flow. Notice the rubbly nature of the aa flow.

» Critical Thinking Question
What controls whether a lava flow is aa or pahoehoe?

Figure 5.6 **Pillow Lava** Much of the upper part of the oceanic crust is made up of pillow lava that formed when lava erupted underwater.

NOAA

a Pillow lava on the seafloor in the Pacific Ocean about 240 km west of Oregon that formed about five years before the photo was taken.

James S. Monroe

b Ancient pillow lava now on land in Kenai Fjords National Park, Alaska.

» Critical Thinking Question
What inferences can you make about where this pillow lava formed?

5.3 Types of Volcanoes

Although volcanoes vary in size and shape, all have a conduit or conduits leading to a magma chamber beneath the surface. Vulcan, the Roman deity of fire, was the inspiration for calling these mountains volcanoes, and because of their danger and obvious connection to Earth's interior, they have been held in awe by many cultures.

In Hawaiian legends, the volcano goddess Pele resides in the crater of Kilauea on Hawaii. During one of her frequent rages, Pele causes earthquakes and lava flows, and she may hurl flaming boulders at those who offend her. Native Americans in the Pacific Northwest tell of a titanic battle between the volcano gods Skel and Llao to account for huge eruptions that took place about 7,700 years ago in Oregon and California.

Most volcanoes have a circular depression known as a **crater** at their summit, or on their flanks, that forms by

Figure 5.7 **Columnar Jointing** Columnar jointing is seen mostly in mafic lava flows and related intrusive rocks.

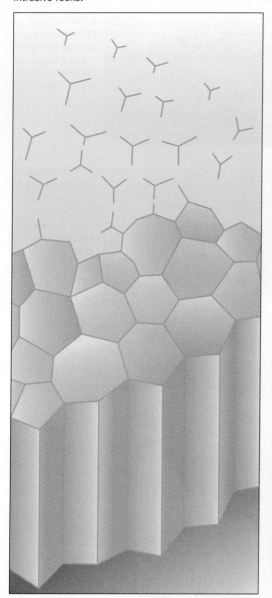

a As lava cools and contracts, three-pronged cracks form that grow and intersect to form four- to seven-sided columns, most of which are six-sided.

b Columnar joints in a basalt lava flow at Devil's Postpile National Monument in California. The rubble in the foreground is collapsed columns.

c Surface view of the columns from (b). The straight lines and polish resulted from abrasion by a glacier that moved over this surface.

explosions or collapse. Craters are generally less than 1 km across, whereas much larger rimmed depressions on volcanoes are **calderas.** In fact, some volcanoes have a summit crater within a caldera. Calderas are huge structures that form following voluminous eruptions during which part of a magma chamber drains and the mountain's summit collapses into the vacated space below. An excellent example is misnamed Crater Lake in Oregon (**I** Figure 5.8). Crater Lake is actually a steep-rimmed caldera that formed about 7,700 years ago in the manner just described; it is more than 1,200 m deep and measures 9.7 by 6.5 km. As impressive as Crater Lake is, it is not

nearly as large as some other calderas, such as the Toba caldera in Sumatra, which is 100 km long and 30 km wide.

Geologists recognize several major types of volcanoes, but one must realize that each volcano is unique in its history of eruptions and development. For instance, the frequency of eruptions varies considerably; the Hawaiian volcanoes and Mount Etna on Sicily have erupted repeatedly, whereas Pinatubo in the Philippines erupted in 1991 for the first time in 600 years. Furthermore, some volcanoes are complex and have the characteristics of more than one type of volcano.

Figure 5.8 **The Origin of Crater Lake, Oregon** Remember, Crater Lake is actually a caldera that formed by partial draining of a magma chamber.

a Eruption begins as huge quantities of ash are ejected from the volcano.

c The collapse of the summit into the partially drained magma chamber forms a huge caldera.

b The eruption continues as more ash and pumice are ejected into the air and pyroclastic flows move down the flanks of the mountain.

d Postcaldera eruptions partly cover the caldera floor, and the small cinder cone called Wizard Island forms.

From Howell Williams, Crate Lake: The Story of Its Origin (Berkeley, Calif. University of California Press): Illustrations from p. 84 (c) 1941 Regents of the University of California, (c) renewed 1969, Howell Williams.

e View from the rim of Crater Lake showing Wizard Island. The lake is 594 m deep, making it the second deepest in North America.

Shield Volcanoes

Volcanoes that look much like the outer surface of a shield laying on the ground with the convex side up are **shield volcanoes** (Figure 5.9). They are composed almost entirely of mafic lava flows that had low viscosity, so the flows spread out and formed thin layers that slope only 2 to 10 degrees. Erupting shield volcanoes, sometimes called *Hawaiian-type eruptions,* are rather quiet compared to eruptions of many other volcanoes, particularly those at convergent plate boundaries. Magma rises to the surface and issues as lava flows that pose little danger to humans. However, jets of incandescent lava may be forcefully but not explosively ejected as lava fountains, some up to 400 m high, when magma reaches the surface and gases expand (Figure 5.10a). These lava fountains

contribute some pyroclastic materials to shield volcanoes, but they are made up mostly of layers of basalt.

Although eruptions of shield volcanoes tend to be rather quiet, some of the Hawaiian volcanoes have, on occasion, produced sizable explosions when groundwater instantly vaporizes as it comes in contact with magma. One such explosion in 1790 killed about 80 warriors in a party headed by Chief Keoua, who was leading them across the summit of Kilauea volcano.

Kilauea volcano is impressive because it has been erupting continuously since January 3, 1983, making it the longest recorded eruption. During these 28 years, more than 2.5 km³ of molten rock has flowed out at the surface, much of it reaching the sea and forming 2.2 km² of new

Figure 5.9 Shield Volcanoes

a Shield volcanoes consist of numerous thin basalt lava flows that build up mountains with slopes rarely exceeding 10 degrees.

b Profile of Mauna Loa in Hawaii. Mauna Loa is one of five huge shield volcanoes that make up the island of Hawaii.

land on the island of Hawaii. Unfortunately, lava flows from Kilauea have also destroyed many homes and caused some $61 million in damages.

Shield volcanoes are most common in the ocean basins, but some are also present on the continents—in East Africa, for instance. The island of Hawaii is made up of five huge shield volcanoes, two of which, Kilauea and Mauna Loa, are active much of the time. Mauna Loa is nearly 100 km across its base and stands more than 9.5 km above the surrounding seafloor. It has a volume estimated at 50,000 km³, making it the world's largest volcano.

Mauna Loa is certainly a large volcano, but the record for size goes to Olympus Mons on Mars, which is by far the largest volcano of any kind known in the solar system (Figure 5.10b). It stands 25 km high, about three times as high as Mount Everest, Earth's highest peak, and measures more than 500 km across. Escarpments 4 km high mark its outer edges. Olympus Mons probably attained its great size because Mars lacks any plate tectonic activity, so prolonged eruptions in the same place simply built up this huge mountain. So far scientists have found no evidence that Olympus Mons is active.

CONNECTION LINK

For more information on Olympus Mons and Mars see page 43 in Chapter 2.

Figure 5.10 Shield Volcanoes in Hawaii and on Mars

a Shield volcanoes are made up mostly of basalt lava flows, but some pyroclastic materials are erupted at lava fountains.

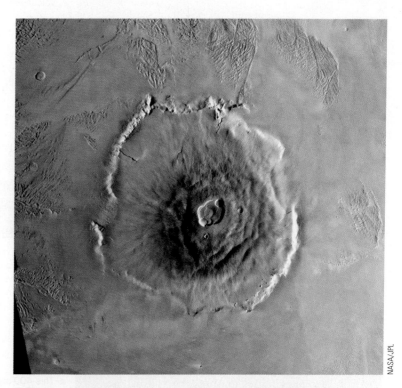

b Olympus Mons on Mars is a huge shield volcano that stands about 25 km high and measures 340 km across. It is encircled by an escarpment 4 km high.

Cinder Cones

Small, steep-sided **cinder cones** made up of particles resembling cinders form when pyroclastic materials accumulate around a vent from which they erupted (❙ Figure 5.11). Cinder cones are small, rarely exceeding 400 m high, with slope angles up to 33 degrees, depending on the angle that can be maintained by the angular pyroclastic materials. Many have a large, bowl-shaped crater, and if they issue any lava flows, they usually break through the base or lower flanks of the mountains. Although all cinder cones are conical, their symmetry varies from those that are almost perfectly symmetrical to those that formed when prevailing winds caused pyroclastic materials to build up higher on the downwind side of the vent.

Many cinder cones form on the flanks or within the calderas of larger volcanoes and represent the final stages of activity, particularly in areas of basaltic volcanism. Wizard Island in Crater Lake, Oregon, is a small cinder cone that formed after the summit of Mount Mazama collapsed to form a caldera (Figure 5.8). Cinder cones are common in the southern Rocky Mountain states, particularly New Mexico and Arizona, and many others are in California, Oregon, Washington, and Hawaii.

In 1973, on the Icelandic island of Heimaey, the town of Vestmannaeyjar was threatened by a new cinder cone. The initial eruption began on January 23, and within two days, a cinder cone, later named Eldfell, rose to about 100 m above the surrounding area (Figure 5.11b). Pyroclastic materials from the volcano buried parts of the town, and by February, a massive aa lava flow was advancing toward the town. The flow's leading edge ranged from 10 to 20 m thick, and its central part was as much as 100 m thick. The residents of Vestmannaeyjar sprayed the leading edge of the flow with seawater in an effort to divert it from the town. The flow was in fact diverted, but how effective the efforts of the townspeople were is not clear. They may have been simply lucky.

Composite Volcanoes (Stratovolcanoes)

Pyroclastic layers, as well as lava flows, both of intermediate composition, are found in **composite volcanoes,** which are also called *stratovolcanoes* (❙ Figure 5.12). As the lava flows cool, they typically form andesite; recall that intermediate lava flows are more viscous than mafic ones. Geologists use the term **lahar** for volcanic mudflows, which are also common on composite volcanoes. A lahar may form when rain falls on unconsolidated pyroclastic materials and creates a muddy slurry that moves downslope (❙ Figure 5.13). On November 13, 1985, a minor eruption of Nevado del Ruiz in Colombia melted snow and ice on the volcano, causing lahars that killed approximately 23,000 people (Table 5.1).

❙ **Figure 5.11** Cinder Cones Cinder cones are small, steep-sided volcanoes made up of pyroclastic materials that resemble cinders.

b Eldfell, a cinder cone in Iceland. Eldfell began erupting in 1973 and in two days grew to 100 m high. The steam visible on the left side of the image resulted from aa lava flowing into the sea. Another cinder cone, known as Helgafel, is also visible.

K. Segerstrom, USGS

a This 400-m-high cinder cone named Paricutín formed in a short time in Mexico in 1943, when pyroclastic materials began to erupt in a farmer's field. Lava flows from the volcano covered two nearby villages, but all activity ceased by 1952.

▌ Figure 5.12 Composite Volcanoes

Pyroclastic layers
Lava flows

a Composite volcanoes, also called stratovolcanoes, are made up mostly of lava flows and pyroclastic materials of intermediate composition, although mudflows (lahars) are also common.

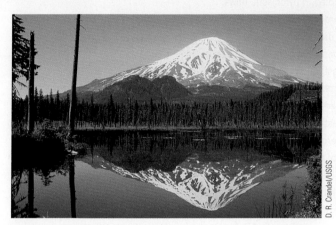

D. R. Crandel/USGS

b Mount St. Helens in Washington State as it appeared from the east in 1978.

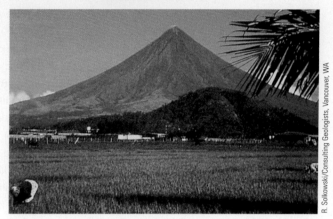

R. Solkowski/Consulting Geologists, Vancouver, WA

c Mayon volcano in the Philippines, a nearly symmetrical composite volcano that last erupted during the 1900s, and also in 2009 and 2010.

Composite volcanoes differ from shield volcanoes and cinder cones in composition and in their overall shape. Remember that shield volcanoes have very low slopes, whereas cinder cones are small, steep-sided, conical mountains. In contrast, composite volcanoes are steep-sided near their summits, perhaps as much as 30 degrees, but the slope decreases toward the base, where it may be no more than 5 degrees (Figure 5.12b). Mayon volcano in the Philippines is one of the most nearly symmetrical composite volcanoes anywhere (Figure 5.12c). It erupted in 1999 for the 13th time during the 1900s.

When most people think of volcanoes, they picture the graceful profiles of composite volcanoes, which are the typical large volcanoes found on the continents and island arcs. And some of these volcanoes are indeed large; Mount Shasta in northern California is made up of about 350 km³ of material and measures 20 km across its base. Remember, though, that Mauna Loa in Hawaii has an estimated volume of 50,000 km³!

Many composite volcanoes erupt regularly and sometimes violently. In 1991, the world's largest volcanic outburst since 1912 took place when Mount Pinatubo in the Philippines discharged huge quantities of gas and an estimated 3 to 5 km³ of ash. Fortunately, warnings of an impending eruption were heeded, and 200,000 people were evacuated from around the volcano. Nevertheless, the eruption was responsible for 722 deaths (Table 5.1).

Lava Domes

Most volcanoes are classified as shield volcanoes, cinder cones, or composite volcanoes. Less common are **lava**

domes, also known as *volcanic domes* and *plug domes,* which are steep-sided, bulbous mountains that form when viscous felsic magma, and occasionally intermediate magma, is forced toward the surface (▌ Figure 5.14). Because felsic magma is so viscous, it moves upward very slowly and only when the pressure from below is great.

Beginning in 1980, a number of lava domes were emplaced in the crater of Mount St. Helens in Washington, most of which were destroyed during subsequent eruptions. Since 1983, Mount St. Helens has been characterized by sporadic dome growth and renewed eruptions in 2004. In June 1991, a lava dome in Japan's Unzen volcano collapsed under its own weight, causing a flow of debris and hot ash that killed 43 people in a nearby town.

USGS/David Johnson

▌ Figure 5.13 **Lahar** Lahars or volcanic mudflows are common on composite volcanoes. Homes partly buried by a lahar on June 15, 1991, following the eruption of Mount Pinatubo in the Philippines.

TABLE 5.1	Some Notable Volcanic Eruptions	
Date Deaths	**Volcano**	**Deaths**
Apr 10, 1815	Tambora, Indonesia	117,000 killed, including deaths from eruption, famine, and disease.
Oct 8, 1822	Galunggung, Java	Pyroclastic flows and mudflows killed 4,011.
Mar 2, 1856	Awu, Indonesia	2,806 died in pyroclastic flows.
Aug 27, 1883	Krakatau, Indonesia	More than 36,000 died, most killed by tsunami.
June 7, 1892	Awu, Indonesia	1,532 died in pyroclastic flows.
May 8, 1902	Mount Pelée, Martinique	Nuée ardente engulfed St. Pierre and killed 28,000.
Oct 24, 1902	Santa Maria, Guatemala	5,000 died during eruption.
May 19, 1919	Kelut, Java	Mudflows devastated 104 villages and killed 5,110.
Jan 21, 1951	Lamington, New Guinea	Pyroclastic flows killed 2,942.
Mar 17, 1963	Agung, Indonesia	1,148 perished during eruption.
May 18, 1980	Mount St. Helens, Washington	63 killed, 600 km^2 of forest devastated.
Mar 28, 1982	El Chichon, Mexico	Pyroclastic flows killed 1,877.
Nov 13, 1985	Nevado del Ruiz, Colombia	Minor eruption triggered mudflows that killed 23,000.
Aug 21, 1986	Oku volcanic field, Cameroon	Cloud of CO_2 released from Lake Nyos killed 1,746.
June 15, 1991	Mount Pinatubo, Philippines	281 killed during eruption, 83 died in later mudflows, 358 died of illness.
July 1999	Soufriére Hills, Montserrat	19 killed, 12,000 evacuated.
Jan 17, 2002	Nyiragongo, Zaire	Lava flow killed 147 in Goma.
Aug 16, 2006	Tungurahua, Ecuador	Explosive eruption killed 7; continuously active since 1999, periodic evacuations and several villages destroyed.
Apr 2010	Eyjafjallajökull, Iceland	No fatalities, but disrupted air traffic over the North Atlantic for several days

▌**Figure 5.14** Lava Domes Lava domes are bulbous masses of magma that are emplaced in the craters of composite volcanoes or stand alone as irregularly shaped mountains flanked by debris shed from the dome.

a Diagram of a mass of viscous magma forming a lava dome.

》 Critical Thinking Question
Why do you think lava domes are so dangerous?

Scott Perry

b The volcano in the distance is Lassen Peak in Lassen Volcanic National Park in California. The dark masses of rock cooled from the viscous magma that was injected in this area. Lassen Peak erupted from 1914 to 1917.

Lava dome eruptions are some of the most violent and destructive. In 1902, viscous magma accumulated beneath the summit of Mount Pelée on the island of Martinique. Eventually, the pressure increased until the side of the mountain blew out in a tremendous explosion, ejecting a mobile, dense cloud of pyroclastic materials and a glowing cloud of gases and dust called a **nuée ardente** (French for "glowing cloud"). The pyroclastic flow followed a valley to the sea, but the nuée ardente jumped a ridge and engulfed the city of St. Pierre (▌Figure 5.15).

▌Figure 5.15 Nuée Ardente

a St. Pierre, Martinique, after it was destroyed by a nuée ardente from Mount Pelée in 1902. Only two of the city's 28,0000 inhabitants survived.

b An April 1986 pyroclastic flow rushing down Augustine volcano in Alaska. This flow is similar to the one that wiped out St. Pierre.

A tremendous blast hit St. Pierre, leveling buildings; hurling boulders, trees, and pieces of masonry down the streets, and moving a 3-ton statue 16 m. Accompanying the blast was a swirling cloud of incandescent ash and gases with an internal temperature of 700°C that incinerated everything in its path. The nuée ardente passed through St. Pierre in two or three minutes, only to be followed by a firestorm as combustible materials burned and casks of rum exploded. But by then most of the 28,000 residents of the city were already dead. In fact, in the area covered by the nuée ardente, only two survived!* One survivor was on the outer edge of the nuée ardente, but even there, he was terribly burned and his family and neighbors were all killed. The other survivor, a stevedore incarcerated the night before for disorderly conduct, was in a windowless cell partly below ground level. He remained in his cell badly burned for four days after the eruption until rescue workers heard his cries for help. He later became an attraction in the Barnum and Bailey Circus where he was advertised as "The only living object that survived in the 'Silent City of Death' where 40,000 beings were suffocated, burned or buried by one belching blast of Mont Pelée's terrible volcanic eruption."**

---

*Although reports commonly claim that only two people survived the eruption, at least 69 and possibly as many as 111 people survived beyond the extreme margins of the nuée ardente and on ships in the harbor. Many, however, were badly injured.

**Quoted from A. Scharth, *Vulcan's Fury: Man Against the Volcano* (New Haven, CT: Yale University Press, 1999), p. 177.

Supervolcano Eruptions

Geologists have no formal definition for *supervolcano eruptions,* but we can take it to mean an explosive eruption of hundreds of cubic kilometers of pyroclastic materials and the origin of a huge caldera. No supervolcano eruptions have occurred in historic times, but geologists know of several that took place during the past two million years—Long Valley in eastern California, Toba in Indonesia, and Taupo in New Zealand, for example.

On three occasions, supervolcano eruptions followed the accumulation of rhyolitic magma beneath Yellowstone National Park, which is mostly in Wyoming, each yielding a widespread blanket of volcanic ash and pumice and gigantic calderas. We can summarize Yellowstone's volcanic history by noting that supervolcano eruptions took place 2 million years ago, 1.3 million years ago, and 600,000 years ago. Then, between 150,000 and 75,000 years ago, an additional 1,000 km³ of pyroclastic materials were erupted within the Yellowstone caldera (▌Figure 5.16).

Geologists think that these huge eruptions were caused by a rising *mantle plume,* a cylindrical mass of magma probably of rhyolitic composition. Because this type of magma is viscous, it triggers explosive eruptions when it nears the surface. Personnel from the U.S. Geological Survey (USGS) and the University of Utah continue to monitor the Yellowstone area for any signs of renewed activity.

GEO-FOCUS   Do Volcanic Gases Cause Ozone Depletion?

Earth supports life because of its distance from the Sun and the fact that it has abundant liquid water and an oxygen-rich atmosphere. Furthermore, an ozone layer (O_3) in the stratosphere (10 to 48 km above the surface) protects Earth because it blocks out most of the harmful ultraviolet radiation that bombards our planet. During the early 1980s, scientists discovered an *ozone hole* over Antarctica that has continued to grow. In fact, depletion of the ozone layer is now also recognized over the Arctic region and elsewhere. Any depletion in ozone levels is viewed with alarm because it would allow more dangerous radiation to reach the surface, increasing the risk of skin cancer, among other effects.

This discovery unleashed a public debate about the primary cause of ozone depletion and how best to combat the problem. Scientists proposed that one cause of ozone depletion is chlorofluorocarbons (CFCs), which are used in various consumer products— for instance, in aerosol cans. According to this theory, CFCs rise into the upper atmosphere where reactions with ultraviolet radiation liberate chlorine, which in turn reacts with and depletes ozone (❙ Figure 1). As a result of this view,

an international agreement called the Montreal Protocol was reached in 1983, limiting the production of CFCs, along with other ozone-depleting substances.

However, during the 1990s this view was challenged by some radio talk show hosts as well as a few government officials. They proposed an alternative idea that ozone depletion occurred because of natural causes rather than commercial products such as CFCs. They pointed out that volcanoes release copious quantities of hydrogen chloride (HCl) gas that rises into the stratosphere and that could be responsible for ozone depletion. Furthermore, they claimed that because CFCs are heavier than air they would not rise into the stratosphere.

It is true that volcanoes release HCl gas as well as several other gases, some of which are quite dangerous— recall the cloud of CO_2 gas in Cameroon that killed 1,746 people and the Blue Haze Famine in Iceland. However, most eruptions are too weak to inject gases of any kind high into the stratosphere. Even when it is released, HCl gas from volcanoes is very soluble and quickly removed from the atmosphere by rain and even by steam (water vapor) from the same eruption that released HCl gas in the first place.

Measurements of chlorine concentrations in the stratosphere show that only temporary increases occur following huge eruptions. For example, the largest volcanic outburst since 1912, the eruption of Mount Pinatubo in the Philippines in 1991, caused little increase in upper atmosphere chlorine. The impact of volcanic eruptions is certainly not enough to cause the average rate of ozone depletion taking place each year.

It is true that CFCs are heavier than air, but this does not mean that they cannot rise into the stratosphere. Earth's surface heats differentially, meaning that more heat may be absorbed in one area than in an adjacent one. The heated air above a warmer area becomes less dense, rises by convection, and carries with it CFCs and other substances that are actually denser than air. Once in the stratosphere, ultraviolet radiation, which is usually absorbed by ozone, breaks up CFC molecules and releases chlorine that reacts with ozone. Indeed, a single chlorine atom can destroy 100,000 ozone molecules (Figure 1). In contrast to the HCl gas produced by volcanoes, CFCs are absolutely insoluble; it is the fact that they are inert that made them so desirable for various uses. Because a

5.4 Other Volcanic Landforms

During *fissure eruptions,* fluid lava pours out and simply builds up rather flat-lying areas, whereas huge explosive eruptions might yield *pyroclastic sheet deposits,* which, as their name implies, have a sheetlike geometry. In both cases, volcanoes fail to develop.

Fissure Eruptions and Basalt Plateaus

Rather than erupting from central vents, the lava flows making up **basalt plateaus** issue from long cracks or fissures during **fissure eruptions.** The lava is so fluid (has such low viscosity) that it spreads out and covers vast areas. A good example is the Columbia River basalt in eastern Washington and parts of Oregon and Idaho. This huge accumulation of 17- to 6-million-year-old overlapping lava

CFC molecule can last for decades, any increase in CFCs is a long-term threat to the ozone layer.

Another indication that CFCs are responsible for the Antarctic ozone hole is that the rate of ozone depletion has slowed since the implementation of the Montreal Protocol. A sound understanding of the science behind these atmospheric processes helped world leaders act quickly to address this issue. Now scientists hope that with continued compliance with the Protocol, the ozone layer will recover by the middle of this century.

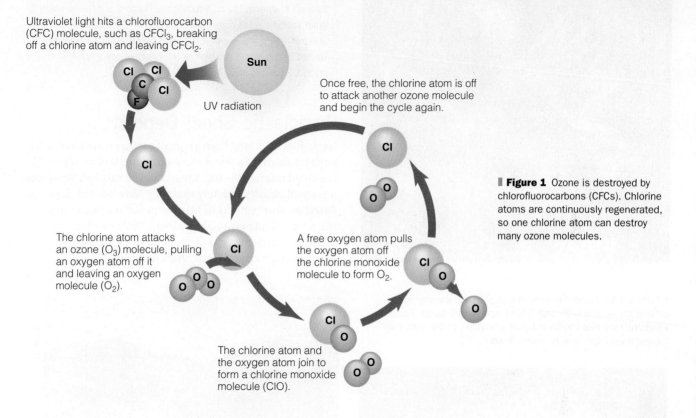

Ultraviolet light hits a chlorofluorocarbon (CFC) molecule, such as CFCl$_3$, breaking off a chlorine atom and leaving CFCl$_2$.

Sun

UV radiation

Once free, the chlorine atom is off to attack another ozone molecule and begin the cycle again.

The chlorine atom attacks an ozone (O$_3$) molecule, pulling an oxygen atom off it and leaving an oxygen molecule (O$_2$).

A free oxygen atom pulls the oxygen atom off the chlorine monoxide molecule to form O$_2$.

The chlorine atom and the oxygen atom join to form a chlorine monoxide molecule (ClO).

Figure 1 Ozone is destroyed by chlorofluorocarbons (CFCs). Chlorine atoms are continuously regenerated, so one chlorine atom can destroy many ozone molecules.

> **CONNECTION LINK**
>
> You can find more information about the Columbia River basalts on pages 618–620 in Chapter 23.

flows covers about 164,000 km^2 (▌ Figure 5.17a and b), and has an aggregate thickness of more than 1,000 m. And some individual flows are enormous—the Roza flow advanced along a front about 100 km wide and covered 40,000 km^2. Geologists have identified 300 huge flows here, one of which flowed 600 km from its source.

Similar accumulations of vast, overlapping lava flows are also found in the Snake River Plain in Idaho (Figure 5.17a and c). These flows are 5.0 to 1.6 million years old, and they represent a style of eruption between fissure eruptions and those of shield volcanoes. In fact, there are small, low shields, as well as fissure flows, in the Snake River Plain.

Currently, fissure eruptions occur only in Iceland. Iceland has a number of volcanoes, but the bulk of the island is composed of basalt lava flows that issued from fissures. In fact, about half of the lava erupted during historic time in Iceland

Figure 5.16 The Yellowstone Tuff The walls of the Grand Canyon of the Yellowstone River are made up of the hydrothermally altered Yellowstone Tuff that partly fills the Yellowstone caldera.

Sue Monroe

Figure 5.17 Basalt Plateaus Basalt plateaus are vast areas of overlapping lava flows that issued from long fissures. Fissure eruptions take place today in Iceland; however, in the past, they formed basalt plateaus in various areas.

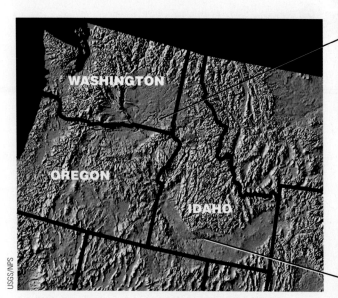

USGS/NPS

a Relief map of the northwestern United States showing the locations of the Columbia River basalt and the Snake River Plain.

came from two fissure eruptions, one in AD 930 and the other in 1783. The 1783 eruption from Laki fissure, which is more than 30 km long, accounted for lava that covered 560 km² and, in one place, filled a valley to a depth of about 200 m.

What Would You Do ?

You are an enthusiast of natural history and would like to share your interests with your family. Accordingly, you plan a vacation to see some of the volcanic features in U.S. national parks and monuments. Let's assume your planned route will take you through Wyoming, Idaho, Washington, Oregon, and California. What specific areas might you visit, and what kinds of volcanic features would you see in these areas? What other parts of the United States might you visit in the future to see additional evidence for volcanism?

Pyroclastic Sheet Deposits

Geologists have long been aware of vast areas covered by felsic volcanic rocks a few meters to hundreds of meters thick. It seemed improbable that these could be vast lava flows, but it seemed equally unlikely that they were ash fall deposits. Based on observations of historic pyroclastic flows, such as the nuée ardente erupted by Mount Pelée in 1902, it seems that these ancient rocks originated as pyroclastic flows—hence the name **pyroclastic sheet deposits.**

Frank Kujawa/University of Central Florida/GeoPhoto

b About 20 lava flows of the Columbia River basalt are exposed in the canyon of the Grand Ronde River in Washington.

Sue Monroe

c Basalt lava flows of the Snake River Plain near Twin Falls, Idaho.

Pyroclastic sheet deposits cover far greater areas than any observed during historic time, however, and apparently erupted from long fissures rather than from a central vent. The pyroclastic materials of many of these flows were so hot that they fused together to form *welded tuff*.

Geologists now think that major pyroclastic flows issue from fissures formed during the origin of calderas. For instance, pyroclastic flows erupted during the formation of a large caldera now occupied by Crater Lake in Oregon (Figure 5.8) and in the Yellowstone caldera in Wyoming.

Similarly, the Bishop Tuff of eastern California erupted shortly before the formation of the Long Valley caldera. Interestingly, earthquake activity in the Long Valley caldera and nearby areas beginning in 1978 may indicate that magma is moving upward beneath part of the caldera. Thus, the possibility of future eruptions in that area cannot be discounted.

5.5 Volcano Belts

Most of the world's active volcanoes are in well-defined zones or belts rather than randomly distributed. The **circum-Pacific belt,** popularly called the Ring of Fire, has more than 60% of all active volcanoes. It includes volcanoes in the Andes of South America; the volcanoes of Central America, Mexico, and the Cascade Range of North America; and the Alaskan volcanoes as well as those in Japan, the Philippines, Indonesia, and New Zealand (▌Figure 5.18). Also in the circum-Pacific belt are the southernmost active volcanoes at Mount Erebus in Antarctica and a large caldera at Deception Island that erupted most recently during 1970.

The second area of active volcanism is the **Mediterranean belt** (Figure 5.18). About 20% of all active volcanism takes place in this belt, where the famous Italian volcanoes such as Mounts Etna and Vesuvius and the Greek volcano Santorini are found. Mount Etna has issued lava flows 190 times since 1500 BC, when activity was first recorded. A particularly violent eruption of Santorini in 1390 BC might be the basis for the myth about the lost continent of Atlantis, and, in AD 79, an eruption of Mount Vesuvius destroyed Pompeii and other nearby cities (see the Introduction).

Nearly all the remaining active volcanoes are at or near mid-oceanic ridges or the extensions of these ridges onto land (Figure 5.18). These include the East Pacific Rise and the longest of all mid-oceanic ridges, the Mid-Atlantic Ridge. The latter is located near the center of the Atlantic Ocean basin, accounting for the volcanism in Iceland and elsewhere. It continues around the southern tip of Africa, where it connects with the Indian Ridge. Branches of the Indian Ridge extend into the Red Sea and East Africa, where such volcanoes as Kilamanjaro in Tanzania, Nyiragongo in Zaire, and Erta Ale in Ethiopia with its continuously active lava lake are found.

Divergent plate boundary
(some transform plate boundaries)

Convergent boundary

▲ Volcano

▌**Figure 5.18 Volcanoes of the World** Most volcanoes are at or near convergent and divergent plate boundaries. The two major volcano belts are the circum-Pacific belt, commonly known as the Ring of Fire, with about 60% of all active volcanoes, and Mediterranean belt, with 20% of active volcanoes. Most of the rest lie near the mid-oceanic ridges.

》》 Critical Thinking Question
Why is magma at divergent plate boundaries mafic, whereas magma at convergent plate boundaries is intermediate or felsic?

5.6 North America's Active Volcanoes

Part of the circum-Pacific belt includes volcanoes in the Pacific Northwest, as well as those in Alaska. Both of these areas of volcanism are at convergent plate boundaries. Of the 80 or so potentially active volcanoes in Alaska, at least half have erupted since 1760. The other active North American volcanoes are in the Cascade Range in the Pacific Northwest where the Juan de Fuca plate is subducted beneath North America. Many of these volcanoes have been historically active, although since 1900, only Lassen Peak in California and Mount St. Helens in Washington have erupted.

Alaska's Volcanoes

Many of the volcanoes in mainland Alaska and in the Aleutian Islands (Figure 5.18) are composite volcanoes, some with huge calderas. Mount Spurr has erupted explosively at least 35 times during the last 5,000 years, but its eruptions pale by comparison with that of Novarupta in 1912. Its defining event was the June 1912 eruption, the largest in the world since the late 1800s. At least 15 km³ of mostly pyroclastic materials erupted during about 60 hours.

When the eruption was over, 120 km² of land was buried beneath pyroclastic deposits as deep as 213 m. In fact, the deposits filled the Valley of Ten Thousand Smokes—so named because of the hundreds of fumaroles where gases vented through the hot deposits for as long as 15 years following the eruption. Fortunately, the eruption took place in a remote area so there were no injuries or fatalities, but enough ash, gases, and pumice were ejected that, for several days, the sky was darkened over much of the Northern Hemisphere.

By the time you read this chapter, several more volcanoes in Alaska will have erupted as the Pacific plate moves relentlessly northward only to be subducted at the Aleutian Trench. The Alaska Volcanoes Observatory in Anchorage, Alaska, continues to monitor these volcanoes and issue warnings about potential eruptions.

The Cascade Range

The **Cascade Range** (❚ Figure 5.19a) stretches from Lassen Peak in northern California north through Oregon and Washington into British Columbia, Canada. Most of the large volcanoes in the range are composite volcanoes, but Lassen Peak in California is the world's largest lava dome (❚ Figure 5.19b). Actually, it is a rather small volcano that developed 27,000 years ago on the flank of a much larger, deeply eroded composite volcano. It erupted from 1914 to 1917, but has since been quiet except for ongoing hydrothermal activity (Figure 5.2).

Two large shield volcanoes lie just to the east of the main Cascade Range volcanoes—Medicine Lake Volcano in California and Newberry Volcano in Oregon. Distinctive features at Newberry Volcano are a 1,600-year-old obsidian flow and casts of trees that formed when lava flowed

around them and solidified. Cinder cones are common throughout the range.

What was once a nearly symmetrical composite volcano changed markedly on May 6, 1980, when Mount St. Helens in Washington erupted explosively, killing 57 people and leveling some 600 km² of forest (❚ Figure 5.19c). A huge lateral blast caused much of the damage and fatalities, but snow and ice on the volcano melted and pyroclastic materials displaced water in lakes and rivers, causing lahars and extensive flooding.

Mount St. Helens's renewed activity, beginning in late September 2004, has resulted in dome growth and small steam and ash explosions. Scientists at the Cascades Volcano Observatory in Vancouver, Washington, issued a low-level alert for an eruption and continue to monitor the volcano.

Several other Cascade Range volcanoes will almost certainly erupt again. Mount Hood in Oregon, one of the most symmetrical volcanoes in the range, last erupted in 1865, but should a large eruption take place, it lies only about 50 km from Portland, an urban area with more than 2 million people (❚ Figure 5.19d). However, the most dangerous is probably Mount Rainier in Washington. The greatest danger from Mount Rainier is volcanic mudflows or huge debris flows. Of the 60 large flows that have occurred during the last 100,000 years, the largest, consisting of 4 km³ of debris, covered an area now occupied by more than 120,000 people.

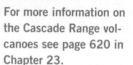

CONNECTION LINK ❮

For more information on the Cascade Range volcanoes see page 620 in Chapter 23.

Indeed, in August 2001, a sizable debris flow took place on the south side of the mountain, but it caused no injuries or fatalities. No one knows when the next flow will take place, but one community has taken the threat seriously enough to formulate an emergency evacuation plan. Unfortunately, the residents would have only one or two hours to evacuate.

What Would You Do ❓

No one doubts that some of the Cascade Range volcanoes will erupt again, but we don't know when or how large these eruptions will be. A job transfer takes you to a community in Oregon that has several nearby large volcanoes. You have some concerns about future eruptions. What kinds of information would you seek before buying a home in this area? In addition, as a concerned citizen, can you make any suggestions about what should be done in the case of a large eruption?

5.7 Plate Tectonics, Volcanoes, and Plutons

In Chapter 4, we discussed the origin and evolution of magma and concluded that (1) mafic magma is generated a Plate tectonic setting for the Pacific Northwest. Suduction of the Juan de Fuca plate accounts for ongoing volcanism in the region.

Figure 5.19
The Cascade Range of the Pacific Northwest

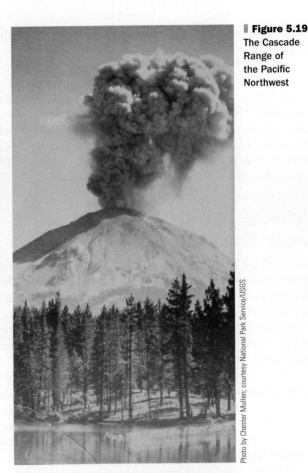

b Lassen Peak in California erupted from 1914 to 1917. This eruption took place in 1915.

c Mount St. Helens, Washington. The lateral blast on May 18, 1980, took place when a bulge on the volcano's north face collapsed, reducing the pressure on gas-charged magma.

d Mount Hood in Oregon as seen from Portland, a city of more than two million people.

beneath spreading ridges, and (2) intermediate or felsic magma forms where an oceanic plate is subducted beneath another oceanic plate or a continental plate. Accordingly, most volcanism and emplacement of plutons take place at or near divergent and convergent plate boundaries.

Igneous Activity at Divergent Plate Boundaries

Much of the mafic magma that originates at spreading ridges is emplaced as vertical dikes and gabbro plutons,

thus composing the lower part of the oceanic crust. However, some rises to the surface and issues forth as submarine lava flows and pillow lava (Figure 5.6), which constitutes the upper part of the oceanic crust. Much of this volcanism goes undetected, but researchers in submersibles have seen the results of recent eruptions.

Mafic lava is very fluid, allowing gases to escape easily, and at great depth in the oceans, the water pressure is so great that explosive volcanism is prevented. In short, pyroclastic materials are rare to absent unless, of course, a volcanic center builds up above sea level. Even if this occurs, however, the mafic magma is so fluid that it forms the gently sloping layers found on shield volcanoes.

Excellent examples of divergent plate boundary volcanism are found along the Mid-Atlantic Ridge, particularly where it rises above sea level as in Iceland (Figure 5.18). In November 1963, a new volcanic island, later named Surtsey, rose from the sea south of Iceland. The East Pacific Rise and the Indian Ridge are areas of similar volcanism. A divergent plate boundary is also present in Africa as the East African Rift system, which is well known for its volcanoes (Figure 5.18).

Igneous Activity at Convergent Plate Boundaries

Nearly all the large active volcanoes in both the circum-Pacific and Mediterranean belts are composite volcanoes near the leading edges of overriding plates at convergent plate boundaries (Figure 5.18). The overriding plate, with its chain of volcanoes, may be oceanic as in the case of the Aleutian Islands, or it may be continental as is, for instance, the South American plate with its chain of volcanoes along its western margin.

As we noted, these volcanoes at convergent plate boundaries consist mostly of lava flows and pyroclastic materials of intermediate to felsic composition. Remember that when mafic oceanic crust partially melts, some of the magma generated is emplaced near plate boundaries as plutons and some is erupted to build up composite volcanoes. More viscous magmas, usually of felsic composition, are emplaced as lava domes, thus accounting for the explosive eruptions that typically occur at convergent plate boundaries.

Good examples of volcanism at convergent plate boundaries are the explosive eruptions of Mount Pinatubo and Mayon volcano in the Philippines; both are near a plate boundary beneath which an oceanic plate is subducted. Mount St. Helens, Washington, is similarly situated, but it is on a continental rather than an oceanic plate.

Intraplate Volcanism

Mauna Loa and Kilauea on the island of Hawaii and Loihi just 32 km to the south are within the interior of a rigid plate far from any divergent or convergent plate boundary (Figure 5.18). The magma is derived from the upper mantle, as it is at spreading ridges, and accordingly is mafic, so it builds up shield volcanoes. Loihi is particularly interesting because it represents an early stage in the origin of a new Hawaiian island. It is a submarine volcano that rises more than 3,000 m above the adjacent seafloor, but its summit is still about 940 m below sea level.

Even though the Hawaiian volcanoes are not at or near a spreading ridge or a subduction zone, their evolution is nevertheless related to plate movements. Notice in Figure 2.23 that the ages of the rocks that make up the Hawaiian islands increase toward the northwest. Kauai formed 5.6 to 3.8 million years ago, whereas Hawaii began forming less than 1 million years ago, and Loihi began to form even more recently. The islands have formed in succession as the Pacific plate moves continuously over a hot spot now beneath Hawaii and just to the south at Loihi.

5.8 Volcanic Hazards, Volcano Monitoring, and Forecasting Eruptions

Undoubtedly you suspect that living near an active volcano poses some risk, and of course this assessment is correct. But what exactly are volcanic hazards, is there any way to anticipate eruptions, and what can we do to minimize the dangers of eruptions? We have already mentioned that lava flows, with few exceptions, pose little threat to humans although they may destroy property. Lava flows, nuée ardentes, and volcanic gases are threats during an eruption (❚ Figure 5.20); however, lahars and landslides may take place even when no eruption occurred for a long time. Certainly, the most vulnerable areas in the United States are Alaska, Hawaii, California, Oregon, and Washington, but some other parts of the West might also experience renewed volcanism.

How Large Is an Eruption, and How Long Do Eruptions Last?

The most widely used indication of the size of a volcanic eruption is the **volcanic explosivity index (VEI)** (❚ Figure 5.21). The VEI ranges from 0 (gentle) to 8 (cataclysmic) and is based on several aspects of an eruption, such as the volume of material explosively ejected and the height of the eruption plume. However, the volume of lava, fatalities, and property damage are not considered. For instance, the 1985 eruption of Nevado del Ruiz in Colombia killed 23,000 people, yet has a VEI value of only 3. In contrast, the huge eruption (VEI = 6) of Novarupta in Alaska in 1912 caused no fatalities or injuries. Since AD 1500, only the 1815 eruption of Tambora had a value of 7; it was both large and deadly (Table 5.1). Of the several eruptions of Eyjafjallajökull in Iceland during April 2010, the largest one had a VEI of no more than 4, but because of its location and the fact that it erupted beneath glacial ice, it caused disruption of air traffic over the North Atlantic.

The duration of eruptions varies considerably. Fully 42% of about 3,300 historic eruptions lasted less than one month. About 33% erupted for one to six months, but some 16 volcanoes have been active more or less continuously for more than 20 years. Stromboli and Mount Etna in Italy and Erta Ale in Ethiopia are good examples. For some explosive volcanoes, the time from the onset of their eruptions to the climactic event is weeks or months. A case in point is the colossal explosive eruption of Mount St. Helens on May 18, 1980, that occurred two months after eruptive activity began (Figure 5.19c). Unfortunately, many volcanoes give little or no warning of such large-scale events; of 252 explosive eruptions, 42% erupted most violently during their first day of activity.

Figure 5.20 Volcanic Hazards A volcanic hazard is any manifestation of volcanism that poses a threat, including lava flows and, more importantly, volcanic gas, ash, and lahars.

a This 2002 lava flow in Goma, Democratic Republic of the Congo, killed 147 people, mostly by causing gasoline storage tanks to explode.

DANGER
CO2 ☠
HAZARD AREA
AHEAD

b This sign at Mammoth Mountain volcano in California warns of the potential danger of CO_2 gas, which has killed 170 acres of trees.

» Critical Thinking Question
Which of the features shown here—lava, gases, and volcanic ash—is most dangerous to humans?

c When Mount Pinatubo in the Philippines erupted on June 15, 1991, this huge cloud of ash and steam formed over the volcano.

Is It Possible to Forecast Eruptions?

Only a few of Earth's potentially dangerous volcanoes are monitored, including some in Japan, Italy, Russia, New Zealand, and the United States. Volcano monitoring involves recording and analyzing physical and chemical changes at volcanoes (Figure 5.22). Tiltmeters detect changes in the slopes of a volcano as it inflates when magma rises beneath it, and a geodimeter uses a laser beam to measure horizontal distances, which change as a volcano inflates. Geologists also monitor gas emissions, changes in groundwater level and temperature, hot springs activity, and changes in the local magnetic and electrical fields. Even the accumulating snow and ice, if any, are evaluated to anticipate hazards from floods should an eruption take place.

Of critical importance in volcano monitoring and warning of an imminent eruption is the detection of **volcanic tremor,** continuous ground motion that lasts for minutes to hours as opposed to the sudden, sharp jolts produced by most earthquakes. Volcanic tremor, also known as *harmonic tremor,* indicates that magma is moving beneath the surface.

To more fully anticipate the future activity of a volcano, its eruptive history must be known. Accordingly, geologists study the record of past eruptions preserved in rocks. Detailed studies before 1980 indicated that Mount St. Helens, Washington, had erupted explosively 14 or 15 times during the last 4,500 years, so geologists concluded that it was one of the most likely Cascade Range volcanoes to erupt again. In fact, maps they prepared showing areas in which damage from an eruption could be expected were helpful in determining which areas should have restricted access and evacuations once an eruption did take place.

Geologists successfully gave timely warnings of impending eruptions of Mount St. Helens in Washington and Mount Pinatubo in the Philippines, but in both cases, the climactic eruptions were preceded by eruptive activity of lesser intensity. In some cases, however, the warning signs are much more subtle and difficult to interpret. Numerous small earthquakes and other warning signs indicated to USGS geologists that magma was moving beneath the surface of the Long Valley caldera in eastern California, so in 1987, they issued a low-level warning, and then nothing happened.

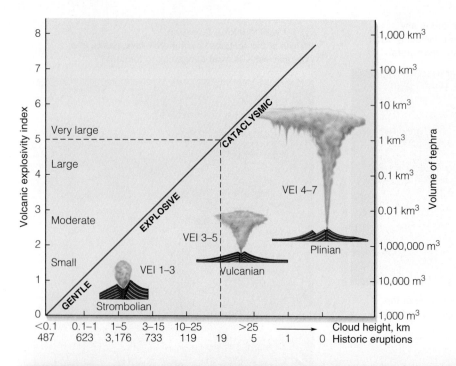

Figure 5.21 The Volcanic Explosivity Index In this example, an eruption with a VEI of 5 has an eruption cloud up to 25 km high and ejects at least 1 km³ of tephra, a collective term for all pyroclastic materials. Geologists characterize eruptions as Hawaiian (nonexplosive), Strombolian, Vulcanian, and Plinian.

>> Critical Thinking Question
Why are numbers of fatalities and property damage not used in assigning a VEI value to an eruption?

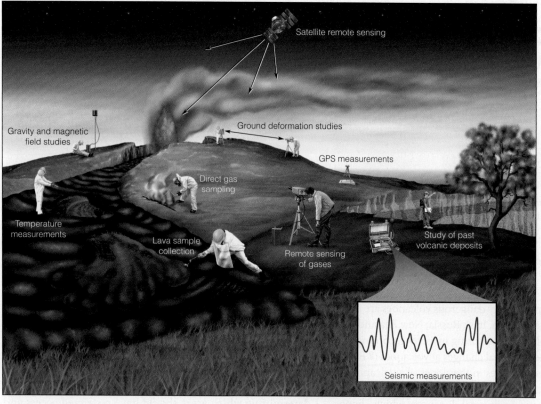

Figure 5.22
Volcanic Monitoring Some important techniques used to monitor volcanoes.

Volcanic activity in the Long Valley caldera occurred as recently as 250 years ago, and there is every reason to think that it will occur again. Unfortunately, the local populace was largely unaware of the geologic history of the region, the USGS did a poor job in communicating its concerns, and premature news releases caused more concern than was justified. In any case, local residents were outraged because the warnings caused a decrease in tourism (Mammoth Mountain on the margins of the caldera is the second-largest ski area in the country) and property values plummeted. Monitoring continues in the Long Valley caldera, and the signs of renewed volcanism, including earthquake swarms, trees being killed by carbon dioxide gas apparently emanating from magma (Figure 5.20b), and hot spring activity, cannot be ignored. In April 2006, three members of a ski patrol were killed by carbon dioxide gas that accumulated in a low area.

GEO-RECAP

» Geo-Image

James S. Monroe

▌ **Figure 1** The Mona Schist at Marquette, Michigan, is basalt that has been changed slightly by metamorphism. How do you think the elliptical features in this ancient lava formed?

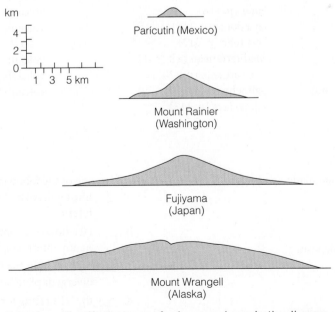

▌ **Figure 2** Identify the types of volcanoes shown in the diagram.

Chapter Summary

- Volcanism encompasses those processes by which magma rises to the surface as lava flows and pyroclastic materials, and gases are released into the atmosphere.

- Gases make up only a few percent by weight of magma. Most is water vapor, but sulfur gases may have far-reaching climatic effects, and carbon dioxide is dangerous.

- Aa lava flows have surfaces of jagged, angular blocks, whereas the surfaces of pahoehoe flows are smoothly wrinkled.

- Other features of lava flows are lava tubes and columnar joints. Lava erupted under water typically forms bulbous masses known as pillow lava.

- Volcanoes are found in various shapes and sizes, but all form where lava and pyroclastic materials are erupted from a vent.

- The summits of volcanoes have either a crater or a much larger caldera. Calderas form following voluminous eruptions, and the volcanic peak collapses into a partially drained magma chamber.

- Shield volcanoes have low, rounded profiles and are composed mostly of mafic flows that cool and form basalt. Small, steep-sided cinder cones form around a vent where pyroclastic materials erupt and accumulate. Composite volcanoes are made up of lava flows and pyroclastic materials of intermediate composition and volcanic mudflows.

- Viscous bulbous masses of lava, mostly of felsic composition, form lava domes, which are dangerous because they erupt explosively.

- Fluid mafic lava from fissure eruptions spreads over large areas to form a basalt plateau.

- Pyroclastic sheet deposits result from huge eruptions of ash and other pyroclastic materials, particularly when calderas form.

- Geologists have devised a volcanic explosivity index (VEI) to give a semiquantitative measure of the size of an eruption. Volume of material erupted and height of the eruption plume are criteria used to determine the VEI; fatalities and property damage are not considered.

- Approximately 80% of all volcanic eruptions take place in the circum-Pacific and the Mediterranean belts, mostly at convergent plate boundaries. Most of the rest of the eruptions occur along mid-oceanic ridges or their extensions onto land.

- The two active volcanoes on the island of Hawaii and one just to the south lie above a hot spot over which the Pacific plate moves.

- To effectively monitor volcanoes, geologists evaluate several physical and chemical aspects of volcanic regions. Of particular importance in monitoring volcanoes and forecasting eruptions is detecting volcanic tremor and determining the eruptive history of a volcano.

Important Terms

aa (p. 113)
basalt plateau (p. 122)
caldera (p. 115)
Cascade Range (p. 126)
cinder cone (p. 118)
circum-Pacific belt (p. 125)
columnar joint (p. 113)
composite volcano (stratovolcano)
 (p. 118)

crater (p. 114)
fissure eruption (p. 122)
lahar (p. 118)
lava dome (p. 119)
lava tube (p. 112)
Mediterranean belt (p. 125)
nuèe ardente (p. 120)
pahoehoe (p. 113)
pillow lava (p. 113)

pyroclastic sheet deposit (p. 124)
shield volcano (p. 116)
volcanic ash (p. 113)
volcanic explosivity index (VEI)
 (p. 128)
volcanic tremor (p. 129)
volcanism (p. 110)
volcano (p. 110)

Review Questions

1. A small, steep-sided volcano made up of pyroclastic materials is a
   a. ____shield volcano.
   b. ____basalt plateau.
   c. ____pyroclastic sheet deposit.
   d. ____lava dome.
   e. ____cinder cone.

2. Which one of the following statements is correct?
   a. ____most volcanism takes place in continental interiors.
   b. ____lava flows on composite volcanoes are predominantly intermediate in composition.
   c. ____volcanism in the Cascade Range occurs at a divergent plate boundary.
   d. ____the VEI rating for an eruption depends on the number of injuries and fatalities.
   e. ____Earth's Moon has many active volcanoes.

3. An aa lava flow is made up of
   a. _____angular blocks and fragments.
   b. _____pyroclastic materials and volcanic mudflows.
   c. _____smooth, ropy lava that resembles taffy.
   d. _____bulbous masses that look like pillows.
   e. _____carbon dioxide and water vapor.
4. The shaking that takes place when magma moves beneath the surface is called
   a. _____columnar jointing.
   b. _____pahoehoe.
   c. _____volcanic tremor.
   d. _____basalt accumulation.
   e. _____cratering.
5. The most common gas emitted by volcanoes is
   a. _____hydrogen sulfide.
   b. _____water vapor.
   c. _____methane.
   d. _____ethane.
   e. _____fluorine.
6. An incandescent cloud of gas and particles erupted by a volcano is a
   a. _____spatter cone.
   b. _____caldera.
   c. _____lapilli.
   d. _____nuée ardente.
   e. _____pressure ridge.
7. When volcanic ash is so hot it fuses it forms
   a. _____andesite.
   b. _____volcanic bombs.
   c. _____a lahar.
   d. _____welded tuff.
   e. _____lava fountain.
8. Most volcanism on Earth takes place in the
   a. _____circum-Pacific belt.
   b. _____Appalachian belt.
   c. _____Indian Ridge-East African belt.
   d. _____Mediterranean belt.
   e. _____mid-oceanic ridges.

9. The Hawaiian volcanoes are made up mostly of
   a. _____cinders.
   b. _____granite.
   c. _____basalt.
   d. _____welded tuff.
   e. _____pyroclastic materials.
10. Water-saturated flows of volcanic debris called _____ are common on _____ volcanoes.
    a. _____nuée ardente/shield.
    b. _____lava flows/lava dome.
    c. _____lahars/composite.
    d. _____pillow lava/submarine.
    e. _____fissure eruptions/cinder cone.
11. How does a caldera form? Where would you go to see one?
12. Why are eruptions of mafic magma rather quiet, whereas those of felsic magma are commonly explosive?
13. Explain how a basalt plateau forms.
14. Why are shield volcanoes most common at divergent plate boundaries, whereas composite volcanoes are found mostly at convergent plate boundaries?
15. What criteria do geologists use and not use to assign a volcanic explosivity index (VEI) value to an eruption?
16. Suppose that you find exposures of rock on land made up of pillow lava overlain by deep-sea sedimentary rocks. Where and how did the pillow lava form, and what type of rock would you expect to find beneath the pillow lava?
17. What are columnar joints and how do they form?
18. Where are Earth's two main volcanic belts, and why do you think they are in these areas?
19. How does a lava tube form?
20. What kinds of information do geologists evaluate when they monitor volcanoes and warn of imminent eruptions?

Differential weathering and erosion in Red Canyon along Utah State Highway 12, which is designated a Scenic Byway. Here the Paleocene to Eocene (40- to 60-million-year-old) Claron Formation has been deeply eroded to form spires and pinnacles because the rock is not structurally and compositionally uniform throughout. The Claron Formation consists of sedimentary rocks composed of gravel, sand, and mud. The rocks are red because they contain small amounts of iron oxide.

CHAPTER 6

Weathering, Erosion, and Soil

OUTLINE

6.1 Introduction

6.2 Alteration of Minerals and Rocks

6.3 Mechanical Weathering—Disaggregation of Earth Materials

6.4 Chemical Weathering—Decomposition of Earth Materials

6.5 Soil and Its Origin

GEO-FOCUS: Industrialization and Acid Rain

6.6 Expansive Soils and Soil Degradation

GEO-INSIGHT: Arches National Park, Utah

6.7 Weathering and Resources

GEO-RECAP

HAVE YOU EVER WONDERED?

- How weathering brings about changes in Earth materials?
- What physical processes break rocks into smaller pieces?
- How chemical reactions decompose rocks?
- What factors control the rate at which Earth materials are altered?
- What soil is, how it forms, and why it varies in fertility?
- Why soil loses its fertility?
- How alteration of rocks is responsible for some important natural resources?

Sue Monroe

6.1 Introduction

Weathering includes several physical and chemical processes that alter surface and near-surface rocks and minerals, whereas **erosion** involves the removal of weathered materials, by running water, wind, glaciers, and waves from the area where weathering takes place. Actually, weathering alters rocks and minerals so that they are more nearly in equilibrium with a new set of environmental conditions. For instance, many igneous and metamorphic rocks form within Earth's crust where pressure and temperature are high and little or no water or oxygen is present. These same rocks at the surface, though, are exposed to low pressure and temperature, water, the atmosphere, and the activities of organisms (▌Figure 6.1). Thus, interactions of Earth materials with the hydrosphere, atmosphere, and biosphere bring about changes as they break down physically (*disintegrate*) and change chemically (*decompose*).

Weathering is such a pervasive phenomenon that it is easy to overlook. Nevertheless, it takes place continuously at variable rates on all surface and near-surface rocks and minerals, including rocklike substances used in construction. Roadways and runways, bricks, and concrete in sidewalks, foundations, and bridges change with time as they are relentlessly attacked by the elements. Obviously, weathering of sidewalks, foundations, and other structures requires costly repairs or replacement.

From the academic point of view, weathering is an important group of processes that once again illustrates the interactions among Earth's systems. But there are other good reasons to study weathering. One reason is so that we can develop construction materials that are more resistant to physical and chemical changes, or develop more effective methods to protect them from the elements. Another reason is that weathering is an essential part of the *rock cycle* (see Figure 1.15). When **parent material**—that is, rocks and minerals exposed to weathering—breaks down into smaller pieces or perhaps dissolves, this weathered material may be eroded from the weathering site and transported elsewhere, by running water or wind, and deposited as *sediment*, the raw materials for sedimentary rocks.

In addition to providing the raw materials for sedimentary rocks, weathering is responsible for the origin of *soils*. Needless to say, we depend on soils directly or indirectly for our existence. In some areas, however, erosion takes place faster than soil-forming processes operate, thereby decreasing

> **CONNECTION LINK** ❮
>
> You can find more information about sediments and sedimentary rocks in Chapter 7.

▌**Figure 6.1** Weathering of Granite

a This exposure of granite has been so thoroughly altered by weathering that only spherical masses of the original rock are visible.

b Close-up view of the weathered granite. Mechanical weathering has predominated, so the particles are mostly small pieces of granite and minerals such as quartz and feldspars.

the amount of productive soil. Of course, erosion is an ongoing natural process, so some soil losses are expected as a normal consequence of the evolution of the land. Unfortunately, some human activities compound the problem, leading to less fertile soil and reduced agricultural production.

Another reason to study weathering is that it accounts for the origin or concentration of some natural resources—the ore of aluminum, for instance. Weathering is also responsible for many clay deposits that are used in ceramics and the manufacture of paper, and weathering coupled with erosion and deposition yields deposits of tin, gold, and diamonds.

6.2 Alteration of Minerals and Rocks

Weathering is a surface or near-surface process, but the rocks it acts on are not structurally and compositionally homogeneous throughout, which accounts for **differential weathering.** That is, weathering takes place at different rates even in the same area, so it commonly results in irregular surfaces. Differential weathering and *differential erosion*—that is, variable rates of erosion—combine to yield some unusual and even bizarre features, such as hoodoos, spires, and arches (see chapter opening photo and the Geo-Insight on pages 150 and 151).

Geologists characterize weathering as *mechanical* and *chemical,* both of which proceed simultaneously on parent material as well as on materials in transport and those deposited as sediment. In short, all surface or near-surface materials weather, although one type of weathering may predominate depending on such variables as climate and rock type.

6.3 Mechanical Weathering— Disaggregation of Earth Materials

Mechanical weathering takes place when physical forces break Earth materials into smaller pieces that retain the composition of the parent material. Granite, for instance, might be mechanically weathered and yield smaller pieces of granite or individual grains of quartz, potassium feldspars, plagioclase feldspars, and biotite (▌Figure 6.1). Several physical processes account for mechanical weathering.

Frost Action

Frost action involving water repeatedly freezing and thawing in cracks and pores in rocks is particularly effective where temperatures commonly fluctuate above and below freezing. Frost action is so effective because water expands by about 9% when it freezes, thus exerting great force on the walls of a crack, widening and extending it by *frost wedging* (▌Figure 6.2a). Repeated freezing and thawing dislodge angular pieces of rock from the parent material that tumble downslope and accumulate as **talus** (▌Figure 6.2b). Frost action is most effective in high mountains, even during the summer months, but it has little or no effect where the temperature rarely drops below freezing, or where Earth materials are permanently frozen.

▌**Figure 6.2** Frost Wedging

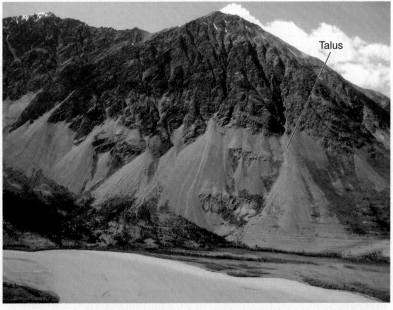

a Frost wedging takes place when water seeps into cracks and expands as it freezes. Angular pieces of rock are pried loose by repeated freezing and thawing.

b Frost wedging and other mechanical weathering processes produced these talus accumulations in Kluane National Park and Reserve in Yukon Territory, Canada.

James S. Monroe

Figure 6.3 Sheet Joints and Exfoliation Domes

Sue Monroe

Mark Gibson/Visual's Unlimited

b Stone Mountain in Georgia is a large exfoliation dome.

a Slabs of granitic rock bounded by sheet joints in the Sierra Nevada of California. The slabs are inclined downward toward the roadway visible at the lower left.

> **≫ Critical Thinking Question**
> Barely visible in the left foreground is a highway. Why do you think these sheet joints might be a problem for highway maintenance crews?

Pressure Release

Some rocks form at depth and are stable under tremendous pressure. Granite crystallizes far below the surface, so when it is uplifted and eroded, its contained energy is released by outward expansion, a phenomenon known as **pressure release.** The outward expansion results in the origin of fractures called **sheet joints** (⏸ Figure 6.3a) that more or less parallel the exposed rock surface. Sheet-joint-bounded slabs of rock slip or slide off the parent rock, leaving large, rounded masses known as **exfoliation domes** (⏸ Figure 6.3b).

That solid rock expands and produces fractures might be counterintuitive, but it is a well-known phenomenon. In deep mines, masses of rock detach from the sides of the excavation, often explosively. These *rock bursts* and less violent *popping* pose a danger to mine workers, and in South Africa they are responsible for approximately 20 deaths per year. In some quarries for building stone, excavations to only 7 or 8 m exposed rocks in which sheet

> **CONNECTION LINK**
>
> For more information on joints see pages 241 and 242 in Chapter 10.

joints formed (⏸ Figure 6.4), in some cases with enough force to throw quarrying machines weighing more than a ton from their tracks.

W.D. Lowry

Figure 6.4 **Sheet Joints in a Quarry** The sheet joint indicated by the hammer formed by expansion in the Mount Airy Granite in North Carolina. The hammer is about 30 cm long.

Figure 6.5 Organisms and Weathering

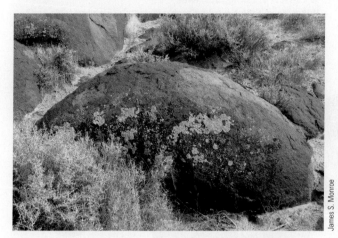

a The orange and gray masses on this rock at Grimes Point Archeological Site in Nevada are lichens, composite organisms made up of fungi and algae. Lichens derive their nutrients from the rock and thus contribute to chemical weathering.

b These trees in the Black Hills of South Dakota contribute to mechanical weathering as they grow in cracks in the rocks, thereby breaking the parent material into smaller pieces.

>> **Critical Thinking Question**
Does vegetation, especially trees, have a detrimental effect on rocklike substances such as foundations and sidewalks?

Thermal Expansion and Contraction

During **thermal expansion and contraction,** the volume of rocks changes as they heat up and then cool down. The temperature may vary as much as 30°C a day in a desert, and rock, being a poor conductor of heat, heats and expands on its outside more than its inside. Even dark minerals absorb heat faster than light-colored ones, so differential expansion takes place between minerals. Surface expansion might generate enough stress to cause fracturing, but experiments in which rocks are heated and cooled repeatedly to simulate years of such activity indicate that thermal expansion and contraction are of minor importance in mechanical weathering.

Growth of Salt Crystals

The formation of salt crystals can exert enough force to widen cracks and dislodge particles in porous, granular rocks such as sandstone. And even in rocks with an interlocking mosaic of crystals, such as granite, **salt crystal growth** pries loose individual minerals. It takes place mostly in hot, arid regions, but also probably affects rocks in some coastal areas.

Activities of Organisms

Animals and plants also participate in the mechanical alteration of rocks (∎ Figure 6.5a). Burrowing animals, such as worms, reptiles, rodents, termites, and ants, constantly mix soil and sediment particles and bring material from depth to the surface where further weathering occurs. The roots of plants, especially large bushes and trees, wedge themselves into cracks in rocks and further widen them (∎ Figure 6.5b).

6.4 Chemical Weathering— Decomposition of Earth Materials

Chemical weathering decomposes rocks and minerals by chemical alteration of the parent material. In contrast to mechanical weathering, chemical weathering changes the composition of weathered materials. For example, several clay minerals (sheet silicates) form by the chemical and structural alteration of other minerals, such as potassium feldspars and plagioclase feldspars, both of which are framework silicates.

CONNECTION LINK ❮

To review the different types of silicate minerals, go to pages 67–69 in Chapter 3.

Figure 6.6 The Solution of Halite

» **Critical Thinking Question**
Which do you think would dissolve fastest, rock salt or table salt? Explain.

a The structure of a water molecule. The asymmetric arrangement of hydrogen atoms causes the molecule to have a slight positive electrical charge at its hydrogen end and a slight negative charge at its oxygen end.

b Solution of sodium chloride (NaCl), the mineral halite, in water. Note that the sodium atoms are attracted to the oxygen end of a water molecule, whereas chloride ions are attracted to the hydrogen end of the molecule.

Other minerals completely decompose during chemical weathering, but some chemically stable minerals are simply liberated from the parent material.

Important agents of chemical weathering include atmospheric gases, especially oxygen, water, and acids. Organisms also play an important role. Rocks with lichens (composite organisms made up of fungi and algae) on their surfaces undergo more rapid chemical alteration than lichen-free rocks (Figure 6.5a). In addition, plants remove ions from soil water and reduce the chemical stability of soil minerals, and plant roots release organic acids.

Solution

When **solution** takes place, the ions of a substance separate in a liquid, and the solid substance dissolves. Water is a remarkable solvent because its molecules have an asymmetric shape, consisting of one oxygen atom with two hydrogen atoms arranged so that the angle between the two hydrogen atoms is about 104 degrees (Figure 6.6a). Because of this asymmetry, the oxygen end of the molecule retains a slight negative electrical charge, whereas the hydrogen end retains a slight positive charge. When a soluble substance such as the mineral halite (NaCl) comes in contact with a water molecule, the positively

charged sodium ions are attracted to the negative end of the water molecule, and the negatively charged chloride ions are attracted to the positively charged end of the molecule (Figure 6.6b). Thus, ions are liberated from the crystal structure, and the solid goes into solution; in other words, it dissolves.

Most minerals are not very soluble in pure water because the attractive forces of water molecules are not sufficient to overcome the forces between particles in minerals. The mineral calcite ($CaCO_3$), the major constituent of the sedimentary rock limestone and the metamorphic rock marble, is practically insoluble in pure water, but it rapidly dissolves if a small amount of acid is present. One way to make water acidic is by dissociating the ions of carbonic acid as follows:

$$H_2O \; + \; CO_2 \; \rightleftharpoons \; H_2CO_3 \; \rightleftharpoons \; H^+ \; + \; HCO_3^-$$

water   carbon   carbonic   hydrogen   bicarbonate
         dioxide   acid   ion   ion

According to this chemical equation, water and carbon dioxide combine to form *carbonic acid*, a small amount of which dissociates to yield hydrogen and bicarbonate ions. The concentration of hydrogen ions determines the acidity of a solution; the more hydrogen ions present, the stronger the acid.

Carbon dioxide from several sources may combine with water and react to form acid solutions. The atmosphere is mostly nitrogen and oxygen, but approximately 0.03% is carbon dioxide, causing rain to be slightly acidic. Decaying organic matter and the respiration of organisms produce carbon dioxide in soils, so groundwater is also usually slightly acidic. Climate also affects the acidity, with arid regions tending to have alkaline groundwater (that is, it has a low concentration of hydrogen ions).

Whatever the source of carbon dioxide, once an acidic solution is present, calcite rapidly dissolves according to the reaction.

$$\underset{\text{calcite}}{CaCO_3} \; + \; \underset{\text{water}}{H_2O} \; + \; \underset{\substack{\text{carbon}\\\text{dioxide}}}{CO_2} \; \rightleftharpoons \; \underset{\substack{\text{calcium}\\\text{ion}}}{Ca^{++}} \; + \; \underset{\substack{\text{bicarbonate}\\\text{ion}}}{2HCO_3^{-}}$$

Oxidation

The term **oxidation** has a variety of meanings for chemists, but in chemical weathering, it refers to reactions with oxygen to form an oxide (one or more metallic elements combined with oxygen) or, if water is present, a hydroxide (a metallic element or radical combined with OH^-). For example, iron rusts when it combines with oxygen to form the iron oxide hematite:

$$\underset{\text{iron}}{4Fe} \; + \; \underset{\text{oxygen}}{3O_2} \; \rightarrow \; \underset{\substack{\text{iron oxide}\\\text{(hematite)}}}{2Fe_2O_3}$$

Atmospheric oxygen is abundantly available for oxidation reactions, but oxidation is generally a slow process unless water is present. Thus, most oxidation is carried out by oxygen dissolved in water.

Oxidation is important in the alteration of ferromagnesian silicates such as olivine, pyroxenes, amphiboles, and biotite. Iron in these minerals combines with oxygen to form the reddish iron oxide hematite (Fe_2O_3) or the yellowish or brown hydroxide limonite [$FeO(OH) \cdot nH_2O$]. The yellow, brown, and red colors of many soils and sedimentary rocks are caused by the presence of small amounts of hematite or limonite.

Hydrolysis

The chemical reaction between the hydrogen (H^+) ions and hydroxyl (OH^-) ions of water and a mineral's ions is known as **hydrolysis.** In hydrolysis, hydrogen ions actually replace positive ions in minerals. Such replacement changes the composition of minerals and liberates iron that then may be oxidized.

The chemical alteration of the potassium feldspar orthoclase provides a good example of hydrolysis. All feldspars are framework silicates, but when altered, they yield soluble salts and clay minerals, such as kaolinite, which are sheet silicates. The chemical weathering of orthoclase by hydrolysis occurs as follows:

$$\underset{\text{orthoclase}}{2KAlSi_3O_8} \; + \; \underset{\substack{\text{hydrogen}\\\text{ion}}}{2H^{+}} \; + \; \underset{\substack{\text{bicarbonate}\\\text{ion}}}{2HCO_3^{-}} \; + \; \underset{\text{water}}{H_2O} \; \rightarrow$$

$$\underset{\text{clay (kaolinite)}}{Al_2Si_2O_5(OH)_4} \; + \; \underset{\substack{\text{potassium}\\\text{ion}}}{2K^{+}} \; + \; \underset{\substack{\text{bicarbonate}\\\text{ion}}}{2HCO_3^{-}} \; + \; \underset{\text{silica}}{4SiO_2}$$

In this reaction, hydrogen ions attack the ions in the orthoclase structure, and some liberated ions are incorporated in a developing clay mineral. The potassium and bicarbonate ions go into solution and combine to form a soluble salt. On the right side of the equation is excess silica that would not fit into the crystal structure of the clay mineral.

The Rate of Chemical Weathering
Chemical weathering operates on the surfaces of particles, so it alters rocks and minerals from the outside inward. In fact, if you break open a weathered stone, you will see a rind of weathering at and near the surface, but the stone is completely unaltered inside. The rate at which chemical weathering proceeds depends on several factors. One is simply the presence or absence of fractures because fluids seep along fractures, and weathering is more intense along these surfaces (▌Figure 6.7). The other factors that control chemical weathering include particle size, climate, and parent material.

▌**Figure 6.7** **Weathering Along Fractures** These granitic rocks in Joshua Tree National Park in California have been chemically weathered more intensely along fractures than in unfractured parts of the same rock outcrop.

Figure 6.8 Particle Size and Chemical Weathering

Surface area = 6 m²

1 m

1 m

a As a rock is divided into smaller particles, its surface area increases, but its volume of 1 m³ remains the same. The surface area is 6 m².

Surface area = 12 m²

0.5 m

0.5 m

b The surface area is 12 m².

Surface area = 24 m²

0.25 m

0.25 m

c The surface area is 24 m², but the volume remains the same at 1 m³. Small particles have more surface area in relation to their volume than do large particles.

Because chemical weathering affects particle surfaces, the greater the surface area, the more effective the weathering. It is important to realize that small particles have larger surface areas compared to their volume than do large particles. Notice in ▌ Figure 6.8 that a block measuring 1 m on a side has a total surface area of 6 m², but when the block is broken into particles measuring 0.5 m on a side, the total surface area increases to 12 m². And if these particles are all reduced to 0.25 m on a side, the total surface area increases to 24 m². Note that although the surface area in this example increases, the total volume remains the same at 1 m³.

We can conclude that mechanical weathering contributes to chemical weathering by yielding smaller particles with greater surface area compared to their volume. Actually, your own experiences with particle size verify our contention about surface area and volume. Because of its very small particle size, powdered sugar gives an intense burst of sweetness as the tiny pieces dissolve rapidly, but otherwise it is the same as the granular sugar we use on our cereal or in our coffee.

It is not surprising that chemical weathering is more effective in the tropics than in arid and arctic regions because temperatures and rainfall are high and evaporation rates are low. In addition, vegetation and animal life are much more abundant. Consequently, the effects of weathering extend to depths of several tens of meters, but they extend only centimeters to a few meters deep in arid and arctic regions.

Some rocks are more resistant to chemical alteration than others, so parent material is another control on the rate of chemical weathering. The metamorphic rock quartzite is an extremely stable substance that alters slowly compared to most other rock types. In contrast, basalt, which contains large amounts of calcium-rich plagioclase and pyroxene minerals, decomposes rapidly because these minerals are chemically unstable. In fact, the stability of common minerals is just the opposite of their order of crystallization in Bowen's reaction series (Table 6.1 and Figure 4.4). The minerals that form last in this series are more stable, whereas those that form early are easily altered because they are most out of equilibrium with their conditions of formation.

One manifestation of chemical weathering is **spheroidal weathering** (▌ Figure 6.9). In spheroidal weathering, a stone, even one that is rectangular to begin with, weathers to form a more spherical shape because that is the most stable shape it can assume. The reason is that on a rectangular stone, the corners are attacked by weathering from three sides, and the edges are attacked from two sides, but the flat surfaces weather more or less uniformly (Figure 6.9). Consequently, the corners and edges are altered more rapidly, the material sloughs off, a more spherical shape develops, and all surfaces weather at the same rate.

TABLE 6.1	Stability of Silicate Minerals	
Ferromagnesian Silicates	**Nonferromagnesian Silicates**	
Olivine	Calcium plagioclase	
Pyroxene		
Amphibole	Sodium plagioclase	
Biotite	Potassium feldspar	
	Muscovite	
	Quartz	

Increasing Stability (arrow pointing down)

6.5 Soil and Its Origin

Much of Earth's land surface is covered by a layer of **regolith** consisting of sediment, pyroclastic materials, and the residue formed in place by weathering. Part of the regolith that contains air, water, and organic matter and supports vegetation is **soil.** Obviously plants grow in soil from which they receive most of their nutrients and water. All land-dwelling animals depend directly or indirectly on soils for their nutrients.

Figure 6.9 Spheroidal Weathering

b Corners and edges weather most rapidly.

c When the blocks are weathered so that they are nearly spherical, their surfaces weather evenly and no further change in shape takes place.

a The rectangular blocks outlined by fractures are attacked by chemical weathering processes.

d An exposure of granite showing spheroidal weathering in Joshua Tree National Park in California.

Sue Monroe

》 Critical Thinking Question
Suppose you place some rectangular ice cubes in the sun. How do you think their shape would change in an hour?

A good soil for farming and gardening is made up of about 45% solid particles derived by weathering of parent material, and most of the rest of its volume is voids filled with air and/or water (■ Figure 6.10a). Another important constituent of soil is *humus* which is carbon that forms by bacterial decay of organic matter and is very resistant to further decay. Even fertile soils may have as little as 5% humus, but it is essential as a source of plant nutrients and it enhances a soil's capacity to retain moisture.

The solid particles in soils may be sand- and silt-sized mineral grains such as quartz, feldspars, and others, which hold soil particles apart and allow oxygen and water to circulate more freely. Clay minerals are also important, because they supply nutrients to plants and aid in soil-water retention. Should excess clay be present, though, a soil drains poorly, and is sticky when wet and hard when dry.

We characterize soils as *residual* or *transported* depending on whether they formed in place or the materials composing them were transported from the weathering site. For example, if a body of granite weathers and the residue that accumulates over the granite is converted to soil, the soil so formed is residual. In contrast, transported soil forms if this same weathering residue is transported elsewhere, deposited, and then converted to soil.

The Soil Profile

Observed in vertical cross section, soil is made up of distinct layers or **soil horizons** that differ in texture, structure, composition, and color (■ Figure 6.10b). From the surface downward, the soil horizons are designed O, A, E, B, and C, although the boundaries between horizons are transitional, and in some cases horizon E is not present. Because soil formation starts at the surface and works downward, horizon A is the most altered from the parent material than the horizons below.

Horizon O is only a few centimeters thick and is composed of organic matter (Figure 6.10b). Plant remains in various states of decomposition are clearly visible in the upper part of this horizon, but its lower part consist of humus. Indeed, the upper and lower parts of horizon O are sometimes referred to as O1 and O2, respectively.

Horizon A, also called *topsoil,* has more organic matter than the horizons below, and is characterized by intense biological activity because plant roots, fungi, bacteria, and worms are abundant (Figure 6.10b). In fact, the earthy aroma of freshly plowed soils comes from thread-like soil bacteria. In soils developed over a long time, horizon A is composed mostly of clays and chemically stable minerals such as quartz.

GEO-FOCUS　Industrialization and Acid Rain

One result of industrialization is atmospheric pollution, which causes smog, possible disruption of the ozone layer, global warming, and acid rain. Acidity, a measure of hydrogen ion concentration, is measured on the pH scale (❚ Figure 1a). A pH value of 7 is neutral, whereas acidic conditions correspond to values less than 7, and values greater than 7 denote alkaline, or basic, conditions. Normal rain has a pH value of about 5.6, making it slightly acidic, but acid rain has a pH of less than 5.0. In addition, some areas experience acid snow and even acid fog with a pH as low as 1.7.

Several natural processes, including soil bacteria metabolism and volcanism, release gases into the atmosphere that contribute to acid rain. Human activities also produce added atmospheric stress, especially burning fossil fuels that release carbon dioxide and nitrogen oxide from internal combustion engines. Both of these gases add to acid rain, but the greatest culprit is sulfur dioxide released mostly by burning coal that contains sulfur that oxidizes to form sulfur dioxide (SO_2) (see Figure 24.2a). As sulfur dioxide rises into the atmosphere, it reacts with oxygen and water droplets

to form sulfuric acid (H_2SO_4), the main component of acid rain.

Robert Angus Smith first recognized acid rain in England in 1872, but not until 1961 did it become an environmental concern when scientists realized that acid rain is corrosive and irritating, kills vegetation, and has a detrimental effect on surface waters. Since then, the effects of acid rain are apparent in Europe (especially in eastern Europe) and the eastern part of North America, where the problem has been getting worse for the last three decades (❚ Figure 1b).

The areas affected by acid rain invariably lie downwind from plants that emit sulfur gases, but the effects of acid rain in these areas may be modified by the local geology. For instance, if the area is underlain by limestone or alkaline soils, acid rain tends to be neutralized, but granite has little or no modifying effect. Small lakes lose their ability to neutralize acid rain and become more and more acidic until various types of organisms disappear, and in some cases all life-forms eventually die.

Acid rain also causes increased chemical weathering of limestone and marble and, to a lesser degree, sandstone. The effects are especially evident

on buildings, monuments, and tombstones, as in Gettysburg National Military Park in Pennsylvania.

The devastation caused by sulfur gases on vegetation near coal-burning plants is apparent, and many forests in the eastern United States show signs of stress that cannot be attributed to other causes (see Figure 24.2b).

Millions of tons of sulfur dioxide are released yearly into the atmosphere in the United States. Power plants built before 1975 have no emission controls, but the problems they pose must be addressed if emissions are to be reduced to an acceptable level. The most effective way to reduce emissions from these older plants is with flue-gas desulfurization, a process that removes up to 90% of the sulfur dioxide from exhaust gases.

Flue-gas desulfurization has some drawbacks. One is that some plants are simply too old to be profitably upgraded. Other problems include disposal of sulfur wastes, the lack of control on nitrogen gas emissions, and reduced efficiency of the power plant, which must burn more coal to make up the difference.

Other ways to control emissions are burning low-sulfur coal, fluidized bed combustion, and conservation of

Below horizon A in some soils is a pale layer with little carbon from which much of the small particles have been removed. This horizon E, as it is called, is present in older, more mature soils, and results from *eluviation,* a process of leaching of minerals by downward moving soil water. Some of this material is then deposited in the horizon below.

Horizon B, or *subsoil,* has fewer organisms and less organic matter than horizon A (Figure 6.10b). This horizon is also called the *zone of accumulation* because soluble materials leached from above accumulate as irregular masses. If horizon A is eroded leaving horizon B exposed, plants do not grow as well, and if it is clayey, it

is harder when dry and stickier when wet than the other soil horizons.

Partially altered bedrock grading down into unaltered bedrock with little organic matter characterizes horizon C (Figure 6.10b). In the horizons above, the parent material has been so thoroughly altered that it is no longer recognizable, but in horizon C, minerals and rock fragments of parent material are easy to identify.

It is worth noting that soils are subdivided, classified, and mapped on the basis of the development and composition of the various soil horizons. In fact, the Soil Survey Division of the Natural Resources Conservation Service (the

▌ Figure 1 Acid Rain

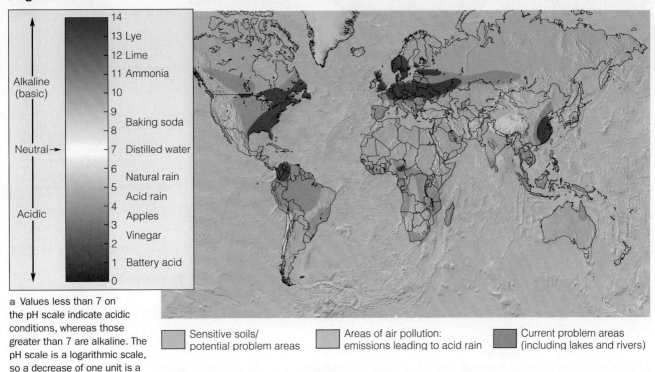

a Values less than 7 on the pH scale indicate acidic conditions, whereas those greater than 7 are alkaline. The pH scale is a logarithmic scale, so a decrease of one unit is a 10-fold increase in acidity.

Sensitive soils/ potential problem areas

Areas of air pollution: emissions leading to acid rain

Current problem areas (including lakes and rivers)

b Areas where acid rain is now a problem, and areas where the problem may develop.

electricity. Natural gas contains practically no sulfur, but converting to this alternative energy source would require the installation of expensive new furnaces in existing plants.

Acid rain, like global warming, is a worldwide problem that knows no national boundaries. Wind may blow pollutants from the source in one country to another where the effects are felt.

For instance, much of the acid rain in eastern Canada actually comes from sources in the United States.

NRCS) divides soils into 12 soil orders, and further subdivides them into small groupings. The 12 soil orders are based on the interaction of such features and processes as parent material, vegetation, and climate. It is beyond the scope of this chapter to go into detail about the 12 soil orders, but information about them is available from the NRCS, a branch of the United States Department of Agriculture.

Factors That Control Soil Formation

Complex interactions among several factors account for soil type, thickness, and fertility, but climate is the single most important factor (▌ Figure 6.11). For example, soils that form in rather humid regions have most of their soluble minerals leached out of horizon A which may be gray but is commonly dark because of abundant organic matter. Aluminum-rich clays and iron oxides tend to accumulate in horizon B (▌ Figure 6.12). Soils that form in semi-arid to arid regions have much less organic matter and more unstable minerals in horizon A because so little water is available to leach them out (Figure 6.12). However, soluble minerals such as calcite ($CaCO_3$) is taken into solution and precipitated in horizon B as irregular masses of *caliche* (Figure 6.12 and ▌ 6.13a). Precipitation of sodium salts in

Figure 6.10 The Composition of Soil and Soil Horizons

O — Loose leaves and organic debris

Partly decomposed organic debris

A — Topsoil; dark in color; rich in organic matter

E — Zone of intense leaching or eluviation

B — Subsoil, zone of accumulation

Transition to C

C — Partly weathered parent material

Parent material

a Soils are made up mostly of minerals and rock fragments derived by weathering, air, water, and organic matter. Most of the organic matter is humus.

b The soil horizons in a fully developed soil. Horizon O is only a few centimeters thick, but it has been exaggerated here to show some detail.

Figure 6.11 Climate and Soil Formation Generalized diagram showing soil formation as a function of the relationships between climate and vegetation, which alter parent material over time.

» Critical Thinking Question
Why does soil extend to much greater depths in the tropics compared to deserts?

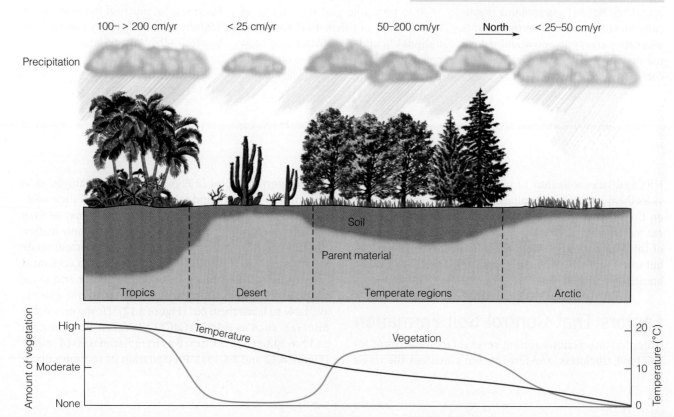

Figure 6.12 Climate, Vegetation, and Soil Formation These idealized soil profiles show the effects of climate and vegetation on soil development.

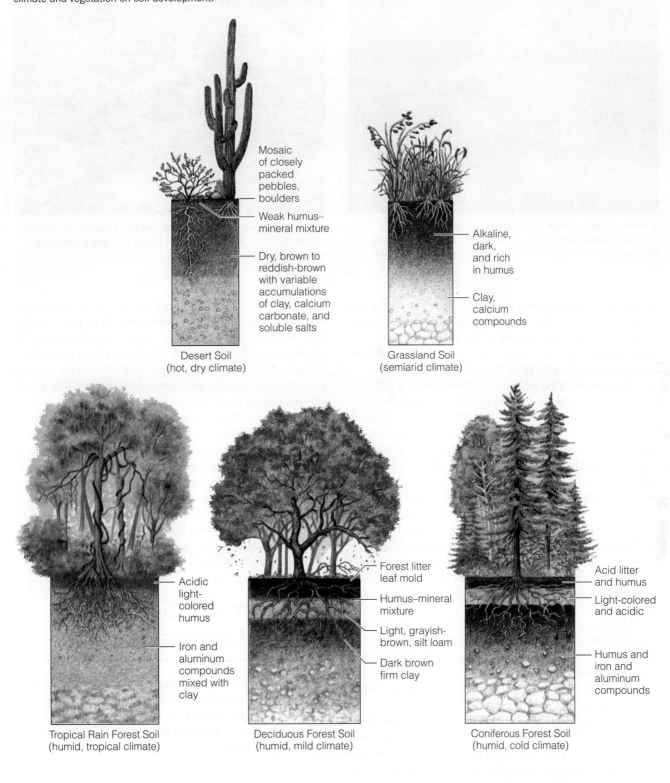

Mosaic of closely packed pebbles, boulders

Weak humus–mineral mixture

Dry, brown to reddish-brown with variable accumulations of clay, calcium carbonate, and soluble salts

Desert Soil (hot, dry climate)

Alkaline, dark, and rich in humus

Clay, calcium compounds

Grassland Soil (semiarid climate)

Acidic light-colored humus

Iron and aluminum compounds mixed with clay

Tropical Rain Forest Soil (humid, tropical climate)

Forest litter leaf mold

Humus–mineral mixture

Light, grayish-brown, silt loam

Dark brown firm clay

Deciduous Forest Soil (humid, mild climate)

Acid litter and humus

Light-colored and acidic

Humus and iron and aluminum compounds

Coniferous Forest Soil (humid, cold climate)

some arid-region soils where soil water evaporates yields *alkali soils* that are so alkaline that they support little or no vegetation (Figure 6.13b).

In the tropics where chemical weathering is intense and leaching of most minerals is complete, a soil termed **laterite**

forms (Figure 6.14a). These red soils extend to depths of many meters and are composed largely of aluminum hydroxides, iron oxides, and clay minerals. Laterite supports lush vegetation but is not very fertile because most plant nutrients have been leached out; the vegetation depends

Figure 6.13 Features of Soils in Arid Regions

a This boulder has been turned over to show the scaley white material known as *caliche*, which formed on its underside. Irregular masses of caliche are common in horizon B of arid region soils.

b An alkali soil near Fallon, Nevada. The white material is sodium carbonate or potassium carbonate. Notice that only a few hardy plants grow in this soil.

mostly on the surface layer of organic matter. In fact, when laterite is cleared of its vegetation and planted for crops, it can sustain farming for only a few years until the soil is depleted, in which case native farmers simply clear another area and repeat the process (Figure 6.14b, c).

The same rock type can yield different soils in different climatic regimes and, in the same climatic regime, the same soils can develop on different rock types. Thus, climate is more important than parent material in determining the type of soil. Nevertheless, rock type does exert some control. For example, the metamorphic rock quartzite will have a thin soil over it because it is chemically stable, whereas an adjacent body of granite will have a much deeper soil.

Soils depend on organisms for their fertility, and in return they provide a suitable habitat for many organisms. Earthworms, ants, sowbugs, termites, centipedes, millipedes, and nematodes, along with fungi, algae, and single-celled organisms, make their homes in soil. All contribute to soil formation and provide humus when they die and decompose by bacterial action.

Much of the humus in soils comes from grasses or leaf litter that microorganisms decompose to obtain food. In so doing, they break down organic compounds in plants and release nutrients back into the soil. In addition, organic acids from decaying soil organisms are important in further weathering of parent materials and soil particles.

Burrowing animals constantly churn and mix soils, and their burrows provide avenues for gases and water. Soil organisms, especially some types of bacteria, are extremely important in changing atmospheric nitrogen into a form of soil nitrogen suitable for use by plants.

The difference in elevation between high and low points in a region is called *relief*. And because climate is such an important factor in soil formation and climate changes with elevation, areas with considerable relief have different soils in mountains and adjacent lowlands. *Slope*, another important control, influences soil formation in two ways. One is

simply *slope angle;* steep slopes have little or no soil because weathered materials are eroded faster than soil-forming processes can operate. The other factor is *slope direction.* In the Northern Hemisphere, north-facing slopes receive less sunlight than south-facing slopes and have cooler internal temperatures, support different vegetation, and if in a cold climate, remain snow covered or frozen longer.

How much time is needed to develop a centimeter of soil or a fully developed soil a meter or so deep? We cannot give a definite answer because weathering proceeds at vastly different rates depending on climate and parent material, but an overall average might be about 2.5 cm per century. Nevertheless, a lava flow a few centuries old in Hawaii may have a well-developed soil on it, whereas a flow the same age in Iceland will have considerably less soil. Given the same climatic conditions, soil develops faster on unconsolidated sediment than it does on bedrock.

6.6 Expansive Soils and Soil Degradation

As you might imagine, soils that expand and contract pose problems for homeowners, developers, and engineers. And given the fact that our survival depends on soils, any loss of soil fertility, contamination of soils, or loss of soils to erosion is viewed with alarm.

Expansive Soils

Some soils contain clay minerals that increase in volume when wet and shrink when they dry out. Any soil that expands by 6% or more is an **expansive soil.** When soil expands and contracts, overlying structures are first uplifted and then subside, thus experiencing forces that usually are not equally applied. About

❙ Figure 6.14 Laterite and Slash-and-Burn Agriculture

a Laterite, shown here in Madagascar, is a deep-red soil that forms in response to intense chemical weathering in the tropics

» Critical Thinking Question
Laterite supports lush vegetation but is not very good for agriculture. Why?

Clearing and burning vegetation

b Indigenous people in some rain forest areas clear and burn the vegetation from a small area where they plant and harvest crops for 2 to 5 years.

Harvesting for 2 to 5 years

c Then the soil fertility is depleted and the farmers move on and repeat the process. An abandoned plot takes 10 to 30 years to completely revegetate and for soil fertility to be restored.

$6 billion in damage to foundations, roadways, sidewalks, and other structures takes place each year in the United States.

Avoiding areas of expansive soils is the best way to prevent damages, but what can be done for structures already sited on these soils? In some cases the soil can be removed, or mixed with chemicals that inhibit expansion, or covered by a layer of nonexpansive fill, all expensive but perhaps necessary remedies. Also, expansive soils should be kept as dry as possible, and specialized construction methods may be used such as placing structures on piers or reinforced foundations.

Soil Degradation

In the context of geologic time, soils form rapidly; however, from the human perspective, the processes are so slow that soil is a nonrenewable resource. So any soil losses that exceed the rate of soil formation and any reduction in soil fertility or

production are causes for concern. Any process that removes soil or makes it less productive is defined as **soil degradation,** a serious problem in many parts of the world that includes erosion, chemical deterioration, and physical changes.

What Would You Do ❓

You have inherited a piece of property ideally located for everything you consider important. Unfortunately, as you prepare to have a house built, your contractor tells you that the soil is rich in clay that expands when wet and contracts when it dries. You nevertheless go ahead with construction but must now decide what measures to take to prevent damage to the structure. Make several proposals that might solve this problem. Which one or ones do you think would be the most cost effective?

GEO-INSIGHT   Arches National Park, Utah

The sandstone arches at Arches National Park near Moab, Utah, are in the Jurassic-age Entrada Sandstone. The park is noted for its arches, spires, balanced rocks, pinnacles, and other features that resulted from differential weathering and erosion.

◄ **1–3.** When the area was uplifted (below), cracks (joints) developed. Differential weathering and erosion enlarged the joints, leaving fins (1, left and bottom left). Continued weathering and erosion of the fins (2) yielded arches, many of which have collapsed (3).

Galen Rowell/Mountain Light

▼ **4.** Delicate Arch shows an advanced stage in arch formation. It measures 9.7 m wide and 14 m high.

Sue Monroe

▶ **5–7.** Probable evolution of Sheep Rock and the fin with Baby Arch or Hole-in-the-Wall (right and above). Close-up view of Baby Arch (below). It measures 7.6 m wide and 4.5 m high.

Sheep Rock

Baby Arch

7

6

▼ **8.** The opening scene in the movie *Indiana Jones and the Last Crusade* (1989), showed a young Indiana Jones, played by River Phoenix, at Double Arch.

◀ **9.** Balanced Rock rises 39 m above its base. Isolated columns and pillars like this one are called *hoodoos*, many of which are in the park.

Figure 6.15 The Dust Bowl of the 1930s was a time of drought, dust storms resulting from wind erosion, and economic hardship.

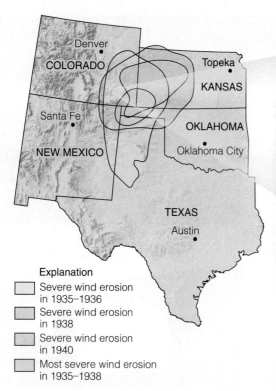

Explanation

Severe wind erosion in 1935–1936

Severe wind erosion in 1938

Severe wind erosion in 1940

Most severe wind erosion in 1935–1938

a Map of the Dust Bowl of the 1930s.

Source: "Map: Extent of Area Subject to Severe Wind Erosion," p. 30, from *Dust Bowl: The Southern Plains in the 1930s,* by Donald Worster. Copyright © 1979, 1982 by Oxford University Press, Inc. Used by permission of Oxford University Press, Inc.

b This huge dust storm was photographed at Lamar, Colorado, in 1934.

c By the mid-1930s, tens of thousands of people were on relief, homeless, or leaving the Dust Bowl. In 1939, Dorothea Lange photographed this homeless family of seven in Pittsburg County, Oklahoma

Erosion, an ongoing natural process, is usually slow enough for soil formation to keep pace, but unfortunately, some human practices add to the problem. Removing natural vegetation by plowing, overgrazing, overexploitation for fire wood, and deforestation all contribute to erosion by wind and running water. The Dust Bowl that developed in several Great Plains states during the 1930s is a poignant example of just how effective wind erosion is on soil pulverized and exposed by plowing (▌Figure 6.15).

Running water removes soil by *sheet erosion,* which involves the removal of thin layers more or less evenly over a broad, sloping surface. *Rill erosion,* in contrast, takes place when running water scours small, troughlike channels. Channels shallow enough to be eliminated by plowing are *rills* (▌Figure 6.16a), but those too deep (about 30 cm) to be plowed over are *gullies* (▌Figure 6.16b). Where gullying is extensive, croplands can no longer be tilled and must be abandoned.

Soil undergoes chemical deterioration when its nutrients are depleted and its productivity decreases. Loss of soil nutrients is most notable in many of the populous developing nations where soils are overused to maintain high levels of agricultural productivity. Chemical deterioration is also caused by insufficient use of fertilizers and by clearing soils of their natural vegetation.

Other types of chemical deterioration are pollution and *salinization,* which occurs when the concentration of salts increases in a soil, making it unfit for agriculture. Improper disposal of domestic and industrial wastes, oil and chemical spills, and the concentration of insecticides and pesticides in soils all cause pollution.

Soil deteriorates physically when it is compacted by the weight of heavy machinery and livestock, especially cattle. Compacted soils are more costly to plow, and plants have a more difficult time emerging from them. Furthermore,

What Would You Do ❓

In the past few years, many gullies have appeared in farmers' fields in your area, and residents of your area are concerned because agriculture is the main source of jobs and tax revenue. Obviously, a decrease in agricultural production would be an economic disaster. You are appointed to a county board charged with making recommendations to prevent, or at least minimize, erosion on local croplands. How would you determine what caused the problem, and what specific recommendations would you make to reduce gullying?

Figure 6.16 Soil Degradation Resulting from Erosion

a Rill erosion in a field in Michigan during a rainstorm. The rill was later plowed over.

b A large gully in the upper basin of the Rio Reventado in Costa Rica. Notice the man at the right for scale.

water does not readily infiltrate, so more runoff occurs; this in turn accelerates the rate of water erosion.

In North America, the rich prairie soils of the midwestern United States and the Great Plains of the United States and Canada are suffering soil degradation. Problems experienced in the past have stimulated the development of methods to minimize soil erosion on agricultural lands. Crop rotation, contour plowing (Figure 6.17), and the construction of terraces have all proved helpful. So has no-till planting, in which the residue from the harvested crop is left on the ground to protect the surface from the ravages of wind and water.

6.7 Weathering and Resources

Soils are certainly one of our most precious natural resources. Indeed, if it were not for soils, food production on Earth would be vastly different and capable of supporting far fewer people. In addition, other aspects of soils are important economically. We discussed the origin of laterite in response to intense chemical weathering in the tropics, and we noted further that laterite is not very productive. If the parent material is rich in aluminum, however, the ore of aluminum called *bauxite* accumulates in horizon B. Some bauxite is found in Arkansas, Alabama, and Georgia, but at present, it is cheaper to import rather than mine these deposits, so both the United States and Canada depend on foreign sources of aluminum ore.

Bauxite and other accumulations of valuable minerals by the selective removal of soluble substances during chemical weathering are known as *residual concentrations.* Certainly, bauxite is a good example, but other deposits that formed in a similar manner are those rich in iron, manganese, clays, nickel, phosphate, tin, diamonds, and gold. Some of the sedimentary iron deposits in the Lake Superior region of the United States and Canada were enriched by chemical weathering when soluble parts of the deposits were carried away.

Figure 6.17 Soil Conservation Practices Contour plowing and strip cropping are two soil conservation practices used on this farm. Contour plowing involves plowing parallel to the contours of the land to inhibit runoff and soil erosion. In strip cropping, row crops such as corn alternate with other crops such as alfalfa.

Some kaolinite deposits in the southern United States formed when chemical weathering altered feldspars in pegmatites or as residual concentrations of clay-rich limestones and dolostones. Kaolinite is a clay mineral used in the manufacture of paper and ceramics.

A *gossan* is a yellow to red deposit made up mostly of hydrated iron oxides that formed by oxidation and leaching of sulfide minerals such as pyrite (FeS_2). The dissolution of pyrite and other sulfides forms sulfuric acid, which causes other metallic minerals to dissolve, and these tend to be carried down toward the groundwater table, where the descending solutions form minerals containing copper, lead, and zinc. Gossans have been mined for iron, but they are far more important as indicators of underlying ore deposits.

CONNECTION LINK

For more about sedimentary iron deposits see page 176 in Chapter 7 and page 489 in Chapter 19.

GEO-RECAP

Geo-Image

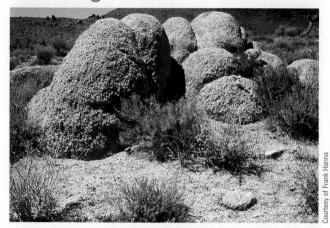

Courtesy of Frank Hanna

▌ **Figure 1**

These two images (▌Figures 1 and 2) show exposures of granite. Explain what weathering phenomena account for their present surface expression.

James S. Monroe

▌ **Figure 2**

Chapter Summary

- Mechanical and chemical weathering disintegrate and decompose parent material so that it is more nearly in equilibrium with new physical and chemical conditions.

- The products of weathering include rock fragments and minerals liberated from parent material as well as soluble compounds and ions in solution.

- Weathering yields materials that may become soil or sedimentary rock.

- Mechanical weathering processes include frost action, pressure release, thermal expansion and contraction, salt crystal growth, and the activities of organisms. The particles yielded retain the composition of the parent material.

- Chemical weathering by solution, hydrolysis, and oxidation results in a chemical change in parent material and proceeds most rapidly in hot, wet environments.

- Mechanical weathering contributes to chemical weathering by breaking parent material into smaller pieces, thereby exposing more surface area.

- Soils possess horizons designated, in descending order, as O, A, E, B, and C, which differ from one another in texture, composition, structure, and color.

- The important factors controlling soil formation are climate, parent material, organic activity, relief and slope, and time.

- Soils in humid regions are darker and more fertile than those of semiarid regions. Laterite is soil that forms in the tropics where chemical weathering is intense.

- Soil degradation results from erosion as well as from physical and chemical deterioration. Human activities such as construction, agriculture, deforestation, waste disposal, and chemical spills contribute to soil degradation.

- Chemical weathering is responsible for the origin of some mineral deposits such as residual concentrations of iron, lead, manganese, and clay.

Important Terms

chemical weathering (p. 139)
differential weathering (p. 137)
erosion (p. 136)
exfoliation dome (p. 138)

expansive soil (p. 148)
frost action (p. 137)
hydrolysis (p. 141)
laterite (p. 147)

mechanical weathering (p. 137)
oxidation (p. 141)
parent material (p. 136)
pressure release (p. 138)

regolith (p. 142)
salt crystal growth (p. 139)
sheet joins (p. 138)
soil (p. 142)

soil degradation (p. 149)
soil horizon (p. 143)
solution (p. 140)
spheroidal weathering (p. 142)

talus (p. 137)
thermal expansion and contraction
    (p. 139)
weathering (p. 136)

Review Questions

1. An essential component of soils is decomposed organic matter called
   a. _____ aluminum oxide.
   b. _____ humus.
   c. _____ laterite.
   d. _____ regolith.
   e. _____ carbonic acid.

2. The accumulation of angular pieces of rock at the base of a slope is known as
   a. _____ talus.
   b. _____ soil.
   c. _____ caliche.
   d. _____ hydrolysis.
   e. _____ oxidation.

3. Spheroidal weathering takes place because
   a. _____ oxidation changes limestone to clay.
   b. _____ aluminum oxides are nearly insoluble.
   c. _____ naturally occurring rocks are spherical to begin with.
   d. _____ thermal expansion and contraction are so effective.
   e. _____ corners and edges of rocks weather faster than flat surfaces.

4. Soil degradation refers to
   a. _____ the depth of a soil in a tropical region.
   b. _____ any loss of soil fertility, contamination, and erosion.
   c. _____ the rate at which feldspars alter to clay by chemical weathering.
   d. _____ how much parent material has been broken down to yield sediment.
   e. _____ the origin of exfoliation domes by pressure release.

5. If a small amount of carbonic acid is present in groundwater, _____ dissolves rapidly.
   a. _____ manganese.
   b. _____ limestone.
   c. _____ talus.
   d. _____ laterite.
   e. _____ granite.

6. An important kind of mechanical weathering is
   a. _____ solution.
   b. _____ leaching.
   c. _____ frost action.
   d. _____ laterization.
   e. _____ gullying.

7. Soils in arid regions commonly contain irregular masses of calcium carbonate called
   a. _____ exfoliation domes.
   b. _____ humus.
   c. _____ caliche.
   d. _____ salt crystals.
   e. _____ spheroids.

8. Horizon C differs from other soil horizons in that it
   a. _____ is the most fertile.
   b. _____ has been weathered the least.
   c. _____ is made up of sodium sulfate.
   d. _____ grades down into parent material.
   e. _____ contains the most humus.

9. Pressure release is the primary process responsible for
   a. _____ residual ores.
   b. _____ exfoliation domes.
   c. _____ soil degradation.
   d. _____ frost heaving.
   e. _____ spheroidal weathering.

10. Soils that when wet expand by at least 6% are _____ soils.
    a. _____ temperamental.
    b. _____ transported.
    c. _____ residual.
    d. _____ expansive.
    e. _____ oxidized.

11. How do parent material, particle size, and climate control the rate of chemical weathering?

12. What are exfoliation domes and how do they form?

13. How does mechanical weathering differ from and contribute to chemical weathering?

14. Why is groundwater in humid regions acidic whereas in arid regions it tends to be alkaline?

15. Describe how limestone goes into solution.

16. Explain what controls the origin, fertility, and depth of soils.

17. What is frost action, and where would you expect it to be most effective?

18. Why do differential weathering and erosion yield uneven surfaces?

19. How does chemical weathering contribute to the origin of and concentration of some natural resources?

20. Diagram and label the horizons found in soil. What are the features of each horizon?

Triassic Peak in Palo Duro Canyon State Park near Amarillo, Texas. The rocks in this view are sedimentary, made up mostly of mud and sand that were deposited in a shallow, retreating sea and in stream channels and on floodplains. Their present surface expression is caused by ongoing erosion, primarily by running water, but mass wasting (gravity-controlled) processes have contributed.

CHAPTER 7

Sue Monroe

Sediment and Sedimentary Rocks

OUTLINE

7.1   Introduction

7.2   Sediment Sources, Transport, and Deposition

7.3   How Does Sediment Become Sedimentary Rock?

7.4   Types of Sedimentary Rocks

7.5   Sedimentary Facies

7.6   Reading the Story in Sedimentary Rocks

GEO-INSIGHT: Fossilization

7.7 Important Resources in Sedimentary Rocks

GEO-FOCUS: Concretions and Geodes

GEO-RECAP

HAVE YOU EVER WONDERED?

- Where gravel, sand, and mud come from?
- How sediments such as gravel, sand, and mud become sedimentary rocks?
- What kinds of sedimentary rocks come from seawater and lake water?
- Which organisms are important in the origin of sedimentary rocks?
- What features in rocks geologists use to determine Earth history?
- How fossils help geologists decipher Earth history?

7.1 Introduction

All **sediment** comes from preexisting rocks and includes *detrital sediment,* the solid particles derived by weathering, and *chemical sediment,* which includes (1) minerals that precipitate from solutions such as sea water, and (2) minerals extracted from seawater by organisms to build their shells. Regardless of how it originates, all sediment is a loose aggregate of solids such as sand on a beach, gravel in a stream channel, or mud on the seafloor. If these loose solids are somehow bound together to form a solid aggregate, they become **sedimentary rock,** which by definition is composed of sediment.

Earth's crust is made up mostly of *crystalline rock,* a term that refers loosely to metamorphic rocks and igneous rocks with the exception of those composed of pyroclastic materials. Nevertheless, sediments and sedimentary rocks, making up only perhaps 5% of the crust, are the most commonly encountered Earth materials. Indeed, they cover about two-thirds of the continents and most of the seafloor, except spreading ridges. This should not be surprising because new oceanic crust forms at spreading ridges and thus has had little time for sediment to accumulate, whereas the seafloor distant from ridges has had sediment deposited on it for a long time.

All rocks are essential in our investigations of Earth history, but sedimentary rocks are particularly significant because they preserve evidence of surface processes such as running water, wind, and glaciers. For example, the rocks present in Palo Duro Canyon State Park, near Amarillo, Texas, are sedimentary, made up mostly of mud and sand (see the chapter opening photo). These rocks from the Permian and Triassic periods date from about 254 to 216 million years ago, although there are some gaps in this record. Detailed studies indicate that the rocks represent deposits from a shallow sea that withdrew from this area, followed by deposits that accumulated in stream channels and on floodplains. Furthermore, some of the rocks contain fossils of large reptiles known as phytosaurs. No dinosaurs have been found in the Palo Duro Canyon rocks, but in a nearby area, rocks of the same age have yielded some of the earliest dinosaurs.

> **CONNECTION LINK**
>
> You can find more information on Earth history in Chapters 19 through 23.

Once again we point out the complex interactions among Earth's systems, all of which are involved in the origin of sediment, its subsequent transport and deposition, and its transformation into sedimentary rock. Sedimentary rocks constitute the last major family of rocks (see Figure 1.16). However, as we mentioned in Chapter 6, some sediment becomes soil rather than sedimentary rock, and sedimentary rocks, or any other rock type, may be weathered to yield soil. The importance of sedimentary rocks in deciphering Earth history is obviously one good reason to study them, but another is that sediments and sedimentary rocks may be resources in their own right, or they may be the host for resources such as coal, petroleum, natural gas, and uranium.

7.2 Sediment Sources, Transport, and Deposition

Weathering is fundamental to the origin of sediment and sedimentary rocks, and so are *erosion* and *deposition*—that is, the movement of sediment from the weathering site by processes such as running water, wind, and glaciers, and its accumulation in some area (❚ Figure 7.1). Because glaciers are moving solids, they can transport sediment of any size, whereas wind transports only sand and smaller particles. Waves and marine currents also transport sediment along shorelines, but running water is by far the most common way to transport sediment from its source to other locations.

One important criterion for classifying detrital sediment is particle size (Table 7.1). Particles described as *gravel* measure more than 2 mm, and *sand* measures between 1/16 and 2 mm. *Silt* is made up of particles from 1/256 to 1/16 mm. None of these designations implies anything about composition; most gravel is made up of rock fragments—that is, small pieces of granite, basalt, or any other type of rock—but most sand and silt grains are single minerals. Particles smaller than 1/256 mm are termed *clay,* but *clay* has two meanings. One is size designation, but the term also refers to certain sheet silicates called *clay minerals.* However, most clay minerals are also clay-sized. Mixtures of silt and clay are called *mud* (Table 7.1).

Remember that detrital sediment is solid particles derived from other rocks. Chemical sediment, in contrast, comes from chemicals that were derived from other rocks that were then extracted from solution by inorganic chemical processes, such as evaporating seawater, or by the activities of organisms. Clams, oysters, corals, and some plants make their skeletons of minerals, especially aragonite or calcite ($CaCO_3$) or silica (SiO_2). In any case, minerals form that may be converted into sedimentary rock.

All detrital sediment is transported some distance from its source, but chemical sediment forms in the area where it is deposited. During transport of detrital sediment, sand

TABLE 7.1	Size Classification of Sedimentary Particles	
Size		**Sediment Name**
>2 mm		Gravel
1/16–2 mm		Sand
1/256–1/16 mm		Silt ⎫
<1/256 mm		Clay ⎭ Mud*

*Mud is a mixture of silt and clay-sized particles

and gravel particles collide, and *abrasion* wears away the sharp corners and edges, causing **rounding** (❙ Figure 7.2a). Transport also results in **sorting,** which refers to the size distribution of particles in sediment or sedimentary rocks that are poorly sorted if a wide range of size is present and well sorted if all the particles are about the same size (❙ Figure 7.2b, c). Rounding and sorting may seem unimportant, but both influence how readily groundwater, petroleum, and gas move through sediments and sedimentary rocks, which is essential to our efforts to recover these materials. They are also useful for determining how sediment deposition occurred, a topic covered more fully in a later section.

Regardless of how detrital sediment is transported or how chemical sediment forms, it is eventually deposited in some geographic area known as a **depositional environment** (❙ Figure 7.3). Deposition may take place in a stream channel or on its floodplain, in a lake, on a beach, or on the deep seafloor where physical and biological processes impart distinctive characteristics to the accumulating sediment. The three broad depositional settings geologists recognize are continental (on the land), transitional (on or near seashores), and marine (in the seas), each with several specific depositional environments (Figure 7.3).

SOURCE OF SEDIMENTARY MATERIALS

MECHANICAL WEATHERING

(gravel, sand, silt, clay–sized particles)

TRANSPORT

CHEMICAL WEATHERING

(clay minerals and ions, compounds in solution)

TO SITES OF DEPOSITION

❙ **Figure 7.1 Origin and Transport of Sediment** Whether derived from preexisting rocks by mechanical or chemical weathering, solid particles and ions and compounds in solution are transported and deposited elsewhere. If they are lithified, they become detrital and chemical sedimentary rocks.

❙ **Figure 7.2** Rounding and Sorting in Sediments

Sue Monroe

a Beginning students mistake rounding to mean ball-shaped or spherical. These three stones are all rounded, but only the one at the upper left is spherical.

James S. Monroe

b Deposit of very poorly sorted but well-rounded gravel. Note the quarter for scale.

James S. Monroe

c Angular, poorly sorted gravel. Note the quarter for scale.

≫ Critical Thinking Question
Do you think that if the gravel in (c) was transported for 10 km in a stream that it would remain angular?

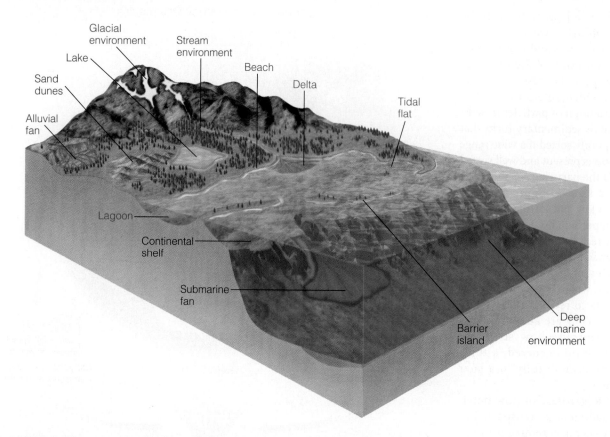

Glacial environment
Stream environment
Lake
Beach
Sand dunes
Delta
Alluvial fan
Tidal flat
Lagoon
Continental shelf
Submarine fan
Barrier island
Deep marine environment

▌**Figure 7.3** **Depositional Environments** Continental environments are shown in red type. The environments along the shoreline, shown in blue type, are transitional between continental and marine. The others, shown in black type, are marine environments.

7.3 How Does Sediment Become Sedimentary Rock?

A deposit of detrital sediment consists of a loose aggregate of particles: Mud in lakes, and sand and gravel in stream channels or on beaches, are good examples. To convert these aggregates of particles into sedimentary rocks requires **lithification** by compaction, cementation, or both (▌Figure 7.4).

To illustrate the relative importance of compaction and cementation, consider a detrital deposit made up of mud and another composed of sand. In both cases, the sediment consists of solid particles and *pore spaces,* the voids between particles. These deposits are subjected to **compaction** from their own weight and the weight of any additional sediment deposited on top of them, thereby reducing the amount of pore space and the volume of the deposit. Our hypothetical mud deposit may have 80% water-filled pore space, but after compaction, its volume is reduced by as much as 40% (Figure 7.4). The sand deposit, with as much as 50% pore space, is also compacted, but far less than the mud deposit, so that the grains fit more tightly together (Figure 7.4).

Compaction alone is sufficient for lithification of mud, but for sand and gravel, **cementation** involving the precipitation of minerals in pore spaces is also necessary. The two most common chemical cements are calcium carbonate ($CaCO_3$) and silicon dioxide (SiO_2), but iron oxide and hydroxide cement, such as hematite (Fe_2O_3) and limonite [$FeO(OH) \cdot nH_2O$], are found in some sedimentary rocks. Recall that calcium carbonate readily dissolves in water that contains a small amount of carbonic acid, and chemical weathering of feldspars and other minerals yields silica in solution. Cementation takes place when minerals precipitate in the pore spaces of sediment from circulating water, thereby binding the loose particles together. Iron oxide and hydroxide cements account for the red, yellow, and brown sedimentary rocks found in many areas (see the chapter opening photograph).

We have explained lithification of detrital sediments, but we have not yet considered this process in chemical sediments. By far the most common chemical sediments are calcium carbonate mud and sand- and gravel-sized accumulations of calcium carbonate grains, such as shells and shell fragments. Compaction and cementation also take place in these sediments, converting them into types of limestone, but compaction is less effective because cementation takes place soon after deposition. In any case, the cement is calcium carbonate derived by partial solution of some of the particles in the deposit.

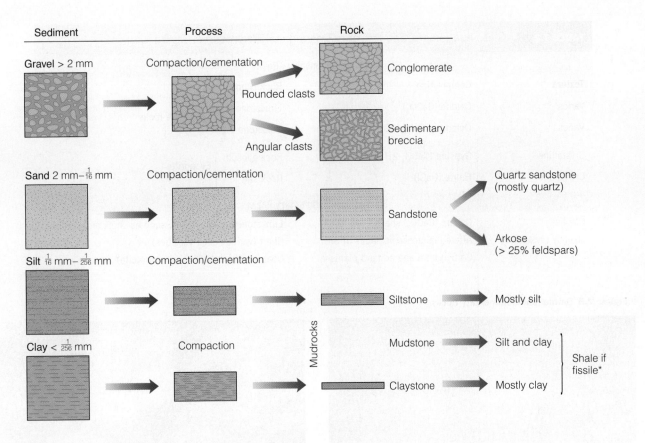

*Fissile refers to rocks that split along closely spaced planes.

❚ **Figure 7.4** Lithification and Classification of Detrital Sedimentary Rocks Notice that little compaction takes place in gravel and sand.

7.4 Types of Sedimentary Rocks

Thus far, we have considered the origin of sediment, its transport, deposition, and lithification. We now turn to the types of sedimentary rocks and how they are classified. The two broad classes or types of sedimentary rocks are *detrital* and *chemical,* although the latter has a subcategory known as *biochemical* (Table 7.2).

Detrital Sedimentary Rocks

Detrital sedimentary rocks are made up of solid particles, gravel, sand, silt, and clay, derived from parent material. All detrital sedimentary rocks have a *clastic texture,* meaning they are composed of particles or fragments known as *clasts.* The several varieties of detrital rocks are classified by the size of their constituent particles, although composition is used to modify some rock names.

Both *conglomerate* and *sedimentary breccia* are composed of gravel-sized particles (Figure 7.4 and ❚ Figure 7.5a, b); but conglomerate has rounded gravel, whereas sedimentary breccia has angular gravel. Conglomerate is common, but sedimentary breccia is not because gravel becomes rounded very

quickly during transport. Thus, if you encounter sedimentary breccia, you can conclude that its angular gravel has experienced little transport, probably less than a kilometer. Considerable energy is needed to transport gravel, so conglomerate is usually found in high-energy environments such as stream channels and beaches.

Sand is a size designation for particles between 1/16 and 2 mm, so any mineral or rock fragment in that size range can be in *sandstone.* Geologists recognize varieties of sandstone based on mineral content (Figure 7.4). *Quartz sandstone* is the most common and, as the name implies, is made up mostly of quartz sand. Another variety of sandstone called *arkose* contains at least 25% feldspar minerals. Sandstone is found in many depositional environments, including stream channels, sand dunes, beaches, barrier islands, deltas, and the continental shelf.

Mudrock is a general term that encompasses all detrital sedimentary rocks composed of silt- and clay-sized particles (Figure 7.4). Varieties include *siltstone* (mostly silt-sized particles), *mudstone* (a mixture of silt and clay), and *claystone* (primarily clay-sized particles). Some mudstones and claystones are designated *shale* if they are fissile, meaning that they break along closely spaced parallel planes (❚ Figure 7.5c). Even weak currents transport silt- and clay-sized particles, and deposition takes place only where currents and fluid turbulence are minimal, as in the quiet offshore waters of lakes or in lagoons.

TABLE 7.2	Classification of Chemical and Biochemical Sedimentary Rocks	
Chemical Sedimentary Rocks		
Texture	**Composition**	**Rock Name**
Varies	Calcite ($CaCO_3$)	Limestone ⎫ Carbonate rocks
Varies	Dolomite [$CaMg(CO_3)_2$]	Dolostone ⎭
Crystalline	Gypsum ($CaSO_4 \cdot 2H_2O$)	Rock gypsum ⎫ Evaporites
Crystalline	Halite (NaCl)	Rock salt ⎭
Biochemical Sedimentary Rocks		
Clastic	Calcite ($CaCO_3$) shells	Limestone (various types, such as chalk and coquina)
Usually crystalline	Altered microscopic shells of SiO_2	Chert (various color varieties)
	Carbon from altered land plants	Coal (lignite, bituminous, anthracite)

▌**Figure 7.5** Detrital and Sedimentary Rocks

James S. Monroe

a Sandstone overlain by conglomerate at Lillooet, British Columbia, Canada.

Sue Monroe

b Sedimentary breccias in Death Valley, California. Notice the angular gravel-sized particles. The largest clast is about 12 cm across.

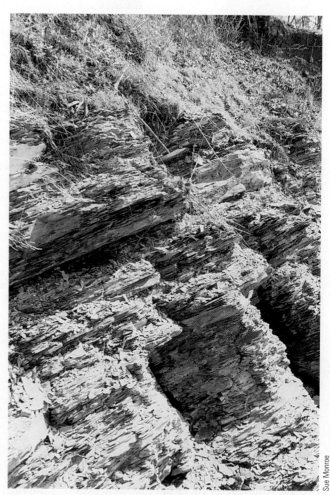

Sue Monroe

c Exposure of shale in Tennessee.

Chemical and Biochemical Sedimentary Rocks

During chemical weathering several compounds and ions go into solution and provide the raw materials for chemical and biochemical sedimentary rocks. For example, seawater contains silica (SiO_2), calcium (Ca), carbonate (CO_3), sulfate (SO_4), potassium (K), sodium (Na) and chlorine (Cl), and many other substances that under certain conditions that are extracted from the water to form minerals which make up **chemical sedimentary rocks.** Organisms play an important role in the origin of some of these rocks, which are designated **biochemical sedimentary rocks** (Table 7.2). Some chemical sedimentary rocks have a crystalline texture, meaning that they are made up of a mosaic of interlocking mineral crystals as in rock salt. Others have a clastic texture as in some limestones composed of fragmented shells.

By far the most common chemical sedimentary rocks are the **carbonates,** so-called because they contain the carbonate radical $(CO_3)^{-2}$. Several rocks meet this criterion but only two are common: limestone, composed of calcite $(CaCO_3)$, and dolostone, composed of dolomite $[CaMg(CO_3)_2]$ (Table 7.2). The origin of limestone is fairly straightforward. Recall from Chapter 6 that calcite in the presence of acidic groundwater rapidly goes into solution, but the chemical reaction leading to solution is reversible, so under some conditions calcite can precipitate from solution. Thus, some limestone forms by inorganic chemical precipitation from seawater or, more rarely, lake water.

Most limestone is biochemical because organisms are so important in its origin. Indeed, the skeletons of sea-dwelling animals are common in many varieties of limestone

(Figure 7.6a). Coquina is a type of limestone composed almost entirely of fragmented seashells (Figure 7.6b), and chalk is a soft type of limestone that consists of microscopic shells. One distinctive variety of limestone contains small spherical grains called *ooids* that have a small nucleus around which concentric layers of calcite precipitate. Lithified deposits of ooids form *oolitic limestone* (Figure 7.6c).

Dolostone is similar to limestone, but it forms mostly by the alteration of limestone when magnesium replaces some of the calcium in calcite, thereby converting it to dolomite. This process may occur in a lagoon where evaporation of seawater takes place, enriching the remaining seawater in magnesium, which permeates limestone and brings about a chemical change.

Most of you are aware that seawater is salty; that is, it contains sodium and chlorine and several other compounds and ions in solution. If you were to take a glass of seawater and let it evaporate completely, you would find a layer of minerals on the bottom of the glass. Obviously evaporation is involved in the origin of these minerals and their corresponding rocks,

▌ Figure 7.6 Chemical Sedimentary Rocks—Limestone

Sue Monroe

a Limestone with numerous fossil shells is called fossiliferous limestone.

Sue Monroe

b Coquina is limestone made up entirely of broken shells.

c This oolitic limestone is made up partly of ooids (see inset), which are rather spherical grains of calcium carbonate.

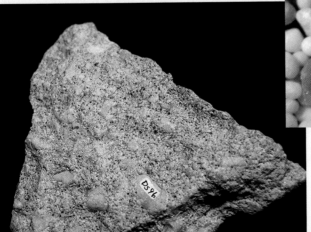

Stan Celestian/Glendale Community College

Sue Monroe

>> **Critical Thinking Question**
Coquina is made up of fragmented seashells, so would you expect it to accumulate on a Florida beach, on the deep seafloor, in a lake, or on a river's floodplain?

Figure 7.7 Chemical Sedimentary Rocks—Rock Salt and Rock Gypsum

a This cylindrical core of rock salt was taken from an oil well in Michigan.

>> **Critical Thinking Question**
Do you know of any commercial uses of rock salt and rock gypsum?

b Rock gypsum. When deeply buried, gypsum ($CaSO_4 \cdot 2H_2O$) loses its water and is converted to anhydrite ($CaSO_4$).

Figure 7.8 Chemical Sedimentary Rocks—Chert Nodules and Bedded Chert

Chert nodule

Chert nodule

a These chert nodules are in limestone at the Valley of the Kings in Egypt. Part of the lower nodule has broken away, but notice that it is not a continuous layer.

b Bedded chert exposed in Marin County, California. Most of the layers are about 5 cm thick.

which are collectively called **evaporites** (Table 7.2). The most familiar evaporites are *rock salt,* composed of the mineral halite (NaCl), and *rock gypsum,* made up of the mineral gypsum ($CaSO_4 \cdot 2H_2O$) (Figure 7.7a, b). Compared with sandstone, mudrocks, and limestone, evaporites are not very common, and yet they are significant deposits in areas such as Michigan, Ohio, New York, and the Gulf Coast region.

Chert is hard rock consisting of microscopic crystals of silica (SiO_2) (Table 7.2, Figure 7.8). Perhaps you have heard of *flint,* which is simply chert colored black by inclusions of organic matter, or *jasper,* which is red or brown chert because of its iron oxide content. Because chert is hard and lacks cleavage, it can be shaped to form sharp cutting edges for tools, spear points, and arrowheads. Some chert

Figure 7.9 The Origin of Coal

Sue Monroe

a Peat is partly decomposed plant material. It represents the first stage in the origin of coal.

Sue Monroe

b Lignite is a dull variety of coal in which plant remains are still visible.

Sue Monroe

c Bituminous coal is shinier and darker than lignite, and only rarely are plant remains visible.

that accumulates in bogs and swamps where the water is deficient in oxygen. Bacteria that decompose vegetation can live without oxygen, but their wastes must be oxidized, and because little or no oxygen is present, wastes build up and kill the bacteria. Decomposition ceases, and the vegetation forms organic muck, which if buried and compressed becomes *peat,* the first step in forming coal (Figure 7.9a).

Where peat is abundant, as in Ireland and Scotland, it is used for fuel, but if it is altered further by deeper burial, and especially if it is also heated, it becomes dull-black coal called *lignite* (Figure 7.9b). During the change from peat to lignite, the volatile elements are driven off, increasing the amount of carbon; peat has about 50% carbon whereas lignite has about 70%. Further changes yield *bituminous coal,* with about 80% carbon, which is dense, black, and so thoroughly altered that plant remains are rarely seen (Figure 7.9c). The highest grade of coal is *anthracite,* a metamorphic type of coal with up to 98% carbon.

> **CONNECTION LINK** ◀
>
> You can find more information on anthracite on page 193 in Chapter 8.

7.5 Sedimentary Facies

Long ago, geologists realized that when they traced a layer of sediment or sedimentary rock laterally, the layers changed in composition, texture, or both. These changes resulted from the simultaneous operation of different processes in adjacent depositional environments. For example, sand may be deposited in a high-energy nearshore marine environment, whereas mud and carbonate sediments accumulate simultaneously in the laterally adjacent low-energy offshore environments (▌Figure 7.10). Deposition in each environment produces **sedimentary facies,** bodies of sediment each possessing distinctive physical, chemical, and biological attributes. Figure 7.10 illustrates three sedimentary facies: a sand facies, a mud facies, and a carbonate facies. If these sediments become lithified, they are sandstone, mudstone (or shale), and limestone facies, respectively.

Many sedimentary rocks in the interiors of continents show clear evidence of deposition in marine environments. The rock layers in ▌Figure 7.11, for example, consist of a sandstone facies that was deposited in a nearshore marine environment overlain by shale and limestone facies deposited in offshore environments. Geologists explain this vertical sequence of facies by deposition occurring during a time when sea level rose with respect to the continents. As sea level rises, the shoreline moves inland, giving rise to a **marine transgression** (Figure 7.10), and the depositional environments parallel to the shoreline migrate landward. As a result, offshore facies are superimposed over nearshore facies, thus accounting for the vertical succession of

is found as irregular masses in other rocks, especially in limestone, whereas other chert is in distinct layers of *bedded chert* made up of tiny shells of silica-secreting organisms and thus is biochemical (Figure 7.8).

We previously mentioned that *coal* consists of altered organic matter but is nevertheless a biochemical sedimentary rock (Table 7.2, ▌Figure 7.9). It forms from vegetation

Figure 7.10 Marine Transgressions and Regressions

Three Stages of Marine Transgression

Three Stages of Marine Regression

sedimentary facies. Even though the nearshore environment is long and narrow at any particular time, deposition takes place continuously as the environment migrates landward. The sand deposit may be tens to hundreds of meters thick but have horizontal dimensions of length and width measured in hundreds of kilometers.

The opposite of a marine transgression is a **marine regression** (Figure 7.10). If sea level falls with respect to a continent, the shoreline and environments that parallel the shoreline move seaward. The vertical sequence produced by a marine regression has facies of the nearshore environment superposed over facies of offshore environments.

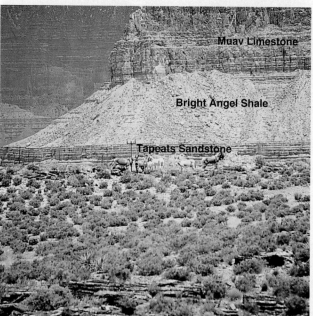

Figure 7.11 Sedimentary Rocks in the Grand Canyon of Arizona View of the Tapeats Sandstone, Bright Angel Shale, and Muav Limestone. The three rock layers are about 300 m thick, although part of the Tapeats Sandstone is buried beneath debris. The rocks were deposited in the sea during a marine transgression. Also notice the layered aspect of these rocks, or what geologists call bedding or stratification.

7.6 Reading the Story in Sedimentary Rocks

No one was present when ancient sediments were deposited, so geologists must evaluate those aspects of sedimentary rocks that allow them to make inferences about the original depositional environment. And making such determinations is of more than academic interest. For instance, barrier island sand deposits make good reservoirs for hydrocarbons, so knowing the environment of deposition and the geometry of these deposits is helpful in exploration for resources.

Sedimentary textures such as sorting and rounding can give clues to depositional processes. Windblown sand dunes tend to be well sorted and well rounded, but poor sorting is typical of glacial deposits. The geometry or three-dimensional shape is another important aspect of sedimentary rock bodies. Marine transgressions and regressions yield sediment bodies with a blanket or sheetlike geometry, but sand deposits in stream channels are long and narrow and are described as having a shoestring geometry. Sedimentary textures and geometry alone are usually insufficient to determine depositional environment, but when considered with other sedimentary rock properties, especially *sedimentary structures* and *fossils,* they enable geologists to reliably determine the history of a deposit.

Sedimentary Structures

Physical and biological processes operating in depositional environments are responsible for a variety of features known as **sedimentary structures** (see Geo-Focus on pages 174 and 175). One of the most common is distinct layers known as **strata** or **beds** (Figure 7.11), with individual layers of less than a millimeter up to many meters thick. These strata or beds are separated from one another by surfaces above and below in which the rocks differ in composition, texture, color, or a combination of features. Layering of some kind is present in almost all sedimentary rocks; however, a few, such as limestone that formed in coral reefs, lack this feature.

Many sedimentary rocks possess **cross-bedding,** in which layers are arranged at an angle to the surface on which they are deposited (Figure 7.12). Cross-beds are found in many depositional environments such as sand dunes in deserts and along shorelines, as well as in stream-channel deposits and shallow marine sediments. Invariably, cross-beds result from transport and deposition by wind or water currents, and the cross-beds are inclined downward in the same direction that the current flowed. Thus, ancient deposits with cross-beds inclined down toward the south, for example, indicate that the currents responsible for them flowed from north to south.

Some individual sedimentary rock layers show an upward decrease in grain size, termed **graded bedding,** and mostly formed by turbidity current deposition. A *turbidity current* is an underwater flow of sediment and water with a greater density than sediment-free water. Because of its greater density, a turbidity current flows downslope until it reaches the relatively flat seafloor, or lakefloor, where it slows and begins

Figure 7.12 Cross-Bedding Cross-bedding takes place when individual beds or strata are deposited at an angle to the surface upon which they accumulate.

Wind or current direction

Sand moves up this slope

Deposition

a Origin of cross-bedding by deposition on the sloping surface of a desert dune. Cross-bedding is also common in dunelike structures in stream and river channels.

James S. Monroe

b Cross-bedding in sandstone at Natural Bridges National Monument in Utah.

» Critical Thinking Question
Can you determine the current direction from the cross-beds?

Figure 7.13 Turbidity Currents and the Origin of Graded Bedding

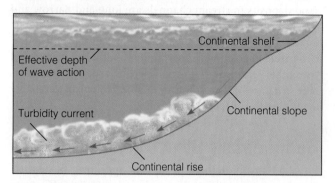

a A turbidity current flows downslope along the seafloor (or a lake bottom) because it is denser than sediment-free water.

b The flow slows and deposits progressively smaller particles, thus forming a graded bed.

depositing large particles followed by progressively smaller ones (▌Figure 7.13).

The surfaces that separate layers in sand deposits commonly have **ripple marks,** small ridges with intervening troughs, giving them a somewhat corrugated appearance. Some ripple marks are asymmetrical in cross section, with a gentle slope on one side and a steep slope on the other. Currents that flow in one direction, as in stream channels, generate these so-called *current ripple*

marks (▌Figure 7.14a,b). And because the steep slope of these ripples is on the downstream side, they are good indications of ancient current directions. In contrast, *wave-formed ripple marks* tend to be symmetrical in cross section and, as their name implies, are generated by the to-and-fro motion of waves (▌Figure 7.14c,d).

When clay-rich sediment dries, it shrinks and develops intersecting fractures called **mud cracks** (▌Figure 7.15). Mud cracks in ancient sedimentary rocks indicate

Figure 7.14 Current and Wave-Formed Ripple Marks

a Current ripple marks form in response to currents that flow in one direction, as in a stream. The enlargement shows the cross-beds in an individual ripple.

c The to-and-fro motion of waves in shallow water yields wave-formed ripple marks.

b Current ripple marks that formed in a stream channel.

d Wave-formed ripple marks in sand in shallow seawater.

》 Critical Thinking Question
How can you determine whether the flow that formed the ripples shown in part (b) was from right to left or from left to right?

James S. Monroe

Figure 7.15 Mud Cracks Mud cracks form in clay-rich sediments when they dry and contract.

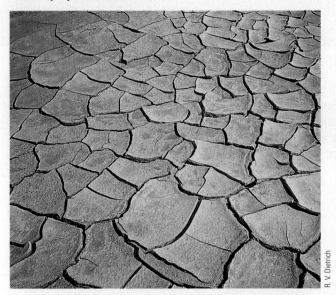

R. V. Dietrich

a Mud cracks in a present-day environment.

» Critical Thinking Question
What inferences can you make about the environment in which the mud in the image below was deposited.

Jeff Mayo/Geophoto Publishing

b Ancient mud cracks in Glacier National Park in Montana. Note that the cracks have been filled in with sediment.

that the sediment was deposited in an environment where periodic drying took place, such as on a river floodplain, near a lakeshore, or where muddy deposits are exposed along seacoasts at low tide. In addition to the sedimentary structures just discussed, there are several others that form long after deposition and accordingly tell nothing about the original deposition processes (see Geo-Focus on pages 174 and 175).

Fossils—Remains and Traces of Ancient Life

Fossils, the remains or traces of ancient organisms, are interesting as evidence of prehistoric life (▌ Figure 7.16), and are also important for determining depositional environments

(see the Geo-Insight on pages 170 and 171). Most people are familiar with fossils of dinosaurs and some other land-dwelling animals, but are unaware that fossils of invertebrates, animals lacking a segmented vertebral column, such as corals, clams, oysters, and a variety of microorganisms, are much more useful because they are so common.

Clams with heavily constructed shells typically live in shallow turbulent seawater, whereas organisms living in low-energy environments commonly have thin, fragile shells. Marine organisms that carry on photosynthesis are restricted to the zone of sunlight penetration, which is usually less than 200 m. The amount of sediment is also a limiting factor; for example, many corals live in shallow, clear seawater because suspended sediment clogs their respiratory and food-gathering organs, and some have photosynthesizing algae living in their tissues.

▌ Figure 7.16 Fossils

Sue Monroe

a Skull of the dinosaur *Allosaurus* on display at the Natural History Museum in Vienna, Austria.

James S. Monroe

b These fossil trilobites are on display in the Mesalands Community College's Dinosaur Museum in Tucumcari, New Mexico. Although distantly related to today's insects, spiders, crabs, and lobsters, trilobites went extinct at the end of the Paleozoic Era, 251 million years ago.

GEO-INSIGHT Fossilization

Although the chance that any one organism will be preserved in the fossil record is slight, fossils are nevertheless common because so many billions of organisms lived during the past several hundred million years. Hard skeletal parts of organisms living where burial was likely are most common.

▼ **1.** ▶ **2.** The bones of this dinosaur and the shells of these marine animals called ammonites have had minerals added to their pores, making them more durable.

Sue Monroe

Sue Monroe

▲ **3.** A coprolite (fossilized feces) from a carnivorous mammal. The coprolite is about 5.5 cm long.

▶ **4.** Trace fossils do not include actual remains—only burrows, nests, droppings, and tracks such as these tiny amphibian tracks.

Sue Monroe

Sue Monroe

◀ **5.** This object looks like a clam, but it is simply sediment that filled a space formed when a clam shell dissolved.

▼ **6.** Fossil insect replaced by silicon dioxide (SiO_2).

Reed Wicander

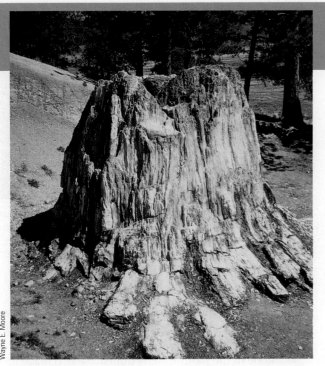

Wayne E. Moore

◀ **7.** Fossilized tree stump at Florissant Fossil Beds National Monument, Colorado. The woody tissue has been replaced by silicon dioxide. Mudflows 3 to 6 m deep buried the lower parts of many trees at this site.

Sue Monroe

▶ **8.** A palm frond and an insect preserved by carbonization.

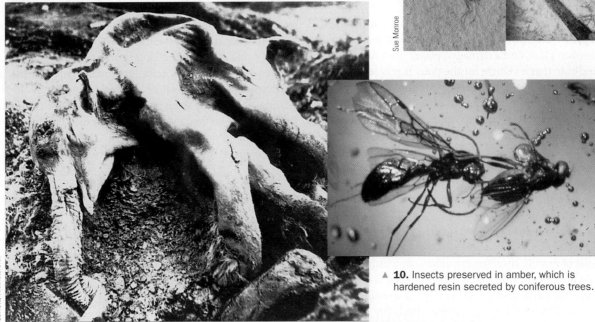

Sovfoto/V.Khristoforov

John Koivula/Photo Researchers, Inc.

▲ **10.** Insects preserved in amber, which is hardened resin secreted by coniferous trees.

▲ **9.** This 6- or 7-month-old frozen baby mammoth was found in Siberia in 1977. It measures 1.15 m long and 1 m high. Most of its hair has fallen out except around the feet. The carcass is about 40,000 years old.

Sue Monroe

◀ **11.** This insect from the La Brea Tar Pits in Los Angeles is just one of hundreds of species of animals found in the asphalt-like substance in the pits.

Microfossils are particularly useful for environmental interpretations because hundreds or even thousands are recovered from small rock samples. When drilling for oil, geologists recover small rock chips called *well cuttings* that may contain numerous fossils of tiny organisms. These fossils are routinely used to determine depositional environments and to match up or correlate rocks of the same relative age in different areas. In addition, fossils provide some of the evidence for organic evolution.

> **CONNECTION LINK**

You can find more information on using fossils for correlation on page 429 in Chapter 17, and the use of fossils as evidence for organic evolution on pages 467 and 468 in Chapter 18.

Sue Monroe

▌**Figure 7.17 Ancient Sedimentary Rocks and Their Interpretation** The Jurassic-age Navajo Sandstone in Zion National Park, Utah, is a wind-blown dune deposit. Notice the well-developed cross-bedding in this image. This rock exposure, or what geologists call an outcrop, measures about 30 m high.

Determining the Environment of Deposition

What kinds of evidence would allow you to determine how a layer of sandstone was deposited? Certainly you would consider texture—that is, rounding and sorting—and also the kinds of sedimentary structures and fossils, if any. You might also compare features in the sandstone with those seen in sand deposits forming today. But are you justified in using present-day processes and deposits to infer what happened when no human observers were present?

Actually you are familiar with the reasoning used to interpret events you did not witness. Skid marks on a street, broken glass, and a damaged power pole almost certainly indicate that a vehicle hit the pole. If you see a badly burned, shattered tree in the forest, you could conclude that a bomb damaged it, but in the absence of any bomb fragments or residue, you would no doubt decide that the tree was hit by lightning. Geologists use exactly the same kind of reasoning—that is, their understanding of natural processes—when they evaluate evidence preserved in sedimentary rocks; when the rocks formed is irrelevant. Geologists simply rely on the principle of uniformitarianism in making these interpretations (see Chapter 1).

So what about that sandstone we mentioned at the beginning of this section? Suppose that it has symmetrical ripples and fossils of marine-dwelling organisms, in which case you would no doubt conclude that it was deposited in a shallow marine environment. If, on the other hand, it contained dinosaur and land-plant fossils, you would probably conclude that it was deposited near a lakeshore. Many other features in sedimentary rocks are used in a similar fashion. Ooids (Figure 7.6c) form today in shallow marine environments with vigorous currents, and we have every reason to think that ancient ones formed in the same way. Glacial deposits are typically poorly sorted, show little stratification, and have other features that indicate glacial transport and deposition.

> **CONNECTION LINK**

You can find more information on glacial transport and deposition on pages 350 and 359 in Chapter 14.

The evidence indicates that the rocks in the lower part of the Grand Canyon (Figure 7.11) were deposited in shallow seas during a marine transgression. The Navajo Sandstone in the southwestern United States is made up of well-sorted sand measuring 0.2 to 0.5 mm in diameter, and it has large cross-beds and the footprints of land-dwelling animals (▌Figure 7.17). Long ago, geologists concluded that the Navajo Sandstone formed as ancient sand dunes. In fact, the inclination of the cross-beds indicates that the wind blew mostly from the northeast.

In some of the later chapters on Earth history, we refer to river deposits, ancient deltas, carbonate shelf deposits, and transgressions and regressions. We cannot include all the supporting evidence for these interpretations, but we can say that they are based on the kinds of criteria discussed in this chapter.

What Would You Do ?

No one was present millions of years ago to record data about the climate, the fauna and flora, the geography, and geologic processes. So how is it possible to decipher unobserved past events? In other words, what features in sedimentary rocks would you look for to determine what happened during the distant past? Can you think of any economic reasons to interpret Earth history from the record preserved in rocks?

7.7 Important Resources in Sedimentary Rocks

Sand and gravel are essential to the construction industry; pure clay deposits are used for ceramics, and limestone is used in the manufacture of cement and in blast furnaces where iron ore is refined to make steel. Evaporites are the source of table salt as well as chemical compounds, and rock gypsum is used to manufacture wallboard. Phosphate-bearing sedimentary rock is used in fertilizers and animal feed supplements.

Some valuable sedimentary deposits are found in streams and on beaches where minerals were concentrated during transport and deposition. These *placer deposits* are surface accumulations resulting from the separation and concentration of materials of greater density from those of lesser density. Much of the gold recovered during the initial stages of the California gold rush (1849–1853) was mined from placer deposits, and placers of other minerals such as diamonds and tin are important.

Historically, most coal mined in the United States has been bituminous coal from the Appalachian region that formed in coastal swamps during the Pennsylvanian Period (299–318 million years ago). Huge lignite and subbituminous coal deposits in the western United States are becoming increasingly important. During 2009, more than 975 million metric tons of coal were mined in this country, more than 70% of it from mines in Wyoming, West Virginia, and Kentucky.

> **CONNECTION LINK**
>
> You can find more information on Pennsylvanian-age coal deposits on pages 534 and 535 in Chapter 20.

Anthracite coal is especially desirable because it burns more efficiently than other types of coal. Unfortunately, it is the least common variety, so most coal used for heating buildings and generating electricity is bituminous (Figure 7.9c). *Coke,* a hard, gray substance consisting of the fused ash of bituminous coal, is used in blast furnaces where steel is produced. Synthetic oil and gas and a number of other products are also made from bituminous coal and lignite.

Petroleum and Natural Gas

Petroleum and natural gas are *hydrocarbons,* meaning that they are composed of hydrogen and carbon. The remains of microscopic organisms settle to the seafloor, or lakefloor in some cases, where little oxygen is present to decompose them. If buried beneath layers of sediment, they are heated and transformed into petroleum and natural gas. The rock in which hydrocarbons form is known as *source rock,* but for hydrocarbons to accumulate in economic quantities, they must migrate from the source rock into some kind of *reservoir rock.* And finally, the reservoir rock must have an overlying, nearly impervious *cap rock;* otherwise, the hydrocarbons would eventually reach the surface and escape (Figure 7.18a, b). Effective reservoir rocks must have appreciable pore space and good permeability, the capacity to transmit fluids; otherwise, hydrocarbons cannot be extracted from them in reasonable quantities.

Figure 7.18 Oil and Natural Gas The arrows in **(a)** and **(b)** indicate the direction of migration of hydrocarbons.

a Two examples of stratigraphic traps: one in sand within shale and the other in a buried reef.

b Two examples of structural traps: one formed by folding and the other by faulting.

c The sedimentary rock oil shale (left) and oil extracted from it. The United States has vast oil shale deposits.

GEO-FOCUS   Concretions and Geodes

Ripple marks and cross-bedding are *primary sedimentary structures,* meaning that they formed at the time of deposition or shortly thereafter. In contrast, *secondary sedimentary structures* form long after deposition and tell nothing about how the sediment was deposited in the first place. Geologists know of many types of these structures, some of which are very attractive, but here we will discuss only concretions and geodes.

Concretions are made up of any mass of material that can easily be separated from the enclosing rock, including shale, sandstone, and limestone. Probably the most common are sandstone concretions resulting from the more thorough cementation in restricted parts of a sand deposit during lithification. Many are only

a few centimeters across and include so-called *imatra stones* or *marlekor,* which are thin, disk-shaped concretions made up of calcium carbonate ($CaCO_3$) (Figure 1). Some of these have such regular geometric shapes that amateur rock and fossil collectors mistake them for manufactured objects.

Some of the more attractive concretions are large spherical objects called *septaria* or *septarian nodules.* These concretions have a series of cracks that widen toward the center, with smaller cracks that more or less parallel the margin of the concretion and intersect the larger cracks (Figure 2). Apparently, the cracks formed when the original concretion shrank as it dried, and the cracks then filled with mineral

crystals, mostly calcite. Septaria that are released from the host rock during weathering so that the cracks are revealed have the appearance of a turtle shell (Figure 2b).

Some oddly shaped sandstone concretions in Oklahoma have been mistakenly identified as human shoe prints in ancient rocks (Figure 3). And because dinosaur fossils come from the same age rocks, some people claim these "shoe prints" constitute evidence that humans and dinosaurs coexisted. However, even a superficial examination of the rocks reveals that the "shoe prints" are nothing more than concretions exposed by differential weathering of the host rock. Their resemblance to shoe prints is purely coincidental.

Figure 2

a Cross section showing the internal structure of a septarian nodule.

b Surface view of a sepatarian nodule resembling a turtle shell.

Figure 1 These disk-shaped concretions, called *imatra stones* or *marlekor,* are from glacial lake deposits in Connecticut. They are composed of calcium carbonate and measure about 4 cm across.

Figure 3 Concretions eroded from Cretaceous rocks in Oklahoma. These were mistakenly identified as human shoe prints. The specimens measure about 20 to 25 cm long.

Figure 4

a A geode formed by the partial filling of a cavity by color-banded agate, which conforms to the walls of the cavity, and by inward-pointing quartz crystals.

b A thunder egg from Oregon.

Sawed specimens of hollow, globular objects called *geodes* are available at just about any rock and mineral shop. Many are used as decorative stones because their internal structure of alternating bands of colored agate and inward-pointing crystals is so attractive. Most are small, measuring 30 cm in diameter or less, and form when minerals grow along the margins of a cavity, partially filling it (Figure 4a). The inward-pointing crystals are quartz or calcite, or more rarely barite ($BaSO_4$) and celestite ($SrSO_4$), all of which precipitated from solution.

Geode-like objects completely filled with minerals are called *thunder eggs* and are found in some igneous rocks, especially welded tuffs and rhyolite (Figure 4b). They form when a gas cavity fills with minerals, especially chalcedony, opal, or agate, which are all varieties of silicon dioxide (SiO_2). One legend holds that the spirits occupying nearby volcanoes in Oregon hurled spherical rocks at one another which they had stolen from the thunderbird, hence the name *thunder egg*. The knobby, ribbed exterior of a thunder egg is not very impressive, but inside it is filled with a dazzling array of minerals.

Many hydrocarbon reservoirs are nearshore marine sandstones with nearby fine-grained, organic-rich source rocks. Such oil and gas traps are called *stratigraphic traps* because they owe their existence to variations in the strata (Figure 7.18a). Indeed, some of the oil in the Persian Gulf region and Michigan is trapped in ancient reefs that are also good stratigraphic traps. *Structural traps* result when rocks are deformed by folding, fracturing, or both. In sedimentary rocks that have been deformed into a series of folds, hydrocarbons migrate to the high parts of these structures. Displacement of rocks along faults (fractures along which movement has occurred) also yields traps for hydrocarbons (■ Figure 7.18b).

Other sources of petroleum that will probably become increasingly important in the future include *oil shales* and *tar sands*. The United States has about two-thirds of all known oil shales, although large deposits are in South America, and all continents have some oil shale. The richest deposits in the United States are in the Green River Formation of Colorado, Utah, and Wyoming. When the appropriate extraction processes are used, liquid oil and combustible gases can be produced from an organic substance called *kerogen*, which is found in oil shale (■ Figure 7.18c). Oil shale in the Green River Formation yields between 10 and 140 gallons of oil per ton of rock processed, and the total amount of oil recoverable with present processes is estimated at 80 billion barrels. Currently, no oil is produced from oil shale in the United States because conventional drilling and pumping are less expensive.

Tar sand is a type of sandstone in which viscous, asphalt-like hydrocarbons fill the pore spaces. This substance is the sticky residue of once-liquid petroleum from which the volatile constituents have been lost. Liquid petroleum can be recovered from tar sand, but for this to happen, large quantities of rock must be mined and processed. Because the United States has few tar-sand deposits, it cannot look to this source as a significant future energy resource. The Athabaska tar sands in Alberta, Canada, however, are one of the largest deposits of this type. These deposits are currently being mined, and it is estimated that they contain several hundred billion barrels of recoverable petroleum.

Uranium

Most of the uranium used in nuclear reactors in North America comes from the complex potassium-, uranium-, vanadium-bearing mineral *carnotite*, which is found in some sedimentary rocks. Some uranium is also derived from *uraninite* (UO_2), a uranium oxide in granitic rocks and hydrothermal veins. Uraninite is easily oxidized and dissolved in groundwater, transported elsewhere, and chemically reduced and precipitated in the presence of organic matter.

The richest uranium ores in the United States are widespread in the Colorado Plateau area of Colorado and adjoining parts of Wyoming, Utah, Arizona, and New Mexico. These ores, consisting of fairly pure masses and encrustations of carnotite, are associated with plant remains in sandstones that formed in ancient stream channels. Although most of these ores are associated with fragmentary plant remains, some petrified trees also contain large quantities of uranium.

Large reserves of low-grade uranium ore also are found in the Chattanooga Shale. The uranium is finely disseminated in this black, organic-rich mudrock that underlies large parts of several states, including Illinois, Indiana, Ohio, Kentucky, and Tennessee. Canada is the world's largest producer and exporter of uranium.

Banded Iron Formation

The chemical sedimentary rock known as *banded iron formation* consists of alternating thin layers of chert and iron minerals, mostly the iron oxides hematite and magnetite (■ Figure 7.19). Banded iron formations are present on all continents and account for most of the iron ore mined in the world today. Vast banded iron formations are present in the Lake Superior region of the United States and Canada and in the Labrador Trough of eastern Canada.

> **CONNECTION LINK** ◀
>
> You can find more information about banded iron formation on page 489 in Chapter 19.

■ **Figure 7.19** Banded Iron Formation Is an Important Sedimentary Rock Because It Is the Main Source of Iron Ore

a The banded iron formation at Ishpeming, Michigan, consists of alternating layers of red chert and silver-colored iron minerals.

b Much of the iron ore in the Great Lakes region is processed and shaped into pellets about 1 cm across, and then either shipped directly to steel mills or stored for later shipment.

GEO-RECAP

» Geo-Image

▌ Figure 1 The cross section shows layers of sedimentary rocks that are no longer in their original horizontal position. See if you can determine which layer is youngest (place a large X on it). Also, indicate on the cross section what the *original* current direction was for the currents that generated the ripple marks.

▌ Figure 2 Suppose you encounter the sedimentary rocks shown in the cross section. Interpret the geologic history of this area as thoroughly as possible.

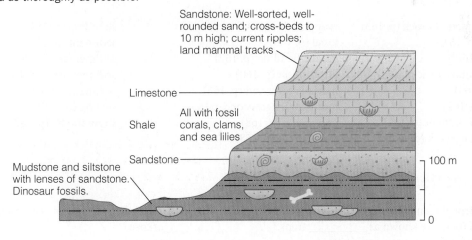

Chapter Summary

- Detrital sediment consists of weathered solid particles, whereas chemical sediment consists of minerals extracted from solution by inorganic chemical processes and the activities of organisms.

- Sedimentary particles are designated in order of decreasing size as gravel, sand, silt, and clay.

- During transport, sedimentary particles are rounded and sorted.

- Major depositional settings are continental, transitional, and marine, each of which includes several specific depositional environments.

- Lithification takes place when sediments are compacted and cemented and thus converted into sedimentary rock. Silica and calcium carbonate are the most common chemical cements.

- Sedimentary rocks are classified as detrital or chemical:

    a. Detrital sedimentary rocks consist of solid particles derived preexisting rocks.

    b. Chemical sedimentary rocks are derived from substances in solution by inorganic chemical processes or the activities of organisms. A subcategory called biochemical sedimentary rocks is recognized.

- Carbonate rocks contain minerals with the carbonate radical $(CO_3)^{-2}$ as in limestone and dolostone.

- Evaporites include rock salt and rock gypsum, both of which form by inorganic precipitation of minerals from evaporating water.

- Coal is a type of biochemical sedimentary rock composed of the altered remains of land plants.

- Sedimentary facies are bodies of sediment or sedimentary rock that are recognizably different from adjacent sediments or rocks.

- Vertical sequences of rocks with offshore facies overlying nearshore facies form when sea level rises with respect to the land, causing a marine transgression. A rise in the land relative to sea level causes a marine regression, which results in nearshore facies overlying offshore facies.

- Sedimentary structures such as bedding, cross-bedding, and ripple marks help geologists determine ancient current directions and depositional environments.

- Fossils provide the only record of prehistoric life and are useful for correlation and environmental interpretations.

- Depositional environments of ancient sedimentary rocks are determined by studying all aspects of the rocks and making comparisons with present-day sediments deposited by known processes.

- Many sediments and sedimentary rocks, including sand, gravel, evaporates, coal, and banded iron formations, are important natural resources. Most oil and natural gas are found in sedimentary rocks.

Important Terms

biochemical sedimentary rock (p. 163)
carbonate rock (p. 163)
cementation (p. 160)
chemical sedimentary rock (p. 162)
compaction (p. 160)
cross-bedding (p. 167)
depositional environment (p. 159)
detrital sedimentary rock (p. 161)

evaporite (p. 164)
fossil (p. 169)
graded bedding (p. 167)
lithification (p. 160)
marine regression (p. 166)
marine transgression (p. 165)
mud crack (p. 168)
ripple mark (p. 168)

rounding (p. 159)
sediment (p. 158)
sedimentary facies (p. 165)
sedimentary rock (p. 158)
sedimentary structure (p. 167)
sorting (p. 159)
strata (beds) (p. 167)

Review Questions

1. Traps for petroleum and natural gas by folding and fracturing of rocks are known as _____ traps.
    a. ____compaction.
    b. ____lithologic.
    c. ____structural.
    d. ____compositional.
    e. ____biochemical.

2. Which one of the following is a common evaporite rock?
    a. ____arkose.
    b. ____chert.
    c. ____conglomerate.
    d. ____rock gypsum.
    e. ____lignite.

3. A deposit of detrital sediment is poorly sorted if it has
   a. _____a large amount of calcium carbonate.
   b. _____particles of very different sizes.
   c. _____mostly ferromagnesian minerals.
   d. _____cross-bedding and symmetrical ripples.
   e. _____iron oxide cement.

4. Lithification involves cementation and
   a. _____compaction.
   b. _____inversion.
   c. _____conduction.
   d. _____stoping.
   e. _____replacement.

5. Cross-bedding preserved in sedimentary rocks is a good indication of
   a. _____the intensity of organic activity.
   b. _____how old the rocks are.
   c. _____the amount of silica cement.
   d. _____how much magnesium has replaced calcium.
   e. _____ancient current directions.

6. A vertical sequence of sedimentary rocks in which marine nearshore facies overlie offshore facies results from
   a. _____deposition by turbidity currents.
   b. _____a marine regression.
   c. _____granitization.
   d. _____deposition in a river system.
   e. _____changing limestone to dolostone.

7. Chemical sedimentary rocks such as limestone and dolostone are called _____ rocks.
   a. _____carbonate.
   b. _____detrital.
   c. _____evaporite.
   d. _____continental.
   e. _____fossil.

8. Turbidity currents in oceans and lakes are responsible for the deposition of
   a. _____graded bedding.
   b. _____mud cracks.
   c. _____chemical cement.
   d. _____evaporites.
   e. _____inverted strata.

9. Arkose is a type of sedimentary rock
   a. _____in which calcium has replaced sodium.
   b. _____with at least 25% feldspar minerals.
   c. _____composed mostly of biotite.
   d. _____with a high percentage of fossil shells.
   e. _____made up of a mixture of silt and clay.

10. Organisms play an important role in the origin of _____ sedimentary rocks.
    a. _____lithified.
    b. _____biochemical.
    c. _____detrital.
    d. _____sorted.
    e. _____compacted.

11. How is it possible to determine how ancient sedimentary rocks were deposited given that no human observers were present to witness their deposition?

12. In your travels you encounter a vertical sequence of rocks with sandstone at the bottom followed upward by shale and then limestone, each of which contains fossil clams and corals. Give an account of the history of these rocks. That is, how were they deposited, and how did they come to be superposed in the sequence you observed?

13. Describe how deposits of mud and sand are lithified.

14. How does coal form and what are the types of coal? Which type makes the best fuel?

15. What distinction(s) do geologists make to differentiate siltstone, mudstone, and claystone, and why are some mudstones and claystones called shale?

16. Illustrate and describe two sedimentary structures that are used to determine ancient current directions.

17. What are structural and stratigraphic traps?

18. What are carbonate rocks and how do they form?

19. The United States uses about 1,020,000,000 metric tons of coal per year from a recoverable reserve base of 250,000,000,000 metric tons. How long will this coal last at the current rate of consumption? Do you see any problems with projections such as this?

20. In what fundamental way(s) do detrital sedimentary rocks differ from chemical sedimentary rocks?

The Taj Mahal in India is constructed mostly of marble. In addition to its main use as a building material, marble was used throughout the structure in artworks and intricately carved marble flowers. It took more than 20,000 workers 17 years to build the Taj Mahal from 1631 to 1648.

CHAPTER 8

Metamorphism and Metamorphic Rocks

OUTLINE

8.1 Introduction

GEO-FOCUS: Asbestos: Good or Bad?

8.2 The Agents of Metamorphism

8.3 The Three Types of Metamorphism

8.4 How Are Metamorphic Rocks Classified?

8.5 Metamorphic Zones and Facies

8.6 Plate Tectonics and Metamorphism

8.7 Metamorphism and Natural Resources

GEO-RECAP

HAVE YOU EVER WONDERED?

- What causes metamorphism?
- How metamorphic rocks form?
- Where metamorphic rocks are most common and why?
- How metamorphic rocks are classified?
- Why marble is such a common rock in sculptures and structures?
- Why asbestos is an important industrial mineral for fireproofing and insulation in buildings and building materials, yet is also considered very dangerous to human health?

8.1 Introduction

Marble is a remarkable stone that has a variety of uses. Formed from limestone or dolostone by the metamorphic processes of heat and pressure, marble comes in a variety of colors and textures and has been used by sculptors and architects for many centuries in statuary and monuments and as a facing and main stone in buildings and structures, as well as in floor tiling and other ornamental and structural uses. For example, the *Aphrodite of Melos*, also known as *Venus de Milo*, is one of the most recognizable works of art in the world. The Taj Mahal in India is constructed mostly of Makrana marble quarried from hills just southwest of Jaipur in Rajasthan. The Peace Monument at Pennsylvania Avenue on the west side of the Capitol in Washington, DC, is constructed from white marble from Carrara, Italy, a locality famous for its marble. These are but a few of the more famous examples of marble used in statues and buildings throughout the world. In addition, ground marble can be found in toothpaste and is a source of lime in agricultural fertilizers.

Metamorphic rocks (from the Greek *meta*, "change," and *morpho*, "shape") are the third major group of rocks we will be examining. They result from the transformation of other rocks by processes that typically occur beneath Earth's surface (see Figure 1.15). During **metamorphism**, rocks are subjected to sufficient heat, pressure, and fluid activity to change their mineral composition, texture, or both, thus forming new rocks that usually do not look anything like the original rock before it was metamorphosed. These transformations take place below the melting temperature of the rock; otherwise, an igneous rock would result.

A useful analogy for metamorphism is baking a cake. Just like a metamorphic rock, the resulting cake depends on the ingredients, their proportions, how they are mixed together, how much water or milk is added, and the temperature and length of time used for baking the cake.

Except for marble and slate, most people are not familiar with metamorphic rocks. We are frequently asked by students why it is important to study metamorphic rocks and processes. Our answer is always, "Just look around you."

A large portion of Earth's continental crust is composed of metamorphic and igneous rocks. Together, they form the crystalline basement rocks underlying the sedimentary rocks of a continent's surface. These basement rocks are widely exposed in regions of the continents known as *shields*, which have been very stable during the past 600 million years (see Figure 19.5). Some of the oldest known rocks, dated at about 4 billion years (▌ Figure 8.1) are metamorphic, which means that they formed from even older rocks!

> **CONNECTION LINK**

To learn more about shields, go to pages 478–480 and Figure 19.5 in Chapter 19.

Copyright and Photograph by Dr. Parvinder S. Sethi

▌ **Figure 8.1 Acasta Gneiss** This metamorphic rock, found in Canada, is estimated to be about 4 billion years old, making it one of the oldest known rocks on Earth.

> **Critical Thinking Question**
> If the Acasta Gneiss (a metamorphic rock) is one of the oldest known rocks on Earth, why does Earth have to be older than 4 billion years?

Metamorphic rocks such as marble and slate are used as building materials, and some metamorphic minerals, such as garnets (used as gemstones or abrasives) and talc (used in cosmetics, in manufacturing paint, and as a lubricant), are economically important. Asbestos, another valuable metamorphic mineral, is used for insulation and fireproofing and has been the subject of much debate over the danger it poses to the public's health (see Geo-Focus on page 183).

What Would You Do ❓

The problem of removing asbestos from public buildings is an important national health and political issue. The current policy of the U.S. Environmental Protection Agency (EPA) mandates that all forms of asbestos are treated as identical hazards. Yet studies indicate that only one form of asbestos is a known health hazard. Because the cost of asbestos removal has been estimated to be as high as $100 billion, many people are questioning whether it is cost effective to remove asbestos from all public buildings where it has been installed.

As a leading researcher on the health hazards of asbestos, you have been asked to testify before a congressional committee on whether it is worthwhile to spend so much money for asbestos removal. How would you address this issue in terms of formulating a policy that balances the risks and benefits of removing asbestos from public buildings? What role would geologists play in formulating this policy?

GEO-FOCUS   Asbestos: Good or Bad?

Asbestos (from the Latin, meaning "unquenchable") is a general term applied to any silicate mineral that easily separates into flexible fibers. The combination of such features as fire resistance and flexibility makes asbestos an important industrial material of considerable value. In fact, asbestos has more than 3,000 known uses, including brake linings, fireproof fabrics, and heat insulators.

Asbestos is divided into two broad groups: *serpentine asbestos* and *amphibole asbestos*. *Chrysotile* is the fibrous form of serpentine asbestos (▌ Figure 1); it is the most valuable type and constitutes the bulk of all commercial asbestos. Its strong, silky fibers are easily spun and can withstand temperatures as high as 2,750°C.

The vast majority of chrysotile asbestos is in serpentine, a type of rock formed by the alteration of ultramafic igneous rocks such as peridotite under low- and intermediate-grade metamorphic conditions. Other chrysotile results when the metamorphism of magnesium limestone or dolostone produces discontinuous serpentine bands within the carbonate beds.

Among the varieties of amphibole asbestos, *crocidolite* is the most common. Also known as blue asbestos, crocidolite is a long, coarse, spinning fiber that is stronger but more brittle than chrysotile and less resistant to heat. Crocidolite is found in such metamorphic rocks as slates and schists and is thought to form by the solid-state alteration of other minerals as a result of deep burial.

Despite the widespread use of asbestos, the U.S. Environmental Protection Agency (EPA) instituted a gradual ban on all new asbestos products. The ban was imposed because some forms of asbestos can cause lung cancer and scarring of the lungs if fibers are inhaled.

Because the EPA apparently paid little attention to the issue of risks versus benefits when it enacted this rule, the U.S. Fifth Circuit Court of Appeals overturned the EPA ban on asbestos in 1991.

The threat of lung cancer has also resulted in legislation mandating the removal of asbestos already in place in all public buildings, including all public and private schools. However, important questions have been raised concerning the threat posed by asbestos and the additional potential hazards that may arise from its improper removal.

Current EPA policy mandates that all forms of asbestos are to be treated as identical hazards. Yet studies indicate that only the amphibole forms constitute a known health hazard. Chrysotile, whose fibers tend to be curly, does not become lodged in the lungs. Furthermore, its fibers are generally soluble and disappear in tissue. In contrast, crocidolite has long, straight, thin fibers that penetrate the lungs and stay there. These fibers irritate the lung tissue and over a long period of time can lead to lung cancer (▌ Figure 2). Thus, crocidolite, and not chrysotile, is overwhelmingly responsible for asbestos-related lung cancer. Because approximately 95% of the asbestos in place in the United States is chrysotile, many people question whether the dangers from asbestos are exaggerated.

▌ **Figure 1** Specimen of Chrysotile from Arizona
Chrysotile is the fibrous form of serpentine asbestos and the most commonly used in buildings and other structures.

Copyright and Photograph by Dr. Parvinder S. Sethi

Removing asbestos from buildings where it has been installed could cost as much as $100 billion. Unless the material containing the asbestos is disturbed, asbestos does not shed fibers and thus does not contribute to airborne asbestos that can be inhaled. Furthermore, improper removal of asbestos can lead to contamination. In most cases of improper removal, the concentration of airborne asbestos fibers is far higher than if the asbestos had been left in place.

The problem of asbestos contamination is a good example of how geology affects our lives and why we should have a basic knowledge of science before making decisions that could have broad economic and societal impacts.

Du Cane Medical Imaging Ltd./Photo Researchers, Inc.

▌ **Figure 2** Lung Cancer
Computerized tomography (CT) scan of an axial section through the chest of a patient with a mesothelioma cancer (light red). It is surrounding and constricting the lung at right (pink). The other lung (dark blue) has a healthy pleura (dark red). The spine (lower center, light blue), the descending aorta (green), and the heart (dark green between lungs) are also seen. Mesothelioma is a malignant cancer of the pleura, the membrane lining the chest cavity and lungs. It is usually caused by asbestos exposure. It often reaches a large size, as here, before diagnosis, and the prognosis is then poor.

8.2 The Agents of Metamorphism

The three principal agents of metamorphism are *heat*, *pressure*, and *fluid activity*. Time is also important to the metamorphic process, because chemical reactions proceed at different rates and thus require different amounts of time to complete. Reactions involving silicate compounds are particularly slow, and because most metamorphic rocks are composed of silicate minerals, it is thought that metamorphism is a very slow geologic process.

During metamorphism, the original rock, which was in equilibrium with its environment, meaning that it was chemically and physically stable under those conditions, undergoes changes to achieve equilibrium with its new environment. These changes may result in the formation of new minerals, a change in the texture of the rock, or both. In some instances, the change is minor, and features of the original rock can still be recognized. In other cases, the rock changes so much that the identity of the original rock can be determined only with great difficulty, if at all.

Heat

Heat is an important agent of metamorphism because it increases the rate of chemical reactions that may produce minerals different from those in the original rock. Heat may come from lava, magma, or as a result of deep burial in the crust due to subduction along a convergent plate boundary.

When rocks are intruded by bodies of magma, they are subjected to intense heat that affects the surrounding rock; the most intense heating usually occurs adjacent to the magma body and gradually decreases with distance from the intrusion. The zone of metamorphosed rocks that forms in the country rock adjacent to an intrusive igneous body is usually distinct and easy to recognize.

It is known that temperature increases with depth. Some rocks that form at the surface may be transported to great depths by subduction along a convergent plate boundary and thus subjected to increasing temperature and pressure. During subduction, some minerals may be transformed into other minerals that are more stable under the higher temperature and pressure conditions.

Pressure

During burial, rocks are subjected to increasingly greater pressure, just as you feel greater pressure the deeper you dive in a body of water. Whereas the pressure you feel is known as *hydrostatic pressure*, because it comes from the water surrounding you, rocks undergo **lithostatic pressure**, which means that the stress (force per unit area) on a rock in Earth's crust is the same in all directions (▌Figure 8.2a). A similar situation occurs when an object is immersed in water. For example, the deeper a cup composed of Styrofoam™ is submerged in the ocean, the smaller it gets because pressure increases with depth and is exerted on the cup equally in all directions, thereby compressing the cup (▌Figure 8.2b).

CONNECTION LINK ◀

To learn more about stress and strain, and how they relate to the deformation of rocks, see pages 235 and 236 in Chapter 10.

▌**Figure 8.2** Lithostatic Pressure

1 kilobar (kbar) = 1,000 bars
Atmospheric pressure at sea level = 1 bar

a Lithostatic pressure is applied equally in all directions in Earth's crust due to the weight of overlying rocks. Thus, pressure increases with depth, as indicated by the sloping black line. From C. Gillen, *Metamorphic Geology*, Figure 4.4, p. 73. Copyright 1982 Kluwer Academic Publishers. Reprinted by permission of the author.

b A similar situation occurs when 200-ml cups composed of Styrofoam™ are lowered to ocean depths of approximately 750 m and 1,500 m. Increased water pressure is exerted equally in all directions on the cups, and they consequently decrease in volume while maintaining their general shape.

Along with lithostatic pressure resulting from burial, rocks may experience **differential pressure** (❙ Figure 8.3). In this case, the stresses are not equal in all directions, but are stronger from some directions than from others. Differential pressures typically occur when two plates collide, thus producing distinctive metamorphic textures and features.

Fluid Activity

In almost every region of metamorphism, water and carbon dioxide (CO_2) are present in varying amounts along mineral grain boundaries or in the pore spaces of rocks. These fluids, which may contain ions in solution, enhance metamorphism by increasing the rate of chemical reactions. Under dry conditions, most minerals react very slowly, but when even small amounts of fluid are introduced, reaction rates increase, mainly because ions can move readily through the fluid and thus enhance chemical reactions and the formation of new minerals.

The following reaction provides a good example of how new minerals can be formed by **fluid activity**. Seawater moving through hot basaltic rock in the oceanic crust transforms olivine into the metamorphic mineral serpentine:

$$2Mg_2SiO4 + 2H_2O \rightarrow (Mg_3Si_2O_5(OH)_4 + MgO$$

olivine   water   serpentine   carried away
                                              in solution

The chemically active fluids important in the metamorphic process come primarily from three sources: (1) water trapped in the pore spaces of sedimentary rocks as they form, (2) the volatile fluid within magma, and (3) the dehydration of water-bearing minerals such as gypsum ($CaSO_4 \cdot 2H_2O$) and some clays.

8.3 The Three Types of Metamorphism

Geologists recognize three major types of metamorphism: (1) *contact (thermal) metamorphism*, in which magmatic heat and fluids act to produce change; (2) *dynamic metamorphism*, which is principally the result of high differential pressures associated with intense deformation; and (3) *regional metamorphism*, which occurs within a large area and is associated with major mountain-building episodes. Even though we will discuss each type of metamorphism separately, the boundary between them is not always distinct and depends largely on which of the three metamorphic agents was dominant (❙ Figure 8.4).

Contact Metamorphism

Contact (thermal) metamorphism takes place when a body of magma alters the surrounding country rock. At shallow depths, intruding magma raises the temperature of the

Courtesy of Eric Johnson

❙ **Figure 8.3 Differential Pressure** Differential pressure results from stress that is unequally applied to an object. Rotated garnets are a good example of the effects of differential pressure applied to a rock during metamorphism. In this example from a schist in northeast Sardinia, stress was applied in opposite directions on the left and right sides of the garnet (center), causing it to rotate.

surrounding rock, causing thermal alteration. Furthermore, the release of hot fluids into the country rock by the cooling intrusion can aid in the formation of new minerals.

Important factors in contact metamorphism are the initial temperature, the size of the intrusion, and the fluid content of the magma and the country rock, or all of these. The initial temperature of an intrusion depends, in part, on its composition; mafic magmas are hotter than felsic magmas (see Chapter 4) and hence have a greater thermal effect on the rocks surrounding them. The size of the intrusion is also important. In the case of small intrusions, such as dikes and sills, usually only those rocks in immediate contact with the intrusion are affected. Because large intrusions, such as batholiths, take a long time to cool, the increased temperature in the surrounding rock may last long enough for a larger area to be affected.

The area of metamorphism surrounding an intrusion is an **aureole**, and the boundary between an intrusion and its aureole may be either sharp or transitional (❙ Figure 8.5). Metamorphic aureoles vary in width depending on the size, temperature, and composition of the intruding magma, as well as the mineralogy of the surrounding country rock. Aureoles range from a few centimeters wide bordering small dikes and sills to several hundred meters or even several kilometers wide around large plutons.

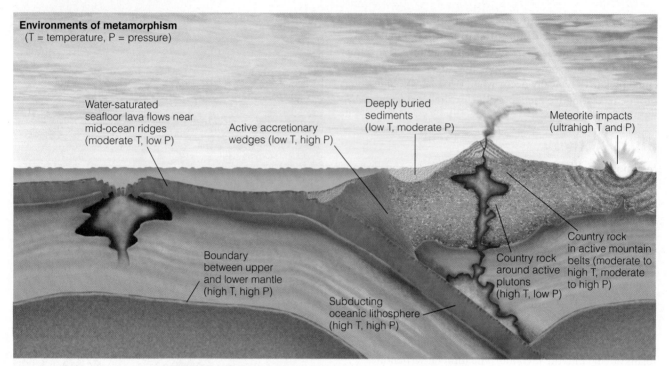

Environments of metamorphism
(T = temperature, P = pressure)

Water-saturated seafloor lava flows near mid-ocean ridges (moderate T, low P)

Active accretionary wedges (low T, high P)

Deeply buried sediments (low T, moderate P)

Meteorite impacts (ultrahigh T and P)

Boundary between upper and lower mantle (high T, high P)

Country rock around active plutons (high T, low P)

Country rock in active mountain belts (moderate to high T, moderate to high P)

Subducting oceanic lithosphere (high T, high P)

▌ **Figure 8.4 Environments of Metamorphism** The type of metamorphism that results depends largely on which of the three metamorphic agents was dominant. Illustrated here are some of the common metamorphic environments associated with plate movement, and whether the temperature and pressure in this environment is considered low, moderate, or high. The third agent of metamorphism, fluid activity, although playing an important role in metamorphism, isn't shown here, but is discussed in the text.

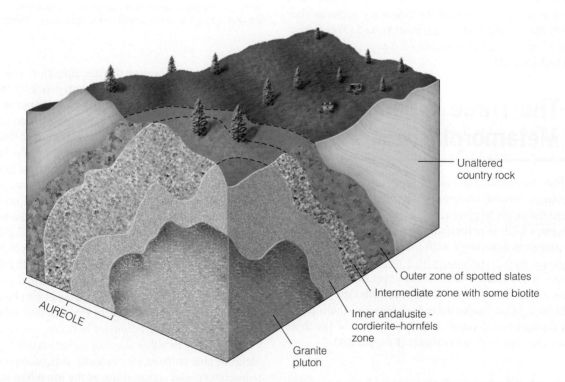

Unaltered country rock

Outer zone of spotted slates

Intermediate zone with some biotite

Inner andalusite - cordierite–hornfels zone

Granite pluton

AUREOLE

▌ **Figure 8.5 Metamorphic Aureole** A metamorphic aureole, the area surrounding an intrusion, consists of zones that reflect the degree of metamorphism. The metamorphic aureole associated with this idealized granite pluton contains three zones of mineral assemblages reflecting the decreases in temperature with distance from the intrusion. An inner andalusite–cordierite hornfels zone forms adjacent to the pluton and indicates the high temperatures near the intrusion. This zone is followed by an intermediate zone of extensive recrystallization in which some biotite develops, and farthest from the intrusion is an outer zone characterized by spotted slates.

The degree of metamorphic change within an aureole generally decreases with distance from the intrusion, reflecting the decrease in temperature from the original heat source. The region or zone closest to the intrusion, and hence subject to the highest temperatures, commonly contains high-temperature metamorphic minerals (that is, minerals in equilibrium with the higher-temperature environment) such as sillimanite. The outer zones—that is, those farthest from the intrusion—are typically characterized by lower-temperature metamorphic minerals such as chlorite, talc, and epidote.

Contact metamorphism can result not only from igneous intrusions but also from lava flows, either along mid-oceanic ridges (Figure 8.4) or from lava flowing over land and thermally altering the underlying rocks (⏸ Figure 8.6). Whereas recognizing a recent lava flow and the resulting contact metamorphism of the rocks below is easy, it is less obvious whether an igneous body is intrusive or extrusive in a rock outcrop where sedimentary rocks occur above and below the igneous body. Recognizing which sedimentary rock units have been metamorphosed enables geologists to determine whether the igneous body is intrusive (such as a sill or dike) or extrusive (lava flow). Such a determination is critical in reconstructing the geologic history of an area and may have important economic implications as well.

Fluids also play an important role in contact metamorphism. Magma is usually wet and contains hot, chemically active fluids that may emanate into the surrounding rock. These fluids can react with the rock and aid in the formation of new minerals. In addition, the country rock may contain pore fluids that, when heated by magma, increase reaction rates.

Because heat and fluids are the primary agents of contact metamorphism, two types of contact metamorphic rocks are generally recognized: those resulting from baking of country rock and those altered by hot solutions. Many of the rocks that result from contact metamorphism have the texture of porcelain; that is, they are hard and fine-grained. This is particularly true for rocks with a high clay content, such as shale. Such texture results because the clay minerals in the rock are baked, just as a clay pot is baked when fired in a kiln.

During the final stages of cooling, when an intruding magma begins to crystallize, large amounts of hot, watery solutions are often released. These solutions may react with the country rock and produce new metamorphic minerals. This process, which usually occurs near Earth's surface, is called *hydrothermal alteration* (from the Greek *hydro*, "water," and *therme*, "heat") and may result in valuable mineral deposits.

Geologists think that many of the world's ore deposits result from the migration of metallic ions in hydrothermal solutions. Examples are copper, gold, iron ores, tin, and zinc in various localities including Australia, Canada, China, Cyprus, Finland, Russia, and the western United States.

Dynamic Metamorphism

Most **dynamic metamorphism** is associated with faults (or fault zones), fractures along which movement has occurred, and where rocks are subjected to high levels of differential pressure. The metamorphic rocks that result from pure dynamic metamorphism are called *mylonites*, and typically they are restricted to narrow zones adjacent to faults. Mylonites are hard, dense, fine-grained rocks, many of which are characterized by thin laminations (⏸ Figure 8.7). Tectonic settings where mylonites occur include the Moine

> **CONNECTION LINK**
>
> To learn more about how to differentiate between an intrusive and extrusive igneous body in an outcrop, and about its importance in working out the correct sequence of events, go to pages 420–428 and Figure 17.4 in Chapter 17.

Lava flow

Baked zone

Ash layer

James S. Moore

⏸ **Figure 8.6 Contact Metamorphism from a Lava Flow** A highly weathered basaltic lava flow near Susanville, California, has altered an underlying rhyolitic volcanic ash by contact metamorphism. The red zone below the lava flow has been baked by the heat of the lava when it flowed over the ash layer. The lava flow displays spheroidal weathering, a type of weathering common in fractured rocks (see Chapter 6).

» **Critical Thinking Question** What criteria would you use to determine that what you see is indeed spheroidal weathering and not weathered pillow lavas?

> **CONNECTION LINK**

To learn more about the San Andreas fault, see the Geo-Insight on pages 214 and 215 in Chapter 9.

Thrust Zone in northwest Scotland, the Adirondack Highlands in New York, portions of the San Andreas fault in California, and parts of the northeastern United States.

Figure 8.7 Mylonite An outcrop of mylonite from the Adirondack Highlands, New York. Mylonites result from dynamic metamorphism, where rocks are subjected to high levels of differential pressure. Note the thin laminations that are characteristic of many mylonites.

Regional Metamorphism

Most metamorphic rocks result from **regional metamorphism**, which occurs over a large area and is usually caused by tremendous temperatures, pressures, and deformation all occurring together within the deeper portions of the crust. Regional metamorphism is most obvious along convergent plate boundaries where rocks are intensely deformed and recrystallized during convergence and subduction (Figure 8.4). Within these metamorphic rocks there is usually a gradation of metamorphic intensity from areas that were subjected to the most intense pressures, highest temperatures, or both, to areas of lower pressures and temperatures. Such a gradation in metamorphism can be recognized by the metamorphic minerals that are present.

Regional metamorphism is not, however, confined only to convergent margins. It can also occur in areas where plates diverge, although usually at much shallower depths because of the high geothermal gradient associated with these areas (Figure 8.4).

Index Minerals and Metamorphic Grade

From field studies and laboratory experiments, certain minerals are known to form only within specific temperature and pressure ranges. Such minerals are known as **index minerals** because their presence allows geologists to recognize low-, intermediate-, and high-grade metamorphism (Figure 8.8).

Metamorphic grade is a term that generally characterizes the degree to which a rock has undergone metamorphic change (Figure 8.8). Although the boundaries between the different metamorphic grades are not sharp, the distinction is nonetheless useful for communicating in a general way the degree to which rocks have been metamorphosed. The presence of index minerals thus helps determine metamorphic grade. For example, when a clay-rich rock such as shale

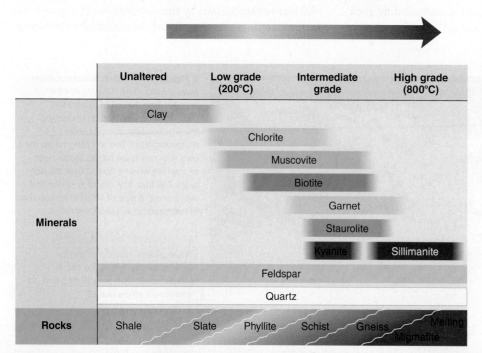

Figure 8.8 Metamorphic Grade Change in mineral assemblage and rock type with increasing metamorphism of shale. When a clay-rich rock such as shale is subjected to increasing metamorphism, new minerals form, as shown by the colored bars. The progressive appearance of certain minerals, known as index minerals, allows geologists to recognize low-, intermediate-, and high-grade metamorphism.

undergoes regional metamorphism, the mineral chlorite first begins to crystallize under relatively low temperatures of about 200°C. Its presence in these rocks thus indicates low-grade metamorphism. If temperatures and pressures continue to increase, new minerals form to replace chlorite, because they are more stable under those new conditions. Thus, there is a progression in the appearance of new minerals from chlorite, whose presence indicates low-grade metamorphism, to biotite and garnet, which are good index minerals for intermediate-grade metamorphism, to sillimanite, whose presence indicates high-grade metamorphism and temperatures exceeding 500°C (Figure 8.8).

Different rock compositions develop different sets of index minerals. For example, clay-rich rocks such as shale will develop the index minerals shown in Figure 8.8, whereas a sandy dolomite will produce a different set of index minerals as metamorphism progresses, because it has a different mineral composition than shale. Thus, a particular set of index minerals will form based on the original composition of the parent rock undergoing metamorphism.

Although such common minerals as mica, quartz, and feldspar can occur in both igneous and metamorphic rocks, other minerals such as andalusite, sillimanite, and kyanite generally occur only in metamorphic rocks derived from clay-rich rocks such as shale. Whereas these three minerals all have the same chemical formula (Al_2SiO_5), they differ in crystal structure and other physical properties because each forms under a different range of pressures and temperatures and is thus reflective of a different metamorphic grade. For this reason, they are useful index minerals for metamorphic rocks that formed from clay-rich rocks.

8.4 How Are Metamorphic Rocks Classified?

For purposes of classification, metamorphic rocks are commonly divided into two groups: those exhibiting a *foliated texture* (from the Latin *folium*, "leaf") and those with a *nonfoliated texture* (Table 8.1).

Foliated Metamorphic Rocks

Rocks subjected to heat and differential pressure during metamorphism typically have minerals arranged in a parallel fashion, giving them a **foliated texture** (❙ Figure 8.9). Low-grade metamorphic rocks, such as slate, have a finely foliated texture in which the mineral grains are so small that they cannot be distinguished without magnification. High-grade foliated rocks, such as gneiss, are coarse-grained, such that the individual grains can easily be seen with the unaided eye. Foliated metamorphic rocks can be arranged in order of increasingly coarse grain size and perfection of foliation.

Slate is a very fine-grained foliated metamorphic rock that commonly exhibits *slaty cleavage* (❙ Figure 8.10). It results from regional metamorphism of shale or, more rarely, volcanic ash. Because it can easily be split along cleavage planes into flat pieces, slate is an excellent rock for roofing and floor tiles, billiard and pool table tops, and blackboards. The different colors of slate are caused by minute amounts of graphite (black), iron oxide (red and purple), and chlorite (green).

❙ **Figure 8.9** Foliated Texture

Random arrangement of elongated minerals before pressure is applied to two sides

Elongated minerals arranged in a parallel manner as a result of pressure applied to two sides

Pressure

Pressure

Pressure

Pressure

a When rocks are subjected to differential pressure, the mineral grains are typically arranged in a parallel manner, producing a foliated texture.

Elongated minerals arranged in a parallel manner as a result of shear

b Photomicrograph of a metamorphic rock with a foliated texture showing the parallel arrangement of mineral grains.

Reed Wicander

TABLE 8.1	Classification of Common Metamorphic Rocks				
Texture	**Metamorphic Rock**	**Typical Mineral**	**Metamorphic Grade**	**Characteristics of Rocks**	**Parent Rock**
Foliated	Slate	Clays, micas, chlorite	Low	Fine-grained, splits easily into flat pieces	Mudrocks, volcanic ash
	Phyllite	Fine-grained quartz, micas, chlorite	Low to medium	Fine-grained, glossy or lustrous sheen	Mudrocks
	Schist	Micas, chlorite, quartz, talc, hornblende, garnet, staurolite, graphite	Low to high	Distinct foliation, minerals visible	Mudrocks, carbonates, mafic igneous rocks
	Gneiss	Quartz, feldspars, hornblende, micas	High	Segregated light and dark bands visible	Mudrocks, sandstones, felsic igneous rocks
	Amphibolite	Hornblende, plagioclase	Medium to high	Dark, weakly foliated	Mafic igneous rocks
	Migmatite	Quartz, feldspar, hornblende, micas	High	Streaks or lenses of granite intermixed with gneiss	Felsic igneous rocks mixed with sedimentary rocks
Nonfoliated	Marble	Calcite, dolomite	Low to high	Interlocking grains of calcite or dolomite, reacts with HCl	Limestone or dolostone
	Quartzite	Quartz	Medium to high	Interlocking quartz grains, hard, dense	Quartz sandstone
	Greenstone	Chlorite, epidote, hornblende	Lot to high	Fine-grained, green	Mafic igneous rocks
	Hornfels	Micas, garnets, andalusite, cordierite, quartz	Low to medium	Fine-grained, equidimensional grains, hard, dense	Mudrocks
	Anthracite	Carbon	High	Black, lustrous, subconcoidal fracture	Coal

❚ **Figure 8.10** Slate

Copyright and Photograph by Dr. Parvinder S. Sethi

a Hand specimen of red slate.

James S. Monroe

b Slate roof of Chalet Enzian, Switzerland.

Phyllite is similar in composition to slate but is coarser grained. The minerals, however, are still too small to be identified without magnification. Phyllite can be distinguished from slate by its glossy or lustrous sheen (❚ Figure 8.11) and represents an intermediate grain size between slate and schist.

Schist is most commonly produced by regional metamorphism. The type of schist formed depends on the intensity of metamorphism and the character of the original rock (❚ Figure 8.12). Metamorphism of many rock types can yield schist, but most schist appears to have formed from clay-rich sedimentary rocks (Table 8.1).

Figure 8.12 Schist

Figure 8.11 **Phyllite** Hand specimen of phyllite. Note the lustrous sheen as well as the bedding (upper left to lower right) at an angle to the cleavage of the specimen.

» Critical Thinking Question
How do you distinguish between bedding and cleavage in a metamorphic rock?

Copyright and Photograph by Dr. Parvinder S. Sethi

a Almandine garnet crystals in a mica schist.

Sue Monroe

b Hornblende–mica schist.

Figure 8.13 **Gneiss** Gneiss is characterized by segregated bands of light and dark minerals. This folded gneiss is exposed at Wawa, Ontario, Canada.

Reed Wicander

All schists contain more than 50% platy and elongated minerals, all of which are large enough to be clearly visible. Their mineral composition imparts a *schistosity* or *schistose foliation* to the rock that usually produces a wavy type of parting when split. Schistosity is common in low- to high-grade metamorphic environments, and each type of schist is known by its most conspicuous mineral or minerals, such as mica schist, chlorite schist, or garnet–mica schist (Figure 8.12).

Gneiss is a high-grade metamorphic rock that is streaked or has segregated bands of light and dark minerals. Gneisses are composed of granular minerals such as quartz, feldspar, or both, with lesser percentages of platy or elongated minerals such as micas or amphiboles (Figure 8.13). Quartz and feldspar are the principal minerals composing the light-colored bands of minerals, whereas biotite and hornblende comprise the dark mineral bands. Gneiss typically breaks in an irregular manner, much like coarsely crystalline nonfoliated rocks.

Most gneiss probably results from recrystallization of clay-rich sedimentary rocks during regional metamorphism (Table 8.1). Gneiss also can form from igneous rocks such as granite or older metamorphic rocks.

Another fairly common foliated metamorphic rock is *amphibolite*. A dark rock, it is composed mainly of hornblende and plagioclase. The alignment of the hornblende crystals produces a slightly foliated texture. Many amphibolites result from intermediate- to high-grade metamorphism of basalt and ferromagnesian-rich mafic rocks.

In some areas of regional metamorphism, exposures of "mixed rocks" called *migmatites*, having both igneous and high-grade metamorphic characteristics, are present (Figure 8.14). Migmatites are thought to result from the extremely high temperatures produced during metamorphism. However, part of the problem in determining the origin of migmatites is explaining how the granitic component formed. According to one model, the granitic magma formed in place by the partial melting of rock during intense metamorphism. Such an origin is possible provided that the host rocks contained quartz and feldspars and that water was present.

Others argue that the characteristic layering or wavy appearance of migmatites arises by the redistribution of minerals during recrystallization in the solid state—that is, through purely metamorphic processes.

Figure 8.14 Migmatite A migmatite boulder in Rocky Mountain National Park, near Estes Park, Colorado. Migmatites consist of high-grade metamorphic rock intermixed with streaks or lenses of granite.

Nonfoliated Metamorphic Rocks

In some metamorphic rocks, the mineral grains do not show a discernible preferred orientation. Instead, these rocks consist of a mosaic of roughly equidimensional minerals and are characterized as having a **nonfoliated texture** (Figure 8.15). Most nonfoliated metamorphic rocks result from contact or regional metamorphism of rocks with no platy or elongate minerals. Frequently, the only indication that a granular rock has been metamorphosed is the large grain size resulting from recrystallization.

Figure 8.15 Nonfoliated Texture Nonfoliated textures are characterized by a mosaic of roughly equidimensional minerals, as in this photomicrograph of marble.

Nonfoliated metamorphic rocks are generally of two types: those composed of mainly one mineral—for example, marble or quartzite—and those in which the different mineral grains are too small to be seen without magnification, such as greenstone and hornfels.

Marble is a well-known metamorphic rock composed predominantly of calcite or dolomite; its grain size ranges from fine to coarsely granular. Marble results from either contact or regional metamorphism of limestones or dolostones (Figure 8.16 and Table 8.1). Pure marble is snowy white or bluish; however, many color varieties exist because of the presence of mineral impurities in the original sedimentary rock. The softness of marble, its uniform texture, and its varying colors have made it the favorite rock of builders and sculptors throughout history.

Quartzite is a hard, compact rock typically formed from quartz sandstone under intermediate- to high-grade metamorphic conditions during contact or regional metamorphism (Figure 8.17). Because recrystallization is so complete, metamorphic quartzite is of uniform strength and therefore usually breaks across the component quartz grains rather than around them when it is struck. Pure quartzite is white; however, iron and other impurities commonly impart a pinkish-red or other color to it. Quartzite is commonly used as foundation material for road and railway beds.

The name *greenstone* is applied to any compact, dark-green, altered, mafic igneous rock that formed under low- to high-grade metamorphic conditions. The green color results from the presence of chlorite, epidote, and hornblende.

Hornfels is a common, fine-grained, nonfoliated metamorphic rock resulting from contact metamorphism and composed of various equidimensional mineral grains. The

Copyright and Photograph by Dr. Parvinder S. Sethi

Metamorphism

Limestone

Marble

Copyright and Photograph by Dr. Parvinder S. Sethi

▮ **Figure 8.16 Marble** Metamorphism of the sedimentary rock limestone or dolostone yields marble.

Copyright and Photograph by Dr. Parvinder S. Sethi

Metamorphism

Quartz sandstone

Quartzite

Copyright and Photograph by Dr. Parvinder S. Sethi

▮ **Figure 8.17 Quartzite** Metamorphism of the sedimentary rock quartz sandstone yields quartzite.

composition of hornfels depends directly on the composition of the original rock, and many compositional varieties are known. The majority of hornfels, however, is apparently derived from contact metamorphism of clay-rich sedimentary rocks or impure dolostones.

Anthracite is a black, lustrous, hard coal that contains a high percentage of fixed carbon and a low percentage of volatile matter and is highly valued by people who burn coal for heating and power. Anthracite usually forms from the metamorphism of lower-grade coals by heat and pressure, and many geologists consider it to be a metamorphic rock.

> **CONNECTION LINK**

To learn more about the formation of coal and its different varieties, see page 165 in Chapter 7. More can be learned about the distribution of coal in Chapter 20 on pages 534 and 535 and Figure 20.29.

8.5 Metamorphic Zones and Facies

While mapping the 440- to 400-million-year-old Dalradian schists of Scotland in the late 1800s, George Barrow and other British geologists made the first systematic study of metamorphic zones. Here, clay-rich sedimentary rocks have been subjected to regional metamorphism, and the resulting metamorphic rocks can be divided into different zones based on the presence of distinctive silicate mineral assemblages. These mineral assemblages, each recognized by the presence of one or more index minerals, indicate different degrees of metamorphism. The index minerals that Barrow and his associates chose to represent increasing metamorphic intensity were chlorite, biotite, garnet, staurolite, kyanite, and sillimanite (Figure 8.8), which we now know all result from the recrystallization of clay-rich sedimentary rocks. As previously mentioned, other mineral assemblages and index minerals are produced from rocks with different original compositions.

The successive appearance of metamorphic index minerals indicates gradually increasing or decreasing intensity of metamorphism. Going from lower- to higher-grade metamorphic zones, the first appearance of a particular index mineral indicates the location of the minimum temperature and pressure conditions needed for the formation of that mineral. When the locations of the first appearances of that index mineral are connected on a map, the result is a line of equal metamorphic intensity or an *isograd*. The region between two adjacent isograds makes up a single **metamorphic zone**—a belt of rocks displaying the same general degree of metamorphism. By mapping adjoining metamorphic zones, geologists can reconstruct metamorphic conditions throughout an entire region (▮ Figure 8.18).

Not long after Barrow and his coworkers completed their work, geologists in Norway and Finland came up with a different method of mapping metamorphism that was even more useful than the metamorphic zone approach. A **metamorphic facies** is defined as a group of metamorphic rocks characterized by particular mineral assemblages formed under broadly similar temperature and pressure conditions (❚ Figure 8.19). Each facies is named after its most characteristic rock or mineral. For example, the green metamorphic mineral chlorite, which forms under relatively low temperatures and pressures, yields rocks belonging to the *greenschist facies*. Under increasingly higher temperatures and pressures, mineral assemblages indicative of the *amphibolite* and *granulite facies* develop.

Although usually applied to areas where the original rocks were clay-rich, the concept of metamorphic facies can also be used with modification in other situations. It cannot, however, be used in areas where the original rocks were pure quartz sandstone or pure limestone or dolostone. Such rocks, regardless of the imposed temperature and pressure conditions, will yield only quartzite and marble, respectively. In such cases, all one can say is that "metamorphism has happened."

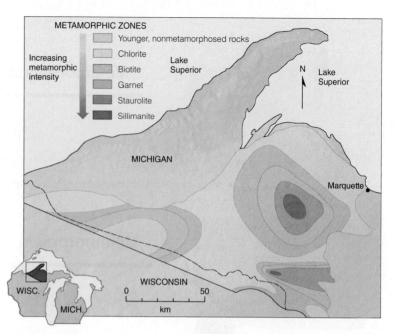

❚ **Figure 8.18** Metamorphic Zones in the Upper Peninsula of Michigan The zones in this region are based on the presence of distinctive silicate mineral assemblages resulting from the metamorphism of sedimentary rocks during an interval of mountain building and minor granitic intrusion during the Proterozoic Eon, about 1.5 billion years ago. The lines separating the different metamorphic zones are isograds.

8.6 Plate Tectonics and Metamorphism

Although metamorphism is associated with all three types of plate boundaries (see Figure 1.14), it is most common along convergent plate margins. Metamorphic rocks form at convergent plate boundaries because temperature and pressure increase as a result of plate collisions.

❚ Figure 8.20 illustrates the various metamorphic facies produced along a typical oceanic–continental convergent plate boundary. When an oceanic plate collides with a continental plate, tremendous pressure is generated as the oceanic plate is subducted. Because rock is a poor heat conductor, the cold, descending oceanic plate heats slowly, and metamorphism is caused mostly by increasing pressure with depth. Metamorphism in such an environment produces rocks typical of the *blueschist facies* (low temperature, high pressure). Geologists use the presence of blueschist facies rocks as evidence of ancient subduction zones.

As subduction along the oceanic–continental convergent plate boundary continues, both temperature and pressure increase with depth and yield high-grade metamorphic rocks.

Eventually, the descending plate begins to melt and generate magma that moves upward. This rising magma may alter the surrounding rock by contact metamorphism, producing migmatites in the deeper portions of the crust and hornfels at shallower depths. High temperatures and low to medium pressures characterize such an environment.

Although metamorphism is most common along convergent plate margins, many divergent plate boundaries are characterized by contact metamorphism. Rising magma at mid-oceanic ridges heats the adjacent rocks, producing contact metamorphic minerals and textures. In

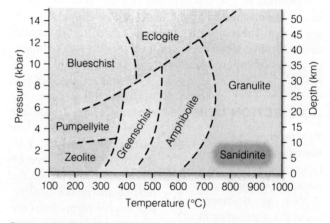

❚ **Figure 8.19** Metamorphic Facies and Their Associated Temperature–Pressure Conditions A temperature–pressure diagram showing under what conditions various metamorphic facies occur. A metamorphic facies is characterized by a particular mineral assemblage that formed under the same broad temperature–pressure conditions. Each facies is named after its most characteristic rock or mineral.

》》 Critical Thinking Question
Go to a point that is represented by 200°C and 2 kbar of pressure. What metamorphic facies is represented by those conditions? If the pressure is raised to 12 kbar, what facies is represented by the new conditions? What change in depth of burial is required to effect the pressure change from 2 to 12 kbar?

addition, fluids emanating from the rising magma—and its reaction with seawater—very commonly produce metal-bearing hydrothermal solutions that may precipitate minerals of economic value, such as the copper ores of Cyprus (see Chapter 2).

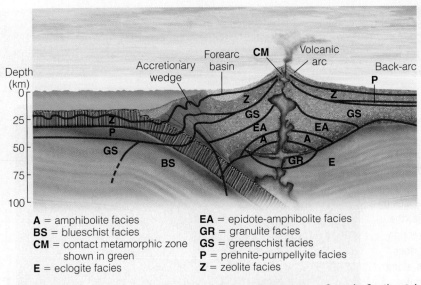

A = amphibolite facies
BS = blueschist facies
CM = contact metamorphic zone
       shown in green
E = eclogite facies

EA = epidote-amphibolite facies
GR = granulite facies
GS = greenschist facies
P = prehnite-pumpellyite facies
Z = zeolite facies

Figure 8.20 Relationship of Facies to Major Tectonic Features at an Oceanic–Continental Convergent Plate Boundary

8.7 Metamorphism and Natural Resources

Many metamorphic minerals and rocks are valuable natural resources. Although these resources include various types of ore deposits, the two most familiar and widely used metamorphic rocks are marble and slate, which have been used for centuries in a variety of ways (Figure 8.21).

Many ore deposits result from contact metamorphism during which hot, ion-rich fluids migrate from igneous intrusions into the surrounding rock, thereby producing rich ore deposits. The most common sulfide ore minerals associated with contact metamorphism are bornite, chalcopyrite, galena (lead), pyrite, and sphalerite (zinc); two common iron oxide ore minerals are hematite and magnetite. Tin and tungsten are also important ores associated with contact metamorphism (Table 8.2).

Other economically important metamorphic minerals include asbestos, used for insulation and fire-proofing in buildings and building materials (see Geo-Focus on page 183), talc for talcum powder, graphite for pencils and dry lubricants, and garnets and corundum, which are used as abrasives or gemstones, depending on their quality. In addition, andalusite, kyanite, and sillimanite, which, as we mentioned earlier, all have the same chemical composition but differ in crystal structure, are used in manufacturing high-temperature porcelains and temperature-resistant materials for products such as sparkplugs and furnace linings.

© Peter Hulme/Ecoscene/CORBIS

Figure 8.21 Slate Quarry in Wales Slate, which has a variety of uses, is the result of low-grade regional metamorphism of shale. These high-quality slates were formed by a mountain-building episode that took place approximately 400 to 440 million years ago in the present-day countries of Ireland, Scotland, Wales, and Norway.

TABLE 8.2	The Main Ore Deposits Resulting from Contact Matamorphism		
Ore Deposit	**Major Mineral**	**Formula**	**Use**
Copper	Bornite, Chalcopyrite	Cu_5FeS_4, $CuFeS_2$	Important sources of copper, which is used in manufacturing, transportation, communications, and construction
Iron	Hematite, Magnetite	Fe_2O_3, Fe_3O_4	Major sources of iron for manufacture of steel, which is used in nearly every form of construction, manufacturing, transportation, and communications
Lead	Galena	PbS	Chief source of lead, which is used in batteries, pipes, solder, and elsewhere resistance to corrosion is required
Tin	Cassiterite	SnO_2	Principal source of tin, which is used for tin plating, solder, alloys, and chemicals
Tungsten	Scheelite, Wolframite	$CaWO_4$, $(FeMn)WO_4$	Chief sources of tungsten, which is used in hardening metals and manufacturing carbides
Zinc	Sphalerite	(Zn,Fe)S	Major source of zinc, which is used in batteries and in galvanizing iron and making brass

GEO-RECAP

❯❯ Geo-Image

▌Figure 1 Foliated metamorphic rocks are characterized by having their minerals arranged in a parallel fashion, such that they can be split along foliation planes (Figure 8.9). In this photo of 500 million to 1 billion year old schists and gneisses of the Blue Ridge Belt in southwestern Virginia, the foliation planes are dipping in the same direction as the slope of the hillside. What problems does this present in terms of potential landslides along the roadway? Hint: Note the several large rocks along the base of the hillside.

Copyright and Photograph by Dr. Parvinder S. Sethi

Chapter Summary

■ Metamorphic rocks result from the transformation of other rocks, usually beneath Earth's surface, as a consequence of one or a combination of three agents: heat, pressure, and fluid activity.

■ Heat for metamorphism comes from intrusive magmas, extrusive lava flows, or deep burial. Pressure is either lithostatic (uniformly applied stress) or differential (stress unequally applied from different directions). Fluids trapped in sedimentary rocks or emanating from intruding magmas can enhance chemical changes and the formation of new minerals.

■ The three major types of metamorphism are contact, dynamic, and regional.

■ Contact metamorphism results when a magma or lava alters the surrounding country rock.

■ Dynamic metamorphism is associated with fault zones where rocks are subjected to high differential pressure.

■ Most metamorphic rocks result from regional metamorphism, which occurs over a large area and is usually caused by tremendous temperatures, pressures, and deformation within the deeper portions of the crust.

■ Metamorphic grade generally characterizes the degree to which a rock has undergone metamorphic change.

■ Index minerals—minerals that form only within specific temperature and pressure ranges—allow geologists to recognize low-, intermediate-, and high-grade metamorphism.

■ Metamorphic rocks are primarily classified according to their texture. In a foliated texture, platy and elongate minerals have a preferred orientation. A nonfoliated texture does not exhibit any discernible preferred orientation of the mineral grains.

■ Foliated metamorphic rocks can be arranged in order of increasing grain size, perfection of their foliation, or both. Slate is fine-grained, followed by (in increasingly larger grain size) phyllite and schist; gneiss displays segregated bands of minerals. Amphibolite is another fairly common foliated metamorphic rock. Migmatites have both igneous and high-grade metamorphic characteristics.

■ Marble, quartzite, greenstone, hornfels, and anthracite are common nonfoliated metamorphic rocks.

■ Metamorphic zones are based on index minerals and are areas of rock that all have similar grades of metamorphism; that is, they have all experienced the same intensity of metamorphism.

■ A metamorphic facies is a group of metamorphic rocks whose minerals all formed under a particular range of temperatures and pressures. Each facies is named after its most characteristic rock or mineral.

■ Metamorphism occurs along all three types of plate boundaries, but is most common at convergent plate margins.

■ Many metamorphic rocks and minerals, such as marble, slate, graphite, talc, and asbestos, are valuable natural resources. In addition, many ore deposits are the result of metamorphism and include copper, tin, tungsten, lead, iron, and zinc.

Important Terms

aureole (p. 185)
contact (thermal) metamorphism (p. 185)

differential pressure (p. 185)
dynamic metamorphism (p. 187)
fluid activity (p. 185)

foliated texture (p. 189)
heat (p. 184)
index mineral (p. 188)

lithostatic pressure (p. 184)
metamorphic facies (p. 193)
metamorphic grade (p. 188)

metamorphic rock (p. 182)
metamorphic zone (p. 193)
metamorphism (p. 182)

nonfoliated texture (p. 192)
regional metamorphism
(p. 188)

Review Questions

1. Which of the following metamorphic rocks displays a nonfoliated texture?
   a. ___ phyllite.
   b. ___ marble.
   c. ___ gneiss.
   d. ___ amphibolite.
   e. ___ schist.

2. Along what type of plate boundary is metamorphism most common?
   a. ___ divergent.
   b. ___ transform.
   c. ___ aseismic.
   d. ___ convergent.
   e. ___ lithospheric.

3. Which is the correct metamorphic sequence of increasingly coarser grain size?
   a. ___ gneiss—> schist—> phyllite—> slate.
   b. ___ phyllite—> slate—> schist—> gneiss.
   c. ___ schist—> slate—> gneiss—> phyllite.
   d. ___ slate—> phyllite—> schist—> gneiss.
   e. ___ slate—> schist—> phyllite—> gneiss.

4. Which of the following are considered the three principal agents of metamorphism?
   a. ___ heat, pressure, fluid activity.
   b. ___ heat, pressure, gravity.
   c. ___ gravity, fluid activity, pressure.
   d. ___ gravity, heat, fluid activity.
   e. ___ none of these.

5. To which metamorphic facies do metamorphic rocks formed under the lowest-temperature and lowest-pressure conditions belong?
   a. ___ blueschist.
   b. ___ pumpellyite.
   c. ___ zeolite.
   d. ___ greenschist.
   e. ___ granulite.

6. What type of metamorphism is associated with fault zones in the shallow crust where rocks are subjected to concentrated high levels of differential pressure?
   a. ___ thermodynamic.
   b. ___ lithostatic.
   c. ___ dynamic.
   d. ___ contact.
   e. ___ regional.

7. From which of the following rock groups can metamorphic rocks form?
   a. ___ plutonic.
   b. ___ sedimentary.
   c. ___ metamorphic.
   d. ___ volcanic.
   e. ___ all of these.

8. The metamorphic rock formed from coal is called
   a. ___ anthracite.
   b. ___ marble.
   c. ___ hornfels.
   d. ___ greenstone.
   e. ___ amphibolite.

9. What type of pressure results from stress that is unequally applied, such as along a convergent plate boundary?
   a. ___ directional.
   b. ___ differential.
   c. ___ lithostatic.
   d. ___ monolithic.
   e. ___ unilateral.

10. Concentric zones surrounding an igneous intrusion and characterized by distinctive mineral assemblages are
    a. ___ thermodynamic rings.
    b. ___ hydrothermal regions.
    c. ___ aureoles.
    d. ___ regional facies.
    e. ___ metamorphic zones.

11. Why is it important that the average citizen know something about metamorphism, metamorphic rocks, and how they form?

12. Describe the two types of metamorphic texture, and discuss how they are produced.

13. Why is metamorphism more widespread along convergent plate boundaries than along any other type of plate boundary?

14. What specific features about foliated metamorphic rocks would make them unsuitable as foundations for dams? Are there any metamorphic rocks that would make good foundations? Explain your answer.

15. Do you think metamorphic rocks exist on other planets in our solar system, and if so, why?

16. Discuss the role each of the three major agents of metamorphism plays in transforming any rock into a metamorphic rock.

17. How can aureoles be used to determine the effects of metamorphism?

18. Where does contact metamorphism occur, and what types of changes does it produce?

19. If plate tectonic movement did not exist, could there be metamorphism?

20. How do metamorphic rocks record the influence of differential pressure in their structures and mineral textures?

People walking over the rubble of destroyed buildings in downtown Port-au-Prince, Haiti, following the 7.0-magnitude earthquake that struck the island nation on January 12, 2010, and left more than 222,500 dead.

CHAPTER 9

Nick Kozak/Sipa Press/haiti_kozak 017/10011191829 (Sipa via AP images)

Earthquakes and Earth's Interior

OUTLINE

9.1 Introduction

9.2 Elastic Rebound Theory

9.3 Seismology

9.4 Where Do Earthquakes Occur, and How Often?

9.5 Seismic Waves

9.6 Locating an Earthquake

9.7 Measuring the Strength of an Earthquake

9.8 What Are the Destructive Effects of Earthquakes?

GEO-INSIGHT: The San Andreas Fault

9.9 Earthquake Prediction

9.10 Earthquake Control

GEO-FOCUS: Paleoseismology

9.11 What Is Earth's Interior Like?

9.12 The Core

9.13 Earth's Mantle

9.14 Seismic Tomography

9.15 Earth's Internal Heat

9.16 Earth's Crust

GEO-RECAP

HAVE YOU EVER WONDERED?

- What causes earthquakes?
- Why earthquakes occur where they do?
- How an earthquake's epicenter is located?
- How the strength of an earthquake is measured?
- Whether earthquakes can be predicted or even controlled?
- What you can do to protect yourself during an earthquake?
- How geologists know what Earth's internal structure is like?

9.1 Introduction

In the afternoon of January 12, 2010, a magnitude 7.0 earthquake struck the island nation of Haiti. According to official estimates, 222,570 people died, at least 300,000 were injured, and more than 285,000 residences and businesses were destroyed or severely damaged. Widespread devastation occurred in the capital of Port-au-Prince and elsewhere throughout the region, exacerbated by an almost total collapse of the vital infrastructure needed to respond to such a disaster, including medical, transportation, and communication systems. This was a disaster of epic proportions, and one whose repercussions are being felt more than a year later. Yet it was not the first nor will it be the last major devastating earthquake in this region or other parts of the world. Most recently, a 9.0-magnitude earthquake and tsunami struck Japan, causing more than 13,000 deaths and tremendous destruction (see Chapter 2 opening photo, page 26). The cause of the Haiti earthquake was movement along a strike-slip fault that is part of the boundary separating the Caribbean plate from the North American plate, whereas the Japanese earthquake resulted from movement along an oceanic-oceanic convergent plate boundary off the eastern coast of Japan (see Chapter 2).

Earthquakes, along with volcanic eruptions, are manifestations of Earth's dynamic and active makeup. As one of nature's most frightening and destructive phenomena, earthquakes have always aroused feelings of fear and have been the subject of myths and legends. An earthquake is so frightening because when it begins, there is no way to tell how long it will last or how violent it will be. Approximately 13 million people have died in earthquakes during the past 4,000 years, with about 3 million of these deaths occurring during the last century (Table 9.1).

Geologists define an **earthquake** as the shaking or trembling of the ground caused by the sudden release of energy, usually as a result of faulting, which involves the displacement of rocks along fractures (we discuss the different types of faults in Chapter 10). After an earthquake, continuing adjustments along a fault may generate a series of earthquakes known as *aftershocks*. Most aftershocks are smaller than the main shock, but they can still cause considerable damage to already weakened structures.

Although the geologic definition of an earthquake is accurate, it is not nearly as imaginative or colorful as the explanations many people had in the past. Many cultures attributed the cause of earthquakes to movements of some kind of animal on which Earth rested. A legend from Mexico holds that earthquakes occur when the devil, El Diablo, rips open the crust so that he and his friends can reach the surface.

If earthquakes are not the result of animal movement or the devil ripping open the crust, what does cause earthquakes? Geologists know that most earthquakes result from energy released along plate boundaries, and as such, earthquakes are a manifestation of Earth's dynamic nature and the fact that Earth is an internally active planet.

Why should you study earthquakes? The obvious reason is that they are destructive and cause many deaths and injuries to

TABLE 9.1	Some Significant Earthquakes		
Year	**Location**	**Magnitude (estimated before 1935)**	**Deaths (estimated**
1556	China (Shanxi Province)	8	1,000,000
1755	Portugal (Lisbon)	8.6	70,000
1906	U.S.A. (San Francisco, California)	8.3	3,000
1923	Japan (Tokyo)	8.3	143,000
1960	Chile	9.5	5,700
1964	U.S.A (Anchorage, Alaska)	8.6	131
1976	China (Tangshan)	8	242,000
1985	Mexico (Mexico City)	8.1	9,500
1988	Armenia	6.9	25,000
1990	Iran	7.3	50,000
1993	India	6.4	30,000
1995	Japan (Kobe)	7.2	6,000+
1999	Turkey	7.4	17,000
2001	India	7.9	14,000+
2003	Iran	6.6	43,000
2004	Indonesia	9	>220,000
2005	Pakistan	7.6	>86,000
2006	Indonesia	6.3	>6,200
2008	China (Sichuan Province)	7.9	>69,000
2010	Haiti	7.0	>220,000
2011	New Zealand	6.3	181
2011	Japan	9.0	>13,000

the people living in earthquake-prone areas. Earthquakes also affect the economies of many countries in terms of cleanup costs, lost jobs, and lost business revenues. From a purely personal standpoint, you someday may be caught in an earthquake. Even if you don't live in an area subject to earthquakes, you may travel to places where the threat of earthquakes exists, and you should know what to do if you experience one. Such knowledge may help you avoid serious injury or even death.

9.2 Elastic Rebound Theory

Based on studies conducted after the 1906 San Francisco earthquake, H. F. Reid of The Johns Hopkins University proposed the **elastic rebound theory** to explain how energy is released during earthquakes. Reid studied three sets of measurements taken across a portion of the San Andreas fault that had broken during the 1906 earthquake. The measurements revealed that points on opposite sides of the fault had moved 3.2 m during the 50-year period prior to breakage in 1906, with the west side moving northward (■ Figure 9.1).

According to Reid, rocks on opposite sides of the San Andreas fault had been storing energy and bending slightly for at least 50 years before the 1906 earthquake. Any straight line, such as a fence or road that crossed the San Andreas fault, was gradually bent because rocks on one side of the fault moved relative to rocks on the other side (Figure 9.1). Eventually, the strength of the rocks was exceeded and then fractured. The rocks on opposite sides of the fault rebounded or "snapped back" to their former undeformed shape, and the energy stored was released as earthquake waves radiating outward from the break (Figure 9.1). Additional field and laboratory studies conducted by Reid and others have confirmed that elastic rebound is the mechanism by which energy is released during earthquakes.

A useful analogy is that of bending a long, straight stick over your knee. As the stick bends, it deforms and eventually reaches the point at which it breaks. When this happens, the two pieces of the original stick snap back into their original straight position. Likewise, rocks subjected to intense forces bend until they break and then return to their original position, releasing energy in the process.

■ **Figure 9.1** The Elastic Rebound Theory

Original position

Deformation

a According to the elastic rebound theory, rocks experiencing deformation store energy and bend. When the internal strength of the rocks is exceeded, they rupture, releasing their accumulated energy, and "snap back" or rebound to their former undeformed shape. This sudden release of energy is what causes an earthquake.

Rupture and release of energy

Rocks rebound to original undeformed shape

b During the 1906 San Francisco earthquake, this fence in Marin County was displaced by almost 5 m. Whereas many people would see a broken fence, a geologist sees that the fence has moved or been displaced and would look for evidence of a fault. A geologist would also notice that the ground has been displaced toward the right side, relative to his or her view. Regardless of what side of the fault you stand on, you must look to the right to see the other part of the fence. Try it!

9.3 Seismology

Seismology, the study of earthquakes, emerged as a true science during the 1880s with the development of **seismographs**, instruments that detect, record, and measure the vibrations produced by an earthquake (Figure 9.2). The record made by a seismograph is called a *seismogram*. Modern seismographs have electronic sensors and record movements precisely using computers rather than simply relying on the drum strip charts commonly used on older seismographs.

When an earthquake occurs, energy in the form of *seismic waves* radiates out from the point of release (Figure 9.3a). These waves are somewhat analogous to the ripples that move out concentrically from the point where a stone is thrown into a pond. Unlike waves on a pond, however, seismic waves move outward in all directions from their source.

Earthquakes take place because rocks are capable of storing energy; however, their strength is limited, so if enough force is present, they rupture and thus release their stored energy. In other words, most earthquakes result when movement occurs along faults, most of which are related, at least indirectly, to plate movements. Once rupturing begins, it moves along the fault at several kilometers per second for as long as conditions for failure exist. The longer the fault along which movement occurs, the more time it takes for the stored energy to be released, and therefore the longer the ground will shake. During some very large earthquakes, the ground might shake for 3 minutes, a seemingly brief time, but interminable if you are experiencing the earthquake firsthand!

The Focus and Epicenter of an Earthquake

The location within Earth's lithosphere where rupturing begins—that is, the point at which energy is first released—is an earthquake's **focus**, or *hypocenter*. What we usually hear in news reports, however, is the location of the **epicenter**, the point on Earth's surface directly above the focus (Figure 9.3a).

Seismologists recognize three categories of earthquakes based on focal depth. *Shallow-focus* earthquakes have focal depths of less than 70 km from the surface, whereas those with foci between 70 and 300 km are *intermediate-focus*, and the foci of those characterized as *deep-focus* are more than 300 km deep. However, earthquakes are not evenly distributed among these three categories. Approximately 90% of all earthquake foci are at depths of less than 100 km, whereas only about 3% of all earthquakes are deep. Shallow-focus earthquakes are, with few exceptions, the most destructive, because the energy they release has little time to dissipate before reaching the surface.

Figure 9.2 Seismographs

a Seismographs record ground motion during an earthquake. The record produced is a seismogram. This seismograph records earthquakes on a strip of paper attached to a rotating drum.

Reed Wicander

b A horizontal-motion seismograph. Because of its inertia, the suspended mass that contains the marker remains stationary, while the rest of the structure moves along with the ground during an earthquake. As long as the length of the arm is not parallel to the direction of ground movement, the marker will record the earthquake waves on the rotating drum. This seismograph would record waves from west or east, but to record waves from the north or south, another seismograph at right angles to this one is needed.

c A vertical-motion seismograph. This seismograph operates on the same principle as the horizontal-motion instrument and records vertical ground movement.

Figure 9.3 The Focus and Epicenter of an Earthquake

Epicenter

Focus

Wave ray

Wavefront   Fault

a The focus of an earthquake is the location where the rupture begins and energy is released. The place on the surface vertically above the focus is the epicenter. Seismic wavefronts move out in all directions from their source, the focus of an earthquake.

North America

Haiti

South America

b The epicenter of the magnitude 7.0 January 12, 2010, earthquake that devastated Haiti was located approximately 25 km west-southwest of Port-au-Prince, and had a focal depth of 13 km.

≫ Critical Thinking Question
Why isn't the epicenter located where the fault emerges at Earth's surface?

> **CONNECTION LINK**

To review the three different types of plate boundaries, and their relationship to surface features, see Figure 2.7, page 33, and pages 40, 41, and 44–47 in Chapter 2.

A definite relationship exists between earthquake foci and plate boundaries. Earthquakes generated along divergent or transform plate boundaries are invariably shallow-focus, whereas many shallow-focus earthquakes and nearly all intermediate- and deep-focus earthquakes occur along convergent margins (Figure 9.4). Furthermore, a pattern emerges when the focal depths of earthquakes near island arcs and their adjacent ocean trenches are plotted. Notice in Figure 9.5 that the focal depth increases beneath the Tonga Trench in a narrow, well-defined zone that dips approximately 45 degrees. Dipping seismic zones, called *Benioff* or *Benioff–Wadati zones*, are common to convergent plate boundaries where one plate is subducted beneath another. Such dipping seismic zones indicate the angle of plate descent along a convergent plate boundary.

9.4 Where Do Earthquakes Occur, and How Often?

No place on Earth is immune to earthquakes, but almost 95% take place in seismic belts corresponding to plate boundaries where plates converge, diverge, and slide past each other. The relationship between plate margins and the distribution of earthquakes is readily apparent when the locations of earthquake epicenters are superimposed on a map showing the boundaries of Earth's plates (Figure 9.4).

The majority of all earthquakes (approximately 80%) occur in the *circum-Pacific belt*, a zone of seismic activity nearly encircling the Pacific Ocean basin. Most of these earthquakes result from convergence along plate margins, as in the case of the 1995 Kobe, Japan, earthquake (Figure 9.6a). The earthquakes along the North American Pacific Coast, especially in California, are also in this belt, but here plates slide past one another rather than converge. The October 17, 1989, Loma Prieta earthquake in the San Francisco area and the January 17, 1994,

• Deep-focus earthquake • Intermediate-focus earthquake • Shallow-focus earthquake

▌**Figure 9.4** **Earthquake Epicenters and Plate Boundaries** This map of earthquake epicenters shows that most earthquakes occur within seismic zones that correspond closely to plate boundaries. Approximately 80% of earthquakes occur within the circum-Pacific belt, 15% within the Mediterranean–Asiatic belt, and the remaining 5% within plate interiors and along oceanic spreading ridges. The dots represent earthquake epicenters and are divided into shallow-, intermediate-, and deep-focus earthquakes. Along with shallow-focus earthquakes, nearly all intermediate- and deep-focus earthquakes occur along convergent plate boundaries.

»Critical Thinking Question
Why are nearly all intermediate- and deep-focus earthquakes associated with convergent plate boundaries?

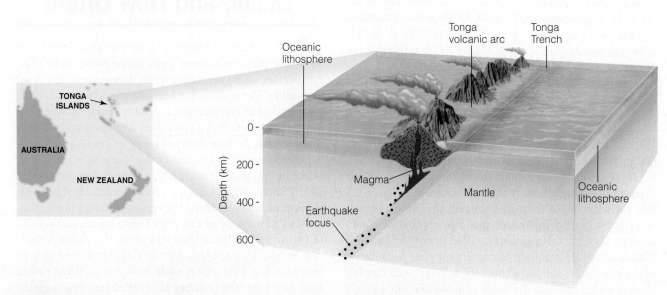

▌**Figure 9.5** **Benioff Zones** Focal depth increases in a well-defined zone that dips approximately 45 degrees beneath the Tonga volcanic arc in the South Pacific. Dipping seismic zones are called *Benioff* or *Benioff-Wadati* zones.

▌Figure 9.6 Earthquake Damage in the Circum-Pacific Belt

a Some of the damage in Kobe, Japan, caused by the January 1995 earthquake in which more than 5,000 people died.

b One of several elevated freeway spans that collapsed during the January 1994 Northridge, California, earthquake in which 61 people were killed.

Northridge earthquake occurred along this plate boundary (▌Figure 9.6b).

The second major seismic belt, accounting for 15% of all earthquakes, is the *Mediterranean–Asiatic belt*. This belt extends westward from Indonesia through the Himalayas, across Iran and Turkey, and westward through the Mediterranean region of Europe. The devastating 2005 earthquake in Pakistan that killed more than 86,000 people is a recent example of the destructive earthquakes that strike this region (Table 9.1).

The remaining 5% of earthquakes occur mostly in the interiors of plates and along oceanic spreading-ridge systems. Most of these earthquakes are not strong, although there have been several major intraplate earthquakes of note. For example, the 1811 and 1812 earthquakes near New Madrid, Missouri, killed approximately 20 people and nearly destroyed the town. So strong were these earthquakes that they were felt from the Rocky Mountains to the Atlantic Ocean and from the Canadian border to the Gulf of Mexico. Within the immediate area, numerous buildings were destroyed and forests were flattened. The land sank several meters in some areas, causing flooding, and reportedly the Mississippi River reversed its flow during the shaking and changed its course slightly.

The cause of intraplate earthquakes is not well understood, but geologists think that they arise from localized stresses caused by the compression that most plates experience along their margins. A useful analogy is moving a house. Regardless of how careful the movers are, moving something so large without its internal parts shifting slightly is impossible. Similarly, plates are not likely to move without some internal stresses that occasionally cause earthquakes.

More than 900,000 earthquakes are recorded annually by the worldwide network of seismograph stations. Many of these are too small to be felt, but are nonetheless recorded. These small earthquakes result from the energy released as continual adjustments take place between the various plates. However, on average, more than 31,000 earthquakes per year are strong enough to be felt and can cause various amounts of damage, depending on how strong they are and where they occur.

9.5 Seismic Waves

Many people have experienced an earthquake, but most are probably unaware that the shaking they feel and the damage to structures are caused by the arrival of *seismic waves*, a general term encompassing all waves generated by an earthquake. When movement on a fault takes place, energy is released in the form of two kinds of seismic waves that radiate outward in all directions from an earthquake's focus. *Body waves*, so called because they travel through the solid body of Earth, are somewhat like sound waves, and *surface waves*, which travel along the ground surface, are analogous to undulations or waves on water surfaces.

Body Waves

An earthquake generates two types of body waves: P-waves and S-waves (▮ Figure 9.7). **P-waves**, or *primary waves* are the fastest seismic waves and can travel through solids, liquids, and gases. P-waves are compressional, or push-pull, waves and are similar to sound waves in that they move material forward and backward along a line in the same direction that the waves themselves are moving (▮ Figure 9.7b). Thus, the material through which a P-wave travels is expanded and compressed as the wave moves through it and returns to its original size and shape after the wave passes by.

S-waves, or *secondary waves* are somewhat slower than P-waves and can travel only through solids. S-waves are *shear waves* because they move the material perpendicular to the direction of travel, thereby producing shear stresses in the material they move through (▮ Figure 9.7c). Because liquids (as well as gases) are not rigid, they have no shear strength, and S-waves cannot be transmitted through them.

The velocities of P- and S-waves are determined by the density and elasticity of the materials through which they travel. For example, seismic waves travel more slowly through rocks of greater density, but more rapidly through rocks with greater elasticity. *Elasticity* is a property of solids, such as rocks, and means that once they have been deformed by an applied force, they return to their original shape when the force is no longer present. Because P-wave velocity is greater than S-wave velocity in all materials, P-waves always arrive at seismic stations first.

Surface Waves

Surface waves travel along the surface of the ground, or just below it, and are slower than body waves (▮ Figure 9.8). Unlike the sharp jolting and shaking that body waves cause, surface waves generally produce a rolling or swaying motion, much like the experience of being on a boat.

Seismologists recognize several types of surface waves. The two most important are Rayleigh waves and Love waves, named after the British scientists who discovered them, Lord Rayleigh and A. E. H. Love. **Rayleigh waves (R-waves)** are generally the slower of the two and behave like water waves in that they move forward while the individual particles of material move in an elliptical path within a vertical plane oriented in the direction of wave movement (▮ Figure 9.8b).

The motion of a **Love wave (L-wave)** is similar to that of an S-wave, but the individual particles of the material move only back and forth in a horizontal plane perpendicular to the direction of wave travel (▮ Figure 9.8c).

▮ **Figure 9.7 Primary and Secondary Seismic Body Waves** Body waves travel through Earth.

a Undisturbed material for reference

Undisturbed material

d The effect of P- and S-waves on a surface structure.

Surface

b Primary waves (P-waves) compress and expand material in the same direction that they travel.

Compression Expansion Compression Expansion Compression

Undisturbed material

Primary wave (P-wave)

Direction of wave movement

c Secondary waves (S-waves) move material perpendicular to the direction of wave movement.

|← Wavelength →|

Secondary wave (S-wave)

Focus

▌ Figure 9.8 Rayleigh and Love Seismic Surface Waves Surface waves travel along Earth's surface or just below it.

Undisturbed material

a Undisturbed material for reference.

Rayleigh wave (R-wave)

b Rayleigh waves (R-waves) move material in an elliptical path in a plan oriented parallel to the direction of wave movement.

Love wave (L-wave)

c Love waves (L-waves) move material back and forth in a horizontal plane perpendicular to the direction of wave movement.

Rayleigh wave   Love wave

d The arrival of R- and L-waves causes the surface to undulate and shake from side to side.

9.6 Locating an Earthquake

We mentioned that news articles commonly report an earthquake's epicenter, but just how is the location of an epicenter determined? Once again, geologists rely on the study of seismic waves. We know that P-waves travel faster than S-waves, nearly twice as fast in all substances, so P-waves arrive at a seismograph station first, followed some time later by S-waves. Both P- and S-waves travel directly from the focus to the seismograph station through Earth's interior, but L- and R-waves arrive last because they are the slowest, and they also travel the longest route along the surface (▌ Figure 9.9a). However, only the P- and S-waves need concern us here because they are the ones important in finding an epicenter.

Seismologists have accumulated a tremendous amount of data over the years and now know the average speeds of P- and S-waves for any specific distance from their source. These P- and S-wave travel times are published in *time–distance graphs* that illustrate the difference between the arrival times of the two waves as a function of the distance between a seismograph and an earthquake's focus (▌ Figure 9.9b). That is, the farther the waves travel, the greater the *P–S time interval*, which is simply the time difference between the arrivals of P- and S-waves (Figure 9.9a, b).

If the P–S time intervals are known from at least three seismograph stations, then the epicenter of any earthquake

can be determined (▌ Figure 9.10). Here is how it works: Subtracting the arrival time of the first P-wave from the arrival time of the first S-wave gives the P–S time interval for each seismic station. Each of these time intervals is plotted on a time–distance graph, and a line is drawn straight down to the distance axis of the graph, thus giving the distance from the focus to each seismic station (Figure 9.9b). Next, a circle whose radius equals the distance shown on the time–distance graph from each of the seismic stations is drawn on a map (Figure 9.10). The intersection of the three circles is the location of the earthquake's epicenter. It should be obvious from Figure 9.10 that P–S time intervals from at least three seismic stations are needed. If only one were used, the epicenter could be at any location on the circle drawn around that station, and using two stations would give two possible locations for the epicenter.

Determining the focal depth of an earthquake is much more difficult and considerably less precise than finding its epicenter. The focal depth is usually found by making computations based on several assumptions, comparing the results with those obtained at other seismic stations, and then recalculating and approximating the depth as closely as possible. Although the results are not highly accurate, they do tell us that most earthquakes, probably 75%, have foci no deeper than 10 to 15 km and that a few are as deep as 680 km.

▌Figure 9.9 Determining the Distance from an Earthquake

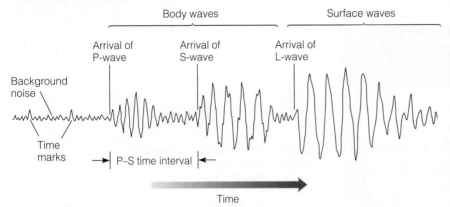

a A schematic seismogram showing the arrival order and pattern produced by P-, S-, and L-waves. When an earthquake occurs, body and surface waves radiate out from the focus at the same time. Because P-waves are the fastest, they arrive at a seismograph first, followed by S-waves and then by surface waves, which are the slowest waves. The difference between the arrival times of the P- and S-waves is the P–S time interval; it is a function of the distance the seismograph station is from the focus.

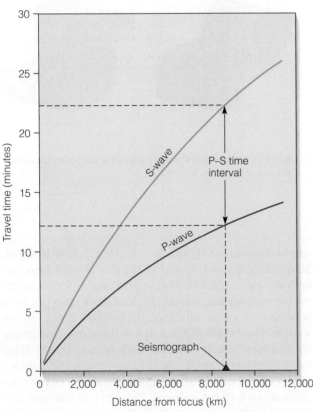

b A time–distance graph showing the average travel times for P- and S-waves. The farther away a seismograph station is from the focus of an earthquake, the longer the interval between the arrival of the P- and S-waves, and hence the greater the distance between the P- and S-wave curves on the time–distance graph as indicated by the P–S time interval. For example, let's assume the difference in arrival times between the P- and S-waves is 10 minutes (P–S time interval). Using the Travel time (minutes) scale, measure how long 10 minutes is (P–S time interval), and move that distance between the S-wave curve and the P-wave curve until the line touches both curves as shown. Then draw a line straight down to the Distance from focus (km) scale. That number is the distance the seismograph is from the earth's focus. In this example, the distance is almost 9,000 km.

9.7 Measuring the Strength of an Earthquake

Following any earthquake that causes extensive damage, fatalities, and injuries, graphic reports of the quake's violence and human suffering are common. Headlines tell us that thousands died, many more were injured or left homeless, and property damage is in the millions and possibly billions of dollars. Although such descriptions of fatalities and damage give some indication of the size of an earthquake, geologists are interested in more reliable methods for determining an earthquake's size.

Two measures of an earthquake's strength are commonly used. One is *intensity*, a qualitative assessment of the kinds of damage done by an earthquake. The other, *magnitude*, is a quantitative measure of the amount of energy released by an earthquake. Each method provides important information that can be used to prepare for future earthquakes.

Intensity

Intensity is a subjective or qualitative measure of the kind of damage done by an earthquake as well as people's reaction to it. Since the mid-19th century, geologists have used intensity as a rough approximation of the size and strength of an earthquake. The most common intensity scale used in the United States is the **Modified Mercalli Intensity Scale**, which has values ranging from I to XII (Table 9.2).

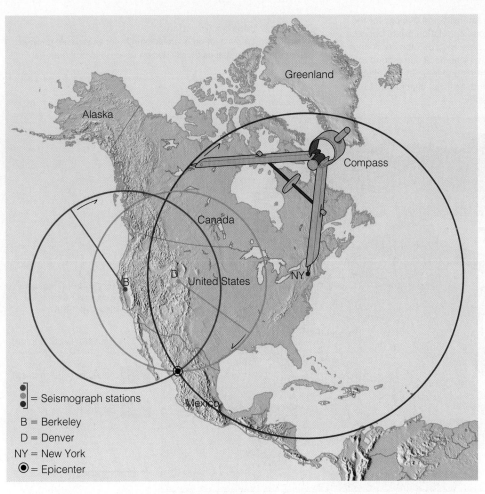

Figure 9.10 Determining the Epicenter of an Earthquake Three seismograph stations are needed to locate the epicenter of an earthquake. The P–S time interval is plotted on a time–distance graph for each seismograph station to determine the distance that station is from the epicenter. A circle with that radius is drawn from each station, and the intersection of the three circles is the epicenter of the earthquake.

= Seismograph stations

B = Berkeley

D = Denver

NY = New York

◉ = Epicenter

TABLE 9.2	Modified Mercalli Intensity Scale

I Not felt except by a very few under especially favorable circumstances.

II Felt by only a few people at rest, especially on upper floors of buildings.

III Felt quite noticeably indoors, especially on upper floors of buildings, but many people do not recognize it as an earthquake. Standing automobiles may rock slightly.

IV During the day felt indoors by many, outdoors by few. At night some awakened. Sensation like heavy truck striking building, standing automobiles rocked noticeably.

V Felt by nearly everyone, many awakened. Some dishes, windows, etc. broken, a few instances of cracked plaster. Disturbance of trees, poles, and other tall objects sometimes noticed.

VI Felt by all, many frightened and run outdoors. Some heavy furniture moved. A few instances of fallen plaster or damaged chimneys. Damage slight.

VII Everybody runs outdoors. Damage negligible in buildings of good design and construction; slight to moderate in well-built ordinary structures; considerable in poorly built or badly designed structures; some chimneys broken. Noticed by people driving automobiles.

VIII Damage slight in specially designed structures; considerable in normally constructed buildings with possible partial collapse; great in poorly built structures. Fall of chimneys, monuments, walls. Heavy furniture overturned. Sand and mud ejected in small amounts.

IX Damage considerable in specially designed structures. Buildings shifted off foundations. Ground noticeably cracked. Underground pipes broken.

X Some well-built wooden structures destroyed; most masonry and frame structures with foundations destroyed; ground badly cracked. Rails bent. Landslides considerable from river banks and steep slopes. Water splashed over river banks.

XI Few, if any (masonry) structures remain standing. Bridges destroyed. Broad fissures in ground. Underground pipelines completely out of service.

XII Damage total. Waves seen on ground surface. Objects thrown upward into the air.

Source: U.S. Geological Survey.

Figure 9.11 **Relationship between Intensity and Geology for the 1906 San Francisco Earthquake** A close correlation exists between the **(a)** geology of the San Francisco area and **(b)** intensity during the 1906 earthquake. Areas underlain by bedrock correspond to the lowest intensity values, followed by areas underlain by thin alluvium (sediment) and thick alluvium. Bay mud, artificial fill, or both lie beneath the areas shaken most violently.

» Critical Thinking Question
Why are structures built on bedrock usually not as severely damaged during an earthquake as those sited on unconsolidated material?

Bay mud (in places covered by artificial fill as of 1906)

Alluvium (>30 m thick)

Alluvium (<30 m thick)

Bedrock

0   1   2   3
km

a Geology of the San Francisco area.

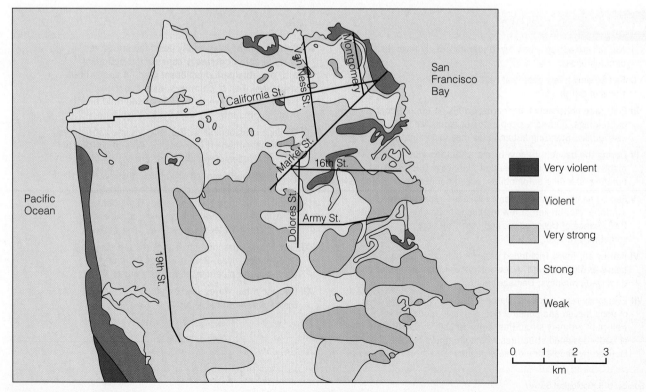

Very violent

Violent

Very strong

Strong

Weak

0   1   2   3
km

b Intensity of ground shaking in the San Francisco area during the 1906 earthquake.

Intensity maps can be constructed for regions hit by earthquakes by dividing the affected region into various intensity zones. The intensity value given for each zone is the maximum intensity that the earthquake produced for that zone. Even though intensity maps are not precise because of the subjective nature of the measurements, they do provide geologists with a rough approximation of the location of the earthquake, the kind and extent of the damage done, and the effects of local geology on different types of building construction (▮ Figure 9.11). Because intensity is a measure of the kind of damage done by an earthquake, insurance companies still classify earthquakes on the basis of intensity.

Generally, a large earthquake will produce higher intensity values than a small earthquake, but many other factors besides the amount of energy released by an earthquake also affect its intensity. These factors include distance from the epicenter, focal depth of the earthquake, population density and geology of the area, type of building construction employed, and duration of shaking.

A comparison of the intensity map for the 1906 San Francisco earthquake and a geologic map of the area shows a strong correlation between the amount of damage done and the underlying rock and soil conditions (Figure 9.11). Damage was greatest in those areas underlain by poorly consolidated material or artificial fill because the effects of shaking are amplified in these materials, whereas damage was less in areas of solid bedrock. The correlation between the geology and the amount of damage done by an earthquake was further reinforced by the 1989 Loma Prieta earthquake, when many of the same areas that were extensively damaged in the 1906 earthquake were once again heavily damaged.

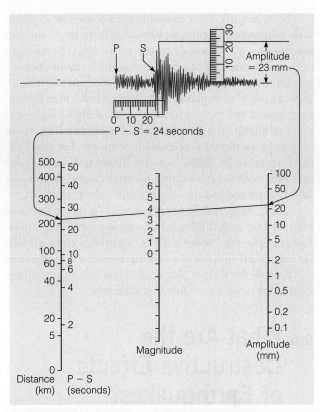

▮ **Figure 9.12 Richter Magnitude Scale** The Richter Magnitude Scale measures the total amount of energy released by an earthquake at its source. The magnitude is determined by measuring the maximum amplitude of the largest seismic wave and marking it on the right-hand scale. The difference between the arrival times of the P- and S-waves (recorded in seconds) is marked on the left-hand scale. When a line is drawn between the two points, the magnitude of the earthquake is the point at which the line crosses the center scale.

Magnitude

If earthquakes are to be compared quantitatively, we must use a scale that measures the amount of energy released and is independent of intensity. Charles F. Richter, a seismologist at the California Institute of Technology, developed such a scale in 1935. The **Richter Magnitude Scale** measures earthquake **magnitude**, which is the total amount of energy released by an earthquake at its source. It is an open-ended scale with values beginning at zero. The largest magnitude recorded was a magnitude 9.5 earthquake in Chile on May 22, 1960 (Table 9.1).

Scientists determine the magnitude of an earthquake by measuring the amplitude of the largest seismic wave as recorded on a seismogram (▮ Figure 9.12). To avoid large numbers, Richter used a conventional base-10 logarithmic scale to convert the amplitude of the largest recorded seismic wave to a numerical magnitude value. Therefore, each whole-number increase in magnitude represents a 10-fold increase in wave amplitude. For example, the amplitude of the largest seismic wave for an earthquake of magnitude 6 is 10 times that produced by an earthquake of magnitude 5, 100 times as large as a magnitude 4 earthquake, and 1,000 times that of an earthquake of magnitude 3 ($10 \times 10 \times 10 = 1,000$).

A common misconception about the size of earthquakes is that an increase of one unit on the Richter Magnitude Scale—a 7 versus a 6, for instance—means a 10-fold increase in size. It is true that each whole-number increase in magnitude represents a 10-fold increase in the wave amplitude, but each magnitude increase of one unit corresponds to a roughly 30-fold increase in the amount of energy released (actually, it is 31.5, but 30 is close enough for our purposes). This means, for example, that the 1964 Alaska earthquake with a magnitude of 8.6 released nearly 900 times more energy than the 1994 Northridge, California, earthquake of magnitude 6.7!

The Richter Magnitude Scale was devised to measure earthquake waves on a particular seismograph and at a specific distance from an earthquake. One of its limitations is that it underestimates the energy of very large earthquakes because it measures the highest peak on a seismogram, which represents only an instant during an earthquake. For large earthquakes, though, the energy might be released over several minutes and along hundreds of kilometers of a fault. For example, during the 1857 Fort Tejon, California, earthquake, the ground shook for longer than 2 minutes, and energy was released for 360 km along the fault.

Seismologists now commonly use a somewhat different scale to measure magnitude. Known as the *seismic-moment magnitude scale*, this scale takes into account the strength of the rocks, the area of a fault along which rupture occurs, and the amount of movement of rocks adjacent to the fault. Because larger earthquakes rupture more rocks than smaller earthquakes and rupture usually occurs along a longer segment of a fault and therefore for a longer duration, these very large earthquakes release more energy. For example, the December 26, 2004, Sumatra, Indonesia, earthquake that generated the devastating tsunami created the longest fault rupture and had the longest duration ever recorded.

Thus, magnitude is now frequently given in terms of both Richter magnitude and seismic-moment magnitude. For example, the 1964 Alaska earthquake is given a Richter magnitude of 8.6 and a seismic-moment magnitude of 9.2. Because the Richter Magnitude Scale is most commonly used in the news, we will use that scale here.

9.8 What Are the Destructive Effects of Earthquakes?

The number of deaths and injuries, as well as the amount of property damage, depends on several factors. Generally speaking, earthquakes that occur during working and school hours in densely populated urban areas are the most destructive and cause the most fatalities and injuries. However, magnitude, duration of shaking, distance from the epicenter, geology of the affected region, and the type of structures are also important considerations. Given these variables, it should not be surprising that a comparatively small earthquake can have disastrous effects, whereas a much larger one might go largely unnoticed, except perhaps by seismologists.

The destructive effects of earthquakes include ground shaking, fire, seismic sea waves, and landslides, as well as panic, disruption of vital services, and psychological shock. In some cases, rescue attempts are hampered by inadequate resources or planning, conditions of civil unrest, or simply the magnitude of the disaster.

Ground Shaking

Ground shaking, the most obvious and immediate effect of an earthquake, varies depending on the earthquake's magnitude, distance from the epicenter, and type of underlying materials in the area—unconsolidated sediment or fill versus bedrock, for instance. Certainly, ground shaking is terrifying, and it may be violent enough for fissures to open in the ground. Nevertheless, contrary to popular myth, fissures do not swallow up people and buildings and then close on them. And although California will no doubt have big earthquakes in the future, rocks cannot store enough energy to displace a landmass as large as California into the Pacific Ocean, as some alarmists claim.

Figure 9.13 Relationship between Seismic Wave Amplitude and Underlying Geology The amplitude and duration of seismic waves generally increase as the waves pass from bedrock into poorly consolidated or water-saturated material. Thus, structures built on weaker material typically suffer greater damage than similar structures built on bedrock because the shaking lasts longer.

Figure 9.14 Liquefaction The effects of ground shaking on water-saturated soil are dramatically illustrated by the collapse of these buildings in Niigata, Japan, during a 1964 earthquake. The buildings were designed to be earthquake resistant and fell over on their sides intact when the ground below them underwent liquefaction.

The effects of ground shaking, such as collapsing buildings, falling building facades and window glass, and toppling monuments and statues, cause more damage and result in more loss of life and injuries than any other earthquake hazard. Structures built on solid bedrock generally suffer less damage than those built on poorly consolidated material such as water-saturated sediments or artificial fill.

Structures built on poorly consolidated or water-saturated material are subjected to ground shaking of longer duration and greater S-wave amplitude than structures built on bedrock (Figure 9.13). In addition, fill and water-saturated sediments tend to liquefy, or behave as a fluid, a process known as *liquefaction*. When shaken, the individual grains lose cohesion, and the ground flows. Two dramatic examples of damage resulting from liquefaction are Niigata, Japan, and Turnagain Heights, Alaska. In Niigata, Japan, large apartment buildings were tipped to their sides after the water-saturated soil of the hillside collapsed (Figure 9.14).

In Turnagain Heights, Alaska, many homes were destroyed when the Bootlegger Cove Clay lost all its strength when it was shaken by the 1964 earthquake (see Figure 11.18).

Besides the magnitude of an earthquake and the underlying geology, the material used and the type of construction affect the amount of damage done. Adobe and mud-walled structures are the weakest and almost always collapse during an earthquake. Unreinforced brick structures and poorly built concrete structures are also particularly susceptible to collapse, as was the case in the 1999 Turkey earthquake in which an estimated 17,000 people died (Figure 9.15). The 1976 earthquake in Tangshan, China, completely leveled the city because hardly any structures were built to resist seismic forces. In fact, most had unreinforced brick walls, which have no flexibility, and consequently they collapsed during the shaking.

Fire

In many earthquakes, particularly in urban areas, fire is a major hazard. Nearly 90% of the damage done in the 1906 San Francisco earthquake was caused by fire. The shaking severed many of the electrical and gas lines, which touched off flames and started fires throughout the city. Because the earthquake ruptured water mains, there was no effective way to fight the fires that raged out of control for three days, destroying much of the city.

Eighty-three years later, during the 1989 Loma Prieta earthquake, a fire broke out in the Marina district of San Francisco. This time, however, the fire was contained within a small area because San Francisco had a system of valves throughout its water and gas pipeline system so that lines could be isolated from breaks (see the Geo-Insight on pages 214 and 215).

During the September 1, 1923, earthquake in Japan, fires destroyed 71% of the houses in Tokyo and practically all the houses in Yokohama. In all, 576,262 houses were destroyed by fire, and 143,000 people died, many as a result of fire. A horrible example occurred in Tokyo where thousands of people gathered along the banks of the Sumida River to escape the raging fires. Suddenly, a firestorm swept over the area, killing more than 38,000 people. The fires from this earthquake were particularly devastating because most of the buildings were constructed of wood and were fanned by 20-km/hr winds.

Tsunami: Killer Waves

On December 26, 2004, a magnitude 9.0 earthquake struck 160 km off the west coast of northern Sumatra, Indonesia, generating the deadliest tsunami in history. Within hours, walls of water as high as 10.5 m pounded the coasts of Indonesia, Sri Lanka, India, Thailand, Somalia, Myanmar, Malaysia, and the Maldives, killing more than 220,000 people and causing billions of dollars in damage.

Following the 9.0-magnitude earthquake that struck Japan on March 11, 2011, a massive tsunami was generated that resulted in tremendous property damage and loss of life along the coastline of Japan (Figure 9.16). Within minutes after the earthquake, walls of water, some as much as up to 37 m high, inundated low-lying areas along the Japanese coast and extended as far as 10 km inland. Boats, vehicles, and structures were destroyed or swept aside as if they were toys, leaving a wake of destruction as the waters finally receded. Of concern also, is the fact that three nuclear power plants suffered damage from the tsunami, and at least one has leaked radiation into the surrounding area as well as the Pacific Ocean.

Yann Arhus-Betrand/Peter Arnold, Inc./PhotoLibrary

 Figure 9.15 Ground Shaking Most of the buildings collapsed or were severely damaged as a result of ground shaking during the August 17, 1999 Turkey earthquake, which killed more than 17,000 people.

GEO-INSIGHT   The San Andreas Fault

The circum-Pacific belt is well known for its volcanic activity and earthquakes. Indeed, about 60% of all volcanic eruptions and 80% of all earthquakes take place in this belt that nearly encircles the Pacific Ocean basin (Figure 9.4).

One well-known and well-studied segment of the circum-Pacific belt is the 1,300-km-long San Andreas fault extending from the Gulf of California north through coastal California until it terminates at the Mendocino fracture zone off California's north coast. In plate tectonic terminology, it marks a transform plate boundary between the North American and Pacific plates (see Chapter 2).

Earthquakes along the San Andreas and related faults will continue to occur. But other segments of the circum-Pacific belt, as well as the Mediterranean–Asiatic belt, are also quite active and will continue to experience earthquakes.

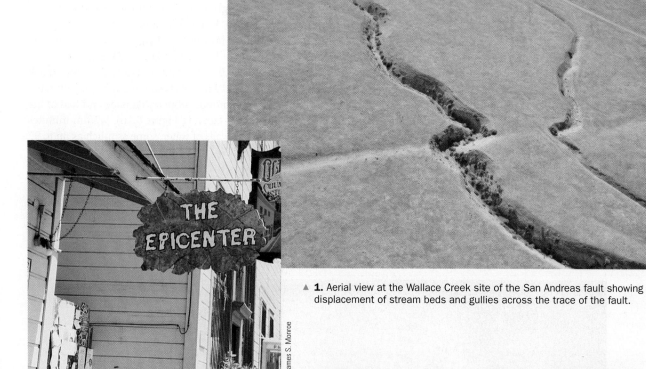

USGS photograph by David K. Lynch, Kenneth W. Hudnut and David S. P. Dearborn (2009)

▲ **1.** Aerial view at the Wallace Creek site of the San Andreas fault showing displacement of stream beds and gullies across the trace of the fault.

James S. Monroe

▲ **2.** This shop in Olema, California, is rather whimsically called The Epicenter, alluding to the fact that it is in the San Andreas fault zone.

▶ **3.** Rocks on opposite sides of the San Andreas fault periodically lurch past one another, generating large earthquakes. The most famous one destroyed San Francisco on April 18, 1906. It resulted when 465 km of the fault ruptured, causing about 6 m of horizontal displacement in some areas (Figure 9.1b). It is estimated that 3,000 people died. The shaking lasted nearly 1 minute and caused property damage estimated at $400 million in 1906 dollars! Approximately 28,000 buildings were destroyed, many of them by the three-day fire that raged out of control and devastated about 12 km^2 of the city.

Steinbrugge Collection, Earthquake Engineering Research Center, University of California Berkeley

San Francisco following the 1906 earthquake. This view along Sacramento Street shows damaged buildings and the approaching fire.

▶ **4.** Since 1906, the San Andreas fault and its subsidiary faults have spawned many more earthquakes; one of the most tragic was centered at Northridge, California, a small community north of Los Angeles. During the early morning hours of January 17, 1994, Northridge and surrounding areas were shaken for 40 seconds. When it was over, 61 people were dead and thousands injured; an oil main and at least 250 gas lines had ruptured, igniting numerous fires; nine freeways were destroyed; and thousands of homes and other buildings were damaged or destroyed by ground shaking. The nearly total destruction of this apartment complex resulted in 16 deaths.

Tim Clary/AFP/Getty Images

▶ **5.** A spectacular fire on Balboa Boulevard, Northridge, was caused by a gas-main explosion during the earthquake.

Ted Soqui/Sygma/Corbis

▌Figure 9.16 2011 Japanese Tsunami

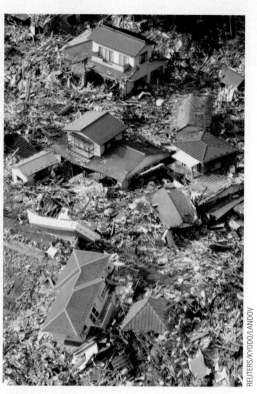

a Map showing the location of the epicenter (pink circled area - 2011 Eq) of the 9.0-magnitude earthquake that struck Japan on March 11, 2011 causing tremendous destruction and loss of life from both the earthquake and subsequent tsunami. The red lines are plate boundaries. Volcanoes are shown as red triangles. Cities with populations of greater than 250,000 are shown as red squares. The two yellow circled areas show previous large earthquakes (1896 and 1933) in the epicentral area.

b Aerial view of Minamisanriku, Miyagi Prefecture in northeastern Japan, showing houses clogged with debris and a large boat that had been swept inland as a result of the tsunami.

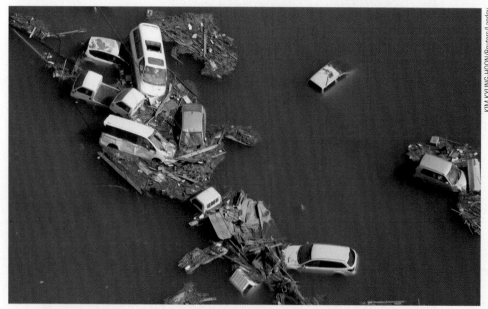

c Vehicles, tossed like toys, mix with the rubble in floodwaters near the coastal town of Sendai, following the tsunami that ravaged the northeastern portion of Japan following the devastating 9.0-magnitude earthquake on March 11, 2011.

Both of these earthquakes generated what is popularly called a "tidal wave," but is more correctly termed a *seismic sea wave* or **tsunami**, a Japanese term meaning "harbor wave." The term *tidal wave* nevertheless persists in popular literature and some news accounts, but these waves are not caused by or related to tides. Indeed, tsunami are destructive sea waves generated when the seafloor undergoes sudden, vertical movements. Many result from submarine earthquakes, but volcanoes at sea or submarine landslides can also cause them. For example, the 1883 eruption of Krakatau between Java and Sumatra generated a large sea wave that killed 36,000 on nearby islands.

Once a tsunami is generated, it can travel across an entire ocean and cause devastation far from its source. In the open sea, tsunami travel at several hundred kilometers per hour and commonly go unnoticed as they pass beneath

CONNECTION LINK

To learn more about how waves form and their terminology, see pages 395 and 396 in Chapter 16.

ships because they are usually less than 1 m high and the distance between wave crests is typically hundreds of kilometers. When they enter shallow water, however, the waves slow down, and water piles up to heights anywhere from a meter or two to many meters high. The 1946 tsunami that struck Hilo, Hawaii, was 16.5 m high! In any case, the tremendous energy possessed by a tsunami is concentrated on a shoreline when it hits either as a large breaking wave or, in some cases, as what appears to be a very rapidly rising tide.

A common popular belief is that a tsunami is a single large wave that crashes onto a shoreline. Any tsunami consists of a series of waves that pour onshore for as long as 30 minutes followed by an equal time during which water rushes back to sea. Furthermore, after the first wave hits, more waves follow at 20- to 60-minute intervals. Approximately 80 minutes after the 1755 Lisbon, Portugal, earthquake, the first of three tsunami, the largest more than 12 m high, destroyed the waterfront area and killed thousands of people. Following the arrival of a 2-m-high tsunami in Crescent City, California, in 1964, curious people went to the waterfront to inspect the damage. Unfortunately, 10 were killed by a following 4-m-high wave!

One of nature's warning signs of an approaching tsunami is a sudden withdrawal of the sea from a coastal region. In fact, the sea might withdraw so far that it cannot be seen and the seafloor is laid bare over a huge area. On more than one occasion, people have rushed out to inspect exposed reefs or to collect fish and shells only to be swept away when the tsunami arrived.

During the December 2004 Indian Ocean tsunami, a 10-year-old British girl saved numerous lives because she recognized the warning signs that she had learned in a school lesson on tsunami only two weeks before. While vacationing with her mother on the island of Phuket, Thailand, a popular resort area, the girl noticed the water quickly receding from the beach. She immediately told her mother that she thought a tsunami was coming, and her mother, along with the resort staff, quickly warned everyone standing around watching the water recede to clear the beach area. Their quick action resulted in many lives being saved.

Following the tragic 1946 tsunami that hit Hilo, Hawaii, the U.S. Coast and Geodetic Survey established a Pacific Tsunami Early Warning System in Ewa Beach, Hawaii. This system combines seismographs and instruments that detect earthquake-generated sea waves. Whenever a strong earthquake takes place anywhere within the Pacific basin, its location is determined, and instruments are checked to see whether a tsunami has been generated. If it has, a warning is sent out to evacuate people from low-lying areas that may be affected.

Nevertheless, tsunami remain a threat to people in coastal areas, especially around the Pacific Ocean. Unfortunately, no such warning system exists for the Indian Ocean.

If one had been in place, it is possible that the death toll from the December 26, 2004, tsunami would have been significantly lower.

Ground Failure

Earthquake-triggered landslides are particularly dangerous in mountainous regions and have been responsible for tremendous amounts of damage and many deaths. The 1959 earthquake in Madison Canyon, Montana, for example, caused a huge rock slide (❚ Figure 9.17), and the 1970 Peru earthquake caused an avalanche that destroyed the town of Yungay and killed an estimated 25,000 people (see Figure 11.23). Most of the 100,000 deaths from the 1920 earthquake in Gansu, China, resulted when cliffs composed of loess (wind-deposited silt) collapsed. More than 2,000 people were killed when two-thirds of the town of Port Royal, Jamaica, slid into the sea following an earthquake on June 7, 1692.

9.9 Earthquake Prediction

A successful prediction must include a time frame for the occurrence of an earthquake, its location, and its strength. Despite the tremendous amount of information geologists have gathered about the cause of earthquakes, successful predictions are still rare. Nevertheless, if reliable predictions can be made, they can greatly reduce the number of deaths and injuries.

From an analysis of historic records and the distribution of known faults, geologists construct *seismic risk maps* that indicate the likelihood and potential severity of future earthquakes based on the intensity of past earthquakes. An international effort by scientists from several countries resulted in the publication of the first Global Seismic Hazard Assessment Map in December 1999 (❚ Figure 9.18). Although such maps cannot be used to predict when an earthquake will take place in any particular area, they are useful in anticipating future earthquakes and helping people plan and prepare for them (see Geo-Focus on page 222).

Earthquake Precursors

Studies conducted during the past several decades indicate that most earthquakes are preceded by both short-term and long-term changes within Earth. Such changes are called *precursors* and may be useful in earthquake prediction.

One long-range prediction technique used in seismically active areas involves plotting the location of major earthquakes and their aftershocks to detect areas that have had major earthquakes in the past, but are currently inactive. Such regions are said to be locked and not releasing energy. Nevertheless, pressure is continuing to accumulate in these regions because of plate motions, making these *seismic gaps* prime locations for future earthquakes. Several seismic gaps along the San Andreas fault have the

Figure 9.17 Ground Failure On August 17, 1959, an earthquake with a Richter magnitude of 7.3 shook southwestern Montana and a large area in adjacent states.

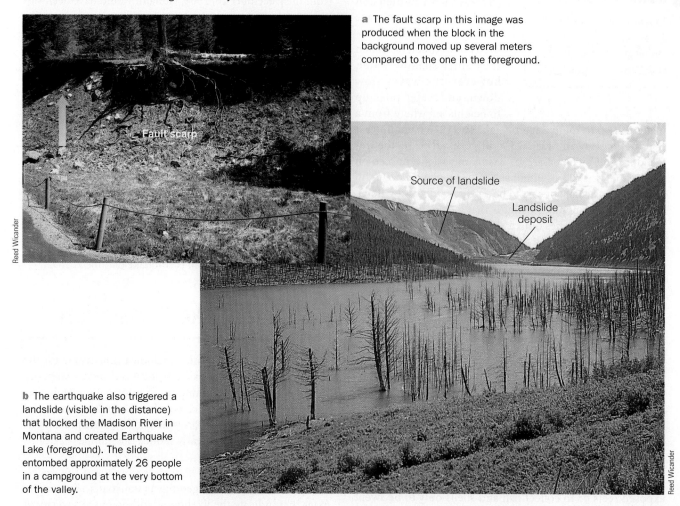

a The fault scarp in this image was produced when the block in the background moved up several meters compared to the one in the foreground.

Source of landslide

Landslide deposit

Fault scarp

b The earthquake also triggered a landslide (visible in the distance) that blocked the Madison River in Montana and created Earthquake Lake (foreground). The slide entombed approximately 26 people in a campground at the very bottom of the valley.

Reed Wicander

Reed Wicander

Figure 9.18 Global Seismic Hazard Assessment Map The Global Seismic Hazard Assessment Program published this seismic hazard map showing peak ground accelerations. The values are based on a 90% probability that the indicated horizontal ground acceleration during an earthquake is not likely to be exceeded in 50 years. The higher the number, the greater the hazard. As expected, the greatest seismic risks are in the circum-Pacific belt and the Mediterranean–Asiatic belt.

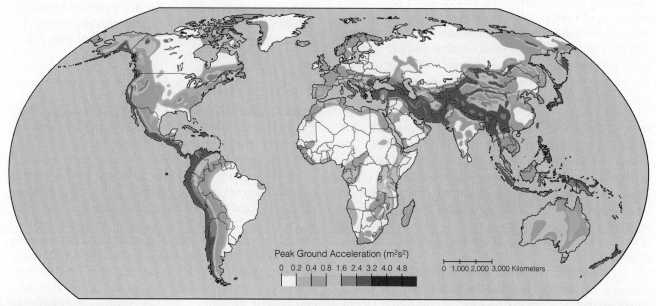

Peak Ground Acceleration (m²s²)

0   0.2 0.4 0.8 1.6 2.4 3.2 4.0 4.8

0 1,000 2,000 3,000 Kilometers

potential for future major earthquakes (■ Figure 9.19). A major earthquake that damaged Mexico City in 1985 occurred along a seismic gap in the convergence zone along the west coast of Mexico.

Earthquake precursors that may be useful in making short-term predictions include slight changes in elevation and tilting of the land surface, fluctuations in the water level in wells, changes in Earth's magnetic field, and the electrical resistance of the ground.

Earthquake Prediction Programs

Currently, only a handful of nations—such as the United States, Japan, Russia, and China—have government-sponsored earthquake prediction programs. These programs include laboratory and field studies of rock behavior before, during, and after large earthquakes, as well as monitoring activity along major active faults. Most earthquake prediction work in the United States is done by the U.S. Geological Survey (USGS) and involves research into all aspects of earthquake-related phenomena.

The Chinese have perhaps the most ambitious earthquake prediction program in the world, which is understandable considering their long history of destructive earthquakes. Their earthquake prediction program was initiated soon after two large earthquakes occurred at Xingtai (300 km southwest of Beijing) in 1966. This program includes extensive study and monitoring of all possible earthquake precursors. Chinese seismologists successfully predicted the 1975 Haicheng earthquake, but failed to predict the devastating 1976 Tangshan earthquake that killed at least 242,000 people.

Progress is being made toward dependable, accurate earthquake predictions, and studies are under way to assess public reactions to long-, medium-, and short-term earthquake warnings. However, unless short-term warnings are actually followed by an earthquake, most people will probably ignore the warnings as they frequently do now for hurricanes, tornadoes, and tsunami. Perhaps the best we can hope for is that people in seismically active areas will take measures to minimize their risk from the next major earthquake (Table 9.3).

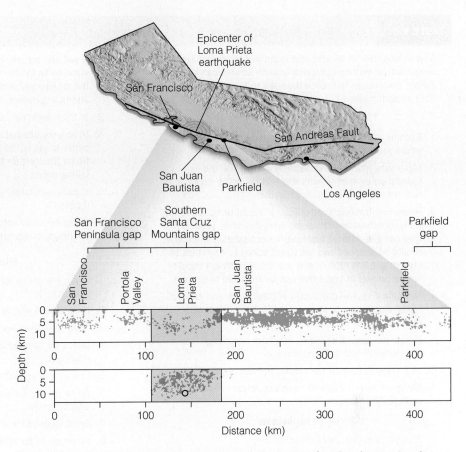

■ Figure 9.19 Earthquake Precursors Seismic gaps are one type of earthquake precursor that can indicate a potential earthquake in the future. Seismic gaps are regions along a fault that are locked; that is, they are not moving and releasing energy. Three seismic gaps are evident in this cross section along the San Andreas fault from north of San Francisco to south of Parkfield. The first is between San Francisco and Portola Valley, the second is near Loma Prieta Mountain, and the third is southeast of Parkfield. The top section shows the epicenters of earthquakes between January 1969 and July 1989. The bottom section shows the southern Santa Cruz Mountains gap after it was filled by the October 17, 1989, Loma Prieta earthquake (open circle) and its aftershocks.

9.10 Earthquake Control

Reliable earthquake prediction is still in the future, but can anything be done to control or at least partly control earthquakes? Because of the tremendous energy involved, it seems unlikely that humans will ever be able to prevent earthquakes. However, it may be possible to gradually release the energy stored in rocks, thus decreasing the probability of a large earthquake and extensive damage.

During the early- to mid-1960s, Denver, Colorado, experienced numerous small earthquakes, which was surprising because Denver had not been prone to earthquakes in the past. Geologist David M. Evans suggested that the earthquakes were directly related to the injection of contaminated wastewater into a 3,674 m disposal well at the Rocky Mountain Arsenal, northeast of Denver (■ Figure 9.20a). A USGS study concluded that the pumping of waste fluids into the fractured rocks beneath the disposal well decreased the friction on opposite sides of fractures—in effect, lubricating them so that movement occurred, thus causing the earthquakes that Denver experienced.

TABLE 9.3	What You Can Do to Prepare for an Earthquake

Anyone who lives in an area that is subject to earthquakes or who will be visiting or moving to such an area can take certain precautions to reduce the risks and losses resulting from an earthquake.

Before an earthquake:

1. Become familiar with the geologic hazards of the area where you live and work.

2. Make sure your house is securely attached to the foundation by anchor bolts and that the walls, floors, and roof are all firmly connected together.

3. Heavy furniture such as bookcases should be bolted to the walls; semiflexible natural gas lines should be used so that they can give without breaking; water heaters and furnaces should be strapped and the straps bolted to wall studs to prevent gas-line rupture and fire. Brick chimneys should have a bracket or brace that can be anchored to the roof.

4. Maintain a several-day supply of fresh water and canned foods, and keep a fresh supply of flashlight and radio batteries, as well as a fire extinguisher.

5. Maintain a basic first-aid kit and have a working knowledge of first-aid procedures.

6. Learn how to turn off the various utilities at your house.

7. Above all, have a planned course of action for when an earthquake strikes.

During an earthquake:

1. Remain calm and avoid panic.

2. If you are indoors, get under a desk or table if possible, or stand in an interior doorway or room corner as these are the structurally strongest parts of a room; avoid windows and falling debris.

3. In a tall building, do not rush for the stairwells or elevators.

4. In an unreinforced or other hazardous building, it may be better to get out of the building rather than to stay in it. Be on the alert for fallen power lines and the possibility of falling debris.

5. If you are outside, get to an open area away from buildings if possible.

6. If you are in an automobile, stay in the car, and avoid tall buildings, overpasses, and bridges if possible.

After an earthquake:

1. If you are uninjured, remain calm and assess the situation.

2. Help anyone who is injured.

3. Make sure there are no fires or fire hazards.

4. Check for damage to utilities and turn off gas valves if you smell gas.

5. Use your telephone only for emergencies.

6. Do not go sightseeing or move around the streets unnecessarily.

7. Avoid landslide and beach areas.

8. Be prepared for aftershocks.

Interestingly, a high degree of correlation was found when comparing the number of earthquakes in Denver against the average amount of contaminated fluids injected into the disposal well per month (▮ Figure 9.20b). Furthermore, when no waste fluids were injected, earthquake activity decreased dramatically.

Experiments conducted in 1969 at an abandoned oil field near Rangely, Colorado, confirmed the hypothesis that fluids injected into the fractured bedrock at the Rocky Mountain Arsenal was causing Denver's earthquakes. When geologists pumped water into abandoned oil wells, seismographs recorded small earthquakes in the area. When fluids were pumped out, earthquake activity declined. Geologists were thus starting and stopping earthquakes at will and validating the relationship between pore-water pressures and earthquakes.

In a situation similar to that in Denver, the number of earthquakes in the Dallas–Ft. Worth area of Texas has dramatically increased, resulting in more earthquakes during 2008 and 2009 than in the previous 30 years combined. Beginning in 2004, thousands of gas wells have been drilled in the area, and almost all of them have undergone hydraulic fracturing, in which large amounts of a high-pressure water mix are injected into the wells to open up the preexisting fractures to allow gas to flow into the wells. Although no one is claiming there is a link between the hydraulic fracturing associated with the drilling and the onset of small earthquakes, geologists are looking closely at the situation to see if there is a cause-and-effect relationship between the two events.

Based on these results, some geologists have proposed that fluids be pumped into the locked segments, or seismic gaps, of active faults to cause small- to moderate-sized earthquakes. They think that this action would relieve the pressure on the fault and prevent a major earthquake from occurring.

Although this plan is intriguing, it also has many potential problems. For instance, there is no guarantee that only a small earthquake might result. Instead, a major earthquake might occur, causing tremendous property damage and loss of life, especially in a densely populated area (▮ Figure 9.21). Who would be responsible? Certainly, a great deal more research is needed before such an experiment is performed, even in an area of low population density.

What Would You Do ❓

Some geologists think that by pumping liquids into locked segments of active faults, they can generate small- to moderate-sized earthquakes. These earthquakes would relieve the buildup of pressure along a fault and thus prevent very large earthquakes from taking place. What do you think of this proposal? What kind of social, political, and economic consequences would there be? Do you think such an effort will ever actually reduce the threat of earthquakes?

Figure 9.20 Controlling Earthquakes

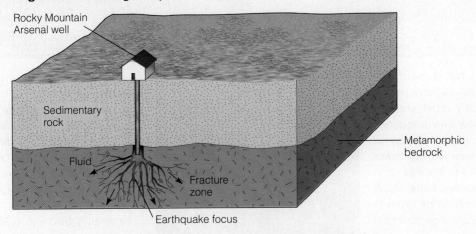

a A block diagram of the Rocky Mountain Arsenal well and the underlying geology.

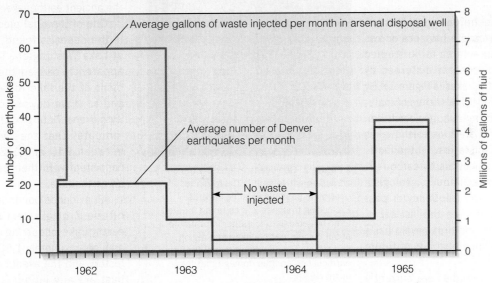

b A graph showing the relationship between the amount of wastewater injected into the well per month and the average number of Denver earthquakes per month. There have been no significant earthquakes in Denver since injection of wastewater into the disposal well ceased in 1965.

Figure 9.21 **Population Density of San Francisco** Downtown San Francisco sits near the active plate boundary between the Pacific plate and the North American plate. The high population density poses a risk to residents in the event of an earthquake. Although pumping fluids into a fault zone may relieve the pressure on a fault and prevent major earthquakes from occurring, people in populated areas are reluctant to risk earthquake control, lest a major earthquake be initiated by the process.

GEO-FOCUS Paleoseismology

Paleoseismology is the study of prehistoric earthquakes. As more people move into seismically active areas, it is important to know how frequently earthquakes in the area have occurred in the past, and how strong those earthquakes were. In this way, prudent decisions can be made about what precautions need to be taken in developing an area and how stringent the building codes for a region need to be.

A typical technique in paleoseismology is to excavate trenches across active faults in an area to be studied and date the sediments disturbed by prehistoric earthquakes (■ Figure 1). By exposing the upper few meters of material along an active fault, geologists can find evidence of previous earthquakes in the ancient soil layers. Furthermore, by dating the paleosoils by carbon-14 or other dating techniques, geologists can determine the frequency of past earthquakes and when the last earthquake occurred, and thus have a basis for estimating the probability of future earthquakes.

Paleoseismic studies are currently underway in many areas of North America, particularly along the San Andreas fault in California and in the coastal regions of Washington. An interesting case in point concerns an ancient earthquake in what is now Seattle, Washington.

Data from a variety of sources have convinced many geologists that a shallow-focus earthquake of at least magnitude 7 occurred beneath Seattle, Washington, less than 1,100 years ago. In a point not lost on officials, they noted the catastrophic effects that a similar-sized earthquake would have if it occurred in the same area today.

The first link in the chain of evidence for a paleoearthquake came from the discovery of a marine terrace that had been uplifted some

▌ Figure 1 Geologists examine a trench across an active fault in California to determine possible seismic hazards. Excavating trenches is a common method used by geologists to gather information about ancient earthquakes in a region and to help assess the potential for future earthquakes and the damage that they might cause.

© John Karachewski, Geoscapes Photography

7 m at Restoration Point, 5 km west of Seattle. Carbon-14 analysis of peat within sediments of the terrace indicates that uplift occurred between 500 and 1,700 years ago. Carbon-14 dating of other sites to the north, south, and east also indicates a sudden uplift in the area within the same time period. The amount of uplift suggests to geologists a magnitude-7 or greater earthquake.

Because many earthquakes in and around the Pacific Ocean basin can cause tsunami, geologists looked for evidence that a tsunami occurred, and they found it in the form of unusual sand layers in nearby tidal marsh deposits. Carbon-14 dating of organic

matter associated with the sands yielded an age of 850 to 1,250 years ago, well within the time period during which the terraces were uplifted.

Geologists also found evidence of rock avalanches in the Olympic Mountains that dammed streams, thereby forming lakes. Drowned trees in the lakes were dated as having died between 1,000 and 1,300 years ago, again fitting in nicely with the date of the ancient earthquake.

One of the final pieces of evidence is the deposits found on the bottom of Lake Washington. The earthquake apparently caused the bottom sediment of the lake to be resuspended and to move downslope as a turbidity current. Dating of these sediments indicates that they were deposited between 940 and 1,280 years ago, consistent with their being caused by an earthquake.

All evidence points to a large (magnitude 7 or greater) shallow-focus earthquake occurring in the Seattle area approximately 1,000 years ago. If history and the events of the geologic past are any guide, it is very likely that another large earthquake will hit the Seattle area in the future. When this will occur can't yet be predicted, but it would be wise to plan for such an eventuality. After all, metropolitan Seattle has a population of more than 2.5 million people, and its entire port area is built on fill that would probably be hard hit by an earthquake. Furthermore, most of Seattle's schools, hospitals, utilities, and fire and police stations are not built to withstand a strong earthquake. What the future holds for Seattle has yet to be determined. Geologists have provided a window on what has happened seismically in the past, and it is up to today's government and its various agencies to decide how they want to use this information.

9.11 What Is Earth's Interior Like?

During most of historic time, Earth's interior was perceived as an underground world of vast caverns, heat, and sulfur gases, populated by demons. By the 1860s, scientists knew the average density of Earth and knew that pressure and temperature increased with depth. And even though Earth's interior is hidden from direct observation, scientists now have a reasonably good idea of its internal structure and composition.

Earth is generally depicted as consisting of concentric layers that differ in composition and density and that are separated from adjacent layers by rather distinct boundaries (❚ Figure 9.22). Recall that the outermost layer, or *crust*, is Earth's thin skin. Below the crust and extending about halfway to Earth's center is the *mantle*, which comprises more than 80% of Earth's volume. The central part of Earth consists of a *core*, which is divided into a solid inner portion and a liquid outer part (Figure 9.22).

The behavior and travel times of P- and S-waves provide geologists with information about Earth's internal structure. Seismic waves travel outward as wavefronts from their source areas, although it is most convenient to depict them as *wave rays*, which are lines showing the direction of movement of small parts of wavefronts (Figure 9.3a).

As we noted earlier, P- and S-wave velocity is determined by the density and elasticity of the materials they travel through, both of which increase with depth. Wave velocity is slowed by increasing density, but increases in materials with greater elasticity. Because elasticity increases with depth faster than density, a general increase in seismic wave velocity takes place as the waves penetrate to greater depths. P-waves travel faster than S-waves under all circumstances, but unlike P-waves, S-waves are not transmitted through liquids because liquids have no shear strength (rigidity); liquids simply flow in response to shear stress.

Because Earth is not a homogeneous body, seismic waves travel from one material into another of different density and elasticity, and thus their velocity and direction of travel change. That is, the waves are bent, a phenomenon known as *refraction*, in much the same way as light waves are refracted as they pass from air into more dense water. Because seismic waves pass through materials of differing density and elasticity, they are continually refracted so that their paths are curved; wave rays travel in a straight line only when their direction of travel is perpendicular to a boundary (❚ Figure 9.23).

In addition to refraction, seismic rays are *reflected*, much as light is reflected from a mirror. When seismic rays encounter a boundary separating materials of different density or elasticity, some of a wave's energy is reflected back to the surface (Figure 9.23). If we know the wave velocity and the time required for the wave to travel from its source to the boundary and back to the surface, we can calculate the depth of the reflecting boundary. Such information is useful in determining not only Earth's internal structure but also

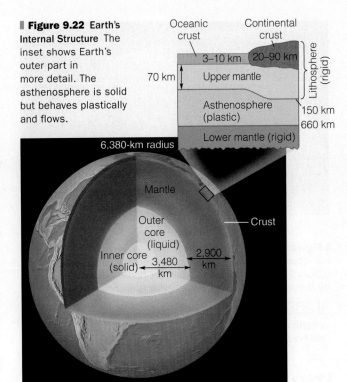

❚ **Figure 9.22** Earth's **Internal Structure** The inset shows Earth's outer part in more detail. The asthenosphere is solid but behaves plastically and flows.

Earth's Composition and Density		
	Composition	Density (g/cm³)
Continental crust	Average composition of granodiorite	≈2.7
Oceanic crust	Upper part basalt, lower part gabbro	≈3.0
Mantle	Peridotite (made up of ferromagnesian silicates)	3.3–5.7
Outer core	Iron with perhaps 12% sulfur, silicon, oxygen, nickel, and potassium	9.9–12.2
Inner core	Iron with 10–20% nickel	12.6–13.0
Earth		5.5

the depths of sedimentary rocks that may contain petroleum. Seismic reflection is a common tool used in petroleum exploration.

Although changes in seismic wave velocity occur continuously with depth, P-wave velocity increases suddenly at the base of the crust and decreases abruptly at a depth of approximately 2,900 km (❚ Figure 9.24). These marked changes in seismic wave velocity indicate a boundary called a **discontinuity** across which a significant change in Earth materials or their properties occurs. Discontinuities are the basis for subdividing Earth's interior into concentric layers.

CONNECTION LINK

To review how petroleum forms and is trapped beneath Earth's surface, see pages 173 and 176 in Chapter 7.

Figure 9.23 Refraction and Reflection of Seismic Waves Refraction and reflection of P-waves as they encounter boundaries separating materials of different density or elasticity. Notice that the only wave ray not refracted is the one perpendicular to boundaries.

>> Critical Thinking Question

If Earth were a homogeneous body, how would seismic waves behave as they moved through Earth?

Figure 9.24 Seismic Wave Velocities Profiles showing seismic wave velocities versus depth. Several discontinuities are shown across which seismic wave velocities change rapidly.

9.12 The Core

In 1906, R. D. Oldham of the Geological Survey of India realized that seismic waves arrived later than expected at seismic stations more than 130 degrees from an earthquake focus. He postulated that Earth has a core that transmits seismic waves more slowly than shallower Earth materials. We now know that P-wave velocity decreases markedly at a depth of 2,900 km, which indicates an important discontinuity now recognized as the core–mantle boundary (Figure 9.24).

Because of the sudden decrease in P-wave velocity at the core–mantle boundary, P-waves are refracted in the core, so that little P-wave energy reaches the surface in the area between 103 degrees and 143 degrees from an earthquake focus (Figure 9.25a). This **P-wave shadow zone**, as it is called, is an area in which little P-wave energy is recorded by seismographs.

The P-wave shadow zone is not a perfect shadow zone because some weak P-wave energy is recorded within it. Scientists proposed several hypotheses to account for this observation, but all were rejected by the Danish seismologist Inge Lehmann, who in 1936 postulated that the core is not entirely liquid as previously thought. She proposed that seismic wave reflection from a solid inner core accounts for the arrival of weak P-wave energy in the P-wave shadow zone, a proposal that was quickly accepted by seismologists.

In 1926, the British physicist Harold Jeffreys realized that S-waves were not simply slowed by the core, but were completely blocked by it. So, besides a P-wave shadow zone, a much larger and more complete **S-wave shadow zone** also exists (Figure 9.25b). At locations greater than 103 degrees from an earthquake focus, no S-waves are recorded, which indicates that S-waves cannot be transmitted through the core. S-waves will not pass through a liquid, so it seems that the outer core must be liquid or behave as a liquid. The inner core, however, is thought to be solid because P-wave velocity increases at the base of the outer core.

Density and Composition of the Core

The core constitutes 16.4% of Earth's volume and nearly one-third of its mass. Geologists can estimate the core's density and composition by using seismic evidence and laboratory experiments. Furthermore, meteorites, which are thought to represent remnants of the material from which the solar system formed, are used to make estimates of density and composition. From these studies, the density of the outer core has been determined to vary from 9.9 to 12.2 g/cm³. At Earth's center, the pressure is equivalent to approximately 3.5 million times normal atmospheric pressure.

The core cannot be composed of minerals common at the surface because, even under the tremendous pressures at great depth, they would still not be dense enough to yield an average density of 5.5 g/cm³ for Earth. Both the outer

Figure 9.25 P-Wave and S-Wave Shadow Zones

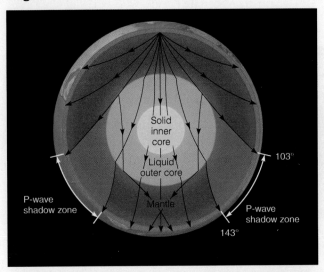

a P-waves are refracted so that no direct P-wave energy reaches the surface in the P-wave shadow zone.

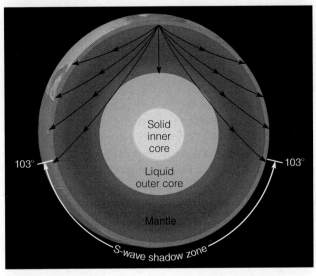

b The presence of an S-wave shadow zone indicates that S-waves are being blocked within Earth.

its interior has crystallized. Indeed, the inner core continues to grow as Earth slowly cools, and the liquid of the outer core crystallizes as iron. Recent evidence also indicates that at present the inner core rotates faster than the outer core, moving approximately 20 km/yr relative to the outer core.

CONNECTION LINK

To review the early history of Earth and its place in our solar system, see page 12 and Figure 1.10 in Chapter 1.

9.13 Earth's Mantle

Another significant discovery about Earth's interior was made in 1909 when the Yugoslavian seismologist Andrija Mohorovičić detected a seismic discontinuity at a depth of about 30 km. While studying the arrival times of seismic waves from Balkan (part of southeastern Europe) earthquakes, Mohorovičić noticed that seismic stations a few hundred kilometers from an earthquake's epicenter were recording two distinct sets of P- and S-waves.

From his observations, Mohorovičić concluded that a sharp boundary separates rocks with different properties at a depth of about 30 km. He postulated that P-waves below this boundary travel at 8 km/sec, whereas those above the boundary travel at 6.75 km/sec. When an earthquake occurs, some waves travel directly from the focus to a seismic station, whereas others travel through the deeper layer and some of their energy is refracted back to the surface (Figure 9.26). The waves traveling through the deeper layer (the mantle) travel farther to a seismic station, but they do so more rapidly and arrive before those that travel more slowly in the shallower layer.

The boundary identified by Mohorovičić separates the crust from the mantle and is now called the **Mohorovičić discontinuity**, or simply the **Moho**. It is present everywhere except beneath spreading ridges. However, its depth varies: Beneath continents, it ranges from 20 to 90 km, with an average of 35 km; beneath the seafloor, it is 5 to 10 km deep.

The Mantle's Structure, Density, and Composition

Although seismic wave velocity in the mantle increases with depth, several discontinuities exist. Between depths of 100 and 250 km, both P- and S-wave velocities decrease markedly (Figure 9.27). This 100- to 250-km-deep layer is the *low-velocity zone*, which corresponds closely to the *asthenosphere* (Figure 9.22), a layer in which the rocks are close to their melting point and are less elastic, accounting for the observed decrease in seismic wave velocity. The asthenosphere is an important zone because it is where most magma is generated, especially under the ocean

and inner cores are thought to be composed largely of iron, but pure iron is too dense to be the sole constituent of the outer core. It must be "diluted" with elements of lesser density. Laboratory experiments and comparisons with iron meteorites indicate that perhaps 12% of the outer core consists of sulfur and possibly some silicon, oxygen, nickel, and potassium (Figure 9.22).

In contrast, pure iron is not dense enough to account for the estimated density of the inner core, so perhaps 10–20% of the inner core consists of nickel. These metals form an iron–nickel alloy thought to be sufficiently dense under the pressure at that depth to account for the density of the inner core.

When the core formed during early Earth history, it was probably entirely molten and has since cooled so that

Figure 9.26 Seismic Discontinuity Andrija Mohorovičić studied seismic waves and detected a seismic discontinuity at a depth of about 30 km. The deeper, faster seismic waves arrive at seismic stations first, even though they travel farther. This discontinuity, now known as the Moho, is between the crust and mantle.

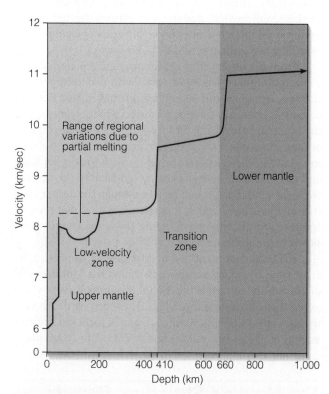

Figure 9.27 Variation in P-wave Velocity Variations in P-wave velocity in the upper mantle and transition zone.

represent structural changes in minerals rather than compositional changes. In other words, geologists think that the mantle is composed of the same material throughout, but the structural states of minerals such as olivine change with depth.

At a depth of 410 km, seismic wave velocity increases slightly as a consequence of such changes in mineral structure (Figure 9.27). Another velocity increase occurs at about 660 km, where the minerals break down into metal oxides, such as FeO (iron oxide) and MgO (magnesium oxide), and silicon dioxide (SiO_2). These two discontinuities define the top and base of a *transition zone* separating the upper mantle from the lower mantle (Figure 9.27).

Although the mantle's density, which varies from 3.3 to 5.7 g/cm³, can be inferred rather accurately from seismic waves, its composition is less certain. The igneous rock *peridotite*, containing mostly ferromagnesian silicates, is considered the most likely component of the upper mantle. Laboratory experiments indicate that it possesses physical properties that account for the mantle's density and observed rates of seismic wave transmissions. Peridotite also forms the lower parts of igneous rock sequences thought to be fragments of the oceanic crust and upper mantle emplaced on land (see Figure 2.20).

basins. Furthermore, it lacks strength, flows plastically, and is thought to be the layer over which the plates of the outer, rigid *lithosphere* move.

Other discontinuities are also present at deeper levels within the mantle. But unlike those between the crust and mantle or between the mantle and core, these probably

9.14 Seismic Tomography

The model of Earth's interior consisting of a core and a mantle is probably accurate, but not very precise. In recent years, geophysicists have developed a technique called *seismic tomography* that allows them to develop more accurate models of Earth's interior. In seismic tomography, numerous crossing seismic waves are analyzed much as CAT (computerized axial tomography) scans are analyzed. In CAT scans, X-rays penetrate the body and a two-dimensional image of its interior is formed. Repeated CAT scans from slightly different angles are stacked to produce a three-dimensional image.

In a similar manner, geophysicists use seismic waves to probe Earth's interior. In seismic tomography, the average velocities of numerous crossing seismic waves are analyzed so that "slow" and "fast" areas of wave travel are detected. Remember that seismic wave velocity depends partly on elasticity; cold rocks have greater elasticity and therefore transmit seismic waves faster than hotter rocks.

As a result of studies in seismic tomography, a much clearer picture of Earth's interior is emerging. It has already given us a better understanding of complex convection within the mantle and a clearer picture of the nature of the core–mantle boundary.

9.15 Earth's Internal Heat

During the 19th century, scientists realized that the temperature in deep mines increases with depth. More recently, the same trend has been observed in deep drill holes. This temperature increase with depth, or **geothermal gradient**, is approximately 25°C/km near the surface. In areas of active or recently active volcanism, the geothermal gradient is greater than in adjacent nonvolcanic areas, and temperature rises faster beneath spreading ridges than elsewhere beneath the seafloor.

> **CONNECTION LINK**
>
> To learn more about radioactive decay and the different radioactive elements, see pages 432 and 433, and Table 17.1 on page 434 in Chapter 17.

Much of Earth's internal heat is generated by radioactive decay, especially the decay of isotopes of uranium and thorium and, to a lesser degree, potassium-40. When these isotopes decay, they emit energetic particles and gamma rays that heat surrounding rocks. Because rock is such a poor conductor of heat, it takes little radioactive decay to build up considerable heat, given enough time.

Unfortunately, the geothermal gradient is not useful for estimating temperatures at great depth. If we were simply to extrapolate from the surface downward, the temperature at 100 km would be so high that, despite the great pressure, all known rocks would melt. Yet except for pockets of magma, it appears that the mantle is solid rather than liquid because it transmits S-waves. Accordingly, the geothermal gradient must decrease markedly.

Current estimates of the temperature at the base of the crust are 800 to 1,200°C. The latter figure seems to be an upper limit; if it were any higher, melting would be expected. Furthermore, fragments of mantle rock thought to have come from depths of 100–300 km appear to have reached equilibrium at these depths at a temperature of approximately 1,200°C. At the core–mantle boundary, the temperature is probably between 2,500 and 5,000°C; the wide range of values indicates the uncertainties of such estimates. If these figures are reasonably accurate, the geothermal gradient in the mantle is only about 1°C/km.

9.16 Earth's Crust

Our main concern in the latter part of this chapter is Earth's interior; however, to be complete, we must briefly discuss the crust, which along with the upper mantle constitutes the lithosphere.

Continental crust is complex, consisting of all rock types, but it is usually described as "granitic," meaning that its overall composition is similar to that of granitic rocks. With the exception of metal-rich rocks such as iron ore deposits, most rocks of the continental crust have densities between 2.5 and 3.0 g/cm^3, with the average density of the crust being about 2.7 g/cm^3. P-wave velocity in continental crust is about 6.75 km/sec, but at the base of the crust, P-wave velocity abruptly increases to about 8 km/sec. Continental crust averages 35 km thick, but its thickness varies from 20 to 90 km. Beneath mountain ranges such as the Rocky Mountains, the Alps in Europe, and the Himalayas in Asia, continental crust is much thicker than it is in adjacent areas. In contrast, continental crust is much thinner than average beneath the Rift Valleys of East Africa and in a large area called the Basin and Range Province in the western United States and northern Mexico. The crust in these areas has been stretched and thinned in what appear to be the initial stages of rifting (see Chapter 2).

In contrast to continental crust, oceanic crust is simpler, consisting of gabbro in its lower part overlain by basalt. It is thinnest, about 5 km, at spreading ridges, and nowhere is it thicker than 10 km. Its average density of 3.0 g/cm^3 accounts for the fact that it transmits P-waves at about 7 km/sec. In fact, this P-wave velocity is what one would expect if oceanic crust is composed of basalt and gabbro.

GEO-RECAP

❱❱ Geo-Image

At 3:02 a.m., on August 17, 1999, violent shaking from an earthquake awakened millions of people in Turkey. Unfortunately for many, their houses or apartment buildings collapsed, causing an estimated 17,000 deaths and at least 50,000 injuries and leaving more than 150,000 buildings moderately to heavily damaged.

What makes Turkey such an earthquake-prone area, and why are there so many deaths and so much destruction when earthquakes of similar size along the San Andreas fault usually result in far fewer casualties and much less damage? To answer this question, look at ▌Figure 1a in terms of Turkey's plate tectonic setting. For the second part of the question, look at ▌Figure 1b and c and think about the types of construction, the population, and other factors that lead to large numbers of deaths and massive destruction in earthquake-prone regions.

▌**Figure 1** The Izmit, Turkey, earthquake of August 17, 1999

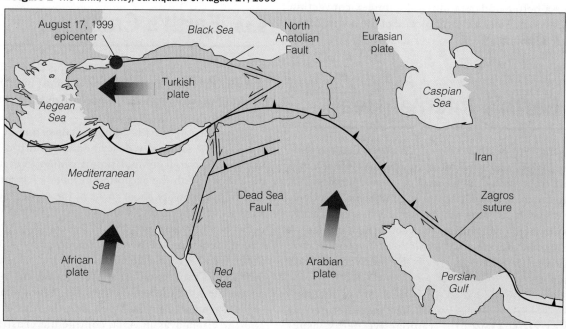

a Map showing the plate tectonic relationships in the region. The earthquake took place along the North Anatolian Fault.

Mehmet Celebi, U.S. Geological Survey

b Some of the numerous severely damaged structures in Izmit, Turkey.

Charles Meuller, U.S. Geological Survey

c Damage ranged from minimal to complete in this area of Turkey following the 7.4-magnitude earthquake on Auguest 17, 1999.

Chapter Summary

- Earthquakes are vibrations caused by the sudden release of energy, usually along a fault.

- The elastic rebound theory is an explanation for how energy is released during earthquakes. As rocks on opposite sides of a fault are subjected to force, they accumulate energy and slowly deform until their internal strength is exceeded. At that time, a sudden movement occurs along the fault, releasing the accumulated energy, and the rocks snap back to their original, undeformed shape.

- Seismology is the study of earthquakes. Earthquakes are recorded on seismographs, and the record of an earthquake is a seismogram.

- An earthquake's focus is the location where rupture within Earth's lithosphere occurs and energy is released. The epicenter is the point on Earth's surface directly above the focus.

- Approximately 80% of all earthquakes occur in the circum-Pacific belt, 15% within the Mediterranean–Asiatic belt, and the remaining 5% mostly in the interior of the plates and along oceanic spreading ridges.

- The two types of body waves are P-waves (primary waves), which are compressional (expanding and compressing) and the fastest seismic waves, traveling through all material, and S-waves (secondary waves), which are shear (moving material perpendicular to the direction of travel) and slower than P-waves and can travel only through solids.

- Rayleigh waves (R-waves) and Love waves (L-waves) move along or just below Earth's surface.

- An earthquake's epicenter is determined using a time–distance graph of the P- and S-waves to calculate how far away a seismic station is from an earthquake. The greater the difference in arrival times between the two waves, the farther away the seismic station is from the earthquake. Three seismographs are needed to locate the epicenter.

- Intensity is a subjective, or qualitative, measure of the kind of damage done by an earthquake. It is expressed in values from I to XII in the Modified Mercalli Intensity Scale.

- The Richter Magnitude Scale measures an earthquake's magnitude, which is the total amount of energy released by an earthquake at its source. It is an open-ended scale with values beginning at 1. Each increase in magnitude number represents about a 30-fold increase in energy released.

- The seismic-moment magnitude scale more accurately measures the total energy released by very large earthquakes.

- The destructive effects of earthquakes include ground shaking, fire, tsunami, landslides, and disruption of vital services.

- Seismic risk maps help geologists in determining the likelihood and potential severity of future earthquakes based on the intensity of past earthquakes.

- Earthquake precursors are changes preceding an earthquake and include seismic gaps, changes in surface elevations, and fluctuations of water levels in wells.

- A variety of earthquake research programs are under way in various countries. Studies indicate that most people would probably not heed a short-term earthquake warning.

- Although it is unlikely that earthquakes can ever be prevented, it might be possible to release small amounts of the energy stored in rocks and thus avoid a large, devastating earthquake.

- Various studies indicate that Earth has an outer layer of oceanic and continental crust below which lies a rocky mantle and an iron-rich core with a solid inner part and a liquid outer part.

- Density and elasticity of Earth materials determine the velocity of seismic waves. Seismic waves are refracted when their direction of travel changes. Wave reflection occurs at boundaries across which the properties of rocks change.

- Geologists use the behavior of P- and S-waves and the presence of the P- and S-wave shadow zones to estimate the density and composition of Earth's interior, as well as to estimate the size and depth of the core and mantle.

- Earth's inner core is probably made up of iron and nickel, whereas the outer core is mostly iron with 10–20% other substances.

- Peridotite, an igneous rock composed mostly of ferromagnesian silicates, is the most likely rock making up Earth's mantle.

- Oceanic crust is composed of basalt and gabbro, whereas continental crust has an overall composition similar to that of granite. The Moho is the boundary between the crust and the mantle.

- The geothermal gradient of 25°C/km cannot continue to great depths; within the mantle and core, it is probably about 1°C/km. The temperature at Earth's center is estimated to be 6,500°C.

Important Terms

discontinuity (p. 223)
earthquake (p. 200)
elastic rebound theory (p. 201)
epicenter (p. 202)
focus (p. 202)
geothermal gradient (p. 227)
intensity (p. 208)
Love wave (L-wave) (p. 206)

magnitude (p. 211)
Modified Mercalli Intensity Scale
  (p. 208)
Mohorovičić discontinuity (Moho)
  (p. 225)
P-wave (p. 206)
P-wave shadow zone (p. 224)
Rayleigh wave (R-wave) (p. 206)

Richter Magnitude Scale (p. 211)
seismograph (p. 202)
seismology (p. 202)
S-wave (p. 206)
S-wave shadow zone (p. 224)
tsunami (p. 216)

Review Questions

1. The focus of an earthquake is
   a. ____ the term for how destructive an earthquake is.
   b. ____ the point on Earth's surface vertically above where rupture occurs.
   c. ____ the opposite of the hypocenter.
   d. ____ the location where energy is released.
   e. ____ none of these.

2. The majority of all earthquakes take place in the
   a. ____ spreading-ridge zone.
   b. ____ Mediterranean–Asiatic belt.
   c. ____ rifts in continental interiors.
   d. ____ circum-Pacific belt.
   e. ____ Appalachian fault zone.

3. The seismic discontinuity at the base of the crust is known as the
   a. ____ transition zone.
   b. ____ magnetic reflection point.
   c. ____ low-velocity zone.
   d. ____ Moho.
   e. ____ high-velocity zone.

4. The minimum number of seismographs needed to determine an earthquake's epicenter is
   a. ____ 1.
   b. ____ 2.
   c. ____ 3.
   d. ____ 4.
   e. ____ 5.

5. The total amount of energy released by an earthquake at its source is its
   a. ____ intensity.
   b. ____ dilatancy.
   c. ____ seismicity.
   d. ____ magnitude.
   e. ____ liquefaction.

6. An S-wave is one in which
   a. ____ material moves perpendicular to the direction of wave travel.
   b. ____ Earth's surface moves as a series of waves.
   c. ____ material is expanded and compressed as the wave moves through it.
   d. ____ large waves crash onto a shoreline following a submarine earthquake.
   e. ____ none of these.

7. In urban areas in particular, what generally causes the most damage in an earthquake?
   a. ____ ground shaking.
   b. ____ fire.
   c. ____ tsunami.
   d. ____ ground failure.
   e. ____ flooding.

8. How much more energy is released by a magnitude 5 earthquake than a magnitude 2 earthquake?
   a. ____ 4.
   b. ____ 810,000.
   c. ____ 27,000.
   d. ____ 90.
   e. ____ 2,500,000.

9. Regions along a major fault that are locked and not releasing energy are prime sites for future earthquakes and are known as
   a. ____ epicenters.
   b. ____ foci.
   c. ____ hypocenters.
   d. ____ discontinuities.
   e. ____ seismic gaps.

10. The geothermal gradient is Earth's
    a. ____ capacity to reflect and refract seismic waves.
    b. ____ most destructive aspect of earthquakes.
    c. ____ temperature increase with depth.
    d. ____ average rate of seismic wave velocity in the mantle.
    e. ____ elastic rebound potential.

11. What are precursors, and how can they be used to predict earthquakes?

12. Why do insurance companies use the qualitative Modified Mercalli Intensity Scale instead of the quantitative Richter Magnitude Scale in classifying earthquakes?

13. How are plate boundaries and earthquake foci related?

14. Refer to the graph in Figure 9.9b. A seismograph in Berkeley, California, recorded the arrival time of an earthquake's P-waves as 6:59:54 p.m. and the S-waves as 7:00:02 p.m. The maximum amplitude of the S-waves as recorded on the seismogram was 75 mm. What was the magnitude of the earthquake, and how far away from Berkeley did it occur?

15. Why do scientists think that the inner core is solid and the outer core is liquid? What is the core composed of?

16. What is the likely explanation for changes with depth in the mantle?

17. How does the elastic rebound theory account for the energy released during an earthquake?

18. What is the difference between an earthquake's focus and its epicenter? Why is an earthquake's epicenter the location that is usually reported in the news?

19. Describe how a tsunami is generated, how it travels, and what impact it has on shorelines.

20. From the arrival times of P- and S-waves shown in the accompanying chart and the graph in Figure 9.9b, calculate how far away from each seismograph station the earthquake occurred. How would you determine the epicenter of this earthquake?

	Arrival Time of P-Wave	Arrival Time of S-Wave
Station A	2:59:03 p.m.	3:04:03 p.m.
Station B	2:51:16 p.m.	3:01:16 p.m.
Station C	2:48:25 p.m.	2:55:55 p.m.

View from Pakistan of the Karakoram Range, which is part of the Himalayan mountain system. This range lies on the borders of Pakistan, India, and China. The Karakoram Range has the most peaks greater than 8,000 m high, including K2, which at 8,611 m is the second-highest peak on Earth. In addition, the range has a large concentration of glaciers.

CHAPTER 10

Persis Sturges

Deformation, Mountain Building, and Earth's Crust

OUTLINE

10.1 Introduction

10.2 Rock Deformation–How Does It Occur?

10.3 Strike and Dip–The Orientation of Deformed Rock Layers

10.4 Deformation and Geologic Structures

10.5 Deformation and the Origin of Mountains

GEO-FOCUS: Engineering and Geology

10.6 Earth's Crust

GEO-RECAP

HAVE YOU EVER WONDERED?

- What kinds of forces deform rocks?
- How geologists interpret rocks that have been crumpled and fractured?
- Whether deformed rocks are a problem in planning and building highways, dams, and other structures?
- How mountains differ from other parts of Earth's crust?
- If there is a relationship between plate tectonics and the origin of mountains?
- Why the continents stand higher than the ocean basins, and why mountains are higher than adjacent parts of continents?

10.1 Introduction

"Solid as a rock" implies permanence and durability, but you already know that when rocks weather they disaggregate and decompose (see Chapter 6), and that they behave very differently during metamorphism (see Chapter 8). Under the tremendous pressure and high temperature at several kilometers below the surface, rock layers yield to forces and actually crumple or fold, yet remain solid. And at shallower depths they may fracture or fold and fracture. In any case, dynamic forces within Earth cause **deformation,** a general term for all changes in the shape or volume of rocks (❚ Figure 10.1).

The fact that dynamic forces continue to operate within Earth is obvious from seismic activity, volcanism, plate movements, and the evolution of mountains in South America, Asia, and elsewhere. In short, Earth is an active planet with several processes driven by internal heat, especially plate movements; much of Earth's seismic activity, volcanism, deformation, and mountain building take place at convergent plate boundaries. The other terrestrial planets, with the possible exception of Venus, and Earth's Moon show little evidence of ongoing deformation, volcanism, and so on.

The origin of Earth's truly large mountain ranges on the continents involves tremendous deformation, usually accompanied by emplacement of plutons, volcanism, and metamorphism, at convergent plate boundaries (see the chapter opening photo). The Appalachians of North America, the Alps in Europe, the Himalayas of Asia, and the Andes in South America all owe their existence to deformation at convergent plate boundaries. And, in some cases, this activity continues even now. Thus, deformation and mountain building are closely related topics, and, accordingly, we consider both in this chapter.

> **CONNECTION LINK**
>
> You can find more information about continental accretion on pages 478–480 in Chapter 19.

The past and continuing evolution of continents involves not only deformation at continental margins but also additions of new material to existing continents, a phenomenon known as *continental accretion.* North America, for instance, has not always had its present shape and area. Indeed, it began evolving during the Archean Eon (4.6–2.5 billion years ago) as new material was added to the continent at deformation belts along its margins.

Much of this chapter is devoted to a review of *geologic structures,* such as folded and fractured rock layers resulting from deformation, their descriptive terminology, and the forces responsible for them. There are several practical reasons to study deformation and mountain building. For one thing, crumpled and fractured rock layers provide a record of the kinds and intensities of forces that operated during the past. Thus, interpretations of these structures allow us to satisfy our curiosity about Earth history, and, in addition, such studies are essential in engineering endeavors such as choosing sites for dams, bridges, and nuclear power plants, especially if they are in areas of ongoing deformation. Also, many aspects of mining and exploration for petroleum and natural gas rely on correctly identifying geologic structures.

❚ **Figure 10.1** Many Rocks Show the Effects of Deformation

a These rock layers in the Balkan Mountains of Serbia have been folded. The vertical dimension of this image is about 10 m.

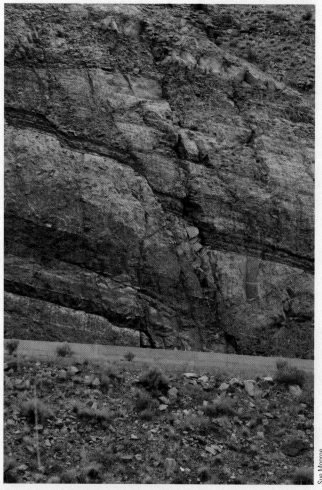

b Deformation of rocks in Utah by fracturing. Notice that the continuity of the rock layers has been disrupted along fractures or what geologists call faults.

10.2 Rock Deformation— How Does It Occur?

We defined the term *deformation* as any change in the volume or shape of rocks, but our reference to rocks means rock layers as depicted in many images in this text, not isolate stones you might find in your driveway or in a streambed. The type of rock is irrelevant, although layered rocks show the effects of deformation most clearly. In any case, rock layers may be crumpled into folds or fractured as a result of **stress,** which results from force applied to a given area of rock. The rock's internal strength resists stress, but if the stress is great enough the rock undergoes **strain,** which is simply deformation caused by stress. The terminology is a little confusing at first, but the following discussion and reference to Figure 10.2 will help you understand these terms.

Stress and Strain

Remember that stress is the force applied to a given area of rock, usually expressed in kilograms per square centimeter (kg/cm^2). For example, the stress, or force, exerted by a person walking on an ice-covered pond is a function of the person's weight and the area beneath her or his feet. The ice's internal strength resists the stress unless the stress is too great, in which case the ice may bend or crack as it is strained (deformed) (❙ Figure 10.2). To avoid breaking through the ice, the person may lie down; this does not reduce the weight of the person on the ice, but it does distribute it over a larger area, thus reducing the stress per unit area.

Although stress is force per unit area, it comes in three varieties: *compression, tension,* and *shear,* depending on the direction of the applied forces. In **compression,** rocks or any other object are squeezed or compressed by forces directed toward one another along the same line, as when you squeeze a rubber ball in your hand. Rock layers in compression tend to be shortened in the direction of stress by either folding or fracturing (❙ Figure 10.3a). **Tension** results from forces acting along the same line, but in opposite directions. Tension tends to lengthen rocks or pull them apart (❙ Figure 10.3b). Incidentally, rocks are much stronger in compression than they are in tension. In **shear stress,** forces act parallel to one another, but in opposite directions, resulting in deformation by displacement along closely spaced planes (❙ Figure 10.3c).

Types of Strain

Geologists characterize strain as **elastic strain** if deformed rocks return to their original shape when the deforming forces are relaxed. In Figure 10.2, the ice on the pond may bend under a person's weight, but return to its original shape once the person leaves. As you might expect, rocks are not very elastic, but Earth's crust behaves elastically when loaded by glacial-ice and depressed into the mantle.

❙ **Figure 10.2** Stress and Strain Exerted on an Ice-Covered Pond

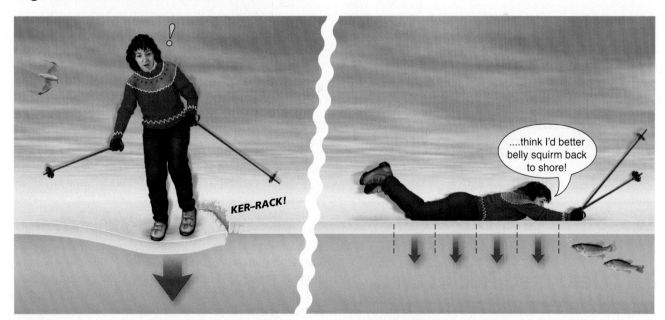

a The woman weighs 65 kg (6,500 g). Her weight is imparted to the ice through her feet, which have a contact area of 120 cm². The stress she exerts on the ice (6,500 g/120 cm²) is 54 g/cm³. This is now sufficient stress to cause the ice to crack.

b To avoid plunging into the freezing water, the woman lays out flat, thereby decreasing the stress she exerts on the ice. Her weight remains the same, but her contact area with the ice is now 3,150 cm², so the stress is now only about 2 g/cm² (6,500 g/3,150 cm², which is well below the threshold needed to crack the ice.

As stress is applied, rocks respond first by elastic strain, but when strained beyond their elastic limit, they undergo **plastic strain** as when they yield by folding, or they behave like brittle solids and **fracture** (❙ Figure 10.4). In either folding or fracturing, the strain is permanent; that is, the rocks do not recover their original shape or volume even if the stress is removed.

Whether strain is elastic, plastic, or fracture depends on the kind of stress applied, pressure and temperature, rock type, and the length of time rocks are subjected to stress. A small stress applied over a long period, as on a mantelpiece supported only at its ends, will cause the rock to sag; that is, the rock deforms plastically (Figure 10.4). By contrast, a large stress applied rapidly to the same object, as when struck by a hammer, results in fracture. Rock type is important because not all rocks have the same internal strength and thus respond to stress differently. Some rocks are *ductile,* whereas others are *brittle,* depending on the amount of plastic strain they exhibit. Brittle rocks show little or no plastic strain before they fracture, but ductile rocks exhibit a great deal (Figure 10.4).

Many rocks show the effects of plastic deformation that must have taken place deep within the crust. At or near the surface, rocks commonly behave like brittle solids and fracture, but at depth, they more often yield by plastic deformation; they become more ductile with increasing pressure and temperature.

❙ **Figure 10.3** Stress and Possible Types of Resulting Deformation

a Compression causes shortening of rock layers by folding or faulting.

b Tension lengthens rock layers and causes faulting.

c Shear stress causes deformation by displacement along closely spaced planes.

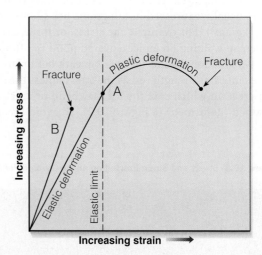

❙ **Figure 10.4** Rock Responses to Stress Rocks initially respond to stress by elastic deformation and then return to their original shape when the stress is released. If the elastic limit is exceeded, as in curve A, rocks deform plastically, which is permanent deformation. The amount of plastic deformation rocks exhibit before fracturing depends on their ductility. If they are ductile, they show considerable plastic deformation (curve A), but if they are brittle, they show little or no plastic deformation before failing by fracture (curve B).

What Would You Do ?

The types of stresses, as well as elastic versus plastic strain might seem rather esoteric, but perhaps understanding these concepts has some practical applications. What relevance do you think knowing about stress and strain has to some professions, other than geology, and what professions might these be? Can you think of stresses and strain that we contend with in our daily lives? As an example, what happens when a car smashes into a tree?

10.3 Strike and Dip— The Orientation of Deformed Rock Layers

During the 1660s, Nicholas Steno, a Danish anatomist, proposed several principles essential for deciphering Earth history from the record preserved in rocks. One is the *principle of original horizontality,* meaning that sediments accumulate in horizontal or nearly horizontal layers. Thus, if we observe steeply inclined sedimentary rocks, we are justified in inferring that they were deposited nearly horizontally, lithified, and then tilted into their present position (❙ Figure 10.5). Rock layers deformed by folding, faulting, or both are no longer in their original position, so geologists use *strike* and *dip* to describe their orientation with respect to a horizontal plane.

By definition, **strike** is the direction of a line formed by the intersection of a horizontal plane and an inclined plane.

Figure 10.5 Strike and Dip of Deformed Rock Layers We can infer that the sedimentary rocks in this illustration were deposited horizontally or nearly so, lithified, and then deformed. To describe their orientation, geologists use the terms *strike* and *dip*. Strike is the intersection of a horizontal plane with an inclined plane. Dip is the maximum angular deviation of the inclined layers from horizontal. Notice the symbol that shows strike and dip.

>> **Critical Thinking Question**
Suppose it is 1 km from the right side of the diagram to the point where the uppermost sandstone layer plunges beneath the surface. Can you predict the depth at which this sandstone would be encountered if you drilled a hole at the right side of the diagram? Explain.

> **CONNECTION LINK**
>
> You can find more information about Nicholas Steno and his principle of original horizontality on pages 418 and 419 in Chapter 17.

The surfaces of the rock layers in Figure 10.5 are good examples of inclined planes, whereas the ground surface is a horizontal plane. The direction of the line formed at the intersection of these planes is the strike of the rock layers. The strike line's orientation is determined by using a compass to measure its angle with respect to north. **Dip** is a measure of an inclined plane's deviation from horizontal, so it must be measured at right angles to strike direction (Figure 10.5).

Geologic maps showing the age, aerial distribution, and geologic structures in an area use a special symbol to indicate strike and dip. A long line oriented in the appropriate compass direction indicates strike, and a short line perpendicular to the strike line shows the direction of dip (Figure 10.5). Adjacent to the strike and dip symbol is a number corresponding to the dip angle. The usefulness of strike and dip symbols will become apparent in the sections on folds and faults.

10.4 Deformation and Geologic Structures

Remember that deformation and its synonym strain refer to changes in the shape or volume of rocks. During deformation, rocks might be crumpled into folds, or they might be fractured, or perhaps folded and fractured (Figure 10.1). Any of these features resulting from deformation is referred

to as a **geologic structure.** Geologic structures are present almost everywhere that rock exposures are present, and many are detected far below the surface by drilling and several geophysical techniques.

Folded Rock Layers

If you place your hands on a tablecloth and move them toward one another, the tablecloth would be crumpled into a series of up- and down-arched folds. Rock layers behave similarly when in compression as they deform into **folds,** but in this case the folding is permanent. That is, plastic strain has taken place, so even if the stress is removed, the rock layers remain folded. Most folding probably occurs deep in the crust where pressure and temperature are high and rocks are more ductile than they are at or near the surface. The configuration of folds and the intensity of folding vary, but there are only three basic types of folds: *monoclines, anticlines,* and *synclines.*

Monoclines A simple bend or flexure in otherwise horizontal or uniformly dipping rock layers is a **monocline** (Figure 10.6a). The large monocline in Figure 10.6b formed when the Bighorn Mountains in Wyoming rose vertically along a fracture. The fracture did not penetrate to the surface, so as uplift of the mountains proceeded, the near-surface rocks were bent so that they now appear to be draped over the margin of the uplifted block. In a manner of speaking, a monocline is simply one-half of an anticline or syncline.

Anticlines and Synclines An **anticline** is an up-arched or convex upward fold with the oldest rock layers in its core, whereas a **syncline** is a down-arched or concave downward fold in which the youngest rock layers are in its core (Figure 10.7). Anticlines and synclines have an axial plane connecting the points of maximum curvature of each folded layer (Figure 10.8); the axial plane divides folds into halves, each half being a *limb.* Because folds are most often found in a series of anticlines alternating with synclines, an anticline and adjacent syncline share a limb. It is important to remember that anticlines and synclines are simply folded rock layers and do not necessarily correspond to high and low areas at the surface (Figure 10.9).

Folds are commonly exposed to view in areas of deep erosion, but even where eroded, strike and dip and the relative ages of the folded rock layers easily distinguish anticlines from synclines. Notice in Figure 10.9 that in the surface view of the anticline, each limb dips outward or

▌Figure 10.6 Monocline

a A monocline. Notice the strike and dip symbol and the circled cross, which is the symbol for horizontal layers.

b A monocline in the Bighorn Mountains in Wyoming.

away from the center of the fold, and the oldest exposed rocks are in the fold's core. In an eroded syncline, though, each limb dips inward toward the fold's center, where the youngest exposed rocks are found.

The folds described so far are *upright,* meaning that their axial planes are vertical and both fold limbs dip at the same angle (Figure 10.9). In many folds, the axial plane is not vertical, the limbs dip at different angles, and the folds are characterized as *inclined* (▌ Figure 10.10a). If both limbs dip in the same direction, the fold is *overturned.* That is, one limb has been rotated more than 90 degrees from its original position so that it is now upside down (▌ Figure 10.10b). In some areas, deformation has been so intense that axial planes of folds are now horizontal, giving rise to what geologists call *recumbent folds* (▌ Figure 10.10c). Overturned and recumbent folds are particularly common in mountains

resulting from compression at convergent plate boundaries (discussed later in this chapter).

For upright folds, the distinction between anticlines and synclines is straightforward, but interpreting complex folds that have been tipped on their sides or turned completely upside down is more difficult. Can you determine which of the two folds shown in Figure 10.10c is an anticline? Even if strike and dip symbols were shown, you would still not be able to resolve this question, but knowing the relative ages of the folded rock layers provides a solution. Remember that an anticline has the oldest rock layers in its core, so the fold nearest the surface is an anticline and the lower fold is a syncline.

▌Figure 10.7 Anticlines and Synclines Rock layers folded into anticlines and synclines in the Calico Mountains of southeastern California. Compression caused the folds, which are, from left to right, a syncline, an anticline, and a syncline.

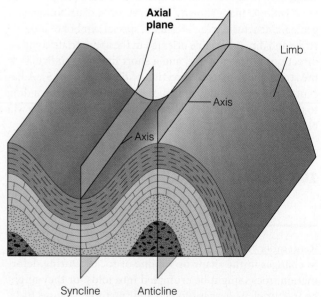

▌Figure 10.8 Syncline and Anticline Axial Planes Syncline and anticline showing the axial plane, axis, and fold limbs.

❚ **Figure 10.9 Eroded Anticlines and Synclines** Geologists identify eroded anticlines and synclines by strike and dip and the relative ages of the folded rock layers.

>> **Critical Thinking Question**
Show on this illustration the direction of the forces that caused deformation.

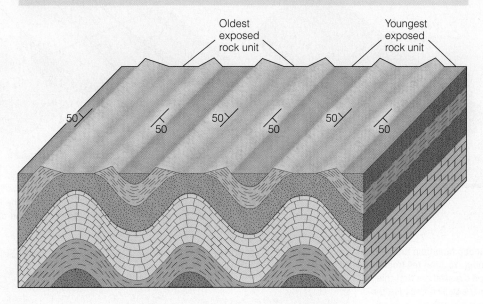

Oldest exposed rock unit

Youngest exposed rock unit

50 50 50 50 50 50

❚ **Figure 10.10** Inclined, Overturned, and Recumbent Folds

Inclined

Axial plane

20 55

a An inclined fold. The axial plane is not vertical, and the fold limbs dip at different angles.

Overturned

Axis

Axial plane

75 25 75 25

b Overturned folds. Both fold limbs dip in the same direction, but one limb is inverted. Notice the special strike-and-dip symbol to indicate overturned beds.

Recumbent

Oldest rocks

Youngest rocks

Axial plane

c Recumbent folds.

❚ **Figure 10.11** Plunging Folds

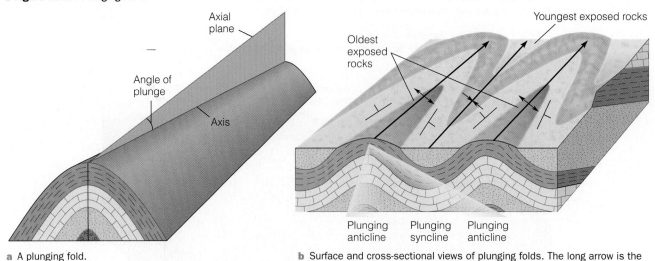

a A plunging fold.

b Surface and cross-sectional views of plunging folds. The long arrow is the geologic symbol for a plunging fold; it shows the direction of plunge.

c The Sheep Mountain anticline in Wyoming. You can tell from the strike-and-dip symbols that this is a plunging anticline because the rock layers dip outward or away from the fold axis.

>> Critical Thinking Question
Notice the small fold on the right side of this large plunging anticline. What kind of fold is it and how do you know?

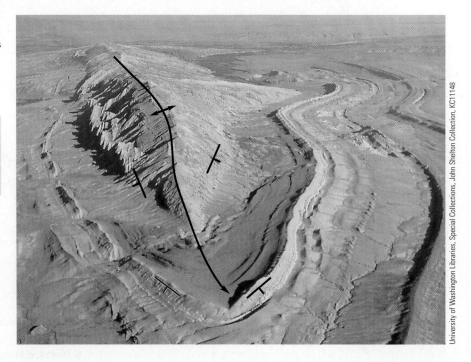

Plunging Folds To complicate matters further, folds are also characterized as *nonplunging* or *plunging*. In some folds, the fold axis, a line formed by the intersection of the axial plane with the folded layers, is horizontal and the folds are nonplunging (Figure 10.9). Much more commonly, though, fold axes are inclined so that they appear to plunge beneath adjacent rocks, and the folds are said to be plunging (❚ Figure 10.11a).

It might seem that with this additional complication, differentiating plunging anticlines from plunging synclines would be much more difficult; however, geologists use exactly the same criteria that they use for nonplunging folds. Therefore, all rock layers dip away from the fold axis

in plunging anticlines and toward the axis in plunging synclines. The oldest exposed rocks are in the core of an eroded plunging anticline, whereas the youngest exposed rock layers are found in the core of an eroded plunging syncline (❚ Figure 10.11b).

In Chapter 7, we noted that anticlines form one type of structural trap in which petroleum and natural gas might accumulate (see Figure 7.18b). As a matter of fact, most of the world's petroleum production comes from anticlines, although other geologic structures and stratigraphic traps are important, too. Accordingly, geologists are particularly interested in correctly identifying geologic structures in areas of potential hydrocarbon production.

Domes and Basins   Anticlines and synclines are elongate structures, meaning that their length greatly exceeds their width. In contrast, folds that are nearly equidimensional—that is, oval to circular—are *domes* and *basins*. In a **dome,** all of the folded strata dip outward from a central point (as opposed to outward from a line as in an anticline), and the oldest exposed rocks are at the center of the fold (❚ Figure 10.12a), so we characterize a dome as the circular equivalent of an anticline. In contrast, a **basin,** the circular counterpart of a syncline, has all strata dipping inward toward a central point and the youngest exposed rocks are at the fold's center (❚ Figure 10.12b). Many domes and basins are so large that they can be visualized only on geologic maps or aerial photographs.

Unfortunately, the terms *dome* and *basin* are also used to distinguish high and low areas of Earth's surface, but as with anticlines and synclines, domes and basins resulting from deformation do not necessarily correspond with mountains or valleys. In some of the following discussions, we will have occasion to use these terms in other contexts, but we will try to be clear when we refer to surface elevations as opposed to geologic structures.

❚ **Figure 10.12** Domes and Basins

Dome

a Notice that in a dome, the oldest exposed rocks are in the center and all rocks dip outward from a central point.

Basin

b In a basin, the youngest exposed rocks are in the center and all rocks dip inward toward a central point.

Joints

Besides folding, rocks are also permanently deformed by fracturing. **Joints** are fractures along which no movement has taken place parallel with the fracture surface (❚ Figure 10.13), although they may open up; that is, joints may show movement perpendicular to the fracture. Coal miners used the term *joint* long ago for cracks in rocks that they thought were surfaces where the adjacent blocks were "joined" together.

Remember that rocks near the surface are brittle and therefore commonly fail by fracturing when subjected to stress. In fact, almost all near-surface rocks have joints that form in response to compression, tension, and shearing. They vary from minute fractures to those extending for many kilometers and are often arranged in two or perhaps three

❚ **Figure 10.13** Joints Fractures along which no movement has taken place parallel with the fracture surface are called joints.

a Joints intersecting at right angles yield this rectangular pattern in Wales.

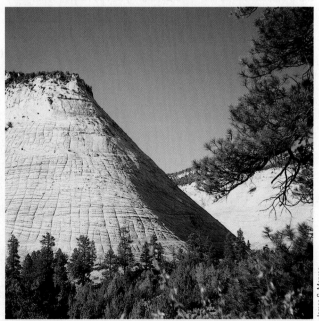

b The Jurassic-age Navajo Sandstone in Zion National Park, Utah. Vertical joints intersect cross-beds, hence the name Checkerboard Mesa.

prominent sets. Regional mapping reveals that joints and sets of joints are usually related to other geologic structures such as faults and folds. In Figures 1–3, Geo-Insight on page 150, for example, the rock layers were broadly folded and parallel fractures developed. In this example, compression caused the folding, but as folding progressed the upper part of the folded layer was stretched (experienced tension) and fractured.

Other types of joints that we have discussed previously include columnar joints that form when lava or magma in some shallow plutons cools and contracts (see Figure 5.7), and sheet jointing that forms in response to pressure release (see Figure 6.3).

Faults

Joints and faults are both fractures, but on joints no movement occurs parallel with the fracture surface, whereas on **faults**, blocks on opposite sides of the fracture move parallel with the fracture surface, which is a **fault plane** (Figure 10.14a). Faults that penetrate to the surface might show a *fault scarp,* a bluff or cliff formed by vertical movement (Figure 10.14b). In some cases, the fault plane is scratched and polished, but in others the movement of the blocks on opposite sides of the fault grinds and pulverizes the rock into a zone of *fault breccia* (Figure 10.14c).

Figure 10.14 Faults Fractures along which movement has occurred parallel to the fracture surface are called faults.

a Terms used to describe the orientation of a fault plane. Striae are scratch marks that form when one block slides past another. You can measure offset or displacement on a fault wherever the truncated end of one feature (point A) can be related to its equivalent across the fault (B).

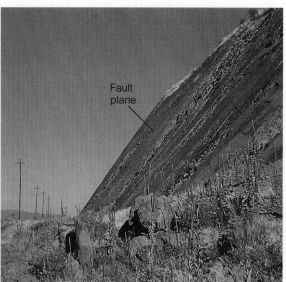

b A polished, scratched fault plane and fault scarp near Klamath Falls, Oregon

c Fault breccias, the zone of rubble along a fault in the Bighorn Mountains, Wyoming. The arrows show the direction of movement along the fault.

» Critical Thinking Question
If the arrows were not shown in Figure 10.14c, could you figure out the relative movement on this fault?

Notice in Figure 10.14a that the rocks overlying the fault make up the **hanging wall block,** whereas those beneath the fault constitute the **footwall block.** You can recognize these two blocks on any fault with the exception of one that is vertical—that is, one that dips at 90 degrees. In order to identify some faults you must identify these two blocks and determine the direction of relative movement of the blocks. By relative movement, we mean which one appears to have moved up or down the fault plane. For example, in Figure 10.14a, the footwall block may have moved up, the hanging wall may have moved down, or both blocks may have moved. Nevertheless, the hanging wall block appears to have moved down relative to the footwall block.

Remember that geologists use strike and dip to define the orientation of dipping rock layers. Fault planes are also dipping planes, so this same concept applies to them. As a matter of fact, two types of faults are recognized based on whether the blocks on opposite sides of the fault moved parallel to the direction of dip (dip-slip faults) or along the direction of strike (strike-slip faults).

Dip-Slip Faults

All movement on **dip-slip faults** takes place parallel with the fault's dip; that is, movement is vertical, either up or down the fault plane. In ▌Figure 10.15a, for example, the hanging wall block moved down relative to the footwall block, giving rise to a **normal fault** (▌Figure 10.16a). In contrast, in

▌**Figure 10.15** Types of Faults

a Normal fault—hanging wall block (HW) moves down relative to the footwall block (FW).

b Reverse fault—hanging wall block moves up relative to the footwall block.

c A thrust is a type of reverse fault with a fault plane dipping at less than 45 degrees.

d Left-lateral strike-slip fault.

e An oblique-slip fault involves a combination of dip-slip and strike-slip movements.

Figure 10.16 Dip-Slip Faults

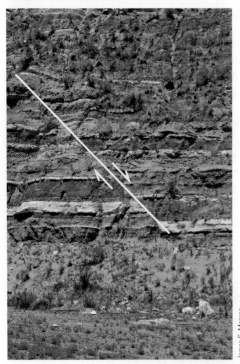

James S. Monroe

a A normal fault at Mt. Carmel Junction, Utah. The hanging wall block (on the right) has moved down about 2 m relative to the footwall block.

Sue Monroe

b A small reverse fault. Notice that the hanging wall block (on the right) has moved up relative to the footwall block.

>> **Critical Thinking Question**
Was compression or tension responsible for these faults?

a **reverse fault,** the hanging wall block moves up relative to the footwall block (█ Figures 10.15b and 10.16b). In █ Figure 10.15c, the hanging wall block also moved up relative to the footwall block, but the fault has a dip of less than 45 degrees and is a special variety of reverse fault known as a **thrust fault.**

Normal faults result from tension. Numerous normal faults are present along one or both sides of mountain ranges in the Basin and Range Province of the western United States where the crust is being stretched and thinned. The Sierra Nevada at the western margin of the Basin and Range is bounded by normal faults, and the range has risen along these faults so that it now stands more than 3,000 m above the lowlands to the east. Reverse and thrust faults result from compression (Figure 10.15b, c). Large-scale examples of both are found in mountain ranges that formed at convergent plate margins, where one would expect compression (discussed later in this chapter). A well-known thrust fault is the Lewis overthrust of Montana. (An overthrust is a low-angle thrust fault with movement measured in kilometers.) On this fault, a huge slab of Precambrian-age rocks moved at least 75 km eastward and now rests upon much younger Cretaceous-age rocks.

> **CONNECTION LINK**
>
> You can find more about the Sierra Nevada and the Basin and Range on pages 620 and 621 in Chapter 23.

Strike-Slip Faults **Strike-slip faults,** resulting from shear stresses, show horizontal movement with blocks on opposite sides of the fault sliding past one another (█ Figure 10.15d). In other words, all movement is in the direction of the fault plane's strike—hence the name *strike-slip* fault. Several large strike-slip faults are known, but the best studied is the San Andreas fault, which cuts through coastal California. Recall from Chapter 2 that the San Andreas fault is called a *transform fault* in plate tectonics terminology.

Strike-slip faults are characterized as right-lateral or left-lateral, depending on the apparent direction of offset. In Figure 10.15d, for example, observers looking at the block on the opposite side of the fault from their location notice that it appears to have moved to the left. Accordingly, this is a *left-lateral strike-slip fault.* If it had been a *right-lateral strike-slip fault,* the block across the fault from the observers would appear to have moved to the right. The San Andreas fault in California is a right-lateral strike-slip fault (█ Figure 10.17a), whereas the Great Glen fault in Scotland is a left-lateral strike-slip fault.

Oblique-Slip Faults The movement on most faults is primarily dip-slip or strike-slip, but on **oblique-slip faults,** both types of movement take place. Strike-slip movement might be accompanied by a component of dip-slip, giving rise to a combined movement that includes left-lateral and reverse, or right-lateral and normal (█ Figures 10.15e and 10.17b).

■ Figure 10.17 Strike-Slip and Oblique-Slip Faults

University of Washington Libraries, Special Collections, John Shelton Collection, KC6977

a Right-lateral offset of a gully by the San Andreas fault in central California. The gully is offset about 21 m.

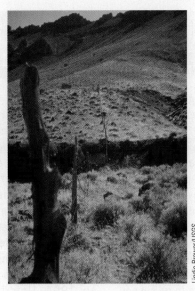

Lindie Brewer/USGS

b Oblique slip took place on this fault in central Nevada during a 1915 earthquake. Notice the fence that shows right-lateral displacement and dip-slip displacement.

As we mentioned in the Introduction, the study of geologic structures is important in the exploration for minerals, oil, and natural gas. In addition, geologic engineers evaluate geologic structures in the planning stages for many projects, especially if they are in tectonically active areas (see Geo-Focus on pages 246 and 247).

What Would You Do ?

As a member of a planning commission, you are charged with developing zoning regulations and building codes for an area with known active faults, steep hills, and deep soils. A number of contractors, as well as developers and citizens in your community, are demanding action because they want to begin several badly needed housing developments. How might geologic maps and an appreciation of geologic structures influence you in this endeavor? Do you think, considering the probable economic gains for your community, that the regulations you draft should be rather lenient or very strict? If the latter, how would you explain why you favored regulations that would involve additional cost for houses?

10.5 Deformation and the Origin of Mountains

Mountains form in several ways, but the truly large mountains on continents result mostly from compression-induced deformation at convergent plate boundaries. Before discussing mountain building, though, we should

define what we mean by the term *mountain* and briefly discuss the types of mountains. *Mountain* is a designation for any area of land that stands significantly higher, at least 300 m, than the surrounding country and has a restricted summit area. Some mountains are single, isolated peaks, but more commonly they are parts of linear associations of peaks and ridges known as *mountain ranges* that are related in age and origin. A *mountain system,* a complex linear zone of deformation and crustal thickening, on the other hand, consists of several or many mountain ranges. The Bitterroot Mountains of Idaho and Montana, the Teton Range in Wyoming, the Wasatch Mountains of Utah, and the Sangre de Cristo Mountains of Colorado and New Mexico are a few of the many ranges in the Rocky Mountain system. The Appalachian Mountains of the eastern United States and Canada are a mountain system with ranges such as the Great Smoky Mountains of North Carolina and Tennessee, the Adirondack Mountains of New York, and the Green Mountains of Vermont.

Mountain Building

Mountains form in several ways, some involving little or no deformation. For example, differential weathering and erosion yield high areas with adjacent lowlands in the southwestern United States, but these erosional remnants are flat-topped or pinnacle-shaped and go by the names *mesa* and *butte,* and most are less than 300 m high. *Block-faulting* is another way that mountains form, but this is caused by deformation of the crust. It involves movement on normal faults so that one or more blocks are elevated relative to adjacent blocks

CONNECTION LINK

You can find out more about mesas and buttes on page 382 in Chapter 15.

GEO-FOCUS   Engineering and Geology

As you might expect, engineering geologists apply the principles and concepts of geology to engineering practices. Geologists with this specialized training may be involved in slope stability studies and in studies of acceptable areas for power plants, for highways in mountainous regions, for tunnels and canals, and for structures to protect riverbank and seashore communities.

A good example is the concern prior to building the Mackinac (pronounced "mack-in-aw") Bridge, a huge suspension bridge that connects the Upper and Lower Peninsulas of Michigan (❙ Figure 1). Geologists and engineers were aware that some of the rock in the area, called Mackinac Breccia, was a collapse breccia, or rubble that formed when caverns collapsed. The concern was whether the breccia or any uncollapsed caverns beneath the area would support the weight of the huge piers and abutments for the bridge. Obviously, the project was completed successfully, but detailed studies were done before construction began.

Geologic engineers are invariably involved in planning for large-scale structures such as bridges, dams, power plants, and highways, especially in tectonically active regions. For example, the bridge in ❙ Figure 2 is only a short distance from the San Andreas fault, so engineers had to take into account the near certainty that it would be badly shaken during an earthquake. Many

other structures on or near the San Andreas fault were constructed when codes were much less stringent, and now they are being retrofitted to make them safer during earthquakes.

Being aware of a problem and taking remedial action sometimes come too late. For instance, engineers were aware that the Santa Monica Freeway in the Los Angeles area would likely be damaged during an earthquake and retrofitting was scheduled for February 1994. Unfortunately, the Northridge earthquake struck on January 17, 1994, and part of the freeway collapsed. Of course, this and other similar events provide important information that can be incorporated into engineering practice to make freeways, buildings, and bridges safer during earthquakes.

Natural hazards include wildfires, swarms of insects, and severe weather phenomena, as well as geologic hazards such as flooding, landslides, volcanic eruptions, earthquakes, tsunami, land subsidence, soil creep, and radon gas. Geologic hazards account for thousands of fatalities and billions of dollars in property damage every year, and whereas the incidence of hazards has not increased, fatalities and damages have grown because more and more people live in disaster-prone areas.

We cannot eliminate geologic hazards, but we can better understand these events, design structures to protect

shoreline communities and withstand shaking during earthquakes, enact zoning and land-use regulations, and, at the very least, decrease the amount of damage and human suffering.

Unfortunately, geologic information that is readily available is often ignored or overlooked. A case in point is the Turnagain Heights subdivision in Anchorage, Alaska, that was so heavily damaged when the soil liquefied during the 1964 earthquake (see Figure 11.18). Not only were reports on soil stability ignored or overlooked before homes were built there, but since 1964, new homes were built on part of the same site!

Along rugged seacoasts and in mountainous areas, highways are notoriously unstable and slump or slide from hillsides (❙ Figure 3). When this happens, engineering geologists are consulted for their recommendations for stabilizing slopes. They may suggest building retaining walls or drainage systems to keep the slopes dry, planting vegetation, or, in some cases, simply rerouting the highway if it is too costly to maintain in its present position (see Chapter 11).

As you read the following chapters on surface processes, keep in mind that engineering geologists are involved in many aspects of stabilizing slopes, designing and constructing dams and flood-control projects, and building structures to protect seaside communities.

(❙ Figure 10.18). A classic example is the Basin and Range Province, which is centered on Nevada, but extends into adjacent areas. Differential movement on faults has produced uplifted blocks called *horsts* and down-dropped blocks called *grabens*. Erosion of the horsts has produced mountainous topography.

Volcanic outpourings form chains of volcanic mountains such as the Hawaiian Islands, where a plate moves over a hot spot (see Figure 2.23). Some mountains such as the Cascade Range of the Pacific Northwest are made up almost entirely of volcanic rocks (Figure 5.19), and the mid-ocean

ridges are also mountains. However, most mountains on land are composed of all rock types and show clear evidence of deformation by compression.

Plate Tectonics and Mountain Building

Geologists define the term **orogeny** as an episode of mountain building during which intense deformation takes place, generally accompanied by metamorphism, the emplacement

Figure 1 Before the Mackinac Bridge, that connects the Lower and Upper Peninsulas of Michigan, could be built, geologists and engineers had to determine whether the Mackinac Breccia could support such a large structure.

Figure 2 This freeway crosses a small valley a short distance from the San Andreas fault in California. Engineers had to take into account the near certainty that this structure would be badly shaken during an earthquake.

Figure 3 The Devil's Slide area along the Pacific Coast of California. The thin line on the hillside is California Highway 1, which is periodically closed due to slides. Not only are the rocks here weak and susceptible to sliding and slumping, but the San Andreas fault is nearby. The state began building a tunnel to bypass this area. It should be completed in 2011.

of plutons, especially batholiths, and thickening of Earth's crust. The processes responsible for an orogeny are still not fully understood, but it is known that mountain building is related to plate movements.

Any theory that accounts for mountain building must adequately explain the characteristics of mountain ranges, such as their geometry and location; they tend to be long and narrow and at or near plate margins. Mountains also show intense deformation, especially compression-induced overturned and recumbent folds, as well as reverse and thrust faults. Furthermore, granitic plutons and regional metamorphism characterize the interiors or cores of mountain ranges. Another feature is sedimentary rocks now far above sea level that were deposited in shallow and deep marine environments.

Deformation and associated activities at convergent plate boundaries are certainly important processes in mountain building. They account for a mountain system's location and geometry, as well as complex geologic structures, plutons, and metamorphism. Yet, the present-day topographic expression of mountains is also related to several surface processes, such as mass wasting (gravity-driven processes

❚ **Figure 10.18** Origins of Horsts and Grabens

a Block-faulting and the origin of horsts and grabens. Many of the mountain ranges in the Basin and Range Province of the western United States and northern Mexico formed in this manner.

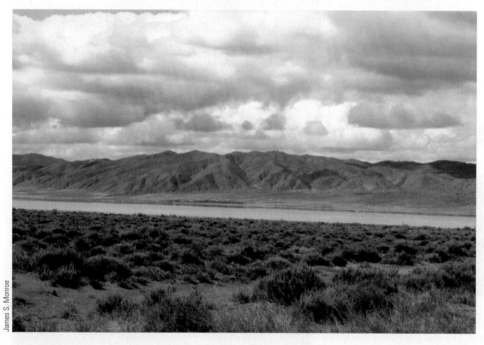

b The Stillwater Range in Nevada is a horst bounded by normal faults.

including landslides), glaciers, and running water. In other words, erosion also plays an important role in the evolution of mountains.

Orogenies at Oceanic–Oceanic Plate Boundaries

Deformation, igneous activity, and the origin of a volcanic island arc characterize orogenies that take place where oceanic lithosphere is subducted beneath oceanic lithosphere. Sediments derived from the island arc are deposited in an adjacent oceanic trench, and then deformed and scraped off against the landward side of the trench (❚ Figure 10.19). These deformed sediments are part of a subduction complex, or an *accretionary wedge,* of intricately folded rocks cut by numerous thrust faults, resulting from compression. In addition, orogenies in this setting are characterized by low-temperature, high-pressure metamorphism of the blueschist facies (see Figure 8.19).

Deformation caused largely by the emplacement of plutons also takes place in the island arc system where many rocks show evidence of high-temperature, low-pressure metamorphism. The overall effect of an island arc orogeny is the origin of two more or less parallel orogenic belts consisting of a landward volcanic island arc underlain by batholiths and a seaward belt of deformed trench rocks (Figure 10.19). The Japanese Islands are a good example.

In the area between an island arc and its nearby continent, the back-arc basin, volcanic rocks and sediments derived from the island arc and the adjacent continent are also deformed as the plates continue to converge. The sediments are intensely folded and displaced toward the continent along low-angle thrust faults. Eventually, the entire island arc complex is fused to the edge of the continent, and the back-arc basin sediments are thrust onto the continent, forming a thick stack of thrust sheets (Figure 10.19).

Figure 10.19 Orogeny and the Origin of a Volcanic Island Arc at an Oceanic–Oceanic Plate Boundary

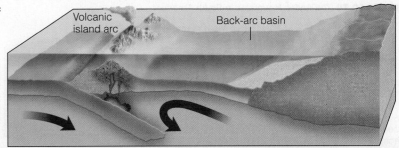

a Subduction of an oceanic plate and the origin of a volcanic island arc and a back-arc basin.

b Continued subduction and back-arc spreading.

c Back-arc basin begins to close, resulting in deformation of back-arc basin and continental margin deposits.

d Thrusting of back-arc sediments onto the adjacent continent and suturing of the island arc to the continent.

Orogenies at Oceanic–Continental Plate Boundaries

The Andes of South America are the best example of continuing orogeny at an oceanic–continental plate boundary. Among the ranges of the Andes are the highest mountain peaks in the Americas and many active volcanoes. Furthermore, the west coast of South America is an extremely active segment of the circum-Pacific earthquake belt, and one of Earth's great oceanic trench systems, the Peru–Chile Trench, lies just off the coast.

Prior to 200 million years ago, the western margin of South America was a broad continental margin where sediments accumulated much as they do now along the East Coast of North America. However, when Pangaea split apart along what is now the Mid-Atlantic Ridge, the South American plate moved westward. As a consequence, the oceanic lithosphere west of South America began subducting beneath the continent (Figure 10.20).

Figure 10.20 The Andes Mountains in South America

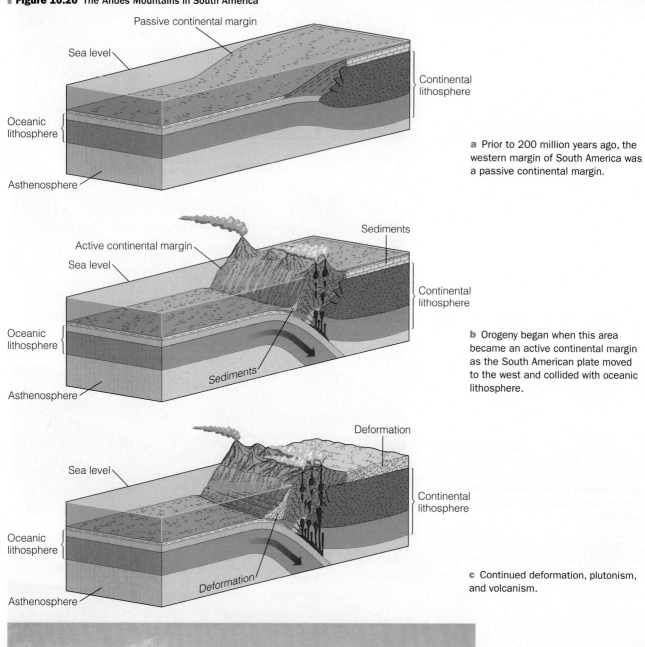

Passive continental margin

Sea level

Continental lithosphere

Oceanic lithosphere

Asthenosphere

a Prior to 200 million years ago, the western margin of South America was a passive continental margin.

Active continental margin

Sea level

Sediments

Continental lithosphere

Oceanic lithosphere

Sediments

Asthenosphere

b Orogeny began when this area became an active continental margin as the South American plate moved to the west and collided with oceanic lithosphere.

Deformation

Sea level

Continental lithosphere

Oceanic lithosphere

Deformation

Deformation

Asthenosphere

c Continued deformation, plutonism, and volcanism.

Tibor Bognar/Corbis

d View of the Andes in Chile.

▌**Figure 10.21** Orogeny at a Continental–Continental Plate Boundary and the Origin of the Himalayas of Asia

a During its long journey north, India moved 15 to 20 cm per year; however, beginning 40 to 50 million years ago, its rate of movement decreased markedly as it collided with the Eurasian plate.

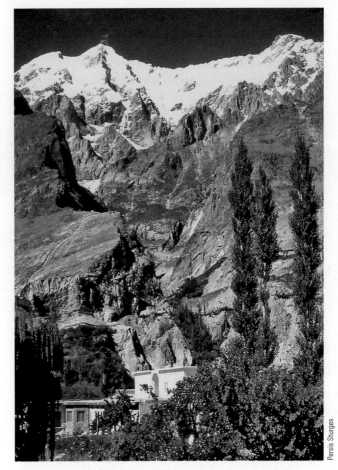

b The Karakoram Range seen here from Karimabad, Pakistan, is within the Himalayan origin. The range lies on the border of Pakistan, China, and India.

Subduction resulted in partial melting of the descending plate, which produced the andesitic volcanic arc of composite volcanoes, and felsic magmas, mostly of granitic composition, were emplaced as large plutons beneath the arc (Figure 10.20).

As a result of the events just described, the Andes Mountains consist of a central core of granitic rocks capped by andesitic volcanoes. To the west of this central core along the coast are the deformed rocks of the accretionary wedge. And to the east of the central core are intensely folded sedimentary rocks that were thrust eastward onto the continent (Figure 10.20). Present-day subduction, volcanism, and seismicity along South America's west coast indicate that the Andes Mountains are still forming.

Orogenies at Continental–Continental Plate Boundaries

The best example of an orogeny along a continental–continental plate boundary is the Himalayas of Asia. The Himalayas began forming when India collided with Asia about 40 to 50 million years ago. Prior to that time, India was far south of Asia and separated from it by an ocean basin (▌Figure 10.21a). As the Indian plate moved northward, a subduction zone formed along the southern margin

of Asia where oceanic lithosphere was consumed. Partial melting generated magma, which rose to form a volcanic arc, and large granite plutons were emplaced into what is now Tibet. At this stage, the activity along Asia's southern margin was similar to what is now occurring along the west coast of South America.

The ocean separating India from Asia continued to close, and India eventually collided with Asia (Figure 10.21a). As a result, two continental plates became welded, or sutured, together. Thus, the Himalayas are now within a continent rather than along a continental margin. The exact time of India's collision with Asia is uncertain, but between 40 and 50 million years ago, India's rate of northward drift decreased abruptly from about 15–20 cm per year to about 5 cm per year. Because continental lithosphere is not dense enough to be subducted, this decrease seems to mark the time of collision and India's resistance to subduction. Consequently, the leading margin of India was thrust beneath Asia, causing crustal thickening, thrusting, and uplift. Sedimentary rocks that had been deposited in the sea south of Asia were thrust northward, and two major thrust faults carried rocks of Asian origin onto the Indian plate. Rocks deposited in the shallow seas along India's northern margin now form the higher parts of the Himalayas (❙ Figure 10.21b). Since its collision with Asia, India has been thrust horizontally about 2,000 km beneath Asia and now moves north at several centimeters per year.

Other mountain systems also formed as a result of collisions between two continental plates. The Urals in Russia and the Appalachians of North America formed by such collisions. In addition, the Arabian plate is now colliding with Asia along the Zagros Mountains of Iran.

Terranes and the Origin of Mountains

In the preceding section, we discussed orogenies along convergent plate boundaries that result in adding material to a continent, a process termed **continental accretion.** Much of the material added to continental margins is eroded older continental crust, but some plutonic and volcanic rocks are new additions. During the 1970s and 1980s, however, geologists discovered that parts of many mountain systems are also made up of small, accreted lithospheric blocks that clearly originated elsewhere. These **terranes,*** as they are called, are fragments of seamounts, island arcs, and small pieces of continents that were carried on oceanic plates that collided with continental plates, thus adding them to the continental margins.

> **CONNECTION LINK**
>
> You can find more information about terranes and their importance in mountain building on pages 589 and 590 in Chapter 22.

10.6 Earth's Crust

Continental crust stands higher than oceanic crust, but why should this be so? Also, why do mountains stand higher than surrounding areas? To answer these questions, we must examine Earth's crust in more detail. You already know that continental crust is granitic with an overall density of 2.7 g/cm³, whereas oceanic crust is made up of basalt and gabbro and its density is 3.0 g/cm³. In most places, continental crust is about 35 km thick except beneath mountain systems, where it is much thicker. Oceanic crust, in contrast, varies from only 5 to 10 km thick. So, these differences, as well as variations in crustal thickness, account for why mountains stand high and why continents stand higher than ocean basins (see the following section).

---

*Some geologists prefer the terms *suspect terrane, exotic terrane,* or *displaced terrane*. Notice also the spelling of terrane as opposed to the more familiar terrain, the latter a geographic term indicating a particular area of land.

How is it possible for a solid (continental crust) to float in another solid (the mantle)? Floating brings to mind a ship at sea or a block of wood in water; however, continents do not behave in this manner. Or do they? Actually, they do float, in a manner of speaking, but a complete answer requires more discussion on the concept of gravity and on the principle of isostasy.

Isaac Newton formulated the law of universal gravitation in which the force of gravity (F) between two masses (m_1 and m_2) is directly proportional to the products of their masses and inversely proportional to the square of the distance between their centers of mass. This means that an attractive force exists between any two objects, and the magnitude of that force varies depending on the masses of the objects and the distance between their centers. We generally refer to the gravitational force between an object and Earth as its *weight.*

Gravitational attraction would be the same everywhere on the surface if Earth were perfectly spherical, homogeneous throughout, and not rotating. But because Earth varies in all of these aspects, the force of gravity varies from area to area. Geologists use a *gravimeter* to measure gravitational attraction and to detect **gravity anomalies**—that is, departures from the expected force of gravity (❙ Figure 10.22). Gravity anomalies might be *positive,* meaning that an excess of mass is present at some location, or *negative,* when a mass deficiency exists.

Principle of Isostasy

Geologists realized long ago that mountains are not simply piles of materials on Earth's surface, and in 1865 George Airy proposed that, in addition to projecting high above sea level, mountains also project far below the surface and thus have a low density root (Figure 10.22). In effect, he was saying that the thicker crust of mountains floats on denser rock at depth, with their excess mass above sea level compensated for by low-density material at depth. Another explanation was proposed by J. H. Pratt, who thought that mountains were high because they were composed of rocks of lower density than those in adjacent regions.

Actually, both Airy and Pratt were correct, because there are places where density or thickness accounts for differences in the level of the crust. For example, Pratt's hyphothesis was confirmed because (1) continental crust is thicker and less dense than oceanic crust and thus stands high, and (2) the mid-oceanic ridges stand high because the crust there is hot and less dense than cooler crust elsewhere. Airy, on the other hand, was correct in his claim that the crust, continental or oceanic, "floats" on the mantle, which has a density of 3.3 g/cm³ in its upper part. However, we have not yet explained what we mean by one solid floating in another solid.

Figure 10.22 Gravity Anomalies

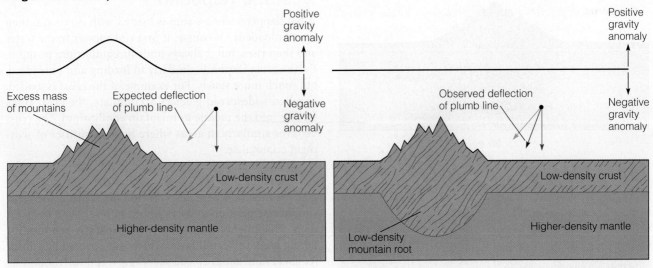

a A plumb line (a cord with a suspended weight) is normally vertical, pointing to Earth's center of gravity. Near a mountain range, the plumb line should be deflected as shown if the mountains are simply thicker, low-density material resting on denser material, and a gravity survey across the mountains would indicate a positive gravity anomaly.

b The actual deflection of the plumb line during the survey in India was less than expected. It was explained by postulating that the Himalayas have a low-density root. A gravity survey, in this case, would show no anomaly because the mass of the mountains above the surface is compensated for at depth by low-density material displacing denser material.

This phenomenon of Earth's crust floating in the denser mantle is now known as the **principle of isostasy,** which is easy to understand by an analogy to an iceberg (‖ Figure 10.23). Ice is slightly less dense than water, so it floats. According to Archimedes' principle of buoyancy, an iceberg sinks in water until it displaces a volume of water whose weight is equal to that of the ice. When the iceberg has sunk to an equilibrium position, only about 10% of its volume is above water level. If some of the ice above water level should melt, the iceberg rises to maintain equilibrium with the same proportion of ice above and below the water.

Earth's crust is similar to the iceberg in that it sinks into the mantle to its equilibrium level. Where the crust is thickest, as beneath mountains, it sinks farther down into the mantle and it also rises higher above the surface. And because continental crust is thicker and less dense than oceanic crust, it stands higher than the ocean basins. Remember, the mantle is hot, yet solid, and under tremendous pressure, so it behaves in a fluid-like manner.

Some of you might realize that crust floating on the mantle raises an apparent contradiction. In Chapter 9, we said that the mantle is a solid because it transmits S-waves, which do not move through a fluid. But according to the principle of isostasy, the mantle behaves as a fluid. When considered in terms of the brief time required for S-waves

to pass through it, the mantle is indeed solid. But when subjected to stress over long periods, it yields by flowage; thus, at this time scale, it is regarded as a viscous fluid.

Figure 10.23 The Principle of Isostasy An iceberg sinks to an equilibrium level with 10% of its mass above water level. The larger iceberg sinks farther below and rises higher above the water surface than the smaller one. If some of the ice above water level melts, the icebergs will rise to maintain the same proportions of ice above and below water level.

(a)

(b)

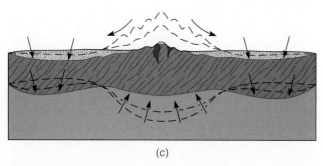

(c)

▌**Figure 10.24** Isostatic Rebound A diagrammatic representation showing the isostatic response of the crust to erosion (unloading) and widespread deposition (loading).

Isostatic Rebound

What happens when a ship is loaded with cargo and then later unloaded? Of course, it first sinks lower in the water and then rises, but it always finds its equilibrium position. Earth's crust responds similarly to loading and unloading, but much more slowly. For example, if the crust is loaded, as when widespread glaciers accumulate, the crust sinks farther into the mantle to maintain equilibrium. The crust behaves similarly in areas where huge quantities of sediment accumulate.

If loading by glacial ice or sediment depresses Earth's crust farther into the mantle, it follows that when vast glaciers melt or where deep erosion takes place, the crust should rise back up to its equilibrium level. And in fact it does. This phenomenon, known as **isostatic rebound,** is taking place in Scandinavia, which was covered by a thick ice sheet until about 10,000 years ago; it is now rebounding at about 1 m per century. In fact, coastal cities in Scandinavia have rebounded rapidly enough that docks constructed several centuries ago are now far from shore. Isostatic rebound has also occurred in eastern Canada where the crust has risen as much as 100 m in the last 6,000 years.

▌Figure 10.24 shows the response of Earth's continental crust to loading and unloading as mountains form and evolve. Recall that during an orogeny, emplacement of plutons, metamorphism, and general thickening of the crust accompany deformation. However, as the mountains erode, isostatic rebound takes place and the mountains rise, whereas adjacent areas of sedimentation subside (Figure 10.24). If continued long enough, the mountains will disappear and then can be detected only by the plutons and metamorphic rocks that show their former existence.

›› Geo-Image

K	Cretaceous	Younger
R	Triassic	
C	Carboniferous	
C	Cambrian	
PC	Precambrian	

Figure 1 The geologic map is a surface view that uses symbols and colors to depict the rocks exposed in the Black Hills of South Dakota. With the information on this map, can you determine whether this is a basin or dome? How do you know?

~~ Wave-formed ripples

VVVV Mud cracks

Graded bedding

Figure 2 The geologic cross section shows a view of several rock layers and a fault. Your task is to decipher the history of the area. What event took place first, second, and so on? What kind of fold is present? What kind of fault? To fully answer these questions, refer to the discussions of sedimentary structures and the principle of superposition in Chapter 7.

Chapter Summary

- Folded and fractured rocks have been deformed or strained by applied stresses.

- Stress is compression, tension, or shear. Elastic strain is not permanent, but plastic strain and fracture are, meaning that rocks do not return to their original shape or volume when the deforming forces are removed.

- Strike and dip are used to define the orientation of deformed rock layers. This same concept applies to other planar features such as fault planes.

- Anticlines and synclines are up- and down-arched folds, respectively. They are identified by strike and dip of the folded rocks and the relative ages of rocks in these folds.

- Domes and basins are the circular to oval equivalents of anticlines and synclines, but they are commonly much larger structures.

- The two structures that result from fracture are joints and faults. Joints show no movement parallel with the fracture surface, whereas faults do.

- On dip-slip faults, all movement is up or down the dip of the fault. If the hanging wall moves relatively down it is a normal fault, but if the hanging wall moves relatively up it is a reverse fault. Normal faults result from tension; reverse faults result from compression.

- In strike-slip faults, all movement is along the strike of the fault. These faults are either right-lateral or left-lateral, depending on the apparent direction of offset of one block relative to the other.

- Oblique-slip faults show components of both dip-slip and strike-slip movement.

- A variety of processes account for the origin of mountains. Some involve little or no deformation, but the large mountain systems on the continents resulted from deformation at convergent plate boundaries.

- Subduction of an oceanic plate beneath another oceanic plate or beneath a continental plate causes an orogeny. At an oceanic–oceanic boundary, a volcanic island arc intruded by plutons forms, whereas at an oceanic–continental boundary a volcanic arc forms on the continental plate. In both cases, deformation and metamorphism occur.

- Some mountain systems are within continents far from a present-day plate boundary. These mountains formed when two continental plates collided and became sutured.

- Geologists now realize that orogenies also involve collisions of terranes with continents.

- Continental crust is characterized as granitic, and it is much thicker and less dense than oceanic crust, which is composed of basalt and gabbro.

- According to the principle of isostasy, Earth's crust floats in equilibrium in the denser mantle below. Continental crust stands higher than oceanic crust because it is thicker and less dense.

Important Terms

anticline (p. 237)
basin (p. 241)
compression (p. 235)
continental accretion (p. 252)
deformation (p. 234)
dip (p. 237)
dip-slip fault (p. 243)
dome (p. 241)
elastic strain (p. 235)
fault (p. 242)
fault plane (p. 242)
fold (p. 237)

footwall block (p. 243)
fracture (p. 236)
geologic structure (p. 237)
gravity anomaly (p. 252)
hanging wall block (p. 243)
isostatic rebound (p. 254)
joint (p. 241)
monocline (p. 237)
normal fault (p. 243)
oblique-slip fault (p. 244)
orogeny (p. 246)
plastic strain (p. 236)

principle of isostasy (p. 253)
reverse fault (p. 244)
shear stress (p. 235)
strain (p. 235)
stress (p. 235)
strike (p. 236)
strike-slip fault (p. 244)
syncline (p. 237)
tension (p. 235)
terrane (p. 252)
thrust fault (p. 244)

Review Questions

1. A fault on which the hanging wall block has moved up relative to the footwall block is a _____ fault.
   a. _____normal.
   b. _____strike-slip.
   c. _____oblique-slip.
   d. _____reverse.
   e. _____overturned.

2. Folding of rock layers into anticlines and synclines results from
   a. _____tension.
   b. _____torque.
   c. _____compression.
   d. _____shear.
   e. _____plunging.

3. A rather circular fold with all strata dipping away from a central point with the oldest exposed rock layers at the center is a/an
   a. ____joint.
   b. ____dome.
   c. ____recumbent fold.
   d. ____syncline.
   e. ____terrane.

4. The line formed by the intersection of a horizontal plane with the surface of an inclined rock layer is called
   a. ____strike.
   b. ____uplift.
   c. ____jointing.
   d. ____stress.
   e. ____brittle behavior.

5. Rocks characterized as ductile
   a. ____show a great amount of plastic strain.
   b. ____are the main rocks in terranes.
   c. ____are found along the crests of anticlines.
   d. ____fracture easily when in compression.
   e. ____are found in the hanging wall blocks of faults.

6. A time of mountain building accompanied by the emplacement of plutons, metamorphism, and thickening of the crust is a/an
   a. ____gravity anomaly.
   b. ____orogeny.
   c. ____magnetic reversal.
   d. ____craton.
   e. ____shield.

7. The maximum inclination of a rock layer from horizontal is its
   a. ____axial plane.
   b. ____joint.
   c. ____clout.
   d. ____limb.
   e. ____dip.

8. A fold in which one limb has been rotated more than 90 degrees from its original position is said to be
   a. ____left-lateral.
   b. ____oblique.
   c. ____anticlinal.
   d. ____overturned.
   e. ____inverted.

9. An ongoing continent–continent collision is responsible for the
   a. ____Cascade Range.
   b. ____Himalaya Mountains.
   c. ____Andes Mountains.
   d. ____Japan Range.
   e. ____Basin and Range.

10. Which one of the following statements is incorrect?
    a. ____most orogenies occur at convergent plate boundaries.
    b. ____shear stress involves forces acting along the same line, but in opposite directions.
    c. ____joints are fractures along which no movement has occurred parallel with the fracture.
    d. ____the dip of a rock layer is its inclination from horizontal.
    e. ____folds, joints, and faults are collectively known as geologic structures.

11. Discuss how time, rock type, pressure, and temperature influence rock deformation.

12. Suppose that rocks were displaced 200 km along a strike-slip fault during a period of 5 million years. What is the average rate of movement per year? Is the average likely to represent the actual rate of displacement on the fault? Explain.

13. How would you explain the concepts of stress and strain to someone unfamiliar with the concepts?

14. What kinds of evidence would indicate that mountain building once took place in an area where mountains are no longer present?

15. How is an eroded anticline different from an eroded syncline?

16. What kinds of dip-slip faults do geologists recognize, and how can you tell one from the other?

17. What is meant by the elastic limit of rocks, and what happens when rocks are strained beyond their elastic limit?

18. How do synclines and basins compare?

19. Is there any connection between mountain building and isostasy? If so, what?

20. What sequence of events takes place when mountains form at an oceanic–continental plate boundary?

A landslide on January 12, 2011, killed and injured dozens of people in Teresopolis near Rio de Janeiro, Brazil. As a result of several days of heavy rains, this hillside gave way, causing a landslide that sent mud, vegetation, and homes sliding down its slope.

CHAPTER 11

Mass Wasting

BRUNO DOMINGOS/Reuters/Landow

OUTLINE

11.1 Introduction

11.2 Factors That Influence Mass Wasting

11.3 Types of Mass Wasting

GEO-FOCUS: Southern California Landslides

11.4 Recognizing and Minimizing the Effects of Mass Wasting

GEO-RECAP

HAVE YOU EVER WONDERED?

- What causes landslides (known as mass movements or mass wasting)?

- Why landslides occur where they do?

- What, if anything, can be done to prevent some of the different types of mass wasting?

- How to minimize the destructive effects of certain types of mass wasting?

- What things you should look for when buying property or building in hilly or coastal areas to avoid, or at least minimize, potential mass-wasting effects to your property?

▌Figure 11.1 Slope Shear Strength

Gravitational Force

Shear Strength

Compression of grains

Cohesiveness of grains

Internal friction between grains

Component of gravitational force acting parallel to slope

a A slope's shear strength depends on the slope material's strength and cohesion, the amount of internal friction between grains, and any external support of the slope. These factors promote slope stability. The force of gravity operates vertically, but also has a component of force acting parallel to the slope. When this force, which promotes instability, exceeds a slope's shear strength, slope failure occurs.

11.1 Introduction

In January 2011, southeast Brazil was struck by devastating floods and landslides triggered by unusually heavy torrential rains that dumped a month's worth of rain in just a few days. The hardest hit area was the mountainous Serrana region north of Rio de Janeiro, where relentless flooding and landslides destroyed hundreds of homes, killing more than 800 people and leaving at least 14,000 people homeless. The death toll easily surpassed that of the 1967 mudslides in Caraguatatuba, making this Brazil's worst natural disaster in more than four decades. Collapsed roads and bridges further exacerbated the situation, hampering rescue operations and emergency aid, and lack of power and telephone service contributed further to the misery of the survivors. Although the floods and landslides swept away homes of both rich and poor alike, the poorer rural areas suffered the most because many structures were built in unstable areas. It is easy to cite the numbers of dead and homeless, but the human side of the disaster was vividly brought home in Terescopolis, where numerous bodies were lined up outside a police station awaiting identification by relatives because there was no more room at the morgue.

This terrible tragedy illustrates how geology affects all of our lives. The underlying causes of the mudslides in Brazil can be found anywhere in the world. In fact, *landslides* (a general term for mass movements of Earth materials) cause, on average, between 25 and 50 deaths and more than $2 billion in damage annually in the United States. By being able to recognize and understand how landslides occur and what the results may be, we can find ways to reduce hazards and minimize damage in terms of both human life and property damage.

Mass wasting (also called *mass movement*) is defined as the downslope movement of material under the direct influence of gravity. Most types of mass wasting are aided by weathering and usually involve surficial material. The material moves at rates ranging from almost imperceptible, as in

b The angle of repose is a function of shear strength. Dry sand has an angle of repose of about 30 degrees. With damp sand, shear strength is increased, and much steeper angles of repose are possible.

the case of creep, to extremely fast, as in a rockfall or slide. Although water can play an important role, the relentless pull of gravity is the major force behind mass wasting.

11.2 Factors That Influence Mass Wasting

Mass wasting is an important geologic process that can occur at any time and almost any place. Although all major landslides have natural causes, many smaller ones are the result of human activity and could have been prevented or their damage minimized.

When the gravitational force acting on a slope exceeds its resisting force, slope failure (mass wasting) occurs. The resisting forces that help maintain slope stability include

❙ Figure 11.2 Undercutting a Slope's Base by Stream Erosion

a Undercutting by stream erosion removes a slope's base,

b which increases the slope angle, and can lead to slope failure.

c Undercutting by stream erosion caused slumping along this stream near Weidman, Michigan. Notice the scarp, which is the exposed surface of the underlying material following slumping.

the slope material's strength and cohesion, the amount of internal friction between grains, and any external support of the slope (❙ Figure 11.1). These factors collectively define a slope's **shear strength**.

Opposing a slope's shear strength is the force of gravity. Gravity operates vertically, but also has a component acting parallel to the slope, thereby causing instability (Figure 11.1a). The steeper a slope's angle, the greater the component of force acting parallel to the slope and the greater the chance for mass wasting. The steepest angle that a slope can maintain without collapsing is its *angle of repose* (Figure 11.1b). At this angle, the shear strength of the slope's material exactly counterbalances the force of gravity. For unconsolidated material, the angle of repose normally ranges from 25 to 40 degrees. Slopes steeper than 40 degrees usually consist of unweathered solid rock.

All slopes are in a state of *dynamic equilibrium*, which means that they are constantly adjusting to new conditions. Although we tend to view mass wasting as a disruptive and usually destructive event, it is one of the ways that a slope adjusts to new conditions. Whenever a building or road is constructed on a hillside, the equilibrium of that slope is affected. The slope must then adjust, perhaps by mass wasting, to this new set of conditions.

Many factors can cause mass wasting: a change in slope angle, weakening of material by weathering, increased water content, changes in the vegetation cover, and overloading. Although most of these are interrelated, we will examine them

separately for ease of discussion, but we will also show how they individually and collectively affect a slope's equilibrium.

Slope Angle

Slope angle is probably the major cause of mass wasting. Generally speaking, the steeper the slope, the less stable it is. Therefore, steep slopes are more likely to experience mass wasting than gentle ones.

A number of processes can oversteepen a slope. One of the most common is undercutting by stream or wave action (❙ Figure 11.2). This process removes the slope's base, increases the slope angle, and thereby increases the gravitational force acting parallel to the slope. Wave action, especially during storms, often results in mass movements along the shores of oceans or large lakes (❙ Figure 11.3).

Excavations for road cuts and hillside building sites are another major cause of slope failure (❙ Figure 11.4). Grading the slope too steeply or cutting into its side increases the stress in the rock or soil until it is no longer strong enough to remain at the steeper angle, and mass movement ensues. Such action is analogous to undercutting by streams or waves and has the same result, thus explaining why so many mountain roads are plagued by frequent mass movements.

Weathering and Climate

Mass wasting is more likely to occur in loose or poorly consolidated slope material than in bedrock. As soon as rock is exposed at Earth's surface, weathering begins to disintegrate and decompose it, reducing its shear strength and increasing its susceptibility to mass wasting. The deeper the

Figure 11.3 Undercutting a Slope's Base by Wave Action This sea cliff north of Bodega Bay, California, was undercut by waves during the winter of 1997–1998. As a result, part of the land slid into the ocean, damaging several houses.

Copyright and Photograph by Dr. Parvinder S. Sethi

Figure 11.4 Highway Excavation

a Highway excavations disturb the equilibrium of a slope by

b removing a portion of its support, as well as oversteepening it at the point of excavation, which can result in

c landslides along the highway.

Courtesy of R. V. Dietrich

d Cutting into the hillside to construct this portion of the Pan-American Highway in Mexico resulted in a rockfall that completely blocked the road.

Figure 11.5 Landslides Caused by Sliding Along Slippery Rock Layers, Point Fermin, California Homes dangerously close to an oversteepened, eroding coastal bluff. Water from rains and groundwater have resulted in slippery layers along which sliding has occurred. Two white drainage pipes that helped drain water out of the ground can be seen dangling from the top of the cliff. It is just a matter of time before these homes are destroyed by the effects of erosion and slumping of the unstable bluff.

weathering zone extends, the greater the likelihood of some type of mass movement.

Recall that some rocks are more susceptible to weathering than others and that climate plays an important role in the rate and type of weathering (see Chapter 6). In the tropics, where temperatures are high and considerable rain falls, the effects of weathering extend to depths of several tens of meters, and mass movements most commonly occur in the deep weathering zone. In arid and semiarid regions, the weathering zone is usually considerably shallower. Nevertheless, intense, localized cloudbursts can drop large quantities of water on an area in a short time. With little vegetation to absorb this water, runoff is rapid and frequently results in mudflows.

An excellent case in point is Point Fermin in Southern California. Dubbed the "sunken city" by residents of the area, Point Fermin is famous for its numerous examples of mass wasting. The area is underlain by fine-grained sedimentary rocks interbedded with diatomite layers (a rock composed of the shells of diatoms, which are siliceous algae) and volcanic ash. When these layers get wet from rains or percolating groundwater, they become slippery and tend to slide easily (Figure 11.5). Furthermore, the rocks also dip slightly toward the ocean and form steep coastal bluffs that are being undercut by constant wave action at their base.

> **CONNECTION LINK**
>
> To learn more about how wave action can result in mass wasting, go to page 398 in Chapter 16.

Water Content

The amount of water in rock or soil influences slope stability. Large quantities of water from melting snow or heavy rainfall greatly increase the likelihood of slope failure. The additional weight that water adds to a slope can be enough to cause mass movement. Furthermore, water percolating through a slope's material helps to decrease friction between grains, contributing to a loss of cohesion. For example, slopes composed of dry clay are usually quite stable, but when wetted, they quickly lose cohesiveness and internal friction and become an unstable slurry. This occurs because clay, which can hold large quantities of water, consists of platy particles that easily slide over each other when wet. For this reason, clay beds are frequently the slippery layer along which overlying rock units slide downslope.

> **CONNECTION LINK**
>
> To learn more about how water moves through various Earth materials, see pages 314 and 315 in Chapter 13.

Vegetation

Vegetation affects slope stability in several ways. By absorbing the water from a rainstorm, vegetation decreases water saturation of a slope's material that would otherwise lead to a loss of shear strength. Vegetation's root system also helps stabilize a slope by binding soil particles together and holding the soil to bedrock.

The removal of vegetation by either natural or human activity is a major cause of many mass movements. Summer brush and forest fires in southern California frequently leave the hillsides bare of vegetation. Fall rainstorms saturate the ground, causing mudslides that do tremendous damage and cost millions of dollars to clean up.

Overloading

Overloading is almost always the result of human activity and typically results from the dumping, filling, or piling up of material. Under natural conditions, a material's load is carried by its grain-to-grain contacts, with the friction

Figure 11.6 Geology, Slope Stability, and Mass Wasting Rocks dipping in the same direction as a hill's slope are particularly susceptible to mass wasting.

1 Water percolates through soil into clay-rich layers ▭ that become slippery, and may swell, weakening the overlying rock ▭.

2 The clay-rich layer dips in the same direction as the even more steeply dipping slope. Gravity can therefore turn it into a skid surface, or potential landslide plane.

3 Undercutting by the stream at the foot of the slope exposes another watery, weak clay layer underlying a heavy, strong limestone bed ▭. The heavy limestone is now prone to slide across the clay, carrying the rest of the overlying slope with it.

4 Layers on this side of the valley dip in an opposite direction from the slope. Thus, gravity cannot easily act to destabilize them, even if water percolation is deep and undercutting occurs.

between the grains maintaining a slope. The additional weight created by overloading increases the water pressure within the material, which in turn decreases its shear strength, thereby weakening the slope material. If enough material is added, the slope will eventually fail, sometimes with tragic consequences.

Geology and Slope Stability

The relationship between the topography and the geology of an area is important in determining slope stability (▌ Figure 11.6). If the rocks underlying a slope dip in the same direction as the slope, mass wasting is more likely to occur than if the rocks are horizontal or dip in the opposite direction. When the rocks dip in the same direction as the slope, water can percolate along the various bedding planes and decrease the cohesiveness and friction between adjacent rock units (Figure 11.5). This is particularly true when clay layers are present because clay becomes slippery when wet.

Even if the rocks are horizontal or dip in a direction opposite to that of the slope, joints may dip in the same direction as the slope. Water migrating through them weathers the rock and expands these openings until the weight of the overlying rock causes it to fall.

Triggering Mechanisms

The factors discussed thus far all contribute to slope instability. Most, though not all, rapid mass movements are triggered by a force that temporarily disturbs slope equilibrium. The most common triggering mechanisms are strong vibrations from earthquakes and excessive amounts of water from a winter snow melt or a heavy rainstorm.

Volcanic eruptions, explosions, and even loud claps of thunder may be enough to trigger a landslide if the slope is sufficiently unstable. Many *avalanches*, which are rapid movements of snow and ice down steep mountain slopes, are triggered by a loud gunshot or, in rare cases, even a person's shout.

11.3 Types of Mass Wasting

Mass movements are generally classified on the basis of three major criteria (Table 11.1): (1) rate of movement (rapid or slow); (2) type of movement (primarily falling, sliding, or flowing); and (3) type of material involved (rock, soil, or debris). Even though many slope failures are combinations of different materials and movements, the resulting mass movements are typically classified according to their dominant behavior.

Rapid mass movements involve a visible movement of material. Such movements usually occur quite suddenly, and the material moves quickly downslope. Rapid mass movements are potentially dangerous and frequently result in loss of life and property damage. Most rapid mass movements occur on relatively steep slopes and can involve rock, soil, or debris.

Slow mass movements advance at an imperceptible rate and are usually detectable only by the effects of their movement, such as tilted trees and power poles or cracked foundations. Although rapid mass movements are more dramatic, slow mass movements are responsible for the downslope transport of a much greater volume of weathered material.

Falls

Rockfalls are a common type of extremely rapid mass movement in which rocks of any size fall through the air (■ Figure 11.7a). Rockfalls occur along steep canyons, cliffs, and road cuts and build up accumulations of loose rocks and rock fragments at their base called *talus* (see Figure 6.2b).

Rockfalls result from failure along joints or bedding planes in the bedrock and are commonly triggered by natural or human undercutting of slopes, or by earthquakes. Many rockfalls in cold climates are the result of frost wedging (see Figure 6.2). Chemical weathering caused by water percolating through the fissures in carbonate rocks (limestone, dolostone, and marble) is also responsible for many rockfalls.

Rockfalls range in size from small rocks falling from a cliff to massive falls involving millions of cubic meters of debris that destroy buildings, bury towns, and block highways. Rockfalls are a particularly common hazard in mountainous areas where roads have been built by blasting and grading through steep hillsides of bedrock (■ Figure 11.7b). Anyone who has ever driven through the Appalachians, the Rocky Mountains, or the Sierra Nevada is familiar with the "Caution Falling Rocks" signs posted to warn drivers of the danger (■ Figure 11.7c). Slopes that are particularly susceptible to rockfalls are sometimes covered with wire mesh in an effort to prevent dislodged rocks from falling to the road below (■ Figure 11.8a). Another tactic is to put up wire mesh fences along the base of the slope to catch or slow down bouncing or rolling rocks (■ Figure 11.8b).

Slides

A **slide** involves movement of material along one or more surfaces of failure. The type of material may be soil, rock,

■ Figure 11.7 Rockfalls

a Rockfalls result from failure along cracks, fractures, or bedding planes in the bedrock and are common features in the areas of steep cliffs.

b A recent rockfall of granite in Yosemite National Park, California.

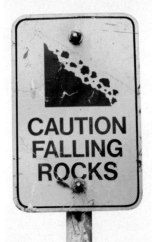

c Warning signs for falling rocks are commonly seen in mountainous areas.

Copyright and Photograph by Dr. Parvinder S. Sethi

© Stephen Finn/Shutterstock

TABLE 11.1	Classification of Mass Movements and Their Characteristics		
Type of Movement	**Subdivision**	**Characteristics**	**Rate of Movement**
Falls	Rockfall	Rocks of any size fall through the air from steep cliffs, canyons, and road cuts	Extremely rapid
Slides	Slump	Movement occurs along a curved surface of rupture; most commonly involves unconsolidated or weakly consolidated material	Extremely slow to moderate
Flows	Rock slide	Movement occurs along a generally planar surface	Rapid to very rapid
	Mudflow	Consists of at least 50% silt- and clay-sized particles and up to 30% water	Very rapid
	Debris flow	Contains larger-sized particles and less water than mudflows	Rapid to very rapid
	Earthflow	Thick, viscous, tongue-shaped mass of wet regolith	Slow to moderate
	Quick clays	Composed of fine silt and clay particles saturated with water; when disturbed by a sudden shock, lose their cohesiveness and flow like a liquid	Rapid to very rapid
	Solifluction	Water-saturated surface sediment	Slow
	Creep	Downslope movement of soil and rock	Extremely slow
Complex movements		Combination of different movement types	Slow to extremely rapid

or a combination of the two, and it may break apart during movement or remain intact. A slide's rate of movement can vary from extremely slow to very rapid (Table 11.1).

Figure 11.8 Minimizing Damage from Rock Falls

Copyright and Photograph by Dr. Parvinder S. Sethi

a Wire mesh has been used to cover this steep slope near Narvik in northern Norway. This is a common practice in mountainous areas to prevent rocks from falling on the road.

Two types of slides are generally recognized: (1) slumps or rotational slides, in which movement occurs along a curved surface; and (2) rock or block slides, which move along a more or less planar surface.

A **slump** involves the downward movement of material along a curved surface of rupture and is characterized by the backward rotation of the slump block (Figure 11.9). Slumps usually occur in unconsolidated or weakly consolidated material and range in size from small individual sets, such as occur along stream banks (Figure 11.2c), to massive, multiple sets that affect large areas and cause considerable damage.

Slumps can be caused by a variety of factors, but the most common is erosion along the base of a slope, which removes support for the overlying material. This local steepening may be caused naturally by stream erosion along its banks (Figure 11.2c) or by wave action at the base of a coastal cliff (Figure 11.10).

James S. Monroe

b A Wire mesh fence along the base of this hillside of Highway 44 in California has caught many boulders and prevented them from rolling onto the highway.

Figure 11.9 Slumping In a slump, material moves downward along the curved surface of a rupture, causing the slump block to rotate backward. Most slumps involve unconsolidated or weakly consolidated material and are typically caused by erosion along the slope's base.

Slope oversteepening can also be caused by human activity, such as the construction of highways and housing developments. Slumps are particularly prevalent along highway cuts, where they are generally the most frequent type of slope failure observed.

Although many slumps are merely a nuisance, large-scale slumps in populated areas and along highways can cause extensive damage. Such is the case in coastal southern California where slumping and sliding have been a constant problem, resulting in the destruction of many homes and the closing and relocation of numerous roads and highways (see Geo-Focus on pages 268 and 269).

Figure 11.10 Slumping in the Pacific Palisades, Southern California Undercutting of steep sea cliffs by wave action resulted in massive slumping in the Pacific Palisades area of southern California on March 31 and April 3, 1958. Highway 1 was completely blocked. Note the heavy earth-moving equipment for scale.

GEO-FOCUS   Southern California Landslides

Southern California is no stranger to landslides. La Conchita, Point Fermin, Pacific Palisades, and Laguna Beach are all locations in southern California that have suffered damaging mass movements during the past 50 years. Two regions in particular, La Conchita and Laguna Beach, have been in the news because of the landslides that destroyed numerous homes.

La Conchita is a small community along the coast at the base of a 100-m-high terrace, 120 km northwest of Los Angeles, California. On March 4, 1995, following a period of heavy rains, some residents of this beach community noticed that the steep slope above their homes was slowly moving and that cracks were appearing in the walls of their houses, indicating that the homes were also moving. Shortly thereafter, a 200,000 m³ slide destroyed or damaged nine homes in its path (Figure 1).

Almost 10 years later, following a week of heavy rainfall in southern California, in which the hillside and previous landslide deposits were saturated with water, another landslide occurred in the same area. This time 10 people were killed and 15 homes were buried under 10 m and 400,000 tons of mud (Figure 2).

What went wrong and why was this situation repeated? The rocks that make up the steep-sloped terrace behind La Conchita consist of soft, weak, and porous sediments that are not well lithified, and thus are easily weathered and susceptible to mass wasting. In addition, an irrigated avocado orchard sits on top of the hill, contributing to the water percolating through the porous sediments and rocks and contributing to the instability of the hillside. Add in heavy rainfall over an extended period, and you have all the ingredients for a landslide in the making. An ancient landslide area to begin with, a steep slope that has been undercut at its base by a road, well-saturated sediments decreasing the cohesion of the sediments that hold the hillside together, and continuing rains all contribute to the making of a landslide. And the potential is still there for another landslide, with no guarantee it won't happen again in the next 10 years!

Farther south in Laguna Beach, another landslide, in this case a rock slide, destroyed 18 expensive hillside homes and severely damaged approximately 20 others on June 1, 2005 (Figure 3). Just as happened in 1978, the main triggering mechanism was probably unusually heavy winter rains, in this case the second-rainiest season

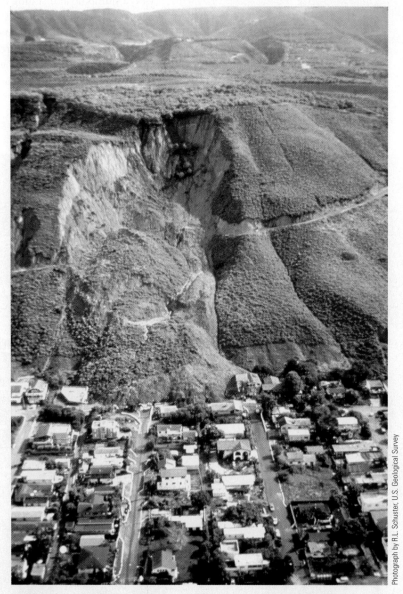

Photograph by R.L. Schuster, U.S. Geological Survey

Figure 1 La Conchita, California, is located at the base of a steep-sloped terrace. Heavy rains and irrigation of an avocado orchard (visible at the top of the terrace) contributed to the landslide that destroyed nine homes in 1995.

on record. In this area of southern California, the rocks dip in the same direction as the slope and contain numerous clay beds interbedded with porous sandstones. Such conditions, when combined with heavy rainfall, are ideal for rock slides. It should come as no surprise that the area where both the 1978 and 2005 rock slides occurred is also part of an ancient slide complex.

Can anything be done to prevent future landslides? The short answer is probably no. Decreasing the slope, benching the hillside, and making sure that there is sufficient drainage and a good cover of vegetation are all steps that can minimize future mass wasting. But the sad fact is that the geologic conditions are such that future landslides are inevitable as the landscape seeks equilibrium conditions by adjusting its slope. Add in the fact that the coastal terraces of Laguna Beach offer some of the most breathtaking views of the Pacific and people are willing to pay a premium to live here, and you have the formula for future landslides, loss of life, and property damage.

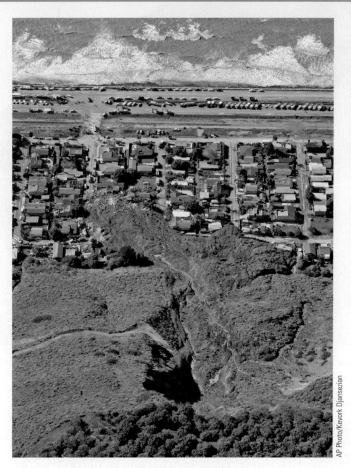

AP Photo/Kevork Djansezian

▌**Figure 2**
La Conchita, California, 10 years later (2005). Similar factors caused another massive landslide in the same area. The landslide can clearly be seen in the center of this photograph, and the scarp and the remains of the 1995 landslide are still visible on the right side, as is the avocado orchard in the foreground.

AP Photo/Nick Ut

▌**Figure 3** A rock slide on June 1, 2005, destroyed 18 expensive homes and damaged at least 20 others in Laguna Beach, California. Heavy rains, combined with unstable underlying geology, contributed to this most recent landslide in this area.

Figure 11.11 Rock Slides Rock slides occur when material moves downslope along a generally planar surface. Most rock slides result when the underlying rocks dip in the same general angle as the slope of the land. Undercutting along the base of the slope and clay layers beneath porous rock or soil layers increase the chance of rock slides.

》Critical Thinking Question
Explain how the geologic planes of weakness in this slope, plus water from rainfall, influenced development of the depicted slide.

A **rock** or *block* **slide** occurs when rocks move downslope along a more or less planar surface. Most rock slides take place because the local slopes and rock layers dip in the same direction (▌ Figure 11.11), although they can also occur along fractures parallel to a slope. Rock slides are also common occurrences along the southern California coast, such as at Point Fermin (Figure 11.5).

Farther south from Point Fermin is the town of Laguna Beach, where residents have been hit by rock slides and mudslides in 1978, 1998, and as recently as 2005 (▌ Figure 11.12). Just as at Point Fermin, the rocks at Laguna Beach dip about 25 degrees in the same direction as the slope of the canyon walls and contain clay beds that "lubricate" the overlying rock layers, causing the rocks and the houses built on them to slide. Percolating water from heavy rains wets subsurface clayey siltstone, thus reducing its shear strength and helping to activate the slide. In addition, these slides are part of a larger ancient slide complex.

Not all rock slides are the result of rocks dipping in the same direction as a hill's slope. The rock slide at Frank, Alberta, Canada, on April 29, 1903, illustrates how nature and human activity can combine to create a situation with tragic results (▌ Figure 11.13).

It would appear at first glance that the coal-mining town of Frank, lying at the base of Turtle Mountain, was in no danger from a landslide (Figure 11.13). After all, many of the rocks dipped away from the mining valley, unlike the situations at Point Fermin and Laguna Beach. The joints in the massive limestone composing Turtle Mountain, however, dip steeply toward the valley and are essentially parallel with the slope of the mountain itself. Furthermore, Turtle Mountain is supported by weak siltstones, shales, and coal layers that underwent slow plastic deformation from the weight of the overlying massive limestone. Coal mining along the base of the valley also contributed to the stress on the rocks by removing some of the underlying support. All of these factors, as well as the frost action and chemical weathering that widened the joints,

Figure 11.12 Rock Slide, Laguna Beach, California A combination of interbedded clay beds that become slippery when wet, rocks dipping in the same direction as the slope of the sea cliffs, and undercutting of the sea cliffs by wave action activated a rock slide at Laguna Beach, California, that destroyed numerous homes and cars on October 2, 1978. This same area was hit by another rock slide in 2005.

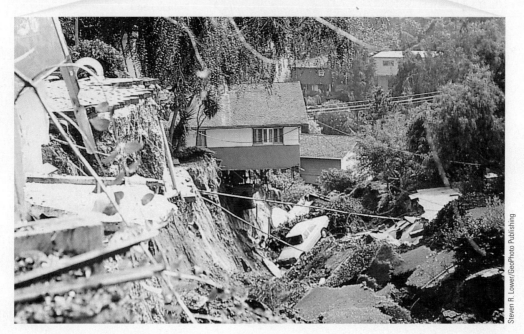

Steven R. Lower/GeoPhoto Publishing

finally resulted in a massive rock slide. Approximately 40 million m³ of rock slid down Turtle Mountain along joint planes, killing 70 people and partially burying the town of Frank.

Flows

Mass movements in which material flows as a viscous fluid or displays plastic movement are termed *flows*. Their rate of movement ranges from extremely slow to extremely rapid (Table 11.1). In many cases, mass movements begin as falls, slumps, or slides and change into flows farther downslope.

Of the major mass movement types, **mudflows** are the most fluid and move most rapidly (at speeds up to 80 km/hr). They consist of at least 50% silt- and clay-sized material combined with a significant amount of water (up to 30%). Mudflows are common in arid and semiarid environments where they are triggered by heavy rainstorms that quickly saturate the regolith, turning it into a raging flow of mud that engulfs everything in its path.

Mudflows can also occur in mountain regions (Figure 11.14) and in areas covered by volcanic ash, where they can be particularly destructive (see Chapter 5). Because mudflows are so fluid, they generally follow preexisting channels until the slope decreases or the channel widens, at which point they fan out.

Debris flows are composed of larger particles than mudflows and do not contain as much water. Consequently, they

CONNECTION LINK

To review the destructive effects of volcanic mudflows (termed *lahars*), see pages 118 and 119 in Chapter 5.

▌ Figure 11.13 Rock Slide, Turtle Mountain, Canada

a The tragic Turtle Mountain rock slide that killed 70 people and partially buried the town of Frank, Alberta, Canada, on April 29, 1903, was caused by a combination of factors. These included joints that dipped in the same direction as the slope of Turtle Mountain, a fault partway down the mountain, weak shale and siltstone beds underlying the base of the mountain, and mined-out coal seams.

b Results of the 1903 rock slide at Frank, Canada.

are usually more viscous than mudflows, typically do not move as rapidly, and rarely are confined to preexisting channels. Debris flows can be just as damaging, though, because they can transport large objects (▌ Figure 11.15).

Earthflows move more slowly than either mudflows or debris flows. An earthflow slumps from the upper part of a hillside, leaving a scarp, and flows slowly downslope as a thick, viscous, tongue-shaped mass of wet regolith (▌ Figure 11.16). Like mudflows and debris flows, earthflows can be of any size and are frequently destructive. They occur most commonly in humid climates on grassy, soil-covered slopes following heavy rains.

Copyright and Photograph by Dr. Parvinder S. Sethi

▌**Figure 11.14** **Mudflow, Estes Park, Colorado** Mudflows move swiftly downslope, engulfing everything in their path. Note how this mudflow in Rocky Mountain National Park has fanned out at the base of the hill. Also note the small lake adjacent to the mudflow that was formed after this mudflow created a dam across the stream.

B. Pipkin, University of Southern California

▌**Figure 11.15** **Debris Flow, Ophir Creek, Nevada** A debris flow and damaged house in lower Ophir Creek, western Nevada. Note the many large boulders that are part of the debris flow. Debris flows do not contain as much water as mudflows and typically are composed of larger particles.

▌ Figure 11.16 Earthflow

Scarp

Scarp

Reed Wicander

a Earthflows form tongue-shaped masses of wet regolith that move slowly downslope. They occur most commonly in humid climates on grassy, soil-covered slopes.

b An earthflow near Baraga, Michigan.

Canada's Air Force

▌ Figure 11.17 **Quick-Clay Slide, Nicolet, Quebec, Canada** The house on the slide (to the right of the bridge and circled in red) traveled several hundred meters with relatively little damage.

Some clays spontaneously liquefy and flow like water when they are disturbed. Such **quick clays** have caused serious damage and loss of lives in Sweden, Norway, eastern Canada (▌ Figure 11.17), and Alaska. Quick clays are composed of fine silt and clay particles made by the grinding action of glaciers. Geologists think that these fine sediments were originally deposited in a marine environment where their pore space was filled with saltwater. The ions in saltwater helped establish strong bonds between the clay particles, thus stabilizing and strengthening the clay. When the clays were subsequently uplifted above sea level, the saltwater was flushed out by fresh groundwater, reducing the effectiveness of the ionic bonds between the clay particles and thereby reducing the overall strength and cohesiveness of the clay. Consequently, when the clay is disturbed by a sudden shock or shaking, it essentially turns to a liquid and flows.

An excellent example of the damage that is done by quick clays occurred in the Turnagain Heights area of Anchorage, Alaska, in 1964 (❙ Figure 11.18a). Underlying most of the Anchorage area is the Bootlegger Cove Clay, a massive clay unit of poor permeability. Because the Bootlegger Cove Clay forms a barrier that prevents groundwater from flowing through the adjacent glacial deposits to the sea, considerable hydraulic pressure builds up on the landward side of the clay. Some of this water has flushed out the saltwater in the clay and has saturated the lenses of sand and silt associated with the clay beds. When the magnitude 8.6 Good Friday earthquake struck on March 27, 1964, the shaking turned parts of the Bootlegger Cove Clay into a quick clay and precipitated a series of massive slides in the coastal bluffs that destroyed most of the homes in the Turnagain Heights subdivision (❙ Figure 11.18b).

> **CONNECTION LINK**
>
> To review the destructive effects of ground shaking by earthquakes, see pages 212 and 213 in Chapter 9.

Solifluction is the slow downslope movement of water-saturated surface sediment. Solifluction can occur in any climate where the ground becomes saturated with water, but it is most common in areas of permafrost.

Permafrost, ground that remains permanently frozen, covers nearly 20% of the world's land surface (❙ Figure 11.19a). During the warmer season when the upper portion of the permafrost thaws, water and surface sediment form a soggy mass that flows by solifluction and produces a characteristic lobate topography (❙ Figure 11.19b).

❙ **Figure 11.18** Quick-Clay Slide, Anchorage, Alaska

a Ground shaking by the 1964 Alaska earthquake turned parts of the Bootlegger Cove Clay into a quick clay, causing numerous slides.

b Low-altitude photograph of the Turnagain Heights subdivision of Anchorage shows some of the numerous landslide fissures that developed, as well as the extensive damage to buildings in the area. The remains of the Four Seasons apartment building can be seen in the background.

A. Grantz/USGS

❙ **Figure 11.19** Permafrost and Solifluction

a Distribution of permafrost areas in the Northern Hemisphere.

b Solifluction flows in Kluane National Park, Yukon Territory, Canada, show the typical lobate topography that is characteristic of solifluction conditions.

James S. Monroe

O.J. Ferrains Jr./USGS

▌Figure 11.20 **Permafrost Damage** This house, south of Fairbanks, Alaska, has settled unevenly because the underlying permafrost in fine-grained silts and sands has thawed.

As might be expected, many problems are associated with construction in a permafrost environment. For example, when an uninsulated building is constructed directly on permafrost, heat escapes through the floor, thaws the ground below, and turns it into a soggy, unstable mush. Because the ground is no longer solid, the building settles unevenly into the ground and numerous structural problems result (▌Figure 11.20).

Creep, the slowest type of flow, is the most widespread and significant mass-wasting process in terms of the total amount of material moved downslope and the monetary damage it does annually. Creep involves extremely slow downhill movement of soil or rock. Although it can occur anywhere and in any climate, it is most effective and significant as a geologic agent in humid regions.

Because the rate of movement is essentially imperceptible, we are frequently unaware of creep's existence until we notice its effects: tilted trees and power poles, broken streets and sidewalks, or cracked retaining walls or foundations (▌Figure 11.21). Creep usually involves the whole hillside and probably occurs, to some extent, on any weathered or soil-covered, sloping surface.

▌Figure 11.21 Creep

B. Bradley and the University of Colorado's Geology Department, National Geophysical Data Center

a Some evidence of creep: (A) curved tree trunks; (B) displaced monuments; (C) tilted power poles; (D) displaced and tilted fences; (E) roadways moved out of alignment; (F) hummocky surface.

b Trees, bent by creep, Wyoming.

B. Bradley and the University of Colorado's Geology Department, National Geophysical Data Center

c Creep has bent these sandstone and shale beds of the Haymond Formation near Marathon, Texas.

David J. Matty

d Stone wall tilted due to creep in Champion, Michigan.

Creep is not only difficult to recognize but also difficult to control. Although engineers can sometimes slow or stabilize creep, many times the only course of action is to simply avoid the area if at all possible or, if the zone of creep is relatively thin, design structures that can be anchored into the bedrock.

Complex Movements

Recall that many mass movements are combinations of different movement types. When one type is dominant, the movement can be classified as one of those described thus far. If several types are more or less equally involved, however, it is called a **complex movement**.

The most common type of complex movement is the *slide-flow*, in which there is sliding at the head and then some type of flowage farther along its course. Most slide-flow landslides involve well-defined slumping at the head, followed by a debris flow or earthflow (❚ Figure 11.22). Any combination of different mass movement types is a complex movement.

A *debris avalanche* is a complex movement that often occurs in very steep mountain ranges. Debris avalanches typically start out as rockfalls when large quantities of rock, ice, and snow are dislodged from a mountainside, frequently as a result of an earthquake. The material then slides or flows down the mountainside, picking up additional surface material and increasing in speed. The 1970 Peru earthquake set in motion the debris avalanche that destroyed the towns of Yungay and Ranrahirca, Peru, and killed more than 25,000 people (❚ Figure 11.23).

11.4 Recognizing and Minimizing the Effects of Mass Wasting

The most important factor in eliminating or minimizing the damaging effects of mass wasting is a thorough geologic investigation of the region in question. In this way, former landslides and areas susceptible to mass movements can be identified and perhaps avoided. By assessing the risks of possible mass wasting before construction begins, engineers can take steps to eliminate or minimize the effects of such events.

What Would You Do ?

You've found your dream parcel of land in the hills of northern Baja California, where you plan to retire someday. Because you want to make sure that the area is safe to build a house, you decide to do your own geologic investigation of the area to make sure that there aren't any obvious geologic hazards. What specific things would you look for that might indicate mass wasting in the past? Even if there is no obvious evidence of rapid mass wasting, what features would you look for that might indicate a problem with slow types of mass wasting such as creep?

Identifying areas with a high potential for slope failure is important in any hazard assessment study; these studies include identifying former landslides, as well as sites of potential mass movement. Scarps, open fissures, displaced or tilted objects, a hummocky surface, and sudden changes in vegetation are some of the features that indicate former landslides or an area susceptible to slope failure. The effects of weathering, erosion, and vegetation may, however, obscure the evidence of previous mass wasting.

The information derived from a hazard assessment study can be used to produce *slope-stability maps* of the area (❚ Figure 11.24). These maps allow planners and developers to make decisions about where to site roads, utility lines, and housing or industrial developments based on the relative stability or instability of a particular location. The maps also

❚ **Figure 11.22** Complex Movement A complex movement is one in which several types of mass wasting are involved. In this example, slumping occurs at the head, followed by an earthflow.

Figure 11.23 Debris Avalanche An earthquake 65 km away triggered this debris avalanche on Nevado Huascarán, Peru, that destroyed the towns of Yungay and Ranrahirca and killed more than 25,000 people.

Horizontal contact between rocks of different stability. Tick marks are toward more easily eroded rocks. Erosional undermining may cause oversteepening of more resistant rocks, leading to landsliding.

Relative areal stability
Decreasing stability

←Relatively stable   Relatively unstable→

0   300   900   1,500 m

N

Figure 11.24 Slope-Stability Map This slope-stability map of part of San Clemente, California, shows areas delineated according to relative stability. Such maps help planners and developers make decisions about where to site roads, utility lines, buildings, and other structures.

» Critical Thinking Question
Locate the line that shows horizontal contact between rocks of different stability. What is the potential for mass wasting along this line, and why?

indicate the extent of an area's landslide problem and the type of mass movement that may occur.

Building codes, which spell out what types of site investigations need to be made and the manner in which structures must be built, also help determine how land will be developed and utilized. Of particular interest to us is the section of the *Uniform Building Code* that deals with slopes and the alteration of the landscape. This chapter deals with such geologic situations as the slope angle and direction of the slope in relation to bedding planes, foliation, faults, and joints. In addition, it specifies surface drainage requirements and compaction of materials in the area under investigation. This section of the *Uniform Building Code* has been widely adopted by many counties and cities in the United States and has led to more stringent local building codes, better land use, and decreased losses to structures built on hillsides and other possibly unstable areas.

Although most large mass movements usually cannot be prevented, geologists and engineers can use various methods to minimize the danger and damage resulting from them. Because water plays such an important role in many landslides, one of the most effective and inexpensive ways to

reduce the potential for slope failure or to increase existing slope stability is surface and subsurface drainage of a hillside. Drainage serves two purposes. It reduces the weight

What Would You Do ?

You are a member of a planning board for your seaside community. A developer wants to rezone some coastal property to build 20 condominiums. This project would be a boon to the local economy because it would provide jobs and increase the tax base. However, because the area is somewhat hilly and fronts the ocean, you are concerned about how safe the buildings would be. What types of studies would need to be done before any rezoning could take place? Is it possible to build safe structures along a hilly coastline? What specifically would you ask the environmental consulting firm that the planning board has hired to look for in terms of actual or potential geologic hazards if the condominiums are built?

▌ Figure 11.25 Using Drainpipes to Remove Subsurface Water

a Driving drainpipes that are perforated on one side into a hillside, with the perforated side up, can remove some subsurface water and help stabilize a hillside.

Flow of groundwater

Reed Wicander

b A drainpipe driven into the hillside at Point Fermin, California, helps to reduce the amount of subsurface water in these porous beds.

》 Critical Thinking Question
What two features in this photo can you identify that indicate mass wasting at this location is both a current and potential problem?

of the material likely to slide, and it increases the shear strength of the slope material by lowering pore pressure.

Surface waters can be drained and diverted by ditches, gutters, or culverts designed to direct water away from slopes. Drainpipes perforated along one surface and driven into a hillside can help remove subsurface water (▌ Figure 11.25). Finally, planting vegetation on hillsides helps stabilize slopes by holding the soil together and reducing the amount of water in the soil.

Another way to help stabilize a hillside is to reduce its slope. Recall that overloading and oversteepening by grading are common causes of slope failure. Reducing the angle of a hillside decreases the potential for slope failure. Two methods are usually employed to reduce a slope's angle. In the *cut-and-fill* method, material is removed from the upper part of the slope and used as fill at the base, thus providing a flat surface for construction and reducing the slope (▌ Figure 11.26).

The second method, which is called *benching*, involves cutting a series of benches or steps into a hillside (▌ Figure 11.27). This process reduces the overall average slope, and the benches serve as collecting sites for small landslides or rockfalls that might occur. Benching is most commonly used on steep hillsides in conjunction with a system of surface drains to divert runoff.

In some situations, retaining walls are constructed to provide support for the base of the slope (▌ Figure 11.28). The walls are usually anchored well into bedrock, backfilled with crushed rock, and provided with drain holes to prevent the buildup of water pressure in the hillside.

Rock bolts, similar to those employed in tunneling and mining, are sometimes used to fasten potentially unstable rock masses into the underlying stable bedrock (▌ Figure 11.29). This technique has been used successfully on the hillsides of Rio de Janeiro, Brazil, and to help secure the slopes at the Glen Canyon Dam on the Colorado River.

Recognition, prevention, and control of landslide-prone areas are expensive, but not nearly as expensive as the damage can be when such warning signs are ignored or not recognized. Unfortunately, there are numerous examples of landfill and dam collapses that serve as tragic reminders of the price paid in loss of lives and property damage when the warning signs of impending disaster are ignored.

▌Figure 11.26 **Stabilizing a Hillside by the Cut-and-Fill Method** One common method used to help stabilize a hillside and reduce its slope is the cut-and-fill method. Material from the steeper upper part of the hillside is removed, thereby decreasing the slope angle, and is used to fill in the base. This provides some additional support at the base of the slope.

Before

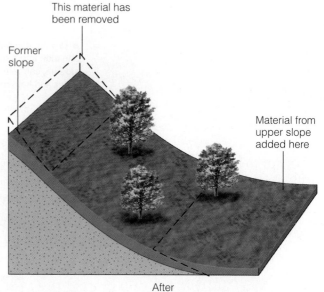

This material has been removed

Former slope

Material from upper slope added here

After

▌Figure 11.27 Stabilizing a Hillside by Benching

Before

Former slope

After

a Another common method used to stabilize a hillside and reduce its slope is benching. This process involves making several cuts along a hillside to reduce the overall slope. Furthermore, individual slope failures are now limited in size, and the material collects on the benches.

© John D. Cunningham/ Visuals Unlimited

b Benching is used in many road cuts and can be clearly seen in this photograph.

⟩⟩ Critical Thinking Question
Given the height of this road cut, how effective do you think benching will be in helping to stabilize this slope? What other measures can be taken to minimize the damage from potential mass wasting?

▌**Figure 11.28** Retaining Walls Help Reduce Landslides

a Retaining walls anchored into bedrock, backfilled with gravel, and provided with drainpipes can support a slope's base and reduce landslides.

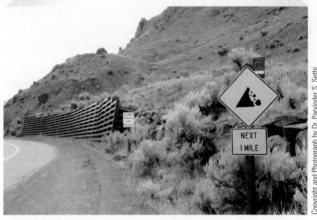

Copyright and Photograph by Dr. Parvinder S. Sethi

b A steel retaining wall built to stabilize the slope and keep falling and sliding rocks off the highway.

▌**Figure 11.29** Rock Bolts and Wire Mesh Help Reduce Landslides

a Rock bolts secured in bedrock can help stabilize a slope and reduce landslides.

Reed Wicander

b Rock bolts and wire mesh are used to secure rock on a steep hillside in Brisbane, Australia.

GEO-RECAP

▶ Geo-Image

© John Karachewski, Geoscapes Photography

▌**Figure 1** What features of slope stabilization do you see in this photograph of a housing development in Concord, California? You should be able to recognize at least three features.

Chapter Summary

- Mass wasting is the downslope movement of material under the direct influence of gravity. It frequently results in loss of life, as well as millions of dollars in damage annually.

- Mass wasting occurs when the gravitational force acting on a slope exceeds the slope's shear strength (the resisting forces that help maintain slope stability).

- The major factors causing mass wasting include slope angle, weathering and climate, water content, overloading, and removal of vegetation. It is usually several of these factors in combination that results in slope failure.

- Mass movements are generally classified on the basis of their rate of movement (rapid versus slow), type of movement (falling, sliding, or flowing), and type of material (rock, soil, or debris).

- Rockfalls are a common mass movement in which rocks free-fall. They are common along steep canyons, cliffs, and road cuts.

- Two types of slides are generally recognized. Slumps are rotational slides that involve movement along a curved surface and are most common in poorly consolidated or unconsolidated material. Rock slides (also known as block slides) occur when movement takes place along a more or less planar surface, and they usually involve solid pieces of rock.

- Several types of flows are recognized on the basis of their rate of movement (rapid versus slow), type of material (rock, sediment, or soil), and amount of water.

- Mudflows consist of mostly clay- and silt-sized particles and contain up to 30% water. They are most common in semiarid and arid environments and generally follow preexisting channels.

- Debris flows are composed of larger particles and contain less water than mudflows.

- Earthflows move more slowly than either debris flows or mudflows and move downslope as thick, viscous, tongue-shaped masses of wet regolith.

- Quick clays are clays that spontaneously liquefy and flow like water when they are disturbed.

- Solifluction is the slow downslope movement of water-saturated surface material and is most common in areas of permafrost.

- Creep, the slowest type of flow, is the imperceptible downslope movement of soil or rock. It is the most widespread of all types of mass wasting.

- Complex movements are combinations of different types of mass movements in which no single type is dominant. Most complex movements involve sliding and flowing.

- The most important factor in reducing or eliminating the damaging effects of mass wasting is a thorough geologic investigation to outline areas susceptible to mass movements.

- Although mass movement cannot be eliminated, its effects can be minimized by building retaining walls, draining excess water, regrading slopes, and planting vegetation.

Important Terms

complex movement (p. 277)
creep (p. 276)
debris flow (p. 271)
earthflow (p. 272)
mass wasting (p. 260)
mudflow (p. 271)

permafrost (p. 275)
quick clay (p. 274)
rapid mass movement (p. 265)
rock slide (p. 270)
rockfall (p. 265)
shear strength (p. 261)

slide (p. 265)
slow mass movement (p. 265)
slump (p. 266)
solifluction (p. 275)

Review Questions

1. A type of mass wasting common in mountainous regions in which talus accumulates is
   a. ____ creep.
   b. ____ solifluction.
   c. ____ rockfalls.
   d. ____ slides.
   e. ____ mudflow.

2. Former landslides and areas currently susceptible to slope failure can be identified by which of the following features?
   a. ____ tilted objects.
   b. ____ open fissures.
   c. ____ scarps.
   d. ____ hummocky surfaces.
   e. ____ all of these.

3. Which of the following factors influence mass wasting?
   a. ___ gravity.
   b. ___ weathering.
   c. ___ slope gradient.
   d. ___ water content.
   e. ___ all of these.

4. The most widespread and costly of all mass-wasting processes is
   a. ___ slumps.
   b. ___ creep.
   c. ___ mudflows.
   d. ___ rockfalls.
   e. ___ quick clays.

5. Which of the following helps reduce the slope angle or provides support at the base of a hillside?
   a. ___ cut and fill.
   b. ___ retaining walls.
   c. ___ benching.
   d. ___ all of these.
   e. ___ none of these.

6. Which of the following factors can actually enhance slope stability?
   a. ___ increasing the slope angle.
   b. ___ vegetation.
   c. ___ overloading.
   d. ___ rocks dipping in the same direction as the slope.
   e. ___ none of these.

7. Mass wasting can occur
   a. ___ on gentle slopes.
   b. ___ on steep slopes.
   c. ___ in flat-lying areas.
   d. ___ all of these.
   e. ___ none of these.

8. Movement of material along a surface or surfaces of failure is a
   a. ___ slide.
   b. ___ fall.
   c. ___ earthflow.
   d. ___ debris flow.
   e. ___ none of these.

9. Which of the following are the most fluid of mass movements?
   a. ___ earthflows.
   b. ___ debris flows.
   c. ___ mudflows.
   d. ___ solifluction.
   e. ___ slumps.

10. Shear strength includes
    a. ___ the strength and cohesion of material.
    b. ___ the amount of internal friction between grains.
    c. ___ gravity.
    d. ___ all of these.
    e. ___ answers a and b.

11. Why is creep so prevalent? Why does it do so much damage? What are some of the ways that creep might be controlled?

12. What roles do climate and weathering play in mass wasting?

13. If an area has a documented history of mass wasting that has endangered or taken human life, how should people and governments prevent such events from happening again? Are most large mass-wasting events preventable or predictable?

14. Discuss how topography and the underlying geology contribute to slope failure.

15. How would removing preexisting vegetation tend to affect most slopes in humid regions with respect to mass wasting?

16. Discuss some of the ways slope stability can be maintained in order to reduce the likelihood of mass movements.

17. Discuss how the different factors that influence mass wasting are interconnected.

18. Why are slumps such a problem along highways and railroad tracks in areas with relief?

19. What features would you look for to determine if the site where you wanted to build a house was safe?

20. How could mass wasting be recognized on other planets or moons? What would that tell us about the geology, and perhaps the atmosphere, of the planet or moon on which it occurred?

Two men paddle through a flooded shopping area in the Rosalie district of Brisbane, Australia on January 13, 2011. Deadly floodwaters that cut a swath across northeastern Australia flowed through the streets of the nation's third-largest city, forcing people to flee suburbs and skyscrapers.

CHAPTER 12

AFP/Stringer/Getty Images

Running Water— Streams and Rivers

OUTLINE

12.1 Introduction

12.2 Water on Earth

GEO-FOCUS: Dams, Reservoirs, and Hydroelectric Power Plants

12.3 Running Water

12.4 Running Water, Erosion, and Sediment Transport

12.5 Deposition by Running Water

12.6 Can Floods Be Predicted and Controlled?

GEO-INSIGHT: Floods and Flooding

12.7 Drainage Systems

12.8 The Evolution of Valleys

GEO-RECAP

HAVE YOU EVER WONDERED?

- How water is continually recycled from the oceans to land and back to the oceans?

- How streams and rivers erode, transport, and deposit sediment?

- What kinds of streams and rivers there are and how they differ?

- Why floods occur and what might be done to predict and control them?

- How streams and rivers adjust to changes in climate?

- How valleys form and evolve?

12.1 Introduction

All the terrestrial planets have a similar early history of accretion, differentiation, and volcanism, but their subsequent histories differ considerably. Indeed, now Earth is the only terrestrial planet with an oxygen-rich atmosphere and abundant surface water. Mercury's small size and its distance from the Sun and Venus's runaway greenhouse effect preclude the possibility of water on those planets. Mars is too small and too cold for liquid water, although it has some frozen water and small amounts of water vapor in its atmosphere. Furthermore, satellite imagery reveals winding valleys and canyons that indicate running water during the planet's early history. In contrast, 71% of Earth's surface is covered by oceans and seas, its atmosphere contains up to 4.0% water vapor, and the continents receive precipitation that feeds streams, rivers, lakes, and the groundwater system.

> **CONNECTION LINK**
>
> You can find more information on the terrestrial planets on pages 14 and 15 in Chapter 1.

The hydrosphere consists of all water on Earth, including water frozen in glaciers, although most of it (97.2%) is in the oceans (Table 12.1). Our main concern here is with that 0.0001% of Earth's water in stream and river channels. Keep in mind, though, that water is continually cycled through channels, and, in fact, running water has a tremendous impact on much of the land surface. Only the most arid deserts and areas covered by glaciers show little or no effects of running water.

You have probably experienced the power of running water if you have ever swum or canoed in a fast-flowing stream or river, but the vivid accounts of floods truly impress us with the energy of moving water. At 4:07 p.m. on May 31, 1889, residents of Johnstown, Pennsylvania, heard "a roar like thunder" and within 10 minutes the town was devastated by an 18-m-high wall of water that tore through the town at 60 km/hr, sweeping up debris, houses, and entire families (▌Figure 12.1a). Heavy rainfall and the failure of a dam upstream from the town caused the flood in which at least 2,200 people were killed, making it the most deadly river flood in U.S. history.

▌**Figure 12.1** Flooding in Pennsylvania in 1889 and in Pakistan in 2010

a Aftermath of the Johnstown, Pennsylvania, flood of May 13, 1889. An 18-m-high wall of water destroyed Johnstown and killed at least 2,200 people.

b As a result of exceptionally heavy monsoon rains, floodwaters from the Indus River covered nearly 20% of Pakistan during July and August 2010. About 2,000 people perished in the flood, and an estimated 18 million were directly affected.

More recently, in July and August 2010, because of exceptionally heavy monsoon rains, the Indus River in Pakistan flooded about 20% of the entire country (▌Figure 12.1b). Although the death toll was less than for some other recent natural disasters, the flood nevertheless was responsible for nearly 2,000 fatalities (see the chapter opening photo), and it had major effects on about 18 million people. Estimates of property damage are in the tens of billions of dollars.

Every year, floods cause extensive property damage and fatalities (see the chapter opening photo), and yet we derive many benefits from running water, even from some floods. Before the completion of the Aswan High Dam in 1970, Egyptian farmers depended on silt deposited on the Nile River floodplain to replenish their croplands. Indeed, in ancient Egypt, taxes were levied based on the level of the Nile River. In addition, running water is one source of freshwater for agriculture, industry, domestic use, and recreation, and about 8% of all electricity used in North America is generated at hydroelectric plants (see Geo-Focus on pages 288 and 289). When Europeans first explored North America, they followed the St. Lawrence, Mississippi, Missouri, and Ohio rivers. These and similar waterways throughout the world are also major avenues of commerce.

TABLE 12.1	Water on Earth	
Location	**Volume (km³)**	**Percentage of Total**
Oceans	1,327,500,000	97.20
Ice caps and glaciers	29,315,000	2.15
Groundwater	8,442,580	0.625
Freshwater and saline lakes and inland seas	230,325	0.017
Atmosphere at sea level	12,982	0.001
Stream channels	1,255	0.0001

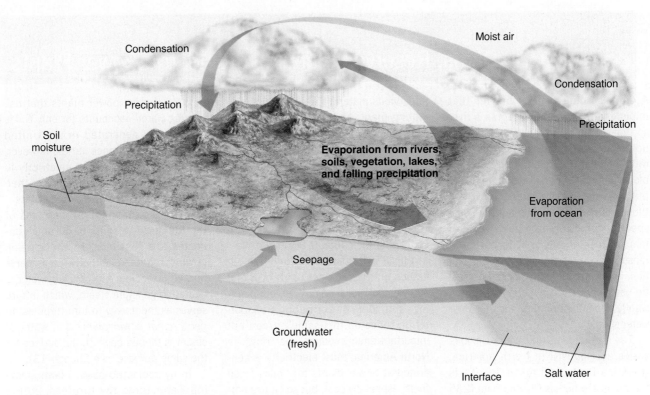

Figure 12.2 The Hydrologic Cycle During the hydrologic cycle, water evaporates from the oceans and rises as water vapor to form clouds that release their precipitation over the oceans or over land. Some of the precipitation falling on land enters the groundwater system, but much of it returns to the oceans by surface runoff, thus completing the cycle.

12.2 Water on Earth

Most of Earth's 1.33 billion km³ of water is in the oceans, and nearly all of the rest is frozen in glaciers on land (Table 12.1). Only 0.65% of all Earth's water is in the atmosphere, groundwater, lakes, swamps, and bogs, and a tiny important amount is in stream and river channels. Nevertheless, the water in channels is, with few exceptions, the most important geologic agent in modifying the land surface. Even in deserts, evidence of running water is common, although channels are dry most of the time.

Much of our discussion of running water is descriptive, but always be aware that streams and rivers are dynamic systems that must continuously respond to change. For example, paving in urban areas increases surface runoff to waterways, and other human activities such as building dams and impounding reservoirs also alter the dynamics of stream and river systems. Natural changes, too, affect the complex interacting parts of stream and river systems.

The Hydrologic Cycle

The connection between precipitation and clouds is obvious, but where does the moisture for rain and snow come from in the first place? In Table 12.1 we noted that 97.2% of all water on Earth is in the oceans, so you might immediately suspect that the oceans are the ultimate source of precipitation. In fact, water is continuously recycled from the oceans, through the atmosphere, to the continents, and back to the oceans. This **hydrologic cycle,** as it is called (Figure 12.2), is powered by solar radiation and is possible because water changes easily from liquid to gas (water vapor) under surface conditions. About 85% of all water entering the atmosphere comes from a layer about 1 m thick that evaporates from the oceans each year. The remaining 15% comes from water on land, but this water originally came from the oceans as well.

Regardless of its source, water vapor rises into the atmosphere where the complex processes of cloud formation and condensation take place. About 80% of the world's precipitation falls directly back into the oceans, in which case the hydrologic cycle is a three-step process of evaporation, condensation, and precipitation. For the 20% of all precipitation that falls on land, the hydrologic cycle is more complex, involving evaporation, condensation, movement of water vapor from the oceans to land, precipitation, and runoff. Some precipitation evaporates as it falls and reenters the cycle, but about 36,000 km³ of the precipitation that falls on land returns to the oceans by **runoff,** the surface flow in streams and rivers.

Not all precipitation returns directly to the oceans. Some is temporarily stored in lakes and swamps, or in snowfields and glaciers, or seeps below the surface where it enters

GEO-FOCUS   Dams, Reservoirs, and Hydroelectric Power Plants

A dam is some kind of barrier that impedes the flow of a stream or a river so that a lake (a reservoir) forms on its upstream side. Some dams form naturally as when a landslide or lava flow blocks a channel or beavers build dams, but the ones we are interested in are those constructed by humans for some specific purpose. These dams come in many sizes and are built for different reasons, including flood control, irrigation, and water for industrial and domestic use. Indeed, many communities are wholly dependent on reservoirs for their water supply.

The size of dams and their reservoirs varies enormously. In North America, the Mica Dam (243 m) in Canada is the highest, whereas Oroville Dam (235 m) in California is the highest in the United States. The latter, an earthfill dam consisting of a compacted core of impervious materials, has a power generating plant, and it supplies water to central and southern California (Figure 12.18a). Hoover Dam (221.4 m) on the Colorado River between Nevada and Arizona is a concrete arch dam that was built between 1931 and 1936. It, too, produces electricity and provides flood control, but it also stores water for irrigation and recreation (▌Figure 1).

Flip a switch and we illuminate our homes and workplaces, or turn a dial and we cook our food or wash our clothing, and some of our public transportation relies on this same unseen but important energy source—electricity. In North America, most electricity is generated at power plants that burn fossil fuels, especially coal, but some use natural gas; most of the rest is produced at nuclear power plants (▌Figure 2). Electricity generated at *hydroelectric power plants*—that is, power plants that use moving water—accounts for only 6.4% of the total generated in the United States. Its importance varies, however, from less than 2% of all electricity in Florida, Ohio, and Texas to more than 80% in Oregon and Washington State.

At all power plants, regardless of the fuel used, a spinning turbine connected to a generator containing a coil of wire produces electricity. So if coal is burned or nuclear fuel is used, water is heated to generate steam, which in turn serves as the energy to turn turbines. In *geothermal power plants,* hot water or steam is tapped beneath the surface for the same purpose (see Chapter 13).

In hydroelectric power plants, moving water turns the turbines, so the water must flow fast enough to accomplish this task. To provide the necessary water, a dam is built impounding a

Lowell Georgia/Photo Researchers, Inc.

▌**Figure 1** Hoover Dam on the Colorado River between Nevada and Arizona. At 221 m, it was the highest dam in the United States when it was completed in 1936. It is now second highest. In addition to electrical power generation, the reservoir functions in flood control, irrigation and recreation.

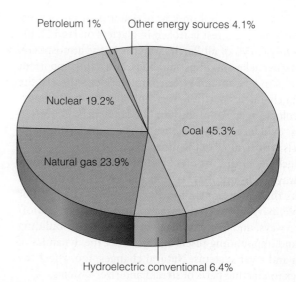

Note: Totals may not equal sum of components because of independent rounding.

▌**Figure 2** Sources of energy to generate electricity in the United States through August 2010. Only 6.4% of all electricity is generated at hydroelectric power plants. Source: U.S. Energy Information Administration

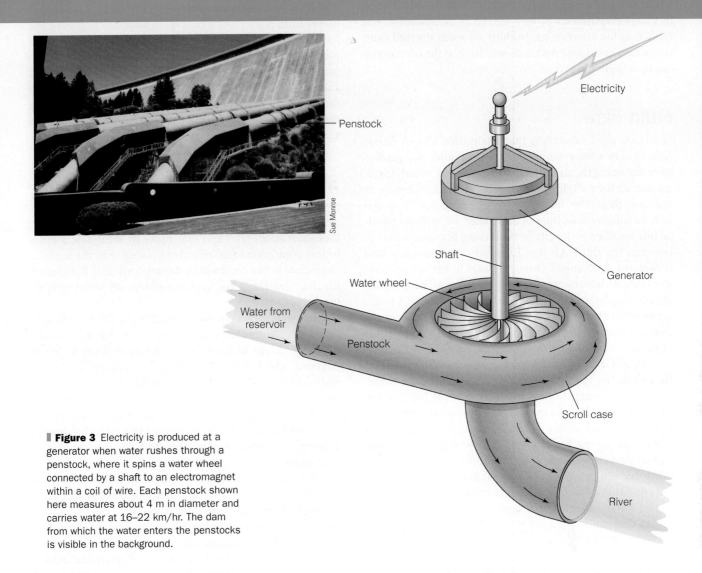

Penstock

Electricity

Shaft

Generator

Water wheel

Water from reservoir

Penstock

Scroll case

River

Sue Monroe

Figure 3 Electricity is produced at a generator when water rushes through a penstock, where it spins a water wheel connected by a shaft to an electromagnet within a coil of wire. Each penstock shown here measures about 4 m in diameter and carries water at 16–22 km/hr. The dam from which the water enters the penstocks is visible in the background.

reservoir where the water is higher than the power plant. Water moves from the reservoir through a large pipe called a penstock and spins a turbine (Figure 3). In other words, the potential energy of the water in the reservoir is converted to electrical energy at the power plant. Falling water was first used to generate electricity in 1892 at Appleton, Wisconsin, and since then hydroelectric power plants have become common on many of the world's waterways. Currently the largest is the Three Gorges Dam on the Yangtze River in the People's Republic of China.

Hydroelectric power has several appealing aspects, not the least of which are that it is a renewable resource and that it causes little pollution. So why don't the states and the federal government simply increase their hydroelectric output? One reason is that not all areas have this potential; suitable sites for dams and reservoirs might not be present, for instance. In addition, dams are very expensive to build and maintain, reservoirs fill with sediment unless dredged, and during droughts enough water may not be available to keep reservoirs sufficiently full.

In addition to construction costs, people must be relocated from areas where reservoirs are impounded and from the discharge areas downstream from dams. Also, remember that dams have failed, causing fatalities and property damage. And as parts of flood control projects, they are designed to retain floods up to a given size, so if a flood of greater size occurs, the dam fills and is of no further use. So although hydroelectric power generating remains important, we cannot realistically look forward to very much additional use of this energy source.

the groundwater system. Water might remain in some of these reservoirs for thousands of years, but eventually glaciers melt, lakes and groundwater feed streams and rivers, and this water returns to the oceans. Even the water used by plants evaporates, a process known as *transpiration,* and returns to the atmosphere. In short, all water derived from the oceans eventually makes its way back to the oceans and can thus begin the hydrologic cycle again.

Fluid Flow

Solids are rigid substances that retain their shapes unless deformed by a force, but fluids—that is, liquids and gases—have no strength, so they flow in response to any force, no matter how slight. Liquid water flows downslope in response to gravity, but its flow may be *laminar* or *turbulent.* In laminar flow, lines of flow called streamlines parallel one another with little or no mixing between adjacent layers in the fluid. All flow is in one direction only, and it remains unchanged through time. In turbulent flow, streamlines intertwine, causing complex mixing within the moving fluid (▌Figure 12.3). If we could trace a single water molecule in turbulent flow, it may move in any direction at a particular time although its overall movement is in the direction of flow.

Runoff during a rainstorm depends on the **infiltration capacity,** the maximum rate at which surface materials absorb water. Several factors control infiltration capacity, including intensity and duration of rainfall. If rain is absorbed as fast as it falls, no surface runoff takes place. Loosely packed dry soil absorbs water faster than tightly packed wet soil, and thus more rain must fall on loose dry soil before runoff begins. Regardless of the initial condition of surface materials, once they are saturated, excess water collects on the surface and, if on a slope, it moves downhill.

▌**Figure 12.3 Turbulent Flow** Running water can move by either laminar flow or turbulent flow, but in streams it is almost always fully turbulent. Laminar flow takes place if water is shallow and flows very slowly. This rapidly flowing stream in Alaska shows turbulent flow.

12.3 Running Water

The term *running water* applies to any surface water that moves from higher to lower areas in response to gravity. We have already noted that running water is very effective in modifying Earth's land surface by erosion, and that it is the primary geologic process responsible for sediment transport and deposition in many areas. Indeed, it is responsible for the tiniest rills in farmer's fields and scenic wonders such as the Grand Canyon in Arizona, as well as vast deposits such as the Mississippi River delta.

Sheet Flow and Channel Flow

Even on steep slopes, flow is initially slow and hence causes little or no erosion. As water moves downslope, though, it accelerates and may move by *sheet flow,* a more or less continuous film of water flowing over the surface. Sheet flow is not confined to depressions, and it accounts for *sheet erosion,* a particular problem on some agricultural lands.

In *channel flow,* surface runoff is confined to trough-like depressions that vary in size from tiny rills with a trickling stream of water to the Amazon River in South America, which is 6,450 km long and at one place 2.4 km wide and 90 m deep. We describe flow in channels with terms such as *rill, brook, creek, stream,* and *river,* most of which are distinguished by size and volume. Here we use the terms *stream* and *river* more or less interchangeably, although the latter usually refers to a larger body of running water.

Streams and rivers receive water from several sources, including sheet flow and rain that falls directly into their channels. Far more important, though, is the water supplied by soil moisture and groundwater, both of which flow downslope and discharge into waterways. In areas where groundwater is plentiful, streams and rivers maintain a fairly stable flow year-round because their water supply is continuous. In contrast, the amount of water in streams and rivers of arid and semiarid regions fluctuates widely because they depend more on infrequent rainstorms and surface runoff for their water.

Gradient, Velocity, and Discharge

Water in any channel flows downhill over a slope known as its **gradient.** Suppose a river has its headwaters (source) 1,000 m above sea level and it flows 500 km to the sea, so it drops vertically 1,000 m over a horizontal distance of 500 km. Its gradient is found by dividing the vertical drop by the horizontal distance, which in this example is 1,000/500 km = 2 m/km (▌Figure 12.4). On average, this river drops vertically 2 m for every kilometer along its course.

In this example, we calculated the average gradient for a hypothetical river, but gradients vary not only among

❚ Figure 12.4 Calculating Gradient The average gradient for this stream is 2 m/km, but the gradient may be calculated for any segment of a stream system. Notice that the gradient is steepest in the headwaters area (where the stream originates) and decreases downstream. In the lower reaches of some large rivers, the gradient is as little as a few centimeters per kilometer.

channels but even along the course of a single channel. Rivers and streams are steeper in their upper reaches (near their headwaters) where they may have gradients of several tens of meters per kilometer, but they have gradients of only a few centimeters per kilometer where they discharge into the sea.

The **velocity** of running water is a measure of the downstream distance water travels in a given time. It is usually expressed in meters per second (m/sec) or feet per second (ft/sec), and it varies across a channel's width as well as along its length. Water moves more slowly and with greater turbulence near a channel's bed and banks because friction is greater there than it is some distance from these boundaries (❚ Figure 12.5a). Channel shape and roughness also influence flow velocity. Broad, shallow channels and narrow, deep channels have proportionately more water in contact with their perimeters than channels with semicircular cross sections (❚ Figure 12.5b). So, if other variables are the same, water flows faster in a semicircular channel because there is less frictional resistance. As one would expect, rough channels, such as those strewn with boulders, offer more frictional resistance to flow than do channels with a bed and banks composed of sand or mud.

❚ Figure 12.5 Velocity and Discharge Flow velocity in rivers and streams varies as a result of friction with their banks and beds.

a The maximum flow velocity is near the center and top of a straight channel where the least friction takes place. The arrows are proportional to velocity.

b These three differently shaped channels have the same cross-sectional area, but the semicircular one has less water in contact with its perimeter and thus less frictional resistance to flow.

	Broad, shallow channel	Narrow, deep channel	Semicircular channel
Cross-sectional area	10 m²	10 m²	10 m²
Perimeter in contact with water	12 m	12 m	7.9 m

Intuitively, you might think that the gradient is the most important control on velocity—the steeper the gradient, the greater the velocity. In fact, a channel's average velocity actually increases downstream even though its gradient decreases! Keep in mind that we are talking about average velocity for a long segment of a channel, not velocity at a single point. Three factors account for this downstream increase in velocity. First, velocity increases even with decreasing gradient in response to the acceleration of gravity. Second, the upstream reaches of channels tend to be boulder-strewn, broad, and shallow, so frictional resistance to flow is high, whereas downstream segments of the same channels are more semicircular and have banks composed of finer materials. And finally, the number of smaller tributaries joining a larger channel increases downstream. Thus, the total volume of water (discharge) increases, and increasing discharge results in greater velocity.

Specifically, **discharge** is the volume of water that passes a particular point in a given period of time. Discharge is found from the dimensions of a water-filled channel—that is, its cross-sectional area (A) and flow velocity (V). Discharge (Q) is then calculated with the formula $Q = VA$ and is expressed in cubic meters per second (m^3/sec) or cubic feet per second (ft^3/sec). The Mississippi River has an average discharge of 18,000 m^3/sec, and the average discharge for the Amazon River in South America is 200,000 m^3/sec.

In most rivers and streams, discharge increases downstream as more and more water enters a channel. However, there are a few exceptions. Because of high evaporation rates and infiltration, the flow in some desert waterways actually decreases downstream until the water disappears. And even in perennial rivers and streams, discharge is obviously highest during times of heavy rainfall and at a minimum during the dry season.

12.4 Running Water, Erosion, and Sediment Transport

Streams and rivers possess two kinds of energy to accomplish their tasks of erosion and transport: potential and kinetic. *Potential energy* is the energy of position, such as the energy of water at high elevation. During stream flow, potential energy is converted to *kinetic energy,* which is the energy of motion. Much of this kinetic energy is used up in fluid turbulence, but some is available for erosion and transport. You already know that erosion involves the removal of weathered materials from their source area, so the materials transported by a stream include a load of solid particles (mud, sand, and gravel) and a dissolved load.

Because the **dissolved load** of a stream is invisible, it is commonly overlooked, but it is an important part of the total sediment load. Some of it is acquired from a stream's bed and banks where soluble rocks such as limestone are present, but much of it is carried into waterways by sheet flow and by groundwater. A stream's solid load is made up of particles ranging from clay sized (>1/256 mm) to huge boulders, much of it supplied by mass wasting, but some is eroded directly from a stream's bed and banks. The direct impact of running water, **hydraulic action,** is sufficient to set particles in motion, just as the stream from a garden hose gouges out a hole in soil (Figure 12.6).

Running water carrying sand and gravel erodes by **abrasion,** as exposed rock is worn and scraped by the impact of these particles. Circular to oval depressions called *potholes* in streambeds are one manifestation of abrasion (Figure 12.7). These depressions form where swirling currents with sand and gravel eroded the rock

Once materials are eroded, they are transported for some distance from their source and eventually deposited. The dissolved load is transported in the water itself, but the load of solid particles moves as *suspended load* or *bed load*. The **suspended load** consists of the smallest particles of silt and clay, which are kept suspended above the channel's bed by fluid turbulence (Figure 12.8). It is the suspended load of streams and rivers that gives the water its murky appearance.

The **bed load** of larger particles, mostly sand and gravel, cannot be kept suspended by fluid turbulence, so it is transported along the bed. However, some of the sand may be temporarily suspended by currents that swirl

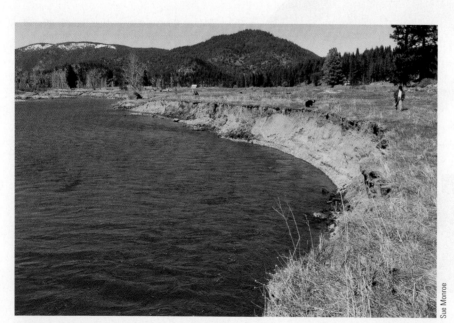

Figure 12.6 Hydraulic Action This stream erodes its bank by the direct impact of water, or what geologists call hydraulic action.

Sue Monroe

Figure 12.7 Abrasion by Running Water Carrying Sand and Gravel

a Potholes in the bed of the McCloud River in California. Also notice the rock surface that has been smoothed by abrasion.

b View into one of the potholes showing the sand and gravel that swirled around to erode the pothole.

across the streambed and lift grains into the water. The grains move forward with the water, but also settle and finally come to rest and then again move by the same process of intermittent bouncing and skipping, a phenomenon known as *saltation* (Figure 12.8). Particles too large to be even temporarily suspended are transported by traction; that is, they simply roll or slide along a channel's bed.

12.5 Deposition by Running Water

Some of the sediment now being deposited in the Gulf of Mexico by the Mississippi River came from sources as distant as Pennsylvania, Minnesota, and Alberta, Canada. Transport might be lengthy, but deposition eventually takes place. Some deposits accumulate along the way in channels, on adjacent floodplains, or where rivers and streams discharge from mountains onto nearby lowlands or where they flow into lakes or seas.

Rivers and streams constantly erode, transport, and deposit sediment, but they do most of their geologic work when they flood. Consequently, their deposits, collectively called **alluvium,** do not represent the day-to-day activities of running water, but rather the periodic sedimentation that takes place during floods. Recall from Chapter 7 that sediments accumulate in *depositional environments* characterized as continental, transitional, and marine (see Figure 7.3). Deposits of rivers and streams are found mostly in the first two of these settings; however, much of the detrital sediment found on continental margins is derived from the land and transported to the oceans by running water.

Figure 12.8 Sediment Transport Sediment transport as bed load, suspended load, and dissolved load. Flow velocity is highest near the surface, but gravel- and sand-sized particles are too large to be lifted far from the streambed so they make up the bed load, whereas silt and clay are in the suspended load.

The Deposits of Braided and Meandering Channels

A **braided stream** has an intricate network of dividing and rejoining channels separated from one another by sand and gravel bars (Figure 12.9). Seen from above, the channels resemble the complex strands of a braid. Braided channels develop when the sediment supply exceeds the transport capacity of running water, resulting in the deposition of sand and gravel bars. During high-water stages, the bars are submerged, but when the water is low, they are exposed and divide a single channel into multiple channels. Braided streams have broad, shallow channels and are characterized as bed-load transport streams because they transport and deposit mostly sand and gravel.

Braided channels are common in arid and semiarid regions with sparse vegetation and surface materials that are

> **CONNECTION LINK**

You can find more about streams that discharge from glaciers on page 358 in Chapter 14.

easily eroded. So much sediment is released from melting glaciers that rivers and streams discharging from them are also commonly braided.

Meandering streams have a single sinuous channel with broadly looping curves known as *meanders* (▮ Figure 12.10). Channels of meandering streams are semicircular in cross section along straight reaches, but markedly asymmetric at meanders, where they vary from shallow to deep across the meander. The deeper side of the channel is known as the *cut bank* because greater velocity and fluid turbulence erode it. In contrast, flow velocity is at a minimum on the opposite bank, which slopes gently into the channel. As a result of this unequal distribution of flow velocity across meanders, the cut bank erodes and a **point bar** is deposited on the gently sloping inner bank. (▮ Figure 12.11).

Meanders commonly become so sinuous that the thin neck of land between adjacent ones is cut off during a flood. Many of the floors of valleys with meandering channels are marked by crescent-shaped **oxbow lakes,** which are simply cutoff meanders (Figure 12.10 and ▮ Figure 12.12). Oxbow

▮ **Figure 12.9** A Braided Stream and Its Deposits This braided stream is in Denali National Park in Alaska. Its deposits are mostly gravel and sand.

>> Critical Thinking Question
What kinds of deposits would you expect to find in a braided stream system?

lakes may persist for a long time, but they eventually fill with organic matter and fine-grained sediments carried by floods.

▮ **Figure 12.10** Diagrammatic View of a Meandering River

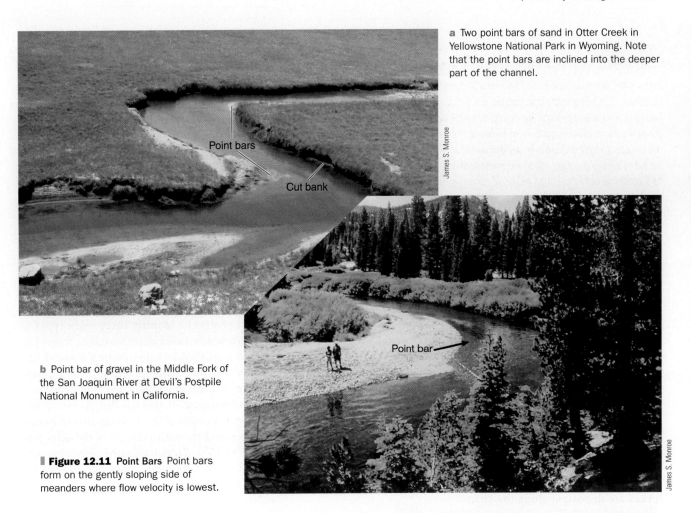

a Two point bars of sand in Otter Creek in Yellowstone National Park in Wyoming. Note that the point bars are inclined into the deeper part of the channel.

b Point bar of gravel in the Middle Fork of the San Joaquin River at Devil's Postpile National Monument in California.

▌ **Figure 12.11** Point Bars Point bars form on the gently sloping side of meanders where flow velocity is lowest.

▌ **Figure 12.12** Oxbow Lakes

a A meandering stream showing various stages in the evolution of oxbow lakes.

≫ Critical Thinking Question
Where do you think the next oxbow will form?

b Oxbow lakes along the Mississippi River in Minnesota.

Floodplain Deposits

Streams and rivers periodically receive more water than their channels can handle, so they overflow their banks and spread across adjacent low-lying, fairly flat **flood-plains** (Figure 12.10). Sand and gravel might be deposited on floodplains where a stream or river bursts from its banks, but much more commonly the deposits are silt and clay, or simply mud. During a flood, a stream or river overtops its banks, and water pours onto the floodplain, but as it does so, its depth and velocity rapidly decrease. As a result, ridges of sandy alluvium known as **natural levees** are deposited along the channel margins, and mud is carried beyond the natural levees into the floodplain where it settles from suspension (Figure 12.13).

Another feature found on floodplains is oxbow lakes; recall that oxbow lakes are cut-off meanders (Figure 12.12a). Once isolated from the main channel, oxbow lakes receive water mostly from periodic floods, although groundwater may also contribute. When they fill in with sediments brought in by floods and accumulating organic matter they are called *meander scars*, some of which are visible in the upper part of the image in Figure 12.12b.

Deltas

Where a river or stream flows into a standing body of water, such as a lake or the ocean, its flow velocity rapidly diminishes and any sediment in transport is deposited. Under some circumstances, this deposition creates a

delta, an alluvial deposit that causes the shoreline to build outward into the lake or sea, a process called *progradation.* The simplest prograding deltas have a characteristic vertical sequence of *bottomset beds* overlain successively by *foreset beds* and *topset beds* (Figure 12.14). This vertical sequence develops when a river or stream enters another body of water where the finest sediment (silt and clay) is carried some distance out into the lake or sea where it settles to form bottomset beds. Nearer shore, foreset beds are deposited as gently inclined layers, and topset beds, consisting of the coarsest sediments, are deposited in a network of *distributary channels* traversing the top of the delta (Figure 12.14).

Small deltas in lakes may have the three-part sequence described above, but deltas deposited along seacoasts are much larger, far more complex, and considerably more important as potential areas of natural resources. In fact, depending on the relative importance of running water, waves, and tides, geologists identify three main types of marine deltas. *Stream-dominated deltas* have long finger-like sand bodies, each deposited in a distributary channel that progrades far seaward. The Mississippi River delta is a good example (Figure 12.15a). In contrast, the Nile River delta in Egypt is *wave dominated.* It also has distributary channels, but the seaward margin of the delta consists of islands reworked by waves, and the entire margin of the delta progrades (Figure 12.15b). *Tide-dominated deltas* are continuously modified into tidal sand bodies that parallel the direction of tidal flow (Figure 12.15c).

Figure 12.13 Floodplain Deposits

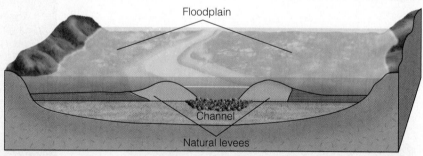

a The origin of floodplain deposits. During floods, streams deposit natural levees, and silt and mud settle from suspension on the floodplain.

b After flooding.

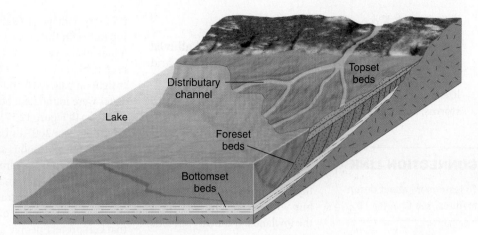

Figure 12.14 Prograding Delta
Internal structure of the simplest type of
prograding delta. Small deltas in lakes
may have this structure, but marine
deltas are much more complex.

Figure 12.15 Stream-, Wave-, and Tide-Dominated Deltas

Image courtesy of the Image Science & Analysis Laboratory, NASA Johnson Space Center
(http://eol.jsc.nasa.gov)

a The Mississippi River delta on the U.S. Gulf Coast is
stream-dominated.

Image courtesy of USGS National Center for EROS and NASA Landsat Project Science Office

c The Ganges-Brahmaputra delta of
Bangladesh is tide-dominated.

Jacques Descloitres, MODIS Rapid Response Team, NASA/GSFC

b The Nile River delta of Egypt is wave-dominated.

Alluvial Fans

Fan-shaped deposits of alluvium on land known as **alluvial fans** form best on lowlands with adjacent highlands in arid and semiarid regions where little vegetation exists to stabilize surface materials (❚ Figure 12.16). During periodic rainstorms, surface materials are quickly saturated, and surface runoff is funneled into a mountain canyon leading to adjacent lowlands. In the mountain canyon, the runoff is confined so that it cannot spread laterally, but when it discharges onto the lowlands, it quickly spreads out, its velocity diminishes, and deposition ensues. Repeated episodes of sedimentation result in the accumulation of a fan-shaped body of alluvium.

> **CONNECTION LINK**
>
> To learn more about desert features, see Chapter 15.

Deposition by running water in the manner just described is responsible for many alluvial fans. In this case, they are composed mostly of sand and gravel, both of which contain a variety of sedimentary structures. In some cases, though, the water flowing through a canyon picks up so much sediment that it becomes a viscous debris flow. Consequently, some alluvial fans consist mostly of debris-flow deposits that show little or no layering. Of course, the dominant type of deposition can change through time, so a particular fan might have both types of deposits.

12.6 Can Floods Be Predicted and Controlled?

When any waterway receives more water than its channel can handle, it floods, occupying part or all of its floodplain. Indeed, floods are so common that unless they cause extensive property damage or fatalities, they rate little more than a passing notice in the news. Dozens of floods occur in the United States every year, but the most extensive river flooding in recent history was the Great Flood of 1993 (see the Geo-Insight on pages 300 and 301). Large parts of nine Midwestern states were inundated, but particularly hard hit were Missouri and Iowa (❚ Figure 12.17). More than 2,100 people died during river floods in 2007 in India and Bangladesh, including dozens killed by poisonous snakes that, like people, sought refuge on high ground. We already mentioned the July and August 2010 flooding of the Indus River in Pakistan, and the more recent flooding in Australia (see the chapter opening photo).

Flooding is a fact of life, but there are several practices that can protect people and their property and minimize the impact of floods. Unfortunately, none of these measures is foolproof, and most are expensive. One solution is to simply move all or parts of a community to higher ground. Indeed, following the Great Flood of 1993, parts of some small communities in Missouri and Illinois were moved to higher ground. Martin, Kentucky, has been flooded 37 times since 1862, and parts of the downtown area are being moved. Moving small communities is one option, but not for cities such as Des Moines, Iowa, which was extensively flooded in 1993.

One common flood control practice is to construct *dams* that impound excess water during floods (❚ Figure 12.18a). Of course dams are expensive, they eventually fill in with sediment, unless dredged, and some fail. In 1976, the Teton Dam in Idaho collapsed, killing 11 people and 13,000 cattle. The total damage was about $2 billion. Another practice is to construct *levees* that raise the banks of streams and rivers, thereby increasing a channel's capacity (❚ Figure 12.18b). Unfortunately, deposition within channels raises streambeds, making the levees less effective unless they are raised, too. Levees along the Huang He in China caused the streambed to rise more than 20 m above its surrounding floodplain in 4,000 years. In 1887 when the river breached its levees, more than 1 million people were killed. Sacramento, California, is a flood-prone city with some levees

❚ **Figure 12.16 Alluvial Fans and Their Deposits**

a These alluvial fans at the base of the Panamint Range in Death Valley, California, were deposited where streams discharged from mountain canvons onto adiacent lowlands.

b Ancient alluvial fan deposits in Montana are made up of sand and gravel.

Figure 12.17 The Great Flood of 1993

Missouri Highway and Transportation Department/USGS

a The flood of '93 was the most extensive river flooding in the U.S. in recent history. It had devastating effects on nine states in the Midwest, but particularly hard hit were Missouri and Iowa. This image shows the Missouri River flooding on July 30, 1993, near the Cedar City and Jefferson City Memorial Airport, which is just north of Jefferson City, Missouri.

> **>> Critical Thinking Question**
> What kinds of deposits would you expect to find on this floodplain when the flood recedes?

Sue Monroe

b Flood waters in Portage des Sioux covered the 5.5-m-high pedestal of this statue on the bank of the Mississippi River.

150 years old and in poor repair. The cost of repairing them has risen to as much as $250,000 per 100 m.

In some areas, state or federal agencies build *floodways,* which are channels used to divert excess water around a community or an area of economic importance (■ Figure 12.18c). Yet another practice is to build *floodwalls* to protect vulnerable areas. These are vertical structures placed along stream or river banks where levees are not practical, such as in cities (■ Figure 12.18d). Reforestation of cleared land also reduces the potential for flooding because vegetated soil absorbs water and decreases runoff.

Flooding streams and rivers account for more than $5 billion in property damage each year in the United States, not including coastal flooding from hurricanes. And even though more and more flood control projects are completed, the damages from flooding are not decreasing. The combination of fertile soils, level surfaces, and proximity to water for agricultural, industrial, and domestic use makes floodplains popular places for development. Unfortunately, urbanization increases

surface runoff because soils are compacted or covered by asphalt or concrete, reducing infiltration capacity. In addition, storm drains in urban areas quickly carry water to nearby channels, many of which flood more often than they did in the past.

As for predicting floods, the best that can be done is to monitor streams, evaluate their past behavior, and anticipate floods of a given size in a specified period. Most people have heard of 10-year floods, 20-year floods, and so on, but how are such determinations made? The U.S. Geological Survey as well as state agencies record and analyze stream behavior through time and anticipate floods of a specified size. So a 20-year flood, for example, is the period during which a flood of a given magnitude can be expected. It does not mean that the river in question will have a flood of that size every 20 years, only that over a long period of time, it will average 20 years. Or we can say that the chances of a 10-year flood taking place in any one year are 1 in 10 (1/10). In fact, it is possible that two 10-year floods could take place in successive years, but then not occur again for several decades.

What Would You Do ?

The largest part of many states' mineral revenues comes from sand and gravel, most of which is used in construction. You become aware of a sand and gravel deposit that you can acquire for a small investment. How would the proximity of the deposit to potential markets influence your decision? Assuming the deposit was stream deposited, would it be important to know whether deposition took place in braided or meandering channels? How could you tell one from the other?

GEO-INSIGHT Floods and Flooding

A flood takes place when water covers an area that is not normally covered. Most floods occur when streams and rivers overtop their banks, but there are other types of floods as well, for example, from failing dams and coastal flooding resulting from hurricanes.

▼ 1. 2. ▶ The St. Francis Dam in southern California in March 1928. On March 13, 1928, only the central part of the dam remained. The catastrophic flooding following the dam's collapse killed at least 450 people. Ironically, the dam was inspected and declared safe not long before its collapse.

Los Angeles Public Library

Los Angeles Public Library

3. ▶ Coastal flooding during hurricanes is caused by exceptionally high waves, called a storm surge, or, in the case of New Orleans, Louisiana, by the failure of levees that were built to protect the city from Lake Pontchartrain and the Mississippi River. This image shows the extensive flooding of New Orleans caused by Hurricane Katrina in August 2005.

Manaham Kahana/AFP/Getty Images

Photo by Merle Agnello of the Tribune-Democrat, Courtesy of the Johnstown Area Heritage Association

4. ▲ Flash floods take place when a channel is filled to overflowing very quickly but only briefly. This flash flood hit Johnstown, Pennsylvania, in 1977 and killed about 85 people. Recall that Johnstown was also hit by a flood from a dam failure in 1889 (see the Introduction and Figure 12.1a).

Gerry Wolfe/NPS

5. ▲ Flash floods often fill gullies in arid areas such as this one that took place in Death Valley, California, in 1976. Nearly 6 cm of rain fell in 4 days, causing a flood that destroyed a paved road.

The images below show the Great Flood of 1993 that inundated parts of nine Midwestern states. It was the most extensive river flooding in the United States in recent history.

Landsat imagery courtesy EOSAT/NASA

◄ **6. 7.** ▼ Satellite images of the Mississippi, Missouri, and Illinois rivers near the juncture of the three rivers during the drought of 1988 (6), and during the flood of 1993 (7).

Landsat Imagery courtesy EOSAT/NASA

Willam Campbell/Pete Arnold Images/ PhotoLibrary

◄ **8.** Breached levee on the Mississippi River near Davenport, Iowa. This is one of 800 levees that failed or were overtopped during the flood.

State	Residential	Agricultural	Other	Total
Illinois	$176,833,000	$166,502,000	$409,020,000	$752,355,000
Iowa	57,827,000	1,030,030,000	334,835,000	1,422,692,000
Kansas	35,829,000	855,849,000	176,162,000	1,067,840,000
Minnesota	16,940,000	694,041,000	286,540,000	997,521,000
Missouri	405,175,000	540,666,000	1,255,191,000	2,201,032,000
Nebraska	18,584,000	120,521,000	67,899,000	207,004,000
North Dakota	10,138,000	35,039,000	25,806,000	70,983,000
South Dakota	21,919,000	276,218,000	195,408,000	493,545,000
Wisconsin	17,747,000	133,835,000	135,917,000	287,499,000
Total—All States	760,992,000	3,852,701,000	2,886,778,000	7,500,471,000

9. ► Damage caused by the Great Flood of '93. Compiled by the U.S. Army Corps of Engineers, figures are rounded to the nearest $1,000.

Figure 12.18 Flood Control   Dams and reservoirs, levees, floodways, and floodwalls are some of the structures used to control floods.

a   Oroville Dam in California, at 235 m high, is the highest dam in the United States. It helps control floods, provides water for irrigation, and produces electricity at its power plant.

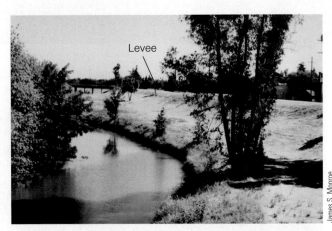

b   This levee, an artificial embankment along a waterway, helps protect nearby areas from floods. A university campus lies out of view just to the right of the levee.

» Critical Thinking Question
What are some of the functions of a dam and its reservoir?

c   This floodway carries excess water from a river (not visible) around a small community.

d   This floodwall on the bank of the Danube River at Mohács, Hungary, helps protect the city from floods.

12.7 Drainage Systems

Thousands of waterways, which are parts of larger drainage systems, flow directly or indirectly into the oceans. The only exceptions are some rivers and streams that flow into desert basins surrounded by higher areas. But even these are parts of larger systems consisting of a main channel with all its tributaries—that is, streams that contribute water to another stream. The Mississippi River and its tributaries such as the Ohio, Missouri, Arkansas, and Red rivers and thousands of smaller ones, or any other drainage system for that matter, carry runoff from an area known as a **drainage basin.** A topographically high area called a **divide** separates a drainage basin from adjoining ones (▮ Figure 12.19). The continental divide along the crest of the Rocky Mountains in North America, for instance, separates drainage in opposite directions; drainage to the west goes to the Pacific, whereas drainage to the east eventually reaches the Gulf of Mexico.

The arrangements of channels within an area are classified as types of **drainage patterns.** The most common is *dendritic drainage,* which consists of a network of channels resembling tree branching (▮ Figure 12.20a). It develops on gently sloping surfaces composed of materials that respond more or less homogeneously to erosion, such as areas underlain by nearly horizontal sedimentary rocks.

In dendritic drainage, tributaries join larger channels at various angles, but *rectangular drainage* is characterized by right-angle bends and tributaries joining larger channels at right angles (▮ Figure 12.20b). Such regularity in channels is strongly controlled by geologic structures, particularly regional joint systems that intersect at right angles.

Trellis drainage, consisting of a network of nearly parallel main streams with tributaries joining them at right angles, is common in some parts of the eastern United States. In Virginia and Pennsylvania, erosion of folded sedimentary rocks developed a landscape of alternating ridges on resistant rocks and valleys underlain by easily eroded rocks. Main waterways follow the valleys, and short

Figure 12.19 Drainage Basins

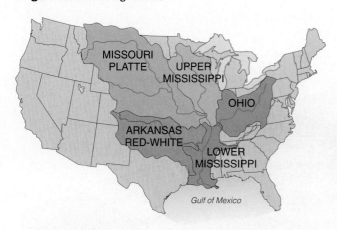

a The drainage basin of the Mississippi River and its main tributaries.

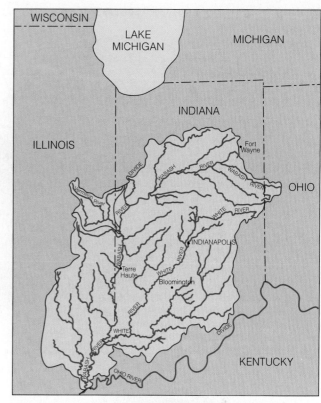

b A detailed view of the Wabash River's drainage basin, a tributary of the Ohio River. All tributary streams within the drainage basin, such as the Vermillion River, have their own smaller drainage basins. Divides are shown by red lines.

tributaries flowing from the nearby ridges join the main channels at nearly right angles (▮ Figure 12.20c).

In *radial drainage,* streams flow outward in all directions from a central high point, such as a large volcano (▮ Figure 12.20d). Many of the volcanoes in the Cascade Range of western North America have radial drainage patterns.

In all of the types of drainage mentioned so far, some kind of pattern is easily recognized. *Deranged drainage,* in contrast, is characterized by irregularity, with streams flowing into and out of swamps and lakes, streams with only a few short tributaries, and vast swampy areas between channels (▮ Figure 12.20e). This kind of drainage developed recently and has not yet formed a fully organized drainage system. In parts of Minnesota, Wisconsin, and Michigan, where glaciers

obliterated the previous drainage, only 10,000 years have elapsed since the glaciers melted. As a result, drainage systems have not fully developed, and large areas remain undrained.

Figure 12.20 Drainage Patterns

a Dendritic drainage.

b Rectangular drainage.

c Trellis drainage.

d Radial drainage.

e Deranged drainage.

Figure 12.21 Base Level and Graded Streams

a Sea level is ultimate base level, but a resistant rock layer over which a waterfall plunges is a local base level. Also, this stream has several irregularities in its profile so it is ungraded.

b Erosion and deposition along its course eliminate irregularities, and the stream becomes graded as it develops a smooth, concave profile of equilibrium.

The Significance of Base Level

Base level is the lowest limit to which a stream or river can erode. With the exception of streams that flow into closed depressions in deserts, all others are restricted ultimately to sea level. That is, they can erode no lower than sea level because they must have some gradient to maintain flow. So *ultimate base level* is sea level, which is simply the lowest level of erosion for any waterway that flows into the sea (▮ Figure 12.21). Ultimate base level applies to an entire stream or river system, but channels may also have a *local* or *temporary base level*. For example, a local base level may be a lake or another stream, or where a stream or river flows across particularly resistant rocks and a waterfall develops (Figure 12.21).

Ultimate base level is sea level, but suppose that sea level dropped or rose with respect to the land, or suppose that the land rose or subsided? In these cases, base level would change and bring about changes in stream and river systems. During the Pleistocene Epoch (Ice Age), sea level was about 130 m lower than it is now, and streams adjusted by eroding deeper valleys and extending well out onto the continental shelves. Rising sea level at the end of the Ice Age accounted for a rising base level, decreased stream gradients, and deposition within channels.

Natural changes, such as fluctuations in sea level during the Pleistocene, alter the dynamics of rivers and streams, but so does human intervention. Geologists and engineers are well aware that building a dam to impound a reservoir creates a local base level. A stream entering a reservoir deposits sediment, so unless dredged, reservoirs eventually fill with sediment. In addition, the water discharged at a dam is largely sediment free, but still possesses energy to carry a sediment load. As a result, a stream may erode vigorously downstream from a dam to acquire a sediment load.

Draining a lake may seem like a small change and well worth the time and expense to expose dry land for agriculture or commercial development. But draining a lake eliminates a local base level, and a stream that originally flowed into the lake responds by rapidly eroding a deeper valley as it adjusts to a new base level.

What Is a Graded Stream?

The *longitudinal profile* of any waterway shows the elevations of a channel along its length as viewed in cross section (Figure 12.21). For some rivers and streams, the longitudinal profile is smooth, but others show irregularities such as lakes and waterfalls, all of which are local base levels (Figure 12.21a). Over time, these irregularities tend to be eliminated because deposition takes place where the gradient is insufficient to maintain sediment transport, and erosion decreases the gradient where it is steep. So, given enough time, rivers and streams develop a smooth, concave longitudinal profile of equilibrium, meaning that all parts of the system dynamically adjust to one another.

A **graded stream** is one with an equilibrium profile in which a delicate balance exists among gradient, discharge, flow velocity, channel shape, and sediment load so that neither significant erosion nor deposition takes place within its channel (Figure 12.21b). Such a delicate balance is rarely attained, so the concept of a graded stream is an ideal. Nevertheless, the graded condition is closely approached in many streams, although only temporarily and not necessarily along their entire lengths.

Even though the concept of a graded stream is an ideal, we can anticipate the response of a graded stream to changes that alter its equilibrium. For instance, a change in base level would cause a stream to adjust as previously discussed. Increased rainfall in a stream's drainage basin would result in greater discharge and flow velocity. In short, the stream would now possess greater energy—energy that must be dissipated within the stream system by, for example, a change from a semicircular to a broad, shallow channel that would dissipate more energy by friction. On the other hand, the stream may respond by eroding a deeper valley, effectively reducing its gradient until it is once again graded.

Vegetation inhibits erosion by stabilizing soil and other loose surface materials. So a decrease in vegetation in a drainage basin might lead to higher erosion rates, causing more sediment to be washed into a stream than it can effectively carry. Accordingly, the stream may respond by deposition within its channel, which increases its gradient until it is sufficiently steep to transport the greater sediment load.

12.8 The Evolution of Valleys

Low areas on land known as **valleys** are bounded by higher land, and most of them have a river or stream running their length, with tributaries draining the nearby high areas. Valleys are common landforms, and with few exceptions, they form and evolve in response to erosion by running water, although other processes, especially mass wasting, contribute. The shapes and sizes of valleys vary from small, steepsided *gullies* to those that are broad with gently sloping valley walls (Figure 12.22). Steep-walled, deep valleys of vast size are *canyons,* and particularly narrow and deep ones are *gorges.*

A valley might start to erode where runoff has sufficient energy to dislodge surface materials and excavate a small rill. Once formed, a rill collects more runoff and becomes deeper and wider and continues to do so until a full-fledged valley develops. Processes that contribute to valley formation include downcutting, lateral erosion, headward erosion, sheetwash, and mass wasting.

Downcutting takes place when a river or stream has more energy than it needs to transport sediment, so some of its excess energy is used to deepen its valley. If downcutting were the only process operating, valleys would be narrow and steep sided. In most cases, though, the valley walls are undercut, a process called *lateral erosion,* creating unstable slopes that may fail by *mass wasting.* Furthermore, erosion by sheetwash and erosion by tributary streams carry materials from the valley walls into the main stream in the valley.

Valleys not only become deeper and wider but also become longer by *headward erosion,* which involves erosion by entering runoff at the upstream end of a valley (Figure 12.23a). Continued headward erosion may result in *stream piracy,* the breaching of a drainage divide and diversion of part of the drainage of another stream (Figure 12.23b,c). Once stream piracy takes place, both drainage systems must adjust to these new conditions; one system now has greater discharge and the potential to do more erosion and sediment transport, whereas the other is diminished in its ability to accomplish these tasks.

According to one concept, stream erosion of an area uplifted above sea level yields a distinctive series of landscapes. When erosion begins, streams erode downward; their valleys are deep, narrow, and V-shaped, and their profiles have a number of irregularities (Figure 12.24a). As streams cease eroding downward, they start eroding laterally, thereby establishing a meandering pattern and a broad floodplain (Figure 12.24b). Finally, with continued erosion, a vast, rather featureless plain develops (Figure 12.24c).

Many streams do indeed show the features typical of these stages. For instance, the Colorado River flows through the Grand Canyon and closely matches the features in the initial stage shown in Figure 12.24a. Streams in many areas approximate the second stage of development, and certainly the lower Mississippi closely resembles the last stage. Nevertheless, the idea of the sequential development of streameroded landscapes has been largely abandoned because there

❚ Figure 12.22 Gullies and Valleys

James S. Monroe

Sue Monroe

a Gullies are small valleys, but they are steep-sided. This gully measures about 15 m across.

b This valley has walls that descend to a narrow valley bottom.

is no reason to think that streams necessarily follow this idealized progression. Indeed, a stream on a gently sloping surface near sea level could develop features of the last stage very early in its history. In addition, as long as the rate of uplift exceeds the rate of downcutting, a stream will continue to erode downward and be confined to a narrow canyon.

Stream Terraces   Rather flat surfaces paralleling a stream or river, but at a higher level than the present-day floodplain, are **stream terraces**. These surfaces represent an older floodplain that was adjacent to the waterway when it flowed at a higher level but subsequently eroded down to a lower level (❚ Figure 12.25). Some streams have several steplike surfaces above their present floodplains, indicating the terraces formed more than once.

The formation of stream terraces is preceded by an episode of deposition followed by erosion as the stream or river begins eroding downward. For example, suppose that a stream is graded so that it is adjusted to its present gradient, volume

of water, and sediment load. If the land over which the stream flows is uplifted or sea level becomes lower, the stream's gradient is steeper, and it has more energy to deepen its valley. When this stream once again reaches a level at which it is graded, downcutting ceases, and it begins to erode laterally, thus establishing a new floodplain at a lower level. Several such episodes of deposition and erosion account for the multiple stream terraces seen in the valleys of some streams.

Although changes in base level probably account for many stream terraces, a change in climate can have the same result. If the amount of precipitation in a stream's drainage basin increases, the stream has a greater volume of water and the capacity to erode a deeper valley, perhaps leaving remnants of an older floodplain as stream terraces.

Incised Meanders

Some streams are restricted to deep, meandering canyons cut into bedrock, where they form features called **incised**

Figure 12.23 The Evolution of a Valley by Headward Erosion, Lateral Erosion, and Stream Piracy

Figure 12.24 Idealized Stages in the Development of a Stream and Its Associated Landform

James S. Monroe

a Headward erosion. The uppermost part, or head, of this small gully will erode at its upper end and extend the gully farther up the slope.

a Initial stage.

b This stream simultaneously extends its valley by headward erosion and widens its valley by lateral erosion.

b Intermediate stage.

c As the larger stream continues to erode headward, stream piracy takes place when it captures some of the drainage of the smaller stream. Notice also that the valley is wider than it was in (b).

c Advanced stage.

laterally; thus, they lack a floodplain and occupy the entire width of the canyon floor.

It is not difficult to understand how a stream can cut downward into rock, but how a stream forms a meandering pattern in bedrock is another matter. Because lateral erosion is inhibited once downcutting begins, one must infer that the meandering course was established when the stream flowed across an area covered by alluvium. For example, suppose that a stream near base level has established a meandering pattern. If the land that the stream flows over is uplifted, then erosion begins and the meanders become incised into the underlying bedrock.

meanders. For example, the Colorado River in Utah occupies a meandering canyon more than 600 m deep (**Figure 12.26**). Streams restricted by rock walls usually cannot erode

Figure 12.25 Origin of Stream Terraces

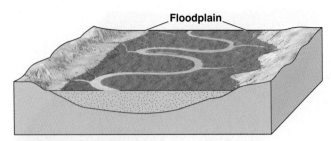

a A stream has a broad floodplain.

c Another level of stream terraces forms as the stream erodes downward again.

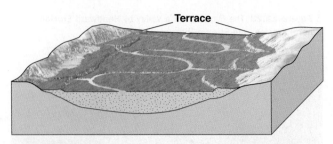

b The stream erodes downward and establishes a new floodplain at a lower level. Remnants of its old, higher floodplain are stream terraces.

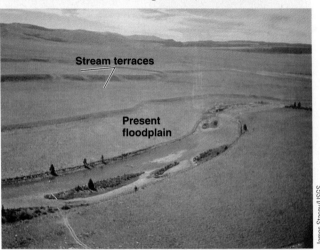

d Stream terraces along the Madison River in Montana.

Figure 12.26 Incised Meanders The Colorado River at Dead Horse State Park in Utah is incised to a depth of 600 m.

Superposed Streams

Water flows downhill in response to gravity, so the direction of flow in streams and rivers is determined by topography. Yet a number of waterways seem, at first glance, to have defied this fundamental control. For instance, the Delaware, Potomac, and Susquehanna rivers in the eastern United States flow in valleys that cut directly through ridges that lie in their paths. These are examples of **superposed streams,** all of which once flowed on a surface at a higher level, but as they eroded downward, they eroded into resistant rocks and cut narrow canyons or what geologists call *water gaps* (**Figure 12.27**).

A water gap has a stream flowing through it, but if the stream is diverted elsewhere, perhaps by stream piracy, the abandoned gap is then called a *wind gap*. The Cumberland Gap in Kentucky is a good example; it was the avenue through which settlers migrated from Virginia to Kentucky from 1790 until well into the 1800s. Furthermore, several water gaps and wind gaps played important strategic roles during the Civil War (1861–1865).

Figure 12.27 Origin of a Superposed Stream

a As a stream erodes down and removes the surface layers of rock, it is lowered onto ridges that form when resistant rocks in the underlying structure are exposed.

Water gap

b The narrow valleys through the ridges are water gaps.

James S. Monroe

c A small stream in Kluane National Park in the Yukon Territory of Canada has been superposed on this block of rock and eroded through it. The small gash through the rock outcrop is a water gap.

GEO-RECAP

≫ Geo-Image

James S. Monroe

Figure 1 The image shows the Sacramento River in California. What is the large deposit indicated by the arrow, and how did it form? Where do you think flow velocity would be greatest in this river?

Chapter Summary

- Water continuously evaporates from the oceans, rises as water vapor, condenses, and falls as precipitation. About 20% of this precipitation falls on land and eventually returns to the oceans, mostly by surface runoff.

- Runoff takes place by sheet flow, a thin, more or less continuous sheet of water, and by channel flow, confined to long, trough-like stream and river channels.

- The vertical drop in a given horizontal distance, or the gradient, for a channel varies from steep in its upper reaches to more gentle in its lower reaches.

- Flow velocity and discharge are related, so that if either changes, the other changes as well.

- Erosion by running water takes place by hydraulic action, abrasion, and solution.

- The bed load in channels is made up of sand and gravel, whereas suspended load consists of silt- and clay-sized particles. Running water also transports a dissolved load.

- Braided streams have a complex of dividing and rejoining channels, and their deposits are mostly sheets of sand and gravel.

- A single sinuous channel is typical of meandering streams that deposit mostly mud, with subordinate point-bar deposits of sand or, more rarely, gravel.

- Broad, flat floodplains adjacent to channels are the sites of oxbow lakes, which are simply abandoned meanders.

- An alluvial deposit at a river's mouth is a delta. Some deltas conform to the three-part division of bottomset, foreset, and topset beds, but large marine deltas are much more complex and are characterized as stream-, wave-, or tide-dominated.

- Alluvial fans are fan-shaped deposits of sand and gravel on land that form best in semiarid regions. They form mostly by deposition from running water, but debris flows are also important.

- Rivers and streams carry runoff from their drainage basins, which are separated from one another by divides.

- Sea level is ultimate base level, the lowest level to which streams or rivers can erode. Local base levels may be lakes or where streams or rivers flow across resistant rocks.

- Graded streams tend to eliminate irregularities in their channels, so they develop a smooth, concave profile of equilibrium.

- A combination of processes, including downcutting, lateral erosion, sheetwash, mass wasting, and headward erosion, are responsible for the origin and evolution of valleys.

- Stream terraces and incised meanders usually form when a stream or river that was formerly in equilibrium, begins a new episode of downcutting.

Important Terms

abrasion (p. 292)
alluvial fan (p. 298)
alluvium (p. 293)
base level (p. 304)
bed load (p. 292)
braided stream (p. 293)
delta (p. 296)
discharge (p. 292)
dissolved load (p. 292)
divide (p. 302)

drainage basin (p. 302)
drainage pattern (p. 302)
floodplain (p. 296)
graded stream (p. 305)
gradient (p. 290)
hydraulic action (p. 292)
hydrologic cycle (p. 287)
incised meanders (p. 307)
infiltration capacity (p. 290)
meandering stream (p. 294)

natural levee (p. 296)
oxbow lake (p. 294)
point bar (p. 294)
runoff (p. 287)
stream terrace (p. 306)
superposed stream (p. 308)
suspended load (p. 292)
valley (p. 305)
velocity (p. 291)

Review Questions

1. A stream or river with multiple channels that divide and rejoin is characterized as
   a. ____braided.
   b. ____sinuous.
   c. ____superposed.
   d. ____meandering.
   e. ____incised.

2. The lowest level to which a stream or river can erode is called
   a. ____point bar.
   b. ____base level.
   c. ____divide.
   d. ____alluvium.
   e. ____infiltration capacity.

3. Dividing the vertical drop of a stream by the horizontal distance it flows gives the
   a. ____velocity.
   b. ____alluvial capacity.
   c. ____discharge.
   d. ____gradient.
   e. ____profile of equilibrium.

4. The silt- and clay-sized particles transported by streams and rivers make up the
   a. ____solution load.
   b. ____sheetwash load.
   c. ____runoff load.
   d. ____bed load.
   e. ____suspended load.

5. The deposit that forms on the gently sloping side of a meander is called a/an
   a. ____point bar.
   b. ____alluvial fan.
   c. ____delta.
   d. ____oxbow lake.
   e. ____alluvial deposit.

6. Drainage basins are separated from one another by a
   a. ____bed load.
   b. ____base level.
   c. ____divide.
   d. ____natural levee.
   e. ____delta.

7. A broad, flat area adjacent to a river is its
   a. ____point bar.
   b. ____natural levee.
   c. ____floodplain.
   d. ____gradient.
   e. ____valley.

8. The recycling of water from the oceans to the land and back to the oceans is known as the
   a. ____alluvial capacity.
   b. ____hydrologic cycle.
   c. ____profile of equilibrium.
   d. ____hydraulic action.
   e. ____drainage pattern.

9. A drainage pattern that looks much like the branching of a tree is called
   a. ____radial.
   b. ____suspended.
   c. ____rectangular.
   d. ____dendritic.
   e. ____irregular.

10. Running water carrying sand and gravel effectively erodes by
    a. ____sheetwash.
    b. ____solution.
    c. ____abrasion.
    d. ____meandering.
    e. ____alluviation.

11. A river with headwaters 2,000 m above sea level flows 1,500 km to the sea. What is its gradient? Do you think that your calculated gradient is valid for all segments of this river? Explain.

12. Explain how the hydrologic cycle works.

13. How do stream terraces form, and what do they indicate about the dynamics of a stream of a river system?

14. How do the deposits of meandering streams compare with those of braided streams?

15. The discharge of most streams and rivers increases downstream, but in a few cases it actually decreases, and they eventually disappear. Why?

16. Calculate the daily discharge of a river 148 m wide and 2.6 m deep, with a flow velocity of 0.3 m/sec.

17. What is a graded stream, and how does it develop?

18. Explain how oxbow lakes form and what eventually happens to them.

19. How are alluvial fans and deltas similar and different?

20. Where and how do point bars form?

A variety of cave deposits such as stalactites, stalagmites, columns, and curtains are present in Mammoth Cave, Kentucky.

Groundwater

OUTLINE

13.1 Introduction

13.2 Groundwater and the Hydrologic Cycle

13.3 Porosity and Permeability

13.4 The Water Table

13.5 Groundwater Movement

13.6 Springs, Water Wells, and Artesian Systems

13.7 Groundwater Erosion and Deposition

13.8 Modification of the Groundwater System and Its Effects

GEO-FOCUS: Arsenic and Old Lace

13.9 Hydrothermal Activity

GEO-RECAP

HAVE YOU EVER WONDERED?

- What percentage of the world's supply of freshwater comes from groundwater?
- What a water table is and where it is found?
- What an artesian system is and why some advertisers tout the superiority of artesian water in their product?
- How caves form?
- Why Old Faithful in Yellowstone National Park, Wyoming, erupts on a regular basis?
- What geothermal energy is and why it is important?

13.1 Introduction

Within the limestone region of western Kentucky lies the largest cave system in the world. In 1941, approximately 51,000 acres were set aside and designated as Mammoth Cave National Park. In 1981, it became a World Heritage Site. From ground level, the topography of the area is unimposing, with gently rolling hills. Beneath the surface, however, are more than 540 km of interconnected passageways whose spectacular geologic features have been enjoyed by millions of cave explorers and tourists.

During the War of 1812, approximately 180 metric tons of saltpeter (potassium nitrate—KNO_3), used in the manufacture of gunpowder, was mined from Mammoth Cave. At the end of the war, the saltpeter market collapsed, and Mammoth Cave was developed as a tourist attraction, easily overshadowing the other caves in the area. During the next 150 years, the discovery of new passageways and links to other caverns helped establish Mammoth Cave as the world's premier cave and the standard against which all others were measured.

The colorful cave deposits are the primary reason that millions of tourists have visited Mammoth Cave over the years. Hanging down from the ceiling and growing up from the floor are spectacular icicle-like structures, as well as columns and curtains in a variety of colors. Moreover, intricate passageways connect rooms of various sizes. The cave is also home to more than 200 species of insects and other animals, including about 45 blind species.

In addition to the beautiful caves, caverns, and cave deposits produced by groundwater movement, groundwater is an important natural resource. Although groundwater constitutes only 0.625% of the world's water, it is, nonetheless, a significant source of freshwater for agricultural, industrial, and domestic use. More than 65% of the groundwater used in the United States each year goes for irrigation, with industrial use second, followed by domestic needs. Furthermore, groundwater provides 80% of the water used for rural livestock and domestic use, as well as providing 40% of public water supplies. In fact, of the largest 100 cities in the United States, 34 depend solely on local groundwater supplies. These demands have severely depleted the groundwater supply in many areas and have led to such problems as ground subsidence and saltwater contamination. In other areas, pollution from landfills, toxic waste, and agriculture has rendered the groundwater supply unsafe.

As the world's population and industrial development expand, the demand for water, particularly groundwater, will increase. Not only must new groundwater sources be located but, once found, these sources must be protected from pollution and managed properly to ensure that users do not withdraw more water than can be replenished.

13.2 Groundwater and the Hydrologic Cycle

Groundwater—water that fills open spaces in rocks, sediment, and soil beneath Earth's surface—is one reservoir in the hydrologic cycle, accounting for approximately 22% (8.4 million km^3) of the world's supply of freshwater (see Table 12.1). Like all other water in the hydrologic cycle, the ultimate source of groundwater is the oceans; however, its more immediate source is the precipitation that infiltrates the ground and seeps down through the voids in soil, sediment, and rocks. Groundwater may also come from water infiltrating from streams, lakes, swamps, artificial recharge ponds, and water-treatment systems.

CONNECTION LINK ❮

To review the hydrologic cycle, see Figure 12.2 and pages 287 and 290 in Chapter 12.

Regardless of its source, groundwater moving through the tiny openings between soil and sediment particles and the spaces in rocks filters out many impurities such as disease-causing microorganisms and many pollutants. However, not all soils and rocks are good filters, and sometimes so much undesirable material may be present that it contaminates the groundwater. Groundwater movement and its recovery at wells depend on two critical aspects of the materials that it moves through: *porosity* and *permeability*.

13.3 Porosity and Permeability

Porosity and permeability are important physical properties of Earth materials and are largely responsible for the amount, availability, and movement of groundwater. Water soaks into the ground because soil, sediment, and rock have open spaces or pores. **Porosity** is the percentage of a material's total volume that is pore space. Porosity most often consists of the spaces between particles in soil, sediment, and sedimentary rocks, but other types of porosity include cracks, fractures, faults, and vesicles in volcanic rocks (❙ Figure 13.1).

Porosity varies among different rock types and is dependent on the size, shape, and arrangement of the material composing the rock (Table 13.1). Most igneous and metamorphic rocks, as well as many limestones and dolostones, have very low porosity because they consist of tightly interlocking crystals. Their porosity can be increased, however, if they have been fractured or weathered by groundwater. This

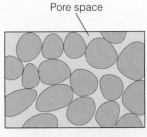

a A well-sorted sedimentary rock has high porosity, whereas

b a poorly sorted one has lower porosity.

c In soluble rocks such as limestone, porosity can be increased by solution, whereas

d crystalline metamorphic and igneous rocks can be rendered porous by fracturing.

▌**Figure 13.1** Porosity A rock's porosity depends on the size, shape, and arrangement of the material composing the rock.

TABLE 13.1 Porosity Values for Different Materials

Material	Percentage Porosity
Unconsolidated sediment	
Soil	55
Gravel	20–40
Sand	25–50
Silt	35–50
Clay	50–70
Rocks	
Sandstone	5–30
Shale	0–10
Solution activity in limestone, dolostone	10–30
Fractured basalt	5–40
Fractured granite	10

Source: U.S. Geological Survey, Water Supply Paper 2220 (1983) and others.

>> Critical Thinking Question
How can some Earth materials be porous, yet not permeable? Give an example.

is particularly true for massive limestone and dolostone, whose fractures can be enlarged by acidic groundwater.

By contrast, detrital sedimentary rocks composed of well-sorted and well-rounded grains can have high porosity because any two grains touch at only a single point, leaving relatively large open spaces between the grains (Figure 13.1a). Poorly sorted sedimentary rocks, on the other hand, typically have low porosity, because smaller grains fill in the spaces between the larger grains, further reducing porosity (Figure 13.1b). In addition, the amount of cement between grains can decrease porosity.

Porosity determines the amount of groundwater that Earth materials can hold, but it does not guarantee that the water can be easily extracted. So, in addition to being porous, Earth materials must have the capacity to transmit fluids, a property known as **permeability**. Thus, both porosity and permeability play important roles in groundwater movement and recovery.

Permeability is dependent not only on porosity but also on the size of the pores or fractures and their interconnections. For example, deposits of silt or clay are typically more porous than sand or gravel, but they have low permeability because the pores between the particles are very small, and molecular attraction between the particles and water is great, thereby preventing movement of the water. In contrast, the pore spaces between grains in sandstone and conglomerate are much larger, and molecular attraction on the water is therefore low. Chemical and biochemical sedimentary rocks, such as limestone and dolostone, and many igneous and metamorphic rocks that are highly fractured can also be very permeable provided that the fractures are interconnected.

The contrasting porosity and permeability of familiar substances are well demonstrated by sand versus clay. Pour some water on sand and it rapidly sinks in, whereas water poured on clay simply remains on the surface. Furthermore, wet sand dries quickly, but once clay absorbs water, it may take days to dry out because of its low permeability. Neither sand nor clay makes a good substance in which to grow crops or gardens, but a mixture of the two, plus some organic matter in the form of humus, makes an excellent soil for farming and gardening (see Chapter 6).

A permeable layer transporting groundwater is an *aquifer*, from the Latin *aqua*, "water." The most effective aquifers are deposits of well-sorted and well-rounded sand and gravel. Limestones in which fractures and bedding planes have been enlarged by solution are also good aquifers. Shales and many igneous and metamorphic rocks make poor aquifers because they are typically impermeable, unless fractured. Rocks such as these and any other materials that prevent the movement of groundwater are *aquicludes*.

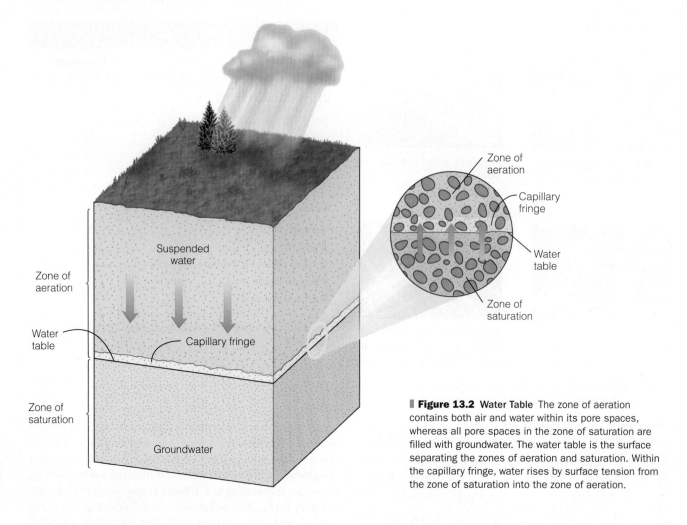

Figure 13.2 Water Table The zone of aeration contains both air and water within its pore spaces, whereas all pore spaces in the zone of saturation are filled with groundwater. The water table is the surface separating the zones of aeration and saturation. Within the capillary fringe, water rises by surface tension from the zone of saturation into the zone of aeration.

13.4 The Water Table

Some of the precipitation on land evaporates, and some enters streams and returns to the oceans by surface runoff; the remainder seeps into the ground. As this water moves down from the surface, a small amount adheres to the material it moves through and halts its downward progress. With the exception of this *suspended water*, however, the rest seeps further downward and collects until it fills all of the available pore spaces. Thus, two zones are defined by whether their pore spaces contain mostly air, the **zone of aeration**, or mostly water, the underlying **zone of saturation**. The surface that separates these two zones is the **water table** (▌ Figure 13.2).

The base of the zone of saturation varies from place to place, but usually extends to a depth where an impermeable layer is encountered or to a depth where confining pressure closes all open space. Extending irregularly upward a few centimeters to several meters from the zone of saturation is the *capillary fringe*. Water moves upward in this region because of surface tension, much as water moves upward through a paper towel.

In general, the configuration of the water table is a sub-dued replica of the overlying land surface; that is, it rises beneath hills and has its lowest elevations beneath valleys.

Several factors contribute to the surface configuration of a region's water table, including regional differences in amount of rainfall, permeability, and rate of groundwater movement. During periods of high rainfall, groundwater tends to rise beneath hills because it cannot flow fast enough into adjacent valleys to maintain a level surface. During droughts, the water table falls and tends to flatten out because it is not being replenished. In arid and semiarid regions, the water table is usually quite flat regardless of the overlying land surface.

13.5 Groundwater Movement

Gravity provides the energy for the downward movement of groundwater. Water entering the ground moves through the zone of aeration to the zone of saturation (▌ Figure 13.3). When water reaches the water table, it continues to move through the zone of saturation from areas where the water table is high toward areas where it is lower, such as streams, lakes, or swamps. Only some of the water follows the direct route along the slope of the water table. Most of it takes longer curving paths down and then enters a stream, lake, or swamp from below, because it moves from areas of high pressure toward areas of lower pressure within the saturated zone.

Figure 13.3 Groundwater Movement Groundwater moves down through the zone of aeration to the zone of saturation. Then some of it moves along the slope of the water table, and the rest moves through the zone of saturation from areas of high pressure toward areas of low pressure. Some water might collect over a local aquiclude, such as a shale layer, thus forming a perched water table.

≫ Critical Thinking Question
If you were drilling a water well on your property and struck water at a considerably shallower depth than your neighbors, should you consider drilling deeper, or just celebrate your good fortune at not having to pay for a deeper well?

Groundwater velocity varies greatly and depends on many factors. Velocities range from 250 m per day in some extremely permeable material to less than a few centimeters per year in nearly impermeable material. In most ordinary aquifers, the average velocity of groundwater is a few centimeters per day.

13.6 Springs, Water Wells, and Artesian Systems

You can think of the water in the zone of saturation much like a reservoir whose surface rises or falls depending on additions as opposed to natural and artificial withdrawals. *Recharge*—that is, additions to the zone of saturation—comes from rainfall or melting snow, or water might be added artificially at wastewater-treatment plants or recharge ponds constructed for just this purpose. But if groundwater is discharged naturally or withdrawn at wells without sufficient recharge, the water table drops—just as a savings account diminishes if withdrawals exceed deposits. Withdrawals from the groundwater system take place where groundwater flows laterally into streams, lakes, or swamps, where it discharges at the surface as *springs*, and where it is withdrawn from the system at water wells.

Springs

Places where groundwater flows or seeps out of the ground as **springs** have always fascinated people. The water flows out of the ground for no apparent reason and from no readily identifiable source. So it is not surprising that springs have long been regarded with superstition and revered for their supposed medicinal value and healing powers. Nevertheless, there is nothing mystical or mysterious about springs.

Although springs can occur under a wide variety of geologic conditions, they all form in basically the same way (▮ Figure 13.4a). When percolating water reaches the water table or an impermeable layer, it flows laterally, and if this flow intersects the surface, the water discharges as a spring (▮ Figure 13.4b). The Mammoth Cave area in Kentucky (see the Introduction) is underlain by fractured limestones whose fractures have been enlarged into caves by solution activity. In this geologic environment, springs occur where the fractures and caves intersect the ground surface, allowing groundwater to exit onto the surface.

Springs can also develop wherever a *perched water table*—a local aquiclude present within a larger aquifer, such as a lens of shale within sandstone—intersects the surface (Figure 13.3). As water migrates through the zone of aeration, it is stopped by the local aquiclude, and a localized zone of saturation "perched" above the main water table forms. Water moving laterally along the perched water table may intersect the surface to produce a spring.

Water Wells

Water wells are openings made by digging or drilling down into the zone of saturation. Once the zone of saturation has been penetrated, water percolates into the well, filling it to the level of the water table. A few wells are free flowing; however, for most, the water must be brought to the surface by pumping.

In some parts of the world, water is raised to the surface with nothing more than a bucket on a rope or a hand-operated pump. In many parts of the United States and Canada, one can see windmills from times past that used wind power to pump water. Most of these are no longer in use, having been replaced by more efficient electric pumps.

CONNECTION LINK ◀

To learn more about windmills and their role in pumping water and generating power, see pages 378 and 379 in the Geo-Focus in Chapter 15.

Figure 13.4 Springs Springs form wherever laterally moving groundwater intersects Earth's surface.

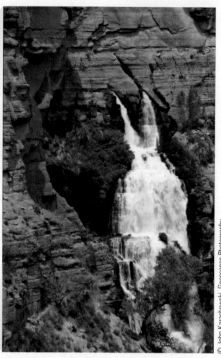

© John Karachewski, Geoscapes Photography

a Most commonly, springs form when percolating water reaches an impermeable layer and migrates laterally until it seeps out at the surface.

b Thunder River Spring in the Grand Canyon, Arizona, issues from rocks along a wall of the Grand Canyon. Water percolating downward through permeable rock is forced to move laterally when it encounters an impermeable zone, and thus gushes out along this cliff. Notice the vegetation parallel to and below the springs, indicating that enough water flows from springs along the cliff wall to support the vegetation.

When groundwater is pumped from a well, the water table in the area around the well is lowered, forming a **cone of depression** (Figure 13.5). A cone of depression forms because the rate of water withdrawal from the well exceeds the rate of water inflow to the well, thus lowering the water table around the well. A cone of depression's gradient—that is, whether it is steep or gentle—depends to a great extent on the permeability of the aquifer being pumped. A highly permeable aquifer produces a gentle gradient in the cone of depression, whereas a low-permeability aquifer results in a steep cone of depression because water cannot easily flow to the well to replace the water being withdrawn.

The formation of a cone of depression does not normally pose a problem for the average domestic well, provided that the well is drilled deep enough into the zone of saturation. However, the tremendous amounts of water used by industry and for irrigation may create a large cone of depression that lowers the water table sufficiently to cause shallow wells in the immediate area to go dry (Figure 13.5). This situation is not uncommon and frequently results in lawsuits by the owners of the shallow dry wells.

Lowering of the regional water table, because more groundwater is being withdrawn than is being replenished, is becoming a serious problem in many areas, particularly in the southwestern

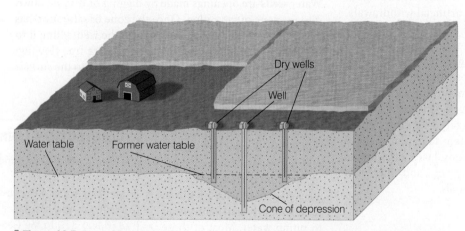

Figure 13.5 Cone of Depression A cone of depression forms whenever water is withdrawn from a well. If water is withdrawn faster than it can be replenished, the cone of depression will grow in depth and circumference, lowering the water table in the area and causing nearby shallow wells to go dry.

Figure 13.6 Artesian System An artesian system must have an aquifer confined above and below by aquicludes, the aquifer must be exposed at the surface, and there must be sufficient precipitation in the recharge area to keep the aquifer filled. The elevation of the water table in the recharge area, which is indicated by a sloping dashed line (the artesian-pressure surface), defines the highest level to which well water can rise.

》Critical Thinking Question
If the elevation of the wellhead is at or above that of the artesian-pressure surface, why then will the well be nonflowing?

United States where rapid growth has placed tremendous demands on the groundwater system. As mentioned earlier, some of the largest cities in the United States depend entirely on groundwater for their municipal needs. Furthermore, some of the largest agricultural states are withdrawing groundwater from regional aquifers that are not being sufficiently replenished. Unrestricted withdrawal of groundwater cannot continue indefinitely, and the rising costs and decreasing supply of groundwater should soon limit the growth of some regions in the United States, such as the Southwest.

Artesian Systems

The word *artesian* comes from the French town and province of Artois (called Artesium during Roman times) near Calais, where the first European artesian well was drilled in 1126 and is still flowing today. The term **artesian system** can be applied to any system in which groundwater is confined and builds up high hydrostatic (fluid) pressure (❙ Figure 13.6). Water in such a system is able to rise above the level of the aquifer if a well is drilled through the confining layer, thereby reducing the pressure and forcing the water

upward. An artesian system can develop when (1) an aquifer is confined above and below by aquicludes; (2) the rock sequence is (usually) tilted to build up hydrostatic pressure; and (3) the aquifer is exposed at the surface, thus enabling it to be recharged.

The elevation of the water table in the recharge area and the distance of the well from the recharge area determine the height to which artesian water rises in a well. The surface defined by the water table in the recharge area, called the *artesian-pressure surface*, is indicated by the sloping dashed line in Figure 13.6. If there were no friction in the aquifer, well water from an artesian aquifer would rise exactly to the elevation of the artesian-pressure surface. Friction, however, slightly reduces the pressure of the aquifer water and consequently the level to which artesian water rises, which is why the pressure surface slopes.

An artesian well will flow freely at the ground surface only if the wellhead is at an elevation below the artesian-pressure surface. In this situation, the water flows out of the well because it rises toward the artesian-pressure surface, which is at a higher elevation than the wellhead. In a nonflowing artesian well, the wellhead is above the artesian-pressure surface, and the water will rise in the well only as high as the artesian-pressure surface.

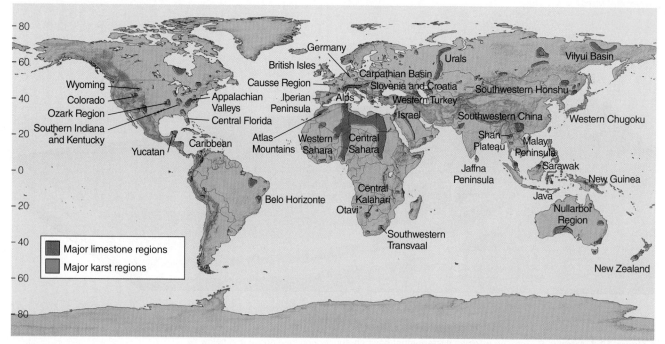

▌Figure 13.7 Distribution of the Major Limestone and Karst Areas of the World Karst topography develops largely by groundwater erosion in areas underlain by soluble rocks.

≫ Critical Thinking Question
Why is karst topography typically restricted to humid and temperate climates?

One of the best-known artesian systems in the United States underlies South Dakota and extends southward to central Texas. The majority of the artesian water from this system is used for irrigation. The aquifer of this artesian system, the Dakota Sandstone, is recharged where it is exposed along the margins of the Black Hills of South Dakota. The hydrostatic pressure in this system was originally great enough to produce free-flowing wells and to operate waterwheels. However, because of the extensive use of this groundwater for irrigation over the years, the hydrostatic pressure in many of the wells is so low that they are no longer free-flowing, and the water must be pumped.

As a final comment on artesian systems, we should mention that it is not unusual for advertisers to tout the quality of artesian water as somehow being superior to other groundwater. Some artesian water might in fact be of excellent quality, but its quality is not dependent on the fact that water rises above the surface of an aquifer. Rather, its quality is a function of dissolved minerals and any introduced substances, so artesian water really is no different from any other groundwater. The myth of its superiority probably arises from the fact that people have always been fascinated by water that flows freely from the ground.

13.7 Groundwater Erosion and Deposition

When rainwater begins to seep into the ground, it immediately starts to react with the minerals it contacts, weathering

them chemically. In an area underlain by soluble rock, groundwater is the principal agent of erosion and is responsible for the formation of many major features of the landscape.

 CONNECTION LINK

To review the process of solution, see pages 140 and 141 in Chapter 6.

Limestone, a common sedimentary rock composed primarily of the mineral calcite ($CaCO_3$), underlies large areas of Earth's surface (▌ Figure 13.7). Although limestone is practically insoluble in pure water, it readily dissolves if a small amount of acid is present. Carbonic acid (H_2CO_3) is a weak acid that forms when carbon dioxide combines with water ($H_2O + CO_2 \rightarrow H_2CO_3$). Because the atmosphere contains a small amount of carbon dioxide (0.03%), and carbon dioxide is also produced in soil by the decay of organic matter, most groundwater is slightly acidic. When groundwater percolates through the various openings in limestone, the slightly acidic water readily reacts with the calcite to dissolve the rock by forming soluble calcium bicarbonate, which is carried away in solution (see Chapter 6).

Sinkholes and Karst Topography

In regions underlain by soluble rock, the ground surface may be pitted with numerous depressions that vary in size and shape. These depressions, called **sinkholes**, or merely *sinks*, mark areas with underlying soluble rock (▌ Figure 13.8). Most sinkholes form in one of two ways. The first is when soluble rock below the soil is dissolved by seeping water, and openings

▌Figure 13.8 Sinkholes

a This sinkhole formed on May 8 and 9, 1981, in Winter Park, Florida. It formed in previously dissolved limestone following a drop in the water table. The 100-m-wide, 35-m-deep sinkhole destroyed a house, numerous cars, and a municipal swimming pool.

Frank Kujawa/University of Central Florida

b A small sinkhole in Montana, now occupied by a lake. The water enters the lake from a hot spring so it remains warm year-round. In fact, tropical fish have been introduced into the lake and thus live in an otherwise inhospitable climate.

James S. Monroe

in the rock are enlarged and filled in by the overlying soil. As the groundwater continues to dissolve the rock, the soil is eventually removed, leaving shallow depressions with gently sloping sides. When adjacent sinkholes merge, they form a network of larger, irregular, closed depressions called *solution valleys.*

Sinkholes also form when a cave's roof collapses, usually producing a steep-sided crater. Sinkholes formed in this way are a serious hazard, particularly in populated areas. In regions prone to sinkhole formation, extensive geologic and hydrogeologic investigation must be performed to determine the depth and extent of underlying cave systems prior to any site development. This examination is necessary to ensure that the underlying rocks are thick enough to support planned structures.

Karst topography, or simply *karst*, develops largely by groundwater erosion in many areas underlain by soluble rocks (▌ Figure 13.9). The name *karst* is derived from the plateau region of the border area of Slovenia, Croatia, and northeastern Italy where this type of topography is well developed. In the United States, regions of karst topography include large areas of southwestern Illinois, southern Indiana, Kentucky, Tennessee, northern Missouri, Alabama, and central and northern Florida (Figure 13.7).

Karst topography is characterized by numerous caves, springs, sinkholes, solution valleys, and disappearing streams (Figure 13.9). *Disappearing streams* are so named because they typically flow only a short distance at the surface and then disappear into a sinkhole. The water continues flowing

▌**Figure 13.9** Features of Karst Topography Erosion of soluble rock by groundwater produces karst topography. Features commonly found include solution valleys, springs, sinkholes, and disappearing streams.

underground through fractures or caves until it surfaces again at a spring or other stream.

Karst topography varies from the spectacular high-relief landscapes of China to the subdued and pockmarked landforms of Kentucky (▌Figure 13.10). Common to all karst topography, though, is the presence of thick-bedded, readily soluble rock at the surface or just below the soil, and enough water for solution activity to occur. Karst topography is therefore typically restricted to humid and temperate climates.

Caves and Cave Deposits

Caves are perhaps the most spectacular examples of the combined effects of weathering and erosion by groundwater. As groundwater percolates through carbonate rocks, it dissolves and enlarges fractures and openings to form a complex, interconnecting system of crevices, caves, caverns, and underground streams. A **cave** is usually defined as a naturally formed, subsurface opening that is generally connected to the surface and is large enough for a person to enter. A *cavern* is a very large cave or a system of interconnected caves.

More than 17,000 caves are known in the United States. Most of them are small, but some are large and spectacular. Some of the more famous ones are Mammoth Cave (Kentucky), Carlsbad Caverns (New Mexico), Lewis and Clark

Caverns (Montana), Lehman Cave (Nevada), and Meramec Caverns (Missouri), which Jesse James and his outlaw band often used as a hideout. Canada also has many famous caves, including the 536-m-deep Arctomys Cave in Mount Robson Provincial Park, British Columbia, the deepest known cave in North America.

Caves and caverns form as a result of the dissolution of carbonate rocks by weakly acidic groundwater (▌Figure 13.11). Groundwater percolating through the zone of aeration slowly dissolves the carbonate rock and enlarges its fractures and bedding planes. On reaching the water table, the groundwater migrates toward the region's surface streams. As the groundwater moves through the zone of saturation, it continues to dissolve the rock and gradually forms a system of horizontal passageways through which the dissolved rock is carried to the streams. As the surface streams erode deeper valleys, the water table drops in response to the lower elevation of the streams. The water that flowed through the system of horizontal passageways now percolates to the lower water table where a new system of passageways begins to form. The abandoned channelways form an interconnecting system of caves and caverns. Caves eventually become unstable and collapse, littering the floor with fallen debris.

When most people think of caves, they think of the seemingly endless variety of colorful and bizarre-shaped deposits found in them. Although a great many different types of cave deposits exist, most form in essentially the same manner and are collectively known as *dripstone*. As water seeps into a cave, some of the dissolved carbon dioxide in the water escapes, and a small amount of calcite is precipitated. In this manner, the various dripstone deposits are formed (Figure 13.11c).

What Would You Do ❓

Your well periodically goes dry and never produces the amount of water you would like or need. Is your well too shallow, is it drilled into Earth materials with low permeability, or are nearby irrigation pumps causing your problems? What can you do to resolve these problems?

Figure 13.10 Karst Landscape in Kunming, China, and Bowling Green, Kentucky

a The Stone Forest, 125 km southeast of Kunming, China, is a high-relief karst landscape formed by the dissolution of carbonate rocks.

b Solution valleys, sinkholes, and sinkhole lakes dominate the subdued karst topography east of Bowling Green, Kentucky.

Stalactites are icicle-shaped structures hanging from cave ceilings that form as a result of precipitation from dripping water (▌Figure 13.12). With each drop of water, a thin layer of calcite is deposited over the previous layer, forming a cone-shaped projection that grows down from the ceiling. The water that drips from a cave's ceiling also precipitates a small amount of calcite when it hits the floor. As additional calcite is deposited, an upward-growing projection called a *stalagmite*

forms (Figure 13.12). If a stalactite and stalagmite meet, they form a *column*. Groundwater seeping from a crack in a cave's ceiling may form a vertical sheet of rock called a *drip curtain*, and water flowing across a cave's floor may produce *travertine terraces* (see Figure 13.11c).

13.8 Modification of the Groundwater System and Its Effects

Groundwater is a valuable natural resource that is rapidly being exploited with seemingly little regard to the effects on it of overuse and misuse. Currently, approximately 20% of all water used in the United States is groundwater. This percentage is rapidly increasing, and unless this resource is used more wisely, sufficient amounts of clean groundwater will not be available in the future. Modification of the groundwater system may have many consequences, including (1) lowering of the water table, causing wells to dry up; (2) salt-water incursion; (3) subsidence; and (4) contamination.

Lowering the Water Table

Withdrawing groundwater at a significantly greater rate than it is replaced by either natural or artificial recharge can have serious effects. For example, the High Plains aquifer is one of the most important aquifers in the United States. It underlies more than 450,000 km², including most of Nebraska, large parts of Colorado and Kansas, portions of South Dakota, Wyoming, and New Mexico, as well as the panhandle regions of Oklahoma and Texas, and accounts for approximately 30% of the groundwater used for irrigation in the United States (▌Figure 13.13).

Significant withdrawal of groundwater from the High Plains aquifer for irrigation began in the 1950s, and by 1980, the water table had dropped an average of 3 meters. Irrigation from the High Plains aquifer is largely responsible for

█ Figure 13.11 Cave Formation

a As groundwater percolates through the zone of aeration and flows through the zone of saturation, it dissolves the carbonate rocks and gradually forms a system of passageways.

b Groundwater moves along the surface of the water table, forming a system of horizontal passageways through which dissolved rock is carried to the surface streams, thus enlarging the passageways.

c As the surface streams erode deeper valleys, the water table drops, and the abandoned channelways form an interconnecting system of caves and caverns.

the region's agricultural productivity, including a significant percentage of the nation's corn, cotton, and wheat, and half of U.S. beef cattle.

Although the High Plains aquifer has contributed to the high productivity of the region, it cannot continue to provide the quantities of water that it has in the past. A 1982 Department of Commerce study estimated that, based on the present rate of groundwater withdrawal, by 2020, about a fourth of the High Plains aquifer water will have been withdrawn. In some parts of the High Plains, from 2 to 100 times more water is being pumped annually than is being recharged, causing a substantial drop in the water table in many areas (Figure 13.13). It must be noted that much of the aquifer's water infiltrated during wetter glacial climates more than 10,000 years ago. Consequently, most of the water being pumped is fossil water that is not being replenished at anywhere near the same rate as when it formed during the Pleistocene Epoch.

What will happen to this region's economy if long-term withdrawal of water from the High Plains aquifer continues to greatly exceed its recharge rate, and the aquifer can no longer supply the quantities of water necessary for irrigation? Solutions range from going back to farming without irrigation to diverting water from other regions such as the Great Lakes. Most users of the aquifer realize that they cannot continue to withdraw the quantities of groundwater that they have in the past, and thus are turning to greater conservation, monitoring of the aquifer, and use of new technologies to try to better balance withdrawal with recharge rates.

Another excellent example of what we might call deficit spending with regard to groundwater took place in California during the drought of 1987–1992. During that time, the state's aquifers were overdrawn at a rate of 10 million acre-feet per year (an acre-foot is the amount of water that covers 1 acre, 1 foot deep). In short, during each year of the drought, California was withdrawing more than 12 km³ of groundwater more than was being replaced. Unfortunately, excessive depletion of the groundwater reservoir has other consequences,

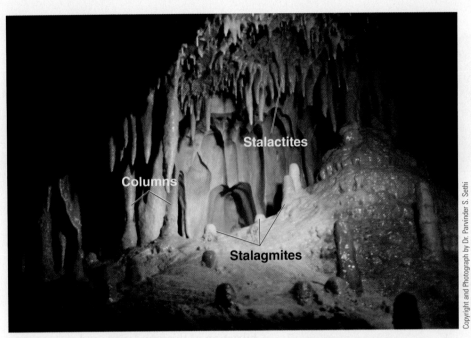

Figure 13.12 Cave Deposits Stalactites are the icicle-shaped structures hanging from the cave's ceiling, whereas the upward-pointing structures on the floor are stalagmites. Columns result when stalactites and stalagmites meet. All three structures are visible in Buchan Cave, Victoria, Australia.

Copyright and Photograph by Dr. Parvinder S. Sethi

such as subsidence involving sinking or settling of the ground surface (discussed in a later section).

Water supply problems certainly exist in many areas, but on the positive side, water use in the United States actually declined during the five years following 1980 and has remained nearly constant since then, even though the population has increased. This downturn in demand resulted largely from improved techniques in irrigation, more efficient industrial water use, and a general public awareness of water problems coupled with conservation practices. Nevertheless, the rates of withdrawal of groundwater from some aquifers still exceed their rates of recharge, and population growth in the arid to semiarid Southwest is continuing to put significant demands on an already limited water supply.

Saltwater Incursion

The excessive pumping of groundwater in coastal areas has resulted in *saltwater incursion*, which has become a major problem in many rapidly growing coastal communities where greater demand for groundwater creates an even greater imbalance between withdrawal and recharge. Along coastlines where permeable rocks or sediments are in contact with the ocean, the fresh groundwater, being less dense than seawater, forms a lens-shaped body above the underlying saltwater (Figure 13.14a). The weight of the freshwater exerts pressure on the underlying saltwater. As long as rates of recharge equal rates of withdrawal, the contact between the fresh groundwater and the seawater remains the same.

If excessive pumping occurs, however, a deep cone of depression forms in the fresh groundwater (Figure 13.14b). Because some of the pressure from the overlying freshwater has been removed, saltwater forms a *cone of ascension* as it rises to fill the pore space that formerly contained freshwater. When this occurs, wells become contaminated with saltwater and remain contaminated until recharge by freshwater restores the former level of the fresh-groundwater water table.

To counteract the effects of saltwater incursion, recharge wells are often drilled to pump water back into the groundwater system (Figure 13.14c). Recharge ponds that allow large quantities of fresh surface water to infiltrate the groundwater supply may also be constructed.

From J. B. Weeks et al., U.S. Geological Survey Professional Paper 1400-A, 1988.

Explanation

- Declines greater than 15 m
- Between 15 m decline and 3 m rise
- Rises greater than 3 m

0 50 100 Kilometers

Figure 13.13 High Plains Aquifer The geographic extent of the High Plains aquifer and changes in water level from predevelopment through 1993. Irrigation from the High Plains aquifer is largely responsible for the region's agricultural productivity.

Subsidence

As excessive amounts of groundwater are withdrawn from poorly consolidated sediments and sedimentary rocks, the

Figure 13.14 Saltwater Incursion

a Because freshwater is not as dense as saltwater, it forms a lens-shaped body above the underlying saltwater.

b If excessive pumping occurs, a cone of depression develops in the fresh groundwater, and a cone of ascension forms in the underlying salty groundwater, which may result in saltwater contamination of the well.

c Pumping water back into the groundwater system through recharge wells can help lower the interface between the fresh groundwater and the salty groundwater and reduce saltwater incursion.

water pressure between grains is reduced, and the weight of the overlying materials causes the grains to pack more closely together, resulting in *subsidence* of the ground. As more and more groundwater is pumped to meet the increasing needs of agriculture, industry, and population growth, subsidence is becoming more prevalent.

The San Joaquin Valley of California is a major agricultural region that relies largely on groundwater for irrigation.

Between 1925 and 1977, groundwater withdrawals in parts of the valley caused subsidence of nearly 9 m (Figure 13.15). Other areas in the United States that have experienced subsidence due to groundwater withdrawal are New Orleans, Louisiana, and Houston, Texas, both of which have subsided more than 2 m, and Las Vegas, Nevada, where 8.5 m of subsidence has taken place.

Looking elsewhere in the world, the tilt of the Leaning Tower of Pisa in Italy is partly due to groundwater withdrawal (Figure 13.16). The tower started tilting soon after construction began in 1173 because of differential compaction of the foundation. During the 1960s, the city of Pisa withdrew ever-greater amounts of groundwater, causing the ground to subside further; as a result, the tilt of the tower increased until it was in danger of falling over. Strict control of groundwater withdrawal, stabilization of the foundation, and recent renovations have reduced the amount of tilting to about 1 mm per year, thus ensuring that the tower should stand for several more centuries.

A spectacular example of continuing subsidence is taking place in Mexico City, which is built on a former lake bed. As groundwater is removed for the increasing needs of the city's 21.2 million people, the water table has been lowered up to 10 m. As a result, the fine-grained lake deposits are compacting, and Mexico City is slowly and unevenly subsiding. Its opera house has settled more than 3 m, and half of the first floor is now below ground level. Other parts of the city have subsided as much as 7.5 m, creating similar problems for other structures. The fact that 72% of the city's water comes from the aquifer beneath the metropolitan area ensures that problems of subsidence will continue.

The extraction of oil can also cause subsidence. Long Beach, California, has subsided 9 m as a result of many decades of oil production. More than $100 million in damages was done to the pumping, transportation, and harbor facilities in this area because of subsidence and encroachment of the sea (Figure 13.17). Once water was pumped back into the oil reservoir, thus stabilizing it, subsidence virtually stopped.

Groundwater Contamination

A major problem facing our society is the safe disposal of the numerous pollutant by-products of an industrialized economy. We are becoming increasingly aware that streams, lakes, and oceans are not unlimited reservoirs for waste and that we must find new, safe ways to dispose of pollutants.

The most common sources of groundwater contamination are sewage, landfills, toxic waste disposal sites, and agriculture. Once pollutants get into the groundwater system, they spread wherever groundwater travels, which can make their containment difficult (see Geo-Focus on pages 328 and 329). Furthermore, because groundwater moves so slowly, it takes a long time to cleanse a groundwater reservoir once it has become contaminated.

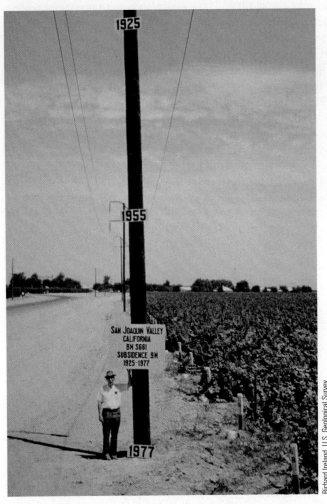

Figure 13.15 Subsidence in the San Joaquin Valley, California The dates on this power pole dramatically illustrate the amount of subsidence in the San Joaquin Valley, California. Because of groundwater withdrawals and subsequent sediment compaction, the ground subsided nearly 9 m between 1925 and 1977. For a time, surface water use reduced subsidence, but during the drought of 1987 to 1992 it started again as more groundwater was withdrawn.

Richard Ireland, U.S. Geological Survey

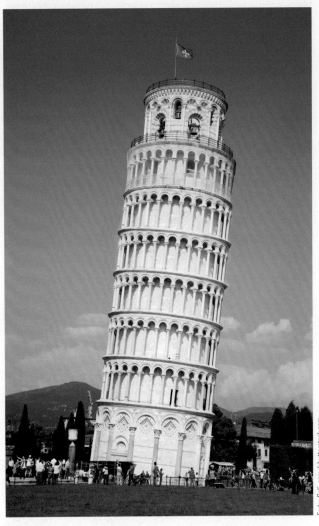

Figure 13.16 The Leaning Tower of Pisa, Italy The tilting is partly the result of subsidence due to the removal of groundwater. Strict control of groundwater withdrawal, recent stabilization of the foundation, and renovation of the structure itself have ensured that the Leaning Tower will continue leaning for many more centuries.

Fedor Selivanov/shutterstock.com

Figure 13.17 Oil Field Subsidence, Long Beach, California The withdrawal of petroleum from the Long Beach, California, oil field resulted in up to 9 m of ground subsidence in some areas because of sediment compaction. In this photograph, note that the ground has settled around the well stems (the white "posts"), leaving the wellheads up above the ground. The levee on the left edge of the photo was built to keep seawater in the adjacent marina from flooding the oil field. It was not until water was pumped back into the reservoir to replace the petroleum that was being extracted that ground subsidence finally ceased.

City of Long Beach Department of Oil Properties

GEO-FOCUS Arsenic and Old Lace

Many people probably learned that arsenic is a poison from either reading or seeing the play *Arsenic and Old Lace*, written by Joseph Kesselring. In the play, the elderly Brewster sisters poison lonely old men by adding a small amount of arsenic to their homemade elderberry wine.

Arsenic is a naturally occurring toxic element found in the environment, and several types of cancer have been linked to arsenic in water. Arsenic also harms the central and peripheral nervous systems and may cause birth defects and reproductive problems. In fact, because of arsenic's prevalence in the environment and its adverse health effects, Congress included it in the amendments to the Safe Drinking Water Act in 1996. Arsenic gets into the groundwater system mainly as arsenic-bearing minerals dissolve in the natural weathering process of rocks and soils.

A map published in 2001 by the U.S. Geological Survey (USGS) shows the extent and concentration of arsenic in the nation's groundwater supply (❙ Figure 1). The highest concentrations of groundwater arsenic were found throughout the West and in parts of the Midwest and Northeast. Although the map is not intended to provide specific information for individual wells or even a locality within a county, it helps researchers and policy makers identify areas of high concentration so that they can make informed decisions about water use. We should point out,

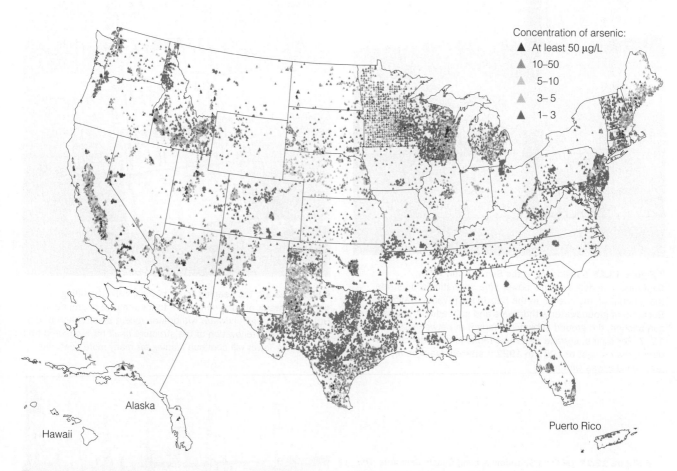

Concentration of arsenic:
▲ At least 50 µg/L
▲ 10–50
▲ 5–10
▲ 3–5
▲ 1–3

Alaska

Hawaii

Puerto Rico

Figure 1 Arsenic concentrations for 31,350 groundwater samples collected from 1973 to 2001.

however, that a high degree of local variability in the amount of arsenic in the groundwater can be caused by local geology, type of aquifer, depth of well, and other factors. The only way to learn the arsenic concentration in any well is to have it tested.

What is considered a safe level of arsenic in drinking water? In 2001, the U.S. Environmental Protection Agency (U.S. EPA, or EPA) lowered the maximum level of arsenic permitted in drinking water to 0.010 ppm (parts per million) of arsenic per liter, with a mandate for all United States drinking water systems to meet this level by 2006.

From the data in Figure 1, additional maps were created. ▮ Figure 2 shows arsenic concentrations found in at least 25% of groundwater samples per county. Based on these data, approximately 10% of the samples in the USGS study exceed the new standard of 10 μg (10 micrograms) of arsenic per liter (10 μg/L) of drinking water.

Public water supply systems that exceed the existing EPA arsenic standard are required to either treat the water to remove the arsenic or find an alternative supply. Although reducing the acceptable level of arsenic in drinking water will surely increase the cost of water to consumers, it will also decrease their exposure to arsenic and the possible adverse health effects associated with this toxic element.

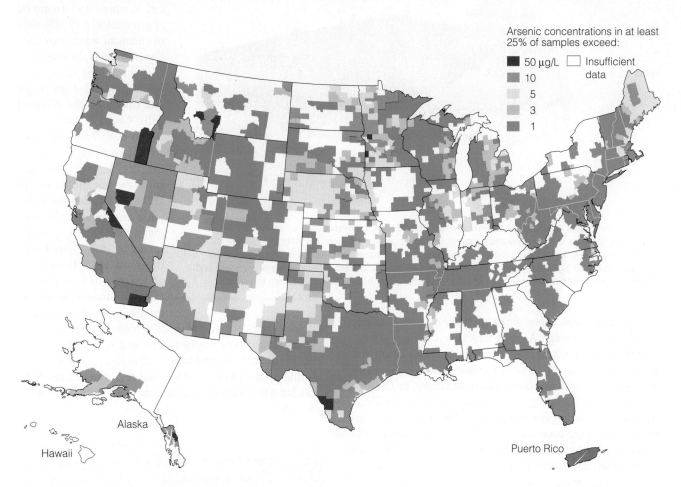

Figure 2 Arsenic concentrations found in at least 25% of groundwater samples per county. The map is based on 31,350 groundwater samples collected between 1973 and 2001.

Figure 13.18 Groundwater Contamination

a A septic system slowly releases sewage into the zone of aeration. Oxidation, bacterial degradation, and filtering usually remove impurities before they reach the water table. However, if the rocks are very permeable or the water table is too close to the septic system, contamination of the groundwater can result.

b Unless there is an impermeable barrier between a landfill and the water table, pollutants can be carried into the zone of saturation and contaminate the groundwater supply: (1) Infiltrating water leaches contaminants from the landfill; (2) the polluted water enters the water table and moves away from the landfill; (3) wells may tap the polluted water and thus contaminate drinking water supplies; and (4) the polluted water may emerge into streams and other water bodies downslope from the landfill.

In many areas, septic tanks are the most common way of disposing of sewage. A septic tank slowly releases sewage into the ground, where it is decomposed by oxidation and microorganisms and filtered by the sediment as it percolates through the zone of aeration. In most situations, by the time the water from the sewage reaches the zone of saturation, it has been cleansed of any impurities and is safe to use (Figure 13.18a). If the water table is close to the surface or if the rocks are very permeable, however, water entering the zone of saturation may still be contaminated and unfit to use.

Landfills are also potential sources of groundwater contamination (Figure 13.18b). Not only does liquid waste seep into the ground but rainwater also carries dissolved chemicals and other pollutants down into the groundwater reservoir. Unless the landfill is carefully designed and lined with an impermeable layer such as clay, many toxic compounds such as paints, solvents, cleansers, pesticides, and battery acid will find their way into the groundwater system.

Toxic waste sites where dangerous chemicals are either buried or pumped underground are an increasing source of groundwater contamination. The United States alone must dispose of several thousand metric tons of hazardous chemical waste per year. Unfortunately, much of this waste has been and still is being improperly dumped and is contaminating the surface water, soil, and groundwater.

Examples of indiscriminate dumping of dangerous and toxic chemicals can be found in every state. Perhaps the most famous is Love Canal, near Niagara Falls, New York. During the 1940s, the Hooker Chemical Company dumped approximately 19,000 tons of chemical waste into Love Canal. In 1953, Hooker covered one of the dump sites with dirt and sold it for one dollar to the Niagara Falls Board of Education, which built an elementary school and playground on the site. Heavy rains and snow during the winter of 1976–1977 raised the water table and turned the area into a muddy swamp in the spring of 1977. Mixed with the mud were thousands of toxic, noxious chemicals that formed puddles in the playground, oozed into people's basements, and covered gardens and lawns. Trees, lawns, and gardens began to die, and many of the residents of the area suffered from serious illnesses. The cost of cleaning up the Love Canal site and relocating its residents exceeded $100 million, and the site and neighborhood are now vacant.

What Would You Do ?

Americans generate tremendous amounts of waste. Some of this waste, such as battery acid, paint, cleaning agents, insecticides, and pesticides, can easily contaminate the groundwater system. Your community is planning to construct a city dump to contain waste products, but it simply wants to dig a hole, dump waste in, and then bury it. What do you think of this plan? Are you skeptical of this plan's merits, and if so, what would you suggest to remedy any potential problems?

Groundwater Quality

Finding groundwater is rather easy because it is present beneath the land surface nearly everywhere, although the depth to the water table varies considerably. But just finding water is not enough. Sufficient amounts in porous and permeable materials must be located if groundwater is to be withdrawn for agricultural, industrial, or domestic use. The availability of groundwater was important in the westward expansion in both Canada and the United States, and now more than one-third of all water for irrigation comes from the groundwater system. As we mentioned in the Introduction, groundwater provides 80% of the water used for rural livestock and domestic use in rural America, and it is the primary source of water for a number of large cities. Furthermore, as one would expect, quality is more important here than it is for most other purposes.

If we discount contamination by humans from landfills, septic systems, toxic waste sites, and industrial effluents, groundwater quality is mostly a function of (1) the kinds of materials that make up an aquifer; (2) the residence time of water in an aquifer; and (3) the solubility of rocks and minerals. These factors account for the amount of dissolved materials in groundwater, such as calcium, iron, fluoride, and several others. Most pose no health problems, but some have undesirable effects such as deposition of minerals in water pipes and water heaters or an offensive taste or smell, or they may stain clothing and fixtures or inhibit the effectiveness of detergents.

A good example that everyone is familiar with is *hard water*, which is a problem in many areas, especially those underlain by limestone and dolostone. Hard water is caused by dissolved calcium (Ca^{+2}) and magnesium (Mg^{+2}) ions. Water containing less than 60 milligrams of Ca^{+2} and Mg^{+2} per liter (mg/L) is considered soft, whereas 61–120 mg/L indicates moderately hard water, values from 121 to 180 mg/L characterize hard water, and any water with more than 180 mg/L is very hard.

One of the negative aspects of hard water is the precipitation of scale (Ca and Mg salts) in water pipes, water heaters, dishwashers, and even on glasses and dinnerware. To remedy this problem, many households have a water softener, whereby calcium and magnesium ions in the water are replaced by sodium (Na^+) ions through the use of an ion exchanger or a mineral sieve. Thus, the amount of calcium and magnesium is reduced, and the water is more desirable for most domestic purposes. However, people on low-sodium diets, such as those with hypertension (high blood pressure), are cautioned not to drink softened water because it contains more sodium.

Not all dissolved materials in groundwater are undesirable, at least in small quantities. Fluoride (F^-), for instance, if present in amounts of 1.0 to 1.5 parts per million (ppm), combines with the calcium phosphate in teeth and makes them more resistant to decay. However, too much fluoride—more than 4.0 ppm—gives children's teeth a dark, blotchy appearance.

Fluoride in natural waters is rare, so few communities benefit from its presence. However, many cities and towns add fluorine to their drinking water, and many dentists routinely use fluoride treatments, so that the population as a whole has seen a reduction in cavities and tooth decay. Despite the beneficial effect of fluoridation in reducing tooth decay, there has been an increase in opposition to fluoridating municipal water supplies because of possible health risks.

13.9 Hydrothermal Activity

Hydrothermal is a term referring to hot water. Some geologists restrict the meaning to include only water heated by magma, but here we use it to refer to any hot subsurface water and the surface activity that results from its discharge. One manifestation of hydrothermal activity in areas of active or recently active volcanism is the discharge of gases, such as steam, at vents known as *fumeroles* (see Figure 5.2). Of more immediate concern here, however, is the groundwater that rises to the surface as *hot springs* or *geysers*. It may be heated by its proximity to magma or by Earth's geothermal gradient because it circulates deeply.

Hot Springs

A **hot spring** (also called a *thermal spring* or *warm spring*) is any spring in which the water temperature is higher than 37°C, the temperature of the human body (❚ Figure 13.19a). Some hot springs are much hotter, with temperatures up to the boiling point in many instances (❚ Figure 13.19b). Another type of hot spring, called a *mud pot*, results when chemically altered rocks yield clays that bubble as hot water and steam rise through them (❚ Figure 13.19c). Of the approximately 1,100 known hot springs in the United States, more than 1,000 are in the far West, with the others in the Black Hills of South Dakota, Georgia, the Ouachita region of Arkansas, and the Appalachian region.

Hot springs are also common in other parts of the world. One of the most famous is in Bath, England, where shortly after the Roman conquest of Britain in AD 43, numerous bathhouses and a temple were built around the hot springs (❚ Figure 13.20).

The heat for most hot springs comes from magma or cooling igneous rocks. The geologically recent igneous activity in the western United States accounts for the large number of hot springs in that region. The water in some hot springs, however, circulates deep into Earth, where it is warmed by the normal increase in temperature, the geothermal gradient. For example, the spring water of Warm Springs, Georgia, is heated in this manner. This hot spring was a health and bathing resort long before the Civil War (1861–1865); later, with the establishment of the Georgia Warm Springs Foundation, it was used to help treat polio victims.

CONNECTION LINK

To review the geothermal gradient, see page 227 in Chapter 9.

▌Figure 13.19 Hot Springs

a One of the more colorful hot springs in Yellowstone National Park, the Morning Glory hot spring, is fringed with multicolored mats of heat-loving cyanobacteria and algal mats. Each color represents a certain temperature range that allows for specific bacterial species to thrive in this extreme environment.

Copyright and Photograph by Dr. Parvinder S. Sethi

James S. Monroe

b The water in this hot spring at Bumpass Hell in Lassen Volcanic National Park, California, is boiling.

James S. Monroe

c Mud pot at the Sulfur Works, also in Lassen Volcanic National Park.

James S. Monroe

d The U.S. National Park Service warns of the dangers in hydrothermal areas, but some people ignore the warnings and are injured or killed.

Geysers

Hot springs that intermittently eject hot water and steam with tremendous force are known as **geysers**. The word comes from the Icelandic *geysir*, "to gush" or "to rush forth." One of the most famous geysers in the world is Old Faithful in Yellowstone National Park in Wyoming (▌Figure 13.21). With a thunderous roar, it erupts a column of hot water and steam every 30 to 90 minutes. Other well-known geyser areas are found in Iceland and New Zealand.

Geysers are the surface expression of an extensive underground system of interconnected fractures within hot igneous rocks (▌Figure 13.22). Groundwater percolating down into the network of fractures is heated as it comes into

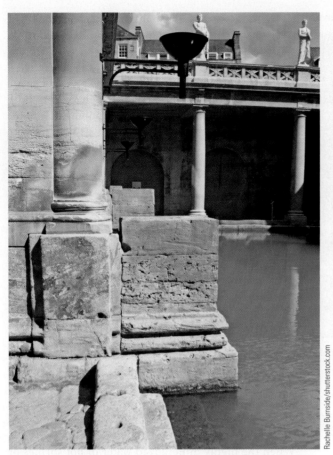

Figure 13.20 **Bath, England** One of the many bathhouses in Bath, England, that were built around hot springs shortly after the Roman conquest in AD 43.

Rachelle Burnside/shutterstock.com

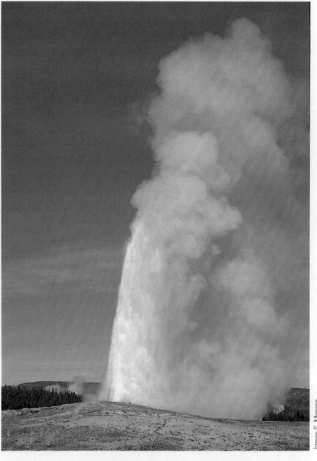

James S. Monroe

Figure 13.21 **Old Faithful Geyser** Old Faithful Geyser in Yellowstone National Park, Wyoming, is one of the world's most famous geysers, erupting faithfully every 30 to 90 minutes and spewing water 32 to 56 m high.

Figure 13.22 Anatomy of a Geyser

a The eruption of a geyser starts when groundwater percolates down into a network of interconnected openings and is heated by the hot igneous rocks. The water near the bottom of the fracture system is under higher pressure than the water near the top and consequently must be heated to a higher temperature before it will boil.

b Any rise in the temperature of the water above its boiling point or a drop in pressure will cause the water to change to steam, which quickly pushes the water above it up and out of the ground, producing a geyser eruption.

Figure 13.23 Hot Spring Deposits in Yellowstone National Park, Wyoming

Copyright and Photograph by Dr. Parvinder S. Sethi

a Minerva Terrace, formed when calcium-carbonate-rich hot-spring water cooled, precipitating travertine.

Reed Wicander

b Liberty Cap is a geyserite mound formed by numerous geyser eruptions of silicon-dioxide-rich hot spring water.

near the boiling point, a slight rise in temperature or a drop in pressure, such as from escaping gas, will instantly change it to steam. The expanding steam quickly pushes the water above it out of the ground and into the air, producing a geyser eruption. After the eruption, relatively cool groundwater starts to seep back into the fracture system where it heats to near its boiling temperature and the eruption cycle begins again. Such a process explains how geysers can erupt with some regularity.

Hot spring and geyser water typically contains large quantities of dissolved minerals because most minerals dissolve more rapidly in warm water than in cold water. Because of this high mineral content, some believe that the waters of many hot springs have medicinal properties. Numerous spas and bathhouses have been built at hot springs throughout the world to take advantage of these supposed healing properties.

When the highly mineralized water of hot springs or geysers cools at the surface, some of the material in solution is precipitated, forming various types of deposits. The amount and type of precipitated minerals depend on the solubility and composition of the material that the groundwater flows through. If the groundwater contains dissolved calcium carbonate ($CaCO_3$), then *travertine* or *calcareous tufa* (both of which are varieties of limestone) are precipitated. Spectacular examples of hot spring travertine deposits are found at Pamukhale in Turkey and at Mammoth Hot Springs in Yellowstone National Park (Figure 13.23a). Groundwater containing dissolved silica will, upon reaching the surface, precipitate a soft, white, hydrated mineral called *siliceous sinter* or *geyserite*, which can accumulate around a geyser's opening (Figure 13.23b).

Geothermal Energy

Geothermal energy is any energy produced from Earth's internal heat. In fact, the term *geothermal* comes from *geo*, "Earth," and *thermal*, "heat." Several forms of internal heat are known, such as hot dry rocks and magma, but so far, only hot water and steam are used.

As oil reserves decline, geothermal energy is becoming an attractive alternative, particularly in parts of the western United States, such as the Salton Sea area of southern California, where geothermal exploration and development have begun. Approximately 1% to 2% of the world's current

contact with the hot rocks. Because the water near the bottom of the fracture system is under higher pressure than the water near the top, it must be heated to a higher temperature before it will boil. Thus, when the deeper water is heated to energy needs could be met by geothermal energy. In those areas where it is plentiful, geothermal energy can supply most, if not all, of the energy needs, sometimes at a fraction of the cost of other types of energy. Some of the countries currently using geothermal energy in one form or another are Iceland, the United States, Mexico, Italy, New Zealand, Japan, the Philippines, and Indonesia.

In the United States, the first commercial geothermal electricity-generating plant was built in 1960 at The Geysers, about 120 km north of San Francisco, California. Here, wells were drilled into the numerous near-vertical fractures underlying the region. As pressure on the rising groundwater decreases, the water changes to steam, which is piped directly to electricity-generating turbines and generators (❙ Figure 13.24).

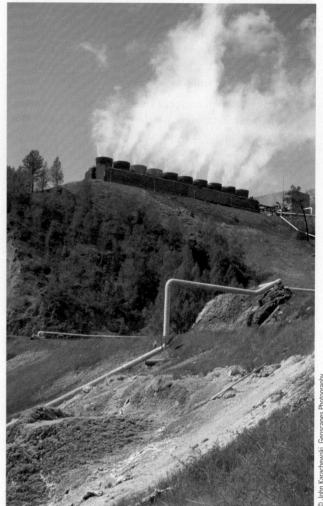

❙ **Figure 13.24 The Geysers, Sonoma County, California** Steam rising from one of the geothermal power plants at The Geysers in Sonoma County, California. Steam from wells drilled into this geothermal region, about 120 km north of San Francisco, is piped directly to electricity-generating turbines to produce electricity that is distributed throughout the area.

≫ Critical Thinking Question
Although geothermal energy is a relatively nonpolluting form of energy used as a source of heat and to generate electricity, why is it typically a more expensive form of energy? (Hint: Which contains more dissolved minerals, hot or cold water?)

© John Karachewski, Geoscapes Photography

GEO-RECAP

≫ Geo-Image

❙ **Figure 1** Withdrawal of large quantities of groundwater in the Las Vegas area has resulted in differential subsidence and damage to roads and buildings. What evidence of subsidence is visible in this photograph?

Marli Bryant Miller

Chapter Summary

- Groundwater is part of the hydrologic cycle and an important natural resource. It consists of all subsurface water trapped in the pores and other open spaces in rocks, sediment, and soil.

- Porosity is the percentage of a material's total volume that is pore space. Permeability is the capacity to transmit fluids. Permeability is dependent on porosity, but also on the size of the pores or fractures and their interconnections.

- The water table is the surface separating the zone of aeration (in which the pores are filled with air and water) from the underlying zone of saturation (in which pores are filled with water). The water table is a subdued replica of the overlying land surface in most places.

- Groundwater moves slowly downward under the influence of gravity through the pore spaces in the zone of aeration to the zone of saturation. Some of it moves along the surface of the water table, and the rest moves from areas of high pressure to areas of low pressure. Generally, the average velocity of groundwater is a few centimeters per day.

- Springs are found wherever the water table intersects the surface. Some springs are the result of a perched water table—that is, a localized aquiclude within an aquifer and above the regional water table.

- Water wells are openings made by digging or drilling down into the zone of saturation. When water is pumped from a well, the water table in the area around the well is lowered, forming a cone of depression.

- In an artesian system, confined groundwater builds up high hydrostatic pressure. For an artesian system to develop, an aquifer must be confined above and below by aquicludes; the aquifer is usually tilted so as to build up hydrostatic pressure; and the aquifer must be exposed at the surface so it can be recharged.

- Karst topography develops by groundwater erosion in many areas underlain by soluble rocks. It is typically characterized by sinkholes, caves, solution valleys, and disappearing streams.

- Caves form when groundwater in the zone of saturation weathers and erodes soluble rock such as limestone. Cave deposits, called dripstone, result from the precipitation of calcite and include such features as stalactites, stalagmites, columns, drip curtains, and travertine terraces.

- Modification of the groundwater system can cause serious problems such as lowering of the water table, saltwater incursion, subsidence, and contamination.

- Groundwater contamination by humans is becoming a serious problem and can result from landfills, septic systems, toxic waste sites, and industrial effluents, all of which affect the quality of the groundwater.

- Hydrothermal refers to hot water, typically heated by magma, but also resulting from Earth's geothermal gradient as it circulates deeply beneath the surface. Manifestations of hydrothermal activity include fumaroles, hot springs, and geysers.

- Geothermal energy is energy produced from Earth's internal heat and comes from the steam and hot water trapped within Earth's crust. It is a relatively nonpolluting form of energy that is used as a source of heat and to generate electricity.

Important Terms

artesian system (p. 319)
cave (p. 322)
cone of depression (p. 318)
geothermal energy (p. 334)
geyser (p. 332)
groundwater (p. 314)

hot spring (p. 331)
hydrothermal (p. 331)
karst topography (p. 321)
permeability (p. 315)
porosity (p. 314)
sinkhole (p. 320)

spring (p. 317)
water table (p. 316)
water well (p. 317)
zone of aeration (p. 316)
zone of saturation (p. 316)

Review Questions

1. The porosity of Earth materials is defined as
   a. ___ their ability to transmit fluids.
   b. ___ the depth of the zone of saturation.
   c. ___ the percentage of void space.
   d. ___ their solubility in the presence of weak acids.
   e. ___ the temperature of groundwater.

2. Two features typical of areas of karst topography are
   a. ___ sinkholes and disappearing streams.
   b. ___ dripstone and a cone of ascension.
   c. ___ saltwater incursion and pollution.
   d. ___ hydrothermal activity and springs.
   e. ___ geysers and hot springs.

3. A cone of depression forms when
   a. ____ a stream flows into a sinkhole.
   b. ____ water in the zone of aeration is replaced by water from the zone of saturation.
   c. ____ a spring forms where a perched water table intersects the surface.
   d. ____ water is withdrawn faster than it can be replaced.
   e. ____ the ceiling of a cave collapses, forming a steep-sided crater.

4. Which of the following statements is correct?
   a. ____ The water table dips beneath hills and rises beneath valleys.
   b. ____ Groundwater erosion yields geysers and hot springs.
   c. ____ The water from an artesian well is better than water from other wells.
   d. ____ A perched water table is found above the main water table.
   e. ____ In the United States, karst topography is best developed in the Pacific Northwest.

5. A hot spring that periodically erupts is known as a
   a. ____ geothermal gradient.
   b. ____ zone of saturation.
   c. ____ geyser.
   d. ____ mud pot.
   e. ____ travertine terrace.

6. Which of the following is a cave deposit?
   a. ____ chamber.
   b. ____ stalactite.
   c. ____ sinkhole.
   d. ____ artesian spring.
   e. ____ aquiclude.

7. The term *hydrothermal* refers to
   a. ____ calcareous tufa.
   b. ____ groundwater contamination.
   c. ____ hot water.
   d. ____ sinkhole formation.
   e. ____ artesian wells.

8. Which of the following conditions must exist for an artesian system to form?
   a. ____ Groundwater must circulate near magma.
   b. ____ The water table must be at or very near the surface.
   c. ____ The rocks below the surface must be especially resistant to solution.
   d. ____ Water must rise very high in the capillary fringe.
   e. ____ An aquifer must be confined above and below by aquicludes.

9. When water is pumped from wells in some coastal areas, a problem arises that is known as
   a. ____ permeability decrease.
   b. ____ saltwater incursion.
   c. ____ geothermal depression.
   d. ____ artesian recharge.
   e. ____ dripstone deposition.

10. A layer of Earth materials that inhibits the movement of groundwater is a(n)
    a. ____ aquiclude.
    b. ____ geyser.
    c. ____ aquifer.
    d. ____ cone of depression.
    e. ____ capillary fringe.

11. Discuss the role of groundwater in the hydrologic cycle.

12. Why should we be concerned about how fast the groundwater supply is being depleted in some areas?

13. What is the difference between an artesian system and a water well? Is there any difference between the water obtained from an artesian system and the water from a water well?

14. Why isn't geothermal energy a virtually unlimited source of energy?

15. Explain how saltwater incursion takes place and why it is a problem in coastal areas.

16. Why does groundwater move so much slower than surface water?

17. Explain how groundwater weathers and erodes Earth materials.

18. Describe three features you might see in an active hydrothermal area. Where in the United States would you go to see such activity?

19. One concern geologists have about burying nuclear waste in present-day arid regions such as Nevada is that the climate may change during the next several thousand years and become more humid, thus allowing more water to percolate through the zone of aeration. Why is this a concern? What would the average rate of groundwater movement have to be during the next 5,000 years to reach buried canisters containing radio-active waste buried at a depth of 400 m?

20. Describe the configuration of the water table beneath a humid area and beneath an arid region. Why are the configurations different?

View of the Jungfrau Firn in Switzerland, which merges with two other valley glaciers to form the Aletsch Glacier, the largest glacier in the Alps of Europe. From this viewpoint, the glacier's terminus is 23 km distant. The Aletsch Glacier covers 120 km², but here its tributary is only about 1.5 km wide. Valley glaciers like this one are present on all the continents except Australia, but during the Ice Age (2.6 million to 10,000 years ago) they were more numerous and

CHAPTER 14

Glaciers and Glaciation

Sue Monroe

OUTLINE

14.1 Introduction

14.2 The Kinds of Glaciers

GEO-INSIGHT: The Little Ice Age

14.3 Glaciers—Moving Bodies of Ice on Land

14.4 The Glacial Budget

GEO-FOCUS: Glaciers and Global Warming

14.5 Erosion and Sediment Transport by Glaciers

14.6 Glacial Deposits

14.7 What Causes Ice Ages?

GEO-RECAP

HAVE YOU EVER WONDERED?

- What glaciers are and where they are found?
- Why glaciers come in different sizes and shapes?
- How glaciers move?
- What controls the length and thickness of glaciers?
- How glaciers transport and deposit sediment?
- What kinds of landscapes result from glacial erosion?
- What the Ice Age was and what caused it?

14.1 Introduction

You have probably heard of the Ice Age, or what geologists call the Pleistocene Epoch, during which time vast glaciers expanded and small glaciers in mountain valleys were much more numerous and larger than they are now (see the chapter opening photo). The Ice Age was brief in the context of geologic time, but it lasted from 2.6 million to 10,000 years ago, although the glaciers waxed and waned several times during this interval. Since then, Earth has gone through several climatic changes. About 6,000 years ago, during the Holocene Maximum, average temperatures were slightly higher than they are now, and some of today's arid regions were much more humid. Today, the only arable land in Egypt is along the Nile River, but during the Holocene Maximum, much of North Africa was covered by grasslands, swamps, and lakes.

Following the Holocene Maximum was a time of cooler temperatures, but from about AD 1000 to 1300, Europe went through the Medieval Warm Period when wine grapes grew 480 km farther north than they do now. Then a cooling trend began in about AD 1300 that led to the **Little Ice Age,** which lasted from 1500 to the middle or late 1800s. During this time glaciers expanded to their greatest historic extent (see the Geo-Insight on pages 342 and 343); summers were cooler and wetter; winters were colder; and sea ice persisted longer around Greenland, Iceland, and the Canadian Arctic islands. The cooler, wetter summers were a problem because the growing seasons were shorter, resulting in several famines.

During the coldest part of the Little Ice Age (1680–1730), the growing season in England was five weeks shorter than during the 1900s. Occasionally, Eskimos following the southern edge of the sea ice paddled their kayaks as far south as Scotland, and the canals in Holland froze over in some winters. In 1607, the first Frost Fair was held in London, England, on the Thames River, which began to freeze over nearly every winter. In the late 1700s, New York Harbor froze over, and 1816 is known as the "year without a summer" when unusually cold temperatures persisted into June and July in New England and northern Europe (the eruption of Tambora in 1815 contributed to the cold spring and summer of 1816).

Many of you have probably heard of the Ice Age and have some idea of what a glacier is, but you may not know about the dynamics of glaciers, how they form, and what may cause ice ages. In any case, *glaciers* are moving bodies of ice on land that are particularly effective at erosion, sediment transport, and deposition. They deeply scour the surfaces they move over, producing many easily recognizable landforms, and they deposit huge amounts of sediment, much of it important sources of sand and gravel. Glaciers today cover about 10% of Earth's land surface, but during the Ice Age they were much more widespread.

Why study glaciers? If for no other reason, glaciers have produced some of the most stunning scenery on Earth, especially in mountains that are or were eroded by glaciers. And of course, glaciers are part of the hydrologic cycle, illustrating again the complex interactions among Earth's systems. In addition, a large part of the water in some countries such as

Nepal, Tibet, and Pakistan comes from melting glaciers. Even in the United States and Canada, some areas depend partly on water stored in glaciers. And finally, glaciers are very sensitive to climatic changes, so scientists are interested in fluctuations in glaciers as possible indications of global warming.

14.2 The Kinds of Glaciers

A **glacier** is a moving body of ice on land that flows downslope or outward from an area of accumulation. We will discuss how glaciers flow in a later section. Our definition of a glacier excludes frozen seawater as in the North Polar region and sea ice that forms yearly adjacent to Greenland and Iceland. Drifting icebergs are not glaciers either, although they may have come from glaciers that flowed into lakes or the sea. The critical points in the definition are *moving* and *on land*. Accordingly, permanent snowfields in high mountains, though on land, are not glaciers because they do not move. All glaciers share several characteristics, but they differ enough in size and location for us to define two specific types—valley glaciers and continental glaciers—and several subvarieties.

Valley Glaciers

Valley glaciers are confined to mountain valleys where they flow from higher to lower elevations (▌ Figure 14.1). Continental glaciers cover vast areas and are not confined by the underlying topography, and they flow outward in all directions from areas of snow and ice accumulation. We use the term **valley glacier,** but some geologists prefer the synonyms *alpine glacier* and *mountain glacier*. Valley glaciers commonly have tributaries, just as streams do, thereby

> **CONNECTION LINK** ◀
>
> To review the hydrologic cycle see page 287 and Figure 12.2 in Chapter 12.

▌ **Figure 14.1** A Valley Glacier in Canada   This glacier is in a mountain valley in the Coast Mountains, Yukon Territory, Canada. Notice the tributaries that join to form a larger glacier.

Sue Monroe

≫ Critical Thinking Question
How are valley glaciers similar to streams and rivers?

Figure 14.2 A Tidewater Glacier in Alaska Wellesley Glacier is a valley glacier that flows into the sea, so it is a tidewater glacier.

forming a network of glaciers in an interconnected system of mountain valleys.

Valley glaciers are common in the mountains of western North America, especially Alaska and Canada, as well as the Andes in South America, the Alps in Europe, the Alps of New Zealand, and the Himalayas in Asia. Even a few of the loftier mountains in Africa, though near the equator, are high enough to support small glaciers. In fact, Australia is the only continent that has no glaciers.

A valley glacier's shape is obviously controlled by the shape of the valley it occupies, so it tends to be a long, narrow tongue of moving ice. Valley glaciers that flow into the ocean are called *tidewater glaciers;* they differ from other valley glaciers only in that their terminus is in the sea rather than on land (Figure 14.2).

Valley glaciers are small compared with the much more extensive continental glaciers, but even so, they may be several kilometers across, 200 km long, and hundreds of meters thick. Some valley glaciers in Alaska are up to 1,500 m thick. Erosion and deposition by valley glaciers were responsible for much of the spectacular scenery in several U.S. and Canadian national parks.

Continental Glaciers

Continental glaciers, also known as *ice sheets,* are vast, covering at least 50,000 km², and they are unconfined by topography; that is, their shape and movement are not controlled by the underlying landscape. Valley glaciers are long, narrow tongues of ice that conform to the shape of the valley they occupy, and the existing slope determines their direction of flow. In contrast, continental glaciers flow outward in all directions from a central area or areas of accumulation in response to variations in ice thickness.

In Earth's two areas of continental glaciation, Greenland and Antarctica, the ice is more than 3,000 m thick and covers all but the highest mountains (Figure 14.3a). The continental glacier in Greenland covers about 1,800,000 km², and in Antarctica the East and West Antarctic Glaciers merge to form a continuous ice sheet covering more than 12,650,000 km². The glaciers in Antarctica flow into the sea, where the buoyant effect of water causes the ice to float in vast *ice shelves;* the Ross Ice Shelf alone covers more than 547,000 km² (Figure 14.3a).

Figure 14.3 Continental Glaciers and Ice Caps

a The West and East Antarctic ice sheets merge to form a nearly continuous ice cover that averages 2,160 m thick. The blue lines are lines of equal thickness.

b View of the Penny Ice Cap on Baffin Island, Canada. It covers about 6,000 km².

GEO-INSIGHT   The Little Ice Age

The Little Ice Age was a time of cooler temperatures, shorter growing seasons, and expansion of glaciers, especially in the northern parts of Europe and North America. It began in about 1300 and ended some time during the middle- to late-1800s.

▶ **1. 2.** The Mer de Glace (Sea of Ice) at present is 7 km long, making it the longest glacier in the French Alps. The painting on the left by Samuel Birmann (1793–1847), a Swiss painter, shows the extent of the glacier not long after the Little Ice Age maximum. Notice the arrows that show the positions of the glacier's margin during the Little Ice Age and in 2000 (right).

▼ **3.** The River Thames froze over regularly during the Little Ice Age. The first recorded Frost Fair was held on the Thames in 1608; the last one was in 1814 as shown in this image. Although the river froze over many times between these dates, the most extensive was during the winter of 1683–84 when it was frozen for two months with ice 28 cm thick.

▲ **4.** Samuel Birmann (1793–1847) painted this view in Switzerland, titled *Unterer Grindelwald,* in 1826 when the glacier extended onto the valley floor in the foreground.

▲ **5.** The same glacier today. Its terminus is hidden behind a rock projection at the lower end of its valley.

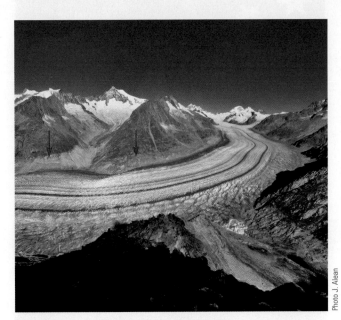

Photo J. Alean

▲ **6.** The 23-km-long Aletsch Glacier in Switzerland is the country's longest. Notice the trim line indicated by the arrows, which show that the glacier was much thicker during the Little Ice Age. Also note the valley directly across the glacier; this once contained a tributary glacier that flowed into the larger glacier. For a view of one of the present-day tributaries to the Aletsche Glacier, see the chapter opening photo.

Photo: Ó. Ingólfsson 2004. http://www3.hi.is/~oi/

▲ **7.** Gigjökull (Crater Glacier) in Iceland was more extensive during the Little Ice Age, but since 1900 its terminus has retreated. The people in the foreground are on the deposit that marks the maximum extent of the glacier, or what geologists call a terminal moraine. This glacier actually is an outlet from an ice cap that covers the active volcano Eyjafjallajökull, which erupted in April 2010 and disrupted air travel over the North Atlantic. During the eruption, Gigjökull's surface was covered by volcanic ash.

Washington Crossing the Delaware River, 25th December 1776, 1851 (oil on canvas) (copy of an original painted in 1848), Leutze, Emanuel Gottlieb (1816-68)/Metropolitan Museum of Art, New York, USA/The Bridgeman Art Library International

◀ **8.** Most of our information on the Little Ice Age comes from Europe, but the North American colonies were impacted by this event, too. In this iconic painting of General George Washington crossing the Delaware River on December 25, 1777, large blocks of ice are shown in the river. On many occasions, the rivers in eastern North America froze over, and sea ice was present in many areas along the coastline.

Although valley glaciers and continental glaciers are easily differentiated by size and location, geologists also recognize an intermediate variety called an **ice cap.** Ice caps are similar to, but smaller than, continental glaciers, covering less than 50,000 km². The 6,000-km² Penny Ice Cap on Baffin Island, Canada (❚ Figure 14.3b), and the Juneau Icefield in Alaska and Canada at about 3,900 km² are good examples. Some ice caps form when valley glaciers grow and overtop the divides and passes between adjacent valleys and coalesce to form a continuous ice cover. They also form on fairly flat terrain in Iceland and some of the islands in the Canadian Arctic.

14.3 Glaciers—Moving Bodies of Ice on Land

We use the term **glaciation** to indicate all glacial activity, including the origin, expansion, and retreat of glaciers, as well as their impact on Earth's surface. Presently, glaciers cover nearly 15 million km², or about 10% of Earth's land surface. As a matter of fact, if all of Earth's glacial ice were in the United States and Canada, it would form a continuous ice cover about 1.5 km thick!

At first glance, glaciers appear static. Even briefly visiting a glacier may not dispel this impression because, although glaciers move, they usually do so slowly. Nevertheless, they do move, and just like other geologic agents such as running water, glaciers are dynamic systems that continuously adjust to changes. For example, a glacier may flow slower or more rapidly depending on decreased or increased amounts of snow or the absence or presence of water at its base. And glaciers may expand or contract depending on climatic changes.

Glaciers—Part of the Hydrologic Cycle

In Chapter 1, you learned that one of Earth's systems, the hydrosphere, consists of all surface water, including water frozen in glaciers. So glaciers make up one reservoir in the hydrologic cycle where it is stored for long periods, but even this water eventually returns to its original source, the oceans (see Figure 12.2). Glaciers at high latitudes, as in Alaska, northern Canada, and Scandinavia, flow directly into the oceans where they melt, or icebergs break off (a process known as calving) and drift out to sea where they eventually melt. At low latitudes or areas remote from the oceans, glaciers flow to lower elevations where they melt and the liquid water enters the groundwater system (another reservoir in the hydrologic cycle) or it returns to the seas by surface runoff.

In addition to melting, glaciers lose water by *sublimation,* when ice changes to water vapor without an intermediate liquid phase. Sublimation is not an exotic process; it occurs in the freezer compartment of a refrigerator. Because

❚ **Figure 14.4** Glacial Ice

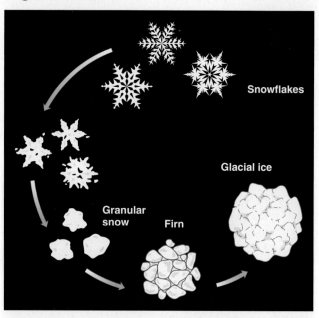

a The conversion of freshly fallen snow to firn and then to glacial ice.

b This iceberg in Portage Lake in Alaska shows the blue color of glacial ice. The longer wavelengths of white light are absorbed by the ice, but blue (short wavelength) is transmitted into the ice and scattered, accounting for the blue color.

of sublimation, the older ice cubes at the bottom of the container are much smaller than the more recently formed ones. In any case, the water vapor so derived enters the atmosphere where it may condense and fall as rain or snow, but in the long run, this water also returns to the oceans.

How Do Glaciers Originate and Move?

Ice is a crystalline solid with characteristic physical properties and a specific chemical composition, and thus is a mineral. Accordingly, glacial ice is a type of metamorphic rock, but one that is easily deformed. Glaciers form in any area where more snow falls than melts during the warmer season and a net accumulation takes place. Freshly fallen snow has about 80% air-filled pore space and 20% solids, but it compacts as it accumulates, partially thaws, and refreezes, converting to a granular type of snow known as **firn.** As more snow accumulates, the firn is buried and further compacted and recrystallized until it is transformed into **glacial ice,** consisting of about 90% solids and 10% air (▌ Figure 14.4).

Now you know how glacial ice forms, but we still have not addressed how glaciers move. At this time, it is useful to recall some terms from Chapter 10. Remember that *stress* is force per unit area and *strain* is a change in the shape or volume of solids. When accumulating snow and ice reach a critical thickness of about 40 m, the stress on the ice at depth is great enough to induce **plastic flow,** a type of permanent deformation involving no fracturing. Glaciers move mostly by plastic flow, but they may also slide over their underlying surface by **basal slip** (▌ Figure 14.5). Liquid water facilitates basal slip because it reduces friction between a glacier and the surface over which it moves.

> **CONNECTION LINK**
>
> For more on stress and strain see pages 235 and 236 in Chapter 10.

The total movement of a glacier in a given time is a consequence of plastic flow and basal slip, although the former occurs continuously, whereas the latter varies depending on the season, latitude, and elevation. Indeed, if a glacier is solidly frozen to the surface below, as in the case of many polar environments, it moves only by plastic flow. Furthermore, basal slip is far more important in valley glaciers as they flow from higher to lower elevations, whereas continental glaciers need no slope for flow.

Although glaciers move by plastic flow, the upper 40 m or so of ice behaves like a brittle solid and fractures if subjected to stress. Large crevasses commonly develop in glaciers where they flow over an increase in slope of the underlying surface or where they flow around a corner (▌ Figure 14.6). In either case, the ice is stretched (subjected to tension) and crevasses open, which extend down to the zone of plastic flow. In some cases, a glacier descends over such a steep precipice that crevasses break up the ice into a jumble of blocks and spires, and an icefall develops.

Distribution of Glaciers

As you might suspect, the amount of snowfall and temperature are important factors in determining where glaciers form. Parts of northern Canada are cold enough to support glaciers but receive too little snowfall, whereas some mountain areas in California receive huge amounts of snow but are too warm for glaciers. Of course, temperature varies with elevation and latitude, so we would expect to find glaciers in high mountains and at high latitudes, if these areas receive enough snow.

Many small glaciers are present in the Sierra Nevada of California, but only at elevations exceeding 3,900 m. In fact, the high mountains in California, Oregon, and Washington all have glaciers because they are high and receive so much snow. Mount Baker in Washington had almost 29 m of snow during

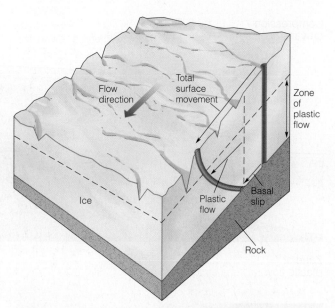

▌ **Figure 14.5 Part of a Glacier Showing Movement by a Combination of Plastic Flow and Basal Slip** Plastic flow involves internal deformation within the ice, whereas basal slip is sliding over the underlying surface. If a glacier is solidly frozen to its bed, it moves only by plastic flow. Notice that the top of the glacier moves farther in a given time than the bottom does.

What Would You Do ❓

Suppose you are a high school earth science teacher trying to explain to students that ice is a mineral and rock, and how a solid like ice can flow like a fluid. Furthermore, you explain that the upper 40 m or so of a glacier is brittle and fractures, whereas ice below that depth simply flows when subjected to stress. Now that your students are thoroughly confused, how will you explain and demonstrate that ice can behave like a solid yet show properties of fluid flow?

the winter of 1998–1999, and average accumulations of 10 m or more are common in many parts of these mountains.

Glaciers are also found in the mountains along the Pacific Coast of Canada, which also receive considerable snowfall, and of course, they are farther north. Some of the higher peaks in the Rocky Mountains in both the United States and Canada also support glaciers. At even higher latitudes, as in Alaska, northern Canada, and Scandinavia, glaciers exist at sea level.

14.4 The Glacial Budget

Just as a savings account grows and shrinks as funds are deposited and withdrawn, a glacier expands and contracts in response to accumulation and wastage. We describe a glacier's behavior in terms of a **glacial budget,** which is essentially a balance sheet of accumulation and wastage. For

Compression–
crevasses close

Tension–
crevasses open

Brittle
zone

Zone of
plastic flow

Figure 14.6 Crevasses Crevasses are common in the upper parts of glaciers when the ice is subjected to tension.

a Crevasses open where the brittle part of a glacier is stretched as it moves over a steeper slope in its valley.

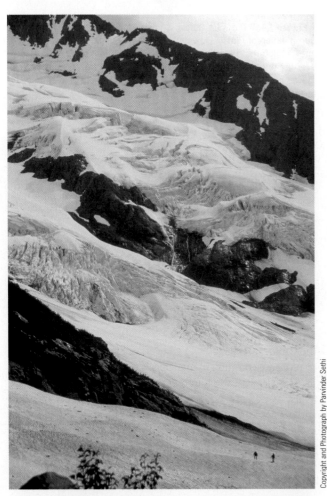

Copyright and Photograph by Parvinder Sethi

b These crevasses are on the Byron Glacier in Kenai Fjords National Park and Preserve, near Seward, Alaska. Note the two hikers for scale.

>> **Critical Thinking Question**
Why do crevasses extend down only to the zone of plastic flow?

instance, the upper part of a valley glacier is a **zone of accumulation,** where additions exceed losses and the surface is perennially snow covered. In contrast, the lower part of the

same glacier is a **zone of wastage,** where losses from melting, sublimation, and calving of icebergs exceed the rate of accumulation (Figure 14.7).

At the end of winter, a glacier's surface is completely covered with the accumulated seasonal snowfall. During the spring and summer, the snow begins to melt, first at lower elevations and then progressively higher up the glacier. The elevation to which snow recedes during a wastage season is the *firn limit* (Figure 14.7). You can easily identify the zones of accumulation and wastage by noting the location of the firn limit.

The firn limit on a glacier may change yearly, but if it does not change or shows only minor fluctuations, the glacier has a balanced budget. That is, additions in the zone of accumulation are exactly balanced by losses in the zone of wastage, and the distal end, or terminus, of the glacier remains stationary (Figure 14.7). If the firn limit moves up the glacier, indicating a negative budget, the glacier's terminus retreats. If the firn limit moves down the glacier, however, the glacier has a positive budget, additions exceed losses, and its terminus advances.

Even though a glacier may have a negative budget and a retreating terminus, the glacial ice continues to move toward the terminus by plastic flow and basal slip. If a negative budget persists long enough, though, the glacier continues to recede, and it thins until it is no longer thick enough to maintain flow. It then ceases moving and becomes a *stagnant glacier;* if wastage continues, the glacier eventually disappears.

We used a valley glacier as an example, but the same budget considerations control the flow of ice caps and continental glaciers as well. The entire Antarctic ice sheet is in the zone of accumulation, but it flows into the ocean where wastage occurs.

We mentioned in the Introduction that glaciers are very sensitive to climate changes and that they may tell us about global warming (see Geo-Focus on pages 348 and 349). One alarming trend in glaciers is that most of those studied are retreating. In fact, glaciers will probably be gone from Glacier National Park in Montana in only a few decades, and similar trends are taking place elsewhere.

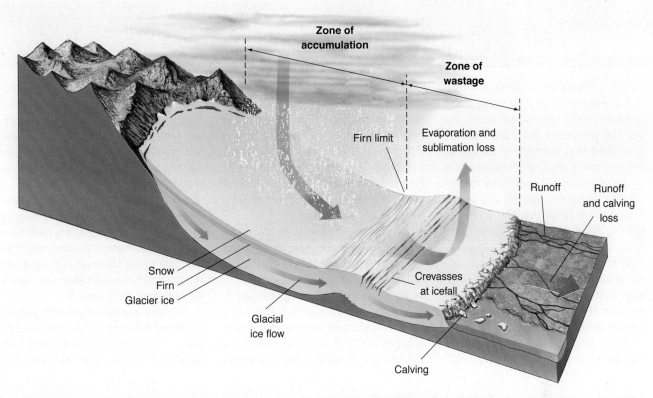

Figure 14.7 The Glacial Budget All glaciers have a zone of accumulation where additions exceed losses and the glacier's surface is perennially covered by snow. They also have a zone of wastage where losses from melting, calving of icebergs, evaporation, and sublimation exceed gains. If a glacier's budget is balanced, its terminus remains in the same position. However, should the budget be positive the terminus advances, and if the budget is negative the terminus retreats.

How Fast Do Glaciers Move?

In general, valley glaciers move more rapidly than continental glaciers, but the rates for both vary from centimeters to tens of meters per day. Valley glaciers moving down steep slopes flow more rapidly than glaciers of comparable size on gentle slopes, assuming that all other variables are the same. The main glacier in a valley glacier system contains a greater volume of ice and thus has a greater discharge and flow velocity than its tributaries (Figure 14.1). Temperature exerts a seasonal control on valley glaciers because, although plastic flow remains rather constant year-round, basal slip is more important during warmer months when meltwater is abundant.

Flow rates also vary within the ice itself. For example, flow velocity increases downslope in the zone of accumulation until the firn limit is reached; from that point, the velocity becomes progressively slower toward the glacier's terminus. Valley glaciers are similar to streams, in that the valley walls and floor cause frictional resistance to flow, so the ice in contact with the walls and floor moves more slowly than the ice some distance away (Figure 14.8a).

Notice in Figure 14.8a that flow velocity in the interior of a glacier increases upward until the top few tens of meters of ice are reached, but little or no additional increase occurs after that point. This upper ice layer constitutes the rigid part of the glacier that is moving as a result of basal slip and plastic flow below.

Continental glaciers ordinarily flow at a rate of centimeters to meters per day. One reason continental glaciers move comparatively slowly is that they exist at higher latitudes and are frozen to the underlying surface most of the time, which limits the amount of basal slip. But some basal slip does occur even beneath the Antarctic ice sheet, although most of its movement is by plastic flow. Nevertheless, some parts of continental glaciers manage to achieve extremely high flow rates. Near the margins of the Greenland ice sheet, the ice is forced between mountains in what are called *outlet glaciers*. In some of these outlets, flow velocities exceed 100 m per day.

In parts of the continental glacier covering West Antarctica, scientists have identified ice streams in which flow rates are considerably higher than in adjacent glacial ice. Drilling has revealed a 5-m-thick layer of water-saturated sediment beneath these ice streams, which acts to facilitate movement of the ice above. Some geologists think that geothermal heat from subglacial volcanism melts the underside of the ice, thus accounting for the layer of water-saturated sediment.

Glacial Surges A **glacial surge** is a short-lived episode of accelerated flow during which the glacier's surface breaks

GEO-FOCUS   Glaciers and Global Warming

As you probably know, global warming is a phenomenon of a warming trend in Earth's lower atmosphere during the last 100 years or so. Many scientists are convinced that warming is caused by the concentration of greenhouse gases in the atmosphere, especially carbon dioxide, methane, and nitrous oxide. They further think that the source of the increased quantities of these gases is the burning of fossil fuels, especially coal and petroleum. Of course, there are dissenters, most of whom acknowledge that Earth's surface temperatures are warmer but attribute the increase to normal climatic variations.

Some ask, "So what if global temperatures increase by a few degrees?"

After all, residents of Fairbanks, Alaska, might welcome slightly warmer winter temperatures, and it is unlikely that the same trend will make the summers in Phoenix, Arizona, or Death Valley, California, much more uncomfortable than they are now. However, temperature increase has far-reaching effects. Remember that during the Little Ice Age, temperatures were only a little cooler than they are now, and yet the cooler weather, especially during the summers, had tragic consequences.

Whatever the cause of global warming, no one doubts that glaciers are good indicators of short-term climatic changes and that with few exceptions glaciers are retreating. No one knows

how many glaciers Earth has; one estimate is 160,000 outside Antarctica and Greenland with Alaska alone having several tens of thousands. So, what do glaciers tell us about climate? Remember that the behavior of glaciers depends on their glacial budget (Figure 14.7), which in turn depends on temperature and precipitation. It is true that not many of these 160,000 or so glaciers have been studied, but those that have reveal an alarming trend: Many are retreating and thinning and in some cases have nearly or completely disappeared.

In Africa, the area covered by glaciers on Mount Kilimanjaro has decreased by about 85%, and a similar trend is seen on Mount Kenya. Glaciers here, and

▌ **Figure 1** Grinnel Glacier in Glacier National Park, Montana.

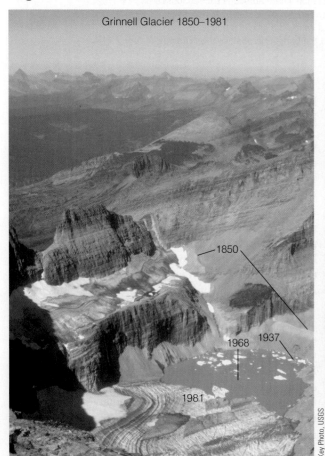

Grinnell Glacier 1850–1981

1850
1937
1968
1981

Key Photo, USGS

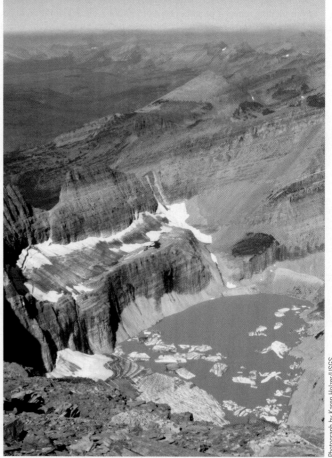

Photograph by Karen Holzer/USGS

a In 1850, at the end of the little Ice Age, the glacier extended much farther and covered about 576 acres. By 1981, its terminus had retreated to the position shown, and in 1993 it covered only 217 acres.

b By 2006, not much of Grinnel Glacier remained. Now, it is one of only about 25 left in the park, all of which are very small.

elsewhere, store water that they slowly release during the dry months and that makes up an important part of the water supply. Of the approximately 150 glaciers in Glacier National Park in Montana in 1850, only perhaps 25 remain, and many are such shrunken remnants that it is difficult to determine if they are still active (▌Figure 1). Indeed, the consensus is that all of them will be gone in a decade or two. The warming in Glacier National Park is slight, but fewer days per year have below freezing temperatures, and there are more warmer days.

The glaciers in Alaska at higher elevations are not showing any significant retreat, but those that flow into the sea, the tidewater glaciers, are almost all retreating and thinning. Many are much smaller than they were only a few decades ago. It is true that some glaciers in Alaska and elsewhere do surge and advance markedly, but these episodes usually do not persist for long (Figure 14.8b). It is also true that a few glaciers are "bucking the trend" and actually growing. A case in point are the glaciers on Mount Shasta in California, but these glaciers are very small (the largest measures 4.8 km long), and they, along with one glacier in Washington State, are the only ones in the continental United States that are growing.

Mount Shasta is one of the dozen or so large composite volcanoes in the Cascade Range of California, Oregon, Washington, and British Columbia, Canada. Many glaciers are present on the higher peaks in these mountains, but only Crater Glacier on Mount St. Helens has grown in recent years (▌Figure 2a). When Mount St. Helens erupted in May 1980, all of its 12 glaciers were destroyed or vastly diminished. By 1982, the lava dome in its crater had cooled enough for snow to accumulate yearly and transform into glacial ice that is now about 190 m thick. In any case, the ice is now thick enough to flow, so a glacier formed where none was present a mere 28 years ago.

Several factors account for the origin of Crater Glacier. First, the Cascade Range receives huge amounts of snowfall: We mentioned earlier in this chapter the annual snowfall in the Cascade Range is measured in the tens of meters. Second, the crater provides protection from the Sun for the accumulating snow and ice. And third, rockfalls from the crater walls help insulate the newly formed glacier. The story for the other Cascade Range volcanoes is not so comforting.

Glacier Peak in Washington, which last erupted in 1880, has more than a dozen glaciers, all of which are retreating.

In fact, the Whitechuck Glacier will soon be inactive (▌Figure 2b). During the Little Ice Age the Whitechuck Glacier had northern and southern branches that merged to form a single glacier. However, a rapid episode of retreat began in 1930, and now the glacier has shrunk and is only about 35 m thick. In fact, by 2002, the northern branch was gone, but the southern branch will probably persist for a few more decades.

The decade from 2000 to 2009 was the hottest on record, and 2010 showed no relief. Indeed, 2010 will likely be among the hottest years ever recorded. One consequence of higher temperatures is, of course, retreating glaciers as we have discussed. But remember that glaciers are one reservoir in the hydrologic cycle, so as they waste away, their water returns to the oceans and sea level rises. We know that during the Pleistocene Epoch (Ice Age) sea level was as much as 130 m lower than it is today, and we know that it has risen on average nearly 2 mm per year for the last 100 years; the current rate is estimated at about 3 mm per year. This increase may not sound like much, but even a slight rise in sea level will eventually be a real problem for many of Earth's coastal regions.

▌**Figure 2**
Glaciers in the Cascade Range in Washington State.

Matt Logan/USGS

a View of a lava dome and the newly formed glacier in the crater of Mount St. Helens on April 19, 2005. Notice the ash on the surface of the glacier, which also shows large crevasses.

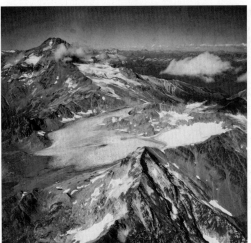

© Bettman/CORBIS

b Whitechuck Glacier on Glacier Peak in Washington State. The south branch of the glacier on the right has a small accumulation area, but the north branch has none.

a Flow velocity in a valley glacier varies horizontally and vertically. Velocity is greatest at the top center because friction with the walls and floor of the trough slows the flow adjacent to these boundaries. The lengths of the arrows are proportional to velocity.

❚ Figure 14.8 Flow Velocity in Valley Glaciers

b The terminus of the Lowell Glacier in Kluane National Park in the Yukon Territory of Canada. The glacier was surging when this image was taken on July 2, 2010. Its surge began probably in October 2009 when its terminus was more than 1.5 km farther up the valley.

>> Critical Thinking Question
How do you think the debris in the foreground was deposited?

into a maze of crevasses and its terminus advances noticeably. These brief episodes are best known in valley glaciers, but they also take place in ice caps and probably even in continental glaciers. In 1995, a huge ice shelf in Antarctica broke apart, and several ice streams from the Antarctic ice sheet surged toward the ocean.

During a surge, a glacier's terminus may advance several tens of meters per day for weeks or months and then return to its previous flow rate. Not many glaciers surge, and none does so in the United States outside Alaska. Even in Canada, they are found only in the Yukon Territory and the Queen Elizabeth Islands. Lowell Glacier in Kluane National Park in the Yukon Territory began surging probably in late 2009, and by May 2010 its terminus had advanced 1.5 km (❚ Figure 14.8b). The fastest surge ever recorded was in 1953 in the Kutiah Glacier in Pakistan; the glacier's terminus advanced 12 km in three months, or 130 m per day on average.

The onset of a glacial surge is heralded by a thickened bulge in the upper part of a glacier that begins to move toward the terminus at several times the glacier's normal velocity. When the bulge reaches the terminus, it causes rapid movement and displacement of the terminus by as much as 20 km. Surges are also probably related to accelerated rates of basal slip rather than more rapid plastic flow. One theory holds that thickening in the zone of accumulation with concurrent thinning in the zone of wastage increases the glacier's slope and accounts for accelerated flow. According to another theory, pressure on soft sediment beneath a glacier squeezes fluids through the sediment, thereby allowing the overlying glacier to slide more effectively.

What Would You Do ❓

After carefully observing the same valley glacier for several years, you conclude (1) that its terminus has retreated at least 1 km and (2) that debris on its surface has obviously moved several hundreds of meters toward the glacier's terminus. Can you think of an explanation for these observations? Also, why do you think studying glaciers might have some implications for the debate on global warming?

14.5 Erosion and Sediment Transport by Glaciers

As moving solids, glaciers erode, transport, and eventually deposit huge quantities of sediment and soil. Indeed, they have the capacity to transport any size of sediment, including boulders the size of houses as well as clay-sized particles. Important processes of erosion include bulldozing, plucking, and abrasion.

Although *bulldozing* is not a formal geologic term, it is fairly self-explanatory; glaciers shove or push unconsolidated materials in their paths. This effective process was aptly described in 1744 during the Little Ice Age by an observer in Norway:

When at times [the glacier] pushes forward a great sound is heard, like that of an organ and it pushes in front of it unmeasurable masses of soil, grit and rocks bigger than any house could be, which it then crushes small like sand.*

Plucking, also called *quarrying*, results when glacial ice freezes in the cracks and crevices of a bedrock projection and eventually pulls it loose. One manifestation of plucking is a landform called a *roche moutonnée*, a French term for "rock sheep." As shown in ▮ Figure 14.9, a glacier smoothes the "upstream" side of an obstacle, such as a small hill, and plucks pieces of rock from the

---

*Quoted in C. Officer and J. Page, *Tales of the Earth* (New York: Oxford University Press, 1993), p. 99.

▮ **Figure 14.9 Roche Moutonée** A French term meaning "rock sheep," a roche moutonée is a bedrock projection that was shaped by glacial abrasion and plucking.

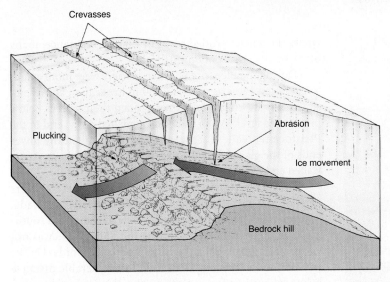

a Origin of a roche moutonée. A glacier abrades and polishes the "upstream" side of a bedrock projection and shapes its "downstream" side by plucking.

b A roche moutonée in Montana. The glacier that formed this feature moved from right to left.

"downstream" side by repeatedly freezing to and pulling away from the obstacle.

Bedrock over which sediment-laden glacial ice moves is effectively eroded by **abrasion** and develops a **glacial polish,** a smooth surface that glistens in reflected light (▮ Figure 14.10a). Abrasion also yields **glacial striations,** rather straight scratches rarely more than a few millimeters deep on rock surfaces. Abrasion thoroughly pulverizes rocks, yielding an aggregate of clay- and silt-sized particles that have the consistency of flour—hence, the name *rock flour*. Rock flour is so common in streams discharging from glaciers that the water has a milky appearance (▮ Figure 14.10b).

Continental glaciers derive sediment from mountains projecting through them, and wind-blown dust settles on their surfaces, but most of their sediment comes from the surface they move over. As a result, most sediment is transported in the lower part of the ice sheet. In contrast, valley glaciers carry sediment in all parts of the ice, but it is concentrated at the base and along the margins (▮ Figure 14.11). Some of the marginal sediment is derived by abrasion and plucking, but much of it is supplied by mass-wasting processes, as when soil, sediment, or rock falls or slides onto the glacier's surface.

Erosion by Valley Glaciers

When mountain ranges are eroded by valley glaciers, they take on a unique appearance of angular ridges and peaks in the midst of broad, smooth valleys with near-vertical walls. The erosional landforms produced by valley glaciers include U-shaped glacial troughs, hanging valleys, cirques, arêtes, and horns, most of which are easily recognized.

U-Shaped Glacial Troughs
A **U-shaped glacial trough** is one of the most distinctive features of valley glaciation. Mountain valleys eroded by running water are typically V-shaped in cross section; that is, they have valley walls that descend to a narrow valley bottom (▮ Figure 14.12a). In contrast, valleys scoured by glaciers are deepened, widened, and straightened so that they have very steep or vertical walls, but broad, rather flat valley floors; thus they exhibit a U-shaped profile (▮ Figures 14.12c, 14.13a).

a Glacial polish and striations, the straight scratches, on basalt at Devils Postpile National Monument in California.

>> **Critical Thinking Question**
Why is the basalt in this image broken into 5- and 6-sided polygons?

▌ **Figure 14.10** Glacial Striations, Polish, and Rock Flour

b The water in this stream in Switzerland is discolored by rock flour, small particles produced by glacial abrasion.

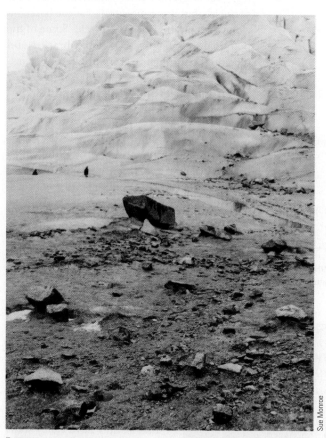

▌ **Figure 14.11** Sediment Transport by Valley Glaciers Debris on the surface of the Mendenhall Glacier in Alaska. The largest boulder is about 2 m across. Notice the icefall in the background. The person left of center provides scale.

Many glacial troughs contain triangular-shaped *truncated spurs,* which are cutoff or truncated ridges that extend into the preglacial valley (Figure 14.12c). Another common feature is a series of steps or rock basins in the valley floor where the glacier eroded rocks of varying resistance; many of the basins now contain small lakes.

During the Pleistocene, when glaciers were more extensive, sea level was as much as 130 m lower than at present, so glaciers flowing into the sea eroded their valleys below present sea level. When the glaciers melted at the end of the Pleistocene, sea level rose, and the ocean filled the lower ends of the glacial troughs so that now they are long, steep-walled embayments called **fiords** (▌Figure 14.13b).

Fiords are restricted to high latitudes where glaciers exist at low elevations, such as Alaska, western Canada, Scandinavia, Greenland, southern New Zealand, and southern Chile. Lower sea level during the Pleistocene was not entirely responsible for the formation of all fiords. Unlike running water, glaciers can erode a considerable distance below sea level. In fact, a glacier 500 m thick can stay in contact with the seafloor and effectively erode it to a depth of about 450 m before the buoyant effects of water cause the glacial ice to float! The depth of some fiords is impressive; some in Norway and southern Chile are about 1,300 m deep.

Hanging Valleys Some of the world's highest and most spectacular waterfalls are found in recently glaciated areas. Bridalveil Falls in Yosemite National Park, California, plunge 188 m from a **hanging valley,** which is a tributary valley whose floor is at a higher level than that of the main valley. Where the two valleys meet, the mouth of the hanging valley is perched far above the main valley's floor (Figures 14.12c, ▌14.14). Accordingly, streams flowing through hanging valleys plunge over vertical or steep precipices.

Although not all hanging valleys form by glacial erosion, many do. As Figure 14.12 shows, the large glacier in the main valley vigorously erodes, whereas the smaller glaciers in tributary valleys are less capable of erosion. When the glaciers disappear, the smaller tributary valleys remain as hanging valleys.

▌Figure 14.12 Erosional Landforms Produced by Valley Glaciers

a A mountain area before glaciation.

b The same area during the maximum extent of valley glaciers.

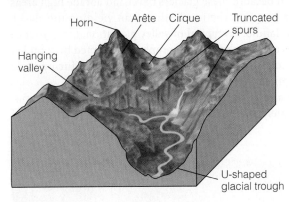

Horn   Arête   Cirque   Truncated spurs

Hanging valley

U-shaped glacial trough

c After glaciation.

Cirques, Arêtes, and Horns Perhaps the most spectacular erosional landforms in areas of valley glaciation are at the upper ends of glacial troughs and along the divides that separate adjacent glacial troughs. Valley glaciers form and move out from steep-walled, bowl-shaped depressions called **cirques** at the upper end of their troughs (▌Figure 14.15). Cirques are typically steep-walled on three sides, but one side is open and leads into the glacial trough. Some cirques slope continuously into the glacial trough, but many have a lip or threshold at their lower end.

The details of cirque origin are not fully understood, but they probably form by erosion of a preexisting depression on a mountainside. As snow and ice accumulate in the depression, frost wedging and plucking, combined with glacial erosion, enlarge and transform the head of a steep

▌Figure 14.13 U-Shaped Glacial Trough and Fiord

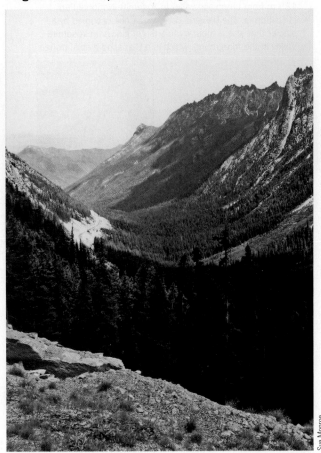

Sue Monroe

a This U-shaped glacial trough is in the Cascade Range in Washington State.

James S. Monroe

b This steep-walled U-shaped glacial trough in Norway extends below sea level, so it is a fiord.

》Critical Thinking Question
How does a U-shaped glacial trough differ from a mountain valley eroded by running water?

mountain valley into a typical amphitheater-shaped cirque. Tension in the upper part of the glacier may reduce the erosive power of the ice on the immediate, downslope side of a cirque, leaving a lip or threshold in the valley floor after the

▌Figure 14.14 Hanging Valley Bridalveil Falls plunge 188 m from a hanging valley in Yosemite National Park in California. The hanging valley was once occupied by a glacier that was a tributary to a larger glacier in Yosemite Valley in the foreground, which is a U-shaped glacial trough.

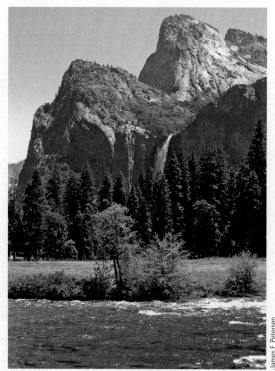

ice melts away. Small lakes of meltwater, called *tarns,* often form on the floors of cirques behind such thresholds.

Cirques become wider and are cut deeper into mountainsides by headward erosion as a result of abrasion,

plucking, and several mass-wasting processes. For example, part of a steep cirque headwall may collapse while frost wedging continues to pry loose rocks that tumble downslope, so a combination of processes erode a small mountainside depression into a large cirque.

Arêtes—narrow, serrated ridges—form in two ways. In many cases, cirques form on opposite sides of a ridge, and headward erosion reduces the ridge until only a thin partition of rock remains (Figure 14.15a). The same effect occurs when erosion in two parallel glacial troughs reduces the intervening ridge to a thin spine of rock.

The most majestic of all mountain peaks are **horns,** steep-walled, pyramidal peaks formed by headward erosion of cirques. For a horn to form, a mountain peak must have at least three cirques on its flanks, all of which erode headward (Figure 14.12). Excellent examples of horns are Mount Assiniboine in the Canadian Rockies, the Grand Teton in Wyoming, and the most famous of all, the Matterhorn in Switzerland (▌Figure 14.16).

Continental Glaciers and Erosional Landforms

Areas eroded by continental glaciers tend to be smooth and rounded because these glaciers bevel and abrade high areas that project into the ice. Rather than yielding the sharp, angular landforms typical of valley glaciation, they produce a landscape of subdued topography interrupted by rounded hills because they bury landscapes entirely during their development.

In a large part of Canada, particularly the vast Canadian Shield region, continental glaciers have stripped off

▌Figure 14.15 Cirques and Arêtes

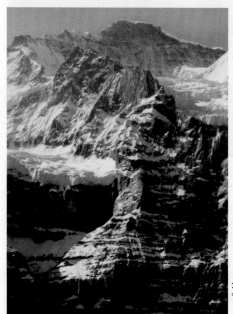

a This view in Switzerland shows two small valley glaciers, the headwall of a cirque, and an arête.

b A cirque in the St. Elias Mountains of the Yukon Territory of Canada.

Figure 14.16 The Matterhorn in Switzerland Is a Well-Known Horn Horns are pyramid-shaped peaks eroded by three or more glaciers on a single mountain.

Figure 14.17 Ice-Scoured Plain This ice-scoured plain in the Northwest Territories of Canada shows little relief, numerous lakes, little or no soil, and extensive bedrock exposures, which is typical of areas eroded by continental glaciers.

the soil and unconsolidated surface sediment, revealing extensive exposures of striated and polished bedrock. In addition to exposed bedrock, these areas, called **ice-scoured plains,** have deranged drainage (Figure 14.17), numerous lakes and swamps, low relief, and little or no soil. Similar though smaller bedrock exposures are also found in the northern United States from Maine through Minnesota.

14.6 Glacial Deposits

Given that glaciers erode and transport gravel, sand, and mud, it follows that they must also deposit sediment. All glacial deposits go by the general term **glacial drift,** but geologists recognize two types of drift—till and stratified drift. **Till** is made up of any sediment deposited directly by glacial ice, as, for example, at the terminus of a glacier. Till is not sorted by particle size, and it shows no stratification. The till of both valley and continental glaciers is similar, but that of continental glaciers is much more extensive and usually has been transported much farther. As opposed to till, **stratified drift,** as the name implies, is layered or stratified, and invariably it shows some degree of sorting by particle size. As a matter of fact, most stratified drift is layers of gravel and sand or mixtures thereof deposited in braided stream channels that discharge from melting glaciers.

The appearance of till and stratified drift may not be as inspiring as some landforms resulting from glacial erosion, but they are important groundwater reservoirs, and they are exploited for their sand and gravel. In fact, sand and gravel, mostly for construction, are the most valuable mineral commodities in many areas. It is true that all glaciers deposit sediment, but as you would expect, the deposits of continental glaciers are far more extensive and more varied (Figure 14.18). In contrast, the deposits of a valley

glacier tend to be restricted to the lower parts of the valley occupied by the glacier.

One conspicuous aspect of glacial drift is rock fragments of various sizes that were obviously not derived from the underlying bedrock. These fragments, called **glacial erratics,** came from some distant source and were transported and then deposited in their present location (Figure 14.19). Some glacial erratics are gigantic—the Madison Boulder in New Hampshire and Daggett Rock in Maine weigh about 4,550 and 7,270 metric tons, respectively.

Landforms Composed of Till

Remember, till is deposited directly by glacial ice, and it shows no sorting or stratification. Till is found in several types of deposits collectively called moraines and in elongated hills known as drumlins.

End Moraines The terminus of any glacier may become stabilized in one position for some period of time, perhaps a few years or even decades. Stabilization of the ice front does not mean that the glacier has ceased to flow, only that it has a balanced budget. When an ice front is stationary, flow within the glacier continues, and any sediment transported within or upon the ice is dumped as a pile of rubble at the glacier's terminus (Figure 14.20). These deposits are **end moraines,** which continue to grow as long as the ice front remains stationary. End moraines of valley glaciers are crescent-shaped ridges of till spanning the valley occupied by the glacier. Those of continental glaciers similarly parallel the ice front but are much more extensive.

Following a period of stabilization, a glacier may advance or retreat, depending on changes in its budget. If it advances, the ice front overrides and modifies its former moraine. If it has a negative budget, though, the ice front retreats toward the zone of accumulation. As the ice front recedes, till is deposited as it is liberated from the melting ice and forms a layer of **ground moraine.** Ground moraine

Figure 14.18 Development of Features Associated with Past Continental Glaciation

Terminal moraine

Emerging subglacial stream

Sediment accumulations in low spots in ice

RETREATING ICE FRONT

a This retreating continental glacier once covered a larger area as indicated by the terminal moraines.

Outwash plain

Buried and isolated ice blocks

Terminal moraine

Recessional moraine

Kames

Esker

Drumlins

ICE GONE

b Moraines, eskers, and drumlins form during glaciation, though eskers and drumlins originate under the ice cover. Kames and kettles develop at the end of glaciation.

Kettle ponds

Abandoned outflow channel

Figure 14.19 Glacial Erratics

a A glacial erratic in the making. This boulder on the Mendenhall Glacier in Alaska will be deposited far from its source.

b Glacial erratic that was deposited by the Exit Glacier in Kenai Fjords National Park near Seward, Alaska.

Figure 14.20 End Moraine End moraines are accumulations of unsorted, nonstratified sediment at the end of a glacier.

a End moraine of the Salmon Glacier in the Coast Mountains of British Columbia, Canada.

b This ridge of gravel is a small end moraine was deposited in 1995 by the Exit Glacier in Kenai Fjords National Park, near Seward, Alaska.

has an irregular, rolling topography, whereas end moraine consists of long, ridgelike accumulations of sediment.

After a glacier has retreated for some time, its terminus may once again stabilize, and it deposits another end moraine. Because the ice front has receded, such moraines are called **recessional moraines** (Figure 14.18b). During the Pleistocene Epoch, continental glaciers in the mid-continent region extended as far south as southern Ohio, Indiana, and Illinois. Their outermost end moraines, marking the greatest extent of the glaciers, go by the special name **terminal moraine** (valley glaciers also deposit terminal moraines). As the glaciers retreated from the positions where their terminal moraines were deposited, they temporarily stopped retreating numerous times and deposited dozens of recessional moraines.

> **CONNECTION LINK**

For more information about Pleistocene glaciation, see pages 626–631 in Chapter 23.

Lateral and Medial Moraines Valley glaciers transport considerable sediment along their margins, much of it abraded and plucked from the valley walls, but a significant amount falls or slides onto the glacier's surface by mass-wasting processes. In any case, this sediment is transported and deposited as long ridges of till called **lateral moraines** along the margin of the glacier (**I** Figure 14.21).

Where two lateral moraines merge, as when a tributary glacier flows into a larger glacier, a **medial moraine** forms (Figure 14.21). A large glacier will often have several dark stripes of sediment on its surface, each of which is a medial moraine. One can determine how many tributaries a valley glacier has by the number of its medial moraines.

Drumlins In many areas where continental glaciers deposited till, the till has been reshaped into elongated hills known as **drumlins** (Figure 14.18b). Some drumlins are as large as 50 m high and 1 km long, but most are much smaller. From the side, a drumlin looks like an inverted spoon, with the steep end on the side from which the glacial ice advanced and the gently sloping end pointing in the direction of ice movement. Drumlins are rarely found as single, isolated hills; instead, they occur in *drumlin fields* that contain hundreds or thousands of drumlins. Drumlin fields are found in several states and Ontario, Canada, but perhaps the finest example is near Palmyra, New York.

Figure 14.21 Types of Moraines A moraine is a mound or ridge of unstratified till. Moraines are called end, lateral, or medial depending on their position. End moraines are deposited at a glacier's terminus (Figure 14.20), but a lateral moraine is found along the margin of a glacier, and a medial moraine, which forms where two lateral moraines merge, is on the more central part of a glacier.

Lateral moraine

Medial moraine

>> Critical Thinking Question
In addition to moraines, what other features of glacial erosion can you identify?

James S. Monroe

▌ **Figure 14.22 Valley Train and Kettles** This deposit of sand and gravel is the valley train deposited by meltwaters from the Salmon Glacier in the Coast Mountains of British Columbia, Canada. Also, notice the circular, water-filled depressions; these are kettle lakes, or simply kettles if not filled with water. Notice in Figure 14.20a, which shows the terminus of the Salmon Glacier, that several blocks of ice are isolated from the glacier. Should these be partially or wholly buried in valley train deposits, they, too, will form kettles.

According to one hypothesis, drumlins form when till beneath a glacier is reshaped into streamlined hills as the ice moves over it by plastic flow. Another hypothesis holds that huge floods of glacial meltwater modify till into drumlins.

Landforms Composed of Stratified Drift

Stratified drift is a glacial deposit that exhibits sorting and layering, both indications that it was deposited by running water. Stratified drift is deposited by streams discharging from both valley and continental glaciers, but as you would expect, it is more extensive in areas of continental glaciation.

Outwash Plains and Valley Trains Glaciers discharge meltwater laden with sediment most of the time, except perhaps during the coldest months. This meltwater forms a series of braided streams that radiate out from the front of continental glaciers over a wide region. So much sediment is supplied to these streams that much of it is deposited within their channels as sand and gravel bars. The vast blanket of sediment so formed is an **outwash plain** (Figure 14.18).

CONNECTION LINK ◄

For more on braided streams and their deposits see pages 293 and 294 in Chapter 12.

Valley glaciers also discharge large amounts of meltwater and, like continental glaciers, have braided streams extending from them. However, these streams are confined to the lower parts of glacial troughs, and their long, narrow deposits of stratified drift are known as **valley trains** (▌ Figure 14.22).

Outwash plains, valley trains, and some moraines commonly contain numerous circular to oval depressions, many of which contain small lakes. These depressions are *kettles* that form when a retreating glacier leaves a block of ice that is subsequently partly or wholly buried (Figure 14.22). When the ice block eventually melts, it leaves a depression; if the depression extends below the water table, it becomes the site of a small lake. Some outwash plains have so many kettles that they are called *pitted outwash plains.*

Kames and Eskers **Kames** are conical hills of stratified drift up to 50 m high (Figure 14.18 and ▌ Figure 14.23a). Many form when a stream deposits sediment in a depression on a glacier's surface; as the ice melts, the deposit is lowered to the land surface. Kames also form in cavities within or beneath stagnant ice.

Long sinuous ridges of stratified drift, many of which meander and have tributaries, are **eskers** (Figures 14.18 and ▌ 14.23b). Most eskers have sharp crests and sides that slope at about 30 degrees. Some are as high as 100 m and can be traced for more than 500 km. The sorting and stratification of the sediments in eskers clearly indicate deposition by running water. The features of ancient eskers and observations of present-day glaciers show that they form in tunnels beneath stagnant ice. Excellent examples of eskers can be seen at Kettle Moraine State Park in Wisconsin and in several other states, but the most extensive eskers in the world are in northern Canada.

▌ **Figure 14.23 Kames and Eskers**

Courtesy of B.M.C. Pape

a This small hill in Wisconsin is a kame.

William W. Shilts/Illinois State Geological Survey. Photograph taken by Christian Zdanovich

b This esker was exposed in 1992 from a melting glacier on Bylot Island, Nunavut, Canada.

Deposits in Glacial Lakes

Some lakes in areas of glaciation formed as a result of glaciers scouring out depressions; others occur where a stream's drainage was blocked; and others are the result of water accumulating behind moraines or in kettles. Regardless of how they formed, glacial lakes, like all lakes, are areas of deposition. Sediment may be carried into them and deposited as small deltas, but of special interest are the fine-grained deposits. Mud deposits in glacial lakes are commonly finely laminated (having layers less than 1 cm thick) and consist of alternating light and dark layers known as *varves* (▌ Figure 14.24a), which represents an annual episode of deposition. The light layer formed during the spring and summer and consists of silt and clay; the dark layer formed during the winter when the smallest particles of clay and organic matter settled from suspension as the lake froze over. The number of varves indicates how many years a glacial lake has existed.

Another distinctive feature of glacial lakes with varves is *dropstones* (▌ Figure 14.24b). These are pieces of gravel, some of boulder size, in otherwise very fine-grained deposits. The presence of varves indicates that currents and turbulence in these lakes were minimal; otherwise, clay and organic matter would not have settled from suspension. How then can we account for dropstones in a low-energy environment? Most of them were probably carried into the lakes by icebergs that eventually melted and released sediment contained in the ice.

▌ Figure 14.24 Varves and a Dropstone in Glacial Deposits

a Each of these pairs of dark and light layers makes up a varve, an annual deposit in a glacial lake.

P. Judson-Rhodes

b These varves have a dropstone that was probably liberated from floating ice.

14.7 What Causes Ice Ages?

We discussed the conditions necessary for a glacier to form earlier in this chapter: More snow falls than melts during the warm season, thus accounting for a net accumulation of snow and ice over the years. But this really does not address the broader question of what causes ice ages—that is, times of much more extensive glaciation. Actually, we need to address not only what causes ice ages but also why there have been so few episodes of widespread glaciation in all of Earth history. Only during the Late Proterozoic Eon, the Late Ordovician, Carboniferous, and Permian periods, and the Pleistocene Epoch has Earth had glaciers on a grand scale. Additionally, widespread glaciation occurred at least four times during the Pleistocene, with each glacial episode separated by a long *interglacial stage* during which glaciers were restricted in their distribution.

For more than a century, scientists have attempted to develop a comprehensive theory explaining all aspects of ice ages, but they have not yet been completely successful. One reason for their lack of success is that the climatic changes responsible for glaciation, the cyclic occurrence of glacial–interglacial episodes, and short-term events such as the Little Ice Age operate on vastly different time scales.

Only a few periods of glaciation are recognized in the geologic record, each separated from the others by long intervals of mild climate. Such long-term climatic changes probably result from slow geographic changes related to plate tectonic activity. Moving plates carry continents to high latitudes where glaciers exist, provided they receive enough precipitation as snow. Plate collisions, the subsequent uplift of vast areas far above sea level, and the changing atmospheric and oceanic circulation patterns caused by the changing shapes and positions of plates also contribute to long-term climate change.

The Milankovitch Theory

Changes in Earth's orbit as a cause of intermediate-term climatic events were first proposed during the mid-1800s, but the idea was made popular during the 1920s by the Serbian astronomer Milutin Milankovitch. He proposed that minor irregularities in Earth's rotation and orbit are sufficient to alter the amount of solar radiation received at any given latitude and hence bring about climate changes. Now called the **Milankovitch theory,** the proposal was initially ignored, but it has received renewed interest since the 1970s and is now widely accepted.

Milankovitch attributed the onset of glacial episodes to variations in three aspects of Earth's orbit. The first is *orbital eccentricity,* which is the degree to which Earth's orbit around the Sun changes over time (▌ Figure 14.25a). When the orbit

> **CONNECTION LINK**

To learn more about Proterozoic and Paleozoic glaciation see pages 488 and 489 in Chapter 19 and pages 506 and 509 in Chapter 20.

is nearly circular, both the Northern and Southern Hemispheres have similar contrasts between the seasons. However, if the orbit is more elliptic, hot summers and cold winters will occur in one hemisphere, whereas warm summers and cool winters will take place in the other hemisphere. Calculations indicate a roughly 100,000-year cycle between times of maximum eccentricity, which corresponds closely to the 20 warm–cold climatic cycles that took place during the Pleistocene.

Milankovitch also pointed out that the angle between Earth's axis and a line perpendicular to the plane of Earth's orbit shifts about 1.5 degrees from its current value of 23.5 degrees during a 41,000-year cycle (▌Figure 14.25b). Although changes in *axial tilt* have little effect on equatorial latitudes, they strongly affect the amount of solar radiation received at high latitudes and the duration of the dark period at and near Earth's poles. Coupled with the third aspect of Earth's orbit, precession of the equinoxes, high latitudes might receive as much as 15% less solar radiation, certainly enough to affect glacial growth and melting.

Precession of the equinoxes, the last aspect of Earth's orbit that Milankovitch cited, refers to a change in the time of the equinoxes. At present, the equinoxes take place on or about March 21 and September 21 when the Sun is directly over the equator. But as Earth rotates on its axis, it also wobbles as its axial tilt varies 1.5 degrees from its current value, thus changing the time of the equinoxes. Taken alone, the time of the equinoxes has little climatic effect, but changes in Earth's axial tilt also change the times of *aphelion* and *perihelion,* which are, respectively, when Earth is farthest from and closest to the Sun during its orbit (▌Figure 14.25c). Earth is now at perihelion, closest to the Sun, during Northern Hemisphere winters, but in about 11,000 years perihelion will be in July. Accordingly, Earth will be at aphelion, farthest from the Sun, in January and have colder winters.

Continuous variations in Earth's orbit and axial tilt cause the amount of solar heat received at any latitude to vary slightly through time. The total heat received by the planet changes little, but according to Milankovitch, and now many scientists agree, these changes cause complex climatic variations and provided the triggering mechanism for the glacial–interglacial episodes of the Pleistocene.

Short-Term Climatic Events

Climatic events with durations of several centuries, such as the Little Ice Age, are too short to be accounted for by plate tectonics or Milankovitch cycles. Several hypotheses have been proposed, including variations in solar energy and volcanism.

Variations in solar energy could result from changes within the Sun itself or from anything that would reduce the amount of energy Earth receives from the Sun. The latter could result from the solar system passing through clouds of interstellar dust and gas or from substances in the atmosphere reflecting solar radiation back into space. Records kept over the past 90 years indicate that during this time the amount of solar radiation has varied only slightly. Although

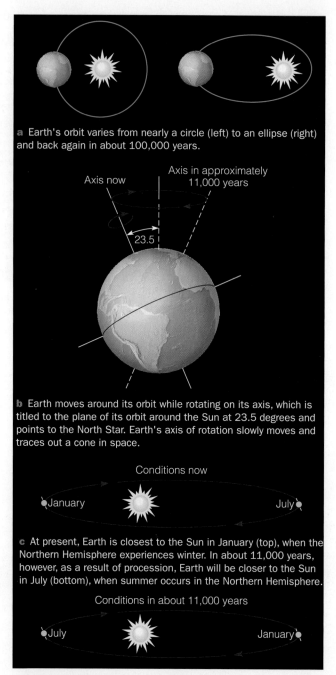

a Earth's orbit varies from nearly a circle (left) to an ellipse (right) and back again in about 100,000 years.

b Earth moves around its orbit while rotating on its axis, which is titled to the plane of its orbit around the Sun at 23.5 degrees and points to the North Star. Earth's axis of rotation slowly moves and traces out a cone in space.

c At present, Earth is closest to the Sun in January (top), when the Northern Hemisphere experiences winter. In about 11,000 years, however, as a result of procession, Earth will be closer to the Sun in July (bottom), when summer occurs in the Northern Hemisphere.

▌**Figure 14.25 The Milankovich Theory** According to the Milankovich theory, minor irregularities in Earth's rotation and orbit may affect climatic changes.

variations in solar energy may influence short-term climatic events, such a correlation has not been demonstrated.

During large volcanic eruptions, tremendous amounts of ash and gases are spewed into the atmosphere, where they reflect incoming solar radiation and thus reduce atmospheric temperatures. Small droplets of sulfur gases remain in the atmosphere for years and can have a significant effect on climate. Several large-scale volcanic events have occurred, such as the 1815 eruption of Tambora, and are known to have had climatic effects. However, no relationship between periods of volcanic activity and periods of glaciation has yet been established.

GEO-RECAP

» Geo-Image

Photograph by Bruce F. Molnia, USGS

Figure 1 The Chugach Mountains in Alaska. Identify the landforms that resulted from valley glaciation.

James S. Monroe

Figure 2 What are the features indicated by the arrows in these Greenland glaciers? Also, explain how they form.

Chapter Summary

- Glaciers currently cover about 10% of the land surface and contain about 2.15% of all water on Earth.

- A glacier forms when winter snowfall exceeds summer melt and accumulates year after year. Snow is compacted and converted to glacial ice, and when the ice is about 40 m thick, pressure causes it to flow.

- Glaciers move by plastic flow and basal slip.

- Valley glaciers are confined to mountain valleys and flow from higher to lower elevations, whereas continental glaciers cover vast areas and flow outward in all directions from a zone of accumulation.

- The behavior of a glacier depends on its budget, which is the relationship between accumulation and wastage. If a glacier has a balanced budget, its terminus remains stationary; a positive or negative budget results in the advance or retreat of the terminus, respectively.

- Glaciers move at varying rates depending on slope, discharge, and season. Valley glaciers tend to flow more rapidly than continental glaciers.

- Glaciers effectively erode and transport because they are solids in motion. They are particularly effective at eroding soil and unconsolidated sediment, and they can transport any size sediment supplied to them.

- Continental glaciers transport most of their sediment in the lower part of the ice, whereas valley glaciers may carry sediment in all parts of the ice.

- Erosion of mountains by valley glaciers yields several sharp, angular landforms including cirques, arêtes, and horns. U-shaped glacial troughs, fiords, and hanging valleys are also products of valley glaciation.

- Continental glaciers abrade and bevel high areas, producing a smooth, rounded landscape known as an ice-scoured plain.

- Depositional landforms include moraines, which are ridgelike accumulations of till. The several types of moraines are terminal, recessional, lateral, and medial.

- Drumlins are composed of till that was apparently reshaped into streamlined hills by continental glaciers or floods.

- Stratified drift in outwash plains and valley trains consists of sand and gravel deposited by meltwater streams issuing from glaciers. Ridges known as eskers, and conical hills called kames are also composed of stratified drift.

- Major glacial intervals separated by tens or hundreds of millions of years probably occur as a result of the changing positions of tectonic plates, which in turn cause changes in oceanic and atmospheric circulation patterns.

- Currently, the Milankovitch theory is widely accepted as the explanation for glacial–interglacial intervals.

- The reasons for short-term climatic changes, such as the Little Ice Age, are not understood. Two proposed causes are changes in the amount of solar energy received by Earth and volcanism.

Important Terms

abrasion (p. 351)
arête (p. 354)
basal slip (p. 345)
cirque (p. 353)
continental glacier (p. 341)
drumlin (p. 357)
end moraine (p. 355)
esker (p. 358)
fiord (p. 352)
firn (p. 345)
glacial budget (p. 345)
glacial drift (p. 355)
glacial erratic (p. 355)
glacial ice (p. 345)

glacial polish (p. 351)
glacial striations (p. 351)
glacial surge (p. 347)
glaciation (p. 344)
glacier (p. 340)
ground moraine (p. 355)
hanging valley (p. 352)
horn (p. 354)
ice cap (p. 344)
ice-scoured plain (p. 355)
kame (p. 358)
lateral moraine (p. 357)
Little Ice Age (p. 340)
medial moraine (p. 357)

Milankovitch theory (p. 359)
outwash plain (p. 358)
plastic flow (p. 345)
recessional moraine (p. 357)
stratified drift (p. 355)
terminal moraine (p. 357)
till (p. 355)
U-shaped glacial trough (p. 351)
valley glacier (p. 340)
valley train (p. 358)
zone of accumulation (p. 346)
zone of wastage (p. 346)

Review Questions

1. Fine-grained, alternating dark and light layers in glacial lakes are
   a. ____ eskers.
   b. ____ varves.
   c. ____ erratics.
   d. ____ horns.
   e. ____ moraines.

2. One of the most distinctive features of an area eroded by a valley glacier is a/an
   a. ____ ice-scoured plain.
   b. ____ valley train.
   c. ____ cirque.
   d. ____ drumlin.
   e. ____ firn.

3. Glacial erratics are
   a. ____ sinuous ridges of stratified drift.
   b. ____ tear-drop shaped hills composed of till.
   c. ____ the deposits laid down by braided streams issuing from continental glaciers.
   d. ____ the features formed where two medial moraines merge.
   e. ____ boulders deposited far from their source by a glacier.

4. A time of glacial advance from about 1500 until the mid- to late-1800s is known as the
   a. ____ Holocene Minimum.
   b. ____ Pleistocene frigid period.
   c. ____ 300 years without a summer.
   d. ____ Little Ice Age.
   e. ____ Medieval cold snap.

5. When freshly fallen snow compacts and partly melts and refreezes, it forms granular ice known as
   a. ____ firn.
   b. ____ till.
   c. ____ outwash.
   d. ____ kame.
   e. ____ glacial frost.

6. A medial moraine forms where
   a. ____ snow is converted to glacial ice.
   b. ____ a valley glacier erodes an arête.
   c. ____ two lateral moraines merge.
   d. ____ isolated blocks of ice are buried in outwash.
   e. ____ a continental glacier abrades its underlying surface.

7. The only two areas where continental glaciers are present today are
   a. ____ Canada and Alaska.
   b. ____ Montana and Wyoming.
   c. ____ Greenland and Antarctica.
   d. ____ Scandinavia and Siberia.
   e. ____ Iceland and Asia.

8. Glaciers move mostly by
   a. ____ elastic rebound.
   b. ____ plastic flow.
   c. ____ lateral compression.
   d. ____ abrasion.
   e. ____ basal surging.

9. Which one of the following features results from erosion by valley glaciers?
   a. ____ outwash plain.
   b. ____ terminal moraine.
   c. ____ valley train.
   d. ____ arête.
   e. ____ ice-scoured plain.

10. The line on a glacier that separates the zone of accumulation from the zone of wastage is the
    a. ____ drift promontory.
    b. ____ lateral moraine.
    c. ____ ice line.
    d. ____ firn limit.
    e. ____ snow elevation.

11. Explain in terms of the glacial budget how a once-active glacier becomes stagnant.

12. How do roche moutonneés and drumlins form? Can you tell the direction of ice flow from these features? Explain.

13. How is it that glaciers can erode below sea level but streams and rivers cannot?

14. What kinds of evidence would tell you that an ice-free mountain area was once glaciated?

15. How does glacial ice originate, and why do geologists consider it to be rock?

16. A valley glacier has a cross-sectional area of 400,000 m^2 and a flow velocity of 2 m per day. How long will it take for 1 km^3 of ice to move past a given point?

17. How do glaciers move?

18. How does the Milankovitch theory account for the onset of glacial episodes?

19. What are hanging valleys, and how do they form?

20. How do lateral, medial, and end moraines form?

A house surrounded by palm trees, nestled in an oasis in the Sahara Desert of Morocco. Large sand dunes can be seen in the background. Contrast this view with the fact that the Sahara was a fertile savannah more than 6,000 years ago, thus illustrating the importance of studying how global climate change affects the various Earth systems and subsystems.

CHAPTER 15

The Work of Wind and Deserts

OUTLINE

15.1 Introduction

15.2 Sediment Transport by Wind

15.3 Wind Erosion

15.4 Wind Deposits

15.5 Air-Pressure Belts and Global Wind Patterns

15.6 The Distribution of Deserts

15.7 Characteristics of Deserts

GEO-FOCUS: Windmills and Wind Power

15.8 Desert Landforms

GEO-RECAP

HAVE YOU EVER WONDERED?

- How wind transports and erodes sediment, thus modifying the landscape?
- How dunes form, and how they are classified?
- What loess is, and why it forms some of the world's most fertile soils?
- Why deserts are so hot and dry, and why they are distributed where they are in the world today and in the past?
- If running water is important in the formation of desert landforms?
- Why buttes and mesas rise pillar-like in deserts?
- What petroglyphs are, and why they are an important cultural resource?

faberfoto/shutterstock.com

15.1 Introduction

During the past several decades, deserts have been advancing across millions of acres of productive land, destroying rangelands, croplands, and even villages. Such expansion, estimated at 70,000 km² per year, has exacted a terrible toll in human suffering. Because of the relentless advance of deserts, hundreds of thousands of people have died of starvation or been forced to migrate as "environmental refugees" from their homelands to camps, where the majority are severely malnourished. This expansion of deserts into formerly productive lands is called **desertification** and is a major problem in many countries.

Most regions undergoing desertification lie along the margins of existing deserts where a delicately balanced ecosystem serves as a buffer between the desert on one side and a more humid environment on the other. These regions have limited potential to adjust to increasing environmental pressures from natural causes as well as human activity. Ordinarily, desert regions expand and contract gradually in response to natural processes such as climatic change, but much of recent desertification has been greatly accelerated by human activities.

In many areas, the natural vegetation has been cleared as crop cultivation has expanded into increasingly drier fringes to support growing populations. Because grasses are the dominant natural vegetation in most desert fringe areas, raising livestock is a common economic activity. However, increasing numbers of livestock in many areas have greatly exceeded the land's ability to support them. Consequently, the vegetation cover that protects the soil has diminished, causing the soil to crumble and be stripped away by wind and water, which results in increased desertification.

One particularly hard-hit area of desertification is the Sahel of Africa (a belt 300–1,100 km wide, lying south of the Sahara). Because drought is common in the Sahel, the region can support only a limited population of livestock and humans. Unfortunately, expanding human and animal populations and more intensive agriculture have increased the demands on the land. Plagued with periodic droughts, this region has suffered tremendously as crops have failed and livestock has overgrazed the natural vegetation, resulting in thousands of deaths, displaced people, and the encroachment of the Sahara.

The tragedy of the Sahel and prolonged droughts in other desert fringe areas remind us of the delicate equilibrium of ecosystems in such regions. Once the fragile soil cover has been removed by erosion, it takes centuries for new soil to form (see Chapter 6).

There are many important reasons to study deserts and the processes that are responsible for their formation. First, deserts cover large regions of Earth's surface. More than 40% of Australia is desert, and the Sahara occupies a vast part of northern Africa. Furthermore, many desert regions are experiencing increases in population growth, such as parts of Nevada and Arizona in the United States. Many of these places already have problems associated with increasing population and the strains it places on the environment, particularly the need for great amounts of groundwater.

CONNECTION LINK ◀

To review some of the problems associated with the increasing withdrawal of groundwater, see pages 323–326 in Chapter 13.

Furthermore, with the current debate about global warming, it is important to understand how desert processes operate and how global climate changes affect the various Earth systems and subsystems. Learning about the underlying causes of climate change by examining ancient desert regions may provide insight into the possible duration and severity of future climatic changes. This understanding can have important ramifications in decisions about whether burying nuclear waste in a desert, such as Yucca Mountain, Nevada, is as safe as some claim and is in our best interests as a society.

As an example, more than 6,000 years ago, the Sahara was a fertile savannah supporting diverse fauna and flora, including humans. Then the climate changed, and the area became a desert. How did this happen? Will this region change back again in the future? These are some of the questions geoscientists hope to answer by studying deserts.

15.2 Sediment Transport by Wind

Wind is a turbulent fluid and therefore transports sediment in much the same way as running water. Although wind typically flows at a greater velocity than water, it has a lower density and thus can carry only clay- and silt-sized particles as *suspended load*. Sand and larger particles are moved along the ground as *bed load*.

Bed Load

Sediments that are too large or heavy to be carried in suspension by water or wind are moved as bed load either by *saltation* or by rolling and sliding. Saltation on land occurs when wind starts sand grains rolling and lifts and carries some grains short distances before they fall back to the surface. As the descending sand grains hit the surface, they strike other grains, causing them to bounce along (▌Figure 15.1). Wind-tunnel experiments show that once sand grains begin moving, they continue to move, even if the wind drops below the speed necessary to start them moving! This happens because once saltation begins, it sets off a chain reaction of collisions between sand grains that keeps the grains in constant motion.

Saltating sand usually moves near the surface, and even when winds are strong, grains are rarely lifted higher than about a meter. If the winds are very strong, these wind-whipped grains can cause extensive abrasion. A car's paint can be removed by sandblasting in a short time, and its windshield will become completely frosted and translucent from pitting.

Suspended Load

Silt- and clay-sized particles constitute most of a wind's suspended load. Even though these particles are much smaller and lighter than sand-sized particles, wind usually starts the latter moving first. The reason for this phenomenon is that a very thin layer of motionless air lies next to the ground where the small silt and clay particles remain undisturbed. The larger sand grains, however, stick up into the turbulent air zone where they can be moved. Unless the stationary air layer is disrupted, the silt and clay particles remain on the ground, providing a smooth surface.

This phenomenon can be observed on a dirt road on a windy day. Unless a vehicle travels over the road, little dust is raised even though it is windy. When a vehicle moves over the road, it breaks the calm boundary layer of air and disturbs the smooth layer of dust, which is picked up by the wind and forms a dust cloud in the vehicle's wake.

In a similar manner, when a sediment layer is disturbed, silt- and clay-sized particles are easily picked up and carried in suspension by the wind, creating clouds of dust or even dust storms. Once these fine particles are lifted into the atmosphere, they may be carried thousands of kilometers from their source. For example, large quantities of fine dust from the southwestern United States were blown eastward and fell on New England during the dust storms of the 1930s (see Figure 6.15).

15.3 Wind Erosion

Although wind action produces many distinctive erosional features and is an extremely efficient sorting agent, running water is responsible for most erosional landforms in arid regions, even though stream channels are typically dry. Wind erodes material in two ways: *abrasion* and *deflation*.

Abrasion

Abrasion involves the impact of saltating sand grains on an object and is analogous to sandblasting. The effects of abrasion are usually minor because sand, the most common agent of abrasion, is rarely carried more than a meter above the surface (Figure 15.2). Rather than creating major erosional features, wind abrasion typically modifies existing features by etching, pitting, smoothing, or polishing. Nonetheless, wind abrasion can produce many strange-looking and bizarre-shaped features.

Ventifacts are a common product of wind abrasion; these are stones whose surfaces have been polished, pitted, grooved, or faceted by the wind (Figure 15.3). If the wind blows from different directions, or if the stone is moved, the ventifact will have multiple facets. Ventifacts are most common in deserts, yet they can form wherever stones are exposed to saltating sand grains—for example, on beaches in humid regions and on some outwash plains.

Yardangs are larger features than ventifacts and also result from wind erosion (Figure 15.4). They are elongated, streamlined ridges that look like an overturned ship's hull.

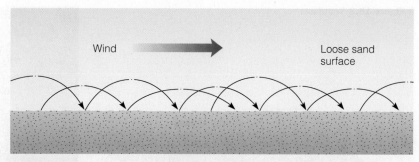

Figure 15.1 **Saltation** Most sand is moved near the ground surface by saltation. Sand grains are picked up by the wind and carried a short distance before falling back to the ground, where they usually hit other grains, causing them to bounce and move in the direction of the wind.

Figure 15.2 **Wind Abrasion** Wind abrasion has formed these bizarre-shaped structures by eroding the lower part of the exposed limestone in the Libyan Desert, Egypt.

>> *Critical Thinking Question*
Why is only the lower part of these outcrops eroded by wind abrasion?

▌Figure 15.3 Ventifacts

Wind

a A ventifact forms when windborne particles (1) abrade the surface of a rock, (2) forming a flat surface. If the rock is moved, (3) additional flat surfaces are formed.

b Numerous ventifacts are visible in this photo, which also shows desert pavement in Death Valley, California. Desert pavement prevents further erosion and transport of a desert's surface materials by forming a protective layer of close-fitting, larger rocks.

Copyright and Photograph by Dr. Parvinder S. Sethi

Marion A. Whitney

▌Figure 15.4 Yardang A profile view of a streamlined yardang in the Roman playa deposits of the Kharga Depression, Egypt. Yardangs form by wind erosion.

These shallow depressions of variable dimensions result from differential erosion of surface materials. Ranging in size from several kilometers in diameter and tens of meters deep to small depressions only a few meters wide and less than a meter deep, deflation hollows are common in the southern Great Plains region of the United States.

In many dry regions, the removal of sand-sized and smaller particles by wind leaves a surface of pebbles, cobbles, and boulders. As the wind removes the fine-grained material from the surface, the effects of gravity and occasional heavy rain, and even the swelling of clay grains, rearrange the remaining coarse particles into a mosaic of close-fitting rocks called **desert pavement** (Figure 15.3b and ▌Figure 15.6). Once desert pavement forms, it protects the underlying material from further deflation.

Yardangs are typically found in clusters aligned parallel to the prevailing winds. They probably form by differential erosion in which depressions, parallel to the direction of wind, are carved out of a rock body, leaving sharp, elongated ridges. These ridges may then be further modified by wind abrasion into their characteristic shape.

Deflation

Another important mechanism of wind erosion is **deflation**, which is the removal of loose surface sediment by the wind. Among the characteristic features of deflation in many arid and semiarid regions are *deflation hollows*, or *blowouts* (▌Figure 15.5).

15.4 Wind Deposits

Although wind is of minor importance as an erosional agent, it is responsible for impressive deposits, which are primarily of two types. The first, *dunes*, occur in several distinctive types, all of which consist of sand-sized particles that are usually deposited near their source. The second is *loess*, which consists of layers of wind-blown silt and clay deposited over large areas downwind and commonly far from their source.

Figure 15.5 Deflation Hollow A deflation hollow, the low area, between two sand dunes in Death Valley, California. Deflation hollows result when loose surface sediment is differentially removed by wind.

Figure 15.6 Desert Pavement

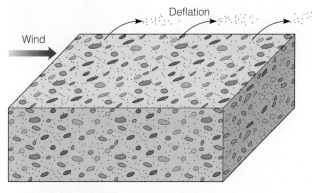

a Fine-grained material is removed by wind,

b leaving a concentration of larger particles that form desert pavement.

>> **Critical Thinking Question**
Why is desert pavement important in desert environments?

The Formation and Migration of Dunes

The most characteristic features in sand-covered regions are **dunes**, which are mounds or ridges of wind-deposited sand (▌Figure 15.7). Dunes form when wind flows over and around an obstruction, resulting in the deposition of sand grains, which accumulate and build up a deposit of sand. As they grow, these sand deposits become self-generating in that they form ever-larger wind barriers that further reduce the wind's velocity, resulting in more sand deposition and growth of the dune.

Most dunes have an asymmetrical profile, with a gentle windward slope and a steeper downwind or leeward slope that is inclined in the direction of the prevailing wind (▌Figure 15.8a). Sand grains move up the gentle windward slope by saltation and accumulate on the leeward side, forming an angle of 30 to 34 degrees from the horizontal, which is the angle of repose of dry sand. When this angle is exceeded by accumulating sand, the slope collapses, and the sand slides down the leeward slope, coming to rest at its base. As sand moves from a dune's windward side and periodically slides down its leeward slope, the dune slowly migrates in the direction of the prevailing wind (▌Figure 15.8b). When preserved in the geologic record, dunes help geologists determine the prevailing direction of ancient winds (▌Figure 15.9).

CONNECTION LINK

To review the sedimentary structure cross-bedding and what it can tell geologists about ancient depositional environments, see page 167 in Chapter 7.

Dune Types

Geologists recognize four major dune types (barchan, longitudinal, transverse, and parabolic), although intermediate forms also exist. The size, shape, and arrangement of dunes result from the interaction of such factors as sand supply, the direction and velocity of the prevailing wind, and the amount of vegetation. Although dunes are usually found in deserts, they can also develop wherever sand is abundant, such as along the upper parts of many beaches.

▌**Figure 15.7** Sand Dunes Large sand dunes in Death Valley, California. The prevailing wind direction is from left to right as indicated by the sand dunes in which the gentle windward side is on the left and the steeper leeward slope is on the right.

▌**Figure 15.8** Dune Migration

Wind

Sand moves by saltation

Windward side

Leeward slope

a Profile of a sand dune.

Wind

Direction of dune migration

b Dunes migrate when sand moves up the windward side and slides down the leeward slope. Such movement of the sand grains produces a series of cross-beds that slope in the direction of wind movement.

Barchan dunes are crescent-shaped dunes whose tips point downwind (▌Figure 15.10). They form in areas that have a generally flat, dry surface with little vegetation, a limited supply of sand, and a nearly constant wind direction. Most barchans are small, with the largest reaching about 30 m high. Barchans are the most mobile of the major dune types, moving at rates that can exceed 10 m per year.

Longitudinal dunes (also called *seif dunes*) are long, parallel ridges of sand aligned generally parallel to the direction of the prevailing winds; they form where the sand supply is somewhat limited (▌Figure 15.11).

▌ Figure 15.9 Ancient Cross-Bedding Ancient cross-bedding in sandstone beds in Zion National Park, Utah, helps geologists determine the prevailing direction of the wind that formed these ancient sand dunes.

» Critical Thinking Question
Did the prevailing direction of wind change during the time the sand forming these sandstone beds was deposited?

▌ Figure 15.10 Barchan Dunes

a Barchan dunes form in areas that have a limited amount of sand, a nearly constant wind direction, and a generally flat, dry surface with little vegetation. The tips of barchan dunes point downward.

Longitudinal dunes result when winds converge from slightly different directions to produce the prevailing wind. They range in height from about 3 m to more than 100 m, and some stretch for more than 100 km. Longitudinal dunes are especially well developed in central Australia, where they cover nearly one-fourth of the continent. They also cover extensive areas in Saudi Arabia, Egypt, and Iran.

Transverse dunes form long ridges perpendicular to the prevailing wind direction in areas that have abundant sand and little or no vegetation (▌ Figure 15.12). When viewed from the air, transverse dunes have a wavelike appearance and are therefore sometimes called *sand seas*. The crests of transverse dunes can be as high as 200 m, and the dunes may be as wide as 3 km.

Parabolic dunes are most common in coastal areas with abundant sand, strong onshore winds, and a partial cover of vegetation (▌ Figure 15.13). Although parabolic dunes have a crescent shape like barchan dunes, their tips point upwind. Parabolic dunes form when the vegetation cover is broken and deflation produces a deflation hollow or blowout. As the wind transports the sand out of the depression, it builds up on the convex downwind dune crest. The central part of the dune is excavated by the wind, while vegetation holds the ends and sides fairly well in place.

Another type of dune commonly found in the deserts of North Africa and Saudi Arabia is the *star dune*, so named because of its resemblance to a multipointed star (▌ Figure 15.14). Star dunes are among the tallest in the world, rising, in some cases, more than 100 m above the surrounding desert plain. They consist of pyramidal hills of sand, from which radiate several ridges of sand, and they develop where the wind direction is variable. Star dunes can remain stationary for centuries, and have served as desert landmarks for nomadic peoples.

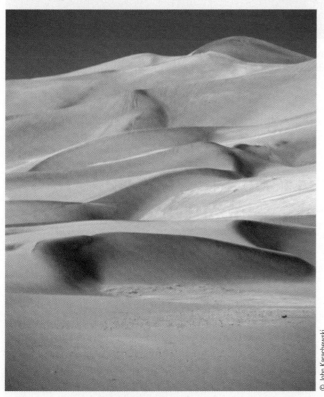

b A ground-level view of several barchan dunes.

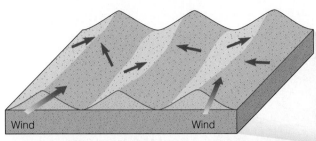

a Longitudinal dunes form long, parallel ridges of sand aligned roughly parallel to the prevailing wind direction. They typically form where sand supplies are limited.

b Longitudinal dunes, 15 m high, in the Gibson Desert, west central Australia. The bright blue areas between the dunes are shallow pools of rainwater, and the darkest patches are areas where the Aborigines have set fires to encourage the growth of spring grasses.

▌ **Figure 15.11** Longitudinal Dunes

▌ **Figure 15.12** Transverse Dunes

a Transverse dunes form long ridges of sand that are perpendicular to the prevailing wind direction in areas of little or no vegetation and abundant sand.

b Transverse dunes, Great Sand Dunes National Monument, Colorado. The prevailing wind direction is from lower left to upper right.

Loess

Wind-blown silt and clay deposits composed of angular quartz grains, feldspar, micas, and calcite are known as **loess**. The distribution of loess shows that it is derived from three main sources: deserts, Pleistocene glacial outwash deposits, and the floodplains of rivers in semiarid regions. Loess must be stabilized by moisture and vegetation in order to accumulate. Consequently, loess is not found in deserts, even though deserts provide much of its material. Because of its unconsolidated nature, loess is easily eroded, and as a result, eroded loess areas are characterized by steep cliffs and rapid lateral and headward stream erosion (▌ Figure 15.15).

Presently, loess deposits cover approximately 10% of Earth's land surface and 30% of the United States. The most extensive and thickest loess deposits are found in northeast China, where accumulations greater than 30 m thick are common. Loess-derived soils are some of the world's most fertile (Figure 15.15). It is therefore not surprising that the world's major grain-producing regions correspond to large loess deposits, such as the North European Plain, Ukraine, and the Great Plains of North America.

15.5 Air-Pressure Belts and Global Wind Patterns

To understand the work of wind and the distribution of deserts, we need to consider the global pattern of air-pressure belts and winds, which are responsible for Earth's atmospheric circulation patterns. Air pressure is the density of air exerted on its surroundings (i.e., its weight). When air is heated, it expands and rises, reducing its mass for a given volume and causing a decrease in air pressure. Conversely, when air is cooled, it contracts and air pressure increases. Therefore, those areas of Earth's surface that receive the most solar radiation, such as the equatorial regions, have low air pressure, whereas the colder areas, such as the polar regions, have high air pressure.

❚ Figure 15.13 Parabolic Dunes

a Parabolic dunes typically form in coastal areas that have a partial cover of vegetation, a strong onshore wind, and abundant sand.

Wind

b A parabolic dune developed along the Lake Michigan shoreline west of St. Ignace, Michigan.

❯❯ Critical Thinking Question
Why do the tips of a parabolic dune point upwind and those of a barchan dune point downwind?

Reed Wicander

❚ Figure 15.14 Star Dunes

a Star dunes are pyramidal hills of sand that develop where the wind direction is variable.

❚ Figure 15.15 Terraced Wheat Fields in the Loess Soil at Tangwa Village, China Because of the unconsolidated nature of loess, many farmers live in hillside caves they carved from the loess.

© Lowell Georgia/CORBIS

© Nigel J. Dennis/Photo Researchers, Inc

b A ground-level view of star dunes in Namib-Naukluft Park, Namibia.

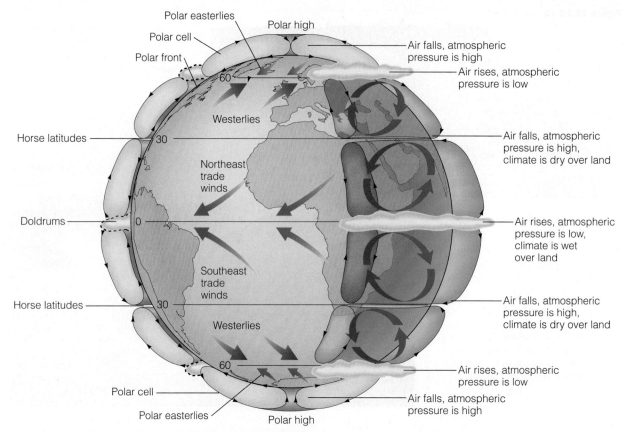

Figure 15.16 The General Circulation Pattern of Earth's Atmosphere Air flows from high-pressure zones to low-pressure zones, and the resulting winds are deflected to the right of their direction of movement (clockwise) in the Northern Hemisphere and to the left of their direction of movement (counterclockwise) in the Southern Hemisphere.

Air flows from high-pressure zones to low-pressure zones. If Earth did not rotate, winds would move in a straight line from one zone to another. Because Earth rotates, however, winds are deflected to the right of their direction of motion (clockwise) in the Northern Hemisphere and to the left of their direction of motion (counterclockwise) in the Southern Hemisphere. This deflection of air between latitudinal zones resulting from Earth's rotation is known as the **Coriolis effect**. The combination of latitudinal pressure differences and the Coriolis effect produces a worldwide pattern of east–west–oriented wind belts (Figure 15.16).

Earth's equatorial zone receives the most solar energy, which heats the surface air and causes it to rise. As the air rises, it cools and releases moisture that falls as rain in the equatorial region (Figure 15.16). The rising air is now much drier as it moves northward and southward toward each pole. By the time it reaches 20 to 30 degrees north and south latitudes, the air has become cooler and denser and begins to descend. Compression of the atmosphere warms the descending air mass and produces a warm, dry, high-pressure area, the perfect conditions for the formation

of the low-latitude deserts of the Northern and Southern Hemispheres (Figure 15.17).

15.6 The Distribution of Deserts

Dry climates occur in the low and middle latitudes where the potential loss of water by evaporation may exceed the yearly precipitation (Figure 15.17). Dry climates cover 30% of Earth's land surface and are subdivided into semiarid and arid regions. *Semiarid regions* receive more precipitation than arid regions, yet are moderately dry. Their soils are usually well developed and fertile, and support a natural grass cover. *Arid regions*, generally described as **deserts**, are dry; they receive less than 25 cm of rain per year, have high evaporation rates, typically have poorly developed soils, and are mostly or completely devoid of vegetation.

The majority of the world's deserts are in the dry climates of the low and middle latitudes (Figure 15.17). The

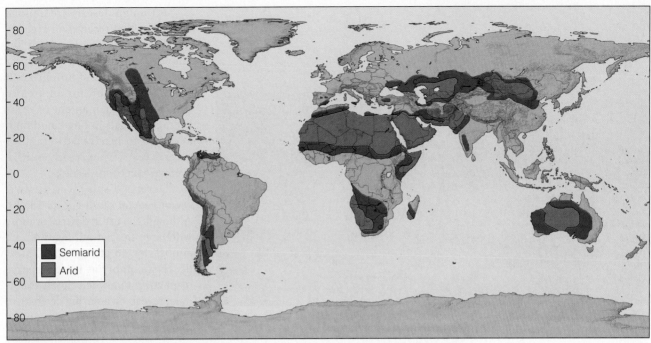

▌Figure 15.17 The Distribution of Earth's Arid and Semiarid Regions Semiarid regions receive more precipitation than arid regions, yet they are still moderately dry. Arid regions, generally described as deserts, are dry and receive less than 25 cm of rain per year. The majority of the world's deserts are located in the dry climates of the low and middle latitudes.

▌Figure 15.18 Rain-Shadow Deserts Many deserts in the middle and high latitudes are rain-shadow deserts, so named because they form on the leeward side of mountain ranges. When moist marine air moving inland meets a mountain range, it is forced upward where it cools and forms clouds that produce rain. This rain falls on the windward side of the mountains. The air descending on the leeward side is much warmer and drier, producing a rain-shadow desert.

remaining dry climates of the world are found in the middle and high latitudes, mostly within continental interiors in the Northern Hemisphere (Figure 15.17). Many of these areas are dry because of their remoteness from moist maritime air and the presence of mountain ranges that produce a **rain-shadow desert** (▌Figure 15.18). When moist marine air moves inland and meets a mountain range, it is forced upward. As it rises, it cools, forming clouds and producing precipitation that falls on the windward side of the mountains. The air that descends on the leeward side of the mountain range is much warmer and drier, producing a rain-shadow desert.

Copyright and Photograph by Dr. Parvinder S. Sethi

▌**Figure 15.19 Desert Vegetation** Desert vegetation is typically sparse, widely spaced, and characterized by slow growth rates. The vegetation shown here in Organ Pipe National Monument, Arizona, includes saguaro and cholla cacti, paloverde trees, and jojoba bushes and is characteristic of the vegetation found in the Sonoran Desert of North America.

15.7 Characteristics of Deserts

To people who live in humid regions, deserts may seem stark and inhospitable. Instead of a landscape of rolling hills and gentle slopes with an almost continuous cover of vegetation, deserts are dry, have little vegetation, and consist of nearly continuous rock exposures, desert pavement, or sand dunes. Yet despite the great contrast between deserts and more humid areas, the same geologic processes are at work, only operating under different climatic conditions.

Temperature, Precipitation, and Vegetation

The heat and dryness of deserts are well known. Many of the deserts of the low latitudes have average summer temperatures that range between 32° and 38°C. It is not uncommon for some low-elevation inland deserts to record daytime highs of 46° to 50°C for weeks at a time. During the winter months, when the Sun's angle is lower and there are fewer daylight hours, daytime temperatures average between 10° and 18°C.

The dryness of the low-latitude deserts results primarily from the year-round dominance of the subtropical high-pressure belt, whereas the dryness of the mid-latitude deserts is due to their isolation from moist marine winds and the rain-shadow effect created by mountain ranges. The dryness of both is accentuated by their high temperatures.

Although deserts are defined as regions that receive, on average, less than 25 cm of rain per year, the amount of rain that falls each year is unpredictable and unreliable. It is not uncommon for an area to receive more than an entire year's average rainfall in one cloudburst and then to receive little rain for several years. Thus, yearly rainfall averages can be misleading.

Deserts display a wide variety of vegetation (▌Figure 15.19). Although the driest deserts, or those with large areas of shifting sand, are almost devoid of vegetation, most deserts support at least a sparse plant cover that when closely examined, reveals an amazing diversity of plants that have evolved the ability to live in the near absence of water.

Desert plants are widely spaced and typically small, and they grow slowly. Their stems and leaves are usually hard and waxy to minimize water loss by evaporation and to protect the plant from sand erosion. Most plants have a widespread shallow root system to absorb the dew that forms each morning in all but the driest deserts and to help anchor the plant in what little soil there may be. In extreme cases, many plants lie dormant during particularly dry years and spring to life after the first rain shower with a beautiful profusion of flowers.

Weathering and Soils

Mechanical weathering is dominant in desert regions. Daily temperature fluctuations and frost wedging are the primary forms of mechanical weathering. The breakdown of rocks by roots and from salt crystal growth is of minor importance. Some chemical weathering does occur, but its rate is greatly reduced by aridity and the scarcity of organic acids produced by the sparse vegetation. Most chemical weathering takes place during the winter months when there is more precipitation, particularly in the mid-latitude deserts.

An interesting feature seen in many deserts is a thin, red, brown, or black, shiny coating on the surface of many rocks. This coating, called *rock varnish*, is composed of iron and manganese oxides. Because many of the varnished rocks contain little or no iron and manganese oxides, the varnish is thought to result either from windblown iron and manganese dust that settles on the ground or from the precipitated waste of microorganisms.

> **CONNECTION LINK** ◀
>
> To review the processes of mechanical and chemical weathering, see pages 137–142 in Chapter 6.

In many desert regions, such as the Southwest and the Great Basin area of North America, a common form of rock art, known as *petroglyphs*, is especially abundant. Petroglyphs (from the Greek *petro*, meaning "rock," and *glyph*, meaning "carving or engraving") are the abraded, pecked, incised, or scratched marks made by humans on boulders, cliffs, and cave walls. Many petroglyphs are the result of removing the thin, brown- or black-colored rock varnish, revealing the underlying lighter-colored natural rock surface, thus providing an excellent contrast for petroglyphs (▮ Figure 15.20). It is important to remember that petroglyphs, which in North America extend back to about 2000 BC, are a fragile and nonrenewable cultural resource that cannot be replaced if they are damaged or destroyed.

Desert soils, if developed, are usually thin and patchy because the limited rainfall and the resultant scarcity of vegetation reduce the efficiency of chemical weathering and hence soil formation. Furthermore, the sparseness of the vegetative cover enhances wind and water erosion of what little soil actually forms.

Mass Wasting, Streams, and Groundwater

› CONNECTION LINK

To review the various ways running water erodes and shapes a landscape, see pages 292 and 293 in Chapter 12.

When traveling through a desert, most people are impressed by such wind-formed features as moving sand, sand dunes, and sand and dust storms. They may also notice the dry washes and dry streambeds. Because of the lack of running water, most people would conclude that wind is the most important erosional agent in deserts. They would be wrong! Running water, even though it occurs infrequently, causes most of the erosion in deserts. The dry conditions and sparse vegetation characteristic of deserts enhance water erosion.

Most of a desert's average annual rainfall of 25 cm or less comes in brief, heavy, localized cloudbursts. During these times, considerable erosion takes place, because the ground cannot absorb all of the rainwater. With so little vegetation to hinder the flow of water, runoff is rapid, especially on moderately to steeply sloping surfaces, resulting in flash floods and sheet flows. Dry stream channels quickly fill with raging torrents of muddy water and mudflows, which carve out steep-sided gullies and overflow their banks. During these times, a tremendous amount of sediment is rapidly transported and deposited far downstream.

Although water is the major erosive agent in deserts today, it was even more important during the Pleistocene Epoch when these regions were more humid. During that time, many of the major topographic features of deserts were forming. Today that topography is modified by wind and infrequently flowing streams.

Reed Wicander

▮ **Figure 15.20** Rock Varnish A human-like petroglyph exposed at an outcrop along Cub Creek Road in Dinosaur National Monument, Utah. Note the contrast between the fresh exposure of the rock where the upper part of the petroglyph's head has been removed, the weathered brown surface of the rest of the petroglyph, and the black rock varnish coating the rock surface.

Most desert streams are poorly integrated and flow only intermittently. Many of them never reach the sea because the water table is usually far deeper than the channels of most streams, so they cannot draw upon groundwater to replace water lost to evaporation and absorption into the ground. This type of drainage in which a stream's load is deposited within the desert is called *internal drainage* and is common in most arid regions.

Although most deserts have internal drainage, some deserts have permanent through-flowing streams, such as the Nile and Niger rivers in Africa, the Rio Grande and Colorado rivers in the southwestern United States, and the Indus River in Asia. These streams can flow through desert regions because their headwaters are well outside the desert and water is plentiful enough to offset losses resulting from evaporation and infiltration.

Wind

Although running water does most of the erosional work in deserts, wind can also be an effective geologic agent capable of producing a variety of distinctive erosional (Figures 15.2 and 15.4) and depositional features (Figures 15.10 through 15.15). Wind is effective in transporting and depositing unconsolidated sand-, silt-, and dust-sized particles. Contrary to popular belief, most deserts are not sand-covered wastelands, but rather vast areas of rock exposures and desert pavement. Sand-covered regions, or sandy deserts, constitute less than 25% of the world's deserts. The sand in these areas has accumulated primarily by the action of wind.

Wind is not only an effective erosional and depositional agent in deserts, it is also becoming an important resource in generating electricity in many parts of the world (see Geo-Focus on pages 378 and 379). The same wind that erodes, transports, and deposits materials is increasingly being harnessed to produce electricity for an energy-hungry world.

GEO-FOCUS   Windmills and Wind Power

Whoosh, whoosh, whoosh. Ah, the gentle sound of a windmill's blades turning in the wind. The image most people associate with windmills is one of a pastoral landscape in Holland dominated by a classic Dutch windmill crafted of wood (Figure 1), or perhaps Don Quixote tilting at windmills in the famous novel *Don Quixote de la Mancha* by Miguel Cervantes. Today, instead of a whoosh, whoosh, whoosh, the sound of modern electricity-generating windmills in a wind farm is more like a woomph, woomph, woomph (Figure 2).

As early as 5000 BC, people began to harness the power of wind to propel boats along the Nile River. The Chinese used windmills to pump water for irrigating crops as long ago as 2000 BC. Wind power was used in the Middle Ages in Europe, particularly in Holland, where windmills have played an important role in society. Windmills were first used to grind corn, which is where the term *windmill* originally came from. Later windmills were used to drain lakes and marshes from the low-lying districts and to saw timber. Settlers in the United States in the late 19th and early 20th centuries used this technology to pump water and generate electricity in the Great Plains.

With the application of steam power and industrialization in Europe and later the United States, the use of windmills rapidly declined. However, industrialization led to the development of larger

Figure 1 Five traditional Dutch windmills lined up along a canal at Kinderdiik, the Netherlands.

Figure 2 A wind farm in the Coachella Valley, California. California has more windmills than any other state in the United States.

What Would You Do ?

As an expert in desert processes, you have been assigned the job of teaching the first astronaut crew that will explore Mars all about deserts and their landforms. The reason is that many Martian features display evidence of having formed as a result of wind processes, and many Martian landforms are the same as those found in deserts on Earth. Describe how you would teach the astronauts to recognize wind-formed features and where on Earth you would take the astronauts to show them the types of landforms they may find on Mars.

15.8 Desert Landforms

Because of differences in temperature, precipitation, and wind, as well as the underlying rocks and recent tectonic events, landforms in arid regions vary considerably. Running water, although infrequent in deserts, is responsible for producing and modifying many distinctive landforms found there.

After an infrequent and particularly intense rainstorm, excess water not absorbed by the ground may accumulate in low areas and form *playa lakes* (Figure 15.21a). These lakes and more efficient windmills exclusively designed to generate electricity. Denmark began using such windmills as early as 1890, and other countries soon followed suit. Interest in electricity-generating windmills has always mirrored the price of fossil fuels. When the price of petroleum and coal is low, it is cheaper to use these fuels to generate electricity. When the price of fossil fuels goes up, interest in wind power also increases. Today, wind power is an important source of electricity, both in the United States and elsewhere around the world, particularly in Europe. As wind turbine technology has increased the efficiency of wind-generated electricity, the price has decreased significantly, so that wind farms can now compete in many areas with traditional fossil fuel–burning power plants.

How do windmills produce electricity? Simply stated, wind turbines (the term commonly used to describe electricity-producing windmills) convert the kinetic energy (the energy of an object due to motion) of the wind into mechanical power, in this case, the generation of electricity. This electricity is sent to the local power grid where it is distributed throughout the area.

In order to be effective, numerous wind turbines are clustered together in wind farms that are located in areas with relatively strong, steady winds. The number of turbines on wind farms can range from several to thousands, as in California (Figure 2). In addition to wind farms on land, wind farms offshore are becoming more popular because they are out of sight and people cannot hear the blades turning. In fact, wind farms in the North Sea already provide approximately 20% of Denmark's energy needs. In the United States, wind power production currently accounts for about 9% of the total power produced by all means.

What are the advantages and disadvantages of wind power? First of all, the wind is a free, renewable energy source, so it cannot be used up. It is also a clean source of energy that doesn't pollute the water or atmosphere, or contribute to greenhouse gases. Thus, it reduces the consumption of fossil fuels. The land on which windmills are sited can still be used for farming and ranching, thereby increasing the productivity of the land and providing an additional source of income to the landowner who leases the land to utilities. In addition, wind farms can benefit the local economy of rural and remote areas by supplying wind energy for local consumption.

There are some disadvantages to wind-generated electricity. The major disadvantage is that wind doesn't always blow with sufficient strength to be totally reliable, thereby necessitating backup generation. Furthermore, good wind sites are frequently located in remote areas, far from the areas where large quantities of electricity are needed, or in coastal areas, where land is expensive and local residents do not want large wind turbines as neighbors. The initial start-up cost of a wind farm is usually higher than the cost of building a conventional power plant. However, as the cost of wind power has decreased because of better technology, wind-generated electricity can compete favorably with traditional power plants in many areas.

The "not-in-my-backyard" opposition to wind farms can make siting a wind farm difficult. The major objection to wind farms is the noise generated by the turbines, although as the windmills are built taller and the turbines are more efficient at noise reduction, that is not the major concern it once was. Of course, aesthetics still play a role in many people's objection to wind farms, and the fact that some birds are killed by the rotating blades is another issue. Studies have indicated, however, that the impact of wind turbines on bird mortality and injury is less than that of many other structures, such as buildings, power lines, and communication towers.

As the price of fossil fuel continues to rise and our dependency on foreign sources of oil increases, the use of a centuries-old staple, the windmill, albeit modernized, will continue to gain in popularity.

---

are temporary, lasting from a few hours to several months. Most of them are shallow and have rapidly shifting boundaries as water flows in or leaves by evaporation and seepage into the ground. The water is often very saline.

When a playa lake evaporates, the dry lake bed is called a **playa** or *salt pan* and is characterized by mud cracks and precipitated salt crystals (▌ Figure 15.21b). Salts in some playas are thick enough to be mined commercially. For example, borates have been mined in Death Valley, California, for more than 100 years.

Other common features of deserts, particularly in the Basin and Range Province of the western United States, are alluvial fans and bajadas. **Alluvial fans** form when sediment-laden streams flowing out from the generally straight, steep mountain fronts deposit their load on the relatively flat desert floor. Once beyond the mountain front where no valley walls confine streams, the sediment spreads out laterally, forming a gently sloping and poorly sorted fan-shaped sedimentary deposit (▌ Figure 15.22).

Alluvial fans are similar in origin and shape to deltas (see Chapter 12), but are formed entirely on land. Alluvial fans may coalesce to form a *bajada*, a broad alluvial apron that typically has an undulating surface resulting from the overlap of adjacent fans.

Figure 15.21 Playas and Playa Lakes

a A playa lake formed after a rainstorm near Badwater, Death Valley National Park, California. Playa lakes are ephemeral features, lasting from a few hours to several months.

b Salt deposits and salt ridges cover the floor of this playa in the Mojave Desert, California. Salt crystals and mud cracks are characteristic features of playas.

Copyright and Photograph by Dr. Parvinder S. Sethi

© John Karachewski

Figure 15.22 Alluvial Fan An alluvial fan, Death Valley, California. Alluvial fans form when sediment-laden streams flowing out from a mountain deposit their load on the relatively flat desert floor, forming a gently sloping, fan-shaped, sedimentary deposit.

» Critical Thinking Question
Why are alluvial fans that form on land the same general shape as deltas that form in water?

Copyright and Photograph by Dr. Parvinder S. Sethi

Large alluvial fans and bajadas are frequently important sources of groundwater for domestic and agricultural use (see Chapter 13). Their outer portions are typically composed of fine-grained sediments suitable for cultivation, and their gentle slopes allow good drainage of water. Many alluvial fans and bajadas are also the sites of large towns and cities, such as San Bernardino, California; Salt Lake City, Utah; and Teheran, Iran.

Most mountains in desert regions, including those of the Basin and Range Province, rise abruptly from gently sloping surfaces called **pediments**. Pediments are erosional bedrock surfaces of low relief that slope gently away from mountain bases (▌Figure 15.23). Most pediments are covered by a thin layer of debris, alluvial fans, or bajadas.

The origin of pediments has been the subject of much controversy. Most geologists agree that they are erosional features developed on bedrock in association with the erosion and retreat of a mountain front (Figure 15.23a). The disagreement concerns how the erosion has occurred. Although not all geologists would agree, it appears that pediments are produced by the combined activities of lateral erosion by streams, sheet flooding, and various weathering processes along the retreating mountain front. Thus, pediments grow at the expense of the mountains, and they will continue to expand as the mountains are eroded away or partially buried.

Rising conspicuously above the flat plains of many deserts are isolated, steep-sided erosional remnants called

▌**Figure 15.23** Pediment

a Pediments are erosional bedrock surfaces formed by erosion along a mountain front.

b A pediment north of Mesquite, Nevada.

inselbergs, a German word meaning "island mountain." Inselbergs have survived for a longer period of time than other mountains because of their greater resistance to weathering. Uluru (formerly known as Ayers Rock) is an excellent example of an inselberg (▌Figure 15.24).

▌**Figure 15.24** Uluru at Sunset Contrary to popular belief, Uluru is not a giant boulder. Rather, it is the exposed portion of the nearly vertically tilted Uluru Arkose, which can clearly be seen in this image. Differential weathering of the sedimentary layers has produced the distinct parallel ridges and other features characteristic of Uluru.

Other easily recognized erosional remnants common to arid and semiarid regions are mesas and buttes (▌Figure 15.25). A **mesa** is a broad, flat-topped erosional remnant bounded on all sides by steep slopes. Continued weathering and stream erosion form isolated, pillar-like structures known as **buttes**. Buttes and mesas consist of relatively easily weathered sedimentary rocks capped by nearly horizontal, resistant rocks such as sandstone, limestone, or basalt. They form when the resistant rock layer is breached, which allows rapid erosion of the less-resistant underlying sediment. One of the best-known areas of mesas and buttes in the United States is Monument Valley on the Arizona–Utah border (Figure 15.25).

▌ **Figure 15.25** Buttes and Mesas

a Several mesas and buttes can be seen in this aerial view of Monument Valley Navajo Tribal Park, Navajo Indian Reservation, Arizona. The mesas are the broad, flat-topped structures, one of which is prominent in the foreground, whereas the buttes are more pillar-like structures.

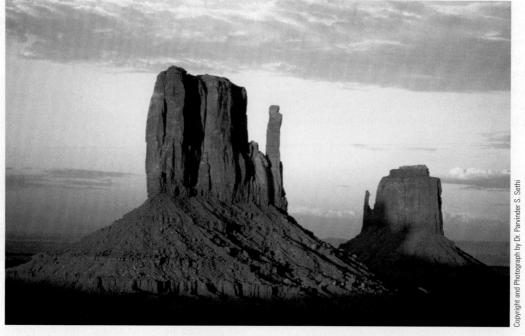

b Left Mitten Butte and Right Mitten Butte in Monument Valley on the border of Arizona and Utah.

GEO-RECAP

» Geo-Image

Figure 1 What desert landforms or features of a desert can you identify in this image of the Mesquite Flat sand dunes in Death Valley, California? What types of sand dunes are present?

Copyright and Photograph by Dr. Parvinder S. Sethi

Chapter Summary

- Desertification is the expansion of deserts into formerly productive lands. It destroys croplands and rangelands, causing massive starvation and forcing hundreds of thousands of people to migrate from their homelands.

- Wind transports sediment in suspension or as bed load. Suspended load is the material that is carried in suspension by water or wind. Silt- and clay-sized particles constitute most of a wind's suspended load. Bed load is the material that is too large or heavy to be carried in suspension, and is thus moved along the surface by saltation, the process of rolling, or sliding.

- Wind erodes material by either abrasion or deflation. Abrasion is the impact of saltating sand grains on an object. Ventifacts are common products of wind abrasion.

- Deflation is the removal of loose surface material by wind. Deflation hollows resulting from differential erosion of surface material are common features of many deserts, as is desert pavement, which effectively protects the underlying surface from additional deflation.

- The two major wind deposits are dunes and loess. Dunes are mounds or ridges of wind-deposited sand that form when wind flows over and around an obstruction, resulting in the deposition of sand grains, which accumulate and build up a deposit of sand.

- The four major dune types are barchan, longitudinal, transverse, and parabolic. The amount of sand available, prevailing wind direction, wind velocity, and amount of vegetation determine which type of dune will form.

- Loess is wind-blown silt and clay deposits composed of angular quartz grains, feldspar, micas, and calcite, and is derived from deserts, Pleistocene glacial outwash deposits, and river floodplains in semiarid regions. Loess covers approximately 10% of Earth's land surface and weathers to a rich, productive soil.

- The winds of the major air-pressure belts, oriented east–west, result from the rising and cooling of air. The winds are deflected clockwise in the Northern Hemisphere and counterclockwise in the Southern Hemisphere by the Coriolis effect to produce Earth's global wind patterns.

- Dry climates, located in the low and middle latitudes where the potential loss of water by evaporation exceeds the yearly precipitation, cover 30% of Earth's land surface and are subdivided into semiarid and arid regions. Semiarid regions receive more precipitation than arid regions, yet are moderately dry. Arid regions, generally described as deserts, are dry and receive less than 25 cm of rain per year.

- The majority of the world's deserts are in the low-latitude, dry-climate zone between 20 and 30 degrees north and south latitudes. Their dry climate results from a high-pressure belt of descending dry air. The remaining deserts are in the middle latitudes, where their distribution is related to the rain-shadow effect, and in the dry polar regions.

- Deserts are characterized by high temperatures, little precipitation, and sparse plant cover. Rainfall is unpredictable and, when it does occur, tends to be intense and of short duration.

- Mechanical weathering is the dominant form of weathering in deserts and, coupled with slow rates of chemical weathering, results in poorly developed soils.

- Running water is the major agent of erosion in deserts and was even more important during the Pleistocene, when wetter climates resulted in humid conditions.

- Wind is also an erosional agent in deserts and is very effective in transporting and depositing unconsolidated, fine-grained sediments.

- Important desert landforms include playas, which are dry lakebeds; when temporarily filled with water, they form playa lakes. Alluvial fans are fan-shaped sedimentary deposits that may coalesce to form bajadas.

- Pediments are erosional bedrock surfaces of low relief that slope gently away from mountain bases.

- Inselbergs are isolated, steep-sided erosional remnants that rise above the surrounding desert plains. Buttes and mesas are, respectively, pinnacle-like and flat-topped erosional remnants with steep sides.

Important Terms

abrasion (p. 367)
alluvial fan (p. 379)
barchan dune (p. 370)
butte (p. 382)
Coriolis effect (p. 374)
deflation (p. 368)
desert (p. 374)

desert pavement (p. 368)
desertification (p. 366)
dune (p. 369)
loess (p. 372)
longitudinal dune (p. 370)
mesa (p. 382)
parabolic dune (p. 371)

pediment (p. 381)
playa (p. 379)
rain-shadow desert (p. 375)
transverse dune (p. 371)
ventifact (p. 367)

Review Questions

1. Which of the following is a feature produced by wind erosion?
   a. ___ playa.
   b. ___ loess.
   c. ___ dune.
   d. ___ yardang.
   e. ___ none of these.

2. Some of the world's most fertile soils are derived from
   a. ___ playas.
   b. ___ dunes.
   c. ___ loess.
   d. ___ alluvial fans.
   e. ___ bajadas.

3. The Coriolis effect causes wind to be deflected
   a. ____ to the right in the Northern Hemisphere and to the left in the Southern Hemisphere.
   b. ____ to the left in the Northern Hemisphere and to the right in the Southern Hemisphere.
   c. ____ only to the left in both hemispheres.
   d. ____ only to the right in both hemispheres.
   e. ____ not at all.

4. The primary process by which bed load is transported is
   a. ____ suspension.
   b. ____ abrasion.
   c. ____ saltation.
   d. ____ precipitation.
   e. ____ answers a and c.

5. Which of the following dunes form long ridges perpendicular to the prevailing wind direction in areas where abundant sand is available and little or no vegetation exists?
   a. ____ barchan.
   b. ____ longitudinal.
   c. ____ star.
   d. ____ transverse.
   e. ____ parabolic.

6. The dominant form of weathering in deserts is _____, and the soils are _____.
   a. ____ mechanical, thick.
   b. ____ mechanical, thin.
   c. ____ mechanical, limited.
   d. ____ chemical, thick.
   e. ____ chemical, thin.

7. Between what latitudes in both hemispheres do the driest deserts in the world occur?
   a. ____ 10° and 20°.
   b. ____ 20° and 30°.
   c. ____ 30° and 40°.
   d. ____ 40° and 60°.
   e. ____ 60° and 80°.

8. Poorly sorted, fan-shaped sedimentary deposits with a gentle slope are
   a. ____ pediments.
   b. ____ inselbergs.
   c. ____ mesas.
   d. ____ alluvial fans.
   e. ____ playas.

9. The major agent of erosion in deserts today is _____, while during the Pleistocene Epoch, it was _____.
   a. ____ wind, running water.
   b. ____ wind, wind.
   c. ____ running water, wind.
   d. ____ running water, running water.
   e. ____ wind, glaciers.

10. The dry lakebeds found in many deserts are
    a. ____ playas.
    b. ____ bajadas.
    c. ____ inselbergs.
    d. ____ pediments.
    e. ____ mesas.

11. Using what you now know about deserts, their location, how they form, and the various landforms found in them, how can you determine where deserts may have existed in the past?

12. If deserts are dry regions in which mechanical weathering predominates, why are so many of their distinctive landforms the result of running water and not of wind?

13. Much of the recent desertification has been greatly accelerated by human activity. Can you think of anything that can be done to slow the process?

14. Why are so many desert rock formations red?

15. Is it possible for Mars to have the same types of sand dunes as Earth? What would that tell us about the climate and geology of Mars?

16. As more people move into arid and semiarid areas such as Las Vegas, Nevada, Phoenix, Arizona, and southern California, an increasing strain is placed on the environment of these areas. What are some of the issues that government entities must face in dealing with these population increases? What are some of the problems that must be dealt with from a geologic perspective?

17. Why are low-latitude deserts so common?

18. Why is desert pavement important in a desert environment?

19. How do dunes form and migrate? Why is dune migration a problem in some areas?

20. One problem of the nuclear age is finding safe storage sites for radioactive waste from nuclear power plants, the manufacture of nuclear weapons, and the radioactive by-products of nuclear medicine. What are the advantages to constructing storage sites, particularly for high-level radioactive waste, in desert areas, such as Nevada's Yucca Mountain? What are the possible disadvantages?

Exposures of sandstone along the shoreline at Shore Acres State Park near Coos Bay, Oregon. This Eocene-age sandstone was deposited in a deltaic depositional environment. Notice that the sandstone layers dip at about 50 degrees, but the overlying layers (below the fence) are nearly horizontal, making this is a good example of an angular unconformity. During storms, waves impact the entire rock face, the top of which is more than 25 m above the sea

Sue Monroe

CHAPTER 16

Oceans, Shorelines, and Shoreline Processes

OUTLINE

16.1 Introduction

16.2 Exploring the Oceans

16.3 Seawater and Oceanic Circulation

16.4 Shorelines and Shoreline Processes

16.5 Erosion and Deposition Along Shorelines

16.6 Types of Coasts

16.7 The Perils of Living Along the Shoreline

16.8 Resources from the Oceans

GEO-FOCUS: The *Deepwater Horizon* Oil Spill

GEO-RECAP

HAVE YOU EVER WONDERED?

- How scientists study the oceans?
- Why saltwater is salty, and whether it is getting saltier with time?
- What causes ocean currents, waves, and tides?
- How waves and currents erode shorelines?
- Why some shorelines have long, sandy beaches and others do not?
- What causes beaches to vary with the seasons?
- How shoreline communities deal with storm waves and coastal flooding?
- What kinds of resources come from the oceans?

16.1 Introduction

In Chapter 2 we discussed several aspects of the seafloor such as oceanic ridges, deep-sea hydrothermal vents (black smokers), seamounts, and guyots. We also discussed the nature of the continental margins, which include the continental shelf, slope, and rise. Our interest here is in the oceanic waters themselves and their impact on shorelines, although we will also briefly discuss some of the sediment found on the deep seafloor.

You already know that the hydrosphere consists of all water on Earth, most of which is in the oceans (see Table 12.1). This vast, interconnected body of saltwater covers 71% of Earth's surface, but parts of it are distinct enough for us to recognize the Pacific, Atlantic, Indian, and Arctic oceans (❙ Figure 16.1). Seas, on the other hand, are mostly marginal parts of oceans, as in the Sea of Japan and the Mediterranean Sea. The oceans and seas are underlain by oceanic crust, but the same is not true of the Dead Sea, the Caspian Sea, or the Salton Sea; these are actually saline lakes on the continents.

What lies beneath the ocean's surface is largely hidden from view, which accounts for why so many sensational

stories about it have persisted for centuries. According to two dialogues written about 350 BC by the Greek Philosopher Plato, a continent called *Atlantis* existed in the Atlantic Ocean west of what we now call the Strait of Gibraltar (❙ Figure 16.2). Plato claimed that following the conquest of Atlantis by Athens, this vast continent sank, and now only "mud shallows" mark its former location. In fact, no geologic evidence intdicates that Atlantis ever existed, not even the "mud shallows" that Plato mentioned, so why has this story persisted for so long? One reason is that stories of lost civilizations are popular, but another is that until fairly recently little was known about what lay beneath the ocean's surface.

In this enormous body of water we call oceans, wave energy is transferred through the water to shorelines where it has a tremendous impact. Accordingly, understanding shoreline processes is important to many people including oceanographers, geologists, and coastal engineers, as well as elected officials and city planners of coastal communities. Indeed, tourism is an important part of the economies of many coastal communities such as Myrtle Beach, South Carolina, Fort Lauderdale, Florida, and Padre Island, Texas. Another important aspect of shorelines is rising sea level, because buildings or even entire communities that were once far inland are now

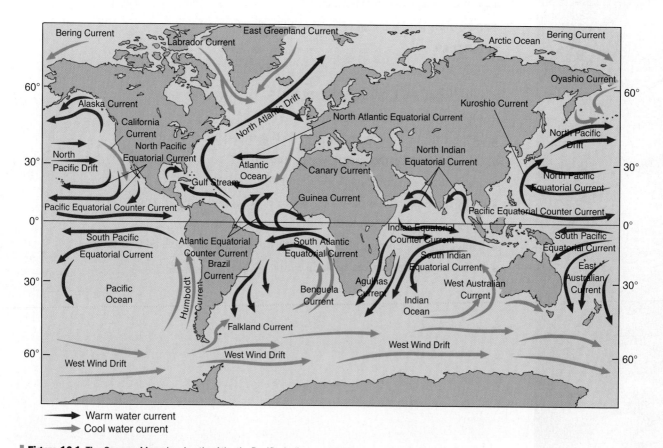

❙ **Figure 16.1** The Oceans Map showing the Atlantic, Pacific, Indian, and Arctic oceans, and the ocean currents.

>> *Critical Thinking Question*
How do ocean currents modify Earth's climate?

Figure 16.2 **Atlantis** According to Plato, Atlantis was a continent west of the Pillars of Hercules, now called the Strait of Gibraltar. In this map from Anthanasium Kircher's *Mundus Subterraneus* (1664), north is toward the bottom of the map. The Strait of Gibraltar is the narrow area between Hispania (Spain) and Africa.

Figure 16.3 **Seismic Profiling** In seismic profiling, the energy generated at a source is reflected from various layers back to the surface where it is detected by hydrophones.

in peril, must be protected or moved, or have already been destroyed. Furthermore, hurricanes expend much of their energy on shorelines resulting in extensive coastal flooding, numerous fatalities, and widespread property damage.

The study of oceans and shorelines provides another excellent example of systems interactions—in this case, between the hydrosphere and the solid Earth. The atmosphere is also involved because energy is transferred from wind to water, thereby causing waves, which in turn generate nearshore currents. And, of course, the gravitational attraction of the Moon and Sun on the ocean waters is responsible for the rhythmic rise and fall of tides. As dynamic systems, shorelines continually adjust to all changes, such as increased wave energy or an increase or a decrease in sediment supply from the land.

16.2 Exploring the Oceans

About 2,200 years ago the Greeks had determined Earth's size and shape rather accurately, but Europeans were unaware of the vastness of the oceans until the 1400s and 1500s, when explorers sought trade routes to the Indies. In 1492 when Christopher Columbus set sail to find a route to the Indies, he greatly underestimated the width of the Atlantic Ocean. He was not, as it is popularly portrayed, attempting to demonstrate that Earth is a sphere; its spherical shape was well accepted by then. The controversy was over Earth's circumference and the shortest route to the Indies; on this point Columbus's critics were correct.

Columbus's voyages and similar ones added considerably to our knowledge of the oceans, but truly scientific investigations did not begin until well into the 1700s. During the 1760s and 1770s the British financed oceanographic expeditions led by Captain James Cook. Charles Darwin was aboard the research vessel HMS *Beagle* as an unpaid naturalist during its around-the-world voyage from 1831 to 1836. As a result of his observations during this journey, he proposed a theory on the evolution of coral reefs. During these 18th- and 19th-century expeditions, researchers determined oceanic depths, sampled seawater, classified marine organisms, sampled the seafloor, and visited many islands previously unknown to Europeans.

CONNECTION LINK

You can learn more about Charles Darwin on pages 451–453 in Chapter 18.

Measuring the length of a weighted line lowered to the seafloor was the first way to determine oceanic depths. Scientists now use an *echo sounder,* which detects sound waves that travel from a ship to the seafloor and back. Depth is calculated by knowing the velocity of sound in water and the time for the waves to reach the seafloor and return to the ship. *Seismic profiling* is even more useful. Strong waves from an energy source reflect from the seafloor or from layers beneath the seafloor (Figure 16.3).

International programs for oceanographic research have been carried out by a succession of ships equipped to drill into the seafloor and retrieve samples of seafloor sediments and oceanic crust. The most recent of these, the R/V *Chikyu* ("Earth"), is a Japanese ship that can drill as deep as 11 kilometers into the seafloor. In addition to surface vessels, submersibles are used in oceanographic research, one of which, the *Trieste,* descended to the greatest known oceanic depth (10,924 m) in 1960. One of the most famous of these submersibles is *Alvin,* which has carried scientists to the seafloor in many places to make observations and collect samples. Other submersibles are remotely controlled or towed by surface ships. In 1985, the *Argo,* equipped with sonar and radar systems, provided the first images of the British ocean liner HMS *Titanic* since it sank in 1912 after hitting an iceberg.

16.3 Seawater and Oceanic Circulation

During its earliest history, Earth was probably hot, airless, and dry, but volcanic eruptions were ubiquitous. Volcanoes emit several gases, but the most abundant one is water vapor, which accumulated in the atmosphere. As Earth cooled, the water vapor condensed, fell as rain, and began collecting on the surface. Geologic evidence indicates that oceans were present by at least 3.5 billion years ago, although their volumes and extent are unknown.

> **CONNECTION LINK**

For more information about the origin and evolution of the oceans, see page 491 in Chapter 19.

Seawater—Its Composition and Color

Seawater contains more than 70 chemical elements in solution, but the most common ones are chloride ions and sodium ions, which make up 85.6% of all dissolved substances and give seawater its most distinctive feature—its saltiness. Seawater's saltiness, or its **salinity**, is a measure of the total quantity of dissolved solids. On average, 1 kg (1,000 g) of seawater contains 35 g of dissolved solids, or 35 parts per thousand, which is symbolized as 35‰. In the open ocean the salinity varies from 32‰ to 37‰, although in some marginal seas, especially in dry, hot areas nearly isolated from the open ocean, values may exceed 40‰.

Most of the chemical elements in seawater come from the continents. During chemical weathering, minerals yield ions in solution, which are transported as the dissolved load of streams and rivers. Runoff from the continents adds about 4 billion tons of dissolved solids to the oceans each year. Another source of elements in seawater is *outgassing,* in which gases from within Earth are released into the oceans and the atmosphere by volcanoes and at deep-sea hydrothermal vents (see Figure 2.8).

The oceans have been salty for at least 1.5 billion years, so seawater is in a state of dynamic equilibrium, meaning that additions are offset by losses. For the salinity of seawater to remain rather constant, continuous recycling of ions in solution must take place, otherwise seawater would get saltier with time. To maintain this balance, ions are removed from seawater when evaporites such as rock salt (NaCl) and rock gypsum ($CaCO_4 \cdot 2H_2O$) are precipitated, when salt spray is blown onshore, when magnesium is used in dolomite and clay minerals, and when organisms use calcium and silica to construct their shells.

When visible light enters seawater, most of it is absorbed in the upper 1 m and elevates the water temperature. However, seawater absorbs the longer wavelengths of visible light, the reds and yellows, more readily than the shorter wavelengths, the greens and blues. The blue color of water in the open ocean is related to the selective absorption of light, but in shallow nearshore waters, the water is green or yellow

■ **Figure 16.4 The Color of Seawater** Notice in this view of the west coast of Florida that the nearshore water is greenish whereas the water offshore is blue.

because suspended fine-grained sediment and organic matter reflect the green and yellow wavelengths (■ Figure 16.4).

Based on the decreasing intensity of light with depth, scientists define two layers in seawater. The upper layer, called the **photic zone,** is usually 100 m or less thick and receives enough light for plants to photosynthesize. Below is the **aphotic zone** where too little light is available for photosynthesis. In this region, most organisms depend directly or indirectly on organic substances that "rain" down from the photic zone. The most notable exception is the organisms living adjacent to deep-sea hydrothermal vents such as black smokers (see Figure 2.8).

CONNECTION LINK ◀

You can find more about black smokers on page 33 and Figure 2.8 in Chapter 2.

Oceanic Circulation

The ocean's surface is nearly always in motion as a result of surface currents, waves, and tides. Even the deep ocean waters are constantly in motion, but unlike wind-driven surface currents and waves, deep-ocean circulation is driven mostly by density differences between adjacent water masses. Much of the circulation in both surface currents and deep-ocean currents is horizontal, but some circulation is vertical when water is transferred from depth to the surface or from the surface to depth. Circulation from depth to the surface, called *upwelling,* has important biological and economic consequences.

As wind blows over a water surface, some of its energy is transferred to the water, which generates surface currents and waves, which we discuss in section 16.4. Figure 16.1 shows the global surface-water current patterns averaged over a long time. Notice in Figure 16.1 that surface currents in the Northern Hemisphere are deflected to the right (clockwise) of their direction of motion and to the left (counterclockwise) of their direction of motion in the Southern Hemisphere. This deflection is the **Coriolis effect,** which results from Earth's rotation. The combination of wind and the Coriolis effect produces large-scale water circulation systems known as **gyres** between the 60° parallels in the Atlantic, Pacific, and Indian oceans (Figure 16.1). One of the best-known ocean currents is the *Gulf Stream,* which is actually part of the much larger North Atlantic gyre.

These wind-driven gyres are important as a world temperature control. Seawater near the equator absorbs huge amounts of heat and transports it to high latitudes in warm currents. Cold currents originate at high latitudes and flow toward the equator. Indeed, the reason that some rather northerly countries, such as Scotland, have mild climates is related to these currents. On the other hand, the current along the northern and central coasts of California is cold, which keeps that area much cooler than the inland parts of the state.

Horizontal circulation of water takes place in the deep ocean basins because of differences in temperature and density of adjacent water masses. Circulation resulting from density differences operates on a simple principle—a water mass of greater density (colder or saltier) will displace and flow beneath a water mass of lesser density. Deep-ocean circulation affects about 90% of all ocean water, but because studying it is expensive and time-consuming, scientists know less about it than about other circulation patterns.

Vertical circulation takes place in the oceans when **upwelling** slowly transfers cold water from depth to the surface and when **downwelling** carries warm water from the surface to depth. Most upwelling takes place along the west coasts of continents where water moves offshore and colder water rises from depth. Upwelling is by far the most important, because it not only transfers water upward but also carries nutrients, particularly nitrates and phosphate, into the photic zone. Here, high concentrations of plankton are sustained that in turn support other organisms. In fact, areas of upwelling are very productive biologically; less than 1% of the ocean surface is in regions of upwelling, yet it supports more than 50% by weight of all fishes. In addition, most of Earth's sedimentary rocks containing phosphate were deposited along continental margins where upwelling takes place (see section 16.8).

Sediments and Sedimentary Rocks on the Seafloor

Much of the sediment eroded from continents is deposited on the continental margins—that is, the continental shelf, slope, and rise—and forms layers of sandstone, shale, and limestone. However, some sediment, mostly silt- and clay-sized particles, is carried into the deep ocean basins where it is deposited. In fact, most of this sediment is *pelagic,* meaning that it settled from suspension far from land. **Pelagic clay** is brown or red and composed mostly of clay-sized particles, whereas **ooze** is made up of the tiny shells of marine organisms. If dominated by calcium carbonate ($CaCO_3$) skeletons, it is *calcarious ooze,* or, if mostly silica (SiO_2) skeletons, it is *siliceous ooze* (Figure 16.5).

Figure 16.5 Sediments on the Deep Seafloor Most of the sediments on the seafloor are pelagic clay and calcarious ooze and siliceous ooze, both of which are composed of the tiny skeletons of single-celled, floating animals and plants.

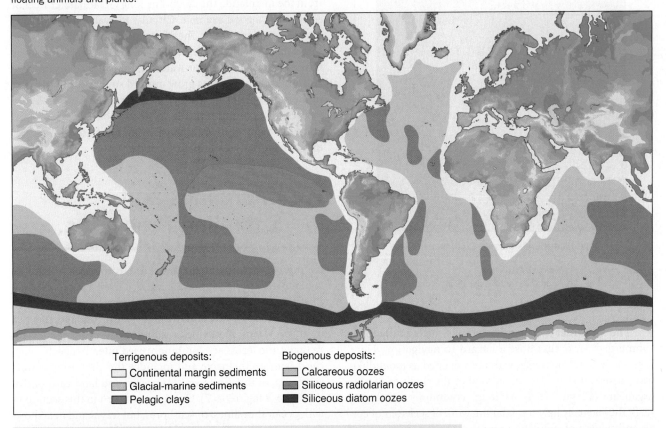

Terrigenous deposits:
☐ Continental margin sediments
▨ Glacial-marine sediments
▧ Pelagic clays

Biogenous deposits:
▨ Calcareous oozes
▨ Siliceous radiolarian oozes
■ Siliceous diatom oozes

» Critical Thinking Question
Why are there no coarse-grained (gravel and sand) sediments on the deep seafloor far from land?

Figure 16.6 The Evolution of Coral Reefs The origin of a fringing reef, barrier reef, and atoll as the plate upon which the fringing reef sits is carried into deeper water.

Fringing reef   Barrier reef   Atoll

a A fringing reef forms around an island in the tropics.

b The island sinks as the oceanic plate on which it rides moves away from a spreading center. In this case, the island does not sink at a rate faster than coral organisms can build upward.

c The island eventually disappears beneath the surface, but the coral remains at the surface as an atoll.

Courtesy of Carl Roassier

d An underwater view of a coral reef in the Red Sea.

» Critical Thinking Question
What kind of sedimentary rock composes coral reefs?

The term **reef** has many meanings, such as shallowly submerged rocks that pose a hazard to navigation. Here, however, we are concerned with reefs defined as moundlike, wave-resistant structures composed of the shells of marine organisms (**Figure 16.6**). Although commonly called coral reefs, they actually have a solid framework of skeletons of corals and clams and encrusting organisms such as sponges and algae. Most reefs are found in tropical seas where the water temperature does not fall below about 20°C, and rarely at depths of more than 50 m because many corals rely on symbiotic algae that must have enough sunlight for photosynthesis.

Most reefs are one of three basic types: fringing, barrier, and atoll. *Fringing reefs* are up to 1 km wide, are solidly attached to a landmass, have a rough, table-like surface, and on their seaward side slope steeply down to the seafloor. *Barrier reefs* are similar except that a lagoon separates them from the mainland. The 2,000-km-long Great Barrier Reef of Australia is a good example. Circular to oval reefs surrounding a lagoon are *atolls*. They form around volcanic islands that subside below sea level as the plate they rest on moves into progressively deeper water. As subsidence occurs, the reef-building organisms grow upward so that the living part of the reef remains in shallow water (Figure 16.6). Atolls are particularly common in the western Pacific Ocean basin, although many of them began as fringing reefs, evolved first into barrier reefs, and finally to atolls.

16.4 Shorelines and Shoreline Processes

No doubt you know that a **shoreline** is the area of land in contact with the ocean or a lake. However, we can expand this definition by noting that ocean shorelines include the land between low tide and the highest level on land affected by storm waves. How does a shoreline differ from a coast? Actually, the terms are commonly used interchangeably, but *coast* is more inclusive and includes the shoreline as well as an area of indefinite width seaward and landward of the shoreline (**Figure 16.7**). Our main concern in this section is with ocean shorelines, or seashores, but waves and nearshore currents are also effective in large lakes. However, waves and

Figure 16.7 Shorelines and Coasts The shoreline of this part of the U.S. Pacific Coast consists of the area from about where the waves break to the base of the sea cliffs. However, the coast extends farther seaward and also includes the sea cliffs and an area some distance inland.

nearshore currents are much more vigorous along seashores, and even the largest lakes have insignificant tides.

Tides, Waves, and Nearshore Currents

In the marine realm, several biological, chemical, and physical processes are operating continuously. Organisms change the local chemistry of seawater and contribute their skeletons to nearshore sediments, and temperature and salinity changes and internal waves occur in the oceans. However, the processes most important for modifying shorelines are purely physical ones, especially waves, tides, and nearshore currents. We cannot totally discount some of these other processes, though; offshore reefs composed of the skeletons of organisms, for instance, may protect a shoreline from most of the energy of waves.

Tides The surface of the oceans rises and falls twice daily in response to the gravitational attraction of the Moon and Sun. These regular fluctuations in the ocean's surface, or **tides,** result in most seashores having two daily high tides and two low tides as sea level rises and falls anywhere from a few centimeters to more than 15 m (Figure 16.8). A complete tidal cycle includes a *flood tide* that progressively covers more and more of a nearshore area until high tide is reached, followed by *ebb tide,* during which the nearshore area is once again exposed (Figure 16.8). These regular fluctuations in sea level constitute one largely untapped source of energy as do waves, ocean currents, and temperature differences in seawater.

Both the Moon and the Sun have sufficient gravitational attraction to exert tide-generating forces strong enough to deform the solid body of Earth, but they have a much greater influence on the oceans. The Sun is 27 million times more massive than the Moon, but it is 390 times as far from Earth, and its tide-generating force is only 46% as strong as that of the Moon. Accordingly, the tides are dominated by the Moon, but the Sun plays an important role as well.

If we consider only the Moon acting on a spherical, water-covered Earth, its tide-generating forces produce two bulges on the ocean surface (Figure 16.9a). One bulge points toward the Moon because it is on the side of Earth where the Moon's gravitational attraction is greatest. The other bulge is on the opposite side of Earth; it points away from the Moon because of centrifugal force due to Earth's rotation, and the Moon's gravitational attraction is less. These two bulges always point toward and away from the Moon (Figure 16.9a), so as Earth rotates and the Moon's position changes, an observer at a particular shoreline location experiences the rhythmic rise and fall of tides twice daily, but the heights of two successive high tides may vary depending on the Moon's inclination with respect to the equator.

Figure 16.8 Low and High Tides Low tide (a) and high tide (b) in Turnagain Arm, part of Cook Inlet in Alaska. The tidal range here is about 10 m. Turnagain Arm is a huge fiord now being filled with sediment carried in by rivers. Notice the mudflats in (a).

a Low tide.

b High tide.

a Tidal bulges if only the Moon caused them.

b When the Moon is new or full, the solar and lunar tides reinforce one another, causing spring tides, the highest high tides and lowest low tides.

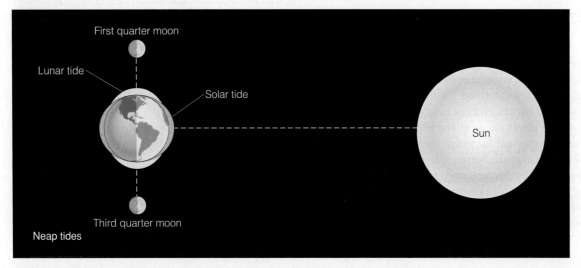

c During the Moon's first and third quarters, the Moon, Sun, and Earth form right angles, causing neap tides, the lowest high tides and highest low tides.

The Moon revolves around Earth every 28 days, so its position with respect to any latitude changes slightly each day. That is, as the Moon moves in its orbit and Earth rotates on its axis, it takes the Moon 50 minutes longer each day to return to the same position it was in the previous day. Thus, an observer would experience a high tide at 1:00 p.m. on one day, for example, and at 1:50 p.m. on the following day.

Even though the Sun's tide-generating force is weaker than the Moon's, when the two are aligned every two weeks, their forces added together generate *spring tides* about 20% higher than average tides (▌Figure 16.9b). When the Moon and Sun are at right angles to each another, also at two-week intervals, the Sun's tide-generating force cancels some of the

Moon's, and *neap tides* about 20% lower than average occur (▌Figure 16.9c).

Tidal ranges are also affected by shoreline configuration. Broad, gently sloping continental shelves as in the Gulf of Mexico have low tidal ranges, whereas steep, irregular shorelines experience much greater rise and fall of tides. Tidal ranges are greatest in some narrow, funnel-shaped bays and inlets. The Bay of Fundy in Nova Scotia has a tidal range of 16.5 m, and ranges greater than 10 m occur in several other areas (Figure 16.8).

Tides have an important impact on shorelines because the area of wave attack constantly shifts onshore and offshore as the tides rise and fall. Tidal currents themselves, however, have little modifying effect on shorelines, except in

Figure 16.10 Waves and Wave Terminology

a Waves and the terminology applied to them. Note that waves are disrupted when they encounter water shallower than wave base, and they eventually form breakers.

b A plunging breaker on the north shore of Oahu, Hawaii.

c A spilling breaker.

narrow passages where tidal current velocity is great enough to erode and transport sediment.

Waves You can see **waves,** or oscillations of a water surface, on all bodies of water, but they are best developed in the oceans. In fact, waves are directly or indirectly responsible for most erosion, sediment transport, and deposition in coastal areas. Wave terminology is illustrated with a typical series of waves in ▌ Figure 16.10a. A *crest,* as you would expect, is the highest part of a wave, whereas the low area between crests is a *trough.* The distance from crest to crest (or trough to trough) is the *wavelength,* and the vertical distance from trough to crest is *wave height.* You can calculate the speed at which a wave advances, called celerity (*C*), by the formula

$$C = L/T$$

where *L* is wavelength and *T* is wave period—that is, the time it takes for two successive wave crests, or troughs, to pass a given point.

The speed of wave advance (*C*) is actually a measure of the velocity of the wave form rather than the speed of the molecules of water in a wave. When waves move across a water surface, the water moves in circular orbits but shows little or no net forward movement (Figure 16.10a). Only the wave form moves forward, and as it does so it transfers energy in the direction of wave movement.

The diameters of the orbits that water follows in waves diminish rapidly with depth, and at a depth of about one-half wavelength (*L*/2), called **wave base,** they are essentially zero. Thus, at a depth exceeding wave base, the water and seafloor or lake floor are unaffected by surface waves (Figure 16.10a). We will explore wave base more fully in later sections of this chapter.

Wave Generation   Several processes generate waves, but most geologic work on shorelines is accomplished by wind-generated waves, especially storm waves. When wind blows over water—that is, one fluid (air) moves over another fluid (water)—friction between the two transfers energy to the water, causing the water surface to oscillate.

In areas where waves are generated, such as beneath a storm center at sea, sharp-crested, irregular waves called *seas* develop. Seas have various heights and lengths, and one wave cannot be easily distinguished from another. But as seas move out from their area of generation, they are sorted into broad *swells* with rounded, long crests and all are about the same size.

The harder and longer the wind blows, the larger are the waves, but these are not the only factors that control wave size. High-velocity wind blowing over a small pond will never generate large waves regardless of how long it blows. In fact, waves on ponds and most lakes appear only while the wind is blowing; once the wind stops, the water quickly smooths out. In contrast, the surface of the ocean is always in motion, and waves with heights of 34 m have been recorded during storms in the open sea.

The reason for the disparity between wave sizes on ponds and lakes and on the oceans is the **fetch,** which is the distance the wind blows over a continuous water surface. So on ponds and lakes the fetch corresponds to their length or width, depending on wind direction. To produce waves of greater length and height, more energy must be transferred from wind to water; hence large waves form beneath large storms at sea.

Waves with different lengths, heights, and periods may merge, making them smaller or larger. Under some circumstances, two wave crests merge to form *rogue waves* that are three or four times higher than the average. These waves can rise unexpectedly out of an otherwise comparatively calm sea and threaten even the largest ships. As recently as April 16, 2005, the Norwegian cruise ship *Dawn* was damaged by a 21-m-high rogue wave that flooded more than 60 cabins and injured four passengers.

Shallow-Water Waves and Breakers   Swells moving out from an area of wave generation lose little energy as they travel long distances across the ocean. In these deepwater swells, the water surface oscillates and water moves in circular orbits, but little net displacement of water takes place in the direction of wave travel (Figure 16.10a). Of course, wind blows some water from wave crests, thus forming whitecaps with foamy white crests, and surface currents transport water great distances; but deepwater waves themselves accomplish little actual water movement. When these waves enter progressively shallower water, however, the wave shape changes, and water is displaced in the direction of wave advance.

Broad, undulating deepwater waves are transformed into sharp-crested waves as they enter shallow water. This transformation begins at a water depth corresponding to wave base—that is, one-half wavelength. At this point, the waves "feel" the seafloor, and the orbital motion of water within the waves is disrupted (Figure 16.10a). As waves continue moving shoreward, the speed of wave advance and wavelength decrease, but wave height increases. Thus, as they enter shallow water, waves become oversteepened as the wave crest advances faster than the wave form, and eventually the crest plunges forward as a **breaker** (Figure 16.10). Breaking waves might be several times higher than their deepwater counterparts, and when they break, they expend their kinetic energy on the shoreline.

The waves just described are the classic *plunging breakers* that crash onto shorelines with steep offshore slopes, such as those on the north shore of Oahu in the Hawaiian Islands (❚ Figure 16.10b). In contrast, shorelines where the offshore slope is more gentle usually have *spilling breakers,* where the waves build up slowly and the wave's crest spills down the wave front (❚ Figure 16.10c). Whether the breakers are plunging or spilling, the water rushes onto the shore and then returns seaward to become part of another breaking wave.

Nearshore Currents   The area extending seaward from the upper limit of the shoreline to just beyond the area of breaking waves is conveniently designated as the *nearshore zone.* Within the nearshore zone are the breaker zone and a surf zone, where water from breaking waves rushes forward and then flows seaward as backwash. The nearshore zone's width varies depending on the length of approaching waves because long waves break at a greater depth, and thus farther offshore, than do short waves. Incoming waves are responsible for two types of currents in the nearshore zone: *longshore currents* and *rip currents.*

Wave Refraction and Longshore Currents   Deepwater waves have long, continuous crests, but rarely are their crests parallel with the shoreline (❚ Figure 16.11). In other words, they seldom approach a shoreline head-on, but rather at some angle. Thus, one part of a wave enters shallow water where it encounters wave base and begins breaking before other parts of the same wave. As a wave begins to break, its velocity diminishes, but the part of the wave still in deep water races ahead until it too encounters wave base. The net effect of this oblique approach is that waves bend so that they more nearly parallel the shoreline, a phenomenon known as **wave refraction** (Figure 16.11).

Even though waves are refracted, they still usually strike the shoreline at some angle, causing the water between the breaker zone and the beach to flow parallel to the shoreline. These **longshore currents,** as they are called, are long and narrow and flow in the same general direction as the approaching waves. These currents are particularly important in transporting and depositing sediment in the nearshore zone.

Rip Currents Waves carry water into the nearshore zone, so there must be a mechanism for mass transfer of water back out to sea. One way in which water moves seaward from the nearshore zone is in **rip currents,** narrow surface currents that flow out to sea through the breaker zone (❙ Figure 16.12). Surfers commonly take advantage of rip currents for an easy ride out beyond the breaker zone, but these currents pose a danger to inexperienced swimmers. Some rip currents flow at several kilometers per hour, so if a swimmer is caught in one, it is useless to try to swim directly back to shore. Instead, because rip currents are narrow and usually nearly perpendicular to the shore, one can swim parallel to the shoreline for a short distance and then turn shoreward with little difficulty.

Rip currents are circulating cells fed by longshore currents that increase in velocity from midway between each rip current (Figure 16.12a). When waves approach a shoreline, the amount of water builds up until the excess moves out to sea through the breaker zone.

Rip currents commonly develop where wave heights are lower than in adjacent areas, and differences in wave height are controlled by variations in water depth. For instance, if

❙ **Figure 16.11** Wave Refraction Waves are refracted as they enter shallow water, but even so they usually come in at an angle to the shoreline.

waves move over a depression, the height of the waves over the depression tends to be less than in adjacent areas, forming the ideal environment for rip currents.

❙ **Figure 16.12** Rip Currents

a Rip currents are fed on each side by currents moving parallel to the shoreline.

Breaker zone

High wave crests

Low wave crests

Rip current

b Suspended sediment, indicated by discolored water, is being carried seaward in these rip currents.

What Would You Do ?

While swimming at your favorite beach, you suddenly notice that you are far down the beach from the point where you entered the water. Furthermore, the size of the waves you were swimming in has diminished considerably. You decide to swim back to shore and then walk back to your original starting place, but no matter how hard you swim, you are carried farther and farther away from the shore. Assuming that you survive this incident, explain what happened and what you did to remedy the situation.

16.5 Erosion and Deposition Along Shorelines

Erosion and deposition by waves and nearshore currents account for many interesting seashore features such as sea stacks, arches, beaches, and spits, all of which are easy to recognize. In fact, many of these features are present along the Pacific, Gulf, and Atlantic coasts as well as along the shorelines of large lakes.

▌ Figure 16.13 Wave Erosion by Abrasion and Hydraulic Action

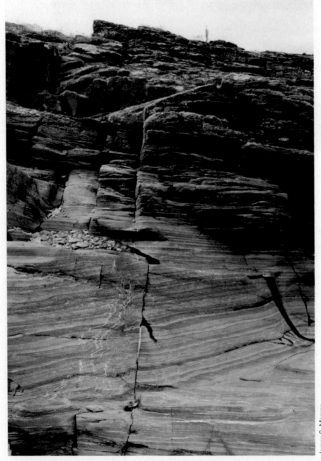

James S. Monroe

Erosion

On many seashores, erosion creates steep or vertical slopes known as *sea cliffs*. During storms, these cliffs are pounded by waves (hydraulic action), worn by the impact of sand and gravel (abrasion) (▌ Figure 16.13), and eroded by dissolution involving the chemical breakdown of rocks by the solvent action of seawater. Tremendous energy from waves is concentrated on the lower parts of sea cliffs and is most effective on those composed of sediments or highly fractured rocks. In any case, the net effect of these processes is erosion of the sea cliff and retreat of the cliff face landward.

Wave-Cut Platforms Wave intensity and the resistance of shoreline materials to erosion determine the rate at which a sea cliff retreats landward. A sea cliff of glacial drift on Cape Cod, Massachusetts, erodes as much as 30 m per century. By comparison, sea cliffs of dense igneous or metamorphic rocks erode and retreat much more slowly.

Sea cliffs erode mostly as a result of hydraulic action and abrasion at their bases. As a sea cliff is undercut by erosion, the upper part is left unsupported and susceptible to mass-wasting processes. Thus, sea cliffs retreat little by little, and as they do, they leave a beveled surface called a **wave-cut platform** that slopes gently seaward (▌ Figure 16.14). Broad wave-cut platforms exist in many areas, but invariably the water over them is shallow because the abrasive planing action of waves is effective to a depth of only about 10 m. The sediment eroded from sea cliffs is transported seaward until it reaches deeper water at the edge of the wave-cut platform. There it is deposited and forms a *wave-built platform,* which is a seaward extension of the wave-cut platform (Figure 16.14). Wave-cut platforms now above sea level are known as **marine terraces.**

a The rocks in the lower part of this image on a small island in the Irish Sea have been smoothed by abrasion, but the rocks higher up are out of the reach of waves.

Sue Monroe

b Hydraulic action and abrasion have undercut these sea cliffs near Bodega Bay, California. Erosion was particularly intense during storms of February 1998. Attempts to stabilize the shoreline have failed and the homes are now gone, and others are now threatened.

Figure 16.14 Origin of a Wave-Cut Platform

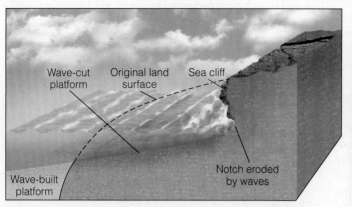

a Erosion causes a sea cliff to migrate landward, leaving a gently sloping surface, called a wave-cut platform. A wave-built platform forms by deposition at the seaward margin of the wave-cut platform.

b A wave-cut platform at Cape Foulweather, Oregon.

Sea Caves, Arches, and Stacks Sea cliffs do not retreat uniformly because some of the materials of which they are composed are more resistant to erosion than others. **Headlands** are seaward-projecting parts of the shoreline that are eroded on both sides by wave refraction (Figure 16.15). *Sea caves* form on opposite sides of a headland, and if these caves join, they form a *sea arch* (Figure 16.15a, b). Continued erosion causes the span of an arch to collapse, creating isolated *sea stacks* on wave-cut platforms (Figure 16.15a, c).

In the long run, shoreline processes tend to straighten an initially irregular shoreline. Wave refraction causes more wave energy to be expended on headlands and less on embayments. Thus, headlands erode, and some of the sediment yielded by erosion is deposited in the embayments.

Figure 16.15 Erosion of a Headland

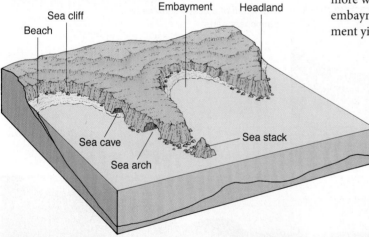

a Erosion of a headland and the origin of a sea cave, a sea arch, and sea stacks.

b A sea arch at Samuel H. Boardman State Park, Oregon.

c Sea stacks at Ruby Beach in Olympic National Park, Washington.

Deposition

In a previous section, we mentioned that longshore currents are effective at transporting sediment. In fact, we can think of the area from the breaker zone to the upper limit of wave swash as a "river" that flows along the shoreline. Unlike rivers on land, though, its direction of flow changes if waves approach from a different direction. Nevertheless, the analogy is apt, and just like rivers on land, a longshore current's capacity for transport varies with flow velocity and water depth.

Wave refraction and the resulting longshore currents are the primary agents of sediment transport and deposition on shorelines, but tides also play a role because, as they rise and fall, the position of wave attack shifts onshore and offshore. Rip currents play no role in shoreline deposition, but they do transport fine-grained sediment (silt and clay) offshore through the breaker zone.

Beaches   By definition, a **beach** is a deposit of unconsolidated sediment extending landward from low tide to a change in topography, such as a line of sand dunes, a sea cliff, or the point where permanent vegetation begins. Typically, a beach has several component parts, including a *backshore* that is usually dry, being covered by water only during storms or exceptionally high tides (Figure 16.16a). The backshore consists of one or more *berms,* platforms composed of sediment deposited by waves; the berms are nearly horizontal or slope gently landward. The sloping area below a berm exposed to wave swash is the *beach face.* The beach face is part of the *foreshore,* an area covered by water during high tide but exposed during low tide.

Depending on shoreline materials and wave intensity, beaches may be discontinuous, existing as only *pocket beaches* in protected areas such as embayments, or they may be continuous for long distances (Figure 16.16b, c).

Some of the sediment on beaches is derived from weathering and wave erosion of the shoreline, but most of it is transported to the coast by streams and redistributed by longshore currents. As we noted, waves usually strike beaches at some angle, causing the sand grains to move up the beach face at a similar angle; as the sand grains are carried seaward in the backwash, however, they move perpendicular to the long axis of the beach. Thus, individual sand grains move in a zigzag pattern in the direction of longshore currents, by what is called *longshore drift*. This movement is not restricted to the beach; it extends seaward to the outer edge of the breaker zone (Figure 16.17a).

Figure 16.16 Beaches

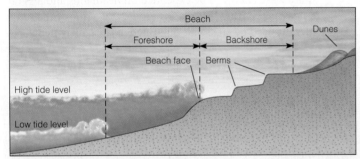

a Diagram of a beach showing its component parts.

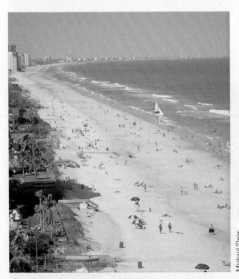

Michael Stear

b The Grand Strand of South Carolina, shown here at Myrtle Beach, is 100 km of nearly continuous beach.

James S. Monroe

c A pocket beach near Seal Rock, Oregon.

Figure 16.17 Longshore Currents and Longshore Drift

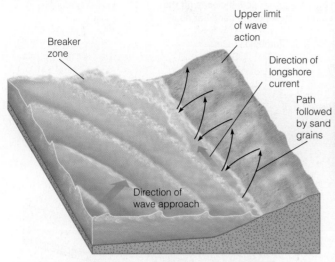

a Longshore currents transport sediment along the shoreline between the breaker zone and the upper limit of wave action.

U.S. Geological Survey Circular 1075

b These groins along the shoreline at Norfolk, Virginia, trap sand to maintain the beach in this area.

» Critical Thinking Question
What is the direction of the longshore currents in this area?

In an attempt to widen a beach or prevent erosion, shoreline residents often build *groins,* structures that project seaward at right angles from the shoreline. Groins interrupt the flow of longshore currents, causing sand deposition on their upcurrent sides and widening of the beach at that location. However, erosion inevitably occurs on the downcurrent side of a groin (Figure 16.17b).

Quartz is the most common mineral in most beach sands, but there are some notable exceptions. For example, the black sand beaches of Hawaii are composed of sand- and gravel-sized basalt rock fragments or small grains of volcanic glass, and some Florida beaches are composed of the fragmented calcium carbonate shells of marine organisms. In short, beaches are composed of whatever material is available.

Seasonal Changes in Beaches
The loose grains on beaches are constantly moved by waves, but the overall configuration of a beach remains unchanged as long as equilibrium conditions persist. We can think of the beach profile consisting of a berm or berms and a beach face as a profile of equilibrium (Figure 16.16a); that is, all parts of the beach are adjusted to the prevailing conditions of wave intensity, nearshore currents, and materials composing the beach.

Tides and longshore currents affect the configuration of beaches to some degree, but storm waves are by far the most important agent modifying their equilibrium profile. In many areas, beach profiles change with the seasons; so we recognize *summer beaches* and *winter beaches,* each of which is adjusted to the conditions prevailing at those times. Summer beaches are sand covered and have a wide berm, a gently sloping beach face, and a smooth offshore profile. Winter beaches, in contrast, tend to be coarser grained and steeper; they have a small berm or none at all, and their offshore profiles reveal sandbars paralleling the shoreline.

Seasonal changes in beach profiles are related to changing wave intensity. During the winter, energetic storm waves erode the sand from beaches and transport it offshore where it is stored in sandbars. The same sand that was eroded from a beach during the winter returns the next summer when it is driven onshore by more gentle swells. The volume of sand in the system remains more or less constant; it simply moves farther offshore or onshore depending on wave energy.

Spits, Baymouth Bars, and Tombolos
Beaches are the most familiar depositional features of coasts, but spits, baymouth bars, and tombolos are common, too. In fact, these features are simply continuations of a beach. A **spit,** for instance, is a finger-like projection of a beach into a body of water such as a bay, and a **baymouth bar** is a spit that has grown until it completely closes off a bay from the open sea (Figure 16.18, 16.19a,b). Both are composed of sand, more rarely gravel, that was transported and deposited by longshore currents where they weakened as they entered the deeper water of a bay's opening. Some spits are modified by waves so that their free ends are curved; they go by the name *hook* or *recurved spit* (Figure 16.18)

A **tombolo** is a type of spit that extends out from the shoreline to an island (Figure 16.18, 16.19c). A tombolo forms on the shoreward side of an island as wave refraction around the island creates converging currents that turn seaward and deposit a sandbar. So, the long axes of tombolos are nearly at right angles to the shoreline.

Spits, baymouth bars, and tombolos are most common along irregular seashores, but they can also be found in large lakes. Regardless of their setting, spits and baymouth bars constitute a continuing problem where bays must be kept open for pleasure boating, commercial shipping, or both. Obviously, a bay closed off by a sandbar is of little use for

❚ **Figure 16.18** Spits, Baymouth Bars, and Tombolos Spits form where longshore currents deposit sand in deeper water, as at the entrance to a bay. A baymouth bar is a spit that has grown until it closes off the mouth of a bay. Wave refraction around an island causes converging currents to deposit a sand bar nearly perpendicular to the shoreline, forming a tombolo.

❚ **Figure 16.19** Spits, Baymouth Bars, and Tombolos

a A spit at the mouth of the Russian River near Jenner, California.

b Rodeo Beach north of San Francisco, California, is a baymouth bar.

either endeavor, so a bay must be regularly dredged or protected from deposition by longshore currents. In some areas, *jetties* are constructed that extend seaward (or lakeward) to interrupt the flow of longshore currents and thus protect the opening to a bay.

Barrier Islands Long, narrow islands of sand lying a short distance offshore from the mainland are **barrier islands** (❚ Figure 16.20). On their seaward sides, they are smoothed by waves, but their landward margins are irregular because storm waves carry sediment over the island and deposit it in a lagoon where it is little modified by further wave activity. The component parts of a barrier island include a beach, windblown sand dunes, and a marshy area on their landward sides.

c A tombolo along the Oregon Coast.

Texas

Corpus Christi

Gulf of Mexico

Padre Island

Laguna Madre

NASA/NOAA

Figure 16.20 **Barrier Islands** View from space of the barrier islands along the Gulf Coast of Texas. Notice that a lagoon up to 20 km wide separates the long, narrow barrier islands from the mainland.

Everyone agrees that barrier islands form on gently sloping continental shelves where abundant sand is available and where both wave energy and the tidal range are low. In fact, these are the reasons that so many are along the United States' Atlantic and Gulf Coasts. But even though it is well known where barrier islands form, the details of their origin are still unresolved. According to one model, they formed as spits that became detached from land, whereas another model holds that they formed as beach ridges that subsequently subsided.

Most barrier islands are migrating landward as a result of erosion on their seaward sides and deposition on their landward sides. This is a natural part of barrier island evolution, and it takes place rather slowly, but fast enough to cause many problems for island residents and communities.

The Nearshore Sediment Budget

We can think of the gains and losses of sediment in the nearshore zone in terms of a **nearshore sediment budget** (Figure 16.21). If a nearshore system has a balanced budget, sediment is supplied to it as fast as it is removed, and the volume of sediment remains more or less constant, although sand may shift offshore and onshore with the changing seasons. A positive budget means that gains exceed losses, whereas a negative budget means that losses exceed gains. If a negative budget prevails long enough, a nearshore system is depleted and beaches may disappear.

Erosion of sea cliffs provides some sediment to beaches, but in most areas probably no more than 5–10% of the total sediment comes from this source. There are exceptions, though; almost all of the sediment on the beaches of Maine is derived from the erosion of shoreline rocks. Most sediment on typical beaches is transported to the shoreline by streams and then redistributed along the shoreline by longshore currents. Thus, longshore currents also play a role in the nearshore sediment budget because they continuously move sediment into and away from beach systems.

The primary ways that a nearshore system loses sediment are offshore transport, wind, and deposition in submarine canyons. Offshore transport mostly involves fine-grained sediment carried seaward where it eventually settles in deeper water. Wind is an important process because it removes sand from beaches and blows it inland where it piles up as sand dunes.

If the heads of submarine canyons are nearshore, huge quantities of sand are funneled into them and deposited in deeper water. La Jolla and Scripps submarine canyons off the coast of southern California funnel off an estimated 2 million m³ of sand each year. In most areas, however, submarine canyons are too far offshore to interrupt the flow of sand in the nearshore zone.

It should be apparent from this discussion that if a nearshore system is in equilibrium, its incoming supply of sediment exactly offsets its losses. Such a delicate balance tends to continue unless the system is somehow disrupted. One change that affects this balance is the construction of dams across the streams that supply sand. Once dams have been built, all sediment from the upper reaches of the drainage systems is trapped in reservoirs and thus cannot reach the shoreline.

What Would You Do ?

While visiting a barrier island, you notice some fine-grained, organic-rich deposits along the beach. In fact, there is so much organic matter that the sediments are black and have a foul odor. Their presence on the beach does not seem to make sense because you know that sediments like these were almost certainly deposited in a marsh on the landward (opposite) side of the island. How can you explain your observations?

16.6 Types of Coasts

Coasts are difficult to classify because of variations in the factors that control their development and variations in their composition and configuration. Rather than attempt to categorize all coasts, we shall simply note that two types of coasts have already been discussed: those dominated by deposition and those dominated by erosion.

In addition, we will examine coasts in terms of their changing relationship to sea level. But note that although

INPUT V^+ Sediment added from erosion by longshore transport onto beach
OUTPUTS V^- Sediment carried down-coast from the beach by longshore transport
  W^- Sediment blown inland by wind
  O^- Sediment cascading down the submarine slope
STABLE BEACH: $(V^+) + (V^- + W^- + O^-) = 0$

Figure 16.21 The Nearshore Sediment Budget The long-term sediment budget can be assessed by considering inputs versus outputs. If inputs and outputs are equal, the system is in a steady state, or equilibrium. If outputs exceed inputs, however, the beach has a negative budget and erosion occurs. Accretion takes place when the beach has a positive budget with inputs exceeding outputs.

>> **Critical Thinking Question**
What will happen to the beaches in this area if a dam is built across the river?

some coasts, such as those in southern California, are described as emergent (uplifted), these same coasts may be erosional as well. In other words, coasts commonly possess features that allow them to be classified in more than one way.

Depositional and Erosional Coasts

Depositional coasts, such as the U.S. Gulf Coast, are characterized by an abundance of detrital sediment and such depositional landforms as wide sandy beaches, deltas, and barrier islands. In contrast, erosional coasts are steep and irregular and typically lack well-developed beaches, except in protected areas (Figure 16.16). They are further characterized by sea cliffs, wave-cut platforms, and sea stacks. Many of the coasts along the West Coast of North America fall into this category.

Submergent and Emergent Coasts

If sea level rises with respect to the land or the land subsides, coastal regions are flooded and said to be **submergent** or *drowned* **coasts** (Figure 16.22). Much of the East Coast of North America from Maine southward through South Carolina was flooded during the rise in sea level following the Pleistocene Epoch, so it is extremely irregular. Recall that during the expansion of glaciers during the Pleistocene, sea level was as much as 130 m lower than at present, and that streams eroded their valleys more deeply and extended across continental shelves. When sea level rose, the lower ends of these valleys were drowned,

CONNECTION LINK

You can review Pleistocene glaciation and its causes on pages 359 and 360 in Chapter 14.

∎ Figure 16.22 Submergent and Emergent Coasts

NASA

a Submergent coasts tend to be extremely irregular, with estuaries such as Chesapeake and Delaware bays. They formed when the East Coast of the United States was flooded as sea level rose following the Pleistocene Epoch.

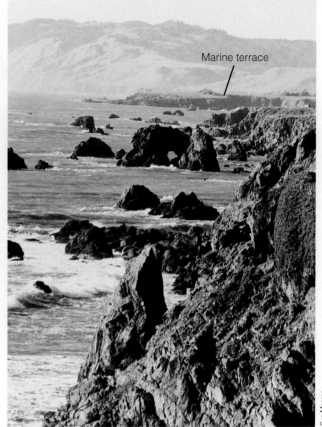

Sue Monroe

b Emergent coasts tend to be steep and straighter than submergent coasts. Notice the several sea stacks and the sea arch. Also, a marine terrace is visible in the distance.

> **» Critical Thinking Question**
> Do you think a wave-cut platform is forming in this area? If so, where?

forming *estuaries* such as Delaware and Chesapeake bays (Figure 16.22a). Estuaries are simply the seaward ends of river valleys where seawater and freshwater mix. The divides between adjacent drainage systems on submergent coasts project seaward as broad headlands or a line of islands.

Emergent coasts are found where the land has risen with respect to sea level (Figure 16.22b). Emergence takes place when water is withdrawn from the oceans, as occurred during the Pleistocene expansion of glaciers. Presently, coasts are emerging as a result of isostasy or tectonism. In northeastern Canada and the Scandinavian countries, for instance, the coasts are irregular because isostatic rebound is elevating formerly glaciated terrain from beneath the sea.

Coasts that form in response to tectonism, on the other hand, tend to be straighter because the seafloor topography being exposed by uplift is smooth. The west coasts of North and South America are rising as a consequence of plate tectonics. Distinctive features of these coasts are marine terraces, which are wave-cut platforms now elevated above sea level (Figure 16.22b). Uplift in these areas appears to be episodic rather than continuous, as indicated by the multiple levels of terraces in some places. In southern California, several terrace levels are

present, each of which probably represents a period of tectonic stability followed by uplift. The highest terrace is now approximately 425 m above sea level.

16.7 The Perils of Living Along a Shoreline

No matter where we live there are risks. Great Plains residents worry about violent storms, whereas those living near active faults or active volcanoes must be aware of the potential danger of earthquakes and eruptions. Likewise, living along a shoreline poses some risk, one of the most obvious being storm waves and coastal flooding, especially during hurricanes. Strong wind is dangerous, too, but most fatalities during hurricanes result from coastal flooding.

Storm Waves and Coastal Flooding

Coastal flooding results when large waves are driven onshore by hurricanes and by heavy rainfall, as much as

Figure 16.23 **Seawall Construction** Construction on this seawall to protect Galveston, Texas, from storm waves began in 1902. Notice that the wall is curved to deflect waves upward.

60 cm in as little as 24 hours. In addition, as a hurricane moves over the ocean, low atmospheric pressure causes the ocean's surface to bulge upward as much as 0.5 m. When the eye of the storm reaches the shoreline, the bulge, coupled with wind-driven waves, piles up in a **storm surge** that may rise several meters above normal high tide and flood areas far inland.

In 1900, hurricane-driven waves surged inland, eventually covering the entire island that Galveston, Texas, was on. As the waves swept into the city, buildings near the shoreline were battered to pieces and "great beams and railway ties were lifted by the [waves] and driven like battering rams into dwellings and business houses."* Between 6,000 and 8,000 people perished. In an effort to protect the city from future storms, a huge seawall was constructed, and the entire city was elevated to the level of the top of the seawall (▮ Figure 16.23).

The seawall has largely protected Galveston from more recent storm surges, but the same is not true for several other areas. Large-scale coastal flooding took place when Hurricane Isabel hit the Outer Banks of North Carolina in 2003, and in 2004, Florida was hit by four hurricanes causing widespread wind damage and coastal flooding. The U.S. Gulf Coast was hardest hit in 2005, first in August by Hurricane Katrina and then again in September by Hurricane Rita. When Hurricane

Katrina roared ashore on August 29, high winds, a huge storm surge, and coastal flooding destroyed nearly everything in an area covering 230,000 km². Parts of Gulfport and Biloxi, Mississippi, were leveled, but most of the public's attention focused on New Orleans, Louisiana (▮ Figure 16.24).

When Hurricane Katrina came ashore, the levees initially held, but on the next day some of the floodwalls were breached and about 80% of the city was flooded. And because New Orleans is mostly below sea level, the floodwaters could not drain out naturally. In fact, the city has 22 pumping stations to remove water from normal rainstorms, but as the city flooded, the pumps were overwhelmed, and when the electricity failed, the pumps were useless. All in all, Hurricane Katrina was the most expensive natural disaster in U.S. history; the property damages exceeded $100 billion and more than 1,800 people died, mostly in Louisiana. Then, in September, Hurricane Rita hit the Gulf Coast of Texas, and caused about 120 more fatalities and nearly $12 billion in damages.

Scientists, engineers, and some politicians had warned of a Katrina-type disaster for New Orleans for many years. They were aware that during a large hurricane the levees or floodwalls would likely fail. "What do we do now?" is a question that will be debated for many years. Most of you reading this book will not become geologists, but perhaps you will be engineers, city planners, members of planning commissions, politicians, or simply concerned citizens of coastal communities and will have to deal with these or similar problems.

---

* L. W. Bates, Jr., "Galveston—A City Built upon Sand," *Scientific American*, 95 (1906), p. 64.

Figure 16.24 Hurricane Katrina, 2005

a Although many areas were hit hard by Katrina, New Orleans was extensively flooded when the floodwalls failed that were built to protect the city from Lake Pontchartrain and the Mississippi River.

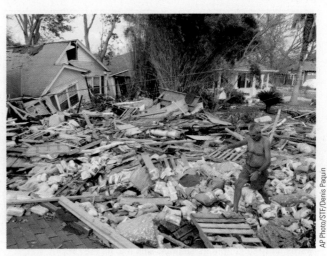

b Destruction caused by the storm surge from Hurricane Katrina. This image shows a man in Biloxi, Mississippi, trying to find his house.

c This is a repaired section of floodwall in New Orleans. This barrier is along one of the canals leading into the city from Lake Pontchartrain. Failure of the floodwalls during and following Hurricane Katrina caused flooding of about 80% of New Orleans.

Coastal Management as Sea Level Rises

During the last century, sea level rose about 12 cm worldwide, and all indications are that it will continue to rise. The absolute rate of sea-level rise in a shoreline region depends on two factors. The first is the volume of water in the ocean basins, which is increasing as a result of glacial ice melting and the thermal expansion of near-surface seawater. Many scientists think that sea level will continue to rise because of global warming caused by increasing concentrations of greenhouse gases in the atmosphere.

The second factor that controls sea level is the rate of uplift or subsidence of a coastal area. In some areas, uplift is occurring fast enough that sea level is actually falling with respect to the land. In other areas, sea level is rising while the coastal region is simultaneously subsiding, resulting in a net change in sea level of as much as 30 cm per century. Perhaps such a "slow" rate of sea level change seems insignificant; after all, it amounts to only a few millimeters per year. But in gently sloping coastal areas, as in the eastern United States from New Jersey southward, even a slight rise in sea level will eventually have widespread effects.

Many of the nearly 300 barrier islands along the East and Gulf coasts of the United States are migrating landward as sea level rises (■ Figure 16.25). Landward migration of barrier islands would pose few problems if it were not for the numerous communities, resorts, and vacation homes located on them. Moreover, barrier islands are not the only threatened areas. For example, Louisiana's coastal wetlands, an important wildlife habitat and seafood-producing area, are currently being lost at a rate of about 90 km² per year. Much of this loss results from sediment compaction, but rising sea level exacerbates the problem.

Rising sea level also directly threatens many beaches upon which communities depend for revenue. The beach at Miami Beach, Florida, for instance, was disappearing at an alarming rate until the Army Corps of Engineers began replacing the eroded beach sand. The problem is even more serious in other countries. A rise in sea level of only 2 m would inundate large areas of the U.S. East and Gulf coasts, but would cover 20% of the entire country of Bangladesh. Other problems associated with rising sea level include increased coastal flooding during storms and saltwater incursions that may threaten groundwater supplies.

Armoring shorelines with *seawalls* (embankments of reinforced concrete or stone) (Figure 16.23) and using *riprap* (piles of stones) (■ Figure 16.26a) protect beachfront structures, but both are initially expensive and during large

❚ Figure 16.25 Barrier Island Migration

a A barrier island.

b A barrier island migrates landward as sea level rises and storm waves carry sand from its seaward side into its lagoon.

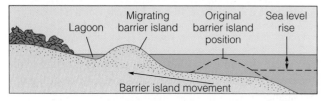

c Over time, the entire island shifts toward the land.

d Jetties were constructed during the 1930s to protect the inlet at Ocean City, Maryland, but they disrupted the net southerly longshore drift and Assateague Island, starved of sediment, has migrated 500 m landward. Beginning in the fall of 2002, beach sand was artificially replenished in an effort to stabilize the island.

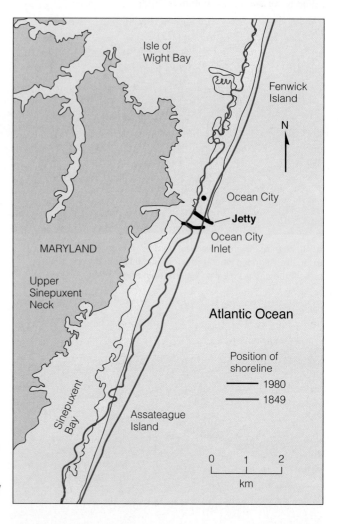

storms are commonly damaged or destroyed. Seawalls do afford some protection and are seen in many coastal areas along the oceans and large lakes, but some states, including North and South Carolina, Rhode Island, Oregon, and Maine, no longer allow their construction. The futility of artificially maintaining beaches is aptly shown by efforts to protect homes on a South Carolina barrier island. After each spring tide, heavy equipment is used to build a sand berm to protect homes from the next spring tide (❚ Figure 16.26b), only to have the berm disappear and then be rebuilt in a never-ending cycle of erosion and expensive artificial replacement.

Because we can do nothing to prevent sea level from rising, engineers, scientists, planners, and political leaders must examine what they can do to prevent or minimize the effects of shoreline erosion. At present, only a few viable options exist. One is to put strict controls on

coastal development. North Carolina, for example, permits large structures no closer to the shoreline than 60 times the annual erosion rate. Although a growing awareness of shoreline processes has resulted in similar legislation elsewhere, some states have virtually no restrictions on coastal development.

Regulating coastal development is commendable, but it has no impact on existing structures and coastal communities. A general retreat from the shoreline may be possible, but expensive, for individual dwellings and small communities, but it is impractical for large population centers. Such communities as Atlantic City, New Jersey, Miami Beach, Florida, and Galveston, Texas, have adopted one of two strategies to combat coastal erosion. One is to build protective barriers such as seawalls. Seawalls, such as the one at Galveston, Texas, are effective, but they are tremendously expensive to construct and maintain. Furthermore, barriers retard erosion only in the area directly behind them; Galveston Island west of its seawall has been eroded back about 45 m.

Another option, adopted by both Atlantic City, New Jersey, and Miami Beach, Florida, is to pump sand onto the beaches to replace that lost to erosion. This, too, is expensive as the sand must be replenished periodically because erosion is a continuing process. In many areas, groins are constructed to preserve beaches, but unless additional sand is artificially supplied to the beaches, longshore currents invariably erode sand from the downcurrent sides of the groins.

16.8 Resources from the Oceans

Seawater contains many elements in solution, the most common being sodium (Na) and chlorine (Cl), which is extracted for salt by evaporation or by mining salt deposits. A large part of the world's supply of magnesium comes from seawater. Seafloor deposits are becoming important, and most nations bordering the ocean claim those resources within their adjacent seafloor. The United States, by presidential proclamation issued in 1983, claims sovereign rights over an area designated the **Exclusive Economic Zone (EEZ)**. The U.S. EEZ extends 200 nautical miles (371 km) seaward and includes areas adjacent to U.S. territories such as Guam and Puerto Rico. Many other nations make similar claims.

Resources within the U.S. EEZ include sand and gravel for construction, and about 34% of all domestic oil production comes from wells on the continental shelf (see Geo-Focus on pages 410 and 411). Spherical objects called manganese nodules are composed of manganese and iron oxides, as well as copper, nickel, and cobalt. The United States imports most of the manganese and cobalt it uses, so these nodules have attracted attention as a potential resource. Another important resource found in shallow marine deposits is phosphate-rich sedimentary rock known as *phorphorite;* it is used for chemical fertilizers and animal feed supplements as well as matches, metallurgy, preserved foods, and ceramics.

A potential resource within the EEZ is methane hydrate, consisting of single methane molecules bound up in networks formed by frozen water. It is also found in permafrost soils. Methane hydrate is stable at water depths of more than 500 m and near-freezing temperatures. According to one estimate, the carbon in these deposits is double that in all coal, oil, and natural gas. However, no one knows if methane hydrate can be effectively recovered and used as an energy source. Another factor to consider is that the volume of methane hydrate is about 3,000 times greater than the volume of methane in the atmosphere—and methane is 10 times more effective than carbon dioxide as a greenhouse gas.

In some coastal locations a dam may be built across an inlet so that it regulates the flow of water landward during flood tides and releases water seaward at ebb tide. The moving water produces electricity by spinning turbines connected to generators, just as at hydroelectric power plants. Unfortunately, not many coastal locations are suitable for tidal power plants. Furthermore, the initial cost of such installations is high, and they can have disastrous effects on the ecology of estuaries.

▌ **Figure 16.26** Riprap and Sand Berm

a Riprap made up of large pieces of basalt was piled on this beach to protect a luxury hotel just to the left of this image.

b Heavy equipment builds a berm, an embankment of sand, on the seaward side of beach homes on the Isle of Palms, South Carolina, to protect them from waves. The berm must be rebuilt every two weeks after each spring tide.

GEO-FOCUS   The *Deepwater Horizon* Oil Spill

In Chapter 7 we noted that hydrocarbons (petroleum and natural gas) are found in stratigraphic traps that result from variations in strata such as buried ancient reefs, and in structural traps that are caused by folding or fracturing of rock layers (see Figures 7.18a, b). We also mentioned that hydrocarbons must be trapped in some permeable rock so that they can be withdrawn in large enough quantities to make a profit. These requirements hold no matter where hydrocarbons are found, on land or on the continental margins, although the latter present challenges not encountered on land.

Offshore oil provides a significant amount of all oil produced in the United States from wells in Alaska, California, and the Gulf Coast. Indeed, wells in the Gulf alone account for about 30% of U.S. oil production, and of that 30% much comes from wells in water more than 300 m deep and some in water more than 1,500 m deep. About 4,000 oil rigs are presently in the Gulf of Mexico in the waters adjacent to the United States. As you would expect, drilling for oil on the continental margins requires specialized technology and equipment, and it is a costly endeavor. Some offshore drill rigs are attached directly to the seafloor, but in deeper waters other platforms are necessary. For instance, ▌ Figure 1 shows the Shell Oil Company's tension-leg platform *Ursa* about 265 km south of New Orleans, Louisiana. It is anchored in 1,160 m of water by steel cables that hold the partly submerged platform in place.

The *Deepwater Horizon* drill rig that sank in the Gulf of Mexico about 66 km off the Louisiana coast was a semi-submersible offshore rig that was equipped to operate in water 2,100 m deep. Unfortunately, as it was drilling in 1,524 m of water, it exploded

on April 20, 2010, killing 11 workers and beginning the largest oil spill in the history of the petroleum industry (▌ Figure 2). Attempts to douse the oil rig fire were unsuccessful, and on the morning of April 22 the rig sank. The well was finally intercepted by a relief well, capped, and declared "dead" on September 19, 2010, after spilling an

estimated 4.9 million barrels of oil into the Gulf. Final estimates of the total damages and cost of cleanup are not in yet; however, preliminary estimates of the economic losses along the Gulf Coast are $30 billion.

The explosion was caused by methane gas under pressure that shot out of the well to the drilling platform where

Courtesy of Shell Oil Company

▌ **Figure 1 Offshore Oil Rig in the Gulf of Mexico** Shell Oil Company's tension-leg platform *Ursa* in the Gulf of Mexico south of New Orleans, Louisiana. It is anchored in 1,160 m of water by steel cables that hold the partly submerged platform in place. It is designed to withstand waves up to 22 m high and hurricane-force winds of 225 km/hr.

it ignited, but as for the flow of oil into the Gulf, a device called a blowout preventer at the wellhead on the sea-floor failed, allowing oil to gush out for three months. Who was responsible is a difficult question to answer because four corporations were involved in the project. BP was the company deemed responsible by the Coast Guard, but the drill rig itself was leased and was owned and operated by Transocean. Haliburton Company was responsible for a cement-ing job to cap the well, and Cameron International manufactured the blow-out preventer. Perhaps you recall the Senate hearings in which executives of these companies shifted blame from one to the other.

In any event, the failure of the blow-out preventer, a special valve to deal with pressure in wells, resulted in the leak-age of as much as 100,000* barrels of oil per day, although the rate decreased with time. Hardest hit was the coastline of Louisiana (❚ Figure 3), but the effects of the oil spill were obvious as far east as Florida. By December 2010, BP had spent billions on cleanup, a process that will continue for some time. And even though the gushing well has been capped since September 2010, the effects of the spill will be felt for years. In addition, the six-month moratorium imposed by the Obama administration on deepwater drilling was lifted in Sep-tember 2010. It is true that oily birds and beaches are not in the news now, and the areas closed to fishing during the leak have been reopened, but much of the spilled oil is still unaccounted for.

Of course the danger of additional oil spills is always present even though blowouts such as the *Deepwater Horizon* event are rare. No one is sure about the long-term effects of the recent spill on the ecology of the Gulf, but of further concern are the approximately 27,000

❚ **Figure 2** The *Deepwater Horizon* drilling platform not long before it sank in the Gulf of Mexico on April 22, 2010.

❚ **Figure 3** *Deepwater Horizon* **Oil Spill** Crews clean oil from a beach in LaFourche Parish, Louisiana. The most extensive damage from the oil spill occurred in Louisiana, but the shorelines of other states were also affected.

abandoned wells in the Gulf's waters, many of which were improperly sealed. Some of these wells were abandoned as far back as the late 1940s, and most are not monitored. We know that some

abandoned wells on land leak, so it is likely that some offshore wells leak, too.

---

* Estimates of the daily flow rate vary considerably, but most reports now put the high rate at about 62,000 barrels per day.

GEO-RECAP

» Geo-Image

Figure 1 The image on the right shows several distinctive shoreline features. What are they, and how do they form?

James S. Monroe

Chapter Summary

- Present-day research vessels are equipped to investigate the seafloor by sampling, drilling, echo sounding, and seismic profiling. Scientists also use submersibles in their studies.

- The upper 100 m or so of the oceans is the photic zone where sunlight is sufficient for photosynthesizing organisms. The aphotic zone lies below.

- Oceanic circulation is mostly horizontal in surface currents and deep-sea currents, but vertical circulation takes place, too.

- The sediments on the seafloor are mostly pelagic clay and ooze consisting of the skeletons of tiny organisms.

- Moundlike, wave-resistant structures consisting of animal skeletons are reefs. Most reefs are fringing reefs, barrier reefs, or atolls.

- Tides are caused by the combined effects of the Moon and the Sun on the oceans.

- As wind-generated waves enter shallow water, they become oversteepened and plunge forward as breakers or spill onto the shoreline, thus expending their kinetic energy.

- Longshore currents resulting from waves approaching a shoreline at an angle erode, transport, and deposit sediment.

- Rip currents carry excess water from the nearshore zone seaward through the breaker zone.

- Erosional coasts have sea cliffs, wave-cut platforms, and sea stacks, whereas depositional coasts have long sandy beaches, deltas, and barrier islands.

- Beaches are continuously modified by waves and nearshore currents, and their profiles usually show seasonal changes.

- Spits, baymouth bars, and tombolos all form and grow as a result of longshore transport and deposition.

- The sediment budget of a nearshore system remains rather constant unless the system is disrupted, as when dams are built across streams supplying sand to the system.

- Submergent and emergent coasts are defined on the basis of their relationship to changes in sea level.

- Coastal flooding during storms by waves and storm surges is an ongoing problem in many areas.

- The United States claims rights to all resources within 200 nautical miles of its shorelines, or what is called its Exclusive Economic Zone (EEZ).

Important Terms

aphotic zone (p. 390)
barrier island (p. 402)
baymouth bar (p. 401)
beach (p. 400)
breaker (p. 396)
Coriolis effect (p. 390)
downwelling (p. 391)

emergent coast (p. 405)
Exclusive Economic Zone (EEZ) (p. 409)
fetch (p. 396)
gyre (p. 391)
headland (p. 399)
longshore current (p. 396)
marine terrace (p. 398)

nearshore sediment budget (p. 403)
ooze (p. 391)
pelagic clay (p. 391)
photic zone (p. 390)
reef (p. 392)
rip current (p. 397)
salinity (p. 390)

shoreline (p. 392)
spit (p. 401)
storm surge (p. 406)
submergent coast (p. 404)

tide (p. 393)
tombolo (p. 401)
upwelling (p. 391)
wave (p. 395)

wave base (p. 395)
wave refraction (p. 396)
wave-cut platform (p. 398)

Review Questions

1. The most common ions in seawater are
   a. ____calcium and iron.
   b. ____silicon and oxygen.
   c. ____manganese and potassium.
   d. ____sodium and chlorine.
   e. ____aluminum and gold.

2. The distance the wind blows over a continuous water surface is the
   a. ____trough.
   b. ____wavelength.
   c. ____fetch.
   d. ____berm.
   e. ____celerity.

3. Longshore currents are generated by
   a. ____water flowing offshore through the breaker zone.
   b. ____the distance wind blows over water.
   c. ____abrasion and hydraulic action.
   d. ____waves approaching a shoreline at an angle.
   e. ____upwelling and downwelling.

4. Large-scale, wind-driven, surface-water currents in the oceans are known as
   a. ____gyres.
   b. ____tombolos.
   c. ____tides.
   d. ____arches.
   e. ____breakers.

5. Waves approaching a shoreline have wavelengths of 15 m, so wave base is at about ____m depth.
   a. ____30.
   b. ____5.0.
   c. ____7.5.
   d. ____45.
   e. ____6.2.

6. A long, narrow, sand deposit separated from the mainland by a lagoon is a/an
   a. ____tidal delta.
   b. ____beach cusp.
   c. ____barrier island.
   d. ____atoll.
   e. ____shelf.

7. A marine terrace is
   a. ____a type of submergent coast.
   b. ____made up of a beach face and a berm.
   c. ____the highest level reached by storm waves.
   d. ____a wave-cut platform above sea level.
   e. ____the area exposed during neap tides.

8. The depth in the oceans below about 100 m where there is too little sunlight for photosynthesis is called the
   a. ____bathosphere.
   b. ____hydrolysis area.
   c. ____aphotic zone.
   d. ____solution depth.
   e. ____wave base.

9. Areas in the oceans where cold water from depth rises and replaces warmer surface water are areas of
   a. ____upwelling.
   b. ____gyres.
   c. ____wave celerity.
   d. ____submergence.
   e. ____lateral mixing.

10. Erosion caused by the wearing action of water carrying sand and gravel is known as
    a. ____cavitation.
    b. ____solution.
    c. ____hydrolysis.
    d. ____emergence.
    e. ____abrasion.

11. Describe the sequence of events leading to the origin of an atoll.

12. How do rising sea level and coastal development, including developments on barrier islands, complicate efforts to control shoreline erosion?

13. Why does an observer at a shoreline location experience two high and two low tides daily?

14. Explain how ocean surface currents are generated and how the Coriolis effect influences them.

15. What is upwelling, and why is it important?

16. What is wave base, and how does it affect waves as they enter shallow water?

17. Explain why waves are so small on ponds and lakes, but much larger ones develop in the oceans.

18. As a member of a planning commission for a coastal community where shoreline erosion is a continuing problem, what recommendations would you make to remedy or at least mitigate the problem?

19. Describe longshore currents, and explain how they transport and deposit sediment.

20. While driving along North America's West Coast you notice a surface above sea level that slopes gently toward the sea and has several masses of rock rising above it. How would you explain the origin of this landform to your children?

The Grand Canyon, Arizona. Major John Wesley Powell led two expeditions down the Colorado River and through the canyon in 1869 and 1871. He was struck by the seemingly limitless time represented by the rocks exposed in the canyon walls and by the recognition that these rock layers, like the pages in a book, contain the geologic history of this region.

Geologic Time: Concepts and Principles

OUTLINE

17.1 Introduction

17.2 How Geologic Time Is Measured

17.3 Early Concepts of Geologic Time and Earth's Age

17.4 James Hutton and the Recognition of Geologic Time

17.5 Relative Dating Methods

17.6 Correlating Rock Units

17.7 Absolute Dating Methods

17.8 Development of the Geologic Time Scale

17.9 Stratigraphy and Stratigraphic Terminology

17.10 Geologic Time and Climate Change

GEO-FOCUS: Denver's Weather—280 Million Years Ago!

GEO-RECAP

HAVE YOU EVER WONDERED?

- How geologic time is measured?
- What the age of Earth is?
- Why it is important to have an appreciation and understanding of geologic time?
- How geologic events in different parts of the world are determined to be synchronous?
- What radioactivity is, and how it can be used to determine the age of different rocks?
- How the geologic time scale came into being?
- How geologists reconstruct past climatic regimes and changes?

17.1 Introduction

In 1869, Major John Wesley Powell, a Civil War veteran who lost his right arm in the battle of Shiloh, led a group of hardy explorers down the uncharted Colorado River through the Grand Canyon. With no maps or other information, Powell and his group ran the many rapids of the Colorado River in fragile wooden boats, hastily recording what they saw. Powell wrote in his diary that "all about me are interesting geologic records. The book is open and I read as I run."

Probably no one has contributed as much to the understanding of the Grand Canyon as Major Powell. In recognition of his contributions, the Powell Memorial was erected on the South Rim of the Grand Canyon in 1969 to commemorate the 100th anniversary of this history-making first expedition.

Most tourists today, like Powell and his fellow explorers in 1869, are astonished by the seemingly limitless time represented by the rocks exposed in the walls of the Grand Canyon. For most visitors, viewing a 1.5-km-deep cut into Earth's crust is the only encounter they'll ever have with the magnitude of geologic time. When standing on the rim and looking down into the Grand Canyon, we are really looking far back in time—all the way back to the early history of our planet. In fact, more than 1 billion years of history are preserved in the rocks of the Grand Canyon.

Vast periods of time set geology apart from most of the other sciences, and an appreciation of the immensity of geologic time is fundamental to understanding the physical and biological history of our planet.

Besides providing an appreciation for the immensity of geologic time, why is the study of geologic time important? One of the most valuable lessons you will learn in this chapter is how to reason and apply some of the fundamental geologic principles to solve basic geologic problems. The logic used in applying these principles to interpret the geologic history of an area involves basic reasoning skills that can be transferred to and used in almost any profession or discipline.

17.2 How Geologic Time Is Measured

In some respects, time is defined by the methods used to measure it. Geologists use two different frames of reference when discussing geologic time. **Relative dating** is placing geologic events in a sequential order as determined from their position in the geologic record. Relative dating will not tell us how long ago a particular event took place, only that one event preceded another.

The principles used to determine relative dating were discovered hundreds of years ago, and since then, they have been used to construct the *relative geologic time scale*

(▌Figure 17.1). Furthermore, these principles are still widely used by geologists today.

Absolute dating provides specific dates for rock units or events expressed in years before the present. Radiometric dating is the most common method of obtaining absolute ages. Dates are calculated from the natural decay rates of various radioactive elements present in trace amounts in some rocks. It was not until the discovery of radioactivity near the end of the 19th century that absolute ages could be accurately applied to the relative geologic time scale.

Today, the geologic time scale is really a dual scale: a relative scale based on rock sequences with radiometric dates expressed as years before the present (Figure 17.1).

17.3 Early Concepts of Geologic Time and Earth's Age

The concept of geologic time and its measurement have changed throughout human history. Some early Christian scholars and clerics tried to establish the date of creation by analyzing historical records and the genealogies found in Scripture. One of the most influential scholars was James Ussher (1581–1656), Archbishop of Armagh, Ireland, who, based upon Old Testament genealogy, asserted that God created Earth on Sunday, October 23, 4004 BC. In 1701, an authorized version of the Bible made this date accepted Church doctrine. For nearly a century thereafter, it was considered heresy to assume that Earth and all of its features were more than about 6,000 years old. Thus, the idea of a very young Earth provided the basis for most Western chronologies of Earth history prior to the 18th century.

During the 18th and 19th centuries, several attempts were made to determine Earth's age on the basis of scientific evidence rather than revelation. The French zoologist Georges Louis de Buffon (1707–1788) assumed that Earth gradually cooled to its present condition from a molten beginning. To simulate this history, he melted iron balls of various diameters and allowed them to cool to the surrounding temperature. By extrapolating their cooling rate to a ball the size of Earth, he determined that Earth was at least 75,000 years old. Although this age was much older than that derived from Scripture, it was vastly younger than we now know our planet to be.

Other scholars were equally ingenious in attempting to calculate Earth's age. For example, if deposition rates could be determined for various sediments, geologists reasoned that they could calculate how long it would take to deposit any rock layer. They could then extrapolate how old Earth was from the total thickness of sedimentary rock in its crust. Rates of deposition vary, however, even for the same type of rock. Furthermore, it is impossible to estimate how much of a rock has been removed by erosion, or how much a rock sequence has been reduced by compaction. As a result of these variables, estimates of Earth's age ranged from younger than 1 million years to older than 2 billion years.

Another attempt to determine Earth's age involved ocean salinity. Scholars assumed that Earth's ocean waters were originally fresh and that their present salinity was the result of dissolved salt being carried into the ocean basins by streams. Knowing the volume of ocean water and its salinity, John Joly, a 19th-century Irish geologist, measured the amount of salt currently in the world's streams. He then calculated that it would have taken at least 90 million years for the oceans to reach their present salinity level. This was still much younger than the now-accepted age of 4.6 billion years for Earth, mainly because Joly had no way to calculate either how much salt had been recycled or the amount of salt stored in continental salt deposits and seafloor clay deposits.

Besides trying to determine Earth's age, the naturalists of the 18th and 19th centuries were formulating some of the fundamental geologic principles that are used in deciphering Earth history. From the evidence preserved in the geologic record, it was clear to them that Earth is very old and that geologic processes have operated over long periods of time.

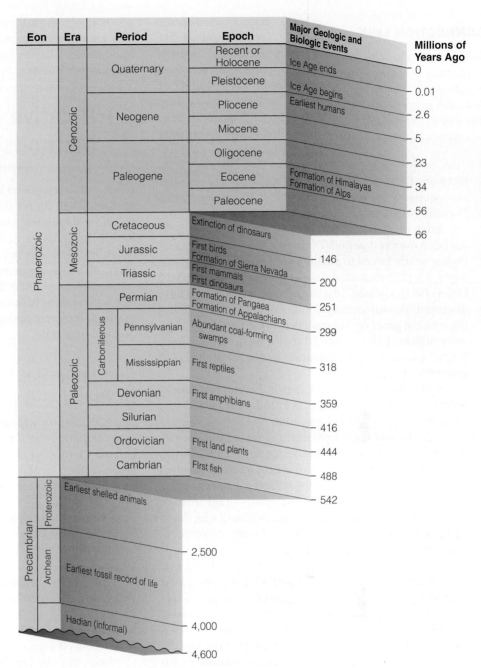

Figure 17.1 The Geologic Time Scale Some of the major geologic and biologic events are indicated for the various eras, periods, and epochs. Dates are from the 2009 International Stratigraphic Chart, © 2009 by the International Commission on Stratigraphy.

17.4 James Hutton and the Recognition of Geologic Time

Many consider the Scottish geologist James Hutton (1726–1797) to be the father of modern geology. His detailed studies and observations of rock exposures and present-day geologic processes were instrumental in establishing the *principle of uniformitarianism* (see Chapter 1), the concept that the same processes seen today have operated over vast amounts of time. Because Hutton relied on known processes to account for Earth history, he concluded that Earth must be very old and wrote that "we find no vestige of a beginning, and no prospect of an end."

In 1830, Charles Lyell published a landmark book, *Principles of Geology*, in which he championed Hutton's concept of uniformitarianism. Instead of relying on catastrophic

CONNECTION LINK

To learn more about Darwin's theory of organic evolution and its implications, see pages 451–453 and 456 in Chapter 18.

events to explain various Earth features, Lyell recognized that imperceptible changes brought about by present-day processes could, over long periods of time, have tremendous cumulative effects. Through his writings, Lyell firmly established uniformitarianism as the guiding principle of geology. Furthermore, the recognition of vastly long periods of time was also necessary for, and instrumental in, the acceptance of Darwin's 1859 theory of organic evolution.

After establishing that present-day processes have operated over vast periods of time, geologists were nevertheless nearly forced to accept a very young age for Earth when a highly respected English physicist, Lord Kelvin (1824–1907), claimed, in a paper written in 1866, to have destroyed the uniformitarian foundation of geology. Starting with the generally accepted belief that Earth was originally molten, Kelvin assumed that it has gradually been losing heat and that, by measuring this heat loss, he could determine its age.

Kelvin knew from deep mines in Europe that Earth's temperature increases with depth, and he reasoned that Earth is losing heat from its interior. By knowing the size of Earth, the melting temperatures of rocks, and the rate of heat loss, Kelvin calculated the age at which Earth was entirely molten. From these calculations, he concluded that Earth could not be older than 400 million years nor younger than 20 million years. This wide discrepancy in age reflected uncertainties in average temperature increases with depth and the various melting points of Earth's constituent materials.

After establishing that Earth was very old and that present-day processes operating over long periods of time account for geologic features, geologists were in a quandary. If they accepted Kelvin's dates, they would have to abandon the concept of seemingly limitless time that was the underpinning of uniformitarian geology and one of the foundations of Darwinian organic evolution and squeeze events into a shorter time frame.

Kelvin's reasoning and calculations were sound, but his basic premises were false, thereby invalidating his conclusions. Kelvin was unaware that Earth has an internal heat source, radioactivity, that has allowed it to maintain a fairly constant temperature through time.* His 40-year campaign for a young Earth ended with the discovery of radioactivity near the end of the 19th century and the insight in 1905 that natural radioactive decay can be used in many cases to date how long ago a rock formed. His calculations were no longer valid, and his proof for a geologically young Earth collapsed.

Although the discovery of radioactivity destroyed Kelvin's arguments, it provided geologists with a clock that could measure Earth's age and validate what geologists had long thought—namely, that Earth was indeed very old!

17.5 Relative Dating Methods

Before the development of radiometric dating techniques, geologists had no reliable means of absolute dating and therefore depended solely on relative dating methods. Recall that relative dating places events in sequential order, but does not tell us how long ago an event took place. Although the principles of relative dating may now seem self-evident, their discovery was an important scientific achievement because they provided geologists with a means to interpret geologic history and develop a relative geologic time scale.

Fundamental Principles of Relative Dating

The 17th century was an important time in the development of geology as a science because of the widely circulated writings of the Danish anatomist Nicolas Steno (1638–1686). Steno observed that when streams flood, they spread out across their floodplains and deposit layers of sediment that bury organisms dwelling on the floodplain. Subsequent floods produce new layers of sediments that are deposited or superposed over previous deposits. When lithified, these layers of sediment become sedimentary rock.

Thus, in an undisturbed succession of sedimentary rock layers, the oldest layer is at the bottom and the youngest layer is at the top. This **principle of superposition** is the basis for relative-age determinations of strata and their contained fossils (▌Figure 17.2a).

Steno also observed that, because sedimentary particles settle from water under the influence of gravity, sediment is deposited in essentially horizontal layers, thus illustrating the **principle of original horizontality** (Figure 17.2a). Therefore, a sequence of sedimentary rock layers that is steeply inclined from the horizontal must have been tilted after deposition and lithification (▌Figure 17.2b).

Steno's third principle, the **principle of lateral continuity**, states that sediment extends laterally in all directions until it thins and pinches out or terminates against the edge of the depositional basin (Figure 17.2a).

James Hutton is credited with formulating the **principle of cross-cutting relationships**. Based on his detailed studies and observations of rock exposures in Scotland, Hutton recognized that an igneous intrusion or fault must be younger than the rocks that it intrudes or displaces (▌Figure 17.3).

---

*Actually, Earth's temperature has decreased through time because the original amount of radioactive materials has been decreasing and thus is not supplying as much heat. However, the temperature is decreasing at a rate considerably slower than would be required to lend any credence to Kelvin's calculations.

Figure 17.2 The Principles of Original Horizontality, Superposition, and Lateral Continuity

a The sedimentary rocks of Bryce Canyon National Park, Utah, were originally deposited horizontally in a variety of continental environments (principle of original horizontality). The oldest rocks are at the bottom of this highly dissected landscape, and the youngest rocks are at the top, forming the rims (principle of superposition). The exposed rock layers extend laterally in all directions for some distance (principle of lateral continuity).

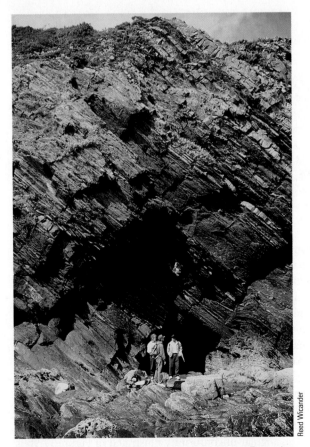

b These shales and limestones of the Postolonnec Formation, at Postolonnec Beach, Crozon Peninsula, France, were originally deposited horizontally, but have been significantly tilted since their formation.

Figure 17.3 The Principle of Cross-Cutting Relationships

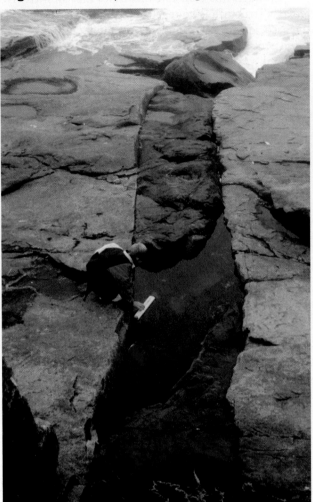

a A dark gabbro dike cuts across granite in Acadia National Park, Maine. The dike is younger than the granite that it intrudes.

b A small fault (arrows show direction of movement) cuts across, and thus displaces, tilted sedimentary beds along the Templin Highway in Castalc, California. The fault is therefore younger than the youngest beds that are displaced.

Although this principle illustrates that an intrusive igneous structure is younger than the rocks it intrudes, the association of sedimentary and igneous rocks may cause problems in relative dating. Buried lava flows and sills look very similar in a sequence of strata (❚ Figure 17.4). A buried lava flow, however, is older than the rocks above it (principle of superposition), whereas a sill, resulting from later igneous intrusion, is younger than all of the beds below it and younger than the immediately overlying bed as well.

To resolve such relative-age problems as these, geologists observe whether the sedimentary rocks in contact with the igneous rocks show signs of baking or alteration by heat (contact metamorphism). A sedimentary rock that shows such effects must be older than the igneous rock with which it is in contact. In Figure 17.4, for example, a sill produces a zone of baking immediately above and below it because it intruded into previously existing sedimentary rocks. A lava flow, in contrast, bakes only those rocks below it.

Another way to determine relative ages is by using the **principle of inclusions**. This principle holds that inclusions, or fragments of one rock contained within a layer of another, are older than the rock layer itself. The batholith shown in ❚ Figure 17.5a contains sandstone inclusions, and the sandstone unit shows the effects of baking. Accordingly, we conclude that the sandstone is older than the batholith. In ❚ Figure 17.5b, however, the sandstone contains granite rock pieces, indicating that the batholith was the source rock for the inclusions and is therefore older than the sandstone.

Fossils have been known for centuries, yet their utility in relative dating and geologic mapping was not fully appreciated until the early 19th century. William Smith (1769–1839), an English civil engineer involved in surveying and building canals in southern England, independently recognized the principle of superposition by reasoning that the fossils at the bottom of a sequence of strata are older than those at the top of the sequence. This recognition served as the basis for the **principle of fossil succession**, or the *principle of faunal and floral succession*, as it is sometimes called (❚ Figure 17.6).

According to this principle, fossil assemblages succeed one another through time in a regular and predictable order. The validity and successful use of this principle depend on three points: (1) life has varied through time, (2) fossil assemblages are recognizably different from one another, and (3) the relative ages of the fossil assemblages can be determined. Observations of fossils in older versus younger strata clearly demonstrate that life-forms have

> **CONNECTION LINK**
>
> To learn more about fossils and their importance in evolution, see pages 467 and 468 in Chapter 18.

changed. Because this is true, fossil assemblages (point 2) are recognizably different. Furthermore, superposition can be used to demonstrate the relative ages of the fossil assemblages.

Unconformities

Our discussion so far has been concerned with conformable strata—sequences in which no deposition breaks of any consequence occur; that is, sedimentation was more or less continuous. A bedding plane between strata may represent a depositional break of anywhere from minutes to tens of years, but it is inconsequential in the context of geologic time. However, in some sequences of strata, surfaces known as **unconformities** may be present, representing times of nondeposition, erosion, or both. Unconformities encompass long periods of geologic time, perhaps millions or tens of millions of years. Accordingly, the geologic record is incomplete wherever an unconformity is present, just as a book with missing pages is incomplete, and the interval of geologic time not represented by strata is called a *hiatus* (❚ Figure 17.7).

Geologists recognize three types of unconformities. First, a **disconformity** is a surface of erosion or nondepo-

> ## What Would You Do ❓
>
> You have been chosen to be part of the first astronaut crew to land on Mars. You were selected because you are a geologist, and therefore your primary responsibility is to map the geology of the landing site area. An important goal of the mission is to work out the geologic history of the area. How will you go about this? Will you be able to use the principles of relative dating? How will you correlate the various rock units? Will you be able to determine absolute ages? How will you do this?

sition separating younger from older rocks, both of which are parallel with one another (❚ Figure 17.8). Unless the erosional surface separating the older from the younger parallel beds is well defined or distinct, the disconformity frequently resembles an ordinary bedding plane. Hence, many disconformities are difficult to recognize and must be identified on the basis of fossil assemblages.

Second, an **angular unconformity** is an erosional surface on tilted or folded strata over which younger strata were deposited (❚ Figure 17.9). The strata below the unconformable surface generally dip more steeply than those above, producing an angular relationship.

The angular unconformity illustrated in Figure 17.9b is probably the most famous in the world. It was here at Siccar

▌**Figure 17.4 Differentiating Between a Buried Lava Flow and a Sill** In (a) and (b) below, the ages of sedimentary strata are shown by numbering from 1–6, with 1 being the oldest sedimentary rock layer in each frame and 6 the youngest. Remember that the lava flow took place during deposition of an ordinary sedimentary sequence, whereas the sill intruded the sedimentary strata after they had accumulated. The notes in both frames highlight the physical features one would look for to distinguish between a lava flow and a sill, which can look quite similar in an outcrop.

Clasts of lava may be present in the overlying, younger layer

Strands and shreds of underlying sedimentary material may be in the base of the flow

Baking zone at bottom of flow

Rubble zones may be present at the top and bottom of the flow

a A buried lava flow has baked underlying bed 2 when it flowed over it. Clasts of the lava were deposited along with other sediments during deposition of bed 3. The lava flow is younger than bed 2 and older than beds 3, 4, and 5.

Baking zones on both sides of sill

Inclusions of rock from both layers above 3 and below 2 may exist in the sill

b The rock units above and below the sill have been baked, indicating that the sill is younger than beds 2 and 3, but its age relative to beds 4–6 cannot be determined.

c This buried lava flow in Yellowstone National Park, Wyoming, displays columnar jointing. A baked zone is present below, and not above, the igneous structure, indicating that it is a buried lava flow and not a sill.

>> **Critical Thinking Question**
In (b), it is stated that the sill is younger than beds 2 and 3, but its age relative to beds 4–6 cannot be determined. Why is this?

Lava flow

Reed Wicander

▌**Figure 17.5** The Principle of Inclusions

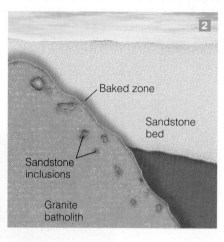

Sandstone bed

Granitic magma

Baked zone

Sandstone bed

Sandstone inclusions

Granite batholith

a The sandstone is older than the granite batholith because there are inclusions of sandstone inside the granite. The sandstone also shows evidence of having been baked along its contact with the granite batholith when the granitic magma intruded the overlying sedimentary beds.

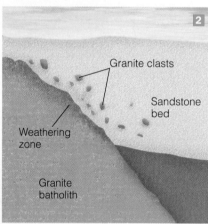

Sandstone deposition

Sandstone bed

Granite batholith

Granite clasts

Weathering zone

Sandstone bed

Granite batholith

b The sandstone is younger than the granite batholith because it contains pieces (clasts) of granite. The granite is also weathered along the contact with the sandstone, indicating that it was the source of the granite clasts, and must therefore be older than the sandstone.

Granite

Basalt inclusion

James S. Monroe

c Outcrop in northern Wisconsin showing basalt inclusions (dark gray) in granite (white). Accordingly, the basalt inclusions are older than the granite.

▌Figure 17.6 The Principle of Fossil Succession This generalized diagram shows how geologists use the principle of fossil succession to identify strata of the same age in different areas. The rocks in the three sections connected by the dashed lines contain similar fossils and are therefore the same age. Note that the youngest rocks in this region are in section B, whereas the oldest rocks are in section C.

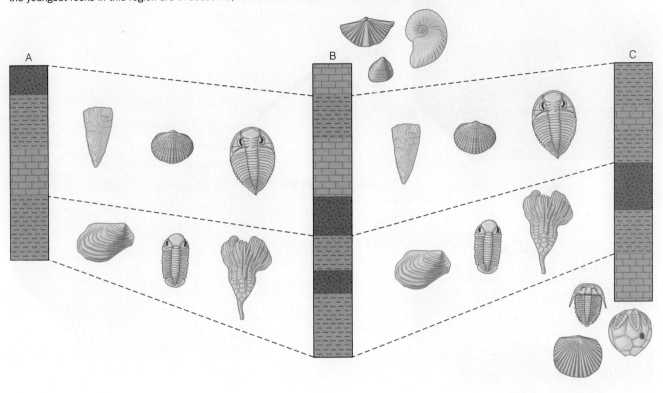

▌Figure 17.7 The Development of a Hiatus and an Unconformity

a Deposition began 12 million years ago (mya) and continued more or less uninterrupted until 4 mya.

b Between 3 and 4 mya, an episode of erosion occurred. During that time, some of the strata deposited earlier was eroded.

c A hiatus of 3 million years thus exists between the older strata and the strata that formed during a renewed episode of deposition that began 3 mya.

d The actual stratigraphic record as seen in an outcrop today. The unconformity is the surface separating the strata and represents a major break in our record of geologic time.

Figure 17.8 Formation of a Disconformity

Uplift and erosion

Deposition

Disconformity

Uplift and erosion

Deposition

a Formation of disconformity.

Jurassic rocks

Mississippian rocks

James S. Monroe

b Disconformity between Mississippian and Jurassic strata in Montana. The geologist at the upper left is sitting on Jurassic strata, and his right foot is resting on Mississippian rocks. This disconformity represents approximately 165 million years.

Point, Scotland, that James Hutton realized that severe upheavals had tilted the lower rocks and formed mountains that were then worn away and covered by younger, flat-lying rocks. The erosional surface between the older tilted rocks and the younger flat-lying strata meant that a significant gap existed in the geologic record. Although Hutton did not use the term

unconformity, he was the first to understand and explain the significance of such discontinuities in the geologic record.

A **nonconformity** is the third type of unconformity. Here, an erosional surface cut into metamorphic or igneous rocks is covered by sedimentary rocks (**Figure 17.10). This type of unconformity closely resembles an intrusive

Figure 17.9 Formation of an Angular Unconformity

Deposition

Uplift and erosion

Erosion

Angular
unconformity

Uplift and tilting

Deposition

Marli Bryant Miller

a Formation of an angular unconformity.

b Angular unconformity at Siccar Point, Scotland. James Hutton first realized the significance of unconformities at this site in 1788.

>> Critical Thinking Question
From what you can see in this photo, which is the same scene Hutton viewed in 1788, what evidence is there that this outcrop represents an ancient mountain range? Were the forces compressional or tensional?

igneous contact with sedimentary rocks. The principle of inclusions (Figure 17.5) is helpful in determining whether the relationship between the underlying igneous rocks and

the overlying sedimentary rocks is the result of an intrusion or erosion. A nonconformity is also marked in many places by an ancient zone of weathering, or even a reddened,

▌**Figure 17.10** Formation of a Nonconformity

Deposition

Uplift and erosion

Nonconformity

Uplift and erosion of overlying sediments

Intrusion of magma

a Formation of a nonconformity.

Paleozoic sedimentary rocks

Nonconformity

Precambrian metamorphic rocks

Marli Bryant Miller

b Nonconformity between Precambrian metamorphic rocks and overlying Paleozoic sedimentary rocks in the Grand Canyon, Arizona.

Figure 17.11 Block Diagram of a Hypothetical Area A block diagram of a hypothetical area in which the various relative dating principles can be applied to determine its geologic history. See Figure 17.12 to learn how the geologic history was determined using relative dating principles.

bricklike soil horizon, or paleosol. In the case of an intrusion, the igneous rocks are younger, whereas in the case of erosion, the sedimentary rocks are younger. Being able to distinguish between a nonconformity and an intrusive contact is important because they represent different sequences of events.

Applying the Principles of Relative Dating

We can decipher the geologic history of the area represented by the block diagram in ▮ Figure 17.11 by applying the various relative-dating principles just discussed. The methods and logic used in this example are the same as those applied by 19th-century geologists in constructing the geologic time scale.

According to the principles of superposition and original horizontality, beds A–G were deposited horizontally; then either they were tilted, faulted (H), and eroded, or after deposition, they were faulted (H), tilted, and then eroded (▮ Figure 17.12a–c). Because the fault cuts beds A–G, it must be younger than the beds according to the principle of cross-cutting relationships.

Beds J–L were then deposited horizontally over this erosional surface, producing an angular unconformity (I)

(▮ Figure 17.12d). Following deposition of these three beds, the entire sequence was intruded by a dike (M), which, according to the principle of cross-cutting relationships, must be younger than all of the rocks that it intrudes (▮ Figure 17.12e).

The entire area was then uplifted and eroded; next, beds P and Q were deposited, producing a disconformity (N) between beds L and P and a nonconformity (O) between the igneous intrusion M and the sedimentary bed P (▮ Figure 17.12f, g). We know that the relationship between igneous intrusion M and the overlying sedimentary bed P is a nonconformity because of the inclusions of M in P (principle of inclusions).

At this point, there are several possibilities for reconstructing the geologic history of this area. According to the principle of cross-cutting relationships, dike R must be younger than bed Q because it intrudes into it. It could have intruded anytime *after* bed Q was deposited; however, we cannot determine whether R was formed right after Q, right after S, or after T was formed. For purposes of this history, we will say that it intruded after the deposition of bed Q (▮ Figure 17.12g, h).

Following the intrusion of dike R, lava S flowed over bed Q, followed by the deposition of bed T (▮ Figure 17.12i, j). Although the lava flow (S) is not a sedimentary unit, the principle of superposition still applies because it flowed onto the surface, just as sediments are deposited on Earth's surface.

a Sediment deposition and lithification.

b Uplift, tilting, and faulting.

c Erosion.

d Sediment deposition and lithification.

e Intrusion.

f Uplift and erosion.

g Sediment deposition and lithification.

h Intrusion.

i Lava flow.

j Sediment deposition.

❚ **Figure 17.12** Using Relative Dating Principles to Interpret the Geologic History of a Hypothetical Area **(a)** Beds A–G are deposited and lithified. **(b)** The preceding beds are tilted and faulted. **(c)** Erosion. **(d)** Beds J–L are deposited and lithified, producing an angular unconformity (I). **(e)** The entire sequence is intruded by a dike (M). **(f)** The entire sequence is uplifted and eroded. **(g)** Beds P and Q are deposited and lithified, producing a disconformity (N) and a nonconformity (O). **(h)** Dike R intrudes. **(i)** Lava (S) flows over bed Q, baking it. **(j)** Bed T is deposited.

We have established a relative chronology for the rocks and events of this area by using the principles of relative dating. Remember, however, that we have no way of knowing how many years ago these events occurred unless we can obtain radiometric dates for the igneous rocks. With these dates, we can establish the range of absolute ages between which the different sedimentary units were deposited and also determine how much time is represented by the unconformities.

17.6 Correlating Rock Units

To decipher Earth history, geologists must demonstrate the time equivalency of rock units in different areas. This process is known as **correlation**.

If surface exposures are adequate, units may simply be traced laterally (principle of lateral continuity), even if occasional gaps exist (❚ Figure 17.13). Other criteria used

Key bed
volcanic ash

Figure 17.13 Correlating Rock Units In areas of adequate exposures, rock units can be traced laterally, even if occasional gaps exist, and correlated on the basis of similarity in rock type and position in a sequence. Rocks can also be correlated by a key bed—in this case, volcanic ash.

>> Critical Thinking Question
Can you give at least one explanation for why the lower tongue of sandstone present in the left and center columns does not reach all the way to the column on the right?

to correlate units are similarity of rock type, position in a sequence, and key beds. *Key beds* are units, such as coal beds or volcanic ash layers, that are sufficiently distinctive to allow identification of the same unit in different areas (Figure 17.13).

Generally, no single location in a region has a geologic record of all events that occurred during its history; therefore, geologists must correlate from one area to another to determine the complete geologic history of the region. An excellent example is the history of the Colorado Plateau (▌ Figure 17.14). This region provides a record of events occurring over approximately 2 billion years. Because of the forces of erosion, the entire record is not preserved at any single location. Within the walls of the Grand Canyon are rocks of the Precambrian and Paleozoic eras, whereas Paleozoic and Mesozoic Era rocks are found in Zion National Park, and Mesozoic and Cenozoic Era rocks are exposed in Bryce Canyon (Figure 17.14). By correlating the uppermost rocks at one location with the lowermost equivalent rocks of another area, geologists can decipher the history of the entire region.

Although geologists match up rocks on the basis of similar rock type and superposition, correlation of this type can be done only in a limited area where beds can be traced from one site to another. To correlate rock units over a large area or to correlate age-equivalent units of different composition, fossils and the principle of fossil succession must be used.

Fossils are useful as relative time indicators because they are the remains of organisms that lived for a certain length of time during the geologic past. Fossils that are easily identified, are geographically widespread, and existed for a rather

short interval of geologic time are particularly useful. Such fossils are **guide fossils**, or *index fossils* (▌ Figure 17.15). The trilobite *Paradoxides* and the brachiopod *Atrypa* meet these criteria and are therefore good guide fossils. In contrast, the brachiopod *Lingula* is easily identified and widespread, but its long geologic range of Ordovician to Recent makes it of little use in correlation.

Because most fossils have fairly long geologic ranges, geologists construct *concurrent range zones* to determine the age of the sedimentary rocks containing the fossils. Concurrent range zones are established by plotting the overlapping geologic ranges of two or more fossils that have different geologic ranges (▌ Figure 17.16). The first and last occurrences of fossils are used to determine zone boundaries. Correlating concurrent range zones is probably the most accurate method of determining time equivalence.

What Would You Do ?

You are a member of a regional planning commission that is considering a plan for constructing what is said to be a much-needed river dam that will create a recreational lake. Opponents of the dam project have come to you with a geologic report and map showing that a fault underlies the area of the proposed dam, and the fault trace can be clearly seen at the surface. The opponents say that the fault may be active, and thus someday it will move suddenly, bursting the dam and sending a wall of water downstream. You seek the advice of a local geologist who has worked in the dam area. She tells you that she found a lava flow covering the fault, less than a kilometer from the proposed dam project site. Can you use this information, along with a radiometric date from the lava flow, to help convince the opponents that the fault has not moved in any direction (vertically or laterally) in the recent past? How would you do this, and what type of reasoning would you use?

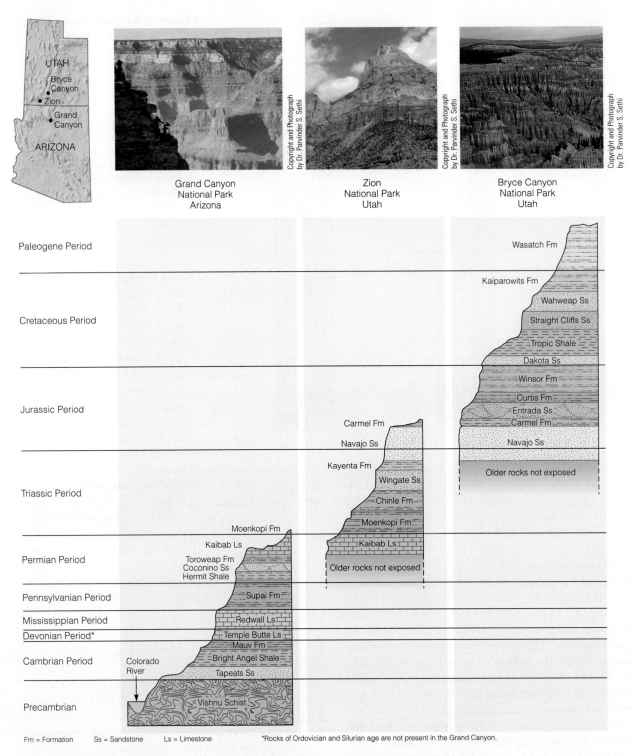

Fm = Formation   Ss = Sandstone   Ls = Limestone   *Rocks of Ordovician and Silurian age are not present in the Grand Canyon.

Figure 17.14 Correlation of Rock Units Within the Colorado Plateau At each location, only a portion of the geologic record of the Colorado Plateau is exposed. By correlating the youngest rocks at one exposure with the oldest rocks at another exposure, geologists can determine the entire history of the region. For example, the rocks forming the rim of the Grand Canyon, Arizona, are the Kaibab Limestone and Moenkopi Formation and are the youngest rocks exposed in the Grand Canyon. The Kaibab Limestone and Moenkopi Formation are the oldest rocks exposed in Zion National Park, Utah, and the youngest rocks are the Navajo Sandstone and Carmel Formation. The Navajo Sandstone and Carmel Formation are the oldest rocks exposed in Bryce Canyon National Park, Utah. By correlating the Kaibab Limestone and Moenkopi Formation between the Grand Canyon and Zion National Park, geologists have extended the geologic history from the Precambrian to the Jurassic. And, by correlating the Navajo Sandstone and Carmel Formation between Zion and Bryce Canyon National Parks, geologists can extend the geologic history through the Paleogene Period. Thus, by correlating the rock exposures between these areas and applying the principle of superposition, geologists can reconstruct the geologic history of the region.

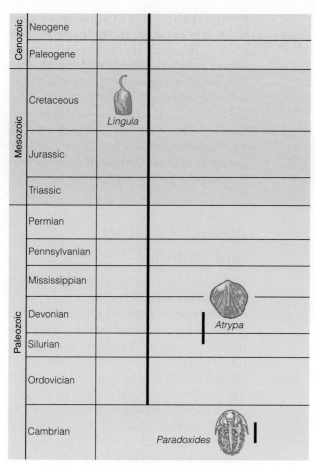

Figure 17.15 Guide Fossils Comparison of the geologic ranges (heavy vertical lines) of three marine invertebrate animals. *Lingula* is of little use in correlation because it has such a long range. But *Atrypa* and *Paradoxides* are good guide fossils because both are widespread, are easily identified, and have short geologic ranges. Thus, both can be used to correlate rock units that are widely separated and to establish the relative age of a rock that contains them.

Figure 17.16 Correlation of Two Sections Using Concurrent Range Zones This concurrent range zone was established by the overlapping geologic ranges of fossils symbolized here by the letters A through E. The concurrent range zone is of shorter duration than any of the individual fossil geologic ranges. Correlating by concurrent range zones is probably the most accurate method of determining time equivalence.

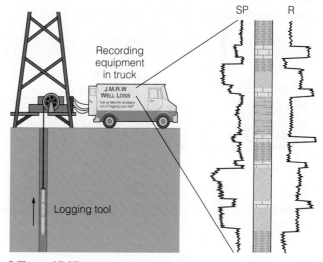

Figure 17.17 Well Logs A schematic diagram showing how well logs are made. As the logging tool is withdrawn from the drill hole, data are transmitted to the surface, where they are recorded and printed as a well log. The curve labeled SP in this diagrammatic electric log is a plot of self-potential (electrical potential caused by different conductors in a solution that conducts electricity) with depth. The curve labeled R is a plot of electrical resistivity with depth. Electric logs yield information about the rock type and fluid content of subsurface formations. Electric logs are also used to correlate from well to well.

Subsurface Correlation

In addition to surface geology, geologists are interested in subsurface geology because it provides additional information about geologic features beneath Earth's surface. A variety of techniques and methods are used to acquire and interpret data about the subsurface geology of an area.

When drilling is done for oil or natural gas, cores or rock chips called *well cuttings* are commonly recovered from the drill hole. These samples are studied under the microscope and reveal such important information as rock type, porosity (the amount of pore space), permeability (the ability to transmit fluids), and the presence of oil stains. In addition, the samples can be processed for a variety of microfossils that aid in determining the geologic age of the rock and the environment of deposition.

Geophysical instruments may be lowered down the drill hole to record such rock properties as electrical resistivity and radioactivity, thus providing a record or *well log* of the rocks penetrated. Cores, well cuttings, and well logs are all extremely useful in making subsurface correlations (Figure 17.17).

> **CONNECTION LINK**
>
> To review what porosity and permeability are and their effect on the transmission of groundwater in different rocks, see pages 314 and 315 in Chapter 13.

Figure 17.18 Three Types of Radioactive Decay

○ Proton ○ Neutron ○ Electron

a Alpha decay, in which an unstable parent nucleus emits 2 protons and 2 neutrons.

○ Proton ○ Neutron ○ Electron

b Beta decay, in which an electron is emitted from the nucleus.

○ Proton ○ Neutron ○ Electron

c Electron capture, in which a protein captures an electron and is thereby converted to a neutron.

Subsurface rock units may also be detected and traced by the study of seismic profiles. Energy pulses, such as those from explosions, travel through rocks at a velocity determined by rock density, and some of this energy is reflected from various horizons (contacts between contrasting layers) back to the surface, where it is recorded (see Figure 16.3). Seismic stratigraphy is particularly useful in tracing units in areas such as the continental shelves, where it is very expensive to drill holes and other techniques have limited use.

17.7 Absolute Dating Methods

Although most of the isotopes of the 92 naturally occurring elements are stable, some are radioactive and

spontaneously decay to other, more stable isotopes of elements, releasing energy in the process. The discovery in 1903 by Pierre and Marie Curie that radioactive decay produces heat meant that geologists finally had a mechanism for explaining Earth's internal heat that did not rely on residual cooling from a molten origin. Furthermore, geologists now had a powerful tool to date geologic events accurately and to verify the long time periods postulated by Hutton and Lyell.

Atoms, Elements, and Isotopes

As discussed in Chapter 3, all matter is made up of chemical elements, each composed of extremely small particles called *atoms*. The *nucleus* of an atom is composed of *protons* (positively charged particles) and *neutrons* (neutral particles), with *electrons* (negatively charged particles) encircling it (see Figure 3.2). The number of protons defines an element's *atomic number* and helps determine its properties and characteristics.

The combined number of protons and neutrons in an atom is its *atomic mass number*. However, not all atoms of the same element have the same number of neutrons in their nuclei. These variable forms of the same element are called *isotopes* (see Figure 3.4). Most isotopes are stable, but some are unstable and spontaneously decay to a more stable form. It is the decay rate of unstable isotopes that geologists measure to determine the absolute ages of rocks.

Radioactive Decay and Half-Lives

Radioactive decay is the process whereby an unstable atomic nucleus is spontaneously transformed into an atomic nucleus of a different element. Three types of radioactive decay are recognized, all of which result in a change of atomic structure (Figure 17.18).

In *alpha decay*, 2 protons and 2 neutrons are emitted from the nucleus, resulting in the loss of 2 atomic numbers and 4 atomic mass numbers. In *beta decay*, a fast-moving electron is emitted from a neutron in the nucleus, changing that neutron to a proton and consequently increasing the atomic number by 1, with no resultant atomic mass number change. *Electron capture* takes place when a proton captures an electron from an electron shell and thereby converts to a neutron, resulting in the loss of 1 atomic number, but not changing the atomic mass number.

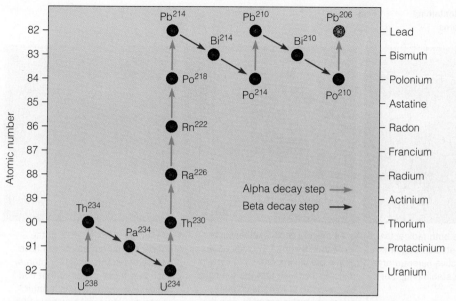

Figure 17.19 Radioactive Decay Series for Uranium 238 to Lead 206 Radioactive uranium 238 decays to its stable daughter product, lead 206, by 8 alpha and 6 beta decay steps. A number of different isotopes are produced as intermediate steps in the decay series.

unstable *parent element* to decay to atoms of a new, more stable *daughter element*. The half-life of a given radioactive element is constant and can be precisely measured. Half-lives of various radioactive elements range from less than a billionth of a second to 49 billion years.

Radioactive decay occurs at a geometric rate rather than a linear rate. Therefore, a graph of the decay rate produces a curve rather than a straight line (❚ Figure 17.20). For example, an element with *1,000,000* parent atoms will have *500,000* parent atoms and *500,000* daughter atoms after one half-life. After two half-lives, it will have *250,000* parent atoms (one-half of the previous parent atoms, which is equivalent to one-fourth of the original parent atoms) and *750,000* daughter atoms. After three half-lives, it will have *125,000* parent atoms (one-half of the previous parent atoms, or one-eighth of the original parent atoms) and *875,000* daughter atoms, and so on, until the number of parent atoms remaining is so few that they cannot be accurately measured by present-day instruments.

By measuring the parent–daughter ratio and knowing the half-life of the parent (which has been determined in the laboratory), geologists can calculate the age of a sample that contains the radioactive element. The parent–daughter ratio is usually determined by a *mass spectrometer*, an instrument that measures the proportions of atoms of different masses.

Some elements undergo only one decay step in the conversion from an unstable form to a stable form. For example, rubidium 87 decays to strontium 87 by a single beta emission, and potassium 40 decays to argon 40 by a single electron capture. Other radioactive elements undergo several decay steps. Uranium 235 decays to lead 207 by 7 alpha and 6 beta steps, whereas uranium 238 decays to lead 206 by 8 alpha and 6 beta steps (❚ Figure 17.19).

When discussing decay rates, it is convenient to refer to them in terms of half-lives. The **half-life** of a radioactive element is the time it takes for half of the atoms of the original

Figure 17.20 Uniform, Linear Change Compared to Geometric Radioactive Decay

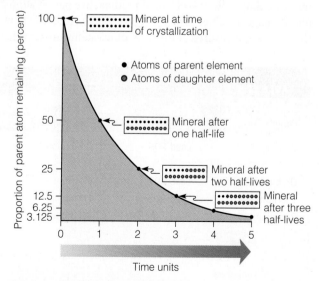

a Uniform, linear change is characteristic of many familiar processes. In this example, water is being added to a glass at a constant rate.

b A geometric radioactive decay curve, in which each time unit represents one half-life, and each half-life is the time it takes for half of the parent element to decay to the daughter element.

▌Figure 17.21 Crystallization of Magma Containing Radioactive Parent and Stable Daughter Atoms

Magma   Mineral crystallizing from magma   Igneous rock

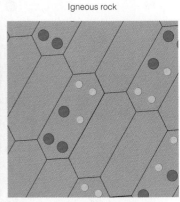

● Radioactive parent atoms
○ Stable daughter atoms

a Magma contains both radioactive parent atoms and stable daughter atoms. The radioactive parent atoms are larger than the stable daughter atoms.

b As magma cools and begins to crystallize, some of the radioactive atoms are incorporated into certain minerals because they are the right size and can fit into the crystal structure. In this example, only the larger radioactive parent atoms fit into the crystal structure. Therefore, at the time of crystallization, minerals in which the radioactive parent atom can fit into the crystal structure will contain 100% radioactive parent atoms and 0% stable daughter atoms.

c After one half-life, 50% of the radioactive parent atoms will have decayed to stable daughter atoms, such that those minerals that had radioactive parent atoms in their crystal structure will now have 50% radioactive parent atoms and 50% stable daughter atoms.

Sources of Uncertainty

The most accurate radiometric dates are obtained from igneous rocks. As magma cools and begins to crystallize, radioactive parent atoms are separated from previously formed daughter atoms. Because they are the right size, some radioactive parent atoms are incorporated into the crystal structure of certain minerals. The stable daughter atoms, though, are a different size from the radioactive parent atoms and consequently cannot fit into the crystal structure of the same mineral as the parent atoms. Therefore, a mineral crystallizing in cooling magma will contain radioactive parent atoms but no stable daughter atoms (▌ Figure 17.21). Thus, the time that is being measured is the time of crystallization of the

mineral that contains the radioactive atoms and not the time of formation of the radioactive atoms.

Except in unusual circumstances, sedimentary rocks cannot be radiometrically dated because one would be measuring the age of a particular mineral rather than the time that it was deposited as a sedimentary particle. One of the few instances in which radiometric dates can be obtained on sedimentary rocks is when the mineral glauconite is present. Glauconite is a greenish mineral containing potassium 40, which decays to argon 40 (Table 17.1). It forms in certain marine environments as a result of chemical reactions with clay minerals during the conversion of sediments to sedimentary rock. Thus, glauconite forms when the sedimentary rock forms, and a radiometric date indicates the time of the

TABLE 17.1	Five of the Principal Long-Lived Radioactive Isotope Pairs Used in Radiometric Dating			
ISOTOPES Parent	Daughter	**Half-Life of Parent (years)**	**Effective Dating Range (years)**	**Minerals and Rocks That Can Be Dated**
Uranium 238	Lead 206	4.5 billion	10 million to 4.6 billion	Zircon Uraninite
Uranium 235	Lead 207	704 million		
Thorium 232	Lead 208	14 billion		
Rubidium 87	Strontium 87	48.8 billion	10 million to 4.6 billion	Muscovite Biotite Potassium feldspar Whole metamorphic or igneous rock
Potassium 40	Argon 40	1.3 billion	100,000 to 4.6 billion	Glauconite Hornblende Muscovite Whole volcanic rock Biotite

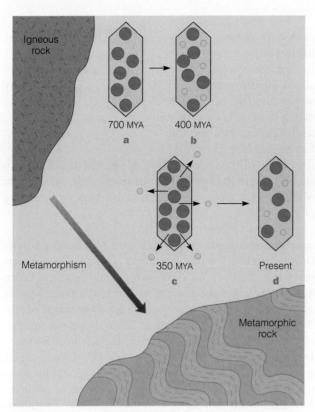

Figure 17.22 Effects of Metamorphism on Radiometric Dating The effect of metamorphism in driving out daughter atoms from a mineral that crystallized 700 million years ago (mya). The mineral is shown immediately after crystallization **(a)**, then at 400 million years **(b)**, when some of the parent atoms had decayed to daughter atoms. Metamorphism at 350 mya **(c)** drives the daughter atoms out of the mineral into the surrounding rock. **(d)** If the rock has remained a closed chemical system throughout its history, dating the mineral today yields the time of metamorphism, whereas dating the whole rock provides the time of its crystallization, 700 mya.

Because heat and pressure affect the parent–daughter ratio, metamorphic rocks are difficult to date accurately. Remember that although the resulting parent–daughter ratio of the sample being analyzed may have been affected by heat, the decay rate of the parent element remains constant, regardless of any physical or chemical changes.

To obtain an accurate radiometric date, geologists must make sure that the sample is fresh and unweathered and that it has not been subjected to high temperature or intense pressures after crystallization. Furthermore, it is sometimes possible to cross-check the radiometric date obtained by measuring the parent–daughter ratio of two different radioactive elements in the same mineral.

For example, naturally occurring uranium consists of both uranium 235 and uranium 238 isotopes. Through various decay steps, uranium 235 decays to lead 207, whereas uranium 238 decays to lead 206 (Figure 17.19). If the minerals that contain both uranium isotopes have remained closed systems, the ages obtained from each parent–daughter ratio should agree closely. If they do, they are said to be *concordant*, thus reflecting the time of crystallization of the magma. If the ages do not closely agree, then they are said to be *discordant*, and other samples must be taken and ratios measured to see which, if either, date is correct.

Recent advances and the development of new techniques and instruments for measuring various isotope ratios have enabled geologists to analyze not only increasingly smaller samples, but with a greater precision than ever before. Presently, the measurement error for many radiometric dates is typically less than 0.5% of the age; in some cases it is even better than 0.1%. Thus, for a rock 540 million years old (near the beginning of the Cambrian Period), the possible error could range from nearly 2.7 million years to less than 540,000 years.

Long-Lived Radioactive Isotope Pairs

Table 17.1 shows the five common, long-lived parent–daughter isotope pairs used in radiometric dating. Long-lived pairs have half-lives of millions or billions of years. All of these were present when Earth formed and are still present in measurable quantities. Other, shorter-lived radioactive isotope pairs have decayed to the point that only small quantities near the limit of detection remain.

The most commonly used isotope pairs are the uranium–lead and thorium–lead series, which are used principally to date ancient igneous intrusives, lunar samples, and some meteorites. The rubidium–strontium pair is also used for very old samples and has been effective in dating the oldest rocks on Earth, as well as meteorites.

CONNECTION LINK ❮

To learn more about Earth's oldest rocks, go to page 480 in Chapter 19.

The potassium–argon method is typically used for dating fine-grained volcanic rocks from which individual crystals cannot be separated; hence, the whole rock is analyzed. Because argon is a gas, great care must be taken to ensure that

sedimentary rock's origin. Being a gas, however, the daughter product argon can easily escape from a mineral. Therefore, any date obtained from glauconite, or any other mineral containing the potassium 40 and argon 40 pair, must be considered a minimum age.

To obtain accurate radiometric dates, geologists must be sure that they are dealing with a *closed system*, meaning that neither parent nor daughter atoms have been added or removed from the system since crystallization, and that the ratio between them results only from radioactive decay. Otherwise, an inaccurate date will result. If daughter atoms have leaked out of the mineral being analyzed, the calculated age will be too young; if parent atoms have been removed, the calculated age will be too old.

Leakage may take place if the rock is heated or subjected to intense pressure as can sometimes occur during metamorphism. If this happens, some of the parent or daughter atoms may be driven from the mineral being analyzed, resulting in an inaccurate age determination. If the daughter product was completely removed, then one would be measuring the time since metamorphism (a useful measurement itself) and not the time since crystallization of the mineral (▌ Figure 17.22).

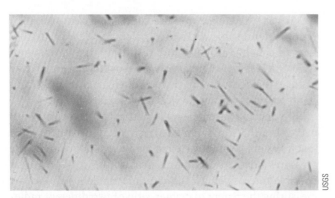

USGS

Figure 17.23 Fission-Track Dating Each fission track (about 16 microns [= 16/1000 mm] long) in this apatite crystal is the result of the radioactive decay of a uranium atom. The apatite crystal, which has been etched with hydrofluoric acid to make the fission tracks visible, comes from one of the dikes at Shiprock, New Mexico, and has a calculated age of 27 million years.

the sample has not been subjected to heat, which would allow argon to escape; such a sample would yield an age that is too young. Other long-lived radioactive isotope pairs exist, but they are rather rare and are used only in special situations.

Fission-Track Dating

The emission of atomic particles that result from the spontaneous decay of uranium within a mineral damages its crystal structure. The damage appears as microscopic linear tracks that are visible only after the mineral has been etched with hydrofluoric acid, an acid so powerful that its vapors can destroy one's sense of smell without careful handling. The age of the sample is determined from the number of fission tracks present and the amount of uranium that the sample contains: the older the sample, the greater the number of tracks (▮ Figure 17.23).

Fission-track dating is of particular interest to archaeologists and geologists because the technique can be used to date samples ranging from only a few hundred to hundreds of millions of years old. It is most useful for dating samples between approximately 40,000 and 1.5 million years ago, a period for which other dating techniques are not always particularly suitable. One of the problems in fission-track dating occurs when the rocks have later been subjected to high temperatures. If this happens, the damaged crystal structures are repaired by annealing, and consequently the tracks disappear. In such instances, the calculated age will be younger than the actual age.

Radiocarbon and Tree-Ring Dating Methods

Carbon is an important element in nature and is one of the basic elements found in all forms of life. It has three isotopes; two of these, carbon 12 and 13, are stable, whereas carbon 14 is radioactive (see Figure 3.4). Carbon 14 has a half-life of 5,730 years plus or minus 30 years. The **carbon-14 dating technique** is based on the ratio of carbon 14 to carbon 12 and is generally used to date formerly living material.

The short half-life of carbon 14 makes this dating technique practical only for specimens younger than about 70,000 years. Consequently, the carbon-14 dating method is especially useful in archaeology and has greatly helped unravel the events of the latter portion of the Pleistocene Epoch such as where the first center for maize domestication in Mesoamerica arose, or when humans began populating North America.

CONNECTION LINK

To learn more about the events of the Pleistocene Epoch, see pages 626–631 in Chapter 23.

Carbon 14 is constantly formed in the upper atmosphere when cosmic rays, which are high-energy particles (mostly protons), strike the atoms of upper-atmospheric gases, splitting their nuclei into protons and neutrons. When a neutron strikes the nucleus of a nitrogen atom (atomic number 7, atomic mass number 14), it may be absorbed into the nucleus and a proton emitted. Thus, the atomic number of the atom decreases by 1, whereas the atomic mass number stays the same. Because the atomic number has changed, a new element, carbon 14 (atomic number 6, atomic mass number 14), is formed. The newly formed carbon 14 is rapidly assimilated into the carbon cycle and, along with carbon 12 and 13, is absorbed in a nearly constant ratio by all living organisms (▮ Figure 17.24). When an organism dies, however, carbon 14 is not replenished, and the ratio of carbon 14 to carbon 12 decreases as carbon 14 decays back to nitrogen by a single beta decay step (Figure 17.24).

Currently, the ratio of carbon 14 to carbon 12 is remarkably constant both in the atmosphere and in living organisms. There is good evidence, however, that the production of carbon 14, and thus the ratio of carbon 14 to carbon 12, has varied somewhat during the past several thousand years. This fact was determined by comparing ages established by carbon-14 dating of wood samples with ages established by counting annual tree rings in the same samples. As a result, carbon-14 ages have been corrected to reflect such variations in the past.

Tree-ring dating is another useful method for dating geologically recent events. The age of a tree can be determined by counting the growth rings in the lower part of the trunk. Each ring represents one year's growth, and the pattern of wide and narrow rings can be compared among trees to establish the exact year in which the rings were formed. The procedure of matching ring patterns from numerous trees and wood fragments in a given area is called *cross-dating*.

By correlating distinctive tree-ring sequences from living and nearby dead trees, scientists can construct a time scale that extends back approximately 14,000 years (▮ Figure 17.25). By matching ring patterns to the composite ring scale, wood samples whose ages are not known can be accurately dated.

The applicability of tree-ring dating is somewhat limited because it can be used only where continuous tree records are found. It is therefore most useful in arid regions, particularly the southwestern United States, where trees live a very long time.

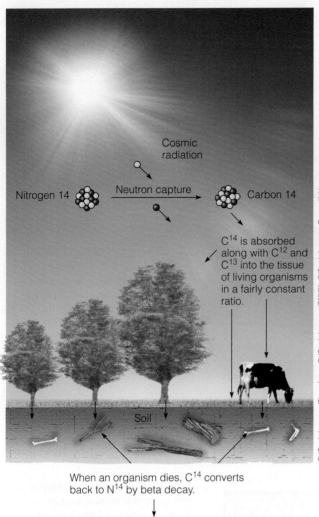

Figure 17.24 Carbon-14 Dating Method The carbon cycle showing the formation of carbon 14 in the upper atmosphere, its dispersal and incorporation into the tissue of all living organisms, and its decay back to nitrogen 14 by beta decay.

» Critical Thinking Question
Why do you think the ratio of carbon 14 to carbon 12 is constant in living organisms?

17.8 Development of the Geologic Time Scale

The geologic time scale is a hierarchical scale in which the 4.6-billion-year history of Earth is divided into time units of varying duration (Figure 17.1). It did not result from the work of any one individual, but rather evolved,

primarily during the 19th century, through the efforts of many people.

By applying relative-dating methods to rock outcrops, geologists in England and western Europe defined the major geologic time units without the benefit of radiometric dating techniques. Using the principles of superposition and fossil succession, they correlated various rock exposures and pieced together a composite geologic section. This composite section is, in effect, a relative time scale because the rocks are arranged in their correct sequential order.

By the beginning of the 20th century, geologists had developed a relative geologic time scale, but did not yet have any absolute dates for the various time-unit boundaries. Following the discovery of radioactivity near the end of the 19th century, radiometric dates were added to the relative geologic time scale (Figure 17.1).

Because sedimentary rocks, with rare exceptions, cannot be radiometrically dated, geologists have had to rely on interbedded volcanic rocks and igneous intrusions to apply absolute dates to the boundaries of the various subdivisions of the geologic time scale (❚ Figure 17.26).

An ashfall or lava flow provides an excellent marker bed that is a time-equivalent surface, supplying a minimum age for the sedimentary rocks below and a maximum age for the rocks above. Ashfalls are particularly useful because they may take place over both marine and nonmarine sedimentary environments and can provide a connection between these different environments.

Thousands of absolute ages are now known for sedimentary rocks of known relative ages, and these absolute dates have been added to the relative time scale. In this way, geologists have been able to determine both the absolute ages of the various geologic periods and their durations (Figure 17.1). In fact, the dates for the era, period, and epoch boundaries of the geologic time scale are still being refined as more accurate dating methods are developed and new exposures dated. The ages shown in Figures 1.18 and 17.1 are the most recently published (2009).

17.9 Stratigraphy and Stratigraphic Terminology

The recognition of a relative geologic time scale brought some order to *stratigraphy* (the study of the composition, origin, areal distribution, and age relationships of layered rocks); however, problems remained because many sedimentary rock units are time transgressive. This means that they were deposited during one geologic period in a particular area and during another period elsewhere (see Figure 7.10). Therefore, modern stratigraphic terminology includes two fundamentally different kinds of units to deal with both rocks and time: those defined by their content and those related to geologic time (Table 17.2).

Figure 17.25 Tree-Ring Dating Method In the cross-dating method, tree-ring patterns from different woods are compared to establish a ring-width chronology backward in time.

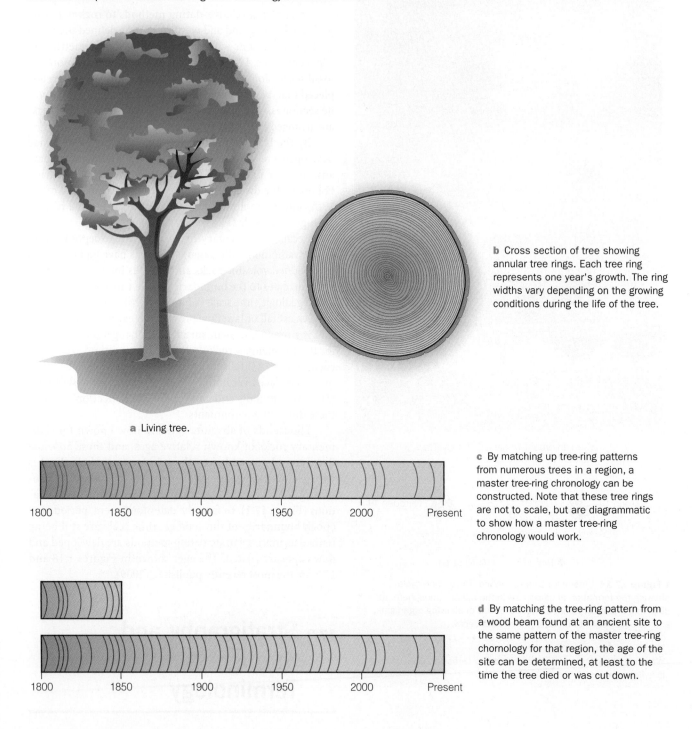

a Living tree.

b Cross section of tree showing annular tree rings. Each tree ring represents one year's growth. The ring widths vary depending on the growing conditions during the life of the tree.

c By matching up tree-ring patterns from numerous trees in a region, a master tree-ring chronology can be constructed. Note that these tree rings are not to scale, but are diagrammatic to show how a master tree-ring chronology would work.

d By matching the tree-ring pattern from a wood beam found at an ancient site to the same pattern of the master tree-ring chornology for that region, the age of the site can be determined, at least to the time the tree died or was cut down.

Units defined by their content include lithostratigraphic and biostratigraphic units. **Lithostratigraphic** (lith- and litho- are prefixes meaning "stone" or "stonelike") **units** are defined by the physical attributes of the rocks, such as rock type (for example, sandstone or limestone), with no consideration of time of origin. The basic lithostratigraphic unit is the *formation*, which is a mapable body of rock with distinctive upper and lower boundaries (Figure 17.27). A formation may consist of a single rock type (e.g., the Redwall Limestone, Figure 17.14), or a variety of related rock types (e.g., the Morrison Formation, see Figure 22.17). Formations are commonly subdivided

Figure 17.26 **Determining Absolute Dates for Sedimentary Rocks** The absolute ages of sedimentary rocks can be determined by dating associated igneous rocks. In **(a)** and **(b)**, sedimentary rocks are bracketed by rock bodies for which absolute ages have been determined.

Figure 17.27 **Lithostratigraphic Units** The Madison Group in Montana consists of two formations, the Lodgepole Formation and the overlying Mission Canyon Formation.

into smaller units known as *members* and *beds*, and they may be parts of larger units known as *groups* and *supergroups* (Table 17.2).

A body of strata recognized only on the basis of its fossil content is a **biostratigraphic unit**, the boundaries of which do not necessarily correspond to those of lithostratigraphic boundaries (▌ Figure 17.28). The fundamental biostratigraphic unit is the *biozone*. Several types of biozones are recognized, one of which, the *concurrent range zone*, was discussed in the section on correlation (Figure 17.16).

The category of units expressing or related to geologic time includes time-stratigraphic units (also known as chronostratigraphic units) and time units (Table 17.2). **Time-stratigraphic units** consist of rocks deposited during a particular interval of geologic time. The *system*, the basic time-stratigraphic unit, is based on a stratotype, which consists of rocks in an area where the system was first described. Systems are recognized beyond their stratotype area by their fossil content.

TABLE 17.2	Classification of Stratigraphic Units			
UNITS DEFINED BY CONTENT		**UNITS EXPRESSING OR RELATED TO GEOLOGIC TIME**		
Lithostratigraphic Units	**Biostratigraphic Units**	**Time-Stratigraphic Units**		**Time Units**
Supergroup	Biozones	Eonothem —————————— Eon		
Group		Erathem —————————— Era		
Formation		System —————————— Period		
Member		Series —————————— Epoch		
Bed		Stage —————————— Age		

Figure 17.28 Rocks and Fossils of the Bearpaw Formation in Saskatchewan, Canada The column on the left shows formation and members that are lithostratigraphic units. Notice that the biozone boundaries do not correspond with lithostratigraphic boundaries. The absolute ages for the two volcanic ash layers indicate that the *Baculites reesidei* zone is about 72 to 73 million years old.

» Critical Thinking Question
Why don't the biozone boundaries correspond with the lithostratigraphic boundaries?

Time units are simply designations for certain parts of geologic time. The basic time unit is the *period*; however, two or more periods may be designated as an *era*, and two or more eras constitute an *eon*. Periods also consist of shorter designations such as *epoch* and *age*. The time units known as period, epoch, and age correspond to the time-stratigraphic units known as system, series, and stage, respectively (Table 17.2). For example, the Cambrian Period is defined as the time during which strata of the Cambrian System were deposited.

17.10 Geologic Time and Climate Change

Given the debate concerning global warming and its possible implications, it is extremely important to be able to reconstruct past climatic regimes as accurately as possible (see Geo-Focus on pages 442 and 443). To model how Earth's climate system has responded to changes in the past

and to use that information for simulations of future climate scenarios, geologists must have a geologic calendar that is as precise and accurate as possible.

New dating techniques with greater precision are providing geologists with more accurate dates for when and how long-past climate changes occurred. The ability to accurately determine when past climate changes occurred helps geologists correlate these changes with regional and global geologic events to see if there are any possible connections.

One interesting method that is becoming more common in reconstructing past climates is to analyze stalagmites from caves. Recall that stalagmites are icicle-shaped structures rising from a cave floor and formed of calcium carbonate precipitated from evaporating water. A stalagmite therefore records a layered history because each newly precipitated layer of calcium carbonate is younger than the previously precipitated layer (▌ Figure 17.29).

Thus, a stalagmite's layers are oldest in the center at its base and progressively younger as they move outward (principle of superposition). Using techniques based on ratios of

▍Figure 17.29 Stalagmites and Climate Change

Stalactites

1 Newly forming layers of calcite in a stalagmite contain U^{234} (substituting for calcium).

3 U^{234} decays to Th^{230} at a predictable, measurable rate.

Stalagmite

$\dfrac{U^{234}}{Th^{230}}$

Time of layer formation

Time

2 The inside of a stalagmite is layered like an onion, showing its incremental growth.

4 The age of each layer can be determined by measuring its ratio of U^{234}/Th^{230}. Layer 5 (to the left) is older than layer 2. It has a lower U^{234}/Th^{230} ratio.

a Stalagmites are icicle-shaped structures rising from the floor of a cave and are formed by the precipitation of calcium carbonate from evaporating water. A stalagmite is thus layered, with the oldest layer in the center and the youngest layers on the outside. Uranium 234 frequently substitutes for the calcium ion in the calcium carbonate of the stalagmite. Uranium 234 decays to thorium 230 at a predictable and measurable rate. Therefore, the age of each layer of the stalagmite can be dated by measuring its ratio of uranium 234 to thorium 230.

3 Water infiltrating the ground carries pollen and a specific climate-related O^{18}/O^{16} ratio with it.

Pollinating plants

1 There is a light isotope of oxygen, O^{16}, and a heavy one, O^{18}.

O^{16} - rich vapor

2 When water (H_2O) evaporates, it is easier for O^{16} to vaporize than O^{18}. Because it is lighter, the unevaporated water becomes O^{18} - enriched. The warmer the climate, the more the evaporation and oxygen isotope separation.

$\dfrac{O^{18}}{O^{16}}$ in water

Temperature

4 The water drips into a cave and is trapped in pores between calcite crystals in the layers of stalagmites. The pollen and O^{18}/O^{16} in the water record climate conditions just before the moment of entrapment.

(Continued on page 444)

b There are two isotopes of oxygen, a light one, oxygen 16, and a heavy one, oxygen 18. Because the oxygen 16 isotope is lighter than the oxygen 18 isotope, it vaporizes more readily than the oxygen 18 isotope when water evaporates. Therefore, as the climate becomes warmer, evaporation increases, and the O^{18}/O^{16} ratio becomes higher in the remaining water. Water in the form of rain or snow percolates into the ground and becomes trapped in the pores between the calcite forming the stalagmites.

GEO-FOCUS Denver's Weather—280 Million Years Ago!

With all of the concern about global climate change, it might be worthwhile to step back a bit and look at climate change from a geologic perspective. We're all aware that some years are hotter than others and some years we have more rain, but generally things tend to average out over time. We know that it will be hot in the summer in Arizona, and it will be very cold in Minnesota in the winter. We also know that scientists, politicians, and concerned people everywhere are debating whether humans are partly responsible for the global warming that Earth seems to be experiencing.

What about long-term climate change? We know that Earth has experienced periods of glaciation in the past—for instance, during the Precambrian, the end of the Ordovician Period, and most recently during the Pleistocene Epoch. Earth has also undergone large-scale periods of aridity, such as during the end of the Permian and beginning of the Triassic periods. Such long-term climatic changes are probably the result of slow geographic changes related to plate tectonic activity. Not only are continents carried into higher and lower latitudes but their movement affects ocean circulation and atmospheric circulation patterns, which in turn affect climate, and result in climate changes.

Even though we can't physically travel back in time, geologists can reconstruct what the climate was like in the past. The distribution of plants and animals is controlled, in part, by climate. Plants are particularly sensitive to climate change, and many can only live in particular environments. The fossils of plants and animals can tell us something about the environment and climate at the time that these organisms were living. Furthermore, climate-sensitive sedimentary rocks can be used to interpret past climatic conditions. Desert dunes are typically well sorted and exhibit large-scale cross-bedding. Coals form in freshwater swamps where climatic conditions promote abundant plant growth. Evaporites such as rock salt result when evaporation exceeds precipitation, such as in desert regions or along hot, dry shorelines. Tillites (glacial sediments) result from glacial activity and indicate cold, wet environments. So, by combining all relevant geologic and paleontologic information, geologists can reconstruct what the climate was like in the past and how it has changed over time at a given locality.

In his book titled *Ancient Denvers: Scenes from the Past 300 Million Years of the Colorado Front Range*, Kirk R. Johnson depicts what Denver, Colorado, looked like at 13 different time periods in the past. The time slices begin during the Pennsylvanian Period, 300 million years ago, and end with a view of the Front Range amid a spreading wave of houses on the southern edge of metropolitan Denver.

The information for piecing together Denver's geologic past was derived mainly from a 688-m-deep well drilled by the Denver Museum of Nature and Science beneath Kiowa, Colorado, in 1999. Using the information gleaned from the rocks recovered from the well, plus additional geologic evidence from other parts of the area, museum scientists and artists were able to reconstruct Denver's geologic past.

Beginning 300 million years ago (Pennsylvanian Period), the Denver area had coastlines on its eastern and western borders and a mountain range (not the Rocky Mountains of today). The climate was mostly temperate with lots of seedless vascular plants, such as ferns, as well as very tall scale trees related to the modern horsetail rush. Huge insects such as millipedes, cockroaches, and dragonflies shared this region with relatively small, fin-backed reptiles and a variety of amphibians.

By 280 million years ago, the area was covered by huge sand seas, much like the Sahara is today (▮ Figure 1). This change in climate and landscape was the result of the formation of Pangaea. As the continents collided, arid and semiarid conditions prevailed over much of the supercontinent, and the Denver area was no exception.

During the Late Jurassic (150 million years ago), herds of plant-eating dinosaurs such as *Apatosaurus* roamed throughout the Denver area, feasting on the succulent and abundant vegetation. Grasses and flowering plants had not yet evolved, so

▮ **Figure 1** Denver as it appeared 280 million years ago. As a result of the collision of continents and the formation of Pangaea, the world's climate was generally arid, and Denver was no exception. Denver was probably covered by great seas of sand, much as the Sahara is today.

the dinosaurs ate the ferns and gymnosperms that were abundant at this time.

As a result of rising sea level, Denver was covered by a warm, shallow sea 70 million years ago (Late Cretaceous). Marine reptiles such as plesiosaurs and mosasaurs ruled these seas, while overhead, pterosaurs soared through the skies, looking for food (Figure 2).

Beginning approximately 66 million years ago, the Rocky Mountains began to form as tectonic forces started a mountain-building episode known as the *Laramide orogeny* that resulted in the present-day Rocky Mountains. Dinosaurs still roamed the land around Denver, and flowering plants began their evolutionary history.

By 55 million years ago (Eocene), the world was in the grip of an intense phase of global warming. A subtropical rainforest with many trees that would be recognizable today filled the landscape. Primitive mammals were becoming more abundant, and many warm-climate–loving animals could be found living north of the Arctic Circle.

Although ice caps still covered portions of North America, mammoths and other mammals wandered among the plains of Denver 16,000 years ago (Figure 3). Mastodons, horses, bison, lions, and giant ground sloths, to name a few, all lived in this region, and their fossils can be found in the sedimentary rocks from this area.

What was once a rainforest, a desert, a warm, shallow sea, and a mountainous region is now home to thousands of people. What the Denver region will be like in the next several million years is anyone's guess. Whereas humans can affect change, what change we will cause is open to debate. Certainly, the same forces that have shaped the Denver area in the past will continue to determine its future. With the rise of humans and technology, we, as a species, will also influence what future Denvers will be like. Let us hope that the choices we make are good ones.

Figure 2 Pterosaurs (flying reptiles) soar over Denver 70 million years ago. At this time, Denver was below a warm, shallow sea that covered much of western North America. Marine reptiles such as plesiosaurs and mosasaurs swam in these seas in search of schools of fish.

Figure 3 Mammoths and camels wander on the prairie during a summer day 16,000 years ago. Whereas much of northern North America was covered by an ice sheet, Denver sported pine trees and prairie grass. This area was home to a large variety of mammals, including mammoths, camels, horses, bison, and giant ground sloths. Humans settled in this area about 11,000 years ago, hunting the plentiful game that was available.

❚ **Figure 17.29** *Continued*

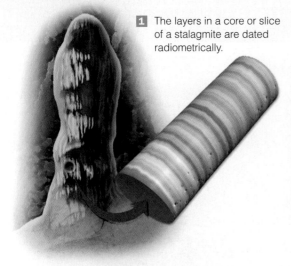

1 The layers in a core or slice of a stalagmite are dated radiometrically.

2 The pore water is analyzed in each layer for O^{18}/O^{16} and species of plants (from the pollen).

3 A record of climatic change is put together for the area of the caves.

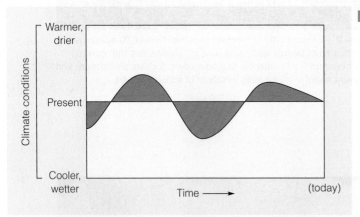

c The layers of a stalagmite can be dated by measuring the U^{234}/Th^{230} ratio, and the O^{18}/O^{16} ratio determined for the pore water trapped in each layer. Thus, a detailed record of climatic change for the area can be determined by correlating the climate of the area as determined by the O^{18}/O^{16} ratio to the time period determined by the U^{234}/Th^{230} ratio.

uranium 234 to thorium 230, and reliable back to approximately 500,000 years, geologists can achieve very precise radiometric dates on individual layers of a stalagmite.

A study of stalagmites from Crevice Cave in Missouri revealed a history of climatic and vegetation change in the mid-continent region of the United States during the interval between 75,000 and 25,000 years ago. Dates obtained from the Crevice Cave stalagmites were correlated with major changes in vegetation and average temperature fluctuations obtained from carbon 13 and oxygen 18 isotope profiles to reconstruct a detailed picture of climate changes during this time period.

As a result, precise dating techniques in stalagmite studies provide an accurate chronology that allows geologists to model climate systems of the past and perhaps to determine what causes global climate changes and their duration. Without these sophisticated dating techniques and others like them, geologists would not be able to make precise correlations and accurately reconstruct past environments and climates. By analyzing past environmental and climate changes and their duration, geologists hope they can use these data, sometime in the near future, to predict and possibly modify regional climate changes.

GEO-RECAP

» Geo-Image

▌**Figure 1** This photograph was taken on Fish Creek Trail in the Sierra Nevada, California, and shows two black metavolcanic erratics on granodiorite bedrock that has been intruded by a white dike. Using your knowledge of igneous rocks and processes, glacial features, and relative dating principles, provide a geologic history of this area based on what is shown in this image.

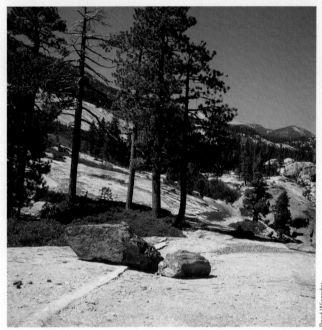

Reed Wicander

Chapter Summary

- Time is defined by the methods used to measure it. Relative dating places geologic events in sequential order as determined from their position in the geologic record. Absolute dating provides specific dates for geologic rock units or events, expressed in years before the present.

- During the 18th and 19th centuries, attempts were made to determine Earth's age based on scientific evidence rather than revelation. Although some attempts were ingenious, they yielded a variety of ages that now are known to be much too young.

- James Hutton, considered by many to be the father of modern geology, thought that present-day processes operating over long periods of time could explain all of the geologic features of Earth. His observations were instrumental in establishing the principle of uniformitarianism and the fact that Earth was much older than earlier scientists thought.

- Uniformitarianism, as articulated by Charles Lyell, soon became the guiding principle of geology. It holds that the laws of nature have been constant through time and that the same processes operating today have operated in the past, although not necessarily at the same rates.

- Besides uniformitarianism, the principles of superposition, original horizontality, lateral continuity, crosscutting relationships, inclusions, and fossil succession are basic for determining relative geologic ages and for interpreting Earth history.

- An unconformity is a surface of erosion, nondeposition, or both, separating younger strata from older strata. These surfaces encompass long periods of geologic time for which there is no geologic record at that location.

- Three types of unconformities are recognized. A disconformity separates younger from older sedimentary strata that are parallel to each other. An angular unconformity is an erosional surface on tilted or folded rocks, over which younger sedimentary rocks were deposited. A nonconformity is an erosional surface cut into igneous or metamorphic rocks and overlain by younger sedimentary rocks.

- Correlation is the demonstration of time equivalency of rock units in different areas. Similarity of rock type, position within a rock sequence, key beds, and fossil assemblages can all be used to correlate rock units.

- Radioactivity was discovered during the late 19th century, and soon afterward, radiometric dating techniques enabled geologists to determine absolute ages for rock units and geologic events.

- Absolute dates for rocks are usually obtained by determining how many half-lives of a radioactive parent element have elapsed since the sample originally crystallized. A half-life is the time it takes for one-half of the original unstable radioactive parent element to decay into a new, more stable daughter element.

- The most accurate radiometric dates are obtained from long-lived radioactive isotope pairs in igneous rocks. The most reliable dates are those obtained by using at least two different radioactive decay series in the same rock.

- Carbon-14 dating can be used only on organic matter such as wood, bones, and shells and is effective back to approximately 70,000 years ago. Unlike the long-lived radioactive isotopic pairs, the carbon-14 dating technique determines age by the ratio of radioactive carbon-14 to stable carbon-12.

- The geologic time scale was developed primarily during the 19th century through the efforts of many people. It was originally a relative geologic time scale, but with the discovery of radioactivity and the development of radiometric dating methods, absolute age dates were added at the beginning of the 20th century. Since then, refinement of the time-unit boundaries has continued.

- Stratigraphic terminology includes two fundamentally different kinds of units: those based on content and those related to geologic time.

- To reconstruct past climate changes and link them to possible causes, geologists must have a geologic calendar that is as precise and accurate as possible. Thus, they must be able to date geologic events and the onset and duration of climate changes as precisely as possible.

Important Terms

absolute dating (p. 416)
angular unconformity (p. 420)
biostratigraphic unit (p. 439)
carbon-14 dating technique (p. 436)
correlation (p. 428)
disconformity (p. 420)
fission-track dating (p. 436)
guide fossil (p. 429)
half-life (p. 433)

lithostratigraphic unit (p. 438)
nonconformity (p. 424)
principle of cross-cutting relationships (p. 418)
principle of fossil succession (p. 420)
principle of inclusions (p. 420)
principle of lateral continuity (p. 418)
principle of original horizontality (p. 418)

principle of superposition (p. 418)
radioactive decay (p. 432)
relative dating (p. 416)
time-stratigraphic unit (p. 439)
time unit (p. 440)
tree-ring dating (p. 436)
unconformity (p. 420)

Review Questions

1. How many half-lives are required to yield a mineral with 625,000,000 atoms of thorium 232 and 19,375,000,000 atoms of lead 208?
   a. ___ 1.
   b. ___ 2.
   c. ___ 3.
   d. ___ 4.
   e. ___ 5.

2. In which type of radioactive decay is an electron emitted from the nucleus?
   a. ___ alpha decay.
   b. ___ beta decay.
   c. ___ electron capture.
   d. ___ fission track.
   e. ___ none of these.

3. Who is generally considered the founder of modern geology?
   a. ___ Abraham Gottlob Werner.
   b. ___ Charles Lyell.
   c. ___ Nicolas Steno.
   d. ___ Georges Cuvier.
   e. ___ James Hutton.

4. If a rock is heated during metamorphism and the parent atoms migrate out of a mineral that is subsequently radiometrically dated, an inaccurate date will be obtained. This date will therefore be _____ the actual date.
   a. ___ the same as.
   b. ___ younger than.
   c. ___ older than.
   d. ___ it cannot be determined.
   e. ___ none of these.

5. What fundamental geologic principle states that the oldest layer is on the bottom of a succession of flat-lying sedimentary rocks and the youngest layer is on top?
   a. ___ lateral continuity.
   b. ___ fossil succession.
   c. ___ original horizontality.
   d. ___ superposition.
   e. ___ cross-cutting relationships.

6. If a flake of biotite within a sedimentary rock (such as a sandstone) is radiometrically dated, the date obtained indicates when
   a. ___ the biotite crystal formed.
   b. ___ the sedimentary rock formed.
   c. ___ the parent radioactive isotope formed.
   d. ___ the daughter radioactive isotope(s) formed.
   e. ___ none of these.

7. Placing geologic events in sequential order as determined by their position in the geologic record is
   a. ___ absolute dating.
   b. ___ correlation.
   c. ___ historical dating.
   d. ___ relative dating.
   e. ___ uniformitarianism.

8. Which of the following methods can be used to demonstrate age equivalency of rock units?
   a. ___ lateral tracing.
   b. ___ radiometric dating.
   c. ___ guide fossils.
   d. ___ position in a sequence.
   e. ___ all of these.

9. As carbon 14 decays back to nitrogen in radiocarbon dating, what isotopic ratio decreases?
   a. ___ nitrogen 14 to carbon 12.
   b. ___ carbon 14 to carbon 12.
   c. ___ carbon 13 to carbon 12.
   d. ___ nitrogen 14 to carbon 14.
   e. ___ nitrogen 14 to carbon 13.

10. The basic lithostratigraphic unit is the
    a. ___ group.
    b. ___ formation.
    c. ___ bed.
    d. ___ member.
    e. ___ supergroup.

11. Can the various principles of relative dating be used to reconstruct the geologic history of Mars? Which principles might not apply to interpreting the geologic history of another planet?

12. What is the major difference between the carbon-14 dating technique and the techniques used for the five common, long-lived radioactive isotope pairs?

13. A volcanic ashfall was radiometrically dated using the potassium 40 to argon 40 and rubidium 87 to strontium 87 isotope pairs. The isotope pairs yielded distinctly different ages. What possible explanation could be offered as to why these two isotope pairs yielded different ages? What would you do to rectify the discrepancy in ages?

14. Why were Lord Kelvin's arguments and calculations so compelling, and what was the basic flaw in his assumption? What do you think the course of geology would have been if radioactivity had not been discovered?

15. Why do igneous rocks yield the most accurate radiometric dates? Why can't sedimentary rocks be dated radiometrically? What problems are encountered in dating metamorphic rocks?

16. How does metamorphism affect the potential for accurate radiometric dating using any and all techniques discussed in this chapter? How would such radiometric dates be affected by metamorphism, and why?

17. When geologists reconstruct the geologic history of an area, why is it important for them to differentiate between a sill and a lava flow? How could you tell the difference between a sill and a lava flow at an outcrop if both structures consisted of basalt? What features would you look for in an outcrop to positively identify the structure as either a sill or a lava flow?

18. In some places where disconformities are particularly difficult to discern from a physical point of view, how could you use the principle of fossil succession to recognize a disconformity? How could the principle of inclusions be used to recognize a nonconformity?

19. Why are the fundamental principles used in relative dating so important in geology?

20. Given the current debate over global warming and the many short-term consequences for humans, can you visualize how the world might look in 10,000 years or even 1 million years from now? Use what you've learned about plate tectonics and the direction and rate of movement of plates, as well as how plate movement and global warming will affect ocean currents, weather patterns, weathering rates, and other factors, to make your prediction. Do you think such short-term changes can be extrapolated to long-term trends in trying to predict what Earth will be like using a geologic time perspective?

M. de Lamarck.

The French botanist–geologist Jean-Baptiste Pierre Antoine de Monet, Chevalier Lamarck (1744–1829) proposed in 1809 the first formal, widely accepted theory for a mechanism to account for evolution. His mechanism, inheritance of acquired characteristics, has since been discredited. Nevertheless, Lamarck made important contributions to our understanding of the natural world in his published works on invertebrate zoology and paleontology.

Organic Evolution— The Theory and Its Supporting Evidence

OUTLINE

18.1 Introduction

18.2 Organic Evolution: What Does It Mean?

GEO-FOCUS: The Tragic Lysenko Affair

GEO-INSIGHT: Artificial Selection, Natural Selection, Fossils, and Evolution

18.3 Mendel and the Birth of Genetics

18.4 The Modern View of Organic Evolution

18.5 Evidence Supporting Evolutionary Theory

GEO-RECAP

HAVE YOU EVER WONDERED?

- What the central claim of evolutionary theory is?

- How the theory of organic evolution developed, and who contributed to it?

- Why inheritance and mutations are important in evolutionary thinking?

- What the modern view of organic evolution is?

- How fossils and studies in biology and biochemistry support evolutionary theory?

- Whether predictive statements can be made from evolutionary theory so that its validity may be assessed?

18.1 Introduction

A rugged group of 13 large islands, 8 smaller ones, and 40 islets, all belonging to Ecuador, lies in the Pacific Ocean about 1,000 km west of South America. Called the Archipelago de Colon after Christopher Columbus, the group is better known as the Galápagos Islands. During Charles Robert Darwin's five-year voyage (1831–1836) as an unpaid naturalist aboard the research vessel HMS *Beagle,* he visited the Galápagos Islands, where he made important observations that changed his ideas about the then widely held concept called *fixity of species.* According to this idea, all present-day species had been created in their present form and had changed little or not at all.

Darwin began his voyage not long after graduating from Christ's College of Cambridge University with a degree in theology, and although he was rather indifferent to religion, he fully accepted the biblical account of creation. During the voyage, though, his ideas began to change. For example, some of the fossil mammals he collected in South America were similar to present-day llamas, sloths, and armadillos yet also differed from them. He realized that these animals had descended with modification from ancestral species, and so he began to question the idea of fixity of species.

Darwin postulated that the 13 species of finches living on the Galápagos Islands had evolved from a common ancestor species that somehow reached the islands as an accidental immigrant from South America. Indeed, their ancestor was very likely a single species resembling the blue-back grassquit finch now living along South America's Pacific Coast. The islands' scarcity of food accounts for the ancestral species evolving different physical characteristics, especially beak shape, to survive (❙ Figure 18.1). And likewise the tortoises, which vary from island to island, still resemble tortoises living along South America's Pacific Coast, even though they differ in subtle ways.

Charles Darwin became convinced that organisms descended with modification from ancestors that lived during the past, which is the central claim of the **theory of evolution.** So why should you study evolution? For one thing, organic evolution involving inheritable changes in organisms through time is fundamental to biology and **paleontology,** the study of life history as revealed by fossils. Furthermore, like plate tectonic theory, evolution is a unifying theory that explains an otherwise encyclopedic collection of facts. And finally, it serves as the framework for discussions of life history in the following chapters.

Unfortunately, evolution is poorly understood by many people, many of whom have misconceptions about it. For instance, some think that (1) evolution precedes strictly by chance; (2) nothing less than fully developed structures such as eyes or wings as they exist now would be of any use; and (3) no transitional fossils (so-called missing links) are found in the fossil record. We address some of these misconceptions in this and the following chapters.

18.2 Organic Evolution: What Does It Mean?

It is important to note that evolution does not explain how life originated, only how life has changed since it first appeared on Earth. There are theories explaining life's origin, but they must be evaluated on their own merits. The idea of evolution is usually attributed solely to Charles Darwin, but it was seriously considered long before he was born, even by some ancient Greeks and by philosophers and theologians during the Middle Ages. Nevertheless, among Europeans, the prevailing belief well into the 1700s was that the works of the Greek philosopher Aristotle (384–322 BC) and the first two chapters of the book of Genesis contained all important knowledge. Literally interpreted, Genesis was taken as the final word on the origin and diversity of life, as well as much of Earth history. To question any aspect of this interpretation was heresy, which was usually dealt with harshly.

> **CONNECTION LINK** ‹
>
> You can find more information on the origin of life on pages 491 and 493 in Chapter 19.

The social and intellectual climate changed in 18th-century Europe, when the absolute authority of the church in all matters was questioned. Ironically, the very naturalists who were trying to find evidence supporting Genesis found more and more evidence that could not be reconciled with a literal reading of Scripture. In this changing intellectual atmosphere, scientists gradually accepted the principle of uniformitarianism and Earth's great age, and the French zoologist Georges Cuvier demonstrated that many plants and animals had become extinct. In view of the accumulating fossil evidence, as well as studies of living organisms, scientists became convinced that organic evolution had occurred, but they lacked a theoretical framework to explain evolution.

Jean-Baptiste de Lamarck and His Ideas on Evolution

Jean-Baptiste de Lamarck (1744–1829) (see the chapter opening photo) was not the first to propose a mechanism to account for evolution, but in 1809 he was the first to be taken seriously. Lamarck contributed greatly to our understanding of the natural world, but unfortunately he is best remembered for his theory of **inheritance of acquired characteristics.** According to this idea, new traits arise in organisms because of their needs and somehow these characteristics are passed on to their descendants. In an ancestral population of short-necked giraffes, for instance, neck stretching to browse in trees gave them the capacity to have offspring with longer necks. In short, Lamarck thought

Figure 18.1 **Darwin's Finches of the Galapagos Islands** These birds probably evolved from accidental immigrants much like the blue-back grassquit finch that lives along the west coast of South America.

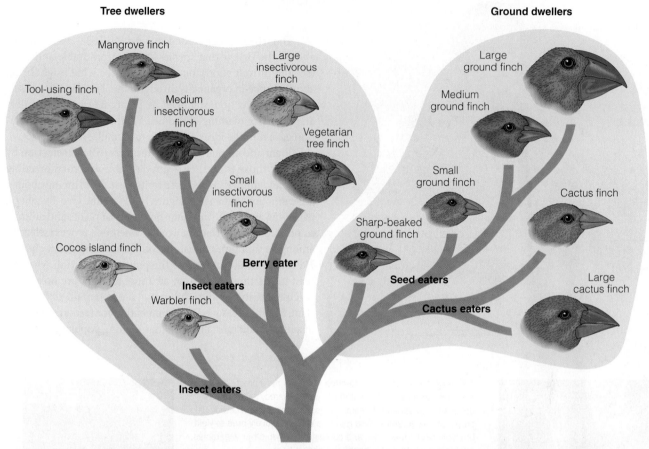

Tree dwellers

Mangrove finch

Tool-using finch

Large insectivorous finch

Medium insectivorous finch

Vegetarian tree finch

Small insectivorous finch

Cocos island finch

Berry eater

Insect eaters

Warbler finch

Ground dwellers

Large ground finch

Medium ground finch

Small ground finch

Cactus finch

Sharp-beaked ground finch

Seed eaters

Large cactus finch

Cactus eaters

Insect eaters

a Darwin's finches arranged to show evolutionary relationships.

b Three of the 13 species of finches on the Galápagos Islands. They are from left to right *Certhidea olivacea,* which has a slender beak to probe for insects in vegetation, *Geospiza scandens,* with a medium-sized beak for eating cactus flowers and fruit, and *Camarhynchus pallidus,* which uses cactus spines to probe for insects.

that characteristics acquired during an individual's lifetime were inheritable.

Given the information available at the time, Lamarck's theory seemed logical and was widely accepted as a viable mechanism for evolution. Indeed, it was not totally refuted until decades later when scientists discovered that the units of heredity known as *genes* cannot be altered by any effort by an organism during its lifetime. Despite many attempts to demonstrate inheritance of acquired characteristics, none has been successful (see Geo-Focus on page 453).

The Contributions of Charles Robert Darwin and Alfred Russel Wallace

In 1859, Charles Robert Darwin (1809–1882) (**Figure 18.2**) published *On the Origin of Species,* in which he detailed his ideas on organic evolution and proposed a mechanism whereby evolution could take place. Although 1859 marks the beginning of modern evolutionary thought, Darwin actually formulated his ideas more than 20 years earlier but, being aware of the furor they would cause, was reluctant to publish them.

Figure 18.2 Charles Robert Darwin in 1840 Darwin was convinced that organisms evolved by natural selection, but he did not publish his ideas until 1859.

American Museum of Natural History #326697

Darwin had concluded during his 1831–1836 voyage aboard the *Beagle* that species are not immutable and fixed (see the Introduction), but he had no idea what might bring about change in organisms through time. However, his observations of selection practiced by plant and animal breeders and a chance reading of Thomas Malthus's essay on population gave him the ideas necessary to formulate his theory.

Plant and animal breeders practice **artificial selection** by selecting those traits they deem desirable and then breeding plants and animals with those traits, thereby bringing about a great amount of change (Figure 18.3) (see the Geo-Insight on pages 454 and 455). The fantastic variety of plants and animals so produced made Darwin wonder whether a process selecting among variant types in nature could also bring about change. He came to fully appreciate the power of selection when he read in Malthus's essay that far more animals are born than reach maturity, yet the adult populations remain rather constant. Malthus reasoned that competition for resources resulted in a high infant mortality rate, thus limiting population size.

Figure 18.3 Artificial Selection Humans have practiced artificial selection for thousands of years, thereby giving rise to dozens of varieties of domestic dogs, pigeons, sheep, cereal crops, and vegetables. Wild mustard was selectively bred to yield broccoli, cauliflower, kale, and cabbage, and two other vegetables, brussels sprouts and kohlrabi, not shown here.

broccoli

© AgStock Images/Corbis

cauliflower

© George D. Lepp/CORBIS

wild mustard

Media Images/SuperStock

kale

© David Cavagnaro/Peter Arnold Images/PhotoLibrary

cabbage

Michael P. Gadomski/Photo Researchers, Inc.

>> Critical Thinking Question
Can you think of other examples of artificial selection in plants and animals?

GEO-FOCUS   The Tragic Lysenko Affair

When Jean-Baptiste de Lamarck (1744–1829) proposed his *theory of inheritance of acquired characteristics,* many scientists accepted it as a viable explanation to account for evolution. In fact, it was not fully refuted for several decades, and even Charles Darwin, at least for a time, accepted some kind of Lamarckian inheritance. Nevertheless, the notion that acquired characteristics could be inherited is now no more than an interesting footnote in the history of science. There was, however, one instance when the idea enjoyed some popularity, but with tragic consequences.

One of the most notable examples of adherence to a disproved scientific theory involved Trofim Denisovich Lysenko (1898–1976), who became president of the Soviet Academy of Agricultural Sciences in 1938. He lost this position in 1953, but in the same year was appointed director of the Institute of Genetics, a post he held until 1965. Lysenko endorsed Lamarck's theory of inheritance

of acquired characteristics. He thought plants and animals could be changed in desirable ways by exposing them to a new environment. Lysenko reasoned that seeds exposed to dry conditions or to cold would acquire a resistance to drought or cold weather, and these traits would be inherited by future generations.

Lysenko accepted inheritance of acquired characteristics because he thought it was compatible with Marxist–Leninist philosophy. Beginning in 1929, Soviet scientists were encouraged to develop concepts consistent with this philosophy. As president of the Academy of Agricultural Sciences and with the endorsement of the Central Committee of the Soviet Union, Lysenko did not allow any other research concerning inheritance. Those who publicly disagreed with him lost their jobs or were sent to labor camps.

Unfortunately for the Soviet people, inheritance of acquired characteristics had been discredited more than 50 years earlier. The results of Lysenko's

belief in the political correctness of this theory and its implementation were widespread crop failures and famine and the thorough dismantling of Soviet genetic research in agriculture. In fact, it took decades for the Soviet genetic research programs to fully recover from this setback, after scientists finally became free to experiment with other theories of inheritance.

Lysenko's ideas on inheritance were not mandated in the Soviet Union because of their scientific merit, but rather because they were deemed compatible with a belief system. The fact that Lysenko retained power and only his type of genetic research was permitted for more than 25 years is a testament to the absurdity of basing scientific theories on philosophic or political beliefs. In other words, a government mandate does not validate a scientific theory, nor does a popular vote or a decree by some ecclesiastic body. Legislating or decreeing that acquired traits are inheritable does not make it so.

In 1858, Darwin received a letter from Alfred Russel Wallace (1823–1913), a naturalist working in southern Asia, who had also read Malthus's essay and came to exactly the same conclusion that a natural process was selecting only a few individuals for survival. Darwin and Wallace presented their idea, called *natural selection,* simultaneously in 1859 to the Linnaean Society in London.

Natural Selection—What Is Its Significance?

We can summarize the salient points of **natural selection,** a mechanism that accounts for evolution, as follows:

1. Organisms in all populations possess heritable variations—size, speed, agility, visual acuity, digestive enzymes, color, and so forth.
2. Some variations are more favorable than others; that is, some variant types have a competitive edge in acquiring resources and/or avoiding predators.
3. Those with favorable variations are *more likely* to survive to reproductive maturity and pass on their favorable variations.

In colloquial usage, natural selection is sometimes expressed as "survival of the fittest," which is misleading because it reduces to "the fittest are those that survive and are thus the fittest." But it actually involves differential rates of survival and reproduction. Therefore, it is largely a matter of reproductive success. Having favorable variations does not guarantee that an individual will live long enough to reproduce, but in a population of perhaps thousands, those with favorable variations are more likely to survive and reproduce (see the Geo-Insight on pages 454 and 455).

Natural selection works on the existing variation in a population, thus simply giving a competitive edge to some individuals. So evolution by natural selection and Lamarck's evolution by inheritance of acquired characteristics are both testable theories, but evidence supports the former, whereas attempts to verify the latter have failed (see Geo-Focus above).

Darwin was not unaware of potential problems for his theory of natural selection. In fact, in *On the Origin of Species,* he wrote:

> If it could be demonstrated that any complex organ existed which could not possibly have been formed by numerous, successive, slight modifications, my theory would absolutely break down. (p. 171)

GEO-INSIGHT   Artificial Selection, Natural Selection, Fossils, and Evolution

Darwin and Wallace used artificial selection as a key element in formulating the theory of natural selection. Fossils provide some of the evidence that organisms have descended with modification (evolved) from ancestors that lived during the past.

▶ **1.** The evolution of corn from teosinte. Wild teosinte has ears 5 to 7 cm long with 5 to 12 kernels per ear. Between 6,000 and 10,000 years ago, Native Americans in Central America selectively bred teosinte, and now a 30-cm ear of corn has about 500 kernels.

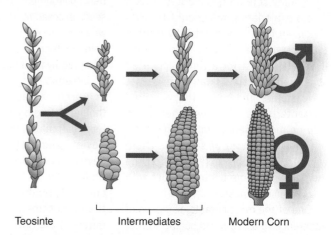

Teosinte   Intermediates   Modern Corn

▼ **2.** Artificial selection of captive populations of rock pigeons has yielded hundreds of varieties of domestic pigeons.

Photographs courtesy Derrell Fowler, Tecumseh, Oklahoma

▼ **3.** Natural selection at work. In preindustrial England, most peppered moths were gray and blended well with the lichens on trees. Black peppered moths were known but rare. With industrialization and pollution, the trees became soot-covered and dark, so dark-colored moths became more common. In rural areas unaffected by pollution, the light–dark frequency did not change.

Michael Willmer Forbes Tweedie/Photo Researchers, Inc.

◀ **4.** Natural selection and the coat color of pocket mice. These mice live in the Sonoran Desert, which has areas of light-brown granite and dark patches of basalt (left). As you might expect, mice living on the light-brown granite have light-brown coats, whereas those living on the basalt have dark-gray coats. The images of the two mice show them on the background that contrasts with their coat color.

© Dr. Hopi Hoekstra, Museum of Comparative Zoology, Harvard University

▶ **5.** Evidence for evolutionary relationships. You would not likely confuse a llama with a dromedary camel, but, in fact, both are camels. They evolved from a common ancestor and even now share several characteristics, although camels live in Asia and llamas live in South America. The animal in the center is a cama, a dromedary camel–llama hybrid. This animal resulted from the artificial insemination of a female llama with sperm from a male dromedary camel.

AP Photo/Kamran Jabreili

Sue Monroe

◀ **6.** Fossils provide some of the evidence for evolution. This duck-billed dinosaur (hadrosaur) measures nearly 15 m long and weighed perhaps 16 metric tons. Found in China, it was on temporary display during 2008 at the Portland Museum of Science and Industry in Portland, Oregon.

▼ **7.** A remarkably well-preserved fossil spider from 35-million-year-old shale in Florissant Fossil Beds National Monument in Colorado.

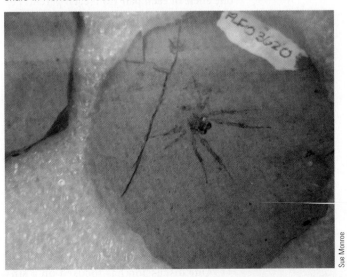

Sue Monroe

James S. Monroe

▲ **8.** About 3 million years ago a blast from an erupting volcano knocked down numerous redwood trees now exposed at The Petrified Forest near Santa Rosa, California. The trees were buried in silica-rich volcanic ash and petrified. The largest part of a tree preserved is about 20 m long.

Critics of natural selection have cited vertebrate eyes and birds' wings that they claim could not have evolved, because, according to them, anything less than eyes and wings as they exist now would be useless. All eyes, no matter if they are nothing more than light-sensitive spots, crude image makers, or image makers as in birds and mammals, are all fully formed and useful. Of what use is half a bird's wing? Of course, this argument presupposes that wings were always used for flying exactly as they are now, which is not necessarily true. The earliest birds may have used rudimentary wings for gliding or for capturing prey. Even today, young partridges have tiny wings of no use for flying and yet to avoid predators they run up trees while flapping their wings. Juvenile partridges with their wings taped down cannot climb trees.

One persistent misconception about natural selection stems from its popularization as "survival of the fittest," a phrase that was never used by Darwin or Wallace. Stated this way, it seems to imply that only the biggest, strongest,

and fastest are likely to survive. In some circumstances these attributes may indeed be favorable, but under different circumstances natural selection may favor the smallest, the most easily concealed, or those having the ability to detoxify some natural or human-made substance.

18.3 Mendel and the Birth of Genetics

The idea of evolution was accepted rather quickly, but the Darwin–Wallace theory of natural selection faced some resistance, mainly because so little was known about inheritance. Critics were quick to point out that Darwin and Wallace could not account for the origin of variations and how variations were maintained in populations. The critics reasoned that should a variant trait arise, it would blend with other traits and be lost. Actually, the answers to these critics existed even then, but would remain in obscurity until 1900.

Mendel's Experiments

During the 1860s, Gregor Mendel, an Austrian monk, performed a series of controlled experiments with true-breeding strains of garden peas (strains that when self-fertilized always display the same trait, such as flower color). He concluded from these experiments that traits are controlled by a pair of factors, or what we now call **genes,** and that these factors (genes) controlling the same trait occur in alternative forms, or **alleles.** He further realized that one allele may be dominant over another and that offspring receive one allele of each pair from each parent. For example, in ▌ Figure 18.4, A represents the allele for red flower color, and a represents white, so any offspring may inherit the combinations of alleles symbolized as AA, Aa, or aa. And since A is dominant over a, only those offspring with the aa combination will have white flowers.

We can summarize the most important aspects of Mendel's work as follows: The factors (genes) that control traits do not blend during inheritance, and even though traits may not be expressed in each generation, they are not lost. Therefore, alternative expressions of genes account for some variation in populations because traits do not blend, as previously thought.

Our discussion has focused on a single gene controlling a trait, but most traits are controlled by many genes, and some genes show incomplete dominance. Mendel

▌ **Figure 18.4** Mendel's Experiment with Flower Color in Garden Peas

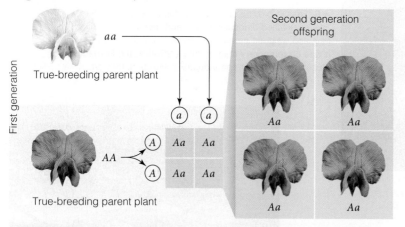

a If true-breeding strains, shown here as the parental generation, are cross-fertilized, they yield offspring with the Aa combination alleles. A is dominant over a, so all offspring in the second generation have red flowers as does the true-breeding parental generation.

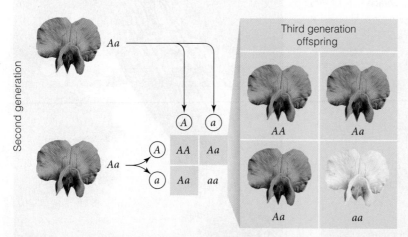

b If the plants from the second generation are cross-fertilized, the third generation has the allele combinations AA, Aa, and aa. Accordingly, this generation has a ratio of three plants with red flowers to one plant with white flowers.

>> **Critical Thinking Question**
Using the third generation, what would be the possible fourth generation offspring if Aa and aa are cross-fertilized?

Figure 18.5 Melosis and Mitosis.

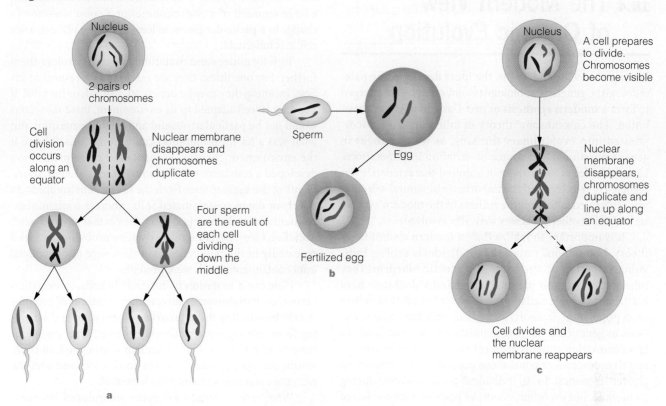

a During melosis, sex cells form that contain one member of each chromosome pair. The formation of sperm Is shown here; eggs form the same way, but only one of the four final eggs is functional.

b The full number of chromosomes is restored when a sperm fertilizes an egg.

c Mitosis results in the complete duplication of a cell. In this example, a cell with four chromosomes (two pairs) produces two cells, each with four chromosomes. Mitosis takes place in all cells except sex cells. Once an egg has been fertilized, the developing embryo grows by mitosis.

was unaware of mutations (changes in the genetic material), chromosomes, and the fact that some genes control the expression of other genes. Nor had he heard of HOX genes that regulate development of major body segments. Nevertheless, Mendel's work provided the answers Darwin and Wallace needed, but his research was published in an obscure journal and went unnoticed until three independent researchers rediscovered it.

Genes and Chromosomes

Complex, double-stranded, helical molecules of **deoxyribonucleic acid (DNA)** called **chromosomes** are found in the cells of all organisms. Specific segments or regions of the DNA molecule are the hereditary units, the genes. The number of chromosomes is specific for a single species, but varies among species. For instance, fruit flies have 8 chromosomes (4 pairs), humans have 46, and domestic horses have 64; chromosomes are always found in pairs carrying genes that control the same traits.

In sexually reproducing organisms, *sex cells* (pollen and ovules in plants and sperm and eggs in animals) are produced when cells undergo a type of cell division known as **meiosis.** This process yields cells with only one

chromosome of each pair, so all sex cells have only half the chromosome number of the parent cell (**Figure 18.5a**). During reproduction, a sperm fertilizes an egg (or pollen fertilizes an ovule), yielding an egg (or ovule) with the full set of chromosomes typical for that species (**Figure 18.5b**).

As Mendel deduced from his experiments, half of the genetic makeup of a fertilized egg comes from each parent. The fertilized egg, however, develops and grows by a cell-division process called **mitosis** during which cells are simply duplicated; that is, there is no reduction in the chromosome number as in meiosis (**Figure 18.5c**).

What Would You Do ?

Suppose that a powerful group in Congress mandated that all future genetic research had to conform to strict guidelines—specifically, that plants and animals should be exposed to particular environments so that they would acquire characteristics that would allow them to live in otherwise inhospitable areas. Furthermore, this group enacted legislation that prohibited any other type of genetic research. Why would it be unwise to implement this research program?

18.4 The Modern View of Organic Evolution

During the 1930s and 1940s, the ideas developed by paleontologists, geneticists, biologists, and others were merged to form a **modern synthesis** or neo-Darwinian view of evolution. The chromosome theory of inheritance was incorporated into evolutionary thinking, as were changes in genes (mutations) as a source of variation in populations. Lamarck's idea of inheritance of acquired characteristics was completely rejected, and the importance of natural selection was reaffirmed, although the authors of the modern synthesis knew that other processes were also involved.

It is important to realize that in modern evolutionary theory, populations rather than individuals evolve. Individuals develop according to the genetic blueprint they inherited from their parents, but an individual may have heritable, favorable variations not present in most members of its population. In short, individuals with favorable variations in genetically varying populations are more likely to leave more descendants than others in their populations, and thus descendent populations possess these variations in greater frequency. So, an individual does not evolve during its lifetime, but evolution does take place in a succession of populations through time.

We emphasize again that the central claim of organic evolution is that organisms have descended with modification from ancestors that lived during the past. Remember, too, that organic evolution involves changes in the genetic makeup of populations through time. It does not encompass changes that take place in organisms as they age or go through some process in their life cycle, such as metamorphosis.

The theory of evolution is not static. New concepts now incorporated into the theory include genetic drift (random changes in populations due to chance), and new ideas on how speciation occurs (see the next sections). Until fairly recently, biologists and paleontologists thought that all genetic transfer was vertical—that is, from one generation to the next. They now know about the processes called *lateral gene transfer* in single-celled organisms such as bacteria, in which one bacterium might incorporate genes from another bacterium but not be the offspring of that bacterium.

What Brings About Variation?

Natural selection works on variations in populations, most of which result from the reshuffling of alleles from generation to generation during sexual reproduction. Given that each of thousands of genes might have several alleles, and that offspring receive half of their genes from each parent, the potential for variation is enormous. Any new variations arise by **mutations** involving some kind of change in the chromosomes or genes—that is, a change in the hereditary information. Whether a *chromosomal mutation* (affecting a large segment of a chromosome) or a *point mutation* (a change in a particular gene), as long as it takes place in a sex cell, it is inheritable.

To fully understand mutations, we must explore them further. For one thing, they are random with respect to fitness, meaning they may be beneficial, neutral, or harmful. If a species is well adapted to its environment, most mutations would not be particularly useful and might be harmful. But what was a harmful mutation can become a useful one if the environment changes. For instance, some plants have developed a resistance to contaminated soils around mines. Plants of the same species from the normal environment do poorly or die in contaminated soils, whereas contaminant-resistant plants do very poorly in the normal environment. Mutations for contaminant resistance probably occurred repeatedly in the population, but they were not beneficial until contaminated soils were present.

How can a mutation be neutral? In cells, information carried on chromosomes directs the formation of proteins in cells by selecting the appropriate amino acids and arranging them into a specific sequence. However, some mutations have no effect on the type of protein synthesized. In other words, the same protein is synthesized before and after the mutation, and thus the mutation is neutral.

What causes mutations? Some are induced by *mutagens,* agents that bring about higher mutation rates. Exposure to some chemicals, ultraviolet radiation, X-rays, and extreme temperature changes might cause mutations. Some mutations are spontaneous, taking place in the absence of any known mutagen.

Sexual reproduction and mutations account for much of the variation in populations, but another factor is *genetic drift,* a random change in the genetic makeup of a population due to chance. Genetic drift is significant in small populations but probably of little importance in large ones. Good examples include evolution of small populations that reach remote areas where only a few individuals with limited genetic diversity rebuild a larger population.

Speciation and the Rate of Evolution

Speciation, the phenomenon of a new species arising from an ancestral species, is well documented, but the rate and ways in which it takes place vary. First, though, let us be clear on what we mean by **species,** a biological term for a population of similar individuals that in nature interbreed and produce fertile offspring. Thus, a species is reproductively isolated from other species. This definition does not apply to organisms such as bacteria that reproduce asexually, but it is nevertheless useful for our discussion of plants, animals, fungi, and single-cell organisms called protistans.

Goats and sheep are distinguished by physical characteristics, and they do not interbreed in nature; thus, they are separate species. Yet in captivity they can produce

■ **Figure 18.6 Allopatric Speciation** An example of allopatric speciation on some remote islands.

a A few individuals of a species on the mainland reach isolated island 1. Speciation follows genetic divergence in a new habitat.

b Later in time, a few individuals of the new species colonize nearby island 2. In this new habitat, speciation follows genetic divergence.

c Speciation may also follow colonization of islands 3 and 4. And it may follow invasion of island 1 by genetically different descendants of the ancestral species.

Akepa (*Loxous coccineus*)

Palila (*Loxioides bailleui*)

Akohekohe (*Palmeria doli*)

Iiwi (*Vestiaria coccinea*)

All © Jack Jeffrey Photography

d More than 20 species of Hawaiian honeycreepers have evolved from a common ancestor, four of which are shown here, as they adapted to diverse resources on the islands.

some species, indicating varying degrees of change from a common ancestral species.

The process of speciation involves a change in the genetic makeup of a population, which also may bring about changes in form and structure. Scientists recognize several ways in which new species may arise, but the most common is **allopatric speciation,** which occurs when a new species evolves from a small population that became isolated from its parent population (■ Figure 18.6). Isolation may result from a marine transgression that effectively separates a once-interbreeding species, or a few individuals may somehow get to a remote area and no longer exchange genes with the parent population (Figure 18.1). Given these conditions and the fact that different selective pressures are likely, they may eventually give rise to a reproductively isolated species.

Widespread agreement exists on allopatric speciation, but scientists disagree on how rapidly a new species may evolve. According to Darwin and reaffirmed by the modern synthesis, the gradual accumulation of minor changes eventually brings about the origin of a new species, a phenomenon called **phyletic gradualism.** Another view, known as **punctuated equilibrium,** holds that little or no change takes place in a species during most of its existence, and then evolution occurs rapidly, giving rise to a new species in perhaps as little as a few thousands of years.

Proponents of punctuated equilibrium argue that few examples of gradual transitions from one species to another are found in the fossil record. Critics, however, point out that neither Darwin nor those who formulated the modern synthesis insisted that all evolutionary change was gradual and continuous, a view shared by many present-day biologists and paleontologists. Furthermore, deposition of sediments in most environments is not continuous; thus, the lack of gradual transitions in many cases is simply an artifact of the fossil record. And finally, despite the incomplete nature of the fossil record, a number of examples of gradual transitions from ancestral to descendant species are well known.

If speciation occurs as we have described, evidence of it taking place now should be available, and in fact, it is. For example, some new plant species have arisen by *polyploidy* when their chromosome count doubled. Speciation, or at least the first steps in speciation, has taken place and continues in populations of mosquitoes, bees, mice, salamanders, fish, and birds that are isolated or partly isolated from one another (■ Figure 18.7).

Divergent, Convergent, and Parallel Evolution

Divergent evolution occurs when an ancestral species gives rise to diverse descendant species. An impressive example involves the mammals whose diversification from a common ancestor during the Late Mesozoic resulted in such varied animals as platypuses, armadillos, rodents, bats, primates, whales, and rhinoceroses (■ Figure 18.8).

fertile offspring. Lions and tigers can also interbreed in captivity, although they do not interbreed in nature and their offspring are sterile, so they too are separate species. Domestic horses that have gone wild can interbreed with zebras to yield a *zebroid,* which is sterile; thus, horses and zebras are separate species. It should be obvious from these examples that reproductive barriers are not complete in

1 A single ancestral species of "greenish warblers" is thought to have originated on the southern edge of the Himalayas 10,000 years ago.

2 Over the years, the warbler established new populations on both sides of the treeless Tibetan Plateau. The birds can interbreed with adjacent populations, indicating that they are still members of the same species.

3 But in Siberia, where birds east and west of the plateau meet, they do not interbreed, they do not recognize each other's songs, their wing coloration changes, and their genes differ. The warblers appear to have become two separate species.

▌**Figure 18.7 Speciation in Songbirds** Two species of Eurasian songbirds appear to have evolved from an ancestral species. Adjacent populations of these birds, a and b, or f and g, for instance, can interbred even though they differ slightly. However, where populations e and h overlap in Siberia, they cannot interbreed.

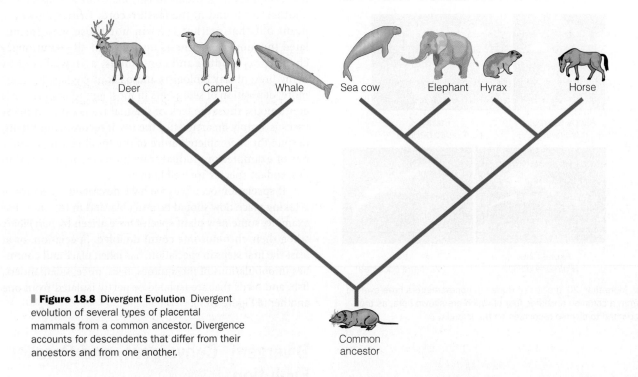

▌**Figure 18.8 Divergent Evolution** Divergent evolution of several types of placental mammals from a common ancestor. Divergence accounts for descendents that differ from their ancestors and from one another.

Divergent evolution leads to descendants that differ markedly from their ancestors. *Convergent evolution* and *parallel evolution* are processes whereby similar adaptations arise in different groups. Unfortunately, they differ in degree and are not always easy to distinguish. **Convergent evolution** involves the development of similar characteristics in distantly related organisms, whereas similar characteristics arising in closely related organisms is **parallel evolution.** In both cases, similar characteristics develop independently because the organisms in question adapt to comparable environments. Perhaps the following examples will clarify the distinction between these two concepts.

❚ Figure 18.9 Convergent and Parallel Evolution

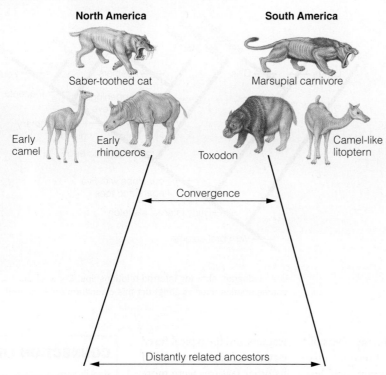

a Convergent evolution takes place when distantly related organisms give rise to species that resemble one another because they adapt in comparable ways.

b Parallel evolution involves the independent origin of similar features in closely related organisms.

During much of the Cenozoic, South America was an island continent with a unique fauna of mammals that evolved in isolation. Nevertheless, several mammals in South and North America adapted in similar ways so that they superficially resembled one another (❚ Figure 18.9a)—a good example of convergent evolution. Likewise, convergence also accounts for the superficial similarities between Australian marsupial (pouched) mammals and placental mammals elsewhere—for example, catlike marsupial carnivores and true cats. Parallel evolution, in contrast, involves closely related organisms, such as jerboas and kangaroo rats, that independently evolved comparable features (❚ Figure 18.9b).

Microevolution and Macroevolution

Micro and *macro* are prefixes that mean "small" and "large," so **microevolution** is any change in the genetic makeup of a species. For example, house sparrows were introduced into North America in 1852 and have evolved so that members of northern populations are larger than those of southern populations, probably a response to climate. Likewise, organisms that develop resistance to insecticides and pesticides, as well as plants that adapt to contaminated soils, are examples of microevolution.

In contrast, **macroevolution** entails changes such as the origin of a new species or changes at even higher levels in the classification of organisms, such as the origin of new genera, families, orders, and classes. The fossil record provides many good examples of macroevolution—the origin of birds from reptiles, the evolution of whales from land-dwelling ancestors, and many others. Although macroevolution encompasses greater changes than microevolution, the cumulative effects of microevolution are responsible for macroevolution. They differ only in the degree of change.

Cladistics and Cladograms

Cladistics is a type of biological analysis in which organisms are grouped together based on derived, as opposed to primitive, characteristics. For instance, all land-dwelling vertebrates have bone and paired limbs, so these characteristics are primitive and of no use in establishing relationships among them. Hair and three middle ear bones, on the other hand, are derived characteristics, sometimes called *evolutionary novelties,* because they served to differentiate mammals from other vertebrates. If we consider only mammals, hair and three middle ear bones are of no further use because all mammals have them, so in this context they are primitive. However, live birth, as opposed to laying eggs, is a derived characteristic that serves to distinguish most mammals from the egg-laying mammals.

Traditionally, scientists have depicted evolutionary relationships with *phylogenetic trees,* in which the horizontal axis shows anatomic differences and the vertical axis denotes time (❚ Figure 18.10a). The patterns of ancestor–descendant relationships are based on shared features, but the ones used are rarely specified. In contrast, a **cladogram** resulting from

❚ Figure 18.10 Phylogenetic Tree and Cladogram

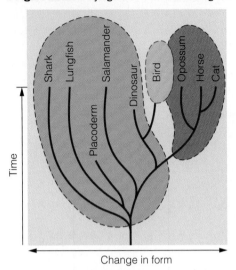

a A phylogenetic tree showing the relationships among various vertebrate animals.

b A cladogram showing inferred relationships. Some of the characteristics used to construct this cladogram are indicated.

cladistic analysis shows the relationships among members of a *clade,* a group of organisms including its most recent common ancestor (❚ Figure 18.10b). Thus, a cladogram includes organisms with shared derived characteristics.

Any number of organisms can be depicted in a cladogram, but the more shown, the more complex and difficult it is to construct. Let's use an example of three animals—bats, dogs, and birds. ❚ Figure 18.11 shows three cladograms, each a different interpretation of their relationship. Bats and birds fly, so we might conclude they are more closely related to each other than to dogs (Figure 18.11a). On the other hand, perhaps birds and dogs are more closely related to each other than either is to bats (Figure 18.11b). But if we concentrate on evolutionary novelties, such as hair and giving birth to live young, we conclude that bats and dogs are more closely related than either is to birds (Figure 18.11c).

Cladistics and cladograms work well for living organisms, but when applied to fossils, care must be taken to determine what are primitive versus derived characteristics, especially in groups with poor fossil records. Furthermore, cladistic analysis depends solely on characteristics inherited from a common ancestor, so paleontologists must be especially careful of characteristics that result from convergent evolution. Nevertheless, cladistics is a powerful tool that has more clearly elucidated the relationships among many fossil lineages and is now used extensively by paleontologists.

Mosaic Evolution and Evolutionary Trends

Evolutionary changes do not involve all aspects of an organism simultaneously because selection pressure is greater on some features than on others. As a result, a key feature we associate with a descendant group may appear before other features typical of that group. For example, the oldest known bird had

feathers and the typical fused clavicles of birds, but most of its other features were more like those of small carnivorous dinosaurs. Accordingly, it represents the concept of **mosaic evolution,** meaning

> **CONNECTION LINK** ◄
>
> You can find more information on the evolution of birds on pages 601 and 602 in Chapter 22.

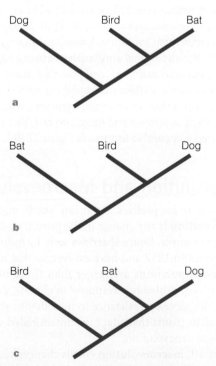

❚ Figure 18.11 Cladograms for Dogs, Birds, and Bats Cladograms showing three hypotheses for the relationships among dogs, birds, and bats. Derived characteristics such as hair and giving birth to live young indicate that dogs and bats are most closely related, as shown in **(c)**.

0   1 m

34 MY

Late

37

Middle

Eocene

49

Early

56

Figure 18.12 Evolutionary Trends in Titanotheres These relatives of horses and rhinoceroses existed for about 20 million years during the Eocene Epoch. During that time, they evolved from small ancestors to giants standing 2.4 m at the shoulder. In addition, they developed large horns, and the shape of the skull changed. Only 4 of the 16 known genera of titanotheres are shown here.

» Critical Thinking Question
Is this sequence of fossil titanotheres consistent with any of the predictions in Table 18.1? If so, which one(s)?

that it had recently evolved characteristics, as well as some features of its ancestral group.

Paleontologists determine in some detail the *phylogeny,* or evolutionary history, and *evolutionary trends* for groups of organisms if sufficient fossil material is available. Size increase is one of the most common evolutionary trends; but trends are extremely complex, they may be reversed, and several trends taking place may not all proceed at the same rate. Abundant fossils show that the Eocene mammals called *titanotheres* not only increased in size but also developed large nasal horns, and the shape of their skull changed (Figure 18.12). The evolution of horses is well documented by fossils that show a general increase in size to living horses, although some fossil species show a size decrease, as well as changes in the skull, teeth, and limbs. However, not all of these changes took place at the same rate.

Isn't evolution by natural selection a random process? If so, how is it possible for a trend to continue long enough to account, just by chance, for such complex structures as eyes, wings, and hands? Actually, evolution by natural selection is a two-step process, and only the first step involves chance. First, variation must be present or arise in a population. Whether or not variations arising by mutations are favorable is indeed a matter of chance, but the second step involving natural selection is not, because only individuals with favorable variations are most likely to survive and reproduce. In one sense, then, natural selection is a process of elimination,

weeding out those individuals not as well adapted to a particular set of environmental circumstances. Of course, such individuals may survive and reproduce just by luck, but in the long run their genes will be eliminated, as selection favors individuals with favorable variations—better visual acuity, appropriate digestive enzymes, more effective camouflage, and so on.

Extinctions

Judging from the fossil record, most organisms that ever existed are now extinct—perhaps as many as 99% of all species. Now, if species actually evolve as natural selection favors certain traits, shouldn't organisms be evolving toward some kind of higher order of perfection or greater complexity? Certainly, vertebrates are more complex, at least in overall organization, than are bacteria, but complexity does not necessarily mean that they are superior in some survival sense—after all, bacteria have persisted for at least 3.5 billion years. Actually, natural selection does not yield some kind of perfect organism, but rather those adapted to a specific set of circumstances at a particular time. Thus, a clam or lizard existing now is not somehow superior to those that lived millions of years ago.

The continual extinction of species is referred to as *background extinction,* to clearly differentiate it from a **mass extinction,** during which accelerated extinction rates sharply reduce Earth's biotic diversity. Extinction is a continuous occurrence, but so is the evolution of new species that usually, but not always, quickly exploit the opportunities created by another species' extinction.

Everyone is familiar with the mass extinction of dinosaurs and other animals at the end of the Mesozoic Era. The greatest extinction, though, during which perhaps more than 90% of all species died out, was at the end of the Paleozoic Era. In some of the following chapters, we discuss those extinctions and their possible causes, as well as other extinctions of lesser magnitude.

CONNECTION LINK ◀

You can find more information on dinosaurs and their extinction in Chapter 22.

18.5 Evidence Supporting Evolutionary Theory

When Charles Darwin proposed his theory of evolution, he cited supporting evidence such as classification, embryology, comparative anatomy, geographic distribution, and, to a limited extent, the fossil record. He had little knowledge of the mechanism of inheritance, and both biochemistry and molecular biology were unknown during his time. Studies in these areas, coupled with a more complete and much better understood fossil record, have convinced scientists that the theory is as well supported by evidence as any other major theory. Of course, scientists disagree on many details, but the central claims of the theory are well established and widely accepted.

But is the theory of evolution truly scientific? That is, can testable predictive statements be made from it? First, we must be clear on what a theory is and what we mean by "predictive." Scientists propose hypotheses to explain natural phenomena, test them, and, in some cases, raise them to the status of a theory—an explanation of some natural phenomenon well supported by evidence from experiments, observations, or both.

Almost everything in the sciences has some kind of theoretical underpinning—optics, the nature of matter, inheritance, the present distribution of continents, diversity in the organic world, and so on. Of course, no theory is ever proven in some final sense, although it might be supported by substantial evidence; all are always open to question, to revision, and occasionally to replacement by a more comprehensive theory. In his book *Why Darwin Matters* (pp. 1–2), Michael Shermer noted that

> a "theory" is not just someone's opinion or a wild guess made by a scientist. A theory is a well-supported and well-tested generalization that explains a set of observations. Science without theory is useless.

Prediction is commonly taken to mean "to foresee an event that has not yet occurred," as in predicting the next solar eclipse. However, not all predictions are about future events. For example, one prediction of seafloor spreading theory is that oceanic crust should be younger at spreading ridges and become progressively older with increasing distance from ridges, which, in fact, it does. Likewise, the theory of evolution allows us to make many predictions about what we should see in the present-day world and in the fossil record if the theory is correct (Table 18.1).

If the theory of evolution is correct, closely related species such as wolves and coyotes should be similar not only in their overall anatomy but also in terms of their biochemistry, genetics, and embryonic development (point 4 in Table 18.1). Suppose they differed in their biochemical mechanisms as well as embryology. Obviously, our prediction would fail, and we would at least have to modify the theory. If other predictions also failed—for example, mammals appeared in the fossil record before fishes—we would have to abandon the theory and find a better explanation for our observations. Accordingly, the theory of evolution is truly scientific because it can, at least in principle, "be falsified"—that is, proven wrong.

Classification—A Nested Pattern of Similarities   Carolus Linnaeus (1707–1778) proposed a classification system in which organisms are given a two-part genus and species name; the coyote, for instance, is *Canis latrans*. Table 18.2 shows Linnaeus's classification, which is a hierarchy of categories that becomes more inclusive as one proceeds up the list. The coyote (*Canis latrans*) and the wolf (*Canis lupus*) share numerous characteristics, so they are members of the same genus, whereas both share some, but fewer, characteristics with the red fox (*Vulpes fulva*), and all three are members of the family Canidae. All canids share some characteristics with cats, bears, and weasels and are grouped together in the order Carnivora, which is one of 18 living orders of the class Mammalia, all of whom are warm-blooded, possess fur or hair, and have mammary glands.

TABLE 18.1	Some Predictions from the Theory of Evolution

1. If evolution has taken place, the oldest fossil-bearing rocks should have remains of organisms very different from those existing now, and more recent rocks should have fossils more similar to today's organisms.

2. If evolution by natural selection actually occurred, Earth must be very old, perhaps many millions of years old.

3. If today's organisms descended with modification from ones in the past, there should be fossils showing characteristics that connect orders, classes, and so on.

4. If evolution is true, closely related species should be similar not only in details of their anatomy but also in their biochemistry, genetics, and embryonic development, whereas distantly related species should show fewer similarities.

5. If the theory of evolution is correct—that is, living organisms descended from a common ancestor—classification of organisms should show a nested pattern of similarities.

6. If evolution actually took place, we would expect cave-dwelling plants and animals to most closely resemble those immediately outside their respective caves rather than being most similar to those in caves elsewhere.

7. If evolution actually took place, we would expect land-dwelling organisms on oceanic islands to most closely resemble those of nearby continents rather than those on other distant islands.

8. If evolution has taken place, a mechanism should exist that accounts for the evolution of one species to another.

9. If evolution occurred, we would expect mammals to appear in the fossil record long after the appearance of the first fish. Likewise, we would expect reptiles to appear before the first mammals or birds.

10. If we examine the fossil record of presumably related organisms such as horses and rhinoceroses, we should find that they were quite similar when they diverged from a common ancestor but became increasingly different as their divergence continued.

TABLE 18.2	Expanded Linnaen Classification Scheme (the animal classified in this example is the coyote, *Canis latrans*)
Kingdom	**Animalia**—Multicelled organisms, cells with nucleus; reproduce sexually; ingest preformed organic molecules
Phylum	**Chordata**—Possess notochord, pharyngeal pouches, dorsal hollow nerve cord at some time during life cycle
Subphylum	**Vertebrata**—Those chordates with a segmented vertebral column
Class	**Mammalia**—Warm-blooded vertebrates with hair or fur and mammary glands
Order	**Carnivora**—Mammals with teeth specialized for a diet of meat
Family	**Canidae**—The doglike carnivores including domestic dogs, wolves, coyotes, and foxes
Genus	***Canis***—Made up only of related species—domestic dogs, wolves, and coyotes
Species	***latrans***—Consists of similar individuals that in nature can interbreed and produce fertile offspring

Linnaeus certainly recognized shared characteristics among organisms, but his intent was simply to categorize species that he thought were specially created and immutable. Following the publication of Darwin's *On the Origin of Species* in 1859, however, scientists quickly realized that shared characteristics constituted a strong argument for evolution.

Linnaeus's classification communicates information about the biological world, but it is arbitrary when assigning organisms to categories above the species level. Also, it does not in all cases reflect evolutionary relationships. As a consequence, scientists now more commonly use evolutionary relationships, or phylogeny, to group organisms. Recall from our discussion of cladistics that a clade is a group of organisms including its most recent common ancestor (Figure 18.10). Accordingly, a well-constructed cladogram shows a nested pattern of similarities (Figure 18.13).

Figure 18.13 Cladograms and Shared Characteristics This example shows the nested pattern of similarities among vertebrate animals—that is, those animals with a segmented vertebral column.

Biological Evidence Supporting Evolution

Life is incredibly diverse—scientists have described at least 1.5 million species and estimate that there may be as many as several tens of millions not yet recognized. And yet, if existing organisms evolved from common ancestors, there should be similarities among all life-forms. As a matter of fact, all living things—be they bacteria, redwood trees, or whales—are composed mostly of carbon, nitrogen, hydrogen, and oxygen. Furthermore, their chromosomes consist of DNA, and all cells synthesize proteins in essentially the same way.

Evidence from biochemistry includes blood proteins that are similar among all mammals, but also indicates that humans are most closely related to great apes, followed in order by Old World monkeys, New World monkeys, and other primates such as lemurs. Biochemical tests support the idea that birds descended from reptiles, a conclusion supported by comparative anatomy and evidence in the fossil record.

The forelimbs of humans, whales, dogs, and birds are superficially dissimilar. Yet all are made up of the same bones; all have basically the same arrangement of muscles, nerves, and blood vessels; all are similarly arranged with respect to other structures; and all have a similar pattern of embryonic development (Figure 18.14). These **homologous structures,** as they are called, are similar even though they may serve different functions—flying versus running, for example—but all were derived from a common ancestor and therefore indicate evolutionary relationships. However, there are some similarities that do not indicate descent from a common ancestor. **Analogous structures** may be superficially similar and serve the same function, as in the wings of insects and birds, but they are dissimilar in structure and development (Figure 18.15).

▌**Figure 18.14** Homologous Structures The forelimbs of several mammals, compared with the forelimb of their ancestors among the reptiles. Homologous structures may serve different functions, but all are composed of the same bones, all have a similar arrangement with respect to other structures, and all undergo a similar embryonic development.

Reptile

Human   Horse   Cat   Bat   Bird   Whale

Bird   Bat

▌**Figure 18.15** Analogous Structures The fly's wings serve the same function as wings of birds and bats, so they are analogous, but they have a different structure and embryologic development.

》Critical Thinking Question
Which of these wings are homologous structures?

Fly

Why do dogs have tiny, functionless toes on their back feet or forefeet or, in some cases, on all feet (▌ Figure 18.16a)? These dewclaws are **vestigial structures**—that is, remnants of structures that were fully functional in their ancestors. Ancestral dogs had five toes on each foot, all of which contacted the ground. But as they evolved, they became toe walkers, with only four digits on the ground, and the big toes and thumbs were lost or reduced to their present state.

Although vestigial structures need not be totally functionless, many are—functionless toes in pigs and deer, remnants of toes in horses, whales and snakes with a pelvis but no rear limbs (▌ Figure 18.16b), and

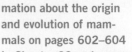

CONNECTION LINK

You can find more information about the origin and evolution of mammals on pages 602–604 in Chapter 22 and pages 635–641 in Chapter 23.

▌Figure 18.16 Vestigial Structures

Sue Monroe

Dewclaw

a Notice the dewclaw, a vestige of the thumb, on the forefoot of the dog.

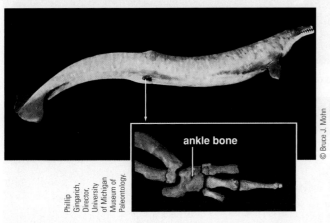

ankle bone

Phillip Gingarich, Director, University of Michigan Museum of Paleontology.

© Bruce J. Mohn

b The Eocene-age whale *Basilosaurus* had tiny vestigial back limbs, but it did not use limbs to support its body weight. Even today, whales have a vestige of a pelvis, and on a few occasions, whales with rear limbs have been caught.

wisdom teeth in humans (many people never have wisdom teeth, and most of those who do, have problems with them). In some cases, vestigial structures are fully functional, but perform a totally different function from what they did in the ancestral condition. For instance, the incus and malleus of the mammalian middle ear were derived from the articular and quadrate bones that formed the joint between the jaw and skull in mammal-like reptiles.

Another type of evidence for evolution is observations of small-scale evolution in living organisms. We have already mentioned one example—the adaptations of some plants to contaminated soils. As a matter of fact, small-scale changes take place rapidly enough that new insecticides and pesticides must be developed continually because insects and rodents develop resistance to existing ones. And development of antibiotic-resistant strains of bacteria is a continuing problem in medicine. Whether the variation in these populations previously existed or was established by

mutations is irrelevant. In either case, some variant types lived and reproduced, bringing about a genetic change.

Fossils: What Do We Learn from Them?

In Chapter 7 we defined **fossils** as the remains or traces of organisms preserved in rocks, and we also noted that they are useful for determining environments of deposition and for relative age determinations. Geologists use the term **body fossil** for remains, mostly shells, bones, and teeth, and rarely mummies and frozen animals (▌Figure 18.17a, b), whereas a **trace fossil** is any indication of organic activity such as burrows and footprints (▌Figure 18.17c). The fossil record is an important repository of life history; in fact, if it were not for fossils we would have no idea that trilobites and dinosaurs ever existed. In addition, fossils provide some of the evidence for organic evolution (see the Geo-Insight on pages 454 and 455).

Some fossils are quite common, especially those of marine-dwelling organisms that have hard skeletons such as corals and clams. After all, so many billions of clams have existed through hundreds of millions of years that if only a tiny fraction of them were preserved as fossils the total number is enormous. On the other hand, the fossil records for worms, jellyfish, and bats are not nearly as good because either the organisms in question did not live where burial in sediment was likely, or they had very delicate skeletons or no skeleton at all.

Fossil marine invertebrates found far from the sea, and even high in mountains, led early naturalists to conclude that the fossil-bearing rocks were deposited during a worldwide flood. In 1508, Leonardo da Vinci realized that the fossil distribution was not what one would expect from a rising flood, but the flood explanation persisted, and John Woodward (1665–1728) proposed a testable hypothesis. According to him, the density of organic remains determined the order in which they settled from floodwaters, so, logically, fossils in the oldest rocks should be denser than those in younger ones. Woodward's hypothesis was quickly rejected because observations did not support it; fossils of various densities are found throughout the fossil record.

The fossil record does show a sequence of different organisms, but not one based on density, size, shape, mobility, or habitat. Rather, the sequence consists of first appearances of groups of organisms through time (▌Figure 18.18). One-celled organisms appeared before multicelled ones, plants before animals, and invertebrates before vertebrates. Among vertebrates, fish appeared first, followed in succession by amphibians, reptiles, mammals, and birds.

If fossils provide evidence for evolution, we might ask, where are the fossils, the so-called missing links, that connect descendants with ancestors? Paleontologists call these "missing links" *transitional fossils* to emphasize the fact that they show characteristics of both ancestors and descendants, and there are, in fact, many of them. In the chapters on Earth and life history we will point out several of these transitional fossils, such as those between amphibians and reptiles, reptiles and birds, and reptiles and mammals. Excellent examples are the Jurassic-age fossils from Germany that have anatomic features much like those of small, carnivorous dinosaurs, and yet they have feathers and the fused clavicles (wishbone) so typical of birds. They have precisely the features we would expect in a transition between reptiles and birds.

Horses, rhinoceroses, and tapirs may seem an odd assortment of mammals, but fossils and studies of living animals clearly show they evolved from a common ancestor (❙ Figure 18.19). If this assessment is correct, we can predict that as we trace these animals back in the fossil record, telling one from the other should become increasingly difficult. And, in fact, the earliest members of each group are remarkably similar, differing mostly in size and minor details of their teeth. As diversification proceeded, though, the differences became more and more apparent.

Of course we will never have enough fossils to document the evolutionary histories of all living organisms, because the remains of some are more likely to be preserved than others, and the accumulation of sediment varies in time and space. Nevertheless, the histories of many groups are well known, and more and more fossils are found that fill in some of the gaps in our knowledge. So even though we may never document the precise history of some organisms, that lack of knowledge has no effect on what we do know about the evolution of trilobites, corals, dinosaurs, and mammals.

❙ **Figure 18.17** Body Fossils and Trace Fossils

a Body fossils are actual remains of organisms, such as these dinosaur bones.

b Body fossils of Oligocene-age clams from Italy.

The Evidence—A Summary

Scientists are convinced that evolutionary theory is as well supported by evidence as any major theory. Indeed, studies as diverse as molecular biology and paleontology all confirm evolutionary theory—that is, organisms living today descended with modification from ancestors that lived during the past. Certainly, scientists disagree on specific issues such as rates of evolution, the significance of some fossils, and precise relationships among evolutionary

▌Figure 18.17 *continued*

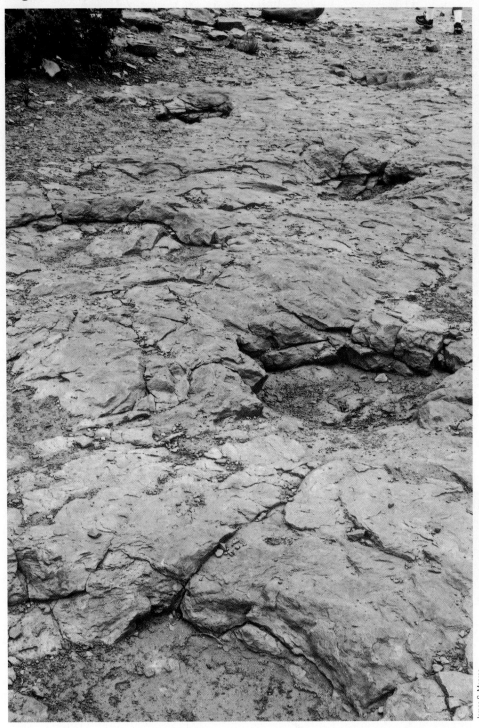

c These circular depressions in sandstone at Copper Ridge, Utah, are trace fossils. They are tracks measuring about 35 cm across that were made by a large dinosaur.

James S. Monroe

lineages, but overall they are overwhelmingly united in their support for the theory. And despite stories holding that scientists are unyieldingly committed to this idea and unwilling to investigate alternative explanations for the same data, just the opposite is true. If another scientific (testable) hypothesis offered a more fruitful approach, scientists would be scrambling to investigate it. There is no better way in the sciences to gain respect and lasting recognition than to modify or replace an existing widely accepted theory.

What Would You Do ❓

Suppose someone told you that evolution is "only a theory that has never been proven" and that "the fossil record shows a sequence of organisms in older to younger rocks that was determined by their density and habitat." Why is the first statement irrelevant to theories in general? What kinds of evidence could you cite to refute the second statement?

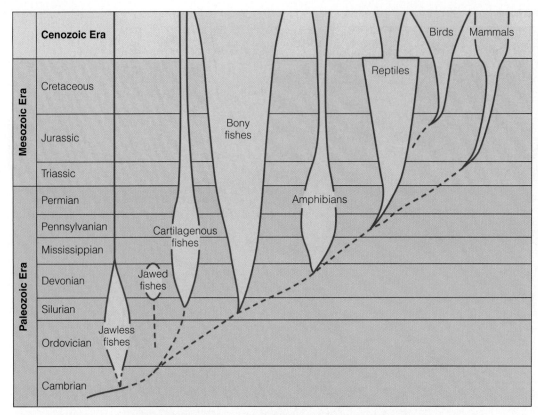

Figure 18.18 **The Fossil Record and Evolution** The spindles indicate the times of appearance and the relative abundance of the members of each group. So, the jawless fishes appeared first, and there are far more species of fishes than there are of amphibians, reptiles, birds, and mammals. A mass extinction at the end of the Mesozoic Era is shown by the marked restriction of the reptile spindle.

Figure 18.19 **Cladogram Showing the Relationships Among Horses, Rhinoceroses, and Tapirs (Order Perissodactyla)** These relationships are well established by fossils and by studies of living animals. When these animals first diverged from a common ancestor, all were remarkably similar, but as divergence continued, they became increasingly different.

GEO-RECAP

» Geo-Image

▌Figure 1 The cladogram below shows that among carnivorous mammals, cats, hyenas, and mongooses are more closely related to one another than to other carnivores. What kinds of evidence from living animals and the fossil record would lend credence to the conclusion that these three animals constitute a clade?

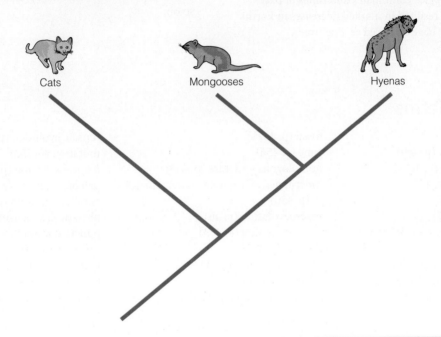

Cats   Mongooses   Hyenas

Chapter Summary

- Jean-Baptiste de Lamarck's proposal of inheritance of acquired characteristics was the first formal explanation for the theory of evolution to be taken seriously.

- In 1859, Charles Robert Darwin and Alfred Russel Wallace published their views on evolution and proposed natural selection as the mechanism for evolutionary change.

- Darwin's observations of variation in natural populations and artificial selection, as well as his reading of Thomas Malthus's essay on population, helped him formulate the idea that natural processes select favorable variants for survival.

- Gregor Mendel's breeding experiments with garden peas provided some of the answers regarding how variation is maintained and passed on.

- Genes are the hereditary determinants in all organisms. This genetic information is carried in the chromosomes of cells, but only the genes in the chromosomes of sex cells are inheritable.

- Sexual reproduction and mutations account for most variation in populations.

- Evolution by natural selection is a two-step process. First, variation must be produced and maintained in interbreeding populations, and second, favorable variants must be selected for survival.

- An important way in which new species evolve is by allopatric speciation. When a group is isolated from its parent population, gene flow is restricted or eliminated, and the isolated group is subjected to different selection pressures.

- Divergent evolution involves an ancestral stock giving rise to diverse species. The development of similar adaptive types in different groups of organisms results from parallel and convergent evolution.

- Microevolution involves changes within a species, whereas macroevolution encompasses all changes above the species level. Macroevolution is simply the outcome of microevolution over time.

- Scientists are increasingly using cladistic analyses to determine relationships among organisms.

- Extinctions take place continually, and times of past mass extinctions resulted in marked decreases in Earth's biologic diversity have occurred several times.

- The theory of evolution is truly scientific because we can think of observations and experiments that could support or falsify it.

- Much of the evidence supporting the theory of evolution comes from classification, embryology, genetics, biochemistry, molecular biology, and present-day small-scale evolution.

- The fossil record also provides evidence for evolution in that it shows a sequence of different groups appearing through time, and some fossils show features that we would expect in the ancestors of birds or mammals, and so on.

Important Terms

allele (p. 456)
allopatric speciation (p. 459)
analogous structure (p. 465)
artificial selection (p. 452)
body fossil (p. 467)
chromosome (p. 457)
cladistics (p. 461)
cladogram (p. 461)
convergent evolution (p. 460)
deoxyribonucleic acid (DNA) (p. 457)
divergent evolution (p. 459)

fossil (p. 467)
gene (p. 456)
homologous structure (p. 465)
inheritance of acquired characteristics (p. 450)
macroevolution (p. 461)
mass extinction (p. 463)
meiosis (p. 457)
microevolution (p. 461)
mitosis (p. 457)
modern synthesis (p. 458)

mosaic evolution (p. 462)
mutation (p. 458)
natural selection (p. 453)
paleontology (p. 450)
parallel evolution (p. 460)
phyletic gradualism (p. 459)
punctuated equilibrium (p. 459)
species (p. 458)
theory of evolution (p. 450)
trace fossil (p. 467)
vestigial structure (p. 466)

Review Questions

1. The theory of natural selection was proposed by Charles Darwin and _____
   a. ____William Smith.
   b. ____Gregor Mendel.
   c. ____Alfred Wallace.
   d. ____Charles Lyell.
   e. ____Jean-Baptiste de Lamarck.

2. The wings of bats and bees serve the same function, but develop differently, so they are _____ structures.
   a. ____analogous.
   b. ____allopatric.
   c. ____divergent.
   d. ____parallel.
   e. ____convergent.

3. Which one of the following is a body fossil?
   a. ____worm burrow.
   b. ____feather impression.
   c. ____dinosaur footprint.
   d. ____mammal tooth.
   e. ____cladogram.

4. The type of cell division that yields eggs and sperm is
   a. ____meiosis.
   b. ____conjugation.
   c. ____chromosomal aberration.
   d. ____allele.
   e. ____mosaic evolution.

5. According to the concept of _____, a species remains unchanged during most of its existence and then gives rise to new species in as little as a few thousands of years.
   a. _____artificial selection.
   b. _____allopatric speciation.
   c. _____punctuated equilibrium.
   d. _____modernization.
   e. _____neo-Darwinism.

6. Convergent evolution accounts for the fact that
   a. _____species evolve very slowly.
   b. _____the fossil record for clams is better than that for birds.
   c. _____body fossils are more useful than trace fossils.
   d. _____many Australian marsupials resemble placental mammals elsewhere.
   e. _____phyletic gradualism takes place in small populations.

7. One prediction of evolutionary theory is
   a. _____humans evolved from monkeys.
   b. _____species are fixed and immutable.
   c. _____features acquired during an organism's lifetime are inheritable.
   d. _____the density of fossils determines where they will be found in the fossil record.
   e. _____mammals should appear in the fossil record long after the first fish.

8. The development of similar features in distantly related species is called
   a. _____fortuitous evolution.
   b. _____microevolution.
   c. _____convergent evolution.
   d. _____cladistic evolution.
   e. _____phyletic evolution.

9. Cell division in which the number of chromosomes is reduced by one-half is
   a. _____divergence.
   b ._____parallelism.
   c. _____meiosis.
   d. _____mutation.
   e. _____lateral gene transfer.

10. Many animals possess _____ that are no longer functional or only partly functional.
    a. _____excess organs.
    b. _____analogous features.
    c. _____phylogenetic atavisms.
    d. _____vestigial structures.
    e. _____cladistic anomalies.

11. Draw three cladograms showing the possible relationships among sharks, whales, and bears. Which cladogram best depicts the relationships among these animals? What criteria did you use to make your decision?

12. Explain what is meant by macroevolution, and give two examples from the fossil record.

13. Does natural selection really mean that only the biggest, strongest, and fastest will survive and reproduce?

14. Give three examples of predictions from evolutionary theory.

15. What criteria are important in defining a vestigial structure? Give some examples.

16. What kinds of evidence should we find in the fossil record if the theory of evolution is correct?

17. How did the experiments carried out by Gregor Mendel help answer the inheritance questions that plagued Darwin and Wallace?

18. What are the concepts of phyletic gradualism and punctuated equilibrium? Is there any evidence supporting either?

19. Explain what mosaic evolution is and give an example.

20. What is/are the difference(s) between inheritance of acquired characteristics and natural selection?

These rocks in the Teton Range in Wyoming are 2.5- to 2.8-billion-year-old, gneiss, schist, and granite. The range itself, however, is only about 10 million years old. The peaks in this image are known as the Cathedral Group.

CHAPTER 19

Sue Monroe

Precambrian Earth and Life History

OUTLINE

19.1 Introduction

19.2 What Happened During the Hadean?

19.3 The Foundations of Continents—Shields, Platforms, and Cratons

19.4 Archean Earth History

19.5 Proterozoic Earth History

19.6 Origin and Evolution of the Atmosphere and Hydrosphere

19.7 Life—Its Origin and Early History

19.8 Resources in Precambrian Rocks

GEO-FOCUS: The Sudbury Meteorite Impact and Its Aftermath

GEO-RECAP

HAVE YOU EVER WONDERED?

- When and how Earth formed?
- What Earth was like during its earliest history?
- How old the oldest known rocks are?
- How the continents formed and evolved?
- Whether Earth's atmosphere has always been like it is now?
- Where the oceans came from and whether they are getting saltier with time?
- What kinds of organisms existed on early Earth?
- Whether Earth's most ancient rocks contain any important resources?

19.1 Introduction

We all know that *time* is used to specify the duration of events and the intervals between events. We are also familiar with time from the human perspective—that is, hours, days, and years—but we have no frame of reference for geologic time. Earth has existed for 4,600,000,000 years, more conveniently stated as 4.6 billion years. So let's suppose than 1 second equals 1 year, and you want to count out Earth's history. Should you take on this task, you and your descendants will be counting for nearly 146 years!

In this chapter we are concerned only with that part of geologic time designated *Precambrian,* 4.6 billion to 542 million years ago. If all geologic time were represented by a 24-hour clock, the Precambrian alone would be more than 21 hours long and constitute 88% of all geologic time (▌ Figure 19.1). And yet we discuss this incredibly long interval of time in a single chapter. Perhaps this treatment seems disproportionate, but we know much more about the Phanerozoic geologic record, which is the more recent 542 million years of geologic time; therefore, we discuss it in more depth.

Precambrian is a widely used term that refers to both time and rocks. As a time term it encompasses all geologic time from Earth's origin until the onset of the Phanerozoic Eon (▌ Figure 19.2). The term also refers to all rocks lying below Cambrian-age rocks, hence the name *Precambrian.* Unfortunately, no rocks are known for the first 600 million years of geologic time, so our geologic record begins about 4.0 billion years ago with the oldest known rocks on Earth except

meteorites. The geologic record we do have for the Precambrian, especially for the part designated Archean (Figure 19.2), is difficult to decipher because (1) many of the rocks have been metamorphosed and complexly deformed; (2) in many places they are deeply buried; and (3) they contain few fossils.

Establishing formal subdivisions of the Precambrian has proved difficult and has varied through the years. However, in 2009, the International Commission on Stratigraphy recommended the use of *Archean Eon* and *Proterozoic Eon* for most of Precambrian time, but it also suggested the informal term *Hadean* for the earliest part of this interval for which there are no known rocks (Figure 19.2). The subdivisions in Figure 19.2 are based on absolute dates rather than time-stratigraphic units, which is a departure from standard practice.

Many of the events that took place during the Precambrian set the stage for further evolution of the planet and life. It was during this time that Earth's systems (see Figure 1.1) became operative, although not all at the same time or in their present form. Earth did not differentiate into a core and mantle until millions of years after it formed, but once it did, internal heat was responsible for plate movements and the evolution of the continents. This was also the time during which the atmosphere formed (although it was very different than it is now), surface waters began to accumulate, and the first organisms appeared at least 3.5 billion years ago. In short, Earth was very different when it first formed, but during the Precambrian it began to evolve and became increasingly more like it is now.

19.2 What Happened During the Hadean?

Hadean is an informal term that encompasses all geologic time from 4.6 to 4.0 billion years ago. Indeed, the oldest known rocks other than meteorites come from the Acasta Gneiss of Canada, which is 3.95 to 4.04 billion years old. So we have no geologic record for the Hadean, but we can make some reasonable inferences about the events that took place then. Of course, Earth accreted as planetesimals collided, it differentiated into a core and mantle (see Figures 1.9 and 1.10), and some continental crust formed, perhaps as much as 4.4 billion years ago.

As the accreting planet grew, it swept up the debris in its vicinity, and just like the other terrestrial planets, it was bombarded by meteorites and comets until about 3.8 billion years ago. Unlike Mercury, Mars, and Earth's Moon, however, the evidence of this period of impacts has been obliterated by weathering, erosion, volcanism, plate movement, and mountain building. In addition to bombardment by meteorites, Earth was probably hit by a Mars-sized planetesimal 4.4 to 4.6 billion years ago, causing the ejection of a huge mass of hot material that coalesced to form the Moon.

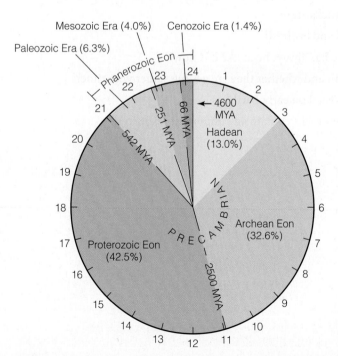

▌ **Figure 19.1** Geologic Time Represented by a 24-Hour Clock If 24 hours represented all geologic time, the Precambrian would be more than 21 hours long, thus more than 88% of the total.

PRECAMBRIAN

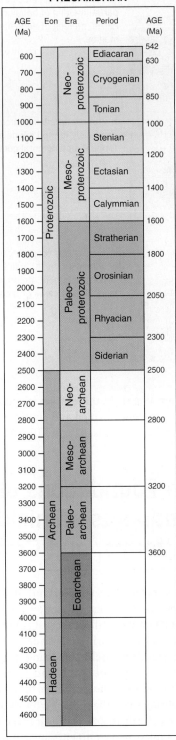

AGE (Ma)	Eon	Era	Period	AGE (Ma)
600	Proterozoic	Neo-proterozoic	Ediacaran	542 / 630
700			Cryogenian	
800				850
900			Tonian	
1000		Meso-proterozoic	Stenian	1000
1100				
1200			Ectasian	1200
1300				
1400			Calymmian	1400
1500				
1600		Paleo-proterozoic	Statherian	1600
1700				
1800				1800
1900			Orosinian	
2000				2050
2100			Rhyacian	
2200				
2300				2300
2400			Siderian	
2500	Archean	Neo-archean		2500
2600				
2700				
2800				2800
2900		Meso-archean		
3000				
3100				
3200		Paleo-archean		3200
3300				
3400				
3500				
3600		Eoarchean		3600
3700				
3800				
3900				
4000	Hadean			
4100				
4200				
4300				
4400				
4500				
4600				

Figure 19.2 The Precambrian Geologic Time Scale This most recent revision of the geologic time scale was published by the International Commission on Stratigraphy (ICS) in 2009. See Figure 1.18 for the complete time scale. Notice the use of the prefixes *eo* (early or dawn), *paleo* (old or ancient), *meso* (middle), and *neo* (new or recent). The age columns on the left and right sides of the time scale are in hundreds and thousands of millions of years (1,800 million years = 1.8 billion years, for example).

Figure 19.3 Earth as it May Have Appeared Soon After it Formed

a No rocks are known from this earliest time in Earth's history, but geologists can make some reasonable inferences about the nature of the newly formed planet.

Copyright and Photograph by Dr. Parvinder S. Sethi

b At about 4.0 billion years old, the Acasta Gneiss from the Northwest Territories in Canada is the oldest known rock on Earth, other than meteorites.

» Critical Thinking Question
Why do you think that so few rocks more than 3.8 billion years old are known on Earth?

After it first formed, Earth retained considerable heat from its origin, and much more heat was generated by radioactive decay; as a result, volcanism was ubiquitous (Figure 19.3a). Gases emitted by volcanoes formed an atmosphere, but it was very unlike the oxygen-rich one present now, and when the planet cooled sufficiently, surface waters began to accumulate. If we could somehow go back and visit early Earth, we would see a rapidly rotating, hot, barren, waterless planet bombarded by meteorites and comets. There were no continents, cosmic radiation would have been intense, and, of course, you would see no organisms.

The age of the oldest continental crust is uncertain, but we can be sure that at least some was present by 3.8 billion years ago, and detrital sedimentary rocks in Australia have zircons ($ZrSiO_4$) 4.4 billion years old, indicating that source rocks that old must have existed. In fact, 3.8-billion-year-old rocks are known from several areas, some of which are metamorphic so their parent rock must be even older.

Figure 19.4 Origin or Granitic Continental Crust

a An andesitic island arc forms by subduction of oceanic lithosphere and partial melting of basaltic oceanic crust. Partial melting of andesite yields granitic magma.

b The island arc in **(a)** collides with a previously formed island arc, thereby forming a continental core.

c The process occurs again when the island arc in **(b)** collides with the evolving continent, thereby forming a craton, the nucleus of a continent.

The friction caused by the Moon on the oceans as well as the continents causes the rate of Earth's rotation to slow very slightly every year. When Earth first formed, it probably rotated in as little as 10 hours so there were many more days in a year; there is no evidence indicating that Earth's orbital period around the Sun has decreased. Another effect of the Earth–Moon tidal interaction is the recession of the Moon from Earth at a few centimeters per year. Thus, during the Hadean, the view of the Moon would have been spectacular!

Geologists agree that when Earth formed, it was exceedingly hot, at least hot enough that it partially melted and differentiated into a core and mantle. Rather than the Earth being a fiery orb for more than a half billion years, as was formerly accepted, some geologists now think that it had cooled enough for surface water to accumulate by 4.4 billion years ago. They base this conclusion on oxygen 18 to oxygen 16 ratios in tiny inclusions in zircon crystals that indicate reactions with surface waters.

When Earth differentiated during the Hadean, the core and mantle formed, but we have made little mention so far of the crust. Remember that we have defined two types of crust, oceanic and continental, which differ in composition, density, and thickness. The first crust probably formed as upwelling mantle currents of mafic magma disrupted the surface, subduction zones formed, and the first island arcs developed (Figure 19.4a). Weathering of these island arcs yielded sediments richer in silica, and partial melting of mafic rocks yielded magma richer in silica. Collisions between island arcs formed a few continental nuclei as silica-rich materials were metamorphosed and intruded by

magma (Figure 19.4b). As these larger island arcs collided, the first protocontinents formed and continued to grow by accretion along their margins (Figure 19.4c).

19.3 The Foundations of Continents—Shields, Platforms, and Cratons

Continents are more than simply land above sea level. Indeed, they consist of rocks with an overall composition similar to granite, and continental crust is thicker and less dense than oceanic crust, which is made up of basalt and gabbro. Furthermore, a **shield** consisting of a vast area or areas of exposed ancient rocks is found on all continents. Continuing outward from the shields are broad **platforms** of buried Precambrian rocks that underlie much of each continent. Collectively, a shield and platform make up a **craton,** which we can think of as a continent's ancient nucleus.

The cratons are the foundations of continents, and along their margins more continental crust was added, a process called **continental accretion,** as they evolved to their present sizes and shapes. Both Archean and Proterozoic rocks are present in cratons, many of which indicate several episodes of deformation accompanied by metamorphism, igneous activity, and mountain building. However, the cratons have experienced remarkably little deformation since the Precambrian.

Figure 19.5 The Distribution of Precambrian Rocks Areas of exposed Precambrian rocks constitute the shields, whereas the platforms consist of buried Precambrian rocks. A shield and its adjoining platform make up a craton.

In North America, the exposed part of the craton is the **Canadian shield,** which occupies most of northeastern Canada, a large part of Greenland, the Adirondack Mountains of New York, and parts of the Lake Superior region in Minnesota, Wisconsin, and Michigan (▮ Figure 19.5). Overall, the Canadian shield is an area of subdued topography, numerous lakes, and exposed Archean and Proterozoic rocks thinly covered in places by Pleistocene glacial deposits. The rocks themselves are plutonic, volcanic, and sedimentary, and metamorphic equivalents of all of these (▮ Figure 19.6).

▮ **Figure 19.6 Archean Rocks in North America** Most exposed Archean rocks in North America are found in the Canadian shield as shown in (**a**), but they are also present in areas of deep erosion and uplift as in the Rocky Mountains (**b**).

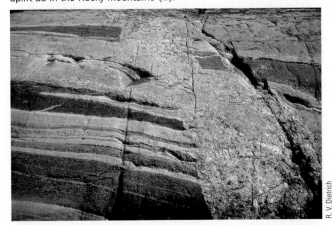

a Outcrop of gneiss cut by a granite dike from a granite-gneiss complex in Ontario, Canada.

>> Critical Thinking Question
Why is it so difficult to study Precambrian rocks, especially those of Archean age, compared to rocks of the Phanerozoic Eon?

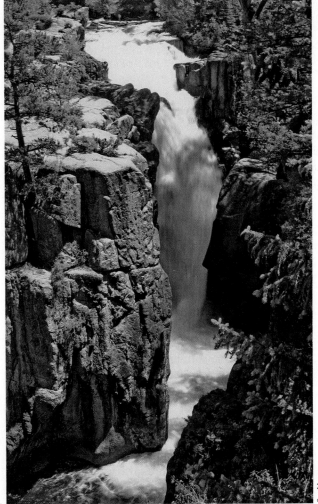

b Shell Creek in the Bighorn Mountains of Wyoming has cut a gorge into this 2.9-billion-year-old (Mesoarchean) granite.

Actually, the Canadian shield, as well as the adjacent platform, are made up of numerous units or smaller cratons that amalgamated along deformation belts during the Paleoproterozoic. Absolute ages and structural trends differentiate these smaller cratons from one another.

Drilling and geophysical evidence indicate that Precambrian rocks underlie much of North America, but beyond the Canadian shield, they are exposed only in areas of deep erosion and uplift. For instance, Archean and Proterozoic rocks are present in the deeper parts of the Grand Canyon of Arizona, as well as in the Appalachian Mountains and the Rocky Mountains.

19.4 Archean Earth History

Geologists place the beginning of the Archean Eon at 4.0 billion years ago (Figure 19.2), which is the age of the oldest known rocks, the Acasta Gneiss in Canada (❙ Figure 19.3b). The end of the Archean Eon and the beginning of the Proterozoic Eon 2.5 billion years ago is arbitrary, but it does correspond to a time when the style of crustal evolution changed and when rock assemblages more like those of the present appeared in abundance.

The Archean Eon alone accounts for 32.3% of all geologic time, yet we review both its physical and biological history in just a few pages. One should not assume, however, that it was an unimportant time in Earth history. The geologic record is more complete for more recent intervals of geologic time, and because Earth is so active, ancient rocks are more likely to have been eroded or changed by metamorphism. In short, older intervals of geologic time are represented by smaller volumes of rock, especially sedimentary rock, and even if preserved, they are more difficult to find and interpret (Table 19.1).

Archean Rocks

Only 22% of Earth's exposed Precambrian crust is Archean, with the most extensive exposures in Africa and North America. Archean crust is made up of many types of rocks, but we characterize most of them as greenstone belts and granite–gneiss complexes, the latter being by far the most common. **Granite–gneiss complexes** are composed of rocks as varied as peridotite and sedimentary rocks; but granitic gneiss and granitic plutonic rocks are the most common, both of which were probably derived from plutons emplaced in volcanic island arcs (❙ Figure 19.7c). Greenstone belts are subordinate, accounting for only 10% of Archean rocks, and yet they are important in unraveling some of the complexities of Archean tectonic events.

A **greenstone belt** has three main components—its lower and middle parts are mostly volcanic rocks, whereas the upper rocks are mostly sedimentary (❙ Figure 19.7a). The origin of the minerals chlorite, actinolite, and epidote during low-grade metamorphism gives the rocks their

greenish color, hence the name *greenstone*. Greenstone belts have a syncline-like structure and measure 40 to 250 km wide and 120 to 800 km long. In addition, they have been intruded by granitic magma and complexly folded and cut by thrust faults.

Much of the volcanism responsible for the lower and middle units in greenstone belts must have been subaqueous because pillow lavas are common (❙ Figure 19.7b), but some large volcanic centers built above sea level. Some of the most interesting volcanic rocks are ultramafic lava flows, or *komatiites*. We noted previously that eruptions producing these flows are rare in rocks younger than Archean because Earth's radiogenic heat production has decreased and near-surface temperatures are not high enough for ultramafic magma to reach the surface (see Chapter 4).

Sedimentary rocks are a minor component in the lower parts of greenstone belts, but they become increasingly abundant toward the top (Figure 19.7a). Associations of graywacke, a sandstone with abundant clay, quartz, feldspars, and rock fragments, and argillite (slightly metamorphosed mudrock) are particularly common. Small-scale graded bedding and cross-bedding indicate that deposition by turbidity currents accounts for the graywacke–argillite successions. Some sedimentary rocks were deposited in delta, tidal flat, barrier island, and shallow marine shelf environments.

Other sedimentary rocks present in Archean greenstone belts include conglomerate, carbonates, and chert. Banded iron formations (BIFs) are also found in Archean greenstone belts, but BIFs are much more common in areas of Proterozoic rocks.

Greenstone belts probably developed in several tectonic settings, but an appealing model for the origin of some relies on Archean plate tectonics and the development of a **back-arc basin** that subsequently closed (❙ Figure 19.8). Recall that a back-arc basin, such as the Sea of Japan, is found between a continent and a volcanic island arc. There is an early stage in which the basin opens, accompanied by volcanism, the emplacement of plutons, and sedimentation, followed by an episode of compression when the basin closes. During this latter phase, the evolving greenstone belt rocks are deformed, metamorphosed, and intruded by granitic magma.

Archean Plate Tectonics and the Origin of Cratons

Certainly, the present plate tectonic regime of opening and closing ocean basins has been a primary agent in Earth's evolution since the Paleoproterozoic. Most geologists are convinced that some kind of plate tectonic activity took place during the Archean as well, but it differed in detail from what is going on now. With more residual heat from Earth's origin and more radiogenic heat, plates must have moved faster and magma was generated more rapidly. As a result, continents no doubt grew more rapidly along their margins as plates collided with island arcs and other plates.

TABLE 19.1 Summary of Archean Geologic and Biologic Events Discussed in the Text

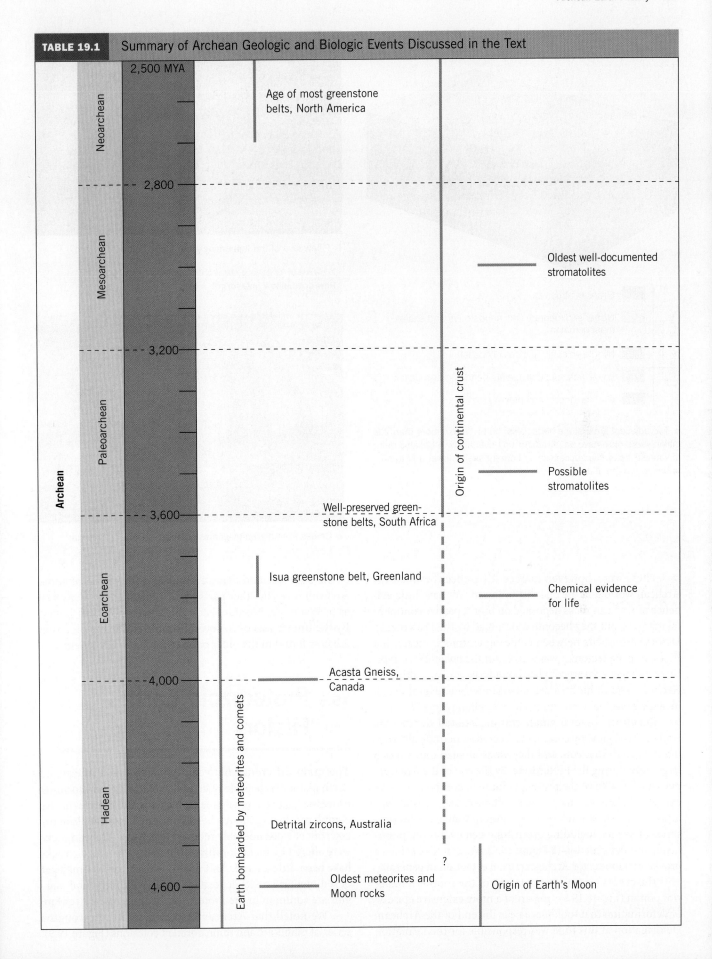

▌Figure 19.7 Greenstone Belts and Granite–Gneiss Complexes

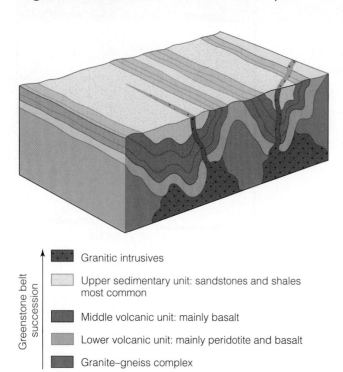

Greenstone belt succession ↑

- **Granitic intrusives**
- **Upper sedimentary unit: sandstones and shales most common**
- **Middle volcanic unit: mainly basalt**
- **Lower volcanic unit: mainly peridotite and basalt**
- **Granite–gneiss complex**

a Two adjacent greenstone belts. Older belts—those more than 2.8 billion years old—have an ultramafic unit overlain by a basaltic unit. In younger belts, the succession is from a basaltic lower unit to an andesite-rhyolite unit.

James S. Monroe

b Pillow lava of the Ispheming greenstone belt in Michigan.

≫ Critical Thinking Question
How does pillow lava form?

R.V. Dietrich

c Gneiss from a granite–gneiss complex in Ontario, Canada.

There were, however, marked differences between the Archean world and the one that followed. We have little evidence of Archean rocks deposited on broad, passive continental margins, but they become widespread by the Proterozoic. Deformation belts between colliding cratons indicate that Archean plate tectonics was active, but the *ophiolites* so typical of younger convergent plate boundaries are rare, although Archean-age ophiolites are now known in several areas, including one in Greenland that is 3.8 billion years old.

Certainly, several small cratons existed during the Archean and grew by continental accretion periodically during the rest of that eon, and they amalgamated into a much larger unit during the Proterozoic. By the end of the Archean, perhaps 30–40% of the present volume of continental crust existed. A plate tectonic model for the Archean crustal evolution of the southern Superior craton of Canada relies on a series of events, including greenstone belt evolution, plutonism, and deformation (▌Figure 19.9). We can take this as a provisional model for Archean crustal evolution in general.

The events leading to the origin of the southern Superior craton (Figure 19.9) are part of a more extensive episode of deformation that took place near the end of the Archean. Deformation at this time was responsible for the formation

of the Superior and Slave cratons and the origin of some Archean rocks in other parts of the Canadian shield, as well as in Wyoming, Montana, and the Mississippi River Valley. By the time it was over, several sizable cratons existed that are now found in the older parts of the Canadian shield.

19.5 Proterozoic Earth History

The main difference between Archean and Proterozoic Earth history is the style of crustal evolution. Crust-forming processes that generated greenstone belts and granite–gneiss complexes during the Archean continued into the Proterozoic, but at a considerably reduced rate. Most Archean rocks were altered by metamorphism, but many Proterozoic rocks have been little altered. And finally, widespread assemblages of sedimentary rocks deposited on passive continental margins are common in the Proterozoic but rare in the Archean.

We noted that Archean cratons assembled through a series of island-arc and minicontinent collisions (Figure 19.9).

Figure 19.8 Origin of a Greenstone Belt in a Back-Arc Basin

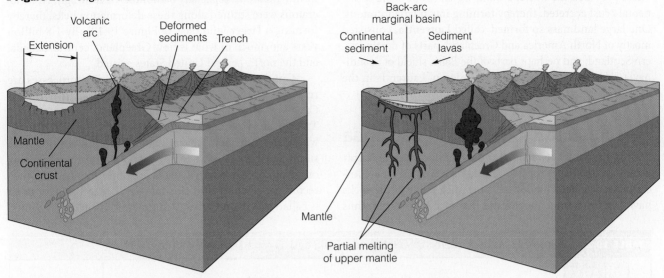

a Rifting on the continental side of the volcanic arc forms a back-arc basin. Partial melting of subducted oceanic crust supplies andesite and dioritic magmas to the island arc.

b Basalt lavas and sediment derived from the continent and island arc fill the back-arc basin.

c Closure of the back-arc basin causes compression and deformation. The evolving greenstone belt is deformed into a syncline-like structure into which granitic magmas intrude.

Figure 19.9 Origin of the Southern Superior Craton

a Geologic map showing greenstone belts (dark green areas) and granite–gneiss subprovinces (light green areas).

b Plate tectonic model for the development of the southern Superior craton. The figure represents a north-south cross section, and the upper diagram is an earlier stage of the lower diagram.

Source: K.C. Condie, *Plate Tectonics and Crustal Evolution,* 2nd edition, p. 218 (Fig. 10.2). Reprinted with permission from the author.

These provided the nuclei around which Proterozoic continental crust accreted, thereby forming much larger cratons. One large landmass so formed, called **Laurentia,** consisted mostly of North America and Greenland, parts of northwestern Scotland, and perhaps parts of the Baltic shield of Scandinavia. We discuss the geologic evolution of Laurentia in the following sections (Table 19.2).

Paleoproterozoic History of Laurentia

During the time between 2.0 and 1.8 billion years ago, collisions among plates formed several **orogens,** which are linear or arcuate deformation belts in which many of the rocks have been metamorphosed and intruded by magma, thus

forming plutons, especially batholiths. Accordingly, Archean cratons were sutured along these deformation belts, thereby forming a larger landmass (▌ Figure 19.10). By 1.8 billion years ago, much of what is now Greenland, central Canada, and the north-central United States existed.

Good examples of these craton-forming processes are recorded in rocks of the Thelon orogen in northwestern Canada where the Slave and Rae cratons collided, and the Trans-Hudson orogen, in Canada and the United States, where the Superior, Hearne, and Wyoming cratons were sutured (Figure 19.10a). Sedimentary rocks of the Wopmay orogen in northwestern Canada are important because they record the opening and closing of an ocean basin, or what is called a **Wilson cycle.** A complete Wilson cycle, named

| TABLE 19.2 | Summary of the Proterozoic Geologic and Biologic Events Discussed in the Text |

Figure 19.10 Proterozoic Evolution of Laurentia These three illustrations show the overall trends in the Proterozoic evolution of Laurentia, but they do not show many of the details in this long, complex episode of Earth history.

a During the Paleoproterozoic, Archean cratons were sutured along deformation belts called orogens.

b Laurentia grew along its southeastern margin by accretion of the Yavapai and Mazatzal provinces.

1.3–1.0 billion ⎱
1.55–1.35 billion ⎰ Mesoproterozoic
1.69–1.65 billion ⎱
1.76–1.72 billion ⎰ Paleoproterozoic
1.9–1.8 billion
2.0–1.8 billion
>2.5 billion ⎱ Archean

c The last episodes in the Proterozoic accretion of Laurentia involved the origin of the Granite-Rhyolite province and the Grenville-Llano provinces.

after the Canadian geologist J. Tuzo Wilson, involves rifting of a continent and the opening of an ocean basin with passive continental margins on both sides. As rifting proceeds, an expansive ocean basin forms, but it eventually begins to close, and subduction zones and volcanic island arcs develop on one or both sides of the ocean basin. The final stage involves a continent–continent collision, during which the rocks that formed along the previous passive continental margins and seafloor are deformed during an orogeny.

Some of the rocks in the Wopmay orogen belong to a suite of sedimentary rocks called a **sandstone–carbonate–shale assemblage** that is deposited on passive continental margins. Paleoproterozoic sandstone–carbonate–shale assemblages are widespread in the Great Lakes region of the United States and Canada (■ Figure 19.11). The sandstones (now quartzites) have sedimentary structures such as ripple marks and cross-beds, whereas some of the carbonate rocks, now mostly dolostone, contain abundant stromatolites (Figure 19.11b, c).

Figure 19.11 Paleoproterozoic Sedimentary Rocks

a Distribution of Paleoproterozoic-age rocks in the Great Lakes region.

b Outcrop of the Mesnard Quartzite. The crests of the ripple marks point toward the observer.

>> **Critical Thinking Question**
In (b), is the layer with the ripples older or younger than the other rocks in the outcrop?

c The Kona Dolomite outcrops. The bulbous structures are stromatolites that resulted from the activities of cyanobacteria (blue-green algae).

Following the initial episode of amalgamation of Archean cratons, accretion took place along Laurentia's southern margin. From 1.8 to 1.6 billion years ago, continental accretion continued in what is now the southwestern and central United States as successively younger belts were

sutured to Laurentia, forming the Yavapai and Mazatzal–Pecos orogens (Figure 19.10b).

The Paleoproterozoic was also the time during which most of Earth's *banded iron formations* (*BIFs*) were deposited. And the first *continental red beds*—sandstone and shale with oxidized iron—were deposited about 1.8 billion years ago. And finally, we should mention that some Paleoproterozoic-age rocks and associated features provide excellent evidence for widespread glaciation.

Paleo- and Mesoproterozoic Igneous Activity

From 1.8 to 1.1 billion years ago, extensive igneous activity took place that was unrelated to orogenic processes. Although quite widespread, much of the magma was either intruded into or erupted onto existing continental crust. These igneous rocks are exposed in eastern Canada, extend across Greenland, and are also found in the Baltic shield of Scandinavia. However, they are deeply buried in most areas.

The origins of these granitic and anorthosite* plutons, calderas and their fill, and vast sheets of rhyolite and ash flows are the subject of debate. According to one hypothesis, large-scale upwelling of magma beneath a Proterozoic supercontinent was responsible for these rocks. According to this hypothesis, the mantle temperature beneath a Proterozoic supercontinent would have been considerably higher than beneath later supercontinents because radiogenic heat production within Earth has decreased. Accordingly, nonorogenic igneous activity would have occurred following the amalgamation of the first supercontinent.

Mesoproterozoic Orogeny and Rifting

The only Mesoproterozoic orogenic event in Laurentia was the 1.3- to 1.0-billion-year-old **Grenville orogeny** (Figure 19.10c). Grenville rocks are well exposed in the present-day northern Appalachian Mountains, as well as in eastern Canada, Greenland, and Scandinavia. Many geologists think that the Grenville orogen resulted from closure of an ocean basin, the final stage in a Wilson cycle. Others disagree, and think that intracontinental deformation or major shearing was responsible for deformation.

Whatever the cause of the Grenville orogeny, it was the final stage in the Proterozoic continental accretion of Laurentia (Figure 19.10c). By then, about 75% of present-day

*Anorthosite is a plutonic rock composed almost entirely of plagioclase feldspars.

Figure 19.12 Proterozoic Rocks in the West

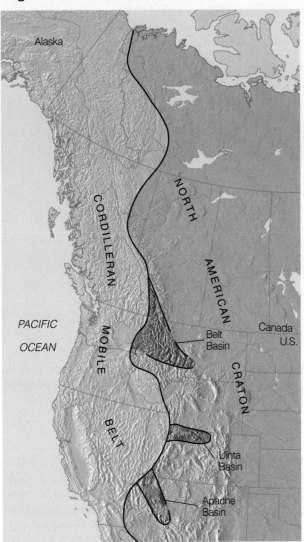

a Meso- to Neoproterozoic sedimentary basins in western North America.

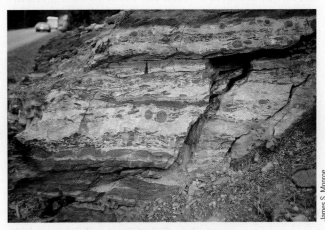

b This 1.0-billion-year-old sandstone in Glacier National Park in Montana has been slightly altered by metamorphism.

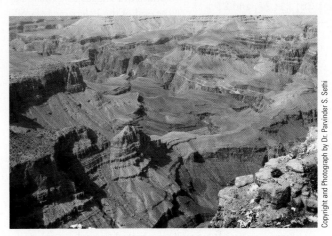

c Sandstone in the Grand Canyon of Arizona. The reddish color results from the presence of small amounts of iron oxides.

North America existed. The remaining 25% accreted along its margins, particularly its eastern and western margins, during the Phanerozoic Eon.

Beginning about 1.1 billion years ago, tensional forces opened the **Midcontinent rift,** a long, narrow trough bounded by faults that outline two branches (Figure 19.10c). Although not all geologists agree, many think that the Midcontinent rift is a failed rift where Laurentia began splitting apart. Had rifting continued, Laurentia would have split into two separate landmasses, but the rifting ceased after about 20 million years.

Most of the Midcontinent rift is buried beneath younger rocks except in the Lake Superior region, where igneous and sedimentary rocks are well exposed. The central part of the rift contains numerous overlapping basalt lava flows, forming a volcanic pile several kilometers thick. Along the rift's margins, coarse-grained sediments were deposited in large alluvial fans that grade into sandstone and shale with increasing distance from the sediment source.

Meso- and Neoproterozoic Sedimentation

The Grenville orogeny between 1.3 and 1.0 billion years ago (Figure 19.10c) was the final episode of continental accretion in Laurentia until the Ordovician Taconic orogeny. Nevertheless, important geologic events were taking place. Meso- and Neoproterozoic-age sedimentary rocks are exceptionally well exposed in the northern Rocky Mountains of Montana, Idaho, and Alberta, Canada (Figure 19.12a,b). Like the rocks in the Great Lakes region, they are mostly sandstones, shales, and stromatolite-bearing carbonates (Figure 19.11).

Sedimentary rocks of Proterozoic age are also found in Utah and the Grand Canyon

CONNECTION LINK

You can find more about the Taconic orogeny on pages 528–530 in Chapter 20.

in Arizona (Figure 19.12c). The rocks, consisting mostly of sandstone, shale, and dolostone, were deposited in shallow-water marine and fluvial environments. The presence of stromatolites and carbonaceous impressions of algae in some of the rocks indicate probable marine deposition.

Proterozoic Supercontinents

A *continent* is one of Earth's major landmasses having an overall composition similar to granite. At present, the continents are mostly above sea level, but at times they have been shallowly submerged during marine transgressions (see Figure 7.10). A *supercontinent* consists of two or more continents sutured together, or, in some cases, most of the continents. But other than its size, a supercontinent is the same as a continent. The supercontinent Pangaea that existed at the end of the Paleozoic is familiar (see Chapter 2), but few are aware of earlier supercontinents.

> **CONNECTION LINK**

For more information about the origin and eventual breakup of the supercontinent Pangaea, see the discussions in Chapters 20, 21, 22, and 23.

Before specifically addressing supercontinents, though, we must note that the present style of plate tectonics involving opening and then closing ocean basins had almost certainly been established by the Paleoproterozoic. In fact, *ophiolites* (rocks of the upper mantle and oceanic crust) (see Figure 2.20) that provide evidence for convergent plate boundaries are known from Neoarchean and Paleoproterozoic rocks from Russia and China, respectively, and recently one dated at 3.8 billion years old (Eoarchean) was discovered in Greenland. Furthermore, these ancient ophiolites compare closely with well-documented younger ones.

Supercontinents may have existed as early as the Late Archean, but if so, we have little evidence of them. The first supercontinent that geologists recognize with some certainty, known as **Rodinia** (Figure 19.13), assembled between 1.3 and 1.0 billion years ago and then began fragmenting 750 million years ago. Judging by the large-scale deformation, the *Pan-African orogeny,* that took place in what are now the Southern Hemisphere continents, geologists conclude that Rodinia's separate pieces reassembled about 650 million years ago and formed another supercontinent, this one known as **Pannotia**. And finally, by the latest Proterozoic, about 550 million years ago, fragmentation was underway again, giving rise to the continental configuration that existed at the onset of the Phanerozoic Eon.

Proterozoic Rocks

The same crust-forming processes that generated Archean granite–gneiss complexes and greenstone belts continued during the Proterozoic, but they did so at a considerably reduced rate. Additionally, many Archean rocks have been thoroughly metamorphosed, whereas vast exposures of Proterozoic rocks show little or no effects of metamorphism.

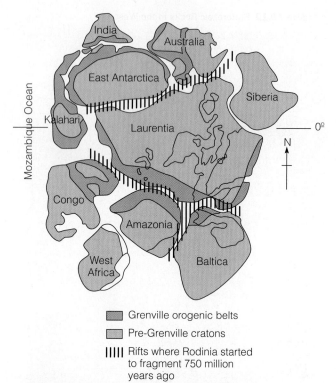

	Grenville orogenic belts
	Pre-Grenville cratons
‖‖‖‖	Rifts where Rodinia started to fragment 750 million years ago

Figure 19.13 Rodinia Possible configuration of the Neoproterozoic supercontinent Rodinia before it began fragmenting about 750 million years ago.

Sandstone–Carbonate–Shale Assemblages Fully 60% of all Proterozoic rocks are sandstone–carbonate–shale assemblages that were deposited along rifted continental margins and in basins within cratons. Their widespread occurrence indicates that large, stable cratons were present with depositional environments much like those of the present. Paleoproterozoic assemblages of these rocks are common in the Great Lakes region of the United States and Canada (Figure 19.11), and in the western part of the continent they are found in three Meso- and Neoproterozoic basins (Figure 19.12).

Proterozoic Glacial Deposits Glacial deposits, as well as striated and polished bedrock, indicate that two major episodes of Proterozoic glaciation took place. North America probably had an extensive ice sheet centered southwest of Hudson Bay during the Paleoproterozoic. Similar deposits of about the same age are present in the United States, Australia, and South Africa, but their ages are not known precisely enough to determine whether there was a single widespread period of glaciation, or a number of glacial events in different areas at different times. Another time of widespread glaciation is indicated by Neoproterozoic-age deposits on all of the continents except Antarctica (Table 19.2; Figure 19.14). It seems that these glaciers existed even in near-equatorial areas.

Based on the fact that Neoproterozoic glacial deposits are so widespread, some geologists think that glaciers covered all land and the seas were frozen—a *snowball Earth,*

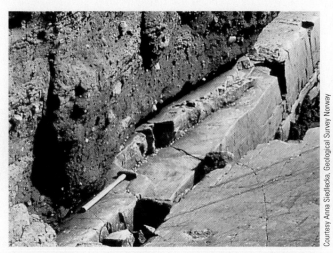

Figure 19.14 Evidence for Proterozoic glaciations in Norway This deposit of glacial till lies upon a striated bedrock surface of sandstone.

Figure 19.15 Proterozoic-Age Banded Iron Formation (BIF) At this outcrop on Jasper Knob in Ishpeming, Michigan, the Negaunee Iron Formation rocks are made up of alternating layers of red chert and silvery iron minerals.

as it has come to be known. The snowball Earth hypothesis is controversial, but proponents claim that the onset of this glacial episode may have been triggered by the near-equatorial location of all continents, and as a result, accelerated weathering would absorb huge quantities of carbon dioxide from the atmosphere. With little CO_2 in the atmosphere, glaciers would form and reflect solar radiation back into space and more glacial ice would form.

So if there actually was a snowball Earth, why wouldn't it stay frozen? Of course, volcanoes would continue to erupt spewing volcanic gases, which includes the greenhouse gases carbon dioxide and methane, which would warm the atmosphere and end the glacial episode. In fact, proponents of this hypothesis note that several such snowball Earths may have occurred until the continents moved into higher latitudes. One criticism of the hypothesis is that if all land was ice covered and the sea froze, how would life survive? Several suggestions have been made to account for this—life persisted at hydrothermal vents on the seafloor; even photosynthesis can take place beneath thin glacial ice; perhaps life persisted in subglacial lakes as it does now in Antarctica; and there may have been pools of liquid water near active volcanoes.

Banded Iron Formation In Chapter 7, we briefly discussed the sedimentary rock known as **banded iron formations (BIFs)** that consist of alternating thin layers of silica (chert) and the iron oxide minerals hematite (Fe_2O_3) and magnetite (Fe_3O_4) (Figure 19.15). Archean-age BIFs are small lens-shaped bodies measuring a few kilometers across and a few meters thick, but those of Proterozoic age are much thicker, cover vast areas, and were probably deposited in shallow marine environments. About 92% of all BIFs formed during the interval from 2.5 to 2.0 billion years ago—that is, during the earlier part of the Paleoproterozoic.

Iron is a highly reactive element that in the presence of oxygen forms rustlike oxides that do not readily dissolve in water. If free oxygen is absent, though, iron goes into solution and accumulates in the oceans. Given that the Archean

atmosphere had little free oxygen, abundant iron was dissolved in seawater. About 2.3 billion years ago, however, there was an increase in the abundance of photosynthesizing bacteria, a corresponding increase in the amount of free oxygen in the atmosphere and oceans, and the precipitation of iron and silica to form BIFs.

Continental Red Beds Obviously, the term **continental red beds** refers to red rocks on the continents; however, more specifically, it means red sandstone or shale colored or stained by iron oxide cement, especially hematite (Fe_2O_3) (Figure 19.12c). These deposits first appear in the geologic record about 1.8 billion years ago, increase in abundance throughout the rest of the Proterozoic, and are quite common in rocks of Phanerozoic age. Iron oxide cement forms under chemically oxidizing conditions, which implies that Earth's atmosphere had some free oxygen by this time, although probably not very much.

19.6 Origin and Evolution of the Atmosphere and Hydrosphere

Early Earth was waterless, its atmosphere lacked free oxygen—that is, oxygen not combined with other elements as in water vapor (H_2O) and carbon dioxide (CO_2)—and no ozone layer was present, so ultraviolet radiation was intense. The atmosphere now contains nearly 21% free oxygen and other important gases in trace amounts, especially carbon dioxide (CO_2) and water vapor. An ozone layer in the upper atmosphere blocks out most ultraviolet radiation, and now 71% of Earth's surface is covered by water. Both the atmosphere and hydrosphere have played important roles in the development of the biosphere. The obvious question is, What brought about such remarkable changes?

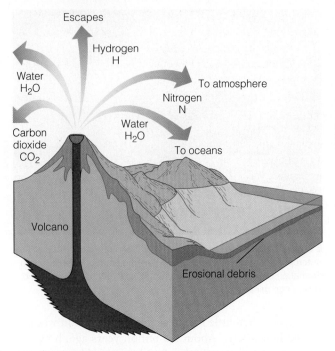

Figure 19.16 Outgassing and Earth's Early Atmosphere. Notice that the atmosphere contains several gases but no free oxygen.

The Atmosphere

Before Earth had a differentiated core, it lacked a magnetic field and a *magnetosphere,* the area around the planet in which the magnetic field is confined. Accordingly, a strong solar wind, an outflow of ions from the Sun, would have swept away any gases that otherwise might have formed

an atmosphere. Once a magnetosphere was established, though, gases from Earth's interior were released during volcanic eruptions and began to accumulate—a phenomenon known as **outgassing** (Figure 19.16).

Archean volcanoes emitted the same gases as today's volcanoes do—mostly water vapor and lesser amounts of carbon dioxide, sulfur dioxide, carbon monoxide, nitrogen, hydrogen, and several others. These gases accumulated rapidly because Earth possessed more residual heat and more radiogenic heat, and as a result, volcanism was common. The atmosphere so formed was rich in carbon dioxide, but contained little or no free oxygen, and with no free oxygen, there was no ozone layer. Furthermore, as these volcanic gases reacted chemically, they very likely yielded atmospheric ammonia (NH_3) and methane (CH_4).

Archean sedimentary deposits with detrital minerals such as pyrite (FeS_2) and uraninite (UO_2) indicate an oxygen-deficient atmosphere because both minerals oxidize rapidly when free oxygen is present. However, oxidized iron is common in Paleoproterozoic-age rocks, so the atmosphere had at least some free oxygen by that time. Two processes account for the introduction of free oxygen into the atmosphere: photochemical dissociation and photosynthesis.

Photochemical dissociation is a process in which water vapor (H_2O) molecules are broken up by ultraviolet radiation in the upper atmosphere to yield hydrogen (H_2) and free oxygen (O_2) (Figure 19.17). However, photochemical dissociation is a self-limiting process and probably accounts for no more than 2% of the present-day level of free oxygen. With this much free oxygen, ozone (O_3) forms a barrier against incoming ultraviolet radiation, so another process must account for most of the atmosphere's free oxygen.

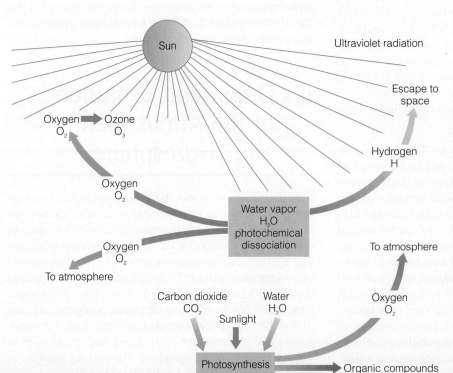

Figure 19.17 Evolution of the Atmosphere Photochemical dissociation occurs when ultraviolet radiation disrupts water molecules that release hydrogen (H_2) and free oxygen (O_2), some of which is converted to ozone (O_3) that blocks most of the ultraviolet radiation. In the presence of sunlight, photosynthesizing organisms use carbon dioxide (CO_2) and water (H_2O) to make organic molecules, and in the processes, release free oxygen as a waste product.

» Critical Thinking Question
Which of these processes, photochemical dissociation or photosynthesis, was the most important in contributing free oxygen to the Precambrian atmosphere?

Photosynthesis is much more important in releasing
free oxygen into the atmosphere, but obviously it could not
take place until organisms practicing this metabolic process had evolved. Organisms that carry out photosynthesis
combine carbon dioxide and water into the organic molecules they need for survival, and they release free oxygen
as a waste product (Figure 19.17). We know from the fossil
record that organisms that almost certainly practiced photosynthesis were present more than 3.0 billion years ago. Even
so, at the end of the Archean Eon, the atmosphere may have
had no more than 1% of its present free oxygen level.

During the Proterozoic Eon, the free oxygen level
increased from 1% to perhaps 10% of its present level, but
probably not until well into the Paleozoic Era, about 400
million years ago, did it reach its current concentration
of 21% of the atmosphere. The cessation of deposition of
banded iron formations and the presence of continental
red beds (Figure 19.12c) provide compelling evidence that
Earth's Proterozoic atmosphere was an oxidizing one.

The Hydrosphere

We have outlined the history of the evolving atmosphere,
and now we turn our attention to the hydrosphere, another
of Earth's major systems. All water on Earth is part of the
hydrosphere, but most of it—more than 97%—is in the
oceans. Where did it come from, and how has it changed?

Certainly, outgassing released water vapor from Earth's
interior, and once the planet cooled sufficiently, water
vapor condensed and surface waters began to accumulate.
Another source of water vapor, and eventually liquid water,
was meteorites and especially icy comets. It is not known
which of these—outgassing or meteorites and comets—was
most important, but we do know that oceans were present
by the Eoarchean, although their volumes and geographic
extent cannot be determined. Nevertheless, we can envision
an early Earth with numerous erupting volcanoes and an
early episode of intense meteorite and comet bombardment
accounting for a rapid rate of surface water accumulation.

Following Earth's early episode of meteorite and comet
bombardment, which ended about 3.8 billion years ago,
these extraterrestrial bodies have added little to the accumulating surface waters. However, volcanoes continue to erupt
and expel water vapor (much of it recycled surface water),
so is the volume of the oceans increasing? Probably it is, but

at a considerably reduced rate. The reason is that much of
Earth's residual heat from when it formed has dissipated and
the amount of radioactive decay to generate internal heat
has diminished, so volcanism is not nearly as commonplace.
Accordingly, the amount of water added to the oceans now
is trivial compared to their volumes.

Recall from Chapter 17 that one early attempt to determine Earth's age was to calculate how long it took for the
oceans to reach their current salinity level—assuming, of
course, that the oceans formed soon after Earth did, that
they were freshwater to begin with, and that their salinity increased at a uniform rate. None of these assumptions
is correct, so the ages determined were vastly different. We
now know that the very early oceans were salty, probably
about as salty as they are now. That is, very early in their
history, the oceans reached chemical equilibrium and have
remained in near-equilibrium conditions ever since.

19.7 Life—Its Origin and Early History

Today, the biosphere is made up of millions of species of
organisms assigned to five kingdoms—animals, plants, protistans, fungi, and monera (bacteria and archaea).* Scientists
have found probable fossils in rocks 3.5 billion years old, and
chemical evidence in 3.8-billion-year-old rocks in Greenland
convinces some that organisms were present then. Certainly
by 3.0 billion years ago (Mesoarchean) organisms were present, but they represent only one of the kingdoms just listed—
monera. In Chapter 18 we discussed the theory of evolution,
which accounts for how organisms have changed through
time, but this theory does not address how life originated in
the first place. Here, we are concerned with **abiogenesis.** that
is, how life formed from nonliving matter. Before proceed-
ing, however, note that abiogenesis does not hold that a living organism such as a bacterium, or even a complex organic
molecule, sprang fully developed from nonliving material.
Rather than one huge step occurring, abiogenesis holds that
several small steps took place, each leading to an increase in
organization and complexity.

But what is living and nonliving? The distinction is clear
in most cases: dogs and trees are alive, rocks and water are
not. A biologist might use several criteria to make the distinction, including growth and reaction to stimuli, but mini-
mally, a living thing must practice some kind of chemical
activity (metabolism) to maintain itself, and it must be capable of reproduction to ensure the long-term survival of the
group to which it belongs. This metabolism–reproduction
criterion might seem sufficient to decide if something is living or not, and yet the distinction is not always easy to make.

Bacteria are living, but under some circumstances, they
can go for long periods during which they show no signs of

---

*Archaea include microscopic organisms that resemble bacteria, but differ
from them genetically and biochemically.

living and then go on living again. Are viruses living? They behave like living organisms in the appropriate host cell, but when outside a host cell, they neither metabolize nor reproduce. Some biologists think that viruses represent another way of living, but others disagree. Comparatively simple organic molecules called *microspheres* form spontaneously and grow and divide in a somewhat organism-like manner, but these processes are more like random chemical reactions, so they are not living.

So what do viruses and microspheres have to do with the origin of life? First, they show that the living versus nonliving distinction is not always clear. And second, if life originated by natural processes, it must have passed through prebiotic stages—that is, stages in which the entities would have shown signs of living organisms but were not truly living.

The Origin of Life

Investigators agree that for life to originate, an energy source must have acted upon the appropriate chemical elements from which organic molecules could synthesize. The early atmosphere, composed of carbon dioxide (CO_2), water vapor (H_2O), nitrogen (N_2), and probably ammonia (NH_3) and methane (CH_4), provided carbon, oxygen, hydrogen, and nitrogen, the primary elements that make up all organisms. Lightning and ultraviolet radiation were two possible sources of the energy necessary for these elements to combine and form rather simple organic molecules known as **monomers,** such as amino acids.

Monomers are needed as basic building blocks of more complex organic molecules, but is it plausible that they formed naturally? Experimental evidence indicates that it is. During the 1950s, Stanley Miller circulated gases approximating Earth's early atmosphere through a closed glass vessel and subjected the mixture to an electric spark to simulate lightning. Within a few days, the mixture became turbid

and an analysis showed that Miller had synthesized several amino acids common in organisms. More recent experiments using different gases have successfully synthesized many of the 20 amino acids in organisms.

Making monomers in a test tube is one thing, but organisms are composed of more complex molecules called **polymers,** such as nucleic acids (DNA and RNA) and proteins consisting of monomers linked together in a specific sequence. So how did this linking of monomers, or polymerization, take place? Researchers have successfully synthesized small molecules called *thermal proteins,* consisting of more than 200 linked amino acids (❚ Figure 19.18). In fact, when heated, dehydrated, and concentrated, amino acids spontaneously polymerize and form thermal proteins.

We call these artificially synthesized thermal proteins *protobionts,* meaning they have characteristics between those of inorganic chemical compounds and living organisms. Suppose that these protobionts came into existence as outlined above. They would have been diluted and would have ceased to exist if they had not developed some kind of outer covering. In other words, they had to be self-contained chemical systems as today's cells are. In the experiments just mentioned, proteinoids have spontaneously aggregated into microspheres that have a cell-like outer covering and grow and divide somewhat like bacteria do (Figure 19.18b).

Perhaps the first steps leading to life took place as monomers formed in great abundance and polymerized, but little is known about how a reproductive mechanism came about. Microspheres divide but do so in a nonbiologic manner. In fact, for some time, researchers were baffled because in present-day organisms either DNA or RNA is necessary for reproduction, but these nucleic acids cannot replicate without protein enzymes, yet protein enzymes cannot be made without nucleic acids. Or so it seemed until a few years ago, when researchers discovered that some small RNA molecules can in fact reproduce without the aid of protein enzymes. So, the first replicating system might have been an RNA molecule.

❚ **Figure 19.18 Experimental Production of Thermal Proteins and Microspheres**

Courtesy of Sidney W. Fox, Distinguished Research Professor, Southern Illinois State University at Carbondale

a Bacteria-like thermal proteins.

Courtesy of Sidney W. Fox, Distinguished Research Professor, Southern Illinois State University at Carbondale

b Thermal proteins aggregate to form microspheres.

VU/WHO/Visuals Unlimited

▌Figure 19.19 Submarine Hydrothermal Vents This black smoker on the East Pacific Rise is at a depth of 2,800 m. The plume of "black smoke" is heated seawater with dissolved minerals. Some scientists think that life may have originated near vents like this one.

Just how these molecules were synthesized under conditions that existed on early Earth has not been resolved.

A common theme is that life originated when organic molecules were synthesized from atmospheric gases, but this idea has been challenged by those who think that the same sequence of events took place near hydrothermal vents (black smokers) on the seafloor (▌Figure 19.19). The necessary elements (carbon, hydrogen, nitrogen, and oxygen) are present in seawater, and heat may have been the energy necessary for monomers to form; in fact, amino acids have been detected in some of these vent waters. Those endorsing this view hold that polymerization took place on the surfaces of clay minerals, and finally protocells were deposited on the seafloor.

Archean Organisms

Prior to the mid-1950s, scientists assumed that the fossils so abundant in Cambrian rocks had a long earlier history, but they had little direct knowledge of Precambrian life. During the early 1900s, Charles Walcott described layered mound-like structures from the Paleoproterozoic-age Gunflint Iron

▌Figure 19.20 Stromatolites and Archean Fossils

Courtesy of Geoffrey Playford, The University of Queensland, Brisbane, Australia

a Present-day stromatolites, Shark Bay, Australia.

Courtesy of J. William Schopf, UCLA

0   10 μm

b Photomicrograph and schematic restoration of fossil prokaryote from the 3.3- to 3.5-billion-year-old Warrawoona Group, Western Australia.

≫ Critical Thinking Question
Stromatolites were widespread during the Proterozoic Eon, but now they have a restricted distribution. Why?

Formation of Canada that he proposed were constructed by algae. We now call these structures **stromatolites,** but not until 1954 did scientists demonstrate that they are actually the result of organic activity. In fact, stromatolites still form in a few areas where they originate by trapping sediment on sticky mats of photosynthesizing cyanobacteria, more commonly known as blue-green algae (▌Figure 19.20a). Although widespread in Proterozoic rocks, they are now restricted to aquatic environments with especially high salinity where snails cannot live and graze on them.

Currently, the oldest known stromatolites are from 3.0-billion-year-old rocks in South Africa, but other probable ones have been discovered in 3.3- to 3.5-billion-year-old rocks near North Pole, Australia. Even more ancient evidence comes from 3.8-billion-year-old rocks in Greenland that have tiny particles of carbon that indicate organic activity, perhaps photosynthesis; but the evidence is not conclusive.

Cyanobacteria, the oldest known fossils, practice photosynthesis, a complex metabolic process that must have been preceded by a simpler process. So it seems reasonable that cyanobacteria were preceded by nonphotosynthesizing organisms for which we so far have no fossils. They probably resembled tiny bacteria, and because the atmosphere lacked free oxygen, they were **anaerobic,** meaning they required no free oxygen. They also were probably **heterotrophic,** dependent on an external source of nutrients, rather than **autotrophic,** capable of manufacturing their own nutrients as in photosynthesis. And finally, they were **prokaryotic cells,** cells lacking a cell nucleus and other internal structures typical of more advanced *eukaryotic cells* (discussed in a later section).

We can characterize these earliest organisms as single-celled, anaerobic, heterotrophic prokaryotes (▌Figure 19.20b). Furthermore, they reproduced asexually. Their energy source was probably adenosine triphosphate (ATP), which can be synthesized from simple gases and phosphate, so it was no doubt available in the early Earth environment. These cells may have simply acquired their ATP directly from their surroundings, but this situation could not have persisted for long because as more and more cells competed for the same resources, the supply should have diminished. Thus, a more sophisticated metabolic process developed, probably *fermentation,* an anaerobic process during which molecules such as sugars are split and release carbon dioxide, alcohol, and energy. As a matter of fact, most living prokaryotes practice fermentation.

Of course, the nature of the earliest cells is informed speculation, but we can say that the most significant event in Archean life history was the development of the autotrophic process of photosynthesis. These more advanced cells were still anaerobic and prokaryotic,

but as autotrophs they no longer depended on an external source of preformed organic molecules for their nutrients. The Archean fossil in Figure 19.20b belongs to the kingdom Monera, which is represented today by bacteria and archaea.

Life of the Proterozoic

The Paleoproterozoic, like the Archean, is characterized mostly by a biota of single-celled, prokaryotic bacteria. No doubt thousands of varieties existed, but none of the more familiar organisms such as plants and animals were present. Before the appearance of cells capable of sexual reproduction, evolution was a comparatively slow process, accounting for the low organic diversity.

A New Type of Cell Appears The origin of **eukaryotic cells** marks one of the most important events in life history (Table 19.2). These cells are much larger than prokaryotic cells; they have a membrane-bounded nucleus containing the genetic material, and most reproduce sexually (▌Figure 19.21). And many eukaryotes—that is, organisms made up of eukaryotic cells—are multicellular and aerobic, so they

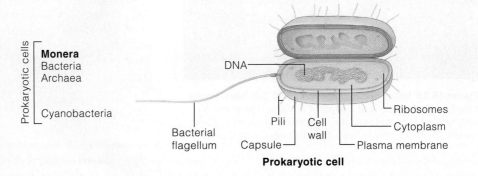

Prokaryotic cells
Monera
Bacteria
Archaea

Cyanobacteria

DNA

Bacterial flagellum | Capsule | Pili | Cell wall | Plasma membrane | Ribosomes | Cytoplasm

Prokaryotic cell

Eukaryotic cells
Protista
Protozoans, algae

Fungi
Spores

Plantae
Plant tissue cells

Animalia
Animal tissue cells

Nuclear envelope / Nucleoplasm + DNA / Nucleolus } Nucleus

Microtubules
Microfilaments
Plasma membrane
Mitochondrion
Pair of centrioles

Vesicle
Lysosome
Rough endoplasmic reticulum
Smooth endoplasmic reticulum
Golgi body
Vesicle

Eukaryotic cell

▌**Figure 19.21 Prokaryotic and Eukaryotic Cells** Eukaryotic cells have a cell nucleus containing the genetic material and organelles such as mitochondria and plastids. In contrast, prokaryotic cells are smaller and not nearly as complex as eukaryotic cells.

Figure 19.22 The Oldest Known Eukaryote and Megafossil

25 μm

N. J. Butterfield

a At 1.2 billion years old, *Bangiomorpha* is the oldest known eukaryotic organism. It is only a few millimeters long.

© Bruce Runnegar

b *Grypania,* at 2.1 billion years old, is the oldest known megafossil. It was probably a bacterium or some kind of algae. It is about 1 cm across.

could not have existed until some free oxygen was present in the atmosphere.

No one doubts that eukaryotes were present by the Mesoproterozoic. Currently, the oldest known eukaryote is *Bangiomorph* from 1.2-billion-year-old rocks in Canada; it was multicelled and reproduced sexually (Figure 19.22a). Other Proterozoic rocks have yielded fossils of unknown affinities. For example, the 2.1-billion-year-old Negaunee Iron Formation of Michigan has fossils known as *Grypania,* which is the oldest known megafossil, but it was probably a single-celled bacterium or some kind of algae (Figure 19.22b).

Cells larger than 60 microns appear in abundance at least 1.4 billion years ago, and many show an increased degree of organizational complexity compared to that of prokaryotic cells. An internal membrane-bounded cell nucleus is present in some, for instance. Furthermore,

microscopic, hollow fossils known as *acritarchs* that probably represent cysts of planktonic algae become common during the Meso- and Neoproterozoic (Figure 19.23a, b). And vase-shaped microfossils from rocks in the Grand Canyon have also been tentatively identified as cysts of some kind of algae (Figure 19.23c).

Endosymbiosis and the Origin of Eukaryotic Cells

According to a widely accepted theory, eukaryotic cells formed from several prokaryotic cells that entered into a symbiotic relationship. Symbiosis, involving a prolonged association of two or more dissimilar organisms, is common today. In many cases, both symbionts benefit from the association, as in lichens, which were once thought to be plants, but actually are symbiotic associations between fungi and algae.

Figure 19.23 Proterozoic Fossils These are probably from eukaryotic organisms and measure only 40 to 50 microns across.

Malgorzata Moczydlowska-Vidal, Uppsala University, Sweden

a The acritarch *Tappania plana* is from Mesoproterozoic rocks in China.

Malgorzata Moczydlowska-Vidal, Uppsala University, Sweden

b This acritarch, known as *Octoedryxium truncatum,* was found in Neoproterozoic rocks in Sweden.

Courtesy of Bonnie Bloeser

c This vase-shaped microfossil from Neoproterozoic rocks in the Grand Canyon in Arizona is probably a cyst from some kind of algae.

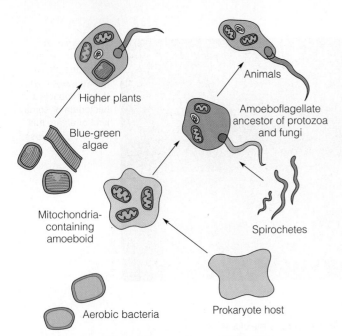

Figure 19.24 Endosymbiosis Theory for the Origin of Eukaryotic Cells An aerobic bacterium and a larger host united to form a mitochondria-containing amoeboid. An amoeboflagellate was formed by a union of the amoebold and a bacterium of the spirochete group; this amoeboflagellate was the direct ancestor of the three kingdoms—protestan, fungi, and animalia. The plant kingdom originated when this amoeboflagellate formed a union with blue-green algae that became plastics.

In a symbiotic relationship, each symbiont must be capable of metabolism and reproduction, but the degree of dependence in some relationships is such that one symbiont cannot live independently. This may have been the case with Proterozoic symbiotic prokaryotes that became increasingly interdependent until the unit could exist only as a whole. In this relationship, though, one symbiont lived within the other, which is a special type of symbiosis called **endosymbiosis** (∥ Figure 19.24).

Supporting evidence for endosymbiosis comes from studies of living eukaryotic cells containing internal structures called organelles, such as mitochondria and plastids, that have their own genetic material. In addition, prokaryotic cells synthesize proteins as a single system, whereas eukaryotic cells are a combination of protein-synthesizing systems. That is, some of the organelles within eukaryotic cells are capable of protein synthesis. These organelles with their own genetic material and protein-synthesizing capabilities are thought to have been free-living bacteria that entered into a symbiotic relationship, eventually giving rise to eukaryotic cells.

The Dawn of Multicelled Organisms **Multicelled organisms** not only are composed of many cells but also have cells specialized to perform specific functions such as reproduction and respiration. Unfortunately, the fossil record does not tell us how multicelled organisms arose from single-celled ancestors. However, studies of present-day organisms give some clues about how this transition might have taken place.

Suppose that a single-celled organism divided and formed a group of cells that did not disperse, but remained together as a colony. The cells in some colonies may have become somewhat specialized, similar to the situation in some living *colonial organisms*. Further specialization might have led to simple multicelled organisms such as sponges, consisting of cells that carry out functions such as reproduction, respiration, and food gathering. Carbonaceous impressions of Proterozoic multicelled algae are known from many areas.

The Ediacaran Fauna In 1947, the Australian geologist R. C. Sprigg discovered a unique assemblage of multicelled, soft-bodied animals preserved as molds and casts on the undersides of sandstone layers (∥ Figure 19.25). Some investigators think that these fossils represent at least three present-day invertebrate phyla: jellyfish and sea pens (phylum Cnidaria), segmented worms (phylum Annelida), and primitive members of the phylum Arthropoda. One wormlike Ediacaran fossil, *Spriggina*, has been cited as a possible ancestor of trilobites, and another, *Tribachidium*, may be a

Figure 19.25 The Ediacaran Fauna of Australia

a The affinities of *Tribachidium* remain uncertain. It may be either a primitive echinoderm or a cnidarian.

b *Spriggina* was originally thought to be a segmented worm (annelid), but now it appears more closely related to arthropods, possibly even an ancestor of trilobites.

c Shield-shaped *Parvanconrina* is perhaps related to the arthropods.

Figure 19.26 Neoproterozoic Fossils

Dr. G. M. Narbonne

a These fossils from Newfoundland have not been given scientific names. The one at the upper center is called a *feather duster,* whereas the other fossils are called *spindles.*

PNAS, December 5, 2000, vol. 97, no. 25, 13684-13689, Shuhai Xiao, Virginia Tech and Andrew Knoll, Harvard University

b These small branching tubes from China may have been made by early relatives of corals. The tubes measure 0.1 to 0.3 mm across.

Ben Waggoner and Mikhail A. Fedarkin

c All agree that *Kimberella* from Russia was an animal; some think it was a mollusk. Specimens measure 3 to 10 cm long.

primitive echinoderm (Figure 19.25a, b). On the other hand, some think Ediacaran fossils represent an early evolutionary radiation distinct from the ancestry of existing invertebrates. These **Ediacaran faunas** existed between 635 and 542 million years ago and are now known on all continents except South America and Antarctica. The animals were widespread, but their fossils are not very common because all of them lacked durable skeletons.

Other Proterozoic Animal Fossils Although scarce, a few animal fossils older than those of the Ediacaran fauna are known from Neoproterozoic rocks (█ Figure 19.26). A jellyfish-like impression is present in rocks 2,000 m below the Pound Quartzite, and in many areas, burrows, presumably made by worms, are found in rocks at least 700 million years old. Wormlike fossils, as well as fossil algae, come from 700- to 900-million-year-old rocks in China, but the identity and age of these "fossils" have been questioned.

All known Proterozoic animals were soft-bodied, but there is some evidence that the earliest stages in the origin of skeletons were under way. Even some Ediacaran animals may have had a chitinous carapace, and others appear to have had areas of calcium carbonate. The odd creature known as *Kimberella,* from the latest Proterozoic of Russia, had a tough outer covering similar to that of some present-day marine invertebrates (Figure 19.26c). Exactly what *Kimberella* was remains uncertain; some think that it was a mollusk.

Neoproterozoic fossils of minute scraps of shell-like material, as well as small toothlike denticles and spicules, presumably from sponges, indicate that several animals with skeletons or at least partial skeletons existed. But more durable skeletons of silica, calcium carbonate, and chitin (a complex organic substance) did not appear in abundance until the beginning of the Phanerozoic Eon 542 million years ago.

CONNECTION LINK

For more information about the emergence of a shelly fauna in the Late Proterozoic and the beginning of the Paleozoic, see pages 543 and 544 in Chapter 21.

19.8 Resources in Precambrian Rocks

Many resources are found in Precambrian rocks, but the one most often associated with rocks of Archean age is gold, whereas iron ore is the one that comes to mind when we discuss Proterozoic resources.

Archean Resources

Archean and Proterozoic rocks near Johannesburg, South Africa, have yielded more than 50% of the world's gold since 1886; but Archean gold is or has also been mined in the Superior craton in Canada and the Homestake Formation in South Dakota. Several Archean-age massive sulfide deposits of zinc, copper, and nickel are known in Australia, Zimbabwe, and Canada, and similar deposits are currently forming adjacent to black smokers on the seafloor.

About one-fourth of the chrome reserves are in Archean rocks, especially in Zimbabwe; they probably formed along with crystals and settled in mafic and ultramafic plutons. The Stillwater Complex in Montana has low-grade chrome and platinum ores that have been mined and

GEO-FOCUS   The Sudbury Meteorite Impact and Its Aftermath

When Earth formed, it swept up the cosmic debris in its orbit, resulting in numerous comet and meteorite impacts. Indeed, this episode in Earth history continued until about 3.8 billion years ago, but even after that Earth was hit by cosmic debris though at a considerably reduced rate. However, because Earth is a dynamic planet the evidence for the earliest episode of impacts has been lost to erosion, plate movements, and mountain building. Scientists are well aware of more recent impacts such as the one 66 million years ago that caused or hastened the extinction of dinosaurs (see Chapter 22). Another impact occurred 35 million years ago where Chesapeake Bay is now located, and about 50,000 years ago a 100-m-diameter meteorite formed Meteor Crater in Arizona.

An especially notable meteor crater, or *astrobleme,* is one measuring about 60 km long and 30 km wide in Ontario, Canada (▌ Figure 1). The impact took place during the Paleoproterozoic, about 1.85 billion years ago, and formed what is now known as the Sudbury Basin. No doubt the impact was spectacular, but the only organisms to witness the event were bacteria and archaea, none of which would have survived within hundreds of kilometers because of the heat and blast. A seismic event would have been recorded around the world if there had been any instruments to record it.

Debris was scattered as far as 800 km from the impact site, thousands of tons of rock were vaporized, releasing copious quantities of gases into the atmosphere, and many hundreds of cubic kilometers of rock melted. Certainly the impact was a catastrophic event, but it was also a fortunate incident for us because it probably accounts for the fact that the Sudbury Basin is one of the richest mining areas in the world.

In 1883, the first metals were discovered in the Sudbury Basin, and since then, 90 or so mines around the basin's margins have produced hundreds of tons of copper, palladium, platinum, nickel, gold, and silver, all commodities essential to industrialized nations. The United States imports some copper, but it also produces copper at mines in several western states. However, much of the platinum-group metals (platinum and palladium) come from South Africa and the United Kingdom, and some is imported from mines in the Sudbury Basin. The United States had no domestic nickel production in 2009 so it depends on imports, 40% of which come from Canada (see Figure 3.20).

One hypothesis for the abundance of metals in the Sudbury Basin is that ores were concentrated from molten rock following the high-speed meteorite impact. In fact, many of the rocks in the basin are mafic intrusives such as gabbro and a similar rock called norite, as well as granophyre and diorite, which are similar to granite. Overlying the igneous rocks are hundreds of meters of impact breccias and younger sedimentary rocks that partly fill the basin.

An interesting note is that the Sudbury Basin has another, much younger meteorite crater lying on its eastern margin. The dark area to the right (east) of the Sudbury Basin in Figure 1 is Lake Wanapitei. This crater is 8.4 km in diameter, and its age is estimated at about 37 million years. It is also interesting to note that Earth continues to be hit by extraterrestrial objects, although none in recent times has produced a large crater. The 1908 Tunguska Event in Siberia leveled hundreds of square kilometers of forest; it was probably a comet that exploded in the atmosphere. And in 1994, a meteorite detonated in the atmosphere over Michigan, and a small piece of it hit a house in the town of Coleman.

NASA WorldWind by User:Vesta using Landsat 7 (Visible Color) satellite image

▌ **Figure 1** The 25-km-wide Sudbury Crater in Ontario, Canada is the oval structure just to the left of the dark circular area, both of which are meteorite impact structures. The Sudbury Crater is about 1.8 billion years old, whereas the younger one, called the Wanapiei Crater is only 34 million years old. When the Sudbury Crater formed it may have been circular and about 250 km across but it was later deformed.

stockpiled during the past, but so far they have not been refined for their chrome or platinum.

Archean-age pegmatites are mostly granitic and of little economic importance. However, some in Manitoba, Canada, and the Rhodesian Province in Africa contain valuable minerals. In addition to gem-quality minerals, a few Archean pegmatites are mined for lithium, beryllium, and rubidium.

Proterozoic Resources

Banded iron formations (BIFs), the world's major source of iron ores, are present on all continents, and 92% of all BIFs were deposited during the Paleoproterozoic (Figure 19.15). The United States must import about 30% of the iron ore it uses, mostly from Canada and Venezuela. Most of the iron ore mined in North America comes from mines in the Lake Superior region and in eastern Canada.

The Sudbury mining district in Ontario, Canada, is an important area of nickel and platinum production (see Geo-Focus on page 498). Nickel is essential for the manufacture of nickel alloys such as stainless steel and Monel metal (nickel plus copper), which are valued for their strength and resistance to corrosion and heat. The United States imports more than 50% of all the nickel it uses, most from the Sudbury mining district in Canada. Some platinum for jewelry, surgical instruments, and chemical and electrical equipment is also exported from Canada to the United States, but the largest exporter is South Africa. The United States also depends on South Africa for much of its chromite, the ore of chromium.

The 2.0-billion-year-old Bushveld complex in South Africa is a huge layered igneous body with rich ores of chromium, platinum, and several other metals. It probably originated as multiple intrusions of magma that slowly crystallized so that different minerals were concentrated in nearly horizontal layers.

Economically recoverable oil and natural gas were discovered in Proterozoic rocks in China and Siberia, arousing interest in the Midcontinent rift as a potential source of hydrocarbons. Some rocks within the rift are known to contain oil, but so far no oil or gas wells are operating.

GEO-RECAP

» Geo-Image

Figure 1 The illustration below shows the 40- to 90-m-thick, 1.2-billion-year-old Purcell Sill in Glacier National Park, Montana. What evidence convinces you that this is a sill rather than a buried lava flow? What can you say about the absolute ages of the rocks below and above the sill? The minerals in the central part of the sill are larger than those near its upper and lower boundaries. Why?

Siyeh Limestone

Purcell Sill (Diorite)

Light-colored Marble

Siyeh Limestone

Chapter Summary

- Geologists use a three-fold division of the Precambrian—the Hadean, Archean, and Proterozoic.

- Each continent has an ancient, stable craton made up of a Precambrian shield and platform. The Canadian shield in North America is made up of several subunits.

- Archean rocks are mostly granite–gneiss complexes and subordinate greenstone belts. One model for the origin of greenstone belts holds that they formed in back-arc marginal basins.

- The amalgamation of Archean cratons and continental accretion along their margins account for the origin of a large landmass known as Laurentia.

- Many geologists think that Archean plates moved faster than plates do now because Earth possessed more radiogenic heat.

- Sandstone–carbonate–shale assemblages deposited on passive continental margins and in intracratonic basins are the most common Proterozoic-age rocks.

- Widespread glaciers were present during the Paleoproterozoic and Neoproterozoic.

- Earth's earliest atmosphere lacked free oxygen, but was rich in carbon dioxide. It was derived by the release of gases during volcanism, a process called outgassing. Meteorite and comet impacts and outgassing yielded the hydrosphere.

- Deposition of widespread banded iron formations between 2.5 and 2.0 billion years ago and the first red beds about 1.8 billion years ago indicate that some free oxygen was present in the atmosphere.

- Energy such as lightning and ultraviolet radiation acting on chemical elements present on early Earth may have yielded the first living things. Some investigators think that RNA molecules were the first molecules capable of reproduction.

- All known Archean fossils represent prokaryotic bacteria. Stromatolites formed by photosynthesizing bacteria may date from 3.5 billion years ago.

- Endosymbiosis practiced by prokaryotic cells was probably responsible for the first eukaryotic cells.

- The oldest megafossils are carbonaceous impressions, probably of algae, in rocks more than 2 billion years old.

- The Neoproterozoic Ediacaran faunas include the oldest well-documented animal fossils. None had durable skeletons, so their fossils are not common.

Important Terms

abiogenesis (p. 491)
anaerobic (p. 494)
autotrophic (p. 494)
back-arc basin (p. 480)
banded iron formation (BIF) (p. 489)
Canadian shield (p. 479)
continental accretion (p. 478)
continental red bed (p. 489)
craton (p. 478)
Ediacaran faunas (p. 497)
endosymbiosis (p. 496)
eukaryotic cell (p. 494)

granite–gneiss complex (p. 480)
greenstone belt (p. 480)
Grenville orogeny (p. 486)
heterotrophic (p. 494)
Laurentia (p. 484)
Midcontinent rift (p. 487)
monomer (p. 492)
multicelled organism (p. 496)
orogen (p. 484)
outgassing (p. 490)
Pannotia (p. 488)
photochemical dissociation (p. 490)

photosynthesis (p. 491)
platform (p. 478)
polymer (p. 492)
prokaryotic cell (p. 494)
Rodinia (p. 488)
sandstone–carbonate–shale
  assemblage (p. 485)
shield (p. 478)
stromatolite (p. 493)
Wilson cycle (p. 484)

Review Questions

1. The uppermost unit in a greenstone belt is characterized by
   a. _____ lava flows.
   b. _____ banded iron formations.
   c. _____ sedimentary rocks.
   d. _____ komatiites.
   e. _____ cratons.

2. Photochemical dissociation is a process whereby
   a. _____ plants synthesize organic molecules.
   b. _____ carbon dioxide is expelled as a waste product of metabolism.
   c. _____ methane and chlorine are emitted during volcanic eruptions.
   d. _____ water molecules are disrupted to yield hydrogen and oxygen.
   e. _____ continents grow along their margins by accretion.

3. The oldest known animal fossils are found in the
   _____ fauna of Australia.
   a. ____Ediacaran.
   b. ____Wendovian.
   c. ____Wilsonian.
   d. ____Rodinian.
   e. ____Laurasian.

4. A large landmass composed mostly of North America
   and Greenland that evolved during the Proterozoic is
   called
   a. ____Romania.
   b. ____Laurentia.
   c. ____Mesoamerica.
   d. ____Grenvillia.
   e. ____Atlantis.

5. The exposed part of the craton in North America is the
   a. ____Wyoming province.
   b. ____Adirondack terrane.
   c. ____Midcontinent platform.
   d. ____Michigan Basin.
   e. ____Canadian shield.

6. Which one of the following sequences of geologic
   time designations is in the correct order from oldest
   to youngest?
   a. ____Phanerozoic-Archean-Proterozoic.
   b. ____Hadean-Archean-Proterozoic.
   c. ____Archean-Hadean-Paleozoic.
   d. ____Mesozoic-Proterozoic-Phanerozoic.
   e. ____Proterozoic-Archean-Phanerozoic.

7. The origin of greenstone belts has not been fully
   resolved, but many geologists think that many of them
   formed in
   a. ____continental shelf environments.
   b. ____carbonate-evaporite depositional areas.
   c. ____back-arc basins.
   d. ____river floodplain environments.
   e. ____transverse shear zones.

8. One type of Proterozoic rock that indicates that at least
   some free oxygen was present is
   a. ____continental red beds.
   b. ____ultramafic lava flows.
   c. ____Wilson cycle deposits.
   d. ____prokaryotic accumulates.
   e. ____glacial deposits.

9. The origin of life from nonliving matter is known as
   a. ____outgassing.
   b. ____abiogenesis.
   c. ____cratonization.
   d. ____biotic accretion.
   e. ____polymerization.

10. Granite–gneiss complexes are
    a. ____the most widespread Archean rocks.
    b. ____found at oceanic spreading ridges.
    c. ____deep-marine deposits of shells of organisms.
    d. ____noted for their fossil plants and animals.
    e. ____greenish because of low-grade metamorphic
       minerals.

11. Explain the two processes that added free oxygen to
    Earth's early atmosphere. Also explain why one process
    was more important than the other.

12. How do stromatolites form, and what is their
    significance?

13. How do continental red beds and banded iron forma-
    tions provide evidence about changes in the Protero-
    zoic atmosphere?

14. What are the first fossil animals recognized in the geo-
    logic record, and why are they so uncommon?

15. Outline the events that led to the origin and growth of
    Laurentia during the Proterozoic.

16. As a high school science teacher you must explain
    that an initially hot Earth has cooled, and at some
    time in the far distant future its internally driven pro-
    cesses will cease. In other words, why will volcanism,
    seismicity, plate movements, and mountain building
    slow down and eventually stop?

17. What are monomers and polymers, and how may they
    have formed naturally?

18. Suppose you were to visit an Earthlike planet that has
    continents and vast oceans. What kinds of evidence
    would indicate that this hypothetical planet's continents
    formed like those on Earth?

19. Proterozoic sedimentary rocks in the northern Rocky
    Mountains are 4,000 m thick and were deposited
    between 1.45 billion and 850 million years ago. What
    was the average rate of sediment accumulation in cen-
    timeters per year? Why is this figure unlikely to repre-
    sent the actual rate of sedimentation?

20. Could you use the principles of superposition, original
    horizontality, and inclusions to decipher Precambrian
    Earth history? Explain.

An aerial view of an open pit coal mine northeast of Somerset, Pennsylvania. The coal being mined is Pennsylvanian in age and formed as a result of tremendous accumulations of plant material deposited in a swampy environment as part of a cyclothem, or repetitive sequence of marine and nonmarine cycles of deposition. Notice the dragline (digging machine) on top of the coal deposit for scale.

CHAPTER 20

Paleozoic Earth History

OUTLINE

20.1 Introduction

20.2 Continental Architecture: Cratons and Mobile Belts

20.3 Paleozoic Paleogeography

20.4 Paleozoic Evolution of North America

20.5 The Sauk Sequence

20.6 The Tippecanoe Sequence

GEO-INSIGHT: The Grand Canyon—A Geologist's Paradise

20.7 The Kaskaskia Sequence

20.8 The Absaroka Sequence

20.9 History of the Paleozoic Mobile Belts

20.10 What Role Did Microplates and Terranes Play in the Formation of Pangaea?

20.11 Paleozoic Mineral Resources

GEO-RECAP

HAVE YOU EVER WONDERED?

- What the geography of Earth looked like during the Paleozoic Era?
- What the climate of Earth was during the Paleozoic Era?
- What North America looked like during the Paleozoic Era, and how it changed from the beginning of the Cambrian Period to the end of the Permian Period?
- How Pangaea came into existence, and how its formation at the end of the Paleozoic Era affected the world's climate?
- How the Appalachian Mountains formed?
- Why the rocks of the Catskill Mountains look the same as many of the rocks that were used to build the castles of England and Scotland?
- How the vast coal deposits found in the midwestern states and the Appalachian region of the United States formed?

20.1 Introduction

August 1, 1815, is an important date in the history of geology. On that date William Smith, a canal builder, published the world's first true geologic map. Measuring more than eight feet high and six feet wide, Smith's hand-painted geologic map of England represented more than 20 years of detailed study of England's rocks and fossils.

England is a country rich in geologic history. Five of the six Paleozoic geologic systems (Cambrian, Ordovician, Silurian, Devonian, and Carboniferous) were described and named for rocks exposed in England. The Carboniferous coal beds of England helped fuel the Industrial Revolution, and the need to transport coal cheaply from where it was mined to where it was used set off a flurry of canal building during the late 1700s and early 1800s. During this time, William Smith, who was mapping various coal mines, first noticed how rocks and fossils repeated themselves in a predictable manner. During the ensuing years, Smith surveyed the English countryside for the most efficient canal routes to bring the coal to market. Much of his success was based on his ability to predict what rocks canal diggers would encounter. Realizing that his observations allowed him to unravel the geologic history of an area and correlate rocks from one region to another, William Smith set out to make the first geologic map of an entire country!

The story of how William Smith came to publish the world's first geologic map is a fascinating tale of determination and perseverance. However, instead of finding fame and success, Smith found himself, slightly less than four years later, in debtors' prison and—upon his release after more than two months—homeless. If such a story can have a happy ending, William Smith at least lived long enough to finally be recognized and honored for the seminal contribution he made to the then fledgling science of geology.

> **CONNECTION LINK**
>
> To review the various principles of relative dating, as well as how rocks are correlated, see pages 418–432 in Chapter 17.

Just as William Smith applied basic geologic principles in deciphering the geology of England, we use these same principles to interpret the geology of the Paleozoic Era. In this chapter, we use the geologic principles and concepts discussed in earlier chapters to help explain how Earth's systems and associated geologic processes interacted during the Paleozoic to lay the groundwork for the distribution of continental landmasses, ocean basins, and the topography we have today.

The Paleozoic history of most continents involves major mountain-building activity along their margins and numerous shallow-water marine transgressions and regressions over their interiors. These transgressions and regressions were caused by global changes in sea level that most probably were related to plate movement and glaciation.

In this chapter, we provide an overview of the geologic history of the world during the Paleozoic Era in order to place into context the geologic events taking place in North America during this time. We then focus our attention on the geologic history of North America—not in a period-by-period chronology, but in terms of the major transgressions and regressions taking place on the continents, as well as the mountain-building activity occurring during this time. Such an approach allows us to place North American geologic events within a global context.

20.2 Continental Architecture: Cratons and Mobile Belts

During the Precambrian, continental accretion and orogenic activity led to the formation of sizable continents. Movement of these continents during the Neoproterozoic led to the first supercontinent known as Rodinia. Following fragmentation of Rodinia sometime around 750 million years ago, a second Pangaea-like supercontinent, referred to as Pannotia, formed. This supercontinent began breaking apart during the latest Proterozoic (550 million years ago), and by the beginning of the Paleozoic Era, six major continents were present. Each continent can be divided into two major components: a craton and one or more mobile belts.

Cratons are the relatively stable and immobile parts of continents and form the foundation on which Phanerozoic sediments were deposited (see Figure 19.5). Cratons typically consist of two parts: a shield and a platform.

Shields are the exposed portions of the crystalline basement rocks of a continent and are composed of Precambrian metamorphic and igneous rocks (see Figure 19.5) that reveal a history of extensive orogenic activity during the Precambrian. During the Phanerozoic, however, shields were extremely stable and formed the foundation of the continents.

Extending outward from the shields are buried Precambrian rocks that constitute a *platform*, another part of the craton. Overlying the platform are flat-lying or gently dipping Phanerozoic detrital and chemical sedimentary rocks that were deposited in widespread shallow seas that transgressed and regressed over the craton. These seas, called **epeiric seas**, were a common feature of most Paleozoic cratonic histories. Changes in sea level, caused primarily by continental glaciation, as well as by plate movement, were responsible for the advance and retreat of these epeiric seas.

Whereas most Paleozoic platform rocks are still essentially flat-lying, in some places they were gently folded into regional arches, domes, and basins (▌Figure 20.1). In many cases, some of these structures stood out as low islands during the Paleozoic Era and supplied sediments to the surrounding epeiric seas.

Mobile belts are elongated areas of mountain-building activity. They are located along the margins of continents where sediments are deposited in the relatively shallow waters of the continental shelf and the deeper waters at the base of the continental slope. During plate convergence along these margins, the sediments are deformed and intruded by magma, creating mountain ranges.

Four mobile belts formed around the margin of the North American craton during the Paleozoic: the Franklin, **Cordilleran**, **Ouachita**, and **Appalachian mobile belts** (Figure 20.1). Each was the site of mountain building in response to compressional forces along a convergent plate boundary and formed mountain ranges such as the Appalachians and Ouachitas.

20.3 Paleozoic Paleogeography

One result of plate tectonics is that Earth's geography is constantly changing. The present-day configuration of the continents and ocean basins is merely a snapshot in time. One of the goals of historical geology is to provide paleogeographic reconstructions of the world during the geologic past. By synthesizing all of the pertinent paleoclimatic, paleomagnetic, paleontologic, sedimentologic, stratigraphic, and tectonic data available, geologists can construct paleogeographic maps. Such maps are simply interpretations of the geography of an area for a particular time in the geologic past and usually show the distribution of land and sea, possible climatic regimes, and such geographic features as mountain ranges, swamps, and glaciers.

The paleogeographic history of the Paleozoic Era, for example, is not as precisely known as that of the Mesozoic and Cenozoic eras, in part because the magnetic anomaly patterns preserved in the oceanic crust were destroyed when much of the Paleozoic oceanic crust was subducted during the formation of Pangaea. Paleozoic paleogeographic reconstructions are therefore based primarily on structural relationships, climate-sensitive sediments such as red beds, evaporites, and coals, as well as the distribution of plants and animals.

At the beginning of the Paleozoic, six major continents were present. Besides these large landmasses, geologists have also identified numerous microcontinents, such as *Avalonia* (composed of parts of present-day Belgium, northern France, England, Wales, Ireland, the Maritime Provinces and Newfoundland of Canada, as well as parts of the New England area of the United States), and various island arcs associated with microplates. We are primarily concerned, however, with the history of the six major continents and their relationships to one another. The six major Paleozoic continents are **Baltica** (Russia west of the Ural Mountains and the major part of northern Europe), **China** (a complex area consisting of at least three Paleozoic continents that were not widely separated and are here considered to include China, Indochina, and the Malay Peninsula), **Gondwana** (Africa, Antarctica, Australia, Florida, India, Madagascar, and parts of the Middle East and southern Europe), **Kazakhstania** (a triangular continent centered on Kazakhstan but considered by some to be an extension of the Paleozoic Siberian continent), **Laurentia** (most of present-day North America, Greenland,

Figure 20.1 Major Cratonic Structures and Mobile Belts The major cratonic structures and mobile belts of North America that formed during the Paleozoic Era.

northwestern Ireland, and Scotland), and **Siberia** (Russia east of the Ural Mountains and Asia north of Kazakhstan and south of Mongolia). The paleogeographic reconstructions that follow (▌ Figures 20.2, 20.3, and 20.4) are based on the methods used to determine and interpret the location, geographic features, and environmental conditions on the paleocontinents.

Early-Middle Paleozoic Global History

In contrast to today's global geography, the Cambrian world consisted of six major continents dispersed around the globe at low tropical latitudes (▌ Figure 20.2a). Water circulated freely among ocean basins, and the polar regions were mostly ice-free. By the Late Cambrian, epeiric seas had covered areas of Laurentia, Baltica, Siberia, Kazakhstania, and China, whereas highlands were present in northeastern Gondwana, eastern Siberia, and central Kazakhstania.

During the Ordovician and Silurian periods, plate movement played a major role in the changing global geography (▌ Figure 20.2b, c). Gondwana moved southward

during the Ordovician and began to cross the South Pole, as indicated by Upper Ordovician tillites (sediments deposited by glaciers and later lithified) found today in the Sahara Desert. During the Early Ordovician, the microcontinent Avalonia separated from Gondwana and began moving northeastward, where it would finally collide with Baltica during the Late Ordovician–Early Silurian. In contrast to the passive continental margin Laurentia exhibited during the Cambrian, an active convergent plate boundary formed along its eastern margin during the Ordovician, as indicated by the Late Ordovician *Taconic orogeny* that occurred in New England.

During the Silurian, Baltica, along with the newly attached Avalonia, moved northwestward relative to Laurentia and collided with it to form the larger continent of **Laurasia**. This collision, which closed the northern Iapetus Ocean, is marked by the *Caledonian orogeny*. After this orogeny, the southern part of the Iapetus Ocean still remained open between Laurentia and Avalonia–Baltica (Figure 20.2c). Siberia and Kazakhstania moved from a southern equatorial position during the Cambrian to north temperate latitudes by the end of the Silurian Period.

▌ **Figure 20.2 Paleozoic Paleogeography** Paleogeography of the world during the **(a)** Late Cambrian Period, **(b)** Late Ordovician Period, and **(c)** Middle Silurian Period.

Shallow sea   Deep ocean   Lowlands   Mountains

a Late Cambrian Period.

Figure 20.2 Paleozoic Paleogeography *(continued)*

Shallow sea Deep ocean Lowlands Mountains Glaciation

b Late Ordovician Period.

Shallow sea Deep ocean Lowlands Mountains

c Middle Silurian Period.

▌Figure 20.3 **Paleozoic Paleogeography** Paleogeography of the world for the **(a)** Late Devonian Period and **(b)** Early Carboniferous Period.

a Late Devonian Period.

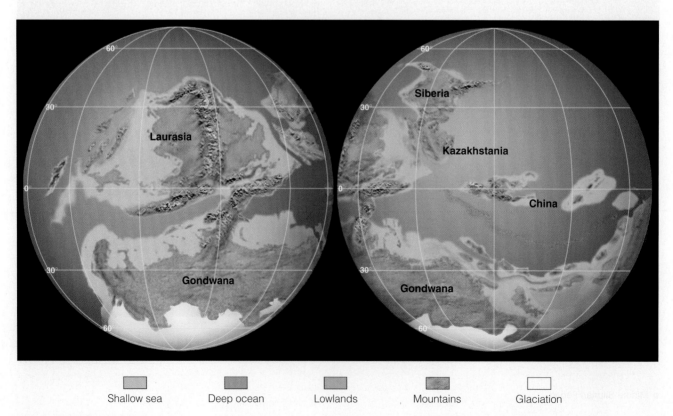

b Early Carboniferous Period.

During the Devonian, as the southern Iapetus Ocean narrowed between Laurasia and Gondwana, mountain building continued along the eastern margin of Laurasia with the *Acadian orogeny* (▌Figure 20.3a). Erosion of the ensuing highlands spread vast amounts of reddish fluvial sediments over large areas of northern Europe (Old Red Sandstone) and eastern North America (the Catskill Delta).

Other Devonian tectonic events, probably related to the collision of Laurentia and Baltica, include the Cordilleran *Antler orogeny,* the *Ellesmere orogeny* along the northern margin of Laurentia (which may reflect the collision of Laurentia with Siberia), and the change from a passive continental margin to an active convergent plate boundary in the Uralian mobile belt of eastern Baltica. The distribution of reefs, evaporites, and red beds, as well as the existence of similar floras throughout the world, suggests a rather uniform global climate during the Devonian Period.

Late Paleozoic Global History

During the Carboniferous Period, southern Gondwana moved over the South Pole, resulting in extensive continental glaciation (▌Figure 20.3b and Figure 20.4a). The advance and retreat of these glaciers produced global changes in sea level that affected sedimentation patterns on the cratons. As Gondwana moved northward, it began colliding with Laurasia during the Early Carboniferous and continued suturing with it during the rest of the Carboniferous (Figures 20.3b and 20.4a). Because Gondwana rotated clockwise relative to Laurasia, deformation generally progressed in a northeast-to-southwest direction along the Hercynian, Appalachian, and Ouachita mobile belts of the two continents. The final phase of collision between Gondwana and Laurasia is indicated by the Ouachita Mountains of Oklahoma, formed by thrusting during the Late Carboniferous and Early Permian.

Elsewhere, Siberia collided with Kazakhstania and moved toward the Uralian margin of Laurasia (Baltica), colliding with it during the Early Permian. By the end of the Carboniferous, the various continental landmasses were fairly close together as Pangaea began taking shape.

The Carboniferous coal basins of eastern North America, western Europe, and the Donets Basin of Ukraine all lay in the equatorial zone, where rainfall was high and temperatures were consistently warm. The absence of strong seasonal growth rings in fossil plants from these coal basins indicates such a climate. The fossil plants found in the coals of Siberia, however, show well-developed growth rings, signifying seasonal growth with abundant rainfall and distinct seasons such as in the temperate zones (latitude 40 degrees to 60 degrees north).

Glacial conditions and the movement of large continental ice sheets in the high southern latitudes are indicated by widespread tillites and glacial striations in southern Gondwana (see Figure 2.5). These ice sheets spread toward the equator and, at their maximum growth, extended well into the middle temperate latitudes.

The assembly of Pangaea was essentially completed during the Permian as a result of the many continental collisions that began during the Carboniferous (▌Figure 20.4b). Although geologists generally agree on the configuration and location of the western half of the supercontinent, there is no consensus on the number or configuration of the various terranes and continental blocks that composed the eastern half of Pangaea. Regardless of the exact configuration of the eastern portion, geologists know that the supercontinent was surrounded by various subduction zones and moved steadily northward during the Permian. Furthermore, an enormous single ocean, **Panthalassa**, surrounded Pangaea and spanned Earth from pole to pole (Figure 20.4b). Waters of this ocean probably circulated more freely than at present, resulting in more equable water temperatures.

The formation of a single large landmass had climatic consequences for the terrestrial environment as well. Terrestrial Permian sediments indicate that arid and semiarid conditions were widespread over Pangaea. The mountain ranges produced by the *Hercynian, Alleghenian,* and *Ouachita orogenies* were high enough to create rain shadows that blocked the moist, subtropical, easterly winds—much as the southern Andes Mountains do in western South America today. This produced very dry conditions in North America and Europe, as evident from the extensive Permian red beds and evaporites found in western North America, central Europe, and parts of Russia. Permian coals, indicating abundant rainfall, were mostly limited to the northern temperate belts (latitude 40 degrees to 60 degrees north), whereas the last remnants of the Carboniferous ice sheets continued their recession.

20.4 Paleozoic Evolution of North America

It is convenient to divide the geologic history of the North American craton into two parts: the first dealing with the relatively stable continental interior over which epeiric seas transgressed and regressed, and the other with the mobile belts where mountain building occurred.

In 1963, the American geologist Laurence Sloss subdivided the sedimentary rock record of North America into six cratonic sequences. A **cratonic sequence** is a large-scale (greater than supergroup) lithostratigraphic unit representing a major transgressive–regressive cycle bounded by craton-wide unconformities (▌Figure 20.5). The transgressive phase, which is usually covered by younger sediments, commonly is well preserved, whereas the regressive phase of each sequence is marked by an unconformity. Where rocks

> **CONNECTION LINK** ◀
>
> To review transgressions and regressions and how they relate to sedimentary facies, see on pages 165 and 166 and Figure 7.10 on page 166 in Chapter 7.

Figure 20.4 Paleozoic Paleogeography Paleogeography of the world for the **(a)** Late Carboniferous Period and **(b)** Late Permian Period.

| Shallow sea | Deep ocean | Lowlands | Mountains | Glaciation |

a Late Carboniferous Period.

| Shallow sea | Deep ocean | Lowlands | Mountains | Glaciation |

b Late Permian Period.

of the appropriate age are preserved, each of the six unconformities can be shown to extend across the various sedimentary basins of the North American craton and into the mobile belts along the cratonic margin.

Geologists have also recognized major unconformity-bounded sequences in cratonic areas outside North America. Such global transgressive–regressive cycles of sea-level changes are thought to result from major tectonic and glacial events.

The realization that rock units can be divided into cratonic sequences and that these sequences can be further subdivided and correlated provides the foundation for an important concept in geology that allows high-resolution analysis of time and facies relationships within sedimentary rocks. **Sequence stratigraphy** is the study of rock relationships within a time-stratigraphic framework of related facies bounded by erosional or nondepositional surfaces. The basic unit of sequence stratigraphy is the *sequence*, which is a succession of rocks bounded by unconformities and their equivalent conformable strata. Sequence boundaries result from a relative drop in sea level. Sequence stratigraphy is an important tool in geology because it allows geologists to subdivide sedimentary rocks into related units that are bounded by stratigraphically significant boundaries. Geologists use sequence stratigraphy for high-resolution correlation and mapping, as well as interpreting and predicting depositional environments.

20.5 The Sauk Sequence

Rocks of the **Sauk Sequence** (Neoproterozoic–Early Ordovician) record the first major transgression onto the North American craton (Figure 20.5). During the Neoproterozoic and Early Cambrian, deposition of marine sediments was limited to the passive shelf areas of the Appalachian

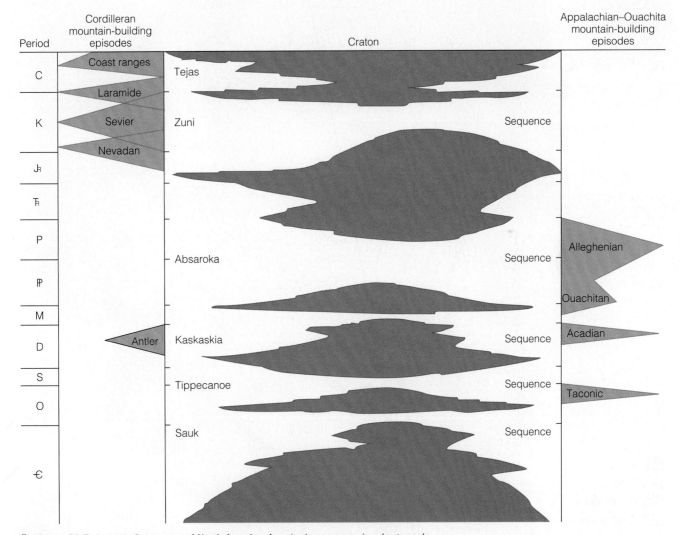

■ **Figure 20.5 Cratonic Sequences of North America** A cratonic sequence is a large-scale lithostratigraphic unit representing a major transgressive–regressive cycle and bounded by craton-wide unconformities. The white areas represent sequences of rocks separated by large-scale unconformities (brown areas). The major Cordilleran orogenies are shown on the left side, and the major Appalachian orogenies are on the right.

and Cordilleran borders of the craton. The craton itself was above sea level and experiencing extensive weathering and erosion. Because North America was located in a tropical climate at this time and there is no evidence of any terrestrial vegetation, weathering and erosion of the exposed Precambrian basement rocks must have proceeded rapidly.

During the Middle Cambrian, the transgressive phase of the Sauk began with epeiric seas encroaching over the craton (see the Geo-Insight on pages 518 and 519). By the Late Cambrian, these epeiric seas had covered most of North America, leaving only a portion of the Canadian shield and a few large islands above sea level (▮ Figure 20.6). These islands, collectively named the **Transcontinental Arch**, extended from New Mexico to Minnesota and the Lake Superior region.

What Would You Do ?

You work for a travel agency and are putting together a raft trip down the Colorado River through the Grand Canyon. In addition to the usual information about such a trip, what kind of geologic information would you include in your brochure to make the trip appealing from an educational standpoint?

▮ **Figure 20.6 Cambrian Paleogeography of North America** Note the position of the Cambrian paleoequator. During this time, North America straddled the equator as indicated in Figure 20.2a.

The sediments deposited both on the craton and along the shelf area of the craton margin show abundant evidence of shallow-water deposition. The only difference between the shelf and craton deposits is that the shelf deposits are thicker. In both areas, the sands are generally clean and well sorted and commonly contain ripple marks and small-scale cross-bedding. Many of the carbonates are bioclastic (composed of fragments of organic remains), contain stromatolites, or have oolitic (small, spherical calcium carbonate grains) textures. Such sedimentary structures and textures indicate shallow-water deposition.

Recall from Chapter 7 that sediments become increasingly finer the farther away from land one goes. Therefore, in a stable environment where sea level remains the same, coarse detrital sediments are typically deposited in the nearshore environment, and finer-grained sediments are deposited in the offshore environment. Carbonates form farthest from land in the area beyond the reach of detrital sediments. During a transgression, these facies (sediments that represent a particular environment) migrate in a landward direction (see Figure 7.10).

This time-transgressive facies relationship occurred not only in the Grand Canyon area (see Figure 7.11 and the Geo-Insight on pages 518 and 519) of the craton during the Sauk Sequence, but elsewhere on the craton as well, as the seas encroached from the Appalachian and Ouachita mobile belts onto the craton interior (▌Figure 20.7). Carbonate deposition dominated on the craton as the Sauk transgression continued during the Early Ordovician, and the advancing Sauk Sea soon covered the islands of the Transcontinental Arch. By the end of Sauk time, much of the craton was submerged beneath a warm, equatorial epeiric sea (Figure 20.2a).

▌**Figure 20.7** Time-Transgressive Cambrian Facies

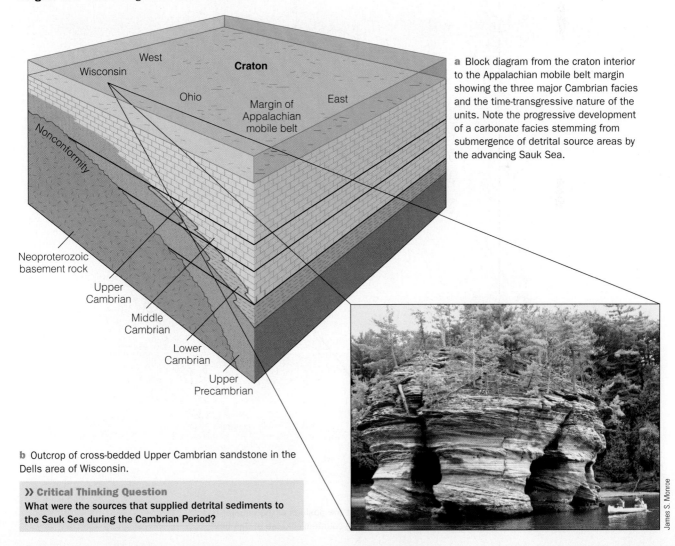

a Block diagram from the craton interior to the Appalachian mobile belt margin showing the three major Cambrian facies and the time-transgressive nature of the units. Note the progressive development of a carbonate facies stemming from submergence of detrital source areas by the advancing Sauk Sea.

b Outcrop of cross-bedded Upper Cambrian sandstone in the Dells area of Wisconsin.

≫ Critical Thinking Question
What were the sources that supplied detrital sediments to the Sauk Sea during the Cambrian Period?

James S. Monroe

20.6 The Tippecanoe Sequence

As the Sauk Sea regressed from the craton during the Early Ordovician, a landscape of low relief emerged. The exposed rocks were predominantly limestones and dolostones deposited earlier as part of the Sauk transgression. Because North America was still located in a tropical environment when the seas regressed, these carbonates experienced extensive erosion at that time (❚ Figure 20.8). The resulting craton-wide unconformity marks the boundary between the Sauk and Tippecanoe sequences.

Like the Sauk Sequence, deposition of the **Tippecanoe Sequence** (Middle Ordovician–Early Devonian) began with a major transgression onto the craton. This transgressing sea deposited clean, well-sorted quartz sands over most of the craton. The best known of the Tippecanoe basal sandstones is the St. Peter Sandstone, an almost-pure quartz sandstone used in manufacturing glass. It occurs throughout much of the mid-continent and resulted from numerous cycles of weathering and erosion of Proterozoic and Cambrian sandstones deposited during the Sauk transgression.

The Tippecanoe basal sandstones were followed by widespread carbonate deposition (Figure 20.8). The limestones were generally the result of deposition by calcium-carbonate–secreting organisms such as corals,

❚ **Figure 20.8 Ordovician Paleogeography of North America** Note that the position of the equator has shifted since the Cambrian, indicating that North America was rotating counterclockwise at this time.

Figure 20.9 Organic Reefs

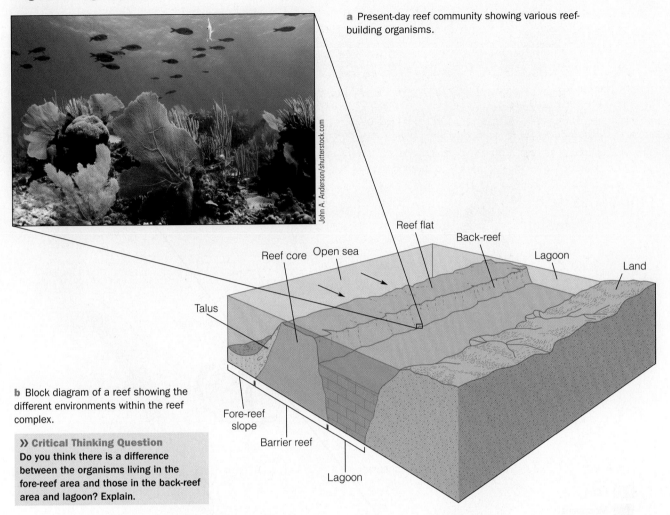

a Present-day reef community showing various reef-building organisms.

b Block diagram of a reef showing the different environments within the reef complex.

>> **Critical Thinking Question**
Do you think there is a difference between the organisms living in the fore-reef area and those in the back-reef area and lagoon? Explain.

brachiopods, stromatoporoids, and bryozoans. Besides the limestones, there were also many dolostones. Most of the dolostones formed as a result of magnesium replacing calcium in calcite, thus converting the limestones into dolostones.

In the eastern portion of the craton, the carbonates grade laterally into shales. These shales mark the farthest extent of detrital sediments derived from weathering and erosion of the Taconic Highlands, which resulted from a tectonic event taking place in the Appalachian mobile belt, and which we will discuss later.

Tippecanoe Reefs and Evaporites

Organic reefs are limestone structures constructed by living organisms, some of which contribute skeletal materials to the reef framework (▮ Figure 20.9). Today corals and calcareous algae are the most prominent reef builders, but in the geologic past, other organisms played a major role.

Regardless of the organisms that dominate reef communities, reefs appear to have occupied the same ecological niche in the geologic past as they do today. Because of the ecological requirements of reef-building organisms, present-day reefs are confined to a narrow latitudinal belt between 30 degrees north and south of the equator. Corals, the major reef-building organisms today, require warm, clear, shallow water of normal salinity for optimal growth.

The size and shape of a reef are mostly the result of the interactions among the reef-building organisms, the bottom topography, wind and wave action, and subsidence of the seafloor. Reefs also alter the area around them by forming barriers to water circulation or wave action.

Reefs have been common features in low latitudes since the Cambrian and have been built by a variety of organisms. The first skeletal builders of reef-like structures were *archaeocyathids*. These conical organisms lived during the Cambrian and had double, perforated, calcareous shell walls. Archaeocyathids built small mounds that have been found on all continents except South America (see Figure 21.5).

CONNECTION LINK

To review the formation and distribution of reefs, see page 392 in Chapter 16.

Figure 20.10 Silurian Paleogeography of North America Note the development of reefs in the Michigan, Ohio, and Indiana–Illinois–Kentucky areas.

Beginning in the Middle Ordovician, stromatoporoid–coral reefs became common in the low latitudes, and similar reefs remained so throughout the rest of the Phanerozoic Eon. The burst of reef building seen in the Late Ordovician through Devonian probably occurred in response to evolutionary changes triggered by the appearance of extensive carbonate seafloors and platforms beyond the influence of detrital sediments.

The Middle Silurian rocks (Tippecanoe Sequence) of the present-day Great Lakes region are world famous for their reef and evaporite deposits (Figure 20.10). The best-known structure in the region, the Michigan Basin, is a

broad, circular basin surrounded by large barrier reefs. No doubt these reefs contributed to increasingly restricted circulation and the precipitation of Upper Silurian evaporites within the basin (Figure 20.11).

Within the rapidly subsiding interior of the basin, other types of reefs are found. *Pinnacle reefs* are tall, spindly structures up to 100 m high. They reflect the rapid upward growth needed to maintain themselves near sea level during subsidence of the basin (Figure 20.11a). Besides the pinnacle reefs, bedded carbonates and thick sequences of salt and anhydrite are also found in the Michigan Basin (Figure 20.11a, b, c).

⬛ Figure 20.11 The Michigan Basin

a Generalized block diagram of the northern Michigan Basin during the Silurian Period.

d Limestone from the carbonate facies.

b Cross section of a stromatoporoid colony from the stromatoporoid barrier reef facies.

c Core of rock salt from the evaporite facies.

As the Tippecanoe Sea gradually regressed from the craton during the Late Silurian, precipitation of evaporite minerals occurred in the Appalachian, Ohio, and Michigan basins (Figure 20.10). In the Michigan Basin alone, approximately 1,500 m of sediments were deposited, nearly half of which are halite and anhydrite. How did such thick sequences of evaporites accumulate? One possibility is that when sea level dropped, the tops of the barrier reefs were as high as, or above, sea level, thus preventing the influx of new seawater into the basin. Evaporation of the basinal seawater would result in the formation of brine, and as the brine became increasingly concentrated, the precipitation of salts would occur. A second possibility is that the reefs grew upward so close to sea level that they formed a sill or barrier that eliminated interior circulation and allowed for the evaporation of the seawater that produced a dense brine that eventually resulted in evaporite deposits (⬛ Figure 20.12).

GEO-INSIGHT   The Grand Canyon—A Geologist's Paradise

"The Grand Canyon is the one great sight which every American should see," declared President Theodore Roosevelt. "We must do nothing to mar its grandeur." And so, in 1908, he named the Grand Canyon a national monument to protect it from exploitation. In 1919, the Grand Canyon National Monument was upgraded to a national park primarily because its scenery and the geology exposed in the canyon are unparalleled.

Located in the Colorado Plateau in northwestern Arizona, Grand Canyon National Park encompasses 1,218,375 acres and contains several major ecosystems (▌Figure 1).

Most of the more than 5,000 km² of the park is maintained as wilderness, with many trails affording visitors both day and overnight backcountry hiking opportunities.

Formed by the erosive power of the Colorado River, the Grand Canyon winds more than 500 km through northwestern Arizona, averages a depth of 1,220 m, and is 1,830 m deep at its deepest point (▌Figure 2). The Colorado River is responsible for carving the Grand Canyon and other canyons during the past 30 million years or so in response to the general uplift of the Colorado Plateau (▌Figure 3).

▶ **1.** Grand Canyon National Park, Arizona, consists of three distinct sections: the South Rim, the North Rim, and the Inner Canyon. Each section has a different climate and vegetation and offers the visitor different experiences.

▲ **2.** The Grand Canyon is world famous for its grandeur and beauty and attracts approximately 5 million visitors a year. Mohave Point, on West Rim Drive, is a popular spot to view the South Rim of the Grand Canyon and the surrounding plateau area.

▲ **3.** The South Rim of the Grand Canyon as viewed from Mohave Point. The Colorado River, which can be seen in the center of this view, is responsible for carving the Grand Canyon and other canyons in the area.

Major John Wesley Powell, a Civil War veteran, led the first geologic expedition down the Colorado River through the Grand Canyon in 1869 (▮ Figure 4). Without any maps or other information, Powell and his group ran the many rapids of the Colorado River in fragile wooden boats, hastily recording what they saw. Powell wrote in his diary that "all about me are interesting geologic records. The book is open and I read as I run." Following this first exploration, he led a second expedition in 1871, which included a photographer, a surveyor, and three topographers. This expedition made detailed topographic and geologic maps of the Grand Canyon area, as well as the first photographic record of the region.

The rocks of the Grand Canyon record more than one billion years of Earth history (▮ Figure 5). The Vishnu Schist represents a major mountain-building episode that occurred during the Precambrian. Following erosion of this mountain range, sediments were deposited in a variety of marine, coastal, and terrestrial settings during the Paleozoic Era and were lithified into the rock formations that are now exposed in the canyon walls (Figures 2 and 3).

Although best known for its geology, Grand Canyon National Park is home to 89 species of mammals, 56 species of reptiles and amphibians, more than 300 species of birds, 17 species of fish, and more than 1,500 plant species (▮ Figure 6). Its great biological diversity can be attributed to the fact that within its boundaries are five of the seven life zones and three of the four desert types in North America.

Library of Congress

▲ **4.** In 1869, Major John Wesley Powell, led the first geologic expedition down the Colorado River through the Grand Canyon in Arizona.

Fm = Formation   Ss = Sandstone   Ls = Limestone

▲ **5.** The rocks of the Grand Canyon preserve more than one billion years of Earth history, including periods of mountain building and erosion, as well as periods of transgressions and regressions of shallow seas over the area.

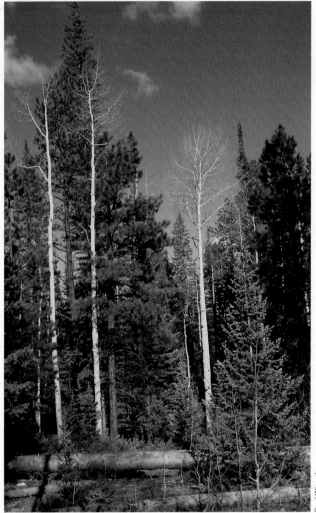

Reed Wicander

▲ **6.** White aspen and pinyon pine, shown here in the DeMotte campground, 28 km north of the North Rim, are typcial tress found in the Grand Canyon.

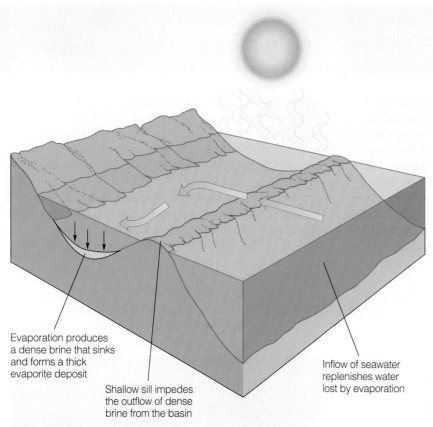

Evaporation produces a dense brine that sinks and forms a thick evaporite deposit

Shallow sill impedes the outflow of dense brine from the basin

Inflow of seawater replenishes water lost by evaporation

▌ **Figure 20.12** **Evaporite Sedimentation** Silled basin model for evaporite sedimentation by direct precipitation from seawater. Vertical scale is greatly exaggerated.

The End of the Tippecanoe Sequence

By the Early Devonian, the regressing Tippecanoe Sea had retreated to the craton margin, exposing extensive lowland topography. During this regression, marine deposition was initially restricted to a few interconnected cratonic basins and, finally, by the end of the Tippecanoe, to only the mobile belts surrounding the craton.

As the Tippecanoe Sea regressed during the Early Devonian, the craton experienced mild deformation, forming many domes, arches, and basins (Figure 20.1). These structures were mostly eroded during the time that the craton was exposed, and deposits from the ensuing and encroaching Kaskaskia Sea eventually covered them.

20.7 The Kaskaskia Sequence

The boundary between the Tippecanoe Sequence and the overlying **Kaskaskia Sequence** (Middle Devonian–Late Mississippian) is marked by a major unconformity. As the Kaskaskia Sea transgressed over the low-relief landscape of the craton, most basal beds deposited consisted of clean, well-sorted quartz sandstones.

The source areas for the basal Kaskaskia sandstones were primarily the eroding highlands of the Appalachian mobile belt area (Acadian Highlands) (▌ Figure 20.13), exhumed Cambrian and Ordovician sandstones cropping out along the flanks of the Ozark Dome, and exposures of the Canadian shield in the Wisconsin area. The lack of similar sands in the Silurian carbonate beds below the Tippecanoe–Kaskaskia unconformity indicates that the source areas of the basal Kaskaskia detrital rocks were submerged when the Tippecanoe Sequence was deposited. Stratigraphic studies indicate that these source areas were uplifted and the Tippecanoe carbonates removed by erosion before the Kaskaskia transgression.

Kaskaskian basal rocks elsewhere on the craton consist of carbonates that are frequently difficult to differentiate from the underlying Tippecanoe carbonates unless they are fossiliferous.

Except for widespread Upper Devonian and Lower Mississippian black shales, the majority of Kaskaskian rocks are carbonates, including reefs, and associated evaporite deposits. In many other parts of the world, such as southern England, Belgium, Central Europe, Australia, and Russia, the Middle and early Late Devonian epochs were times of major reef building.

Reef Development in Western Canada

The Middle and Late Devonian reefs of western Canada contain large reserves of petroleum and have been widely studied from outcrops and in the subsurface (▌ Figure 20.14). These reefs began forming as the Kaskaskia Sea transgressed southward into western Canada. By the end of the Middle Devonian, they had coalesced into a large barrier-reef system that restricted the flow of oceanic water into the back-reef platform, creating conditions for evaporite precipitation (Figure 20.14). In the back-reef area, up to 300 m of evaporites precipitated in much the same way as in the Michigan Basin during the Silurian (Figure 20.11). More than half of the world's potash, which is used in fertilizers, comes from these Devonian evaporites. By the middle of the Late Devonian, reef growth had stopped in western Canada, although non-reef carbonate deposition continued.

Figure 20.13 Devonian Paleogeography of North America

Black Shales

In North America, many areas of carbonate–evaporite deposition gave way to a greater proportion of shales and coarser detrital rocks beginning in the Middle Devonian and continuing into the Late Devonian. This change to detrital deposition resulted from the formation of new source areas brought about by the mountain-building activity associated with the Acadian orogeny in North America (Figure 20.13).

As the Devonian Period ended, a conspicuous change in sedimentation took place over the North American craton with the appearance of widespread black shales. In the eastern United States, these black shales are commonly called the Chattanooga Shale, but they are known by a variety of local names elsewhere (for example, New Albany Shale and Antrim Shale). Although these black shales are best developed from the cratonic margins along the Appalachian mobile belt to the Mississippi Valley, correlative units can also be found in many western states and in western Canada (Figure 20.15a).

The Upper Devonian–Lower Mississippian black shales of North America are typically noncalcareous, thinly bedded, and less than 10 m thick (Figure 20.15b). Fossils are usually rare, but some Upper Devonian black shales do contain rich conodont (microscopic animal) faunas. Because most black shales lack body fossils, they are difficult to date and correlate. However, in places where conodonts,

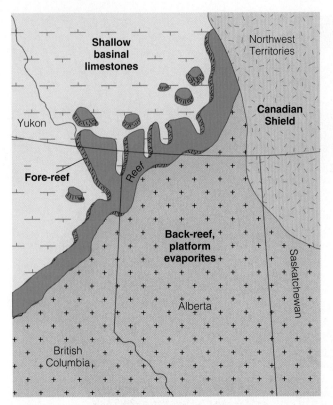

Figure 20.14 Devonian Reef Complex of Western Canada Reconstruction of the extensive Devonian Reef complex of western Canada. These extensive reefs controlled the regional facies of the Devonian epeiric seas.

>> **Critical Thinking Question**
How do you think the reefs controlled the regional facies in this area?

acritarchs (microscopic algae), or plant spores are found, these fossils indicate that the lower beds are Late Devonian and the upper beds are Early Mississippian in age.

Although the origin of these extensive black shales is still being debated, the essential features required to produce them include undisturbed anaerobic bottom water, a reduced supply of coarser detrital sediment, and high organic productivity in the overlying oxygenated waters. High productivity in the surface waters leads to a shower of organic material, which decomposes on the undisturbed seafloor and depletes the dissolved oxygen at the sediment–water interface.

The wide extent of such apparently shallow-water black shales in North America remains puzzling. Nonetheless, these shales are rich in uranium and are an important source rock of oil and gas in the Appalachian region.

The Late Kaskaskia—A Return to Extensive Carbonate Deposition

Following deposition of the widespread Upper Devonian–Lower Mississippian black shales, carbonate sedimentation on the craton dominated the remainder of the Mississippian Period (▌ Figure 20.16). During this time, a variety of carbonate sediments were deposited in the epeiric sea, as indicated by the extensive deposits of crinoidal limestones (rich in crinoid fragments), oolitic limestones, and various other limestones and dolostones. These Mississippian carbonates display cross-bedding, ripple marks, and well-sorted fossil fragments, all of which indicate a shallow-water environment. Analogous features can be observed on the present-day

Figure 20.15 Upper Devonian–Lower Mississippian Black Shales

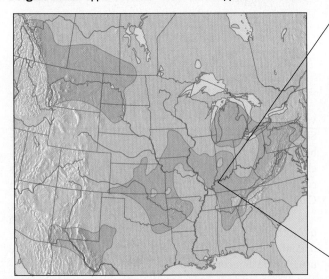

a The extent of the Upper Devonian to Lower Mississippian Chattanooga Shale and its equivalent units (such as the Antrim Shale and New Albany Shale) in North America.

b Upper Devonian New Albany Shale, Button Mold Knob Quarry, Kentucky.

Reed Wicander

Land

Mountains

Evaporites

Barrier reefs

Epeiric sea

Deep ocean

Figure 20.16 Mississippian Paleogeography of North America

Bahama Banks. In addition, numerous small organic reefs occurred throughout the craton during the Mississippian. However, these were all much smaller than the large barrier-reef complexes that dominated the earlier Paleozoic seas.

During the Late Mississippian regression of the Kaskaskia Sea from the craton, vast quantities of detrital sediments replaced carbonate deposition. The resulting sandstones, particularly in the Illinois Basin, have been studied in great detail because they are excellent petroleum reservoirs. Before the end of the Mississippian, the epeiric sea had retreated to the craton margin, once again exposing the craton to widespread weathering and erosion resulting in a craton-wide unconformity at the end of the Kaskaskia Sequence.

20.8 The Absaroka Sequence

The **Absaroka Sequence** includes rocks deposited during the Pennsylvanian through Early Jurassic. In this chapter, however, we are concerned only with the Paleozoic rocks of the Absaroka Sequence. The extensive unconformity separating the Kaskaskia and Absaroka sequences essentially divides the strata into the North American Mississippian and Pennsylvanian systems. These two systems are closely equivalent to the European Lower and Upper Carboniferous

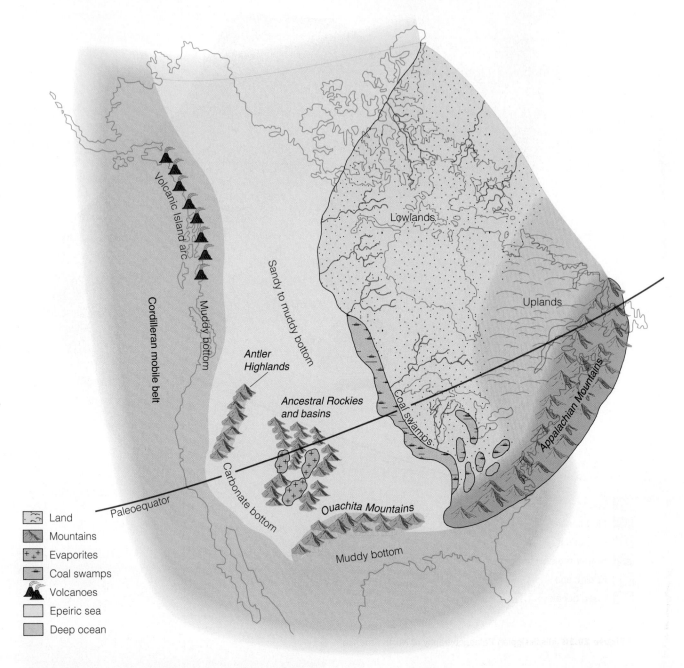

Figure 20.17 Pennsylvanian Paleogeography of North America

systems, respectively. The rocks of the Absaroka Sequence not only differ from those of the Kaskaskia Sequence but also resulted from different tectonic regimes.

The lowermost sediments of the Absaroka Sequence are confined to the margins of the craton. These deposits are generally thickest in the east and southeast, near the emerging highlands of the Appalachian and Ouachita mobile belts, and thin westward onto the craton. The rocks also reveal lateral changes from nonmarine detrital rocks and coals in the east, through transitional marine–nonmarine beds, to largely marine detrital rocks and limestones farther west (Figure 20.17).

What Are Cyclothems, and Why Are They Important?

One characteristic feature of Pennsylvanian rocks is their repetitive pattern of alternating marine and nonmarine strata known as **cyclothems**. They result from repeated alternations of marine and nonmarine environments, usually in areas of low relief. Although seemingly simple, cyclothems reflect a delicate interplay between nonmarine deltaic and shallow-marine interdeltaic and shelf environments.

▌Figure 20.18 Cyclothems

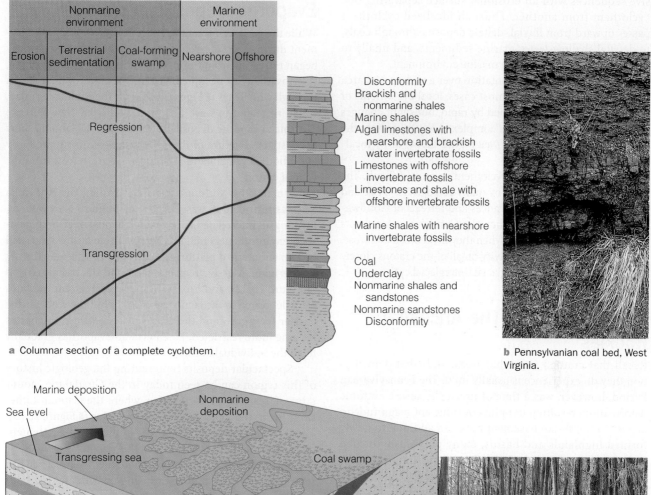

a Columnar section of a complete cyclothem.

Disconformity
Brackish and
  nonmarine shales
Marine shales
Algal limestones with
  nearshore and brackish
  water invertebrate fossils
Limestones with offshore
  invertebrate fossils
Limestones and shale with
  offshore invertebrate fossils

Marine shales with nearshore
  invertebrate fossils

Coal
Underclay
Nonmarine shales and
  sandstones
Nonmarine sandstones
  Disconformity

b Pennsylvanian coal bed, West
Virginia.

Courtesy of Wayne E. Moore

c Reconstruction of the environment of a Pennsylvanian
coal-forming swamp.

d The Okefenokee Swamp, Georgia, is a modern example of a
coal-forming environment, similar to those occurring during the
Pennsylvanian Period.

© Patricia Caulfield/Photo Researchers, Inc.

For illustration, look at a typical coal-bearing cyclothem from the Illinois Basin (▌Figure 20.18a). Such a cyclothem contains nonmarine units, capped by a coal unit and overlain by marine units. Figure 20.18a shows the depositional environments that produced the cyclothem. The initial units represent deltaic and fluvial deposits. Above them is an underclay that frequently contains root casts from the plants and trees that comprise the overlying coal. The coal bed results from accumulations of plant material and is overlain by marine units of alternating limestones and shales, usually with an abundant marine invertebrate fauna. The marine cycle ends with an erosion surface. A new cyclothem begins with a nonmarine deltaic sandstone. All of the beds illustrated in the idealized cyclothem are not always preserved because of abrupt changes from marine to nonmarine conditions or removal of some units by erosion.

Cyclothems represent transgressive and regressive sequences with an erosional surface separating one cyclothem from another. Thus, an idealized cyclothem passes upward from fluvial–deltaic deposits, through coals, to detrital shallow-water marine sediments, and finally to limestones typical of an open marine environment.

Such repetitive sedimentation over a widespread area requires an explanation. In most cases, local cyclothems of limited extent can be explained by rapid, but slight changes in sea level in a swamp–delta complex of low relief near the sea such as by progradation (see Figure 12.14) or by localized crustal movement.

Explaining widespread cyclothems is more difficult. The hypothesis currently favored by many geologists is a rise and fall of sea level related to advances and retreats of Gondwanan continental glaciers. When the Gondwanan ice sheets advanced, sea level dropped; when they melted, sea level rose. Late Paleozoic cyclothem activity on all of the cratons closely corresponds to Gondwanan glacial–interglacial cycles.

Cratonic Uplift—The Ancestral Rockies

Recall that cratons are stable areas, and what deformation they do experience is usually mild. The Pennsylvanian Period, however, was a time of unusually severe cratonic deformation, resulting in uplifts of sufficient magnitude to expose Precambrian basement rocks. In addition to newly formed highlands and basins, many previously formed arches and domes, such as the Cincinnati Arch, Nashville Dome, and Ozark Dome, were also reactivated (Figure 20.1).

During the Late Absaroka (Pennsylvanian), the area of greatest deformation was in the southwestern part of the North American craton, where a series of fault-bounded uplifted blocks formed the **Ancestral Rockies** (▮ Figure 20.19a). These mountain ranges had diverse geologic histories and were not all elevated at the same time. Uplift of these mountains, some of which were elevated more than 2 km along near-vertical faults, resulted in erosion of overlying Paleozoic sediments and exposure of the Precambrian igneous and metamorphic basement rocks (▮ Figure 20.19b). As the mountains eroded, tremendous quantities of coarse, red arkosic sand and conglomerate were deposited in the surrounding basins. These sediments are preserved in many areas, including the rocks of the Garden of the Gods near Colorado Springs (▮ Figure 20.19c) and at the Red Rocks Amphitheatre near Morrison, Colorado.

Intracratonic mountain ranges are unusual, and their cause has long been debated. It is currently thought that the collision of Gondwana with Laurasia along the Ouachita mobile belt (Figure 20.4a) generated great stresses in the southwestern region of the North American craton. These crustal stresses were relieved by faulting. Movement along these faults produced uplifted cratonic blocks and downwarped adjacent basins, forming a series of ranges and basins.

The Late Absaroka—More Evaporite Deposits and Reefs

While the various intracratonic basins were filling with sediment during the Late Pennsylvanian, the epeiric sea slowly began retreating from the craton. During the Early Permian, the Absaroka Sea occupied a narrow region from Nebraska through west Texas (▮ Figure 20.20). By the Middle Permian, it had retreated to west Texas and southern New Mexico. The thick evaporite deposits in Kansas and Oklahoma show the restricted nature of the Absaroka Sea during the Early and Middle Permian and its southwestward retreat from the central craton.

During the Middle and Late Permian, the Absaroka Sea was restricted to west Texas and southern New Mexico, forming an interrelated complex of lagoonal, reef, and open-shelf environments (▮ Figure 20.21). Three basins separated by two submerged platforms developed in this area during the Permian. Massive reefs grew around the basin margins (▮ Figure 20.22), and limestones, evaporites, and red beds were deposited in the lagoonal areas behind the reefs. As the barrier reefs grew and the passageways between the basins became more restricted, Late Permian evaporites gradually filled the individual basins.

Spectacular deposits representing the geologic history of this region can be seen today in the Guadalupe Mountains of Texas and New Mexico, where the Capitan Limestone forms the caprock of these mountains (▮ Figure 20.23). These reefs have been extensively studied because tremendous oil production takes place in this region.

By the end of the Permian Period, the Absaroka Sea had retreated from the craton, exposing continental red beds that had been deposited over most of the southwestern and eastern region (Figure 20.4b).

20.9 History of the Paleozoic Mobile Belts

Having examined the Paleozoic history of the craton, we now turn our attention to the orogenic activity in the mobile belts. The mountain building occurring during the Paleozoic Era had a profound influence on the climate and sedimentary history of the craton. In addition, it was part of the global tectonic regime that sutured the continents together, forming Pangaea by the end of the Paleozoic Era.

Appalachian Mobile Belt

Throughout Sauk time (Neoproterozoic–Early Ordovician), the Appalachian region was a broad, passive, continental margin. Sedimentation was closely balanced by subsidence as extensive carbonate deposits succeeded thick, shallow marine

Figure 20.19 The Ancestral Rockies

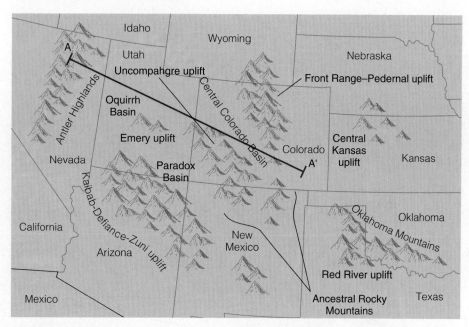

a Location of the principal Pennsylvanian highland areas and basins of the southwestern part of the craton.

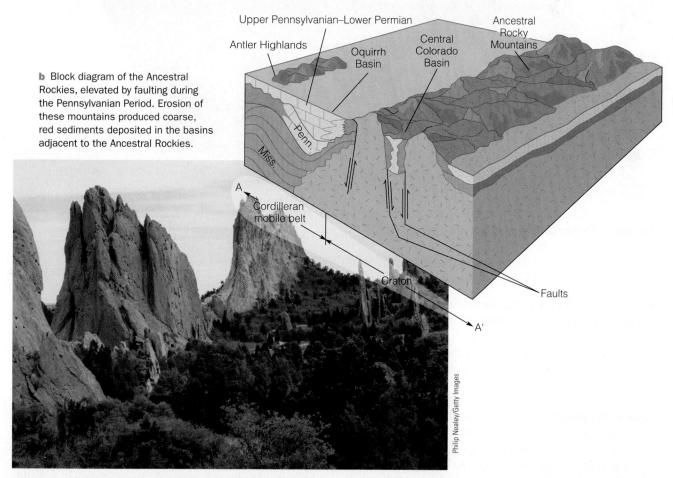

b Block diagram of the Ancestral Rockies, elevated by faulting during the Pennsylvanian Period. Erosion of these mountains produced coarse, red sediments deposited in the basins adjacent to the Ancestral Rockies.

c Garden of the Gods, Storm Sky View from Near Hidden Inn, Colorado Springs, Colorado.

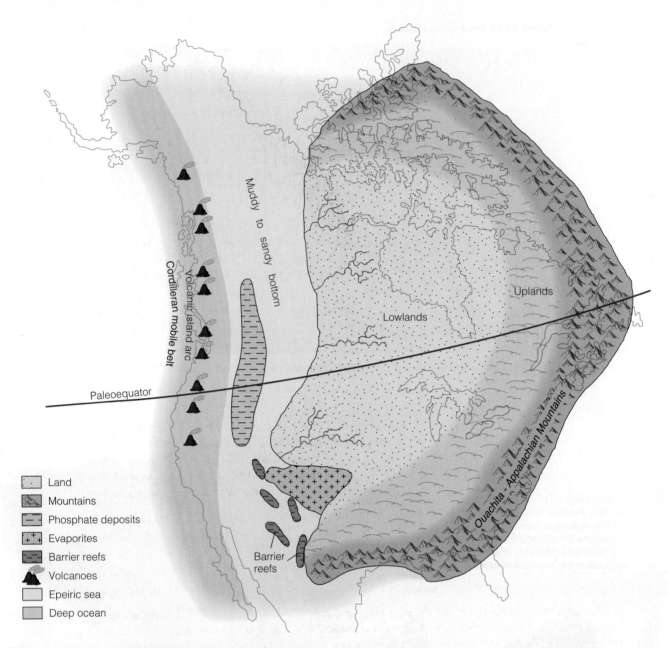

Figure 20.20 Permian Paleogeography of North America

Legend:
- Land
- Mountains
- Phosphate deposits
- Evaporites
- Barrier reefs
- Volcanoes
- Epeiric sea
- Deep ocean

Map labels: Cordilleran mobile belt, Volcanic island arc, Muddy to sandy bottom, Paleoequator, Uplands, Lowlands, Ouachita - Appalachian Mountains, Barrier reefs

sands. During this time, movement along a divergent plate boundary was widening the **Iapetus Ocean** (Figure 20.24a).

Taconic Orogeny Beginning with the subduction of the Iapetus plate beneath Laurentia (an oceanic–continental convergent plate boundary), the Appalachian mobile belt was born (Figure 20.24b). The resulting **Taconic orogeny**— named after the present-day Taconic Mountains of eastern New York, central Massachusetts, and Vermont—was the first of several orogenies to affect the Appalachian region.

The Appalachian mobile belt can be divided into two depositional environments. The first is the extensive, shallow-water carbonate platform (Figure 20.24a) that formed

the broad eastern continental shelf and stretched from Newfoundland to Alabama. It formed during the transgression of the Sauk Sea onto the craton when carbonates were deposited in a vast, shallow sea. Stromatolites, mud cracks, and other sedimentary structures and fossils indicate the shallow water depth on the platform.

Carbonate deposition ceased along the east coast during the Middle Ordovician and was replaced by deepwater deposits characterized by thinly bedded black shales, graded beds, coarse sandstones, graywackes (poorly sorted, immature sandstones), and associated volcanic material. This suite of sediments marks the onset of mountain building, in this case, the Taconic orogeny. The subduction of the

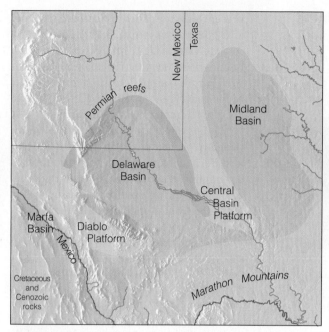

Figure 20.21 **West Texas Permian Basins and Surrounding Reefs** During the Middle and Late Permian, an interrelated complex of lagoonal, barrier-reef, and open-shelf environments formed in the west Texas and southern New Mexico area. Much of the tremendous oil production in this region comes from these reefs.

>> **Critical Thinking Question**
Why is so much oil found in reef structures?

Photo courtesy of the Petroleum Museum, Randolph D. Rubin Collection

Figure 20.22 **Middle Permian Capitan Limestone Reef Environment** A reconstruction of the Middle Permian Capitan Limestone reef environment. Shown are brachiopods, corals, bryozoans, and large glass sponges.

Iapetus plate beneath Laurentia resulted in volcanism and downwarping of the carbonate platform (Figure 20.24b). Throughout the Appalachian mobile belt, facies patterns, paleocurrents, and sedimentary structures all indicate that these deposits were derived from the east, where the Taconic Highlands and associated volcanoes were rising.

The final piece of evidence for the Taconic orogeny is the development of a large **clastic wedge**, an extensive accumulation of mostly detrital sediments deposited adjacent to an uplifted area. These deposits are thickest and coarsest nearest the highland area and become thinner and finer-grained away from the source area, eventually grading into the carbonate cratonic facies (▌Figure 20.25). The clastic wedge resulting from the erosion of the Taconic Highlands is referred to as the **Queenston Delta**. Careful mapping and correlation of these deposits indicate that more than 600,000 km³ of rock were eroded from the Taconic Highlands. Based on this figure, geologists estimate that the Taconic Highlands were at least 4,000 m high.

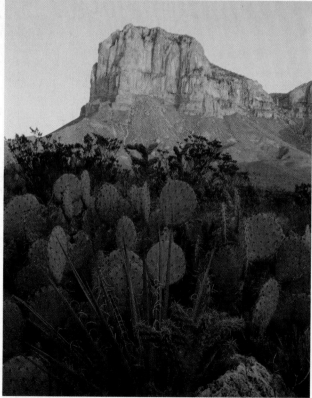

© David Muench/CORBIS

Figure 20.23 **Guadalupe Mountains, Texas** The prominent Capitan Limestone forms the caprock of the Guadalupe Mountains. The Capitan Limestone is rich in fossil corals and associated reef organisms.

■ **Figure 20.24** Neoproterozoic to Late Ordovician Evolution of the Appalachian Mobile Belt

a During the Neoproterozoic to the Early Ordovician, the Iapetus Ocean was opening along a divergent plate boundary. Both the east coast of Laurentia and the west coast of Baltica were passive continental margins with large carbonate platforms.

b Beginning in the Middle Ordovician, the passive margins of Laurentia and Baltica changed to active oceanic–continental plate boundaries, resulting in orogenic activity.

The Taconic orogeny marked the first pulse of mountain building in the Appalachian mobile belt and was a response to the subduction taking place beneath the east coast of Laurentia. As the Iapetus Ocean narrowed and closed, another orogeny occurred in Europe during the Silurian. The Caledonian orogeny was essentially a mirror image of the Taconic and Acadian orogenies and was part of the global mountain-building episode that occurred during the Paleozoic Era.

Caledonian Orogeny The Caledonian mobile belt stretches along the western border of Baltica and includes the present-day countries of Scotland, Ireland, and Norway (Figure 20.2c). During the Middle Ordovician, subduction along the boundary between the Iapetus plate and Baltica (Europe) began, forming a mirror image of the convergent plate boundary off the east coast of Laurentia (North America).

The culmination of the **Caledonian orogeny** occurred during the Late Silurian and Early Devonian with the formation of a mountain range along the western margin of Baltica. Red-colored sediments deposited along the front of the Caledonian Highlands formed a large clastic wedge known as the *Old Red Sandstone*.

Acadian Orogeny The third Paleozoic orogeny to affect Laurentia and Baltica began during the Late Silurian and concluded at the end of the Devonian Period. The **Acadian orogeny** affected the Appalachian mobile belt from Newfoundland to Pennsylvania as sedimentary rocks were folded and thrust against the craton.

As with the preceding Taconic and Caledonian orogenies, the Acadian orogeny occurred along an oceanic–continental convergent plate boundary. As the northern Iapetus Ocean continued to close during the Devonian, the plate carrying Baltica finally collided with Laurentia, forming a continental–continental convergent plate boundary along the zone of collision (Figure 20.3a).

Weathering and erosion of the Acadian Highlands produced the **Catskill Delta**, a thick clastic wedge named for the Catskill Mountains in upstate New York where it is well exposed. The Catskill Delta, composed of red, coarse conglomerates, sandstones, and shales, contains nearly three times as much sediment as the Queenston Delta (■ Figure 20.25).

The Devonian rocks of New York are among the most studied on the continent. A cross section of the Devonian strata clearly reflects an eastern source (Acadian Highlands) for the Catskill facies (■ Figure 20.26). These detrital rocks can be traced from eastern Pennsylvania, where the coarse-grained deposits are approximately 3 km thick, to Ohio, where the deltaic facies are only about 100 m thick and consist of cratonic shales and carbonates.

The red beds of the Catskill Delta derive their color from the hematite in the sediments. Plant fossils and oxidation of the hematite indicate that the beds were deposited in a continental environment. Toward the west, the red beds grade laterally into gray sandstones and shales containing fossil tree trunks, which indicate a swamp or marsh environment.

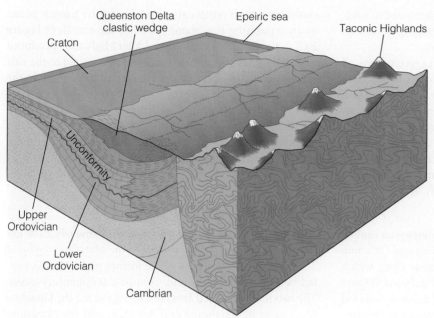

Figure 20.25 Reconstruction of the Taconic Highlands and Queenston Delta Clastic Wedge The Queenston Delta clastic wedge, resulting from the erosion of the Taconic Highlands, consists of thick, coarse-grained detrital sediments nearest the highlands and thins laterally into finer-grained sediments in the epeiric seas covering the craton.

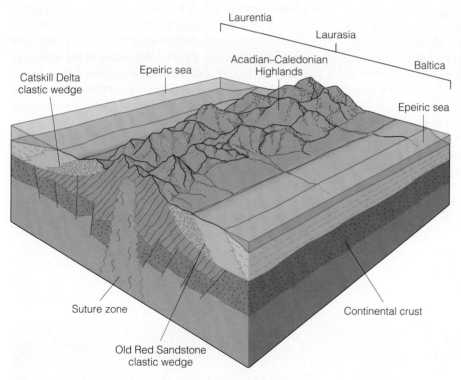

Figure 20.26 Formation of Laurasia Block diagram showing the area of collision between Laurentia and Baltica. Note the bilateral symmetry of the Catskill Delta clastic wedge and the Old Red Sandstone and their relationship to the Acadian and Caledonian Highlands.

The Old Red Sandstone The red beds of the Catskill Delta have a European counterpart in the Devonian Old Red Sandstone of the British Isles (Figure 20.26). The Old Red Sandstone was a Devonian clastic wedge that grew eastward from the Caledonian Highlands onto the Baltica craton. The Old Red Sandstone, just like its North American Catskill counterpart, contains numerous fossils of freshwater fish, early amphibians, and land plants.

By the end of the Devonian Period, Baltica and Laurentia were sutured together, forming Laurasia (Figure 20.26). The red beds of the Catskill Delta can be traced north, through Canada and Greenland, to the Old Red Sandstone of the British Isles

CONNECTION LINK

To learn more about the fish, amphibians, reptiles, and land plants of the Paleozoic, see the respective sections on pages 554–568 in Chapter 21.

and into northern Europe. These beds were deposited in similar environments along the flanks of developing mountain chains formed at convergent plate boundaries.

The Taconic, Caledonian, and Acadian orogenies were all part of the same major orogenic event related to the closing of the Iapetus Ocean (Figures 20.24 and 20.26). This event began with paired oceanic–continental convergent plate boundaries during the Taconic and Caledonian orogenies and culminated along a continental–continental convergent plate boundary during the Acadian orogeny as Laurentia and Baltica became sutured. After this, the Hercynian–Alleghenian orogeny began, followed by orogenic activity in the Ouachita mobile belt.

Hercynian–Alleghenian Orogeny The Hercynian mobile belt of southern Europe and the Appalachian and Ouachita mobile belts of North America mark the zone along which Europe (part of Laurasia) collided with Gondwana (Figure 20.3). While Gondwana and southern Laurasia collided during the Pennsylvanian and Permian periods in the area of the Ouachita mobile belt, eastern Laurasia (Europe and southeastern North America) joined Gondwana (Africa) as part of the **Hercynian–Alleghenian orogeny** (Figure 20.4a).

Initial contact between eastern Laurasia and Gondwana began during the Mississippian Period along the Hercynian mobile belt. The greatest deformation occurred during the Pennsylvanian and Permian periods and is referred to as the *Hercynian orogeny*. The central and southern parts of the Appalachian mobile belt (from New York to Alabama) were folded and thrust toward the craton as eastern Laurasia and Gondwana were sutured. This event in North America is referred to as the *Alleghenian orogeny*.

These three Late Paleozoic orogenies (Hercynian, Alleghenian, and Ouachita) represent the final joining of Laurasia and Gondwana into the supercontinent Pangaea during the Permian.

continental shelf sediments, thus closing the narrow ocean basin separating the island arc from the craton (▌ Figure 20.27). Erosion of the resulting Antler Highlands produced large quantities of sediments that were deposited to the east in the epeiric sea covering the craton and to the west in the deep sea. The Antler orogeny was the first in a series of orogenic events to affect the Cordilleran mobile belt. During the Mesozoic and Cenozoic, this area was the site of major tectonic activity caused by oceanic–continental convergence and accretion of various terranes.

Ouachita Mobile Belt

The Ouachita mobile belt extends for approximately 2,100 km from the subsurface of Mississippi to the Marathon region of Texas. Approximately 80% of the former mobile belt is buried beneath a Mesozoic and Cenozoic sedimentary cover. The two major exposed areas in this region are the Ouachita Mountains of Oklahoma and Arkansas and the Marathon Mountains of Texas.

During the Neoproterozoic to Early Mississippian, shallow-water detrital and carbonate sediments were deposited on a broad continental shelf, and in the deeper-water portion of the adjoining mobile belt, bedded cherts and shales were accumulating (▌ Figure 20.28a). Beginning in the Mississippian Period, the rate of sedimentation increased dramatically as the region changed from a passive continental margin to an active convergent plate boundary, marking the beginning of the **Ouachita orogeny** (▌ Figure 20.28b).

Thrusting of sediments continued throughout the Pennsylvanian and Early Permian, driven by the compressive forces generated along the zone of subduction as Gondwana collided with Laurasia (▌ Figure 20.28c). The collision of Gondwana and Laurasia is marked by the formation of a large mountain range, most of which eroded during the

Cordilleran Mobile Belt

During the Neoproterozoic and Early Paleozoic, the Cordilleran area was a passive continental margin along which extensive continental shelf sediments were deposited. Thick sections of marine sediments graded laterally into thin cratonic units as the Sauk Sea transgressed onto the craton. Beginning in the Middle Paleozoic, an island arc formed off the western margin of the craton. A collision between this eastward-moving island arc and the western border of the craton took place during the Late Devonian and Early Mississippian, producing a highland area.

This orogenic event, the **Antler orogeny**, was caused by subduction, resulting in deep-water deposits and oceanic crustal rocks being thrust eastward over shallow-water

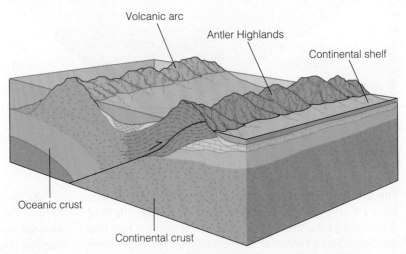

▌ **Figure 20.27** Antler Orogeny Reconstruction of the Cordilleran mobile belt during the Early Mississippian in which deep-water continental slope deposits were thrust eastward over shallow-water continental shelf carbonates, forming the Antler Highlands.

Figure 20.28 Ouachita Mobile Belt Plate tectonic model for deformation of the Ouachita mobile belt.

North America

Continental crust

Oceanic crust

a Depositional environment prior to the beginning of the orogenic activity.

North America

Gondwana

Continental crust

Continental crust

Oceanic crust

b Incipient continental collision between North America and Gondwana began during the Mississippian Period.

North America

Gondwana

c Continental collision continued during the Pennsylvanian and Permian periods.

Mesozoic Era. Only the rejuvenated Ouachita and Marathon Mountains remain of this once lofty mountain range.

The Ouachita deformation was part of the general worldwide tectonic activity that occurred when Gondwana united with Laurasia. The Hercynian, Appalachian, and Ouachita mobile belts were continuous and marked the southern boundary of Laurasia (Figure 20.4). The tectonic activity that uplifted the Ouachita mobile belt was very complex and involved not only the collision of Laurasia and

Gondwana but also several microplates and terranes between the continents that eventually became part of Central America. The compressive forces impinging on the Ouachita mobile belt also affected the craton by broadly uplifting the southwestern part of North America.

20.10 What Role Did Microplates and Terranes Play in the Formation of Pangaea?

We have presented the geologic history of the mobile belts bordering the Paleozoic continents in terms of subduction along convergent plate boundaries. It is becoming increasingly clear, however, that accretion along the continental margins is more complicated than the somewhat simple, large-scale plate interactions that we have described. Geologists now recognize that numerous terranes or microplates existed during the Paleozoic and were involved in the orogenic events that occurred during that time.

In this chapter, we have been concerned only with the six major Paleozoic continents. However, terranes and microplates of varying sizes were present during the Paleozoic and participated in the formation of Pangaea. For example, as we mentioned earlier in this chapter, the microcontinent of Avalonia consisted of some coastal parts of New England, southern New Brunswick, much of Nova Scotia, the Avalon Peninsula of eastern Newfoundland, southeastern Ireland, Wales, England, and parts of Belgium and northern France. This microcontinent separated from Gondwana in the Early Ordovician and existed as a separate continent until it collided with Baltica during the Late Ordovician–Early Silurian and then with Laurentia (as part of Baltica) during the Silurian (Figures 20.2b, c and 20.3a).

Other terranes and microplates include *Iberia–Armorica* (a portion of southern France, Sardinia, and most of the Iberian peninsula), *Perunica* (Bohemia), and numerous Alpine fragments (especially in Austria), as well as many other bits and pieces of island arcs and suture zones. Not only did these terranes and microplates separate and move away from the larger continental landmasses during the Paleozoic, but they usually developed their own unique faunal and floral assemblages.

Thus, although the basic history of the formation of Pangaea during the Paleozoic remains the same, geologists realize that microplates and terranes also played an important role in the formation of Pangaea. Furthermore, they help explain some previously anomalous geologic and paleontologic situations.

20.11 Paleozoic Mineral Resources

Paleozoic-age rocks contain a variety of important mineral resources, including energy resources, metallic and nonmetallic mineral deposits, and sand and gravel for construction. Important sources of industrial or silica sand are the Upper Cambrian Jordan Sandstone of Minnesota and Wisconsin; the Middle Ordovician St. Peter Sandstone; the Lower Silurian Tuscarora Sandstone in Pennsylvania and Virginia; the Devonian Ridgeley Formation in West Virginia, Maryland, and Pennsylvania; and the Devonian Sylvania Sandstone in Michigan.

Silica sand has several uses, including the manufacture of glass, refractory bricks for blast furnaces, and molds for casting iron, aluminum, and copper alloys. Some silica sands, called hydraulic fracturing sands, are pumped into wells to fracture oil- or gas-bearing rocks and provide permeable passageways for the oil or gas to migrate to the well.

Thick deposits of Silurian evaporites, mostly rock salt (NaCl) and rock gypsum ($CaSO_4 \cdot 2H_2O$) altered to rock anhydrite ($CaSO_4$), underlie parts of Michigan, Ohio, New York, and adjacent areas in Ontario, Canada. These rocks are important sources of various salts. In addition, barrier and pinnacle reefs in carbonate rocks associated with these evaporites are reservoirs for oil and gas in Michigan and Ohio.

The Zechstein evaporites of Europe extend from Great Britain across the North Sea and into Denmark, the Netherlands, Germany, and eastern Poland and Lithuania. Besides the evaporites themselves, Zechstein deposits form the caprock for the large reservoirs of the gas fields of the Netherlands and part of the North Sea region.

Other important evaporite mineral resources include those of the Permian Delaware Basin of West Texas and New Mexico, and Devonian evaporites in the Elk Point Basin of Canada. In Michigan, gypsum is mined and used in the production of sheetrock. Late Paleozoic-age limestones from many areas in North America are used in the manufacturing of cement. Limestone is also mined and used in blast furnaces for steel production.

Metallic mineral resources, including tin, copper, gold, and silver, are known from Late Paleozoic-age rocks, especially those deformed during mountain building. The host rocks for deposits of lead and zinc in southeast Missouri are Cambrian dolostones, although some Ordovician rocks contain these

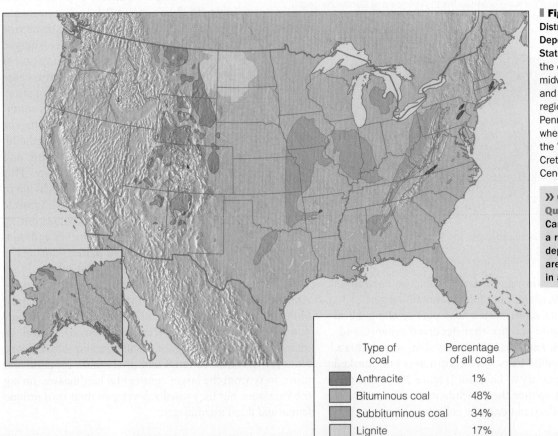

▌Figure 20.29
Distribution of Coal Deposits in the United States The age of the coals in the midwestern states and the Appalachian region are mostly Pennsylvanian, whereas those in the West are mostly Cretaceous and Cenozoic.

≫ Critical Thinking Question
Can you think of a reason why coal deposits in the West are not Paleozoic in age?

Type of coal	Percentage of all coal
Anthracite	1%
Bituminous coal	48%
Subbituminous coal	34%
Lignite	17%

metals as well. These deposits have been mined since 1720, but have been largely depleted. Now most lead and zinc mined in Missouri comes from Mississippian-age sedimentary rocks.

The Silurian Clinton Formation crops out from Alabama north to New York, and equivalent rocks are found in Newfoundland. This formation has been mined for iron in many places. In the United States, the richest ores and most extensive mining occurred near Birmingham, Alabama, but only a small amount of ore is currently produced in that area.

Petroleum and natural gas are recovered in commercial quantities from rocks ranging in age from the Devonian through Permian. For example, Devonian rocks in the Michigan Basin, Illinois Basin, and the Williston Basin of Montana, South Dakota, and adjacent parts of Alberta, Canada, have yielded considerable amounts of hydrocarbons. Permian reefs and other strata in the western United States, particularly Texas, have also been important producers.

Although Permian coal beds are known from several areas, including Asia, Africa, and Australia, much of the coal in North America and Europe comes from Pennsylvanian (Upper Carboniferous) deposits. Large areas in the Appalachian region and the midwestern United States are underlain by vast coal deposits (Figure 20.29). These coal deposits formed from the lush vegetation that flourished in Pennsylvanian coal-forming swamps (Figure 20.18).

Much of this coal is bituminous coal, which contains about 80% carbon. It is a dense, black coal that has been so thoroughly altered that plant remains can be seen only rarely. Bituminous coal is used to make *coke*, a hard, gray substance made up of the fused ash of bituminous coal. Coke is used to fire blast furnaces for steel production.

Some Pennsylvanian coal from North America is *anthracite*, a metamorphic type of coal containing up to 98% carbon. Most anthracite is in the Appalachian region (Figure 20.29). It is especially desirable because it burns with a smokeless flame and yields more heat per unit volume than other types of coal. Unfortunately, it is the least common type—much of the coal used in the United States is bituminous.

GEO-RECAP

≫ Geo-Image

Figure 1 This close-up of a Devonian red rock from a building in Glasgow, Scotland, shows a distinctive sedimentary structure. Identify the sedimentary structure, indicate what type of environment you think it was deposited in, and give the name of the formation this rock comes from.

Chapter Summary

- Tables 20.1 (page 537) and 20.2 (page 538) summarize the Paleozoic geologic history of the North American craton and mobile belts, as well as global events and sea-level changes.

- Most continents consist of two major components: a relatively stable craton over which epeiric seas transgressed and regressed, surrounded by mobile belts in which mountain building took place.

- Six major continents and numerous microcontinents and island arcs existed at the beginning of the Paleozoic Era; all of these were dispersed around the globe at low latitudes during the Cambrian.

- During the Early Paleozoic (Cambrian–Silurian), Laurentia moved northward, while Gondwana moved southward and began to cross the South Pole as indicated by tillite deposits.

- During the Late Paleozoic, Baltica and Laurentia collided, forming Laurasia. Siberia and Kazakhstania also collided and were then sutured to Laurasia. Gondwana continued moving over the South Pole and experienced several glacial–interglacial periods, resulting in global sea-level changes and transgressions and regressions along the low-lying craton margins.

- Laurasia and Gondwana underwent a series of collisions beginning in the Carboniferous, and during the Permian the formation of Pangaea was completed. Surrounding the supercontinent was the global ocean, Panthalassa.

- Geologists divide the geologic history of North America into cratonic sequences that formed as a result of craton-wide transgressions and regressions.

- The Sauk Sequence began with a major marine transgression onto the craton. At its maximum, the Sauk Sea covered the craton except for parts of the Canadian shield and the Transcontinental Arch, a series of large, northeast–southwest trending islands.

- The Tippecanoe Sequence began with deposition of an extensive sand unit over the exposed and eroded Sauk landscape and was followed by extensive carbonate deposition. In addition, large barrier reefs surrounded many cratonic basins, resulting in evaporite deposition within these basins.

- The basal beds of the Kaskaskia Sequence that were deposited on the exposed Tippecanoe surface consisted either of sandstones derived from the eroding Taconic Highlands or of carbonate rocks.

- Most of the Kaskaskia Sequence is dominated by carbonates and associated evaporites. The Devonian Period was a time of major reef building in western Canada, southern England, Belgium, Australia, and Russia.

- Widespread black shales were deposited over large areas of the North American craton during the Late Devonian and Early Mississippian.

- The Mississippian Period was dominated, for the most part, by carbonate deposition.

- Transgressions and regressions over the low-lying North American craton, and probably caused by advancing and retreating Gondwanan ice sheets, resulted in cyclothems and the formation of coals during the Pennsylvanian Period.

- Cratonic mountain building, specifically the Ancestral Rockies, occurred during the Pennsylvanian Period and resulted in thick, nonmarine detrital sediments and evaporites being deposited in the intervening basins.

- By the Early Permian, the Absaroka Sea occupied a narrow zone of the south-central craton. Here, several large reefs and associated evaporites developed. By the end of the Permian Period, this epeiric sea had retreated from the craton.

- The eastern edge of North America was a stable carbonate platform during Sauk time. During Tippecanoe time, an oceanic–continental convergent plate boundary formed, resulting in the Taconic orogeny, the first of three major orogenies to affect the Appalachian mobile belt.

- The newly formed Taconic Highlands shed sediments into the western epeiric sea, producing a clastic wedge that geologists call the Queenston Delta.

- The Caledonian, Acadian, Hercynian, and Alleghenian orogenies were all part of the global tectonic activity that resulted from the assembly of Pangaea.

- The Cordilleran mobile belt was the site of the Antler orogeny, a minor Devonian orogeny during which deep-water sediments were thrust eastward over shallow-water sediments.

- During the Pennsylvanian and Early Permian, mountain building occurred in the Ouachita mobile belt. This tectonic activity was partly responsible for the cratonic uplift in the southwest, resulting in the Ancestral Rockies.

- During the Paleozoic Era, numerous microplates and terranes, such as Avalonia, Iberia–Armorica, and Peru-nica, existed and played an important role in forming Pangaea.

- Paleozoic-age rocks contain a variety of mineral resources, including petroleum, coal, evaporites, silica sand, lead, zinc, and other metallic deposits.

TABLE 20.1	Summary of Early Paleozoic Geologic and Evolutionary Events								
Age (millions of years)	Geologic Period	Sequence	Relative Changes in Sea Level (Rising / Falling)	Cordilleran Mobile Belt	Craton	Ouachita Mobile Belt	Appalachian Mobile Belt	Major Events Outside North America	Major Evolutionary Events
416	Silurian	Tippecanoe	(Present sea level)		Extensive barrier reefs and evaporites common		Acadian orogeny	Caledonian orogeny	First jawed fish evolve / Early land plants—seedless vascular plants
444	Ordovician	Tippecanoe			Queenston Delta clastic wedge / Transgression of Tippecanoe Sea / Regression exposing large areas to erosion		Taconic orogeny	Continental glaciation in Southern Hemisphere	Extinction of many marine invertebrates near end of Ordovician / Plants move to land? / Major adaptive radiation of all invertebrate groups
488	Cambrian	Sauk			Canadian shield and Transcontinental Arch only areas above sea level / Transgression of Sauk Sea				Many trilobites become extinct near end of Cambrian / Earliest vertebrates—jawless fish evolve
542									

Important Terms

Absaroka Sequence (p. 523)
Acadian orogeny (p. 530)
Ancestral Rockies (p. 526)
Antler orogeny (p. 532)
Appalachian mobile belt (p. 505)
Baltica (p. 505)
Caledonian orogeny (p. 530)
Catskill Delta (p. 530)
China (p. 505)
clastic wedge (p. 529)
Cordilleran mobile belt (p. 505)
cratonic sequence (p. 509)

cyclothem (p. 524)
epeiric sea (p. 504)
Gondwana (p. 506)
Hercynian–Alleghenian orogeny (p. 532)
Iapetus Ocean (p. 528)
Kaskaskia Sequence (p. 520)
Kazakhstania (p. 505)
Laurasia (p. 506)
Laurentia (p. 506)
mobile belt (p. 505)
organic reef (p. 515)

Ouachita mobile belt (p. 505)
Ouachita orogeny (p. 532)
Panthalassa Ocean (p. 509)
Queenston Delta (p. 529)
Sauk Sequence (p. 511)
sequence stratigraphy (p. 511)
Siberia (p. 506)
Taconic orogeny (p. 528)
Tippecanoe Sequence (p. 514)
Transcontinental Arch (p. 512)

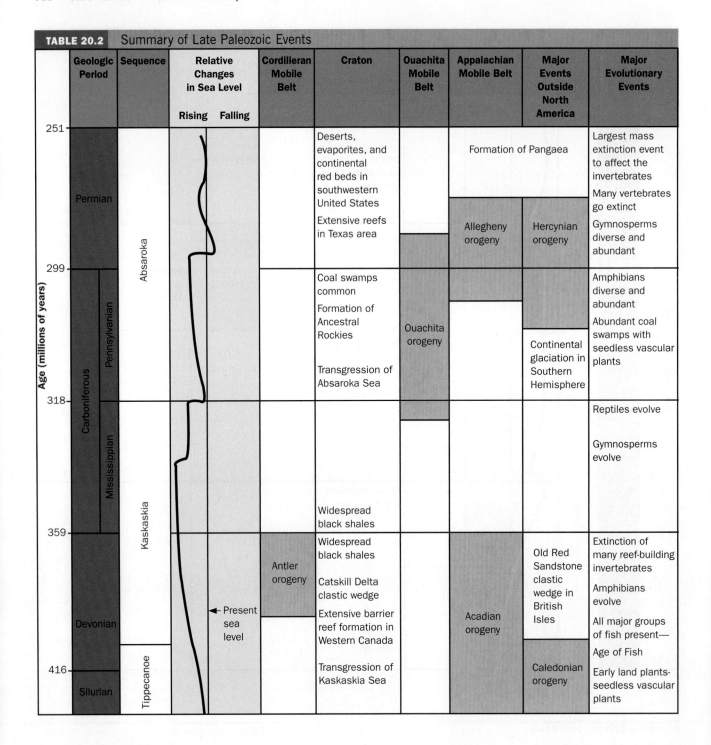

TABLE 20.2　Summary of Late Paleozoic Events

Geologic Period	Sequence	Relative Changes in Sea Level (Rising / Falling)	Cordilleran Mobile Belt	Craton	Ouachita Mobile Belt	Appalachian Mobile Belt	Major Events Outside North America	Major Evolutionary Events
Permian (251)	Absaroka			Deserts, evaporites, and continental red beds in southwestern United States		Formation of Pangaea		Largest mass extinction event to affect the invertebrates
								Many vertebrates go extinct
				Extensive reefs in Texas area		Allegheny orogeny	Hercynian orogeny	Gymnosperms diverse and abundant
Carboniferous — Pennsylvanian (299)				Coal swamps common	Ouachita orogeny			Amphibians diverse and abundant
				Formation of Ancestral Rockies			Continental glaciation in Southern Hemisphere	Abundant coal swamps with seedless vascular plants
				Transgression of Absaroka Sea				
Carboniferous — Mississippian (318)	Kaskaskia							Reptiles evolve
								Gymnosperms evolve
				Widespread black shales				
Devonian (359)		Present sea level	Antler orogeny	Widespread black shales			Old Red Sandstone clastic wedge in British Isles	Extinction of many reef-building invertebrates
				Catskill Delta clastic wedge		Acadian orogeny		Amphibians evolve
				Extensive barrier reef formation in Western Canada				All major groups of fish present— Age of Fish
	Tippecanoe			Transgression of Kaskaskia Sea			Caledonian orogeny	Early land plants— seedless vascular plants
Silurian (416)								

Age (millions of years): 251, 299, 318, 359, 416

Review Questions

1. Weathering of which highlands produced the Queenston Delta clastic wedge?
   a. ____ Transcontinental Arch.
   b. ____ Acadian.
   c. ____ Taconic.
   d. ____ Sevier.
   e. ____ Caledonian.

2. The economically valuable deposit in a cyclothem is
   a. ____ gravel.
   b. ____ metallic ore.
   c. ____ coal.
   d. ____ carbonate.
   e. ____ evaporites.

3. The enormous single ocean surrounding Pangaea and spanning Earth from pole to pole at the end of the Permian is called the
   a. ___ Panthalassa.
   b. ___ Tethys.
   c. ___ Iapetus.
   d. ___ Atlantis.
   e. ___ Avalonia.

4. During which sequence did the eastern margin of Laurentia change from a passive plate margin to an active plate margin?
   a. ___ Zuni.
   b. ___ Tippecanoe.
   c. ___ Sauk.
   d. ___ Kaskaskia.
   e. ___ Absaroka.

5. During which sequence were evaporites and reef carbonates the predominant cratonic rocks?
   a. ___ Kaskaskia.
   b. ___ Zuni.
   c. ___ Sauk.
   d. ___ Absaroka.
   e. ___ Tippecanoe.

6. What type of plate interaction produced the Taconic orogeny?
   a. ___ oceanic–oceanic convergent.
   b. ___ oceanic–continental convergent.
   c. ___ continental–continental convergent.
   d. ___ divergent.
   e. ___ transform.

7. Which of the following resulted from intracratonic deformation?
   a. ___ Antler Highlands.
   b. ___ Acadian Highlands.
   c. ___ Ancestral Rockies.
   d. ___ Caledonian Highlands.
   e. ___ Taconic Highlands.

8. Which orogeny was *not* involved in the closing of the Iapetus Ocean?
   a. ___ Acadian.
   b. ___ Alleghenian.
   c. ___ Antler.
   d. ___ Caledonian.
   e. ___ Taconic.

9. A major transgressive–regressive cycle bounded by craton-wide unconformities is a(n)
   a. ___ cratonic sequence.
   b. ___ biostratigraphic unit.
   c. ___ orogeny.
   d. ___ shallow sea.
   e. ___ cyclothem.

10. The Catskill Delta clastic wedge resulted from weathering and erosion of the _____ highlands.
    a. ___ Taconic.
    b. ___ Antler.
    c. ___ Transcontinental Arch.
    d. ___ Acadian.
    e. ___ Caledonian.

11. Discuss how sequence stratigraphy can be used to make global correlations and why it is so useful in reconstructing past events.

12. Based on the discussion of Milankovitch cycles and their role in causing glacial–interglacial cycles (see Chapter 14), could these cycles be partly responsible for the transgressive–regressive cycles that resulted in cyclothems during the Pennsylvanian Period?

13. Discuss why cratonic sequences are a convenient way to study the geologic history of the Paleozoic Era.

14. How did the formation of Pangaea and Panthalassa affect the world's climate at the end of the Paleozoic Era?

15. What evidence indicates that the Iapetus Ocean began closing during the Middle Ordovician?

16. How are the Caledonian, Acadian, Ouachita, Hercynian, and Alleghenian orogenies related to modern concepts of plate tectonics?

17. What is the economic benefit to the automobile industry in having Paleozoic silica sand deposits near Toledo, Ohio?

18. Paleographic maps of what the world looked like during the Paleozoic Era can be found in almost every Earth history book and in numerous scientific journals. What criteria are used to determine the location of ancient continents and ocean basins, and why are there minor differences in the location and size of these paleocontinents among the various books and articles?

19. How does the origin of evaporite deposits of the Kaskaskia Sequence compare with the origin of evaporites of the Tippecanoe Sequence?

20. According to estimates made from mapping and correlation, the Queenston Delta contains more than 600,000 km³ of rock eroded from the Taconic Highlands. Based on this figure, geologists estimate that the Taconic Highlands were at least 4,000 m high. They also estimate that the Catskill Delta contains three times as much sediment as the Queenston Delta. From what you know about the geographic distribution of the Taconic Highlands and the Acadian Highlands, can you estimate how high the Acadian Highlands might have been?

Diorama of the environment and biota of the Phyllopod bed of the Middle Cambrian Burgess Shale, British Columbia, Canada. In the background is a vertical wall of a submarine escarpment with algae growing on it. The large, cylindrical, ribbed organisms on the muddy bottom in the foreground are sponges.

CHAPTER 21

Paleozoic Life History

Photo by V. Krantz. Courtesy of the Smithsonian Institution. #86-13471A

OUTLINE

21.1 Introduction

21.2 What Was the Cambrian Explosion?

21.3 The Emergence of a Shelly Fauna

21.4 Paleozoic Invertebrate Marine Life

GEO-INSIGHT: Trilobites—Paleozoic Arthropods

21.5 Vertebrate Evolution

21.6 Fish

21.7 Amphibians—Vertebrates Invade the Land

21.8 Evolution of the Reptiles—The Land Is Conquered

21.9 Plant Evolution

GEO-RECAP

HAVE YOU EVER WONDERED?

- Why animals with skeletons appeared abruptly at the beginning of the Paleozoic Era?

- What caused the mass extinction at the end of the Permian Period?

- What the first vertebrates were like and when they evolved?

- What the evolutionary history of fish was, and which group gave rise to the amphibians?

- What problems had to be solved by both animals and plants to make the transition from water onto land, and how they solved these problems?

- What allowed the reptiles to fully exploit the terrestrial environment, and which group of reptiles gave rise to the mammals?

- Why both the amphibians and seedless vascular plants were so abundant during the Pennsylvanian Period?

- What role the onset of arid conditions during the assemblage of Pangaea played in gymnosperms (flowerless seed plants) becoming the dominant element of the world's flora during the Permian Period?

21.1 Introduction

On August 30 and 31, 1909, near the end of the summer field season, Charles D. Walcott, geologist and head of the Smithsonian Institution, was searching for fossils along a trail on Burgess Ridge between Mount Field and Mount Wapta, near Field, British Columbia, Canada. On the west slope of this ridge, he discovered the first soft-bodied fossils from the Burgess Shale, a discovery of immense importance in deciphering the early history of life. During the following week, Walcott and his collecting party split open numerous blocks of shale, many of which yielded carbonized impressions of a number of soft-bodied organisms beautifully preserved on bedding planes. Walcott returned to the site the following summer and located the shale stratum that was the source of his fossil-bearing rocks in the steep slope above the trail. He quarried the site and shipped back thousands of fossil specimens to the United States National Museum of Natural History, where he later cataloged and studied them.

The importance of Walcott's discovery is not that it was just another collection of well-preserved Cambrian fossils, but that it allowed geologists a rare glimpse into a world previously almost unknown: that of the soft-bodied animals that lived during the Middle Cambrian. The beautifully preserved fossils from the Burgess Shale present a much more complete picture of a Middle Cambrian community than do deposits containing only fossils of the hard parts of organisms. In fact, 60% of the total fossil assemblage of more than 100 genera is composed of soft-bodied animals, a percentage comparable to present-day marine communities.

What conditions led to the remarkable preservation of the Burgess Shale fauna? The depositional site of the Burgess Shale lay at the base of a steep submarine escarpment. The animals whose exquisitely preserved fossil remains are found in the Burgess Shale lived in and on mud banks that formed along the top of this escarpment. Periodically, this unstable area would slump and slide down the escarpment as a turbidity current. At the base, the mud and animals carried with it were deposited in a deep-water anaerobic environment devoid of life. In such an environment, bacterial degradation did not destroy the buried animals, and thus they were compressed by the weight of the overlying sediments and eventually preserved as carbonaceous films.

In this chapter, we examine the history of Paleozoic life as a system in which its parts consist of a series of interconnected biologic and geologic events. The underlying processes of evolution and plate tectonics are the forces that drove this system. The opening and closing of ocean basins, transgressions and regressions of epeiric seas, the formation of mountain ranges, and the changing positions of the continents profoundly affected the evolution of the marine and terrestrial communities.

A time of tremendous biologic change began with the appearance of skeletonized animals near the Precambrian–Cambrian boundary. Following this event, marine invertebrates began a period of adaptive radiation and evolution, during which the Paleozoic marine invertebrate community greatly diversified. The earliest fossil record of vertebrates are fish, which evolved during the Cambrian. One group of fish was ancestral to the amphibians, which evolved during the Devonian. Reptiles evolved from amphibians during the Mississippian Period and were the dominant vertebrate animals on land by the end of the Paleozoic. Plants preceded animals onto land, and both plants and animals were confronted with the same basic problems in making the transition from water to land.

The end of the Paleozoic Era witnessed the greatest mass extinction in Earth's history. The marine invertebrate community was decimated, and many amphibians and reptiles also became extinct.

21.2 What Was the Cambrian Explosion?

At the beginning of the Paleozoic Era, animals with skeletons appeared rather abruptly in the fossil record. In fact, their appearance is described as an explosive development of new types of animals and is referred to as the "Cambrian explosion" by most scientists. This sudden appearance of new animals in the fossil record is rapid but only in the context of geologic time, having taken place over millions of years during the Early Cambrian Period.

This seemingly sudden appearance of animals in the fossil record is not a recent discovery. Early geologists observed that the remains of skeletonized animals appeared rather abruptly in the fossil record. Charles Darwin addressed this problem in *On the Origin of Species by Means of Natural Selection* (1859) and observed that, without a convincing explanation, such an event was difficult to reconcile with his newly expounded evolutionary theory.

> **CONNECTION LINK** ◀
>
> To review the contributions of Charles Darwin and the mechanism of natural selection, see pages 451–456 in Chapter 18.

The sudden appearance of shelled animals during the Early Cambrian contrasts sharply with the biota that lived during the preceding Proterozoic Eon. Until the evolution of the Ediacaran fauna, Earth was populated primarily by single-celled organisms. Recall from Chapter 19 that the Ediacaran fauna, which is found on all continents except Antarctica, consists primarily of multicelled, soft-bodied organisms. Microscopic calcareous tubes, presumably housing wormlike suspension-feeding organisms, have also been found at some localities. In addition, trails and burrows, which represent the activities of worms and other sluglike animals, are found associated with Ediacaran faunas throughout the world. The trails and burrows are similar to those made by present-day soft-bodied organisms.

Until recently, it appeared that there was a fairly long period between the extinction of the Ediacaran fauna and the appearance of the first Cambrian fossils. That gap has been considerably narrowed in recent years with the discovery of new Proterozoic fossiliferous localities. Now, Proterozoic fossil assemblages continue right to the base of the Cambrian.

Nonetheless, the cause of the sudden appearance of so many different animal phyla during the Early Cambrian is still hotly debated. Molecular techniques that allow evolutionary biologists to compare molecular sequences of the same gene from different species are being applied to the phylogeny, or evolutionary history, of many organisms. In addition, new fossil sites and detailed stratigraphic studies are shedding light on the early history and ancestry of the various invertebrate phyla.

The Cambrian explosion probably had its roots firmly planted in the Proterozoic. However, the mechanism or mechanisms that triggered this event are still being investigated. Although some would argue for a single causal event, it is more likely that the Cambrian explosion was a combination of factors, both biological and geological. For example, geologic evidence indicates that Earth was glaciated one or more times during the Proterozoic, followed by global warming during the Cambrian. These global environmental changes may have stimulated evolution and contributed to the Cambrian explosion.

Others would argue that a change in the chemistry of the oceans favored the evolution of a mineralized skeleton. In this scenario, an increase in the concentration of calcium from the Neoproterozoic through the Early Cambrian allowed for the precipitation of calcium carbonate and calcium phosphate, compounds that comprise the shells of most invertebrates.

Another hypothesis gaining favor is that the rapid evolution of a skeletonized fauna was a response to the evolution of predators. A shell or mineralized covering would provide protection against predation by the various predators evolving during this time.

A further line of research related to the Cambrian explosion involves *Hox* genes, which are sequences of genes that control the development of individual regions of the body. Studies indicate that the basic body plans for all animals were apparently established by the end of the

> **CONNECTION LINK**
>
> To review Proterozoic glaciation, see pages 488 and 489 in Chapter 19.

Cambrian explosion, and only minor modifications have occurred since then. Whatever the ultimate cause of the Cambrian explosion, the appearance of a skeletonized fauna and the rapid diversification of that fauna during the Early Cambrian were major events in life history.

21.3 The Emergence of a Shelly Fauna

The earliest organisms with hard parts are Proterozoic calcareous tubes found associated with Ediacaran faunas from several locations throughout the world. These are followed by other microscopic skeletonized fossils in Lower Cambrian (▌ Figure 21.1) rocks, and the appearance of large skeletonized animals during the Cambrian explosion. Along with the question of why animals appeared so suddenly in the fossil record is the equally intriguing question of why they initially acquired skeletons and what selective advantage skeletons provided. A variety of explanations about why marine organisms evolved skeletons have been proposed, but none is completely satisfactory or universally accepted.

What Would You Do ❓

Congress has just approved a large cut in funding for the National Science Foundation (NSF). Because of reduced funding, some earth science areas will not be funded this year. One area slated for a major funding reduction or no funding this year is paleontology. As a paleontology professor whose specialty is Paleozoic marine ecosystems and whose research depends on NSF funds, you will be testifying against cutting funding for paleontology.

What arguments would you use to persuade the members of Congress to keep funding paleontologic research? What possible outcomes from your research could you point to that might benefit society today?

The formation of an exoskeleton, or shell, confers advantages on an organism: (1) it provides protection against ultraviolet radiation, allowing animals to move into shallower waters; (2) it helps prevent drying out in an intertidal environment; (3) a supporting skeleton, whether an exo- or endoskeleton, allows animals to increase their size and provides attachment sites for muscles; and (4) it provides protection against predators. Evidence of actual fossils of predators and specimens of damaged prey, as well as anti-predatory adaptations in some animals, indicates that the impact of predation during the Cambrian was great, leading some scientists to hypothesize that the rapid evolution of a shelly invertebrate fauna was a

Figure 21.1 Lower Cambrian Shelly Fossils Two small (several millimeters in size) Lower Cambrian shelly fossils.

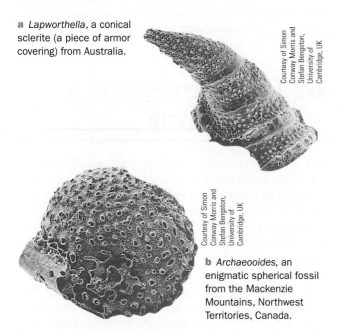

a *Lapworthella*, a conical sclerite (a piece of armor covering) from Australia.

Courtesy of Simon Conway Morris and Stefan Bengston, University of Cambridge, UK

Courtesy of Simon Conway Morris and Stefan Bengston, University of Cambridge, UK

b *Archaeooides,* an enigmatic spherical fossil from the Mackenzie Mountains, Northwest Territories, Canada.

response to the rise of predators. With predators playing an important role in the Cambrian marine ecosystem, any mechanism or feature that protected an animal would certainly be advantageous and confer an adaptive advantage to the organism.

Scientists currently have no clear answer as to why marine organisms evolved mineralized skeletons during the Cambrian explosion and shortly thereafter. They undoubtedly evolved because of a variety of biologic and environmental factors. Whatever the reason, the acquisition of a mineralized skeleton was a major evolutionary innovation

that allowed invertebrates to successfully occupy a wide variety of marine habitats.

21.4 Paleozoic Invertebrate Marine Life

Having considered the origin, differentiation, and evolution of the Proterozoic–Cambrian marine biota, we now examine the changes that occurred in the marine invertebrate community during the Paleozoic Era. Rather than focusing on the history of each invertebrate phylum (Table. 21.1), we will survey the evolution of the Paleozoic marine invertebrate communities through time, concentrating on the major features and changes that took place. To do that, we need to briefly examine the nature and structure of living marine communities so that we can make a reasonable interpretation of the fossil record.

The Present Marine Ecosystem

In order to understand ancient ecosystems, we must analyze the present-day marine ecosystem and look at where organisms live, how they get around, and how they feed (▌ Figure 21.2). Organisms that live in the water column above the seafloor are called *pelagic*. They are divided into two main groups: the floaters, or **plankton**, and the swimmers, or **nekton**.

Plankton are mostly passive and go where currents carry them. Plant plankton, such as diatoms, dinoflagellates, and various algae, are called phytoplankton and are mostly microscopic. Animal plankton are called *zooplankton* and are also mostly microscopic. Examples of zooplankton

TABLE 21.1	The Major Invertebrate Groups and Their Stratigraphic Ranges		
Phylum Protozoa	Cambrian–Recent	**Phylum Mollusca**	Cambrian–Recent
Class Sarcodina	Cambrian–Recent	Class Monoplacophora	Cambrian–Recent
Order Foraminifera	Cambrian–Recent	Class Gastropoda	Cambrian–Recent
Order Radiolaria	Cambrian–Recent	Class Bivalvia	Cambrian–Recent
Phylum Porifera	Cambrian–Recent	Class Cephalopoda	Cambrian–Recent
Class Demospongea	Cambrian–Recent	**Phylum Annelida**	Precambrian–Recent
Order Stromatoporoida	Cambrian–Oligocene	**Phylum Arthropoda**	Cambrian–Recent
Phylum Archaeocyatha	Cambrian	Class Trilobita	Cambrian–Permian
Phylum Cnidaria	Cambrian–Recent	Class Crustacea	Cambrian–Recent
Class Anthozoa	Ordovician–Recent	Class Insecta	Silurian–Recent
Order Tabulata	Ordovician–Permian	**Phylum Enchinodermata**	Cambrian–Recent
Order Rugosa	Ordovician–Permian	Class Blastoidea	Ordovician–Permian
Order Scleractinia	Triassic–Recent	Class Crinoidea	Cambrian–Recent
Phylum Bryozoa	Ordovician–Recent	Class Echinoidea	Ordovician–Recent
Phylum Brachiopoda	Cambrian–Recent	Class Asteroidea	Ordovician–Recent
Class Inarticulata	Cambrian–Recent	**Phylum Hemichordata**	Cambrian–Recent
Class Articulata	Cambrian–Recent	Class Graptolithina	Cambrian–Mississippian

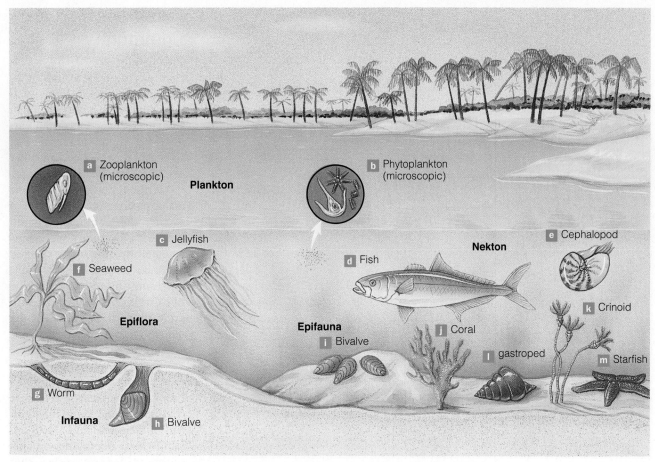

▌Figure 21.2 Marine Ecosystem Where and how animals and plants live in the marine ecosystem. Plankton: (a) through (c). Nekton: (d) and (e). Benthos: (f) through (m). Sessile epiflora: (f). Sessile epifauna: (i), (j), and (k). Mobile epifauna: (l) and (m). Infauna: (g) and (h). Suspension feeders: (i), (j), and (k). Herbivores: (l). Carnivore-scavengers: (m). Sediment-deposit feeders: (g).

include foraminifera, radiolarians, and jellyfish. The nekton are swimmers and are mainly vertebrates such as fish; the invertebrate nekton include cephalopods.

Organisms that live on or in the seafloor make up the **benthos**. They are characterized as *epifauna* (animals) or *epiflora* (plants)—those that live on the seafloor—or as *infauna*—animals that live in and move through the sediments. The benthos are further divided into those organisms that stay in one place, called *sessile*, and those that move around on or in the seafloor, called *mobile*.

The feeding strategies of organisms are also important in terms of their relationships with other organisms in the marine ecosystem. There are basically four feeding groups: **suspension-feeding** animals remove or consume microscopic plants and animals as well as dissolved nutrients from the water; **herbivores** are plant eaters; **carnivore-scavengers** are meat eaters; and **sediment-deposit feeders** ingest sediment and extract the nutrients from it.

We can define an organism's place in the marine ecosystem by where it lives and how it eats. For example, a brachiopod is a benthonic, epifaunal suspension feeder, whereas a cephalopod is a nektonic carnivore.

An ecosystem includes several *trophic levels*, which are tiers of food production and consumption within a feeding

hierarchy. The feeding hierarchy and hence energy flow in an ecosystem comprise a food web of complex interrelationships among the producers, consumers, and decomposers (▌ Figure 21.3). The **primary producers**, or *autotrophs*, are those organisms that manufacture their own food. Virtually all marine primary producers are phytoplankton. Feeding on the primary producers are the primary consumers, which are mostly suspension feeders. Secondary consumers feed on the primary consumers and thus are predators, whereas tertiary consumers, which are also predators, feed on the secondary consumers. Besides the producers and consumers, there are also transformers and decomposers. These are bacteria that break down the dead organisms that have not been consumed into organic compounds, which are then recycled.

When we look at the marine realm today, we see a complex organization of organisms interrelated by trophic interactions and affected by changes in the physical environment. When one part of the system changes, the whole structure changes, sometimes almost insignificantly, other times catastrophically.

As we examine the evolution of the Paleozoic marine ecosystem, keep in mind how geologic and evolutionary changes can have a significant impact on its composition

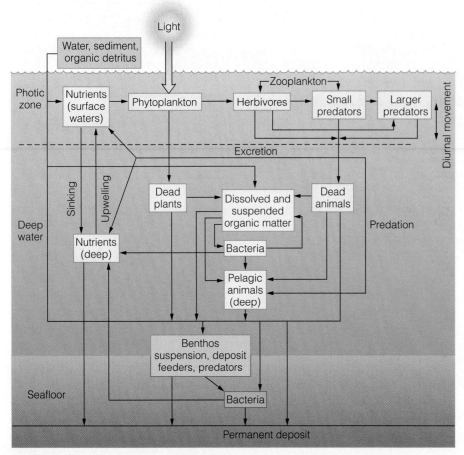

Figure 21.3 Marine Food Web Marine food web showing the relationship between the producers (phytoplankton), consumers (herbivores, small and large predators, pelagic animals, and benthos), and decomposers (bacteria).

Figure 21.4 Cambrian Marine Community Reconstruction of a Cambrian marine community showing floating jellyfish, swimming arthropods, benthonic sponges, and scavenging trilobites.

© Carnegie Museum of Natural History

Cambrian Marine Community

The Cambrian Period was a time during which many new body plans evolved and animals moved into new niches. As might be expected, the Cambrian witnessed a higher percentage of such experiments than any other period of geologic history.

Although almost all of the major invertebrate phyla evolved during the Cambrian Period (Table 21.1), many were represented by only a few species. Whereas trace fossils are common and echinoderms diverse, trilobites, brachiopods, and archaeocyathids made up the majority of Cambrian skeletonized life (❚ Figure 21.4).

Trilobites were by far the most conspicuous element of the Cambrian marine invertebrate community and made up about half of the total fauna. Most trilobites were benthonic, mobile, sediment-deposit feeders that crawled or swam along the seafloor (see the Geo-Insight on pages 548 and 549).

Cambrian *brachiopods* were mostly primitive types and were benthonic, sessile suspension feeders. Brachiopods became an abundant member of the marine invertebrate community during the Ordovician Period.

The third major group of Cambrian organisms was the *archaeocyathids* (❚ Figure 21.5). These organisms were benthonic, sessile suspension feeders that constructed reeflike structures beginning in the Early Cambrian, were abundant during the Early and Middle Cambrian, and went extinct at the end of the Cambrian.

The rest of the Cambrian fauna consisted of representatives of most of the other major phyla, including many organisms

and structure. For example, the major transgressions onto the craton opened up vast areas of shallow seas that could be inhabited. The movement of continents affected oceanic circulation patterns and caused environmental changes.

that were short-lived evolutionary experiments. As might be expected during times of adaptive radiation and evolutionary experimentation, many of the invertebrates that evolved during the Cambrian soon became extinct.

Figure 21.5 **Archaeocyathids** Restoration of a Cambrian reeflike structure built by archaeocyathids.

The Burgess Shale Biota

No discussion of Cambrian life would be complete without mentioning one of the best examples of a preserved soft-bodied fauna and flora: the Burgess Shale biota. As the Sauk Sea transgressed from the Cordilleran shelf onto the western edge of the craton, Early Cambrian-age sands were covered by Middle Cambrian-age black muds that allowed a diverse soft-bodied benthonic community to be preserved. As we discussed in the Introduction, these fossils were discovered in 1909 by Charles D. Walcott near Field, British Columbia, Canada. They represent one of the most significant fossil finds of the 20th century because they consist of impressions of soft-bodied animals and plants that are rarely preserved in the fossil record (**Figure 21.6**).

In recent years, the reconstruction, classification, and interpretation of many of the Burgess Shale fossils have undergone a major change that has led to new theories and explanations of the Cambrian explosion of life. Recall that during the Neoproterozoic, multicelled organisms evolved, and shortly thereafter animals with hard parts made their first appearance. These were followed by an explosion of invertebrate groups during the Cambrian, many of which

Figure 21.6 **Fossils from the Burgess Shale** Some of the fossil animals preserved in the Burgess Shale.

a *Anomalocaris*, a predator from the Early and Middle Cambrian Period, is shown feeding on *Opabinia*, one of many extinct invertebrates found in the Burgess Shale. *Anomalocaris* was approximately 45 cm long and used its gripping appendages to bring its prey to its circular mouth structure.

b *Wiwaxia*, a scaly armored sluglike creature whose affinities remain controversial.

c *Hallucigenia*, a velvet worm.

d *Waptia*, an arthropod.

Trilobites, an extinct class of arthropods, are probably the favorite and most sought after of invertebrate fossils. They lived from the Early Cambrian until the end of the Permian and were most diverse during the Late Cambrian. More than 15,000 species of trilobites have been described, and they are currently grouped into nine orders.

Trilobites had a worldwide distribution throughout the Paleozoic, and they lived in all marine environments, from shallow, nearshore waters to deep oceanic settings. They occupied a wide variety of habitats. Most were bottom-dwellers, crawling around the seafloor and scavenging organic detritus or feeding on microorganisms or algae. Others were free-swimming predators or filter-feeders living throughout the water column.

The trilobite body is divided into three parts (Figure 1); a cephalon (head), containing the eyes, mouth, and sensory organs; the thorax (body), composed of individual segments; and the pygidium (tail). Interestingly, the name *trilobite* does not refer to its three main body parts, but means *three-lobed*, which corresponds to the three longitudinal lobes of the thorax—the axial lobe and the two flanking pleural lobes.

The appendages of trilobites are rarely preserved. However, from specimens in which the soft-part anatomy is preserved as an impression, paleontologists know that beneath each thoracic segment was a two-part appendage consisting of a gill-bearing outer branch used for respiration and an inner branch or walking leg composed of articulating limb segments (Figure 2).

Trilobites range in size from a few millimeters in length, as found in many of the agnostid trilobites (Figure 3), to more than 70 cm, as recorded by the world's largest trilobite (Figure 4). Most trilobites, however, range between 3 and 10 cm long.

Trilobites were either blind, as in the agnostids (Figure 3), or possessed paired eyes. Holochroal eyes are characterized by generally hexagonal-shaped, biconvex lenses that are packed closely together, and covered by a single corneal layer that covers all of the lenses (Figure 5). This type of eye is similar to a modern compound eye and results in mosaic vision.

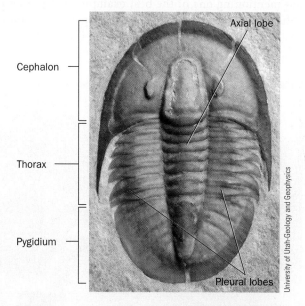

▲ **1.** *Cedaria minor*, from the Cambrian Weeks Formation, Utah, illustrates the major body parts of a trilobite.

▶ **2.** Model of a dorsal **(a)** and ventral **(b)** anatomy of *Triarthus eatoni*, a Late Ordovician trilobite from the Frankfort Shale, New York.

a Dorsal view.

b Ventral view.

Schizochroal eyes are composed of relatively large, individual lenses, each of which is covered by its own cornea and separated from adjacent lenses by a thick dividing wall (❙ Figure 6). Schizochroal lenses are arranged in rows and columns and most likely produced a visual field consisting of shadows.

Although trilobites had a hard dorsal exoskeleton, useful, in part, as a protection against predators, some trilobites also had the ability to enroll, presumably to protect their antennae, limbs, and soft ventral appendages. By doing so, the trilobite could protect its soft ventral anatomy and still view its surroundings (❙ Figure 7).

As mentioned earlier, trilobites were a very successful group. They first appeared in the Early Cambrian, rapidly diversified, reached their maximum diversity in the Late Cambrian, and then suffered major reductions in diversity near the end of the Cambrian, at the end of the Ordovician, and again near the end of the Devonian.

As yet, no consensus exists on what caused the trilobite extinctions, but a combination of factors was likely involved, possibly including a reduction of shelf space, increased competition, and a rise in predators. Some scientists have also suggested that a cooling of the seas may have played a role, particularly for the extinctions at the end of the Ordovician Period. The trilobites, like most of their Permian marine invertebrate contemporaries, finally fell victim to the Permian mass extinction event.

▶ **3.** *Hypagnastus pavifrons,* a small (5–8 mm long) agnostid triolobite from the Cambrian Marjum Formation, Utah.

University of Utah-Geology and Geophysics

▶ **4.** *Isotellus rex,* the world's largest triolbite (more than 70 cm long), from the Ordovician Churchill River Group, Manitoba, Canada.

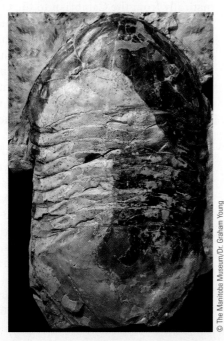

© The Manitoba Museum/Dr. Graham Young

◀ **5.** Holochroal eyes from the trilobite *Scutellum campaniferum.*

Courtesy of E. N. K. Clarkson, used with permission University of Edinburgh

◀ **6.** Schizochroal eyes from the trilobite *Eophacops trapeziceps.*

Courtesy of E. N. K. Clarkson, used with permission

◀ **7.** An enrolled specimen of the Ordovician trilobite *Pilomera fisheri* from Putilowa, Poland

© Sinclair Stammers/SPL/ Photo Researchers, Inc.

are now extinct. These Cambrian organisms represent the root stock and basic body plans from which all present-day invertebrates evolved.

The question that paleontologists are still debating is how many phyla arose during the Cambrian, and at the center of that debate are the Burgess Shale fossils. For years, most paleontologists placed the bulk of the Burgess Shale organisms into existing phyla, with only a few assigned to phyla that are now extinct. Thus, the phyla of the Cambrian world were viewed as being essentially the same in number as the phyla of the present-day world, but with fewer species in each phylum. According to this view, the history of life has been simply a gradual increase in the diversity of species within each phylum through time. The number of basic body plans has therefore remained more or less constant since the initial radiation of multi-celled organisms.

This view, however, has been challenged by other paleontologists who think that the initial explosion of varied life-forms in the Cambrian was promptly followed by a short period of experimentation and then extinction of many phyla. The richness and diversity of modern life-forms are the result of repeated variations of the basic body plans that survived the Cambrian extinctions. In other words, life was much more diverse in terms of phyla during the Cambrian than it is today. The reason why members of the Burgess Shale biota look so strange to us is that no living organisms possess their basic body plans, and therefore many of them have been reassigned to new phyla.

Discoveries of new Cambrian fossils at localities such as Sirius Passet, Greenland, and especially some of the newest finds from beds with exceptional preservation in Yunnan, China, have resulted in reassignment of some Burgess Shale specimens back into extant phyla. If these reassignments to known phyla prove to be correct, then no massive extinction followed the Cambrian explosion, and life has gradually increased in diversity through time. Currently, there is no clear answer to this debate, and the outcome will probably be decided as more fossil discoveries are made.

Ordovician Marine Community

A major transgression that began during the Middle Ordovician (Tippecanoe Sequence) resulted in the most

Figure 21.7 Middle Ordovician Marine Community Reconstruction of a Middle Ordovician seafloor fauna. Cephalopods, crinoids, colonial corals, bryozoans, trilobites, and brachiopods are shown.

» Critical Thinking Question
What role did the transgressing seas during the Tippecanoe Sequence play in terms of opening up new ecologic niches for the marine invertebrate community, and what was the contribution of the invertebrates to the sedimentary rocks of the Tippecanoe Sequence?

widespread inundation of the craton. This vast epeiric sea, which was uniformly warm during this time, opened many new marine habitats that were soon filled by a variety of organisms.

Not only did sedimentation patterns change dramatically from the Cambrian to the Ordovician, but the fauna underwent equally striking changes. Whereas trilobites, brachiopods, and archaeocyathids dominated the Cambrian invertebrate community, the Ordovician was characterized by the continued diversification of brachiopods and by the adaptive radiation of many other animal phyla (such as bryozoans and corals), with a consequent dramatic increase in the diversity of the total shelly fauna (▌ Figure 21.7).

During the Cambrian, archaeocyathids were the main builders of reeflike structures, but bryozoans, stromatoporoids, and tabulate and rugose corals assumed that role beginning in the Middle Ordovician. Many of these reefs were small patch reefs similar in size to those of the Cambrian but of a different composition, whereas others were quite large. As with present-day reefs, Ordovician reefs showed a high diversity of organisms and were dominated by suspension feeders.

The end of the Ordovician was a time of mass extinctions in the marine realm. More than 100 families of marine invertebrates became extinct, and in North America alone, about half of the brachiopods and bryozoans died out. What caused such an event? Many geologists think that these extinctions were the result of the extensive glaciation in Gondwana at the end of the Ordovician Period (see Chapter 20).

Silurian and Devonian Marine Communities

The mass extinction at the end of the Ordovician was followed by rediversification and recovery of many of the decimated groups. Brachiopods, bryozoans, gastropods, bivalves, corals, crinoids, and graptolites were just some of the groups that rediversified beginning during the Silurian.

As discussed in Chapter 20, the Silurian and Devonian were times of major reef building. Whereas most of the Silurian radiations of invertebrates represented the repopulation of niches, organic reef builders diversified in new ways, building massive reefs larger than any produced during the Cambrian or Ordovician. This repopulation was probably caused, in part, by renewed transgressions over the craton, and although a major drop in sea level occurred at the end of the Silurian, the Middle Paleozoic sea level was generally high (see Table 20.1).

The Silurian and Devonian reefs were dominated by tabulate and colonial rugose corals and stromatoporoids (▮ Figure 21.8). Although the fauna of these Silurian and Devonian reefs was somewhat different from that of earlier reefs and reeflike structures, the general composition and structure are the same as in present-day reefs.

The Silurian and Devonian periods were also the time when *eurypterids* (arthropods with scorpion-like bodies and impressive pincers) were abundant, and unlike many other marine invertebrates, eurypterids expanded into brackish and freshwater habitats (▮ Figure 21.9). *Ammonoids*, a subclass of the cephalopods, evolved from nautiloids during the Early Devonian and rapidly diversified.

© The Field Museum, #Ge080821c

▮ **Figure 21.8** Middle Devonian Marine Reef Community Reconstruction of a Middle Devonian reef from the Great Lakes area of North America. Shown are corals, bryozoans, cephalopods, trilobites, crinoids, and brachiopods.

© The Field Museum, #GE080815c

▮ **Figure 21.9** Silurian Brackish Water Community Restoration of a Silurian brackish water scene near Buffalo, New York. Shown are algae, eurypterids, gastropods, worms, and shrimp.

With their distinctive suture patterns, short stratigraphic ranges, and widespread distribution, ammonoids are excellent guide fossils for the Devonian through Cretaceous periods (▮ Figure 21.10).

Near the end of the Devonian, another mass extinction occurred that resulted in a worldwide near-total collapse of the massive reef communities. On land, however, the seedless vascular plants were seemingly unaffected. Thus, extinctions at this time were most extensive among marine life, particularly in the reef and pelagic communities.

Figure 21.10 **Ammonoid Cephalopod** A Late Devonian-age ammonoid cephalopod from Erfoud, Morocco. The distinctive suture pattern, short stratigraphic range, and wide geographic distribution make ammonoids excellent guide fossils.

The demise of the Middle Paleozoic reef communities highlights the geographic aspects of the Late Devonian mass extinction. The tropical groups were most severely affected; in contrast, the higher latitude communities were seemingly little affected. Apparently, an episode of global cooling was largely responsible for the extinctions near the end of the Devonian. During such a cooling, the disappearance of tropical conditions would have had a severe effect on reef and other warm-water organisms. Cool-water species, in contrast, could have simply migrated toward the equator. Although cooling temperatures certainly played an important role in the Late Devonian extinctions, the closing of the Iapetus Ocean and the orogenic events of this time undoubtedly also played a role by reducing the area of shallow shelf environments where many marine invertebrates lived.

Carboniferous and Permian Marine Communities

The Carboniferous invertebrate marine community responded to the Late Devonian extinctions in much the same way that the Silurian invertebrate marine community responded to the Late Ordovician extinctions—that is, by renewed adaptive radiation and rediversification. The brachiopods and ammonoids quickly recovered and again assumed important ecologic roles. Other groups, such as the lacy bryozoans and crinoids, reached their greatest diversity during the Carboniferous. With the decline of the stromatoporoids and the tabulate and rugose corals, large organic reefs such as those existing earlier in the Paleozoic virtually disappeared and were replaced by small patch reefs. These reefs were dominated by crinoids, blastoids, lacy bryozoans, brachiopods, and calcareous algae and flourished during the Late Paleozoic (Figure 21.11). In addition, bryozoans and crinoids contributed large amounts of skeletal debris to the formation of the vast bedded limestones that constitute the majority of Mississippian sedimentary rocks.

The Permian invertebrate marine faunas resembled those of the Carboniferous, but were not as widely distributed because of the restricted size of the shallow seas on the cratons and the reduced shelf space along the continental margins (see Figure 20.20). The spiny and odd-shaped productids dominated the brachiopod assemblage and constituted an important part of the reef complexes that formed in the Texas region during the Permian (Figure 21.12). The fusulinids (spindle-shaped foraminifera), which first evolved during the Late Mississippian and greatly diversified during the Pennsylvanian, experienced a further diversification during the

Figure 21.11 **Late Mississippian Marine Community** Reconstruction of marine life during the Mississippian, based on an Upper Mississippian fossil site at Crawfordsville, Indiana. Invertebrate animals shown include crinoids, blastoids, lacy bryozoans, brachiopods, and small corals.

Figure 21.12 Permian Patch-Reef Marine Community Reconstruction of a Permian patch-reef community from the Glass Mountains of West Texas. Shown are algae, productid brachiopods, cephalopods, sponges, and corals.

1 Late Ordovician	(−12%)
2 Late Devonian	(−14%)
3 Late Permian	(−52%)
4 Late Triassic	(−12%)
5 Late Cretaceous	(−11%)

Geologic time (10⁶ years)

Figure 21.13 Phanerozoic Marine Diversity Phanerozoic diversity for marine invertebrate and vertebrate families. Note the three episodes of Paleozoic mass extinctions, with the greatest occurring at the end of the Permian Period.

≫ Critical Thinking Question
Note that the greatest extinction in Earth history occurred at the end of the Permian, yet the Late Cretaceous mass extinction is the one most people are familiar with. Why do you think most people have heard of the Late Cretaceous mass extinction but not the one at the end of the Permian?

The Permian Marine Invertebrate Mass Extinction

The greatest recorded mass extinction to affect Earth occurred at the end of the Permian Period (Figure 21.13). By the time the Permian ended, roughly 50% of all marine invertebrate families and about 90% of all marine invertebrate species became extinct. Fusulinids, rugose and tabulate corals, and several bryozoan and brachiopod orders, as well as trilobites and blastoids, did not survive the end of the Permian. All of these groups had been very successful during the Paleozoic Era. In addition, more than 65% of all amphibians and reptiles, as well as nearly 33% of insects on land, became extinct.

What caused such a crisis for both marine and land-dwelling organisms? Various hypotheses have been proposed, but no completely satisfactory answer has yet been found. Because the extinction event extended over many millions of years at the end of the Permian, a meteorite impact such as occurred at the end of the Cretaceous Period can be reasonably discounted.

A reduction in the habitable marine shelf area caused by the formation of Pangaea and a widespread marine regression resulting from glaciation can also be rejected as the primary cause of the extinctions. By the end of the Permian, most collisions of the continents had already taken place, such that the reduction in the shelf area had already occurred before the mass extinctions began in earnest. Furthermore, the widespread glaciation that took place during the Carboniferous was now waning in the Permian.

CONNECTION LINK

To learn more about the Cretaceous mass extinction and whether a meteorite impact played a role, see pages 606 and 607 in Chapter 22.

What Would You Do ?

One concern of environmentalists is that environmental degradation is leading to vast reductions in global biodiversity. As a paleontologist, you are aware that mass extinctions have taken place throughout Earth history. What facts and information can you provide from your geologic perspective that will help focus the debate on whether or not Earth's biota is being adversely affected by such human activities as industrialization, and what the possible outcome(s) might be if global biodiversity is severely reduced?

Permian. Because of their abundance, diversity, and worldwide occurrence, fusulinids are important guide fossils for Pennsylvanian and Permian strata. Bryozoans, sponges, and some types of calcareous algae also were common elements of the Permian invertebrate fauna.

Many scientists think that an episode of deep-sea anoxia and increased oceanic CO_2 levels resulted in a highly stratified ocean during the Late Permian. In other words, there was very little, if any, circulation of oxygen-rich surface waters into the deep ocean. During this time, stagnant waters also covered the shelf regions, thus affecting the shallow marine fauna.

During the Late Permian, widespread volcanic and continental fissure eruptions, such as the Siberian Traps, where lava flows covered more than 2 million km² were also taking place. These eruptions released not only large amounts of additional carbon dioxide into the atmosphere but also high amounts of fluorine and chlorine, which could have damaged the ozone layer and helped contribute to increased climatic instability and ecologic collapse.

Several recent studies have examined the role that the Siberian Traps flood basalts may have played in causing widespread ozone depletion, thus allowing an increase in UV-B radiation at the end of the Permian. These studies have included using a two-dimensional atmospheric chemistry-transport model to assess the impact of the Siberian Traps eruption in altering the end-Permian stratospheric ozone layer, as well as an analysis of the relationship of morphologic abnormalities in gymnosperm (flowerless seed plant) pollen grains (see the section "Plant Evolution" later in this chapter) to deteriorating conditions around the Permian–Triassic boundary.

By the end of the Permian, a near collapse of both the marine and terrestrial ecosystems had occurred. Although the ultimate cause of such devastation is still being debated and investigated, it is safe to say that it was probably a combination of interconnected and related geologic and biologic events.

21.5 Vertebrate Evolution

A **chordate** (phylum Chordata) is an animal that has, at least during part of its life cycle, a notochord, a dorsal hollow nerve cord, and pharyngeal slits (▌ Figure 21.14). **Vertebrates**, which are animals with backbones, are simply a subphylum of chordates.

The ancestors and early members of the phylum Chordata were soft-bodied organisms that left few fossils (▌ Figure 21.15). Consequently, we know little about the early evolutionary history of the chordates or vertebrates. Surprisingly, a close relationship exists between echinoderms (Table 21.1) and chordates, and they may even have shared a common ancestor. This is because the development of the embryos of echinoderms and chordates are the same in both groups; that is, the cells divide by radial cleavage so that the cells are aligned directly above each

Dorsal hollow nerve cord · Notochord · Dorsal fin · Tail · Oral tentacles · Velar tentacles · Pharynx · Pharyngeal (gill) slits · Midgut cecum · Atrium · Atrial opening (excurrent) · Ventral fin · Anus · Muscles

▌ **Figure 21.14 Three Characteristics of a Chordate** The structure of the lancelet *Amphioxus* illustrates the three characteristics of a chordate: a notochord, a dorsal hollow nerve cord, and pharyngeal slits.

Courtesy of Dr. Lars Ramsköld

▌ **Figure 21.15** *Yunnanozoon lividum* Found in 525-million-year-old rocks in Yunnan province, China, *Yunnanozoon lividum,* a 5-cm-long animal, is one of the oldest known chordates.

▌ **Figure 21.16 Cell Cleavage**

Radial Cleavage

a Arrangement of cells resulting from radial cleavage is characteristic of chordates and echinoderms. In this configuration, cells are directly above each other.

Spiral Cleavage

b Arrangement of cells resulting from spiral cleavage. In this arrangement, cells in successive rows are nested between each other. Spiral cleavage is characteristic of all invertebrates except echinoderms.

other (▌ Figure 21.16a). In all other invertebrates, cells undergo spiral cleavage, which results in having cells nested between each other in successive rows (▌ Figure 21.16b). Furthermore, the biochemistry of muscle activity

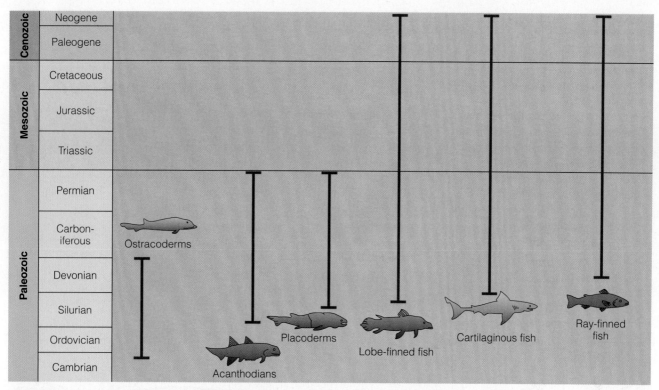

▌Figure 21.17 Geologic Ranges of the Major Fish Groups Ostracoderms are early members of the class Agnatha (jawless fish). Acanthodians (class Acanthodii) are the first fish with jaws. Placoderms (class Placodermii) are armored, jawed fish. Lobe-finned (subclass Sarcopterygii) and ray-finned (subclass Actinopterygii) fish are members of the class Osteichthyes (bony fish), whereas cartilaginous fish belong to the class Chondrichthyes.

and blood proteins, and the larval stages, are similar in both echinoderms and chordates.

The evolutionary pathway to vertebrates thus appears to have taken place much earlier and more rapidly than many scientists have long thought. Based on fossil evidence and recent advances in molecular biology, one scenario suggests that vertebrates evolved shortly after an ancestral chordate, probably resembling *Yunnanozoon*, acquired a second set of genes. According to this hypothesis, a random mutation produced a duplicate set of genes, letting the ancestral vertebrate animal evolve entirely new body structures that proved to be evolutionarily advantageous. Not all scientists accept this hypothesis, and the origin of vertebrates is still hotly debated.

21.6 Fish

The most primitive vertebrates are fish, and some of the oldest fish remains are found in the Upper Cambrian Deadwood Formation in northeastern Wyoming. Here, phosphatic scales and plates of *Anatolepis*, a primitive member of the class Agnatha (jawless fish), have been recovered from marine sediments. All known Cambrian and Ordovician fossil fish have been found in shallow, nearshore marine deposits, whereas the earliest nonmarine (freshwater) fish

remains have been found in Silurian strata. This does not prove that fish originated in the oceans, but it does lend strong support to the idea.

As a group, fish range from the Late Cambrian to the present (▌Figure 21.17). The oldest and most primitive of the class Agnatha are the **ostracoderms**, whose name means "bony skin" (Table 21.2). These are armored, jawless fish that first evolved during the Late Cambrian, reached their zenith during the Silurian and Devonian, and then became extinct.

The majority of ostracoderms lived on the seafloor. *Hemicyclaspis* is a good example of a bottom-dwelling ostracoderm (▌Figure 21.18a). Vertical scales allowed *Hemicyclaspis* to wiggle sideways, propelling itself along the seafloor, and the eyes on top of its head allowed it to see such predators as cephalopods and jawed fish approaching from above. While moving along the sea bottom, it probably sucked up small bits of food and sediment through its jawless mouth.

The evolution of jaws was a major evolutionary advance among primitive vertebrates. Although their jawless ancestors could only feed on detritus, jawed fish could chew food and become active predators, thus opening many new ecological niches.

The vertebrate jaw is an excellent example of evolutionary opportunism. Various studies suggest that the jaw originally evolved from the first two or three anterior gill arches of

TABLE 21.2	Brief Classiciation of Fish Groups Referred to in the Text		
Classification	**Geologic Range**	**Living Example**	
Class Agnatha (jawless fish)	Late Cambrian–Recent	Lampery, hagfish	
Early members of the class are called ostracoderms	Late Cambrian Devonian	No living ostracoderms	
Class Acanthodii (the first fish with jaws)	Early Silurian–Permian	None	
Class Placodermii (armored, jawed fish)	Late Silurian–Permian	None	
Class Chondrichthyes (cartilaginous fish)	Devonian–Recent	Sharks, rays, skates	
Class Osteichthyes (bony fish)	Devonian–Recent	Tuna, perch, bass, pike, catfish, trout, salmon, lungfish, *Latimeria*	
Subclass Actinopterygii (ray-finned fish)	Devonian–Recent	Tuna, perch, bass, pike, catfish, trout, salmon	
Subclass Sarcopterygii (lobe-finned fish)	Devonian–Recent	Lungfish, *Latimeria*	
Order Coelacanthimorpha	Devonian–Recent	*Latimeria*	
Order Dipnoi	Devonian–Recent	Lungfish	
Order Crossopterygii	Devonian–Permian	None	
Suborder Rhipidistia	Devonian–Permian	None	

▌**Figure 21.18 Devonian Seafloor** Recreation of a Devonian seafloor showing (a) an ostracoderm, (b) a placoderm, (c) an acanthodian, and (d) a ray-finned fish.

jawless fish. Because the gills are soft, they are supported by gill arches of bone or cartilage. The evolution of the jaw may thus have been related to respiration rather than to feeding (▌Figure 21.19). By evolving joints in the forward gill arches,

jawless fish could open their mouths wider. Every time a fish opened and closed its mouth, it would pump more water past the gills, thereby increasing the oxygen intake. The modification from rigid to hinged forward gill arches let fish increase

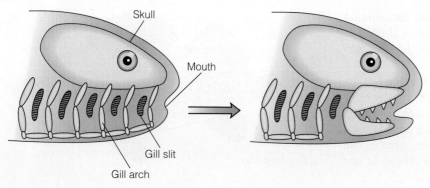

Figure 21.19 Evolution of the Vertebrate Jaw The evolution of the vertebrate jaw is thought to have begun from the modification of the first two or three anterior gill arches. This theory is based on the comparative anatomy of living vertebrates.

Figure 21.20 Late Devonian Seascape A Late Devonian marine scene from the mid-continent of North America. (a) The giant placoderm *Dunkleosteus* (length more than 12 m) is pursuing (b) the shark *Cladoselache* (length up to 1.2 m), a cartilaginous fish. Also shown are (c) the bottom-dwelling placoderm *Bothriolepis* and (d) the swimming ray-finned fish *Cheirolepsis*, both of which reached a length of 40–50 cm.

group included the ancestors of the present-day bony and cartilaginous fish groups.

The other jawed fish, the **placoderms** (class Placodermii), whose name means "plate-skinned," evolved during the Silurian (Table 21.2). Placoderms were heavily armored, jawed fish that lived in both freshwater and the ocean and reached their peak of abundance and diversity during the Devonian.

The placoderms showed considerable variety, including small bottom dwellers (Figure 21.18b) as well as large major predators such as *Dunkleosteus*, a Late Devonian fish that lived in the mid-continental North American epeiric seas (Figure 21.20a). It was by far the largest fish of the time, reaching a length of more than 12 m. It had a heavily armored head and shoulder region, a huge jaw lined with razor-sharp bony teeth, and a flexible tail, all features consistent with its status as a ferocious predator.

Besides the abundant acanthodians, placoderms, and ostracoderms, other fish groups, such as the cartilaginous and bony fish, also evolved during the Devonian Period. Small wonder, then, that the Devonian is informally called the "Age of Fish," because all major fish groups were present during this time period.

The **cartilaginous fish** (class Chondrichthyes) (Table 21.2), represented today by sharks, rays, and skates, first evolved during the Early Devonian, and by the Late Devonian, primitive marine sharks such as *Cladoselache* were abundant (Figure 21.20b). Cartilaginous fish have never been as numerous or as diverse as their cousins, the bony fish, but they were, and still are, important members of the marine vertebrate fauna.

Along with the cartilaginous fish, the **bony fish** (class Osteichthyes) (Table 21.2) also first evolved during the Devonian. Because bony fish are the most varied and numerous of all the fishes, and because the amphibians evolved from them, their evolutionary history is particularly

both their food consumption and oxygen intake, and the evolution of the jaw as a feeding structure rapidly followed.

The fossil remains of the first jawed fish are found in Lower Silurian rocks and belong to the *acanthodians*, a group of small, enigmatic fish characterized by large spines, paired fins, scales covering much of the body, jaws, teeth, and greatly reduced body armor (Figure 21.18c and Table 21.2). Although the relationship of acanthodians to other fish is not yet well established, many scientists think that the important. There are two groups of bony fish: the common *ray-finned fish* (subclass Actinopterygii) (Figures 2l.18d and 21.20d) and the less familiar *lobe-finned fish* (subclass Sarcopterygii) (Table 21.2).

The term *ray-finned* refers to the way that the fins are supported by thin bones that spread away from the body (Figure 21.21a). From a modest freshwater beginning during the Devonian, ray-finned fish, which include most of the familiar fish such as trout, bass, perch, salmon, and tuna, rapidly diversified to dominate the Mesozoic and Cenozoic seas.

Present-day lobe-finned fish are characterized by muscular fins. The fins do not have radiating bones, but, rather, have articulating bones with the fin attached to the body by a fleshy shaft (Figure 21.21b). Such an arrangement allows for a powerful stroke of the fin, making the fish an effective swimmer. Three orders of lobe-finned fish are recognized: *coelacanths*, *lungfish*, and *crossopterygians* (Table 21.2).

Coelacanths (order Coelacanthimorpha) are marine lobe-finned fish that evolved during the Middle Devonian and were thought to have gone extinct at the end of the Cretaceous. In 1938, however, a fisherman caught a coelacanth in the deep waters off Madagascar, and since then, several dozen more have been caught, both there and in Indonesia.

Lungfish (order Dipnoi) were fairly abundant during the Devonian, but today only three freshwater genera exist, one each in South America, Africa, and Australia. Their present-day distribution presumably reflects the Mesozoic breakup of Gondwana.

The "lung" of a modern-day lungfish is actually a modified swim bladder that most fish use for buoyancy in swimming. In lungfish, this structure absorbs oxygen, allowing them to breathe air when the lakes or streams in which they live become stagnant and dry up. During such times, they burrow into the sediment to prevent dehydration and breathe through their swim bladder until the stream begins flowing or the lake that they were living in fills with water. When they are back in the water, lungfish then rely on gill respiration.

The **crossopterygians** (order Crossopterygii) are an important group of lobe-finned fish, because it is probably from them that amphibians evolved. However, the transition between crossopterygians and true amphibians is not as simple as it was once portrayed. Among the crossopterygians, the *rhipidistians* appear to be the ancestral group (Table 21.2). These fish, reaching lengths of more than 2 m, were the dominant freshwater predators during the Late Paleozoic. *Eusthenopteron*, a good example of a rhipidistian crossopterygian and a classic example of a transitional form between fish and amphibians, had an elongate body

> **CONNECTION LINK**

To learn more about the breakup of Gondwana during the Mesozoic, and its effect on global climate and ocean circulation patterns, see pages 576–580 in Chapter 22.

a Ray-finned fish

b Lobe-finned fish

 Figure 21.21 Ray-finned and Lobe-finned Fish Arrangement of fin bones for **(a)** a typical ray-finned fish and **(b)** a lobe-finned fish. The muscles extend into the fin of the lobe-finned fish, allowing greater flexibility of movement than that of the ray-finned fish.

that helped it move swiftly in the water and paired muscular fins that many scientists thought could be used for moving on land (Figure 21.22). The structural similarity between crossopterygian fish and the earliest amphibians is striking, and one of the most widely cited examples of a transition from one major group to another (Figure 21.23). However, recent discoveries of older lobe-finned fish and newly published findings of tetrapod footprints and tetrapod-like fish are filling the gaps in the evolution from fish and tetrapods (animals with four limbs).

Before discussing this transition and the evolution of amphibians, it is useful to place the evolutionary history of Paleozoic fish in the larger context of Paleozoic evolutionary events. Certainly, the evolution and diversification of jawed fish, as well as eurypterids (extinct members of a class of arthropods) and ammonoid cephalopods, had a profound effect on the marine ecosystem. Previously defenseless organisms either evolved defensive mechanisms or suffered great losses, possibly even extinction.

For example, ostracoderms, although armored, would have been easy prey for the swifter jawed fish. Ostracoderms became extinct by the end of the Devonian, a time that coincides with the rapid evolution of jawed fish. Placoderms, like acanthodians, greatly decreased in abundance after the Devonian and became extinct by the end of the Paleozoic Era. In contrast, cartilaginous fish and ray-finned bony fish expanded during the Late Paleozoic, as did the ammonoids (Figure 21.10), the other major predators of the Late Paleozoic seas.

▌Figure 21.22
Rhipidistian
Crossopterygian The
crossopterygians are
the group from which
the amphibians are
thought to have evolved.
Eusthenopteron, a
member of the rhipidistian
crossopterygians, had
an elongated body and
paired fins that could be
used for moving about
on land.

Labyrinthodont

Lobe fin

a Skeletal similarity.

▌Figure 21.23 Similarities Between Crossopterygians and
Labyrinthodonts Similarities between the crossopterygian lobe-
finned fish and the labyrinthodont amphibians.

b Comparison of the limb bones of a crossopterygian (*left*) and
amphibian (*right*); color identifies the bones (u = ulna, shown in
blue, r = radius, mauve, h = humerus, gold) that the two groups have
in common.

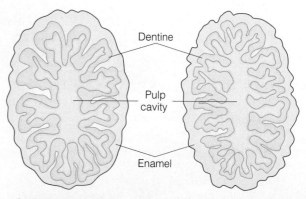

Dentine

Pulp
cavity

Enamel

c Comparison of tooth cross sections shows the complex and
distinctive structure found in both the crossopterygians (*left*) and
labyrinthodont amphibians (*right*).

21.7 Amphibians—Vertebrates Invade the Land

Although amphibians were the first vertebrates to live on land, they were not the first land-living organisms. Land plants, which probably evolved from green algae, first evolved during the Ordovician. Furthermore, insects, millipedes, spiders, and even snails invaded the land before amphibians. Fossil evidence indicates that such land-dwelling arthropods as scorpions and flightless insects had evolved by at least the Devonian.

The transition from water to land required animals to surmount several barriers. The most critical were desiccation, reproduction, the effects of gravity, and the extraction of oxygen from the atmosphere by lungs rather than from water by gills. Until the 1990s, the traditional evolutionary sequence had a rhipidistian crossopterygian, the *Eusthenopteron*, evolving into a primitive amphibian such as *Ichthyostega*. At that time, fossils of those two genera were about all that paleontologists had to work with, and although there were gaps in morphology, the link between crossopterygians and these earliest amphibians was easy to see (Figure 21.23).

Nevertheless, the question remained as to when animals made the transition from water to land, and perhaps even more intriguing is the question of why limbs evolved in the first place. The answer to that question is that it probably was not for walking on land. In fact, the current thinking of many scientists is that aquatic limbs made it easier for animals to move around in streams, lakes, or swamps that were choked with water plants or other debris. The fossil evidence that began to emerge in the 1990s also seems to support this hypothesis.

Fossils of *Acanthostega*, a tetrapod found in the 360-million-year-old Old Red Sandstone from Greenland, reveal an animal that had limbs, but was clearly unable to walk on land. Paleontologist Jennifer Clack, who has recovered and analyzed hundreds of specimens of *Acanthostega*, points out that *Acanthostega's* limbs were not strong enough to support its weight on land, and its rib cage was too small for the necessary muscles needed to hold its body off the ground. In addition, *Acanthostega* had gills and lungs, meaning that it could survive on land, but it was more suited for the water. Clack thinks that *Acanthostega* used its limbs to maneuver around in swampy, plant-filled waters, where swimming would be difficult and limbs were an advantage.

Fragmentary fossils from other tetrapods living at about the same time as *Acanthostega* suggest, however, that some of these early tetrapods may have spent more time on dry land than in the water. These oldest amphibians, many of which are also found in the Upper Devonian Old Red Sandstone of eastern Greenland and belong to genera such as *Ichthyostega*, had streamlined bodies, long tails, and fins along their backs. In addition to four legs, they had a strong backbone, a rib cage, and pelvic and pectoral girdles, all of which were structural adaptations for walking on land (▌Figure 21.24).

These earliest amphibians thus appear to have inherited many characteristics from the crossopterygians with little modification (Figure 21.23). However, with the discovery of such fossils as *Acanthostega* and others like it, the transition between fish and amphibians involves a number of new genera that are intermediary between the two groups and fills in some of the gaps between the earlier postulated Rhipidistian crossopterygian–amphibian phylogeny.

In 2006, an exciting discovery of a 1.2–2.8-m-long, 375-million-year-old (Late Devonian) "fishapod" was announced. Discovered on Ellesmere Island, Canada, *Tiktaalik roseae*, from the Inuktitut meaning "large fish in a stream," was hailed as an intermediary between the lobe-finned fish and the earliest tetrapod, *Acanthostega* (▌Figure 21.25).

Tiktaalik roseae is truly a "fishapod" in that it has a mixture of both fish and tetrapod characteristics (Figure 21.25b). For example, it has gills and fish scales, but also a broad skull, eyes on top of its head, a flexible neck and large rib cage that could support its body on land or in shallow water, and lungs, all of which are tetrapod features. What really excited scientists, however, was that *Tiktaalik roseae* has the beginnings of a true tetrapod forelimb, complete with

▌**Figure 21.24 Late Devonian Landscape** Shown is *Ichthyostega*, both in the water and on land. *Ichthyostega* was an amphibian that grew to a length of about 1 m. The flora at this time was diverse and consisted of a variety of small and large seedless vascular plants.

De Agostini Picture Library/De Agostini/Getty Images

Figure 21.25 *Tiktaalik roseae* *Tiktaalik roseae*, a "fishapod," has been hailed as an intermediary between lobe-finned fish and tetrapods because it has characteristics of both fish and tetrapods.

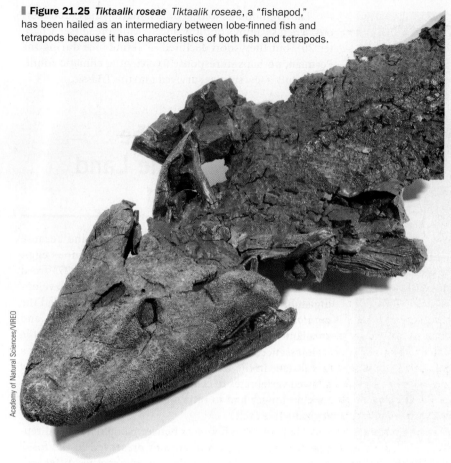

a Skeleton of *Tiktaalik roseae*.

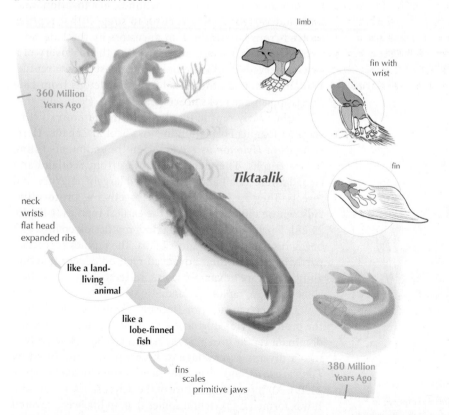

b Diagram illustrating how *Tiktaalik roseae* is a transitional species between lobe-finned fish and tetrapods.

functional wrist bones and five digits, as well as a modified ear region. Sedimentological evidence suggests *Tiktaalik roseae* lived in a shallow-water habitat associated with the Late Devonian floodplains of Laurasia.

Although the transition between fish and the earliest tetrapods is becoming better known with new fossil discoveries, footprints recently discovered from 395-million-year-old rocks in the Holy Cross Mountains in southeast Poland indicate that the lobe-finned fish to tetrapod transition might have taken place even earlier than previously thought. Based on the size of the footprints, it is estimated that the animal that left them was approximately 2.5 m long. If they are shown to be true tetrapod footprints, then the evolutionary transition from fish to amphibians will again have to be revised. Such a discovery once again illustrates how science works, and how theories have to be modified as new information becomes available and new discoveries are made.

As previously mentioned, the oldest known amphibian, *Ichthyostega*, had skeletal features that allowed it to spend its life on land. Because amphibians did not evolve until the Late Devonian, they were a minor element of the Devonian terrestrial ecosystem. Like other groups that moved into new and previously unoccupied niches, amphibians underwent rapid adaptive radiation and became abundant during the Carboniferous and Early Permian.

The Late Paleozoic amphibians did not at all resemble the familiar frogs, toads, newts, and salamanders that make up the modern amphibian fauna. Rather, they displayed a broad spectrum of sizes, shapes, and modes of life (▌Figure 21.26). One group of amphibians was the **labyrinthodonts**, so named for the labyrinthine wrinkling and folding of the chewing surface of their teeth (Figure 21.23). Most labyrinthodonts were large animals, as much as 2 m in length. These typically sluggish creatures lived in swamps and streams, eating fish, vegetation, insects, and other small amphibians (Figure 21.26).

Figure 21.26 Carboniferous Coal Swamp The amphibian fauna during the Carboniferous was varied. Shown is the serpent-like *Dolichosoma* (foreground) and the large labyrinthodont amphibian *Eryops*.

CHARLES R. KNIGHT/National Geographic Stock

>> **Critical Thinking Question**
What were the conditions that made the Late Carboniferous (Pennsylvanian) Period so favorable for the proliferation of amphibians and seedless vascular plants?

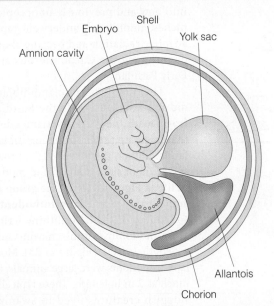

Figure 21.27 Amniote Egg In an amniote egg, the embryo is surrounded by a liquid sac (amnion cavity) and provided with a food source (yolk sac) and waste sac (allantois). The evolution of the amniote egg freed reptiles from having to return to the water for reproduction and let them inhabit all parts of the land.

Labyrinthodonts were abundant during the Carboniferous, when swampy conditions were widespread (see Chapter 20), but they soon declined in abundance during the Permian, perhaps in response to changing climatic conditions. Only a few species survived into the Triassic.

21.8 Evolution of the Reptiles—The Land Is Conquered

Amphibians were limited in colonizing the land because they had to return to water to lay their gelatinous eggs. The evolution of the **amniote egg** (∥ Figure 21.27) freed reptiles from this constraint. In such an egg, the developing embryo is surrounded by a liquid-filled sac called the *amnion* and provided with both a yolk, or food sac, and an allantois, or waste sac. In this way, the emerging reptile is, in essence, a miniature adult, bypassing the need for a larval stage in the water. The evolution of the amniote egg allowed vertebrates to colonize all parts of the land because they no longer had to return to the water as part of their reproductive cycle.

Many of the differences between amphibians and reptiles are physiologic and are not preserved in the fossil record. Nevertheless, amphibians and reptiles differ sufficiently in skull structure, jawbones, ear location, and limb and vertebral construction to suggest that reptiles evolved from labyrinthodont ancestors by the Late Mississippian. This assessment is based on the discovery of a well-preserved fossil skeleton of the oldest known reptile, *Westlothiana*, and other fossil reptile skeletons from Late Mississippian-age rocks in Scotland.

Other early reptile fossils occur in the Lower Pennsylvania Joggins Formation in Nova Scotia, Canada. Here, remains of *Hylonomus* are found in the sediments filling in tree trunks. These earliest reptiles from Scotland and Canada were small and agile and fed largely on grubs and insects. They are loosely grouped together as **protorothyrids**, whose members include the earliest reptiles (∥ Figure 21.28). During the Permian Period, reptiles diversified and began displacing many amphibians. The reptiles succeeded partly because of their advanced method of reproduction and their more advanced jaws and teeth, as well as their ability to move rapidly on land.

The **pelycosaurs**, or finback reptiles, evolved from the protorothyrids during the Pennsylvanian and were the dominant reptile group by the Early Permian. They evolved into a diverse assemblage of herbivores, exemplified by *Edaphosaurus*, and carnivores such as *Dimetrodon* (∥ Figure 21.29). An interesting feature of the pelycosaurs is their sail. It was formed by vertebral spines that, in life, were covered with skin. The sail has been variously explained as a type of sexual display, a means of protection, and a display to look more ferocious. The current consensus seems to be that the sail served as some type of thermoregulatory device, raising the reptile's temperature by catching the sun's rays or

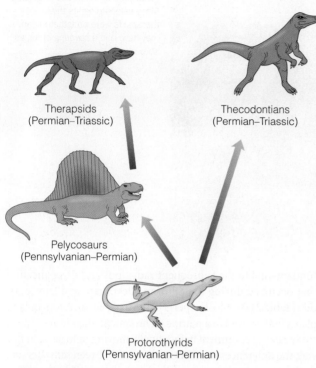

Therapsids
(Permian–Triassic)

Thecodontians
(Permian–Triassic)

Pelycosaurs
(Pennsylvanian–Permian)

Protorothyrids
(Pennsylvanian–Permian)

▌ Figure 21.28 Evolutionary Relationship Among the Paleozoic Reptiles

》 Critical Thinking Question
Although the transition from reptiles to mammals will not be discussed in detail until the next chapter, what is one characteristic in common between pelycosaurs and therapsids that indicates the pelycosaur to therapsid to mammal lineage is a correct evolutionary lineage?

cooling it by facing the wind. Because pelycosaurs are considered to be the group from which therapsids evolved, it is interesting that they may have had some sort of body-temperature control.

The pelycosaurs became extinct during the Permian and were succeeded by the **therapsids**, mammal-like reptiles that evolved from the carnivorous pelycosaur lineage and rapidly diversified into herbivorous and carnivorous lineages (▌ Figure 21.30). Therapsids were small- to medium-sized animals that displayed the beginnings of many mammalian features: fewer bones in the skull because many of the small skull bones were fused; enlarged lower jawbone; differentiation of teeth for various functions such as nipping, tearing, and chewing food; and more vertically placed legs for greater flexibility, as opposed to the way that the legs sprawled out to the side in primitive reptiles.

Furthermore, many paleontologists think that therapsids were *endothermic*, or warm-blooded, enabling them to maintain a constant internal body temperature. This characteristic would have let them expand into a variety of habitats, and indeed the Permian rocks in which their fossil remains are found are distributed not only in low latitudes but in middle and high latitudes as well.

As the Paleozoic Era came to an end, the therapsids constituted about 90% of the known reptile genera and occupied a wide range of ecological niches. The mass extinctions that decimated the marine fauna at the close of the Paleozoic had an equally great effect on the terrestrial population. By the end of the Permian, about 90% of all marine invertebrate species were extinct, compared with more than two-thirds of all amphibians and reptiles. Plants, in contrast, apparently did not experience as great a turnover as animals.

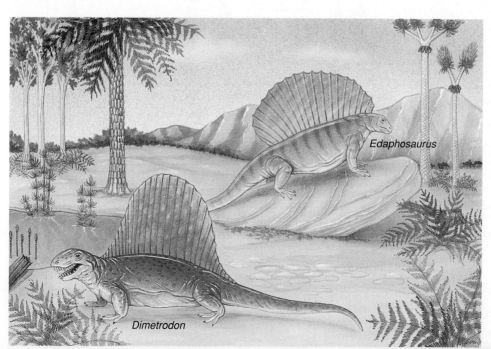

Edaphosaurus

Dimetrodon

▌ Figure 21.29 Pelycosaurs Most pelycosaurs or finback reptiles have a characteristic sail on their back. One hypothesis explains the sail as a rudimentary thermoregulatory device. Other hypotheses are that it was a type of sexual display or a device to make the reptile look more intimidating. Shown here are the carnivore *Dimetrodon* and the herbivore *Edaphosaurus*.

Figure 21.30 Therapsids A Late Permian scene in southern Africa showing various therapsids including *Dicynodon* and *Moschops*. Many paleontologists think therapsids were endothermic and may have had a covering of fur, as shown here.

Moschops

Dicynodon

21.9 Plant Evolution

When plants made the transition from water to land, they had to solve most of the same problems that animals did: desiccation, support, reproduction, and the effects of gravity. Plants did so by evolving a variety of structural adaptations that were

fundamental to the subsequent radiations and diversification that occurred during the Silurian, Devonian, and later periods (Table 21.3). Most experts agree that the ancestors of land plants first evolved in a marine environment, then moved into a freshwater environment before finally moving onto land. In this way, the differences in osmotic pressures between saltwater and freshwater were overcome while the plant was still in the water.

TABLE 21.3	Major Events in the Evolution of Land Plants. The Devonian Period was a time of rapid evolution for the land plants. Major events were the appearance of leaves, heterospory, secondary growth, and the emergence of seeds.

MYA	PERIOD	EPOCH	
2.3	Neogene		Flowering plants
66	Paleogene		
146	Cretaceous		
200	Jurassic		Cycads
251	Triassic		Conifer-type seed plants
299	Permian		Ferns
359	Carboniferous		Seed ferns Lycophytes
	Devonian	Upper	Arborescene Seeds Megaphyllous leaves / Progymnosperms
385			
398	Devonian	Middle	Secondary growth Zosterophyllodphytes
416		Lower	Major diversification of vascular plants Trimerophytes Heterospory Microphyllous leaves
419	Silurian	Pridoli	Rhyniophytes
423		Ludlow	
428		Wenlock	Tracheids Cooksonia
444		Llandovery	
	Ordovician	Upper	First land plants
488		Lower	

The higher land plants are divided into two major groups: nonvascular and vascular. Most land plants are **vascular**, meaning that they have a tissue system of specialized cells for the movement of water and nutrients. The *nonvascular* plants, such as bryophytes (liverworts, hornworts, and mosses) and fungi, do not have these specialized cells and are typically small and live in low, moist areas.

The earliest land plants from the Middle to Late Ordovician were probably small and bryophyte-like in their overall organization (but not necessarily related to bryophytes). The evolution of vascular tissue in plants was an important step because it allowed for the transport of food and water.

Discoveries of probable vascular plant megafossils and characteristic spores indicate to many paleontologists that vascular plants evolved well before the Middle Silurian. Sheets of cuticle-like cells—that is, the cells that cover the surface of present-day land plants—and tetrahedral clusters that closely resemble the spore tetrahedrals of primitive land plants have been reported from Middle to Upper Ordovician rocks from western Libya and elsewhere.

The ancestor of terrestrial vascular plants was probably some type of green alga. Although no fossil record of the transition from green algae to terrestrial vascular plants has been found, comparison of their physiology reveals a strong link. Primitive *seedless vascular plants* (discussed later in this chapter) such as ferns resemble green algae in their pigmentation, important metabolic enzymes, and type of reproductive cycle. Furthermore, green algae are one of the few plant groups to have made the transition from saltwater to freshwater.

The evolution of terrestrial vascular plants from an aquatic, probable green algal ancestry was accompanied by various modifications that let them occupy this new harsh environment. Besides the primary function of transporting water and nutrients throughout a plant, vascular tissue also provides some support for the plant body. Additional strength is derived from the organic compounds *lignin* and *cellulose*, found throughout a plant's walls.

The problem of desiccation was circumvented by the evolution of *cutin*, an organic compound found in the outer-wall layers of plants. Cutin also provides additional resistance to oxidation, the effects of ultraviolet light, and the entry of parasites.

Roots evolved in response to the need to collect water and nutrients from the soil and to help anchor the plant in the ground. The evolution of *leaves* from tiny outgrowths on the stem or from branch systems provided plants with an efficient light-gathering system for photosynthesis.

Silurian and Devonian Floras

The earliest known vascular land plants are small, Y-shaped stems assigned to the genus *Cooksonia* from the Middle Silurian of Wales and Ireland. Together with Upper Silurian and Lower Devonian species from Scotland, New York State, and the Czech Republic, these earliest plants were small, simple, leafless stalks with a spore-producing structure at the tip (▌ Figure 21.31); they are known as **seedless vascular plants** because they did not produce seeds. They also did not have a true root system. A *rhizome*, the underground part

Courtesy of Dianne Edwards, University of Wales College

▌ **Figure 21.31** *Cooksonia* The earliest known fertile land plant was *Cooksonia*, seen in this fossil from the Upper Silurian of South Wales. *Cooksonia* consisted of upright, branched stems terminating in sporangia (spore-producing structures). It also had a resistant cuticle and produced spores typical of a vascular plant. These plants probably lived in moist environments such as mudflats. This specimen is 1.49 cm long.

of the stem, transferred water from the soil to the plant and anchored the plant to the ground. The sedimentary rocks in which these plant fossils are found indicate that they lived in low, wet, marshy, freshwater environments.

An interesting parallel can be seen between seedless vascular plants and amphibians. When they made the transition from water to land, both plants and animals had to overcome the problems that such a transition involved. Both groups, while successful, nevertheless required a source of water in order to reproduce. In the case of amphibians, their gelatinous egg had to remain moist, and the seedless vascular plants required water for the sperm to travel through to reach the egg.

From this simple beginning, the seedless vascular plants evolved many of the major structural features characteristic of modern plants, such as leaves, roots, and secondary growth. These features did not all evolve simultaneously, but rather at different times, a pattern known as *mosaic evolution*. This diversification and adaptive radiation took place during the Late Silurian and Early Devonian and resulted in a tremendous increase in diversity (▌ Figure 21.32). During the Devonian, the number of plant genera remained about the same, yet the composition of the flora changed. Whereas the Early Devonian landscape was dominated by relatively small, low-growing, bog-dwelling types of plants, the Late Devonian witnessed forests of large, tree-sized plants up to 10 m tall.

In addition to the diverse seedless vascular plant flora of the Late Devonian, another significant floral event took

Figure 21.32 Early Devonian Landscape Reconstruction of an Early Devonian landscape showing some of the earliest land plants.

place. The evolution of the seed at this time liberated land plants from their dependence on moist conditions and allowed them to spread over all parts of the land.

Seedless vascular plants require moisture for successful fertilization because the sperm must travel to the egg on the surface of the gamete-bearing plant (gametophyte) to produce a successful spore-generating plant (sporophyte). Without moisture, the sperm would dry out before reaching the egg (Figure 21.33a). In the seed method of reproduction, the spores are not released to the environment, as they are in the seedless vascular plants, but are retained on the spore-bearing plant, where they grow into the male and female forms of the gamete-bearing generation.

In the case of the **gymnosperms**, or flowerless seed plants, a single plant has both male and female cones (Figure 21.33b). The male cone produces pollen, which contains the sperm and has a waxy coating to prevent desiccation, and the egg, or embryonic seed, is contained in the female cone. After fertilization, the seed develops into a mature, cone-bearing plant. In this way, the need for a moist environment for the gametophyte generation is solved. The significance of this development is that seed plants, like reptiles, were no longer restricted to wet areas, but were free to migrate into previously unoccupied dry environments.

Although the seedless vascular plants dominated the flora of the Carboniferous coal-forming swamps, the gymnosperms made up an important element of the Late Paleozoic flora, particularly in the nonswampy areas.

Late Carboniferous and Permian Floras

As discussed earlier, the rocks of the Pennsylvanian Period (Late Carboniferous) are the major source of the world's coal. Coal results from the alteration of plant remains accumulating in low, swampy areas. The geologic and geographic conditions of the Pennsylvanian were ideal for the growth of seedless vascular plants, and consequently, these coal swamps had a very diverse flora (Figure 21.34).

> **CONNECTION LINK** ◀
>
> To review the formation of coal and its distribution, see page 165 in Chapter 7 and pages 534 and 535 in Chapter 20.

It is evident from the fossil record that whereas the Early Carboniferous flora was similar to its Late Devonian counterpart, a great deal of evolutionary experimentation was taking place that would lead to the highly successful Late Paleozoic flora of the coal swamps and adjacent habitats. Among the seedless vascular plants, the *lycopsids* and *sphenopsids* were the most important coal-forming groups of the Pennsylvanian Period.

The lycopsids were present during the Devonian, chiefly as small plants, but by the Pennsylvanian, they were the dominant element of the coal swamps, achieving heights up to 30 m in such genera as *Lepidodendron* and

▌ Figure 21.33 Generalized Life History of a Seedless Vascular Plant and Gymnosperm Plant

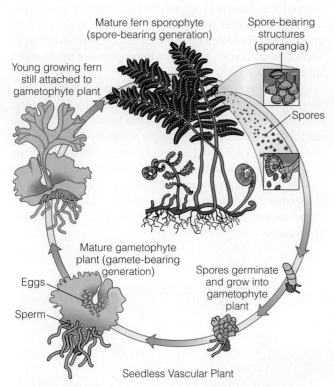

Young growing fern still attached to gametophyte plant

Mature fern sporophyte (spore-bearing generation)

Spore-bearing structures (sporangia)

Spores

Mature gametophyte plant (gamete-bearing generation)

Spores germinate and grow into gametophyte plant

Eggs

Sperm

Seedless Vascular Plant

a Generalized life history of a seedless vascular plant. The mature sporophyte plant produces spores, which, upon germination, grow into small gametophyte plants that produce sperm and eggs. The fertilized eggs grow into the spore-producing mature plant, and the sporophyte–gametophyte life cycle begins again.

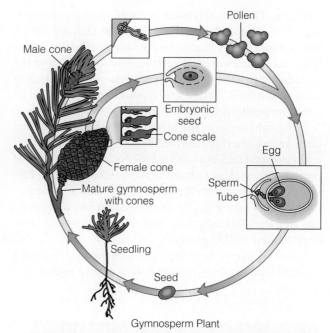

Male cone

Pollen

Embryonic seed

Cone scale

Female cone

Mature gymnosperm with cones

Egg

Sperm

Tube

Seedling

Seed

Gymnosperm Plant

b Generalized life history of a gymnosperm plant. The mature plant bears both male cones that produce sperm-bearing pollen grains and female cones that contain embryonic seeds. Pollen grains are transported to the female cones by the wind. Fertilization occurs when the sperm moves through a moist tube growing from the pollen grain and unites with the embryonic seed, which then grows into a cone-bearing mature plant.

》 Critical Thinking Question
What is it about the life cycle of the gymnosperms that gave them an advantage over the seedless vascular plants in terms of the climatic changes taking place during the Permian Period, when the continents came together to form the supercontinent Pangaea, and the subsequent Triassic Period?

▌ Figure 21.34 Pennsylvanian Coal Swamp Reconstruction of a Pennsylvanian coal swamp with its characteristic vegetation of seedless vascular plants.

Sigillaria. The Pennsylvanian lycopsid trees are interesting because they lacked branches except at their top, which had elongate leaves similar to the individual palm leaf of today. As the trees grew, the leaves were replaced from the top, leaving prominent and characteristic rows or spirals of scars on the trunk. Today, the lycopsids are represented by small, temperate-forest ground pines.

The sphenopsids, the other important coal-forming plant group, are characterized by being jointed and having horizontal underground stem-bearing roots. Many of these plants, such as *Calamites*, average 5 to 6 m tall. Living sphenopsids include the horsetail (*Equisetum*) or scouring rushes (❙ Figure 21.35). Small seedless vascular plants and seed ferns formed a thick undergrowth or ground cover beneath these treelike plants.

Not all plants were restricted to the coal-forming swamps. Among those plants that occupied higher and drier ground were some of the *cordaites*, a group of tall gymnosperm trees that grew up to 50 m high and probably formed vast forests (❙ Figure 21.36). Another important nonswamp dweller was *Glossopteris*, the famous plant so abundant in Gondwana (see Figure 2.1), whose distribution is cited as critical evidence that the continents have moved through time.

The floras that were abundant during the Pennsylvanian persisted into the Permian, but because of climatic and geologic changes resulting from tectonic events, they declined in abundance and importance. By the end of the Permian, the cordaites became extinct, and the lycopsids and sphenopsids were reduced to mostly small, creeping forms. Gymnosperms with lifestyles more suited to the warmer and drier Permian climates diversified and came to dominate the Permian, Triassic, and Jurassic landscapes.

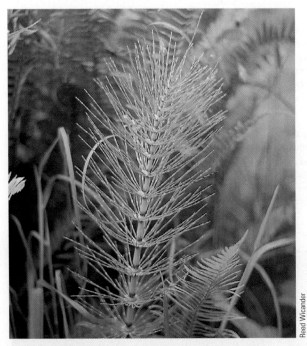

❙ **Figure 21.35 Horsetail *Equisetum*** Living sphenopsids include the horsetail *Equisetum*.

❙ **Figure 21.36 Late Carboniferous Cordaite Forest** Cordaites were a group of gymnosperm trees that grew up to 50 m tall.

GEO-RECAP

» Geo-Image

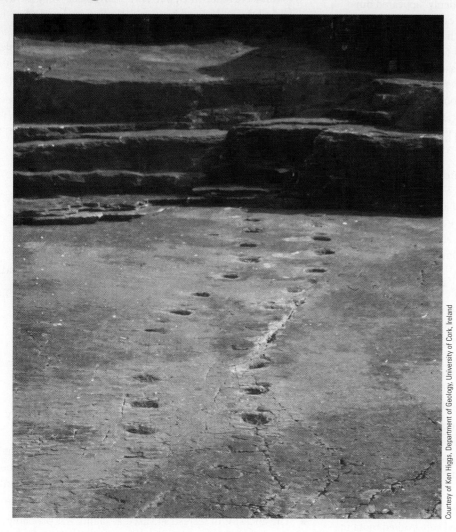

Courtesy of Ken Higgs, Department of Geology, University of Cork, Ireland

Figure 1 These fossilized footprints, which are more than 365 million years old, are part of a tetrapod trackway at Valentia Island, Ireland, and until recently represented evidence of some of the oldest land-dwelling tetrapods. A recent discovery of 395-million-year-old fossilized footprints in the Holy Cross Mountains of Poland indicate tetrapods may have evolved as much as 30 million years earlier, although there is not unanimity that the Poland specimens are indeed tetrapod footprints. The purported footprints in Poland were found in carbonate rocks suggesting an intertidal area, possibly a lagoon. What evidence in this photo indicates the sediments were deposited in a wet environment, and how can you use the footprints to establish that the animal making them was a tetrapod? Also, how do you think scientists estimated the size of the tetrapod that made this trackway based on this outcrop?

Chapter Summary

- Table 21.4 (page 571) summarizes the major evolutionary and geologic events of the Paleozoic Era and shows their relationships to one another.

- Multicelled organisms presumably had a long Precambrian history, during which they lacked hard parts.

- Invertebrates with hard parts "suddenly" appeared during the Early Cambrian in what is called the "Cambrian explosion." Skeletons provided such advantages as protection against predators and support for muscles, enabling organisms to grow large and increase locomotor efficiency. Hard parts probably evolved as a result of various geologic and biologic factors rather than from a single cause.

- Marine organisms are classified as plankton if they are floaters, nekton if they swim, and benthos if they live on or in the seafloor.

- Marine organisms are divided into four basic feeding groups: suspension feeders, which consume microscopic plants and animals as well as dissolved nutrients from water; herbivores, which are plant eaters; carnivore-scavengers, which are meat eaters; and sediment-deposit feeders, which ingest sediment and extract nutrients from it.

- The marine ecosystem consists of various trophic levels of food production and consumption. At the base are primary producers, on which all other organisms are dependent. Feeding on the primary producers are the primary consumers, which in turn are fed on by higher levels of consumers. The decomposers are bacteria that break down the complex organic compounds of dead organisms and recycle them within the ecosystem.

- The Cambrian invertebrate community was dominated by three major groups: the trilobites, brachiopods, and archaeocyathids. Little specialization existed among the invertebrates, and most phyla were represented by only a few species.

- The Middle Cambrian Burgess Shale contains one of the finest examples of a well-preserved soft-bodied biota in the world and provides us with an important glimpse of not only rarely preserved organisms but the soft-part anatomy of many extinct groups.

- The Ordovician marine invertebrate community marked the beginning of the dominance of the shelly fauna and the start of large-scale reef building. The end of the Ordovician Period was a time of major extinctions of many invertebrate phyla.

- The Silurian and Devonian periods were times of diverse faunas dominated by reef-building animals, whereas the Carboniferous and Permian periods saw a great decline in invertebrate diversity.

- Chordates are characterized by a notochord, dorsal hollow nerve cord, and pharyngeal slits. The earliest chordates were soft-bodied organisms that were rarely fossilized. Vertebrates are a subphylum of the chordates.

- Fish are the earliest known vertebrates, with their first fossil occurrence in Upper Cambrian rocks. They have had a long and varied history, including jawless and jawed armored forms (ostracoderms and placoderms), cartilaginous forms, and bony forms. It is from the lobe-finned bony fish that amphibians evolved.

- The link between crossopterygian lobe-finned fish and the earliest amphibians is convincing and includes a close similarity of bone and tooth structures. However, new fossil discoveries show that the transition between the two groups is more complicated than originally hypothesized and includes several intermediate forms.

- Amphibians evolved during the Late Devonian, with labyrinthodont amphibians becoming the dominant terrestrial vertebrate animals during the Carboniferous.

- The Late Mississippian marks the earliest fossil record of reptiles. The evolution of an amniote egg was the critical factor that allowed reptiles to completely colonize the land.

- Pelycosaurs were the dominant reptiles during the Early Permian, whereas therapsids dominated the landscape for the rest of the Permian Period.

- In making the transition from water to land, plants had to overcome the same basic problems as animals—namely, desiccation, reproduction, and gravity.

- The earliest fossil record of land plants is from Middle to Upper Ordovician rocks. These plants were probably small and bryophyte-like in their overall organization.

- The evolution of vascular tissue was an important event in plant evolution as it allowed nutrients and water to be transported throughout the plant and provided the plant with additional support.

- The ancestor of terrestrial vascular plants was probably some type of green algae based on such similarities as pigmentation, metabolic enzymes, and the same type of reproductive cycle.

- The earliest seedless vascular plants were small, leafless stalks, with spore-producing structures on their tips. From this simple beginning, plants evolved many of the major structural features characteristic of today's plants.

- By the end of the Devonian Period, forests with tree-sized plants up to 10 m tall had evolved. The Late Devonian also witnessed the evolution of flowerless seed plants (gymnosperms) whose reproductive style freed them from having to stay near water.

■ The Carboniferous Period was a time of vast coal swamps, where conditions were ideal for the seedless vascular plants. With the onset of more arid conditions during the Permian, the gymnosperms became the dominant element of the world's forests.

■ A major extinction occurred at the end of the Paleozoic Era, affecting the invertebrates as well as the vertebrates. Its cause is still the subject of debate.

TABLE 21.4 Major Evolutionary and Geologic Events of the Paleozoic Era

Age (millions of years ago)	Geologic Period		Invertebrates	Vertebrates	Plants	Major Geologic Events
251 –	Permian		Largest mass extinction to affect the invertebrates	Acanthodians, placoderms, and pelycosaurs become extinct / Therapsids and pelycosaurs are the most abundant reptiles	Gymnosperms diverse and abundant	Formation of Pangaea / Alleghenian orogeny / Hercynian orogeny
299 –	Carboniferous	Pennsylvanian	Fusulinids diversify	Amphibians abundant and diverse	Coal swamps with flora of seedless vascular plants and gymnosperms	Coal-forming swamps common / Formation of Ancestral Rockies / Continental glaciation in Gondwana
318 –	Carboniferous	Mississippian	Crinoids, lacy bryozoans, blastoids become abundant / Renewed adaptive radiation following extinctions of many reef-builders	Reptiles evolve	Gymnosperms appear (may have evolved during Late Devonian)	Ouachita orogeny / Widespread deposition of black shale
359 –	Devonian		Extinctions of many reef-building invertebrates near end of Devonian / Reef building continues / Eurypterids abundant	Amphibians evolve / All major groups of fish present—Age of Fish	First seeds evolve / Seedless vascular plants diversify	Widespread deposition of black shale / Acadian orogeny / Antler orogeny
416 –	Silurian		Major reef building / Diversity of invertebrates remains high	Ostracoderms common / Acanthodians, the first jawed fish evolve	Early land plants— seedless vascular plants	Caledonian orogeny / Extensive barrier reefs and evaporites
444 –	Ordovician		Extinctions of a variety of marine invertebrates near end of Ordovician / Major adaptive radiation of all invertebrate groups / Suspension-feeders dominant	Ostracoderms diversify	Plants move to land?	Continental glaciation in Gondwana / Taconic orogeny
488 –	Cambrian		Many trilobites become extinct near end of Cambrian / Trilobites, brachiopods, and archaeocyathids are most abundant	Earliest vertebrates— jawless fish called ostracoderms		First Phanerozoic transgression (Sauk) onto North American craton
542 –						

Important Terms

amniote egg (p. 562)
benthos (p. 545)
bony fish (p. 557)
carnivore-scavenger (p. 545)
cartilaginous fish (p. 557)
chordate (p. 554)
crossopterygian (p. 558)
gymnosperm (p. 566)

herbivore (p. 545)
labyrinthodont (p. 561)
nekton (p. 544)
ostracoderm (p. 555)
pelycosaur (p. 562)
placoderm (p. 557)
plankton (p. 544)
primary producer (p. 545)

protorothyrids (p. 562)
sediment-deposit feeder (p. 545)
seedless vascular plant (p. 565)
suspension feeder (p. 545)
therapsid (p. 563)
vascular (p. 565)
vertebrate (p. 554)

Review Questions

1. An exoskeleton is advantageous because it
   a. ___ prevents drying out in an intertidal environment.
   b. ___ provides protection against ultraviolet radiation.
   c. ___ provides protection against predators.
   d. ___ provides attachment sites for the development of strong muscles.
   e. ___ all of these.

2. Which was the first plant group that did not require a wet area for the reproductive part of its life cycle?
   a. ___ seedless vascular.
   b. ___ naked seedless.
   c. ___ gymnosperms.
   d. ___ angiosperms.
   e. ___ flowering.

3. An organism must possess which of the following during at least part of its life cycle to be classified a chordate?
   a. ___ notochord, dorsal solid nerve cord, lungs.
   b. ___ vertebrae, dorsal hollow nerve cord, pharyngeal slits.
   c. ___ vertebrae, dorsal hollow nerve cord, lungs.
   d. ___ notochord, ventral solid nerve cord, lungs.
   e. ___ notochord, dorsal hollow nerve cord, pharyngeal slits.

4. Labyrinthodonts are
   a. ___ plants.
   b. ___ fish.
   c. ___ amphibians.
   d. ___ reptiles.
   e. ___ none of these.

5. The greatest recorded mass extinction to affect Earth occurred at the end of which period?
   a. ___ Cambrian.
   b. ___ Ordovician.
   c. ___ Devonian.

   d. ___ Permian.
   e. ___ Cretaceous.

6. Which reptile group gave rise to the mammals?
   a. ___ labyrinthodonts.
   b. ___ acanthodians.
   c. ___ pelycosaurs.
   d. ___ protothyrids.
   e. ___ therapsids.

7. Pelycosaurs are
   a. ___ jawless fish.
   b. ___ jawed, armored fish.
   c. ___ reptiles.
   d. ___ amphibians.
   e. ___ plants.

8. Which evolutionary innovation allowed reptiles to colonize all of the land?
   a. ___ tear ducts.
   b. ___ additional bones in the jaw.
   c. ___ the middle-ear bone.
   d. ___ an egg that contained food and waste sacs and an embryo surrounded by a fluid-filled sac.
   e. ___ limbs and a backbone capable of supporting the animals on land.

9. The oldest and most primitive of the fish are
   a. ___ acanthodians.
   b. ___ cartilaginous.
   c. ___ placoderms.
   d. ___ ostracoderms.
   e. ___ bony.

10. The discovery of *Tiktaalik roseae* is significant because it is
   a. ___ the ancestor of modern reptiles.
   b. ___ an intermediary between lobe-finned fish and amphibians.
   c. ___ the first vascular land plant.
   d. ___ a "missing link" between amphibians and reptiles.
   e. ___ the oldest known fish.

11. If the Cambrian explosion of life was partly the result of filling unoccupied niches, why don't we see such rapid evolution following mass extinctions such as those that occurred at the end of the Permian and Cretaceous periods?

12. What are the major differences between the seedless vascular plants and the gymnosperms, and why are these differences significant in terms of exploiting the terrestrial environment?

13. Discuss how the incompleteness of the fossil record may play a role in such theories as the Cambrian explosion of life.

14. Discuss how changing geologic conditions affected the evolution of invertebrate life during the Paleozoic Era.

15. Discuss some of the possible causes for the Permian mass extinction.

16. Describe the problems that had to be overcome before organisms could inhabit and completely colonize the land.

17. Discuss the significance and possible advantages of the pelycosaur sail.

18. Based on what you know about Pennsylvanian geology (Chapter 20), why was this time period so advantageous to the evolution of both plants and amphibians?

19. Outline the evolutionary history of fish.

20. Why were the reptiles so much more successful at extending their habitat than the amphibians?

View of the Sierra Nevada taken west of Lone Pine, California. The rocks in this image are part of the Sierra Nevada batholith, a huge mass of granite and related rocks made up of many intrusive bodies. The peak toward the right is Mount Whitney, which at 4,421 m is the highest peak in the continental United States.

CHAPTER 22

Sue Monroe

Mesozoic Earth and Life History

OUTLINE

22.1 Introduction

22.2 The Breakup of Pangaea

22.3 Mesozoic History of North America

22.4 What Role Did Accretion of Terranes Play in the Growth of Western North America?

22.5 Mesozoic Mineral Resources

22.6 Life of the Mesozoic Era

GEO-INSIGHT: Dinosaurs

22.7 Mesozoic Climates and Paleogeography

GEO-FOCUS: Mary Anning's Contributions to Paleontology

22.8 Mass Extinctions—A Crisis in the History of Life

GEO-RECAP

HAVE YOU EVER WONDERED?

- What caused the breakup of Pangaea and its effects on the geography, oceanic circulation patterns, climate, and distribution of natural resources during the Mesozoic Era?

- How land-plant communities changed during the Mesozoic Era?

- About the nature of dinosaurs and their relatives such as flying reptiles and marine reptiles?

- When birds first appear in the fossil record?

- How the earliest mammals compare to those living today?

- What happened at the end of the Mesozoic to cause widespread extinctions?

22.1 Introduction

Approximately 150 to 210 million years after the emplacement of massive plutons created the Sierra Nevada (Nevadan orogeny), gold was discovered at Sutter's Mill on the South Fork of the American River at Coloma, California. On January 24, 1848, James Marshall, a carpenter building a sawmill for John Sutter, found bits of the glittering metal in the mill's tailrace. Soon settlements throughout the state were abandoned as word of the chance for instant riches spread throughout California.

Within a year after the news of the gold discovery reached the East Coast, the Sutter's Mill area was swarming with more than 80,000 prospectors, all hoping to make their fortune. At least 250,000 gold seekers prospected the Sutter's Mill area, and although most were Americans, prospectors came from all over the world, even as far away as China.

Although many prospectors searched for the source of the gold, or the mother lode, the gold that they recovered was mostly in the form of placer deposits (deposits of sand and gravel containing gold particles large enough to be recovered by panning). Unfortunately, most thought the gold was simply waiting to be taken and didn't realize that prospecting was hard work. In fact, the shop owners and traders frequently made more money than the prospectors, most of whom barely made a living wage working their claims. Nevertheless, during the five years from 1848 to 1853 that constituted the gold rush proper, more than $200 million in gold was extracted.

The Mesozoic Era (251 to 66 million years ago) was an important time in Earth's history. The major geologic event was the breakup of Pangaea. As a result of plate movement during the Mesozoic, the Atlantic Ocean basin formed, as well as the initial stages in the growth of the Rocky Mountains. The accumulation of vast salt deposits eventually formed salt domes adjacent to which oil and natural gas were trapped, and the emplacement of huge batholiths account for the origin of various mineral resources, including the gold that fueled the California gold rush of the mid-1800s.

The Mesozoic Era is probably best known, however, for dinosaurs that evolved during the Late Triassic and went extinct at the end of the Cretaceous Period. In addition, the Mesozoic was also the time when flying reptiles (pterosaurs), several types of marine reptiles (ichthyosaurs, plesiosaurs, and mosasaurs), and huge crocodiles proliferated. Also present were turtles, lizards, and snakes, although the fossil record for the last two groups is not so good. Birds also first appeared during the Jurassic Period, having evolved from small carnivorous dinosaurs. And mammals evolved from mammal-like reptiles during the Triassic. In fact, mammals were contemporaries with dinosaurs, although they were not nearly as diverse and all were small.

One of the main emphases in this book has been the systems approach to Earth and life history. Remember that just as the formation of Pangaea influenced geologic and biologic events during the Paleozoic, the breakup of this supercontinent had profound geologic and biologic effects during the Mesozoic. The movement of continents affected the global climatic and oceanic regimes, as well as the climates of the individual continents. Populations became isolated or were brought into contact with other populations, leading to evolutionary changes in the biota. So great was the effect of this breakup on the world that it forms the central theme of this chapter. The distribution of land and sea has a profound influence on oceanic circulation, which in turn partly controls climate, and the proximity of continents partly determines the geographic distribution of organisms.

> **CONNECTION LINK** ‹
>
> To review how plate movements affect the distribution of life, see pages 53 and 54 in Chapter 2.

22.2 The Breakup of Pangaea

Pangaea's breakup began with rifting between Laurasia and Gondwana during the Triassic (❚ Figure 22.1a). By the end of the Triassic, the newly formed and expanding Atlantic Ocean separated North America from Africa. This was followed by the rifting of North America from South America, sometime during the Late Triassic and Early Jurassic.

Separation of the continents allowed water from the Tethys Sea to flow into the expanding central Atlantic Ocean, whereas Pacific Ocean waters flowed into the newly formed Gulf of Mexico, which at that time was little more than a restricted bay (❚ Figure 22.2). During that time, these areas were located in the low tropical latitudes, where high temperatures and high rates of evaporation were ideal for the formation of thick evaporite deposits.

During the Late Triassic and Jurassic periods, Antarctica and Australia, which remained sutured together, began separating from South America and Africa, and India began rifting from the Gondwana continent and moved northward.

South America and Africa began rifting apart during the Jurassic (❚ Figure 22.1b), and the subsequent separation of these two continents formed a narrow basin where thick evaporite deposits accumulated from the evaporation of southern ocean waters (Figure 22.2). During this time, the eastern end of the Tethys Sea began closing as a result of the clockwise rotation of Laurasia and the northward movement of Africa. This narrow Late Jurassic and Cretaceous seaway between Africa and Europe was the forerunner of the present Mediterranean Sea.

By the end of the Cretaceous, Australia and Antarctica had detached from each other, and India had moved into the low southern latitudes and was nearly to the equator. South America and Africa were widely separated, and Greenland was essentially an independent landmass with only a shallow sea between it and North America and Europe (❚ Figure 22.1c).

A global rise in sea level during the Cretaceous resulted in worldwide transgressions onto the continents. Higher heat

Figure 22.1 Mesozoic Paleogeography Paleogeography of the world during the **(a)** Triassic Period, **(b)** Jurassic Period, and **(c)** Late Cretaceous Period.

a Triassic Period.

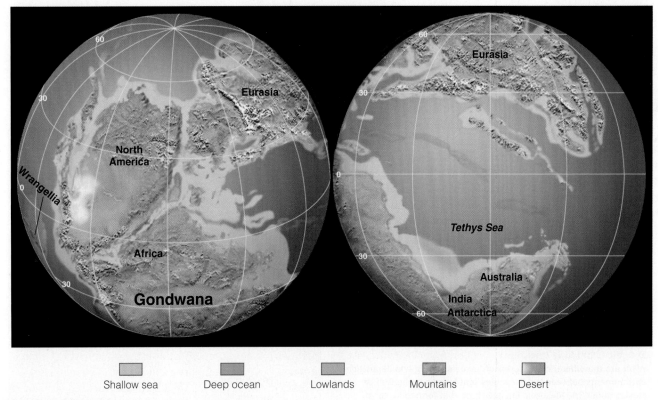

b Jurassic Period.

Figure 22.1 *(Continued)*

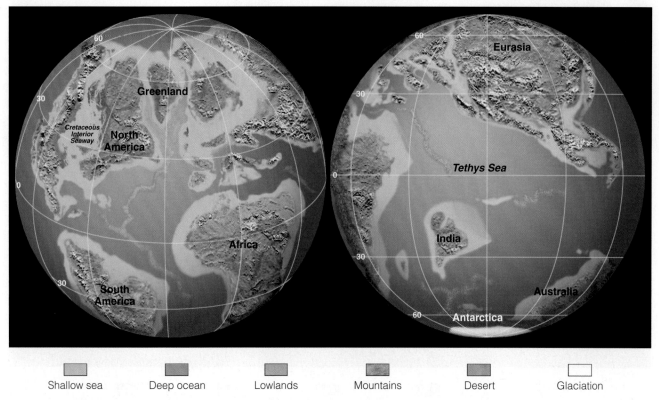

| Shallow sea | Deep ocean | Lowlands | Mountains | Desert | Glaciation |

c Late Cretaceous Period.

>> **Critical Thinking Question**
Why would increased seafloor spreading, resulting in higher heat flow and an increase and rapid
expansion in oceanic ridges during the Cretaceous cause worldwide transgressions onto the continents?

flow and rapid expansion of oceanic ridges was responsible
for these transgressions. By the Middle Cretaceous, sea level
was probably as high as at any time since the Ordovician, and
about one-third of the pres-
ent land area was inundated by
epeiric seas (Figure 22.1c).

The final stage in Pan-
gaea's breakup occurred during the Cenozoic. During this
time, Australia continued moving northward, and Greenland
was completely separated from
Europe and North America and
formed a separate landmass.

> **CONNECTION LINK**

To learn more about plate
tectonics and the paleo-
geography of Earth during
the Cenozoic, go to pages
615 and 616 and Figure
23.2 in Chapter 23.

The Effects of the Breakup of Pangaea on Global Climates and Ocean Circulation Patterns

By the end of the Permian Period, Pangaea extended from
pole to pole, covered about one-fourth of Earth's surface, and

Figure 22.2 Gulf of Mexico Evaporites Evaporites accumulated in
shallow basins as Pangaea broke apart during the Early Mesozoic. Water
from the Tethys Sea flowed into the central Atlantic Ocean, and water from
the Pacific Ocean flowed into the newly formed Gulf of Mexico. Marine
water from the south flowed into the southern Atlantic Ocean

>> **Critical Thinking Question**
What are the differences in terms of tectonic setting and depositional
environments between the evaporites that formed in the Gulf of
Mexico during the Mesozoic Era and those that formed in North
America during the Paleozoic Era?

was surrounded by Panthalassa, a global ocean that encompassed approximately 300 degrees of longitude (see Figure 20.4b). Such a configuration exerted tremendous influence on the world's climate and resulted in generally arid conditions over large parts of Pangaea's interior.

The world's climates result from the complex interaction between wind and ocean currents and the location and topography of the continents. In general, dry climates occur on large landmasses in areas remote from sources of moisture and where barriers to moist air exist, such as mountain ranges. Wet climates occur near large bodies of water or where winds can carry moist air over land.

Past climatic conditions can be inferred from the distribution of climate-sensitive deposits. Evaporite deposits result when evaporation exceeds precipitation. Although sand dunes and red beds may form locally in humid regions, they are characteristic of arid regions. Coal forms in both warm and cool humid climates. Vegetation that is eventually converted into coal requires at least a good seasonal water supply; thus, coal deposits are indicative of humid conditions.

Widespread Triassic evaporites, red beds, and desert dunes in the low and middle latitudes of North and South America, Europe, and Africa indicate dry climates in those regions, whereas coal deposits are found mainly in the high latitudes, indicating humid conditions (Figure 22.1a). These high-latitude coals are analogous to today's Scottish peat bogs or Canadian muskeg. The lands bordering the Tethys Sea were probably dominated by seasonal monsoon rains resulting from the warm, moist winds and warm oceanic currents impinging against the east-facing coast of Pangaea.

The temperature gradient between the tropics and the poles also affects oceanic and atmospheric circulation. The greater the temperature difference between the tropics and the poles, the steeper the temperature gradient, and thus, the faster the circulation of the oceans and atmosphere. Oceans absorb about 90% of the solar radiation they receive, whereas continents absorb only about 50%, even less if they are snow covered. The rest of the solar radiation is reflected back into space. Areas dominated by seas are warmer than those dominated by continents. By knowing the distribution of continents and ocean basins, geologists can generally estimate the average annual temperature for any region on Earth, as well as determine a temperature gradient.

The breakup of Pangaea during the Late Triassic caused the global temperature gradient to increase because the Northern Hemisphere continents moved farther northward, displacing higher-latitude ocean waters. Because of the steeper global temperature gradient produced by a decrease in temperature in the high latitudes and the changing positions of the continents, oceanic and atmospheric circulation patterns greatly accelerated during the Mesozoic (❙ Figure 22.3).

❙ **Figure 22.3** **Mesozoic Oceanic Circulation Patterns** Oceanic circulation evolved from **(a)** a simple pattern in a single ocean (Panthalassa) with a single continent (Pangaea) to **(b)** a more complex pattern in the newly formed oceans of the Cretaceous Period.

a Triassic Period.

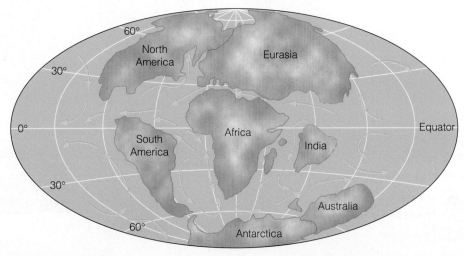

b Cretaceous Period.

Although the temperature gradient and seasonality on land were increasing during the Jurassic and Cretaceous, the middle- and higher-latitude oceans were still quite warm because warm waters from the Tethys Sea were circulating to the higher latitudes. The result was a relatively equable worldwide climate through the end of the Cretaceous.

22.3 Mesozoic History of North America

The beginning of the Mesozoic Era was essentially the same in terms of tectonism and sedimentation as the preceding Permian Period in North America (see Figure 20.20).

Terrestrial sedimentation continued over much of the craton, and block faulting and igneous activity began in the Appalachian region as North America and Africa began separating (▌Figure 22.4). The newly forming Gulf of Mexico experienced extensive evaporite deposition during the Late Triassic and Jurassic as North America separated from South America (Figures 22.2 and ▌22.5).

A global rise in sea level during the Cretaceous resulted in worldwide transgressions onto the continents such that marine deposition was continuous over much of the North America Cordilleran (▌Figure 22.6).

A volcanic island arc system that formed off the western edge of the craton during the Permian was sutured to North America sometime later during the Permian or Triassic. This event is referred to as the *Sonoma orogeny* and will be discussed later in the chapter. During the Jurassic,

▌**Figure 22.4** Triassic Paleogeography of North America

Figure 22.5 Jurassic Paleogeography of North America

the entire Cordilleran area was involved in a series of major mountain-building episodes resulting in the formation of the Sierra Nevada, the Rocky Mountains, and other lesser mountain ranges. Although each orogenic episode has its own name, the entire mountain-building event is simply called the *Cordilleran orogeny* (also discussed later in this chapter). With this simplified overview of the Mesozoic history of North America in mind, we will now examine the specific regions of the continent.

Continental Interior

Recall that the history of the North American craton is divided into unconformity-bound sequences reflecting advances and retreats of epeiric seas over the craton (see Figure 20.5). Although these transgressions and regressions

played a major role in the Paleozoic geologic history of the continent, they were not as important during the Mesozoic. Most of the continental interior during the Mesozoic was well above sea level and was not inundated by epeiric seas. Consequently, the two Mesozoic cratonic sequences, the *Absaroka Sequence* (Late Mississippian to Early Jurassic) and the *Zuni Sequence* (Early Jurassic to Early Paleocene; see Figure 20.5), are not treated separately here; instead, we will examine the Mesozoic history of the three continental margin regions of North America.

Eastern Coastal Region

During the Early and Middle Triassic, coarse detrital sediments derived from erosion of the recently uplifted Appalachians (Alleghenian orogeny) filled the various intermontane

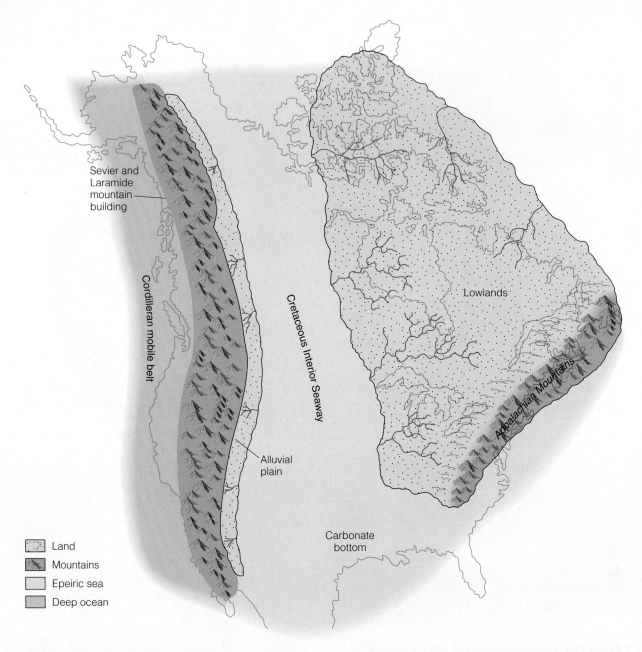

Sevier and
Laramide
mountain
building

Cordilleran mobile belt

Cretaceous Interior Seaway

Alluvial
plain

Lowlands

Appalachian Mountains

Land

Mountains

Epeiric sea

Deep ocean

Carbonate
bottom

▌**Figure 22.6** Cretaceous Paleogeography of North America

basins and spread over the surrounding areas. As weathering and erosion continued during the Mesozoic, this once-lofty mountain system was reduced to a low-lying plain.

During the Late Triassic, the first stage in the breakup of Pangaea began with North America separating from Africa. Fault-block basins developed in response to upwelling magma beneath Pangaea in a zone stretching from present-day Nova Scotia to North Carolina (▌Figure 22.7). Erosion of the fault-block mountains filled the adjacent basins with great quantities (up to 6,000 m) of poorly sorted red, nonmarine detrital sediments known as the *Newark Group*.

Reptiles roamed along the margins of the lakes and streams that formed in these basins, leaving their footprints

and trackways in the soft sediments. Although the Newark rocks contain numerous dinosaur footprints, they are almost completely devoid of dinosaur bones! The Newark Group is mostly Late Triassic in age, but in some areas deposition did not begin until the Early Jurassic.

Concurrent with sedimentation in the fault-block basins were extensive lava flows that blanketed the basin floors, as well as intrusions of numerous dikes and sills. The most famous intrusion is the prominent Palisades sill along the Hudson River in the New York–New Jersey area (Figure 22.7d).

As the Atlantic Ocean grew, rifting ceased along the eastern margin of North America, and this once-active convergent

Figure 22.7 North American Triassic Fault-Block Basins

a Areas where Triassic fault-block basin deposits crop out in eastern North America.

b After the Appalachians were eroded to a low-lying plain by the Middle Triassic, fault-block basins such as this one (shown in cross section) formed as a result of Late Triassic rifting between North America and Africa.

c These valleys accumulated tremendous thickness of sediments and were themselves broken by a complex of normal faults during rifting.

d Palisades of the Hudson River. This sill was one of many intruded into the Newark sediments during the Late Triassic rifting that marked the separation of North America from Africa.

plate margin became a passive, trailing continental margin. The fault-block mountains produced by this rifting continued to erode during the Jurassic and Early Cretaceous until only an area of low relief remained.

The sediments produced by this erosion contributed to the growing eastern continental shelf. During the Cretaceous Period, the Appalachian region was reelevated and once again shed sediments onto the continental shelf, forming a gently dipping, seaward-thickening wedge of rocks up to 3,000 m thick. These rocks are currently exposed in a belt extending from Long Island, New York, to Georgia.

Gulf Coastal Region

The Gulf Coastal region was above sea level until the Late Triassic (Figure 22.4). As North America separated from South America during the Late Triassic and Early Jurassic, the Gulf of Mexico began to form (Figure 22.5). With oceanic waters flowing into this newly formed, shallow, restricted basin, conditions were ideal for evaporite formation. More than 1,000 m of evaporites were precipitated at this time, and most geologists think that these Jurassic evaporites are the source for the Cenozoic salt domes found today in the Gulf of Mexico and southern Louisiana.

By the Late Jurassic, circulation in the Gulf of Mexico was less restricted, and evaporite deposition ended. Normal marine conditions returned to the area with alternating transgressing and regressing seas. The resulting sediments were covered and buried by thousands of meters of Cretaceous and Cenozoic sediments.

During the Cretaceous, the Gulf Coastal region, like the rest of the continental margin, was flooded by northward-transgressing seas (Figure 22.6). As a result, nearshore sandstones are overlain by finer sediments characteristic of deeper waters. Following an extensive regression at the end of the Early Cretaceous, a major transgression began during which a wide seaway extended from the Arctic Ocean to the Gulf of Mexico (Figure 22.6). Sediments deposited in the Gulf Coastal region during the Cretaceous formed a seaward-thickening wedge.

Reefs were also widespread in the Gulf Coastal region during the Cretaceous. Because of their high porosity and permeability, reefs make excellent petroleum reservoirs. The facies patterns of these Cretaceous reefs are as complex as those in the major barrier-reef systems of the Paleozoic Era.

Western Region

Mesozoic Tectonics   The Mesozoic geologic history of the North American Cordilleran mobile belt is very complex, involving the eastward subduction of the oceanic Farallon plate under the continental North American plate. Activity along this oceanic–continental convergent plate boundary resulted in an eastward movement of

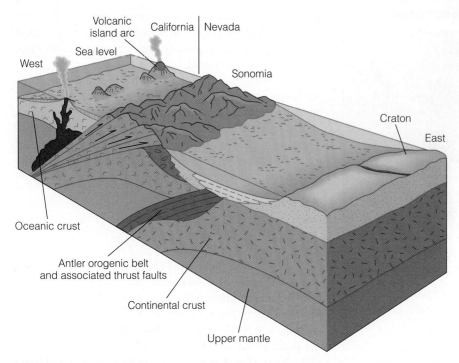

Figure 22.8 Sonoma Orogeny Tectonic activity that culminated in the Permian–Triassic Sonoma orogeny in western North America. The Sonoma orogeny was the result of a collision between the southwestern margin of North America and an island arc system.

deformation. This orogenic activity progressively affected the trench and continental slope, the continental shelf, and the cratonic margin, causing a thickening of the continental crust. In addition, the accretion of terranes and microplates along the western margin of North America also played a significant role in the Mesozoic tectonic history of this area.

Except for the Late Devonian–Early Mississippian Antler orogeny (see Figure 20.27), the Cordilleran region of North America experienced little tectonism during the Paleozoic. However, an island arc and ocean basin formed off the western North American craton during the Permian (Figure 22.4), followed by subduction of an oceanic plate beneath the island arc and the thrusting of oceanic and island arc rocks eastward against the craton margin (Figure 22.8). This event, similar to the preceding Antler orogeny (see Figure 20.27) and known as the **Sonoma orogeny,** occurred at or near the Permian–Triassic boundary and resulted in the suturing of island-arc terranes along the western edge of North America.

Following the Late Paleozoic–Early Mesozoic destruction of the volcanic island arc during the Sonoma orogeny, the western margin of North America became an oceanic–continental convergent plate boundary. During the Late Triassic, a steeply dipping subduction zone developed along the western margin of North America in response to the westward movement of North America over the Farallon plate. This newly created oceanic–continental

plate boundary controlled Cordilleran tectonics for the rest of the Mesozoic and for most of the Cenozoic Era; this subduction zone marks the beginning of the modern circum-Pacific orogenic system (see Chapter 23).

Two subduction zones dipping in opposite directions from each other formed off the west coast of North America during the Middle and early Late Jurassic (Figure 22.9). As the North American plate moved westward, as a result of the opening of the Atlantic Ocean, it soon overrode the westerly dipping subduction zone, destroying it and leaving only the easterly dipping subduction zone along its western periphery. It was also during this time that the *Franciscan Complex* and the *Great Valley Group* were deposited and deformed.

The Franciscan Complex, which is up to 7,000 m thick, is an unusual rock unit consisting of a chaotic mixture of rocks that accumulated during the Late Jurassic and Cretaceous. The various rock types—graywacke, volcanic breccia, siltstone, black shale, chert, pillow basalt, and blueschist metamorphic rocks—suggest that rocks of the continental shelf, slope, and deep-sea environments were brought together in a submarine trench when North America overrode the subducting Farallon plate (Figure 22.10).

The Franciscan Complex and the Great Valley Group, which lies east of it, were both squeezed against the edge of the North American craton as a result of subduction of the Farallon plate beneath the North American plate. The Franciscan Complex and the Great Valley Group are currently separated from each other by a major thrust fault. The Great Valley Group consists of more than 16,000 m of conglomerates, sandstones, siltstones, and shales that were deposited on the continental shelf and slope at the same time that the Franciscan deposits were accumulating in the submarine trench (Figure 22.10).

The general term **Cordilleran orogeny** is applied to the mountain-building activity that began during the Jurassic and continued into the Cenozoic (Figure 22.11). The Cordilleran orogeny consisted of a series of individual mountain-building events, or pulses, that occurred in different regions at different times, but overlapped to some extent. Most of this Cordilleran orogenic activity is related to the continued westward movement of the North American plate as it overrode the Farallon plate, and its history is highly complex.

Figure 22.9 Mesozoic Tectonic Evolution of the Sierra Nevada Interpretation of the tectonic evolution of the Sierra Nevada during the Mesozoic Era.

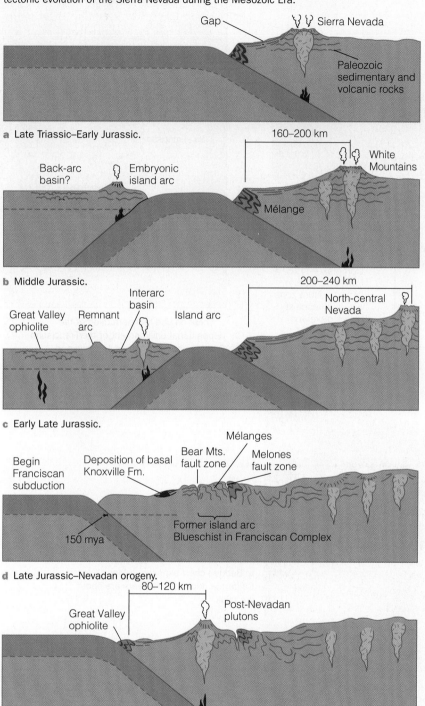

a Late Triassic–Early Jurassic.

b Middle Jurassic.

c Early Late Jurassic.

d Late Jurassic–Nevadan orogeny.

e Latest Jurassic.

The first phase of the Cordilleran orogeny, the **Nevadan orogeny** (Figure 22.11), began during the Late Jurassic and continued into the Cretaceous as large volumes of granitic magma were generated at depth beneath the western edge of North America. These granitic masses were emplaced as

huge batholiths that are now recognized as the Sierra Nevada, Southern California, Idaho, and Coast Range batholiths (Figure 22.12). It was also during this time that the Franciscan Complex and Great Valley Group were being deposited and deformed as part of the Nevadan orogeny within the Cordilleran mobile belt.

By the Late Cretaceous, most of the volcanic and plutonic activity had migrated eastward into Nevada and Idaho. This migration was probably caused by a change from high-angle to low-angle subduction, resulting in the subducting oceanic plate reaching its melting depth farther east (Figure 22.13). Thrusting occurred progressively farther east, so that by the Late Cretaceous, it extended all the way to the Idaho–Washington border.

The second phase of the Cordilleran orogeny, the **Sevier orogeny**, was mostly a Cretaceous event even though it began in the Late Jurassic and is associated with the tectonic activity of the earlier Nevadan orogeny (Figure 22.11). Subduction of the Farallon plate beneath the North American plate continued during this time, resulting in numerous overlapping, low-angle thrust faults in which blocks of older strata were thrust eastward on top of younger strata (Figure 22.14). This deformation produced generally north-south–trending mountain ranges that stretch from Montana to western Canada.

During the Late Cretaceous to Early Cenozoic, the final pulse of the Cordilleran orogeny took place (Figure 22.11). The **Laramide orogeny** developed east of the Sevier orogenic belt in the present-day Rocky Mountain areas of New Mexico, Colorado, and Wyoming. Most features of the present-day Rocky Mountains resulted from the Cenozoic phase of the Laramide orogeny, and for that reason, it will be discussed in Chapter 23.

Mesozoic Sedimentation Concurrent with the tectonism in the Cordilleran mobile belt, Early Triassic sedimentation on the western continental shelf consisted of shallow-water marine sandstones, shales, and limestones. During the Middle and Late Triassic, the western shallow seas regressed

Figure 22.10 The Franciscan Complex

a Location and reconstruction of the depositional environment of the Franciscan Complex during the Late Jurassic and Cretaceous.

b Bedded chert exposed in Marin County, California. Most of the layers are about 5 cm thick.

James S. Monroe

farther west, exposing large areas of former seafloor to erosion. Marginal marine and nonmarine Triassic rocks, particularly red beds, contribute to the spectacular and colorful scenery of the region.

These rocks represent a variety of continental depositional environments. The Upper Triassic *Chinle Formation,* for example, is widely exposed throughout the Colorado Plateau region and is probably most famous for its petrified wood, spectacularly exposed in Petrified Forest National

Triassic formations in the Southwest, also contains the fossilized remains and tracks of various amphibians and reptiles.

Early Jurassic-age deposits in a large part of the western region consist mostly of clean, cross-bedded sandstones indicative of wind-blown deposits. The thickest and most prominent of these is the *Navajo Sandstone,* a widespread cross-bedded sandstone that accumulated in a coastal dune environment along the southwestern margin of the craton. The sandstone's most distinguishing feature is its large-scale cross-beds, some of which are more than 25 m high (Figure 22.16).

▌**Figure 22.11** Mesozoic Cordilleran Orogenies Mesozoic orogenies occurring in the Cordilleran mobile belt.

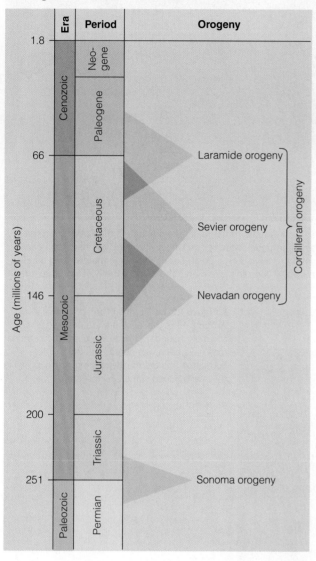

▌**Figure 22.12** Cordilleran Batholiths Location of Jurassic and Cretaceous batholiths in western North America.

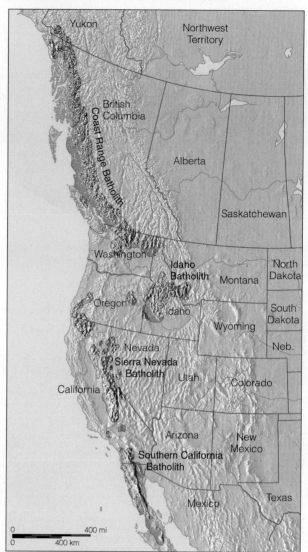

▌**Figure 22.13** Cretaceous Subduction in the Cordilleran Mobile Belt A possible cause for the eastward migration of igneous activity in the Cordilleran region during the Cretaceous Period was a change from (a) high-angle to (b) low-angle subduction. As the subducting plate moved downward at a lower angle, the depth of melting moved farther east.

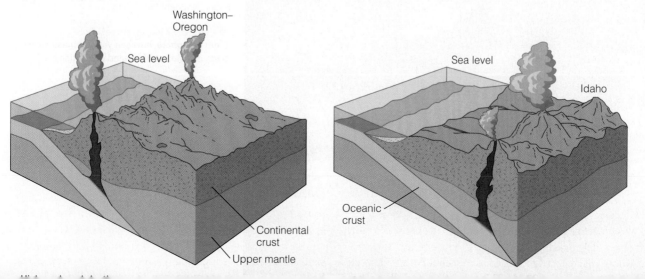

▌ Figure 22.14 Sevier Orogeny

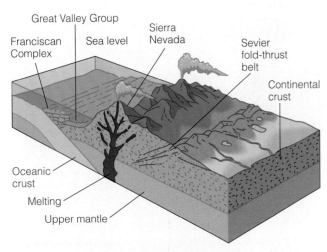

a Restoration showing the associated tectonic features of the Late Cretaceous Sevier orogeny caused by subduction of the Farallon plate under the North American plate.

b The Keystone thrust fault, a major fault in the Sevier overthrust belt, is exposed west of Las Vegas, Nevada. The sharp boundary between the light-colored Mesozoic rocks and the overlying dark-colored Paleozoic rocks marks the trace of the Keystone thrust fault.

▌ Figure 22.15 Petrified Forest National Park, Arizona All of the logs here are *Araucarioxylon*, the most abundant tree in the park. The petrified logs have been weathered from the Chinle Formation and are mostly in the position in which they were buried some 200 million years ago.

▌ Figure 22.16 Navajo Sandstone Large cross-beds of the Jurassic Navajo Sandstone exposed in Zion National Park, Utah.

》 Critical Thinking Question
From this photo, how can you tell these cross-beds are the result of wind deposition?

Figure 22.17 Morrison Formation, Dinosaur National Monument, Utah

a Panoramic view of the Jurassic Morrison Formation as seen from the visitors' center at Dinosaur National Monument, near Vernal, Utah

b North wall of visitors' center showing dinosaur bones in bas relief, just as they were deposited 140 million years ago.

Marine conditions returned to the region during the Middle Jurassic when a seaway called the **Sundance Sea** twice flooded the interior of western North America (Figure 22.5). The resulting deposits, the *Sundance Formation,* were produced from erosion of tectonic highlands to the west that paralleled the shoreline. These highlands resulted from intrusive igneous activity and associated volcanism that began during the Triassic.

During the Late Jurassic, a mountain chain formed in Nevada, Utah, and Idaho as a result of the deformation produced by the Nevadan orogeny. As the mountain chain grew and shed sediments eastward, the Sundance Sea began retreating northward. A large part of the area formerly occupied by the Sundance Sea was then covered by multicolored sandstones, mudstones, shales, and occasional lenses of conglomerates that constitute the world-famous *Morrison Formation* (Figure 22.17a).

The Morrison Formation contains the world's richest assemblage of Jurassic dinosaur remains (Figure 22.17b). Although most of the dinosaur skeletons are broken up, as many as 50 individuals have been found together in a small area. Such a concentration indicates that the skeletons were brought together during times of flooding and deposited on sandbars in stream channels. Soils in the Morrison Formation indicate that the climate was seasonally dry.

Shortly before the end of the Early Cretaceous, Arctic waters spread southward over the craton, forming a large inland sea in the Cordilleran region. Mid-Cretaceous transgressions also occurred on other continents, and all were part of the global mid-Cretaceous rise in sea level that resulted from accelerated seafloor spreading as Pangaea continued to fragment. These Middle Cretaceous transgressions are marked by widespread black shale deposition within the oceanic areas, the shallow sea shelf areas, and the continental regions that were inundated by the transgressions.

By the beginning of the Late Cretaceous, this incursion joined the northward-transgressing waters from the Gulf area to create an enormous **Cretaceous Interior Seaway** that occupied the area east of the Sevier orogenic belt. Extending from the Gulf of Mexico to the Arctic Ocean, and more than 1,500 km wide at its maximum extent, this seaway effectively divided North America into two large landmasses until just before the end of the Late Cretaceous (Figure 22.6).

Cretaceous deposits less than 100 m thick indicate that the eastern margin of the Cretaceous Interior Seaway subsided slowly and received little sediment from the emergent, low-relief craton to the east. The western shoreline, however, shifted back and forth, primarily in response to fluctuations in the supply of sediment from the Cordilleran–Sevier orogenic belt to the west. The facies relationships show lateral changes from conglomerate and coarse sandstone adjacent to the mountain belt through finer sandstones, siltstones, shales, and even limestones and chalks in the east. During times of particularly active mountain building, these coarse clastic wedges of gravel and sand prograded even farther east.

As the Mesozoic Era ended, the Cretaceous Interior Seaway withdrew from the craton. During this regression, marine waters retreated to the north and south, and marginal marine and continental deposition formed widespread coal-bearing deposits on the coastal plain.

22.4 What Role Did Accretion of Terranes Play in the Growth of Western North America?

In the preceding sections, we have discussed orogenies along convergent plate boundaries resulting in continental accretion. Much of the material accreted to continents during such events is simply eroded older continental crust;

however, a significant amount of new material is added to continents as well, such as igneous rocks that formed as a consequence of subduction and partial melting. Although subduction is the predominant influence on the tectonic history in many regions of orogenesis, other processes are also involved in mountain building and continental accretion, especially the accretion of terranes.

Geologists now know that portions of many mountain systems are composed of small accreted lithospheric blocks that are clearly of foreign origin. These **terranes** differ completely in their fossil content, stratigraphy, structural trends, and paleomagnetic properties from the rocks of the surrounding mountain system and adjacent craton. In fact, these terranes are so different from adjacent rocks that most geologists think that they formed elsewhere and were carried great distances as parts of other plates until they collided with other terranes or continents.

Geologic evidence indicates that more than 25% of the entire Pacific Coast from Alaska to Baja California consists of accreted terranes. The accreting terranes are composed of volcanic island arcs, oceanic ridges, seamounts, volcanic plateaus, hot-spot tracks, and small fragments of continents that were scraped off and accreted to the continent's margin as the oceanic plate on which they were carried was subducted under the continent.

Geologists estimate that more than 100 different-sized terranes have been added to the western margin of North America during the last 200 million years (▌Figure 22.18). The Wrangellian terranes (Figure 22.1b) are a good example of terranes that have been accreted to North America's western margin.

The basic plate tectonic reconstruction of orogenies and continental accretion remains unchanged; however, the details of such reconstructions are decidedly different in view of terrane tectonics. For example, growth along active continental margins is faster than along passive continental margins because of the accretion of terranes. Furthermore, these accreted terranes are often new additions to a continent rather than reworked older continental material.

So far, most terranes have been identified in mountains of the North American Pacific Coast region, but a number of such plates are suspected to be present in other mountain systems as well. They are more difficult to recognize in older mountain systems, such as the Appalachians, however, because of greater deformation and erosion. Thus, terranes provide another way of viewing Earth and gaining a better understanding of the geologic history of the continents.

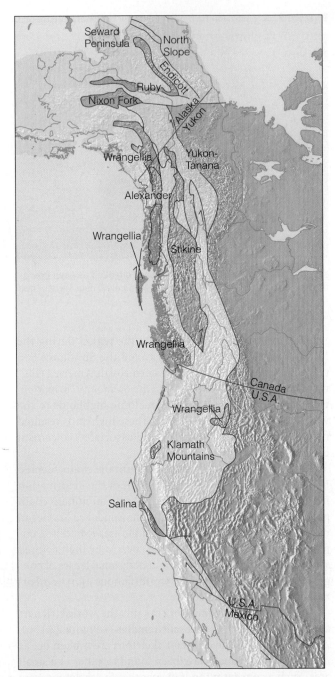

▌**Figure 22.18 Mesozoic Terranes** Some of the accreted lithospheric blocks, called terranes, that form the western margin of the North American craton. The dark-brown blocks probably originated as terranes and were accreted to North America. The light-green blocks are possibly displaced parts of North America. The North American craton is shown in dark green. See Figure 22.1b for the position of the Wrangellian terranes during the Jurassic.

22.5 Mesozoic Mineral Resources

Although much of the coal in North America is Pennsylvanian or Paleogene in age, important Mesozoic coals occur in the Rocky Mountain states. These are mostly lignite and bituminous coals, but some local anthracites are present as well. Particularly widespread in western North America are coals of Cretaceous age. Mesozoic coals are also known from Australia, Russia, and China.

Large concentrations of petroleum occur in many areas of the world, but more than 50% of all proven reserves are in the Persian Gulf region. Conditions similar to those that led to the formation of petroleum in the Persian Gulf region existed in what is now the Gulf Coast region of the United States and Central America. Here, petroleum and natural gas also formed on a broad shelf over which transgressions and regressions occurred. In this region, the hydrocarbons are largely in reservoir rocks that were deposited as distributary channels on deltas and as barrier-island and beach sands. Some of these hydrocarbons are associated with structures formed adjacent to rising salt domes. The salt, called the *Louann Salt,* initially formed in a long, narrow sea when North America separated from Europe and North Africa during the fragmentation of Pangaea (Figure 22.2).

The richest uranium ores in the United States are widespread in Mesozoic rocks of the Colorado Plateau area of Colorado and adjoining parts of Wyoming, Utah, Arizona, and New Mexico. These ores, consisting of fairly pure masses of a complex potassium-, uranium-, and vanadium-bearing mineral called *carnotite,* are associated with plant remains in sandstones that were deposited in ancient stream channels.

As noted in Chapter 19, Proterozoic banded iron formations are the main sources of iron ores. There are, however, some important exceptions. For example, the Jurassic-age "Minette" iron ores of Western Europe, composed of oolitic limonite and hematite, are important ores in France, Germany, Belgium, and Luxembourg. In Great Britain, low-grade iron ores of Jurassic age consist of oolitic siderite, which is an iron carbonate. And in Spain, Cretaceous rocks are the host rocks for iron minerals.

South Africa, the world's leading producer of gem-quality diamonds and among the leaders in industrial diamond production, mines these minerals from kimberlite pipes, conical igneous intrusions of dark gray or blue igneous rock. Diamonds, which form at great depth where pressure and temperature are high, are brought to the surface during the explosive volcanism that forms kimberlite pipes. Although kimberlite pipes have formed throughout geologic time, the most intense episode of such activity in South Africa and adjacent countries was during the Cretaceous Period. Emplacement of Triassic and Jurassic diamond-bearing kimberlites also occurred in Siberia.

In the Introduction we noted that the mother lode or source for the placer deposits mined during the California gold rush is in Jurassic-age intrusive rocks of the Sierra Nevada. Gold placers are also known in Cretaceous-age conglomerates of the Klamath Mountains of California and Oregon.

Porphyry copper was originally named for copper deposits in the western United States mined from porphyritic granodiorite; however, the term now applies to large, low-grade copper deposits disseminated in a variety of rocks. These porphyry copper deposits are an excellent example of the relationship between convergent plate boundaries and the distribution, concentration, and exploitation of valuable metallic ores. Magma generated by partial melting of a subducting plate rises toward the surface, and as it cools, it precipitates and concentrates various metallic ores. The world's largest copper deposits were formed during the Mesozoic and Cenozoic in a belt along the western margins of North and South America (see Figure 2.28).

What Would You Do ?

Because of political events in the Middle East, the oil-producing nations of this region have reduced the amount of petroleum that they export, resulting in shortages in the United States. To alleviate U.S. dependence on overseas oil, the major oil companies want Congress to let them explore for oil in many of our national parks. As director of the National Park Service, you have been called to testify at the congressional hearing addressing this possibility. What arguments would you use to discourage such exploration? Would knowledge of the geology of the area be helpful in your testimony? Explain.

22.6 Life of the Mesozoic Era

Restorations of prehistoric floras and faunas—especially Mesozoic dinosaurs, marine reptiles, and flying reptiles—fascinate many people. Indeed, these animals have been popularized in books, television specials, and movies, including *Jurassic Park* (1993) and its two sequels. Part of the popularity of these creatures is no doubt related to the tremendous increase in knowledge that scientists have gained about the Mesozoic biota during the last few decades.

The Mesozoic Era is commonly called the "Age of Reptiles," calling attention to the fact that reptiles predominated among land-dwelling vertebrate animals. Certainly, the Mesozoic diversification of reptiles was an important event in life history, but other equally important, though not as well known, events took place. Mammals made their appearance during the Triassic, having evolved from mammal-like reptiles, and birds evolved from small carnivorous dinosaurs by the Jurassic. And, of course, there was a variety of reptiles in addition to dinosaurs—flying reptiles and marine reptiles being two of the most notable.

Important changes also took place among plants when the first flowering plants (angiosperms) evolved and soon became the most common plants on land. Even though the major land plants from the Paleozoic Era persisted, and many still exist, they now make up less than 10% of all land plants. Marine invertebrates such as clams, snails, and cephalopods made a remarkable resurgence following the Permian mass extinctions.

CONNECTION LINK

For a review of Paleozoic invertebrates and plants, see pages 544–554 and 564–568 in Chapter 21.

Although Pangaea began fragmenting during the Triassic (Figure 22.2), the proximity of continents and mild, Mesozoic climates made it possible for plants and animals to occupy extensive geographic ranges. But as the fragmentation continued, some continents became isolated, especially Australia and South America, and their faunas, evolving in isolation, became increasingly different from those elsewhere. Indeed, Australia has been a separate landmass since the Mesozoic and even now has a fauna unlike that of any other continent.

Marine Invertebrates and Phytoplankton

Following the Permian mass extinctions, the Mesozoic was a time when marine invertebrates repopulated the seas. The Early Triassic invertebrate fauna was not very diverse, but by the Late Triassic, the seas were once again swarming with invertebrates—from planktonic foraminifera to cephalopods. The brachiopods that had been so abundant during the Paleozoic never completely recovered from their near extinction, and although they still exist, the bivalves have largely taken over their ecologic niche.

Mollusks such as cephalopods, bivalves, and gastropods were the most important members of the Mesozoic marine invertebrate fauna. Their rapid evolution and the fact that many cephalopods were nektonic make them excellent guide fossils (▌ Figure 22.19). The Ammonoidea, cephalopods with wrinkled sutures, constitute three groups: the goniatites, ceratites, and ammonites. The latter, though present during the entire Mesozoic, were most prolific during the Jurassic and Cretaceous. Most ammonites were coiled, some attaining diameters of 2 m, whereas others were uncoiled and led a near benthonic existence. Ammonites became extinct at the end of the Cretaceous, but two related

groups of cephalopods survived into the Cenozoic: the *nautiloids,* including the living pearly nautilus, and the *coleoids,* represented by the extinct belemnoids (Figure 22.19), which are good Jurassic and Cretaceous guide fossils, as well as by the living squid and octopus.

Mesozoic bivalves diversified to inhabit many epifaunal and infaunal niches. Oysters and clams (epifaunal suspension feeders) became particularly diverse and abundant and, despite a reduction in diversity at the end of the Cretaceous, remain important animals in the marine fauna today.

As is true now, where shallow marine waters were warm and clear, coral reefs proliferated. However, reefs did not rebound from the Permian extinctions until the Middle Triassic. An important reef builder throughout the Mesozoic was a group of bivalves known as *rudists.* Rudists are important because they displaced corals as the main reef builders during the later Mesozoic and are good guide fossils for the Late Jurassic and Cretaceous.

A new and familiar type of coral also appeared during the Triassic, the *scleractinians.* Whether scleractinians evolved from rugose corals or from an as yet unknown soft-bodied ancestor with no known fossil record is still unresolved. In addition, another invertebrate group that prospered during the Mesozoic was the echinoids. Echinoids were exclusively epifaunal during the Paleozoic, but branched out into the infaunal habitat during the Mesozoic.

A major difference between Paleozoic and Mesozoic marine invertebrate faunas was the increased abundance and diversity of burrowing organisms. With few exceptions, Paleozoic burrowers were soft-bodied animals such as worms. The bivalves and echinoids, which were epifaunal animals during the Paleozoic, evolved various means of entering infaunal habitats. This trend toward an infaunal existence may have been an adaptive response to increasing predation from the rapidly evolving fish (▌ Figure 22.20)

▌ Figure 22.19 Belemnoids Belemnoids are extinct squidlike cephalopods that were particularly abundant during the Cretaceous. They are excellent guide fossils for the Jurassic and Cretaceous. Shown here are several belemnoids swimming in a Cretaceous sea.

▌ Figure 22.20 The Mesozoic Fish *Leedsichthys* (background) and the Short-Necked Plesiosaur *Liopleurodon* This fish, from the intermediate group of fishes (Holostei), was one of the largest ever, but because some of its spine is missing, its total length is not known; estimates range from 9 to 30 m. It was probably a plankton feeder.

‖ **Figure 22.21** **Primary Producers** Coccolithophores, diatoms, and dinoflagellates were primary producers in the Mesozoic and Cenozoic oceans. **(a)** A Miocene coccolith from the Gulf of Mexico (left): A Miocene-Pliocene coccolith from the Gulf of Mexico (right). Coccoliths are the calcareous plates covering the living coccolithosphere. **(b)** Upper Miocene diatoms from Java (left and right). **(c)** Miocene dinoflagellate from Sale, Morocco (left and right).

inhabit freshwater lakes. *Dinoflagellates,* which are organic-walled phytoplankton, were common during the Mesozoic and today are the major primary producers in warm water (‖ Figure 22.21c).

In general terms, we can think of the Mesozoic as a time of increasing complexity among the marine invertebrate fauna. At the beginning of the Triassic, diversity was low and food chains were short. Near the end of the Cretaceous, though, the marine invertebrate fauna was highly complex, with interrelated food chains. This evolutionary history reflects changing geologic conditions influenced by plate tectonic activity, as discussed in Chapter 2.

Plants—Primary Producers on Land

Just as during the Late Paleozoic, seedless vascular plants and gymnosperms dominated Triassic and Jurassic land-plant communities, and, in fact, representatives of both groups are still common. Among the gymnosperms, the large seed ferns became extinct by the end of the Triassic, but *ginkgos* remained abundant and still exist in isolated regions, and *conifers* continued to diversify and are now widespread in some terrestrial habitats, particularly at high elevations and high latitudes. A new group of gymnosperms known as *cycads* made its appearance during the Triassic. These palmlike plants became widespread and now live in tropical and semitropical areas.

The long dominance of seedless plants and gymnosperms ended during the Early Cretaceous, perhaps the Late Jurassic, when many were replaced by **angiosperms,** or flowering plants (‖ Figure 22.22). Studies of fossils and living gymnosperms show that they are closely related to angiosperms, but unfortunately, the early fossil record of angiosperms is sparse, so their precise ancestry remains obscure.

Since they first evolved, angiosperms have adapted to nearly every terrestrial habitat—from mountains to deserts—and some have even adapted to shallow coastal waters. Several factors account for their phenomenal success, but chief among them is their method of reproduction (Figure 22.22a). Particularly important was the evolution of flowers, which attract animal pollinators, especially insects, and the evolution of enclosed seeds.

Seedless vascular plants and gymnosperms are important and still flourish in many environments. Nevertheless, a measure of the angiosperms' success is that today, with 250,000 to 300,000 species, they account for more than 90% of all land-plant species, and they occupy some habitats in which other land plants do poorly or cannot exist.

The Diversification of Reptiles

Reptiles, birds, and mammals are collectively called **amniotes** because they all produce amniotic eggs, and all evolved from a common ancestor that lived during the Late Paleozoic Era. The first of these ancestors were the so-called protorothyrids that evolved during the Mississippian Period (see Figure 21.28).

and cephalopods. Bivalves, for instance, expanded into the infaunal niche during the Mesozoic, and by burrowing, they escaped predators.

The foraminifera (single-celled consumers) diversified rapidly during the Jurassic and Cretaceous and continued to be diverse and abundant to the present. The planktonic forms, in particular, diversified rapidly, but most genera became extinct at the end of the Cretaceous.

The primary producers in the Mesozoic seas were various types of microorganisms. *Coccolithophores* are an important group of calcareous phytoplankton (‖ Figure 22.21a) that first evolved during the Jurassic and became extremely common during the Cretaceous. *Diatoms* (‖ Figure 22.21b), which build skeletons of silica, made their appearance during the Cretaceous, but they are more important as primary producers during the Cenozoic. Diatoms are presently most abundant in cooler oceanic waters, and some species

Figure 22.22 The Angiosperms or Flowering Plants

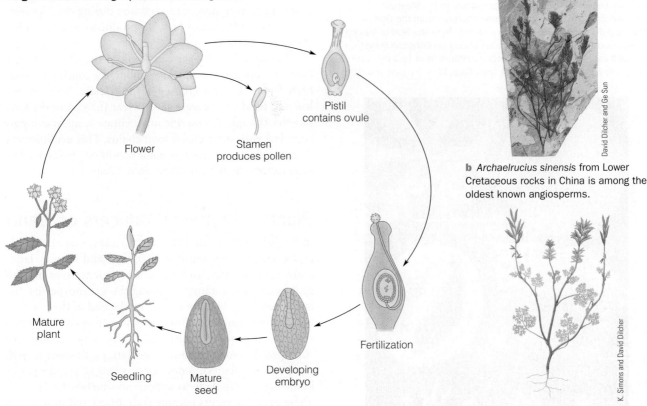

b *Archaelrucius sinensis* from Lower Cretaceous rocks in China is among the oldest known angiosperms.

Pistil contains ovule

Stamen produces pollen

Flower

Fertilization

Mature plant

Seedling

Mature seed

Developing embryo

a The reproductive cycle in angiosperms.

c Restoration of *Archaefructus sinensis*.

From this basal stock of stem reptiles two main branches emerged, one leading to the pelycosaurs and therapsids and eventually to mammals, and the other to all other reptiles and birds. We will consider mammals later in this chapter, but here our main concern is reptiles and their descendants, the birds.

Dinosaurs are included among the reptiles, but they possess several characteristics that set them apart: teeth set in individual sockets, a ball-like head on the upper leg bone (femur), and a fully upright posture with the limbs directly beneath their bodies. Indeed, their upright posture and modifications in their limbs account for more efficient locomotion than in other land-dwelling reptiles, which may have been one factor in their incredible success.

Contrary to popular belief, there were no flying dinosaurs or fully aquatic ones, although there were other Mesozoic reptiles that occupied these niches (▌ Figure 22.23). Nor were all dinosaurs large, although some certainly were. Further, dinosaurs lived only during the Mesozoic Era, unless we consider their evolutionary descendants, the birds.

Archosaurs and the Origin of Dinosaurs Reptiles known as **archosaurs** (*archo* meaning "ruling," and *sauros* meaning "lizard") include crocodiles, flying reptiles (pterosaurs), dinosaurs, and birds (Figure 22.23). The probable ancestors of dinosaurs were archosaurs much like those from Middle and Late Triassic-age rocks in Argentina (▌ Figure 22.24). They were rather small (about

1 m long), long-legged carnivores that walked and ran on their hind limbs, so they were **bipedal,** as opposed to **quadrupedal** animals that moved on all four limbs. In any case, the first dinosaurs evolved by Late Triassic time.

Sir Richard Owen proposed the term **dinosaur** in 1842 to mean "fearfully great lizard," although now "fearfully" has come to mean "terrible," thus the depiction of dinosaurs as "terrible lizards." Of course we have no reason to think they were any more terrible than animals today, and they were not lizards. Nevertheless, these ideas persist and the popularization of dinosaurs in movies and TV shows has often been inaccurate and has contributed to misunderstandings. For example, many people think that all dinosaurs were large and that they must have been poorly adapted because they went extinct. Some were in fact very large, weighing several tens of metric tons, but others weighed no more than 2 or 3 kg. And to consider them poorly adapted is to ignore the fact that they were extremely diverse and existed as a group for more than 140 million years.

Although various media now portray dinosaurs as more active animals, the mistaken belief persists that they were dim-witted, lethargic beasts. Evidence available now indicates that some were more active than formerly thought, some may have been warm-blooded, and a few types cared for their young long after they hatched, a behavioral characteristic more typical of birds and mammals. Much remains to be learned about dinosaurs, but their fossils and the rocks containing them are revealing more and more about their evolution and behavior.

snakes
lizards
"stem" reptiles
tuataras
ichthyosaurs
plesiosaurs
birds
theropod dinosaurs
other dinosaurs
pterosaurs
archosaurs
crocodilians
turtles
therapsids
mammals

CARBONIFEROUS	PERMIAN	TRIASSIC	JURASSIC	CRETACEOUS	
PALEOZOIC ERA		MESOZOIC ERA			CENOZOIC ERA

▌ **Figure 22.23 Evolution of the Amniotes** All of the organisms shown here are amniotes—that is, animals that lay amniote eggs. The living amniotes are snakes, lizards, tuataras, birds, turtles, and mammals.

» Critical Thinking Question
Other than their pelvises, are there other anatomical features you can use to differentiate the two main groups of dinosaurs?

Dinosaurs are united by several shared features, yet they differ enough for paleontologists to recognize two distinct groups based mostly on their type of pelvis—the **Saurischia** and **Ornithischia.** Saurischian dinosaurs have a lizard-like pelvis, whereas ornithischians have a bird-like pelvis, and both groups are further divided into several distinct types (▌ Figure 22.25). It is now clear that both saurischians and ornithischians had a common ancestor much like the Triassic archosaurs that lived in Argentina (Figure 22.24).

Saurischian Dinosaurs Saurischian dinosaurs have a lizard-like pelvis, hence their popularization as lizard-hipped dinosaurs, and include two subgroups, the *theropods* and

▌ **Figure 22.24 Eoraptor** This small, carnivorous biped was one of the first dinosaurs. Populations of these animals, or ones very much like them, are the probable ancestor of all later dinosaurs. *Eoraptor* lived during the Late Triassic (about 230 million years ago) in what is now Argentina. It was only about a meter long and likely weighed no more than 10 kg.

De Agostini Picture Library/De Agostini/ Getty Images

Figure 22.25 Cladogram Showing Relationships Among Dinosaurs Pelvises of ornithischian and saurischian dinosaurs are shown for comparison. All of the dinosaurs shown here were herbivores, except for theropods. Note that bipedal and quadrupedal dinosaurs are found in both ornithischians and saurischians.

sauropods (Figure 22.25). Theropods were bipedal carnivores that varied from tiny *Compsognathus* (2 or 3 kg) to giants such as *Tyrannosaurs* (5 metric tons) (see the Geo-Insight on pages 598 and 599).* Beginning in 1966, Chinese scientists have found several types of theropods with feathers.** No one doubts that these dinosaurs had feathers, and molecular evidence indicates that they were made of the same material as bird feathers.

One of the smaller theropods, the 1.8-m-long predator *Velociraptor,* was popularized in the movie *Jurassic Park.* This animal and its somewhat larger relative, *Deinonychus,* had large, sickle-like claws on the back feet that were probably used in a slashing type of attack. But despite what you might see on TV or in movies, theropods, like predators today, no doubt avoided large, dangerous prey and went for the easy kill, preying on the young, old, or disabled, or they

dined on carrion. And undoubtedly the larger predators simply chased smaller predators away from their kill.

As opposed to theropods, the sauropods were large, quadrupedal herbivores. Among the sauropods were the largest land-dwelling animals ever. *Brachiosaurus,* a giant even by sauropod standards, may have weighted 75 metric tons, and partial remains indicate that even larger sauropods existed. Evidence from trackways shows that sauropods moved in herds.

Sauropods were preceded in the fossil record by the smaller, Late Triassic to Early Jurassic prosauropods. In fact, there is a good record of small, bipedal prosauropods that were ancestors to the larger, quadrupedal sauropods. The true sauropods, represented by *Apatosaurus, Diplodocus,* and *Brachiosaurus,* were common during the Jurassic, but only a few types existed during the Cretaceous.

Ornithischian Dinosaurs In addition to their pelvis, the ornithischians, or so-called bird-hipped dinosaurs, differed from saurischians in other ways. For example, they had no teeth in the front of the mouth, whereas saurischians did, and they had ossified (bonelike) tendons in the back region. Although

---

*One group of theropods, the poorly known therizinosaurs, reverted to a herbivorous diet.

**One feathered specimen purchased in Utah that was reported to have come from China turned out to be a forgery. It even appeared in *National Geographic* before scientists exposed it as a fraud.

you may not know the names of the five subgroups of ornithischians, you surely have seen examples of *ornithopods, pachycephalosaurs, ankylosaurs, stegosaurs,* and *ceratopsians* (Figure
22.25) (see the Geo-Insight on pages 598 and 599).

In 1822, Gideon Mantell and his wife, Mary Ann, discovered some teeth from a dinosaur that he later named
Iguanadon and that proved to be an ornithopod. Another
ornithopod, discovered in North America in 1858, was the
first dinosaur to be assembled and displayed in a museum.
All ornithopods were herbivores and primarily bipedal, but
they had well-developed front limbs and could also walk in
a quadrupedal manner. One group of ornithopods that was
especially numerous during the Cretaceous is called duck-
billed dinosaurs, or hadrosaurs. Some of these animals had
crests on their heads that may have been used to amplify
bellowing, for sexual display, or for species recognition. The
Late Cretaceous ornithopod *Maiasaura* nested in colonies,
used the same nesting area repeatedly, and apparently cared
for its young long after hatching.

The pachycephalosaurs, or dome-headed dinosaurs,
had thickened skulls that may have been used in butting
contests for dominance or mates. Not all agree with this
interpretation and claim that only juveniles had thick skulls.
In any case, pachycephalosaurs were bipedal herbivores that
ranged from 1.0 to 4.5 m long. Their fossils are known only
from Late Cretaceous-age rocks.

Ankylosaurs were quadrupedal herbivores with heavy,
bony armor that protected the back, flanks, and top of the
head. The tail of some species, such as *Ankylosaurus,* ended
in a heavy, bony club that could probably deliver a crippling
blow to an attacking predator. If the tail proved inadequate,
the animal probably simply hunkered down, making it difficult even for *Tyrannosaurs* to flip over a 4.5-ton *Ankylosaurus.*

The most familiar genus of stegosaur is the medium-
sized, herbivorous quadruped known as *Stegosaurus* (see the
Geo-Insight on pages 598 and 599). This dinosaur had distinctive spikes on its tail, almost certainly used for defense,
and bony plates on its back. The arrangement of these plates
is not certainly known, although they are usually depicted
in two rows with the plates on one side offset from those on
the other. Most agree that the plates functioned to absorb
and dissipate heat.

The fossil record of ceratopsians (horned dinosaurs)
shows that small, bipedal, Early Creaceous animals were
ancestors of large, Late Cretaceous, herbivorous quadru-
peds such as *Triceratops. Triceratops* and related genera had
huge heads, a bony frill over the top of the neck, and horn
or horns on the skull. Fossil trackways show that these were
herding animals. A bone bed with numerous fossils of a single species indicates that large numbers of ceratopsians perished quickly, probably during a river crossing.

We have every reason to think that dinosaurs behaved
much as land animals do today. They had to acquire
resources and reproduce, and it seems likely that some of
them had complex social structures. Certainly, some dinosaurs lived in herds and no doubt interacted by bellowing,
snorting, grunting, and foot stomping in defense, territorial
disputes, and attempts to attract mates.

Warm-Blooded Dinosaurs? Were dinosaurs **endotherms** (warm–blooded) like today's mammals and birds,
or were they **ectotherms** (cold–blooded) like today's reptiles? Almost everyone now agrees some compelling evidence exists for dinosaur endothermy, but opinion is still
divided among (1) those holding that all dinosaurs were
endotherms, (2) those who think only some were endotherms, and (3) those proposing that dinosaur metabolism,
and thus the ability to regulate body temperature, changed
as they matured.

Bones of endotherms typically have numerous passageways that, when the animals are alive, contain blood
vessels, but the bones of ectotherms have considerably
fewer passageways. Proponents of dinosaur endothermy
note that dinosaur bones are more similar to the bones of
living endotherms. Crocodiles and turtles have this so-
called endothermic bone, but they are ectotherms, and
some small mammals have bone more typical of ectotherms. Perhaps bone structure is related more to body size
and growth patterns than to endothermy, so this evidence
is not conclusive.

Endotherms must eat more than comparably sized ectotherms because their metabolic rates are so much higher.
Consequently, endothermic predators require large prey populations and thus constitute a much smaller proportion of the
total animal population than their prey, usually only a few percent. In contrast, the proportion of ectothermic predators to
prey may be as high as 50%. Where data are sufficient to allow
an estimate, dinosaur predators made up 3% to 5% of the total
population. Nevertheless, uncertainties in the data make this
argument less than convincing for many paleontologists.

A large brain in comparison to body size requires a rather
constant body temperature and thus implies endothermy. And
some dinosaurs were indeed brainy, especially the small- and
medium-sized theropods, so brain size might be a convincing
argument for these dinosaurs. Even more compelling evidence
for theropod endothermy comes from their probable relationship to birds and from the discoveries in China of dinosaurs
with feathers or a feather-like covering. Today, only endotherms have hair, fur, or feathers for insulation.

Some scientists point out that certain duck-billed dinosaurs grew and reached maturity much more quickly than
would be expected for ectotherms and conclude that they
must have been warm blooded. Furthermore, a fossil ornithopod discovered in 1993 has a preserved four-chambered
heart much like that of living mammals and birds. Three-
dimensional imaging of this structure, now on display at the
North Carolina Museum of Natural Sciences, has convinced
many scientists that this animal was an endotherm.

Good arguments for endothermy exist for several types
of dinosaurs, particularly theropods, although the large sauropods were probably not endothermic, but nevertheless
were capable of maintaining a rather constant body temperature. Large animals heat up and cool down more slowly
than smaller ones because they have a small surface area
compared to their volume. With their comparatively smaller
surface area for heat loss, sauropods probably retained heat
more effectively than their smaller relatives.

GEO-INSIGHT Dinosaurs

Dinosaurs are the best-known extinct animals. Hundreds of dinosaur species existed from the Late Triassic through the Cretaceous. They varied from species that weighted no more than a few kilograms to giants weighing several tens of metric tons.

▲ **1.** Lifelike restoration of the Early Cretaceous 3-m-long theropod *Deinony-chus* "terrible claw" in its probable attack posture. Note the huge curved claws on the back feet and the well-developed forelimbs with sharp claws.

▲ **2.** *Gigantosaurus* was a huge theropod that lived during the Late Cretaceous in Argentina and Brazil. It was nearly 13 m long, weighed about 7.3 metric tons, and lived 25 million years before the more familiar *Tyrannosaurus rex*.

◄ **3.** This Mesozoic scene shows the 10-m-long *Baryonyx* and two large sauropods in the background.

▶ **4.** Skeleton of the rhinoceros-sized ceratopslan dinosaur *Triceratops*. It and other ceratopsian genera were quite common in North America during the Late Cretaceous.

▲ **5.** Restoration of the Early Creta-ceous feathered dinosaur *Caudipteryx*. This turkey-sized animal, along with sev-eral other feathered dinosaurs, comes from the Liaoning Province of northeast-ern China.

© Eivind Bovor

▲ **6.** A Late Cretaceous scene showing the heavily armored dinosaur *Ankylosaurus* defending itself from an attack.

Douglas Henderson

▲▼ **7–9.** The ornithopod dinosaurs were abundant and varied. *Maiasaura* (above) nested in colonies and tended its young long after hatching. It had no crest, but *Corythosaurus* (below left) and *Parasaurolophus* (below) had bony extensions of the skull.

Dave King / Graham High at Centaur Studios - modelmaker
© Dorling Kindersley

▲ **8.** *Corythosaurus.*

Ann Winterbotham © Dorling Kindersley

▲ **9.** *Parasaurolophus.*

© 2011 by Karen Carr and Karen Carr Studio, Inc.

▲ **10.** *Stegosaurus* is noted for the rows of plates on its back and the bony spikes at the end of its tail. Stegosaurs died out by the end of the Jurassic Period.

© CM Studio. All Rights Reserved.

▲ **11.** The pachycephalosaurs, represented here by *Stegoceras,* had dome-shaped skulls from thickening of the bones. *Stegoceras* was small, measuring only 2 m long, but some of its relatives were up to 5 m long.

■ **Figure 22.26** Mesozoic Pterosaurs (Flying Reptiles)

a *Pterodactylus,* a long-tailed pterosaur, is well known from the Late Jurassic. Among the several known species, wingspans ranged from 50 cm to 2.5 m.

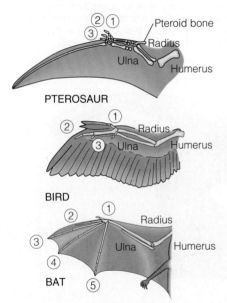

b In all flying vertebrates, the forelimb has been modified into a wing. A long fourth finger supports the pterosaur wing. In birds the second and third fingers are fused together, and in bats fingers 2 through 5 support the wing.

>> **Critical Thinking Question**
Are the forelimbs shown here analogous, homologous, or both? See Figures 18.14 and 18.15.

Flying Reptiles Paleozoic insects were the first flying animals, but the first among vertebrates were **pterosaurs,** which were common from the Late Triassic until their extinction during the Cretaceous (■ Figure 22.26). Many pterosaurs were no longer than today's sparrows, robins, and crows, but there were a few exceptions. One Cretaceous species from Texas had a wingspan of about 12 m, but even the largest species probably weighed no more than few tens of kilograms.

Adaptations for flight include a wing membrane supported by an elongated fourth finger (Figure 22.26b); light, hollow bones; and development of those parts of the brain

c The short-tailed pterosaur known as *Pteranodon* was a large, Cretaceous animal with a wingspan of more than 6 m.

associated with muscular coordination and sight. Experiments and studies of fossils of large pterosaurs, such as *Pteranodon* (Figure 22.26c) indicate that these animals likely relied on thermal updrafts to stay airborne, but occasionally flapped their wings for maneuvering In contrast, small pterosaurs must have flapped their wings vigorously, just as present-day small birds do.

Wing flapping requires a high metabolic rate and efficient respiratory and circulatory systems as in birds today, so it seems likely that pterosaurs were warm-blooded. And, in fact, at least one pterosaur species had a coat of hair or hairlike feathers; a fossil found in China in 2009 has hairlike fibers on its wings. In today's animals, only endotherms have hair or feathers.

Marine Reptiles Mesozoic turtles, some crocodiles, and the Triassic mollusk-crushing placodonts adapted to a marine environment, but here we concentrate on the ichthyosaurs, plesiosaurs, and the less-familiar mosasaurs (Figure 22.23). All were thoroughly aquatic marine predators, only the ichthyosaurs and plesiosaurs were closely related, and none was a dinosaur even though they are commonly depicted as such.

The streamlined, rather porpoise-like **ichthyosaurs** (Greek for "fish lizard") ranged from only 0.7 m to 15 m long (■ Figure 22.27a). Ichthyosaurs used their powerful tail for propulsion and maneuvered with their flipper-like forelimbs, which are simply modified forelimbs of their land-dwelling ancestors. They had numerous sharp teeth, and preserved stomach contents reveal a diet of fish, cephalopods, and other marine organisms. It is doubtful that ichthyosaurs could come onto land, so females must have retained the eggs in their bodies and given birth to live young. Fossils with small ichthyosaurs in the appropriate part of the body cavity support this interpretation. An interesting side note on ichthyosaurs is the story of Mary Anning (see Geo-Focus on page 605), who, when she was only 11 years old, discovered and excavated a nearly complete ichthyosaur in southern England.

Plesiosaurs were also aquatic animals, but they probably came ashore to lay their eggs. It is unlikely that they were graceful on land, but then neither are walruses and seals today, but they certainly manage. Plesiosaurs were either long-necked or short-necked (Figures 22.20 and 22.27b) and most were

Figure 22.27 Mesozoic Marine Reptiles

a Icthyosaur.

b A long-necked plesiosaur.

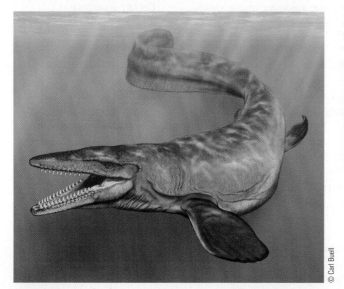

c Mososaur.

modest-sized at 3.6 to 6.0 m long, although one species found in Antarctica measures 15 m. All plesiosaurs were predators.

Mosasaurs were Late Cretaceous marine lizards that measured from 2.5 to 9.0 m long (Figure 22.27c). All were predators, and preserved stomach contents shows that they ate fish, birds, smaller mosasaurs, and several invertebrate animals including amminoids. Mosasaur limbs resembled paddles and were used mostly for maneuvering, whereas their long tail provided propulsion. A possible mosasaur ancestor, found in Texas in 1989, has a mosasaur skeleton, but rather than paddle-like flippers, it retained complete limbs for walking on land.

The Origin and Evolution of Birds

Obviously living birds do not resemble any of today's reptiles, yet the best evidence available indicates that they evolved from theropod dinosaurs. Indeed, both reptiles and birds lay shelled, amniotic eggs, and they share several skeletal features, such as the way the jaw attaches to the skull. Furthermore, since 1860, about 10 fossils have been found in the Solnhofen Limestone in Germany that provide evidence for theropod–bird relationships. These fossils, known as *Archaeopteryx* (from Greek *archaio,* "ancient," and *pteryx,* "feather"), have feathers and a wishbone, the fused clavicles so typical of birds, but numerous skeletal features convince scientists that their ancestors were among the theropods (Figure 22.28). In

Figure 22.28 Fossil of *Archaeopteryx* from the Jurassic-age Solnhofen Limestone of Germany Notice the feather impressions on the wings and the long tail. This animal had feathers and a wishbone, making it a bird, but in most details of its anatomy, it resembled small theropod dinosaurs. For example, it had reptilian teeth, claws on its wings, and a long tail—none of which are found on birds today.

》 Critical Thinking Question
Why do paleontologists think that birds evolved from dinosaurs?

fact, if we could resurrect the oldest known bird, we would be hard-pressed to distinguish it from small theropods.

Even feathers and fused clavicles are no longer sufficient to distinguish birds from theropods, because a wishbone has been found in several theropods, and paleontologists in China have discovered theropods with feathers. The few that oppose the theropod–bird view point to minor skeletal differences and note that theropods are found in Cretaceous-age rocks, but *Archaeopteryx* is Jurassic. However, some of the fossils from China are about the same age as *Archaeopteryx*, narrowing the gap between ancestor and descendant. Indeed, one feathered theropod comes from Late Jurassic-age rocks in China. Furthermore, fossils from China that are slightly younger than *Archaeopteryx* retain ribs in the abdominal region just as in *Archaeopteryx* and small theropods, but they have a reduced tail more like present-day birds.

Unfortunately, the fossil record of early birds is insufficient to resolve whether *Archaeopteryx* was ancestral to all later birds or an early bird that left no evolutionary descendants. Of course, this does not diminish the fact that *Archaeopteryx* had theropod and bird features (recall the concept of *mosaic evolution* from Chapter 18). Another contender for the earliest bird are fossils of two crow-sized individuals called *Protoavis* from Upper Triassic rocks in Texas. However, these fossils are fragmentary and show no feather impressions, so most paleontologists think they are small theropods.

One hypothesis for the origin of bird flight—*from the ground up*—holds that the ancestors of birds were bipedal, fleet-footed ground dwellers that used their wings to leap into the air, at least for short distances, to catch insects or to escape predators. The *from the trees down* hypothesis holds that bird ancestors were bipeds that climbed trees and used their wings for gliding or parachuting. A more recent hypothesis holds that birds may have taken both routes. Researchers know that many of today's bird chicks flap their rudimentary wings to run up steep slopes to avoid predators, and they use the same rudimentary wings to help them descend.

Evolution of Mammals

Recall from Table 18.1 that one prediction of evolutionary theory is, "If today's organisms descended with modification from ones in the past, there should be fossils showing characteristics that connect orders, classes, and so on." *Archaeopteryx* certainly fills the bill, but so do the fossils showing the transition from ancient therapsids (mammal-like reptiles) to mammals. In fact, the transition is so well documented that some fossils are difficult to assign to one group or the other. Among the therapsids, the most mammal-like of all is a group called the **cynodonts,** which by Late Triassic time had given rise to true mammals.

Cynodonts and the Origin of Mammals We can easily recognize living mammals as warm-blooded animals that have hair or fur and mammary glands and, except for the platypus and spiny anteater, give birth to live young. However, these criteria are not sufficient for recognizing fossil mammals; for them, we must rely on skeletal structure only. Several skeletal modifications took place during the transition from mammal-like reptiles to mammals, but

TABLE 22.1	Summary Chart Showing Some Characteristics and How They Changed during the Transition from Reptiles to Mammals		
Features	**Typical Reptile**	**Cynodont**	**Mammal**
Lower Jaw	Dentary and several other bones	Dentary enlarged, other bones reduced	Dentary bone only, except in earliest mammals
Jaw–Skull Joint	Articular quadrate	Articular-quadrate; some advanced cynodonts had both the reptile jaw–skull joint and the mammal jaw–skull joint	Dentary-squamosal
Middle-Ear Bones	Stapes	Stapes	Stapes, Incus, malieus
Secondary Palate	Absent	Partly developed	Well developed
Teeth	No differentiation; chewing teeth single rooted	Some differentiation; chewing teeth partly double rooted	Fully differentiated into incisors, canines, and chewing teeth; chewing teeth double rooted
Tooth Replacement	Teeth replaced continuously	Only two sets of teeth in some advanced cynodonts	Two sets of teeth
Occipital Condyle	Single	Partly divided	Double
Occlusion (chewing teeth meet surface to surface to allow grinding)	No occlusion	Occlusion in some advanced cynodonts	Occlusion
Endothermic vs. Ectothermic	Ectothermic	Probably endothermic	Endothermic
Body Covering	Scales	One fossil shows that it had skin similar to that of mammals	Skin with hair or fur

Figure 22.29 Evolution of the Mammal Jaw and Middle Ear (a) Cynodonts had four bones in the jaw, whereas (b) mammals have only one. Note that the cynodont jaw–skull joint is between the articular and the quadrate bones, but in mammals, it is between the dentary and squamosal bones. (c) Cynodonts had one middle-ear bone—the stapes—but (d) mammals have three: the malleus, the incus, and the stapes. The malleus and incus were derived from the cynodont articular and quadrate bones.

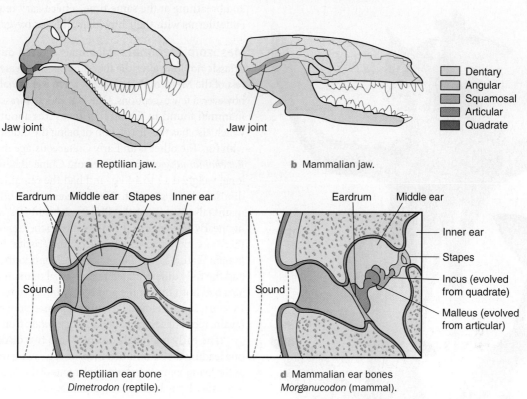

a Reptilian jaw.

b Mammalian jaw.

Dentary
Angular
Squamosal
Articular
Quadrate

c Reptilian ear bone
Dimetrodon (reptile).

d Mammalian ear bones
Morganucodon (mammal).

distinctions between the two groups are based mostly on details of the middle ear, the lower jaw, and the teeth (Table 22.1). Fortunately, the evolution of mammals from cynodonts is well documented by fossils.

Reptiles as well as typical cynodonts have one small bone in the middle ear (the stapes), whereas mammals have three: the incus, the malleus, and the stapes. Also, the lower jaw of a mammal is composed of a single bone called the dentary, but a reptile's jaw is composed of several bones (Figure 22.29). In addition, a reptile's jaw is hinged to the skull at a contact between the articular and quadrate bones, whereas in mammals the dentary contacts the squamosal bone of the skull (Figure 22.29).

During the transition from cynodonts to mammals, the quadrate and articular bones that had formed the joint between the jaw and skull in reptiles were modified into the incus and malleus of the mammalian middle ear (Figure 22.29, Table 22.1). Fossils document the progressive enlargement of the dentary until it became the only element in the mammalian jaw. Likewise, a progressive change from the reptile to mammal jaw joint is documented by fossil evidence. In fact, some of the most advanced cynodonts were truly transitional, because they had a compound jaw joint consisting of (1) the articular and quadrate bones typical of reptiles, and (2) the dentary and squamosal bones as in mammals (Table 22.1).

In Chapter 18, we noted that the study of embryos provides some of the evidence for evolution. Opossum embryos show that the middle-ear bones of mammals were originally part of the jaw. In fact, when opossums are born, the middle ear elements are still attached to the dentary, but as they develop further, these elements migrate to the middle ear, and a typical mammal jaw joint develops.

Several other aspects of cynodonts also indicate they were ancestors of mammals. Their teeth were becoming double rooted as they are in mammals, and they were somewhat differentiated into distinct types that performed specific functions. In mammals, the teeth are fully differentiated into incisors, canines, and chewing teeth (premolars and molars) but typical reptiles do not have differentiated teeth (Figure 22.29a). In addition, mammals have only two sets of teeth during their lifetimes—a set of baby teeth and the permanent adult teeth. Reptiles have teeth replaced continuously throughout their lives, the notable exception being some cynodonts, that in mammal fashion, had only two sets of teeth. Another important feature of mammal teeth is occlusion; that is, the chewing teeth meet surface to surface to allow grinding. Thus, mammals chew their food, but reptiles, amphibians, and fish do not. However, tooth occlusion is known in some advanced cynodonts (Table 22.1).

Reptiles and mammals have a bony protuberance from the skull that fits into a socket in the first vertebra—the atlas. This structure, called the *occipital condyle,* is a single feature in reptiles, but in cynodonts it is partly divided into a double

▌Figure 22.30 Mesozoic Mammals

a Restoration of the oldest known marsupial mammal, *Sinodelphym,* which measures 15 cm long.

b This restoration shows *Eomaia,* the oldest known placental mammal. It was only 12 or 13 cm long. Both fossils come from Early Cretaceous-age rocks in China.

Mark A. Klinger/Carnegie Museum of Natural History

structure typical of mammals. Another mammalian feature, the secondary palate, was partially developed in advanced cynodonts (Table 22.1). This bony shelf separating the nasal passages from the mouth cavity is an adaptation for eating and breathing at the same time, a necessary requirement for endotherms with their high demands for oxygen.

Mesozoic Mammals Mammals evolved during the Late Triassic not long after the first dinosaurs appeared, but for the rest of the Mesozoic Era, most of them were small. There were, however, a few exceptions. One is a Middle Jurassic-age aquatic mammal found in China that measures about 50 cm long, which also has the distinction of being the oldest known fossil with fur. The other is an Early Cretaceous-age mammal called *Repenomamus giganticus,* also from China, that was about 1 m long, weighed 12 to 14 kg, and had the remains of a juvenile dinosaur in its stomach. Nevertheless, most other Mesozoic mammals were about the size of mice and rats, and they were not nearly as diverse as they were during the Cenozoic Era. Furthermore, they retained reptile characteristics, but had mammalian features, too. The Triassic triconodonts, for instance, had the fully differentiated teeth typical of mammals, but they also had both the reptile and the mammal types of jaw joints. In short, some mammal features appeared sooner than others (again, recall the concept of *mosaic evolution* from Chapter 18).

The early mammals diverged into two distinct branches, one leading to the **monotremes,** or egg-laying mammals such as the living spiny anteater and platypus of the Australian region. The other branch includes a group of mammals called eupantotheres that gave rise to both **marsupial** (pouched) **mammals** and the **placental mammals** during the Early Cretaceous (▌Figure 22.30). Most mammals that lived during the Cenozoic Era were placentals, although marsupials were successful in South America and even today make up most of the mammalian fauna of Australia.

> **CONNECTION LINK ◄**
>
> For more information about the diversification of marsupial and placental mammals, see pages 635–637 and 639–641 in Chapter 23.

22.7 Mesozoic Climates and Paleogeography

Fragmentation of the supercontinent Pangaea began by the Late Triassic and continues to the present, but during much of the Mesozoic, close connections existed between the various landmasses. The proximity of these landmasses, however, is not sufficient to explain Mesozoic biogeographic distributions because climates are also effective barriers to wide dispersal. During much of the Mesozoic, though, climates were more equable and lacked the strong north and south zonation characteristic of the present. In short, Mesozoic plants and animals had greater opportunities to occupy much more extensive geographic ranges.

GEO-FOCUS   Mary Anning's Contributions to Paleontology

Paleontologists use fossils to study life of the past, so part of their efforts are spent in finding and collecting fossils. Western European men dominated the early history of this field, but this situation no longer prevails. Indeed, men and women from many countries are now making significant contributions. Perhaps the most notable early exception is Mary Anning (1799–1847), who began a remarkable career as a fossil collector when she was only 11 years old.

Mary Anning was born in Lyme Regis on England's southern coast. When only 15 months old, she survived a lightning strike that, according to one report, killed three girls and according to another, killed a nurse tending her. In 1810, Mary's father, a cabinet maker who also sold fossils part time, died, leaving the family nearly destitute. Mary Anning (▌ Figure 1) expanded the fossil business and became a professional fossil collector

known to the paleontologists of her time, some of whom visited her shop to buy fossils or gather information. She collected fossils from the Dorset coast near Lyme Regis and is reported to have been the inspiration for the tongue twister, "She sell seashells by the seashore."

Soon after her father's death, Mary Anning made her first important discovery—a nearly complete skeleton of a Jurassic ichthyosaur, which was described in 1814 by Sir Everard Home. The sale of this fossil specimen provided considerable financial relief for her family. In 1821, she made a second major discovery and excavated the remains of a plesiosaur. And in 1828, she found the first pterosaur in England, which was sent to the eminent geologist William Buckland at Oxford University.

By 1830, Mary Anning's fortunes began declining as collectors and museums had fewer funds with which

to buy fossils. Indeed, she may once again have become destitute were it not for her geologist friend Henry Thomas de la Beche, also a resident of Lyme Regis. De la Beche drew a fanciful scene called *Duria antiquior,* meaning "An earlier Dorset," in which he brought to life the fossils Mary Anning had collected. The scene was made into a lithograph that was printed and sold widely, and its proceeds went directly to Mary Anning.

Mary Anning died of cancer in 1847, and although only 48 years old, she had a fossil-collecting career that spanned 37 years. Her contributions to paleontology are now widely recognized, but, unfortunately, soon after her death, she was mostly forgotten. Apparently, the people who purchased her fossils were credited with finding them. "It didn't occur to them to credit a woman from the lower classes with such astonishing work. So an uneducated little girl, with a quick mind and an accurate eye, played a key role in setting the course of the 19th century geologic revolution. Then—we simply forgot about her."*

Following Mary Anning's death, her contributions to paleontology were mostly forgotten. Charles Dickens wrote about her life in 1865, but otherwise little notice was taken of her work. However, since 2002, the Palaeontological Association has presented the Mary Anning Award to someone who is "not professionally employed within palaeontology but who made an outstanding contribution to the subject."** She was further recognized in 2010 when the Royal Society of London named her as one of the 10 British women who have most influenced the history of science.

▌ **Figure 1** Mary Anning, who lived in Lyme Regis on England's southern coast, began collecting and selling fossils when she was only 11 years old.

Mary Anning (1799-1847) (oil on canvas), English School, (19th century)/Private Collection/The Bridgeman Art Library International

---

*John H. Lienhard, Professor, College of Engineering, University of Houston.

**The Palaeontological Association, http://www.palass.org/modules.php?name=palaeo&sec=Awards&page=122.

❚ Figure 22.31 Mesozoic Extinctions

a Close-up view of the iridium-rich Cretaceous–Paleogene boundary clay in the Raton Basin, New Mexico.

b Proposed meteorite impact crater centered on Chicxulub on the Yucatán Peninsula of Mexico.

Pangaea persisted as a single unit throughout most of the Triassic (Figure 22.1a), and the Triassic climate was warm-temperate to tropical, although some areas, such as the present southwestern United States, were arid. Mild temperatures extended 50 degrees north and south of the equator, and even the polar regions may have been temperate. The fauna had a truly worldwide distribution. Some dinosaurs had continuous ranges across Laurasia and Gondwana; the peculiar gliding lizards were in New Jersey and England, and reptiles known as phytosaurs lived in North America, Europe, and Madagascar.

By the Late Jurassic, Laurasia had become partly fragmented by the opening North Atlantic, but a connection still existed (Figure 22.1b). The South Atlantic had begun to open so that a long, narrow sea separated the southern parts of Africa and South America. Otherwise, the southern continents were still close together.

The mild Triassic climate persisted into the Jurassic. Ferns, whose living relatives are now restricted to the tropics of southeast Asia, are known from areas as far as 63 degrees south latitude and 75 degrees north latitude. Dinosaurs roamed widely across Laurasia and Gondwana. For example, the giant sauropod *Brachiosaurus* is known from western North America and eastern Africa. Stegosaurs and some families of carnivorous dinosaurs lived throughout Laurasia and in Africa.

By the Late Cretaceous, the North Atlantic had opened further, and Africa and South America were completely separated (Figure 22.1c). South America remained an island continent until late in the Cenozoic and its fauna, evolving in isolation, became increasingly different from faunas elsewhere. Marsupial mammals reached Australia from South America via Antarctica, but the South American connection was eventually severed. Placentals, other than bats and a few rodents, never reached Australia, thus explaining why marsupials continue to dominate the continent's fauna even today.

Cretaceous climates were more strongly zoned by latitude, but they remained warm and equable until the close of that period. Climates then became more seasonal and cooler, a trend that persisted into the Cenozoic. Dinosaur and mammal fossils demonstrate that interchange was still possible, especially between the components of Laurasia.

22.8 Mass Extinctions— A Crisis in the History of Life

The greatest mass extinction took place at the end of the Paleozoic Era, but the one at the close of the Mesozoic has attracted more attention because among its casualties were dinosaurs, flying reptiles, and marine reptiles. Several kinds of marine invertebrates also went extinct, including ammonites, which had been so abundant during the Mesozoic, rudistid bivalves, and some planktonic organisms.

Hypotheses to explain Mesozoic extinctions are numerous, but most have been dismissed as improbable, untestable, or inconsistent with the available data. A proposal that has become popular since 1980 is based on a discovery at the Cretaceous–Paleogene boundary in Italy—a 2.5-cm-thick clay layer with a remarkably high concentration of the platinum-group element iridium (❚ Figure 22.31a). High iridium concentrations have now been identified at many other Cretaceous–Paleogene boundary sites.

The significance of this **iridium anomaly** is that iridium is rare in crustal rocks, but is found in much higher concentrations in some meteorites. Accordingly, some investigators propose a meteorite impact to explain the anomaly and further postulate that the impact of a meteorite, perhaps 10 km in diameter, set in motion a chain of events leading to extinctions. Some Cretaceous–Paleogene boundary sites also contain soot and shock-metamorphosed quartz grains, both of which are cited as additional evidence of an impact.

According to the impact hypothesis, approximately 60 times the mass of the meteorite was blasted from the crust high into the atmosphere, and the heat generated at impact started raging forest fires that added more particulate matter to the atmosphere. Sunlight was blocked for several months, temporarily halting photosynthesis; food chains collapsed; and extinctions followed. Furthermore, with sunlight greatly diminished, Earth's surface temperatures were drastically reduced, adding to the biologic stress. Another consequence of the impact was that vaporized rock and atmospheric gases produced sulfuric acid (H_2SO_4) and nitric acid (HNO_3). Both would have contributed to strongly acid rain that may have had devastating effects on vegetation and marine organisms.

Now some geologists point to a probable impact site centered on the town of Chicxulub on the Yucatán Peninsula of Mexico (❙ Figure 22.31b). The 180-km-diameter structure lies beneath layers of sedimentary rock and appears to be the right age. Evidence that supports the conclusion that the Chicxulub structure is an impact crater includes shocked quartz, the deposits of huge waves, and tektites—small pieces of rock melted during the impact and hurled into the atmosphere.

Even if a meteorite did hit Earth, did it lead to these extinctions? If so, both terrestrial and marine extinctions must have occurred at the same time. To date, strict time equivalence between terrestrial and marine extinctions has not been demonstrated. The selective nature of the extinctions is also a problem. In the terrestrial realm, large animals were the most affected, but not all dinosaurs were large, and crocodiles, close relatives of dinosaurs, survived, although some species died out. Some paleontologists think that dinosaurs, some marine invertebrates, and many plants were already on the decline and headed for extinction before the end of the Cretaceous. A meteorite impact may have simply hastened the process.

In the final analysis, Mesozoic extinctions have not been explained to everyone's satisfaction. Most geologists now concede that a large meteorite impact occurred, but we also know that vast outpourings of lava were taking place in what is now India. Perhaps these brought about detrimental atmospheric changes. Furthermore, the vast shallow seas that covered large parts of the continents had mostly withdrawn by the end of the Cretaceous, and the mild equable Mesozoic climates became harsher and more seasonal by the end of that era. Nevertheless, these extinctions were very selective, and no single explanation accounts for all aspects of this crisis in life history.

GEO-RECAP

❱❱ Geo-Image

❙ **Figure 1** This view in Capitol Reef National Park, Utah, shows excellent exposures of several sedimentary rock formations that were deposited during the Triassic and Jurassic periods.

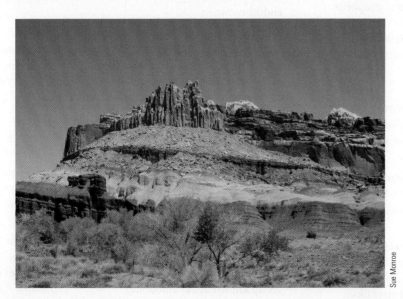

Sue Monroe

a. What accounts for the reddish color of some of the rock layers?

b. The rocks visible on the skyline belong to the Navajo Formation, which is exposed over a large area in the southwest. It is composed of well sorted, well-rounded quartz sandstone, it has tracks of land-dwelling animals, including dinosaurs, and the sandstone has cross-beds up to 30 m high. How do you think it was deposited?

c. Suppose that these rock layers had been deformed and were now upside down. How would you figure out that they had, in fact, been overturned?

Chapter Summary

- Tables 22.2 (page 610) and 22.3 (page 611) summarize many Mesozoic geologic and biologic events.

- We can summarize the breakup of Pangaea as follows:
  1. During the Late Triassic, North America began separating from Africa. This was followed by the rifting of North America from South America.
  2. During the Late Triassic and Jurassic periods, Antarctica and Australia—which remained sutured together—began separating from South America and Africa, and India began rifting from Gondwana.
  3. South America and Africa began separating during the Jurassic, and Europe and Africa began converging during this time.
  4. The final stage in Pangaea's breakup occurred during the Cenozoic, when Greenland completely separated from Europe and North America.

- The breakup of Pangaea influenced global climatic and atmospheric circulation patterns. Although the temperature gradient from the tropics to the poles gradually increased during the Mesozoic, overall global temperatures remained equable.

- An increased rate of seafloor spreading during the Cretaceous Period caused sea level to rise and transgressions to occur.

- Except for incursions along the continental margin and two major transgressions (the Sundance Sea and the Cretaceous Interior Seaway), the North American craton was above sea level during the Mesozoic Era.

- The Eastern Coastal region was the initial site of the separation of North America from Africa that began during the Late Triassic. During the Cretaceous Period, it was inundated by marine transgressions.

- The Gulf Coastal region was the site of major evaporite accumulation during the Jurassic as North America rifted from South America. During the Cretaceous, it was inundated by a transgressing sea, which, at its maximum, connected with a sea transgressing from the north to create the Cretaceous Interior Seaway.

- Mesozoic rocks of the western region of North America were deposited in a variety of continental and marine environments. One of the major controls of sediment distribution patterns was tectonism.

- Western North America was affected by four interrelated orogenies: the Sonoma, Nevadan, Sevier, and Laramide. Each involved igneous intrusions, as well as eastward thrust faulting and folding.

- The cause of the Nevadan, Sevier, and Laramide orogenies was the changing angle of subduction of the oceanic Farallon plate beneath the continental North American plate. The direction and rate of plate movement along the western coast were related to seafloor spreading and the opening of the Atlantic Ocean.

- Although the structural features of North America's western margin are associated with activity along an oceanic–continental convergent plate boundary, it is thought that more than 25% of the western margin originated from the accretion of terranes.

- Mesozoic rocks contain a variety of mineral resources, including coal, petroleum, uranium, gold, and copper.

- The marine invertebrate survivors of the Permian mass extinction diversified and gave rise to increasingly diverse Mesozoic marine invertebrate communities.

- Land-plant communities of the Triassic and Jurassic consisted of seedless vascular plants and gymnosperms. The angiosperms, or flowering plants, evolved during the Early Cretaceous, diversified rapidly, and were soon the most abundant land plants.

- Dinosaurs evolved during the Late Triassic, but were most abundant and diverse during the Jurassic and Cretaceous. The two distinct groups of dinosaurs, based on pelvic structure, are Saurischia (lizard-hipped) and Ornithischia (bird-hipped).

- Bone structure, predator–prey relationships, and other criteria have been cited as evidence of dinosaur endothermy. Although there is still no solid consensus, many paleontologists think that some dinosaurs were endotherms.

- That some theropods had feathers indicates that they were warm-blooded and provides further evidence of their relationship to birds.

- Small pterosaurs were probably active, wing-flapping fliers, whereas large ones may have depended on soaring to stay aloft. At least one pterosaur species had hair or feathers, so it, and perhaps all pterosaurs, were probably endothermic.

- The fish-eating, porpoise-like ichthyosaurs were thoroughly adapted to an aquatic life, whereas the plesiosaurs with their paddle-like limbs could most likely come out of the water to lay their eggs. The marine reptiles known as mosasaurs are most closely related to lizards.

- Birds probably evolved from small theropod dinosaurs. The oldest known bird, *Archaeopteryx*, appeared during the Jurassic; however, few other Mesozoic birds are known.

- The earliest mammals evolved during the Late Triassic, but they are difficult to distinguish from advanced cynodonts. Details of the teeth, middle ear, and lower jaw are used to distinguish the two.

- Several types of Mesozoic mammals existed, but most were small, and their diversity was low. Both marsupials and placentals evolved during the Cretaceous from a group known as eupantotheres.

- Because the continents were close together and climates were mild during much of the Mesozoic, animals and plants occupied much larger geographic ranges than they do now.

- Among the victims of the Mesozoic mass extinctions were dinosaurs, flying reptiles, marine reptiles, and several groups of marine invertebrates. A meteorite impact may have caused these extinctions, but some paleontologists think that other factors were also important.

Important Terms

amniote (p. 593)
angiosperm (p. 593)
archosaur (p. 594)
bipedal (p. 594)
Cordilleran orogeny (p. 584)
Cretaceous Interior Seaway (p. 589)
cynodont (p. 602)
dinosaur (p. 594)
ectotherm (p. 597)

endotherm (p. 597)
ichthyosaur (p. 600)
iridium anomaly (p. 606)
Laramide orogeny (p. 585)
marsupial mammal (p. 604)
monotreme (p. 604)
mosasaur (p. 601)
Nevadan orogeny (p. 585)
Ornithischia (p. 595)

placental mammal (p. 604)
plesiosaur (p. 601)
pterosaur (p. 600)
quadrupedal (p. 594)
Saurischia (p. 595)
Sevier orogeny (p. 585)
Sonoma orogeny (p. 584)
Sundance Sea (p. 589)
terranes (p. 590)

Review Questions

1. Which one of the following animals is not an amniote?
   a. ___ reptile.
   b. ___ mammal.
   c. ___ amphibian.
   d. ___ bird.
   e. ___ dinosaur.

2. A possible cause for the eastward migration of igneous activity in the Cordilleran region during the Cretaceous was a change from
   a. ___ high-angle to low-angle subduction.
   b. ___ divergent plate margin activity to subduction.
   c. ___ subduction to divergent plate margin activity.
   d. ___ oceanic–oceanic convergence to oceanic–continental convergence.
   e. ___ divergent to convergent plate margin activity.

3. The second Mesozoic major seaway to flood North America was the
   a. ___ Cretaceous Interior Seaway.
   b. ___ Sundance Sea.
   c. ___ Cordilleran.
   d. ___ Zuni.
   e. ___ Newark.

4. Among the therapsids, the cynodonts were
   a. ___ the most mammal-like.
   b. ___ the descendants of birds.
   c. ___ the largest dinosaurs.
   d. ___ the ancestors of pterosaurs.
   e. ___ extinct by the end of the Paleozoic Era.

5. The time of greatest post-Paleozoic inundation of the craton occurred during which geologic period?
   a. ___ Triassic.
   b. ___ Jurassic.
   c. ___ Cretaceous.
   d. ___ Paleogene.
   e. ___ Neogene.

6. The jaw–skull joint in mammals is between the
   a. ___ atlas and mandible.
   b. ___ occipital and quadrate.
   c. ___ articular and maxillary.
   d. ___ dentary and squamosal.
   e. ___ occlusal and zygomatic.

7. Which formation or group filled the Late Triassic fault-block basins of the east coast of North America with red, nonmarine sediments?
   a. ___ Franciscan.
   b. ___ Morrison.
   c. ___ Chinle.
   d. ___ Newark.
   e. ___ Navajo.

8. Modification of the hand yielding an elongated fourth finger for wing support is found in
   a. ___ birds.
   b. ___ insects.
   c. ___ pterosaurs.
   d. ___ theropods.
   e. ___ bats.

9. Which orogeny produced the Sierra Nevada, Southern California, Idaho, and Coast Range batholiths?
   a. ____ Laramide.
   b. ____ Sonoma.
   c. ____ Nevadan.
   d. ____ Sevier.
   e. ____ Antler.

10. Stegosaurs were all
   a. ____ small carnivores.
   b. ____ quadrupedal herbivores.
   c. ____ bipedal plant eaters.
   d. ____ omnivorous tree climbers.
   e. ____ placental monotremes.

TABLE 22.2   Mesozoic Geologic Events

Age (millions of years)	Geologic Period	Sequence	Relative Changes in Sea Level (Rising / Falling)	Cordilleran Mobile Belt	Gulf Coast Region	Eastern Coastal Region	Global Plate Tectonic Events
66	Cretaceous	Zuni	Present sea level	Jurassic and Cretaceous tectonism controlled by eastward subduction of the Pacific plate beneath North America and accretion of microplates. Cordilleran orogeny — Laramide orogeny; Sevier orogeny; Nevadan orogeny	Major Late Cretaceous transgression. Reefs particularly abundant. Regression at end of Early Cretaceous. Early Cretaceous transgression and marine sedimentation	Appalachian region uplifted. Erosion of fault-block mountains that formed during the Late Triassic to Early Jurassic	South America and Africa are widely separated. Greenland begins separating from Europe
146	Jurassic				Sandstones, shales, and limestones are deposited in transgressing and regressing seas. Thick evaporites are deposited in newly formed Gulf of Mexico	Fault-block mountains and basins develop in eastern North America	South America and Africa begin separating in the Late Jurassic
200	Triassic	Absaroka		Subduction zone develops as a result of westward movement of North America. Sonoma orogeny	Gulf of Mexico begins forming during Late Triassic	Deposition of Newark Group; lava flows, sills, and dikes	Breakup of Pangaea begins with rifting between Laurasia and Gondwana. Supercontinent Pangaea still in existence
251							

11. Briefly explain some of the evidence geologists use to interpret climatic conditions during the Mesozoic Era.

12. Discuss the details of evolution of the jaw and the jaw–skull joint in the transition from cynodonts to mammals.

13. Why did the angiosperms have so much success in colonizing the land, and what effect did this have on other major groups of plants?

14. From a plate tectonic perspective, how does the orogenic activity that occurred in the Cordilleran mobile belt during the Mesozoic Era differ from that which took place in the Appalachian mobile belt during the Paleozoic Era?

15. What is the most popular hypothesis for the extinctions at the end of the Mesozoic, and what evidence supports this idea?

16. How did the Mesozoic rifting that took place on the eastern coast of North America affect the tectonics in the Cordilleran mobile belt?

17. How does terrane accretion relate to the Mesozoic orogenies that took place on the western margin of North America?

18. What kind of evidence indicates that at least some dinosaurs were endotherms?

19. What modifications took place in marine reptiles as they adapted to an aquatic life?

20. Explain how plate positions and climate influenced the geographic distribution of Mesozoic plants and animals.

TABLE 22.3   Mesozoic Biologic Events

Age (millions of years)	Geologic Period	Invertebrates	Vertebrates	Plants	Climate
66 —	Cretaceous	Extinction of ammonites, rudists, and most planktonic foraminifera at end of Cretaceous Continued diversification of ammonites and belemnoids Rudists become major reef builders	Extinctions of dinosaurs, flying reptiles, marine reptiles, and some marine invertebrates Placental and marsupial mammals diverge	Angiosperms evolve and diversify rapidly Seedless plants and gymnosperms still common but less varied and abundant	Climate becomes more seasonal and cooler at end of Cretaceous North-south zonation of climates more marked but remains equable
146 —	Jurassic	Ammonites and belemnoid cephalopods increase in diversity Scleractinian coral reefs common Appearance of rudist bivalves	First birds (may have evolved in Late Triassic) Time of giant sauropod dinosaurs	Seedless vascular plants and gymnosperms only	Much like Triassic Ferns with living relatives restricted to tropics live at high latitudes, indicating mild climates
200 — 251 —	Triassic	The seas are repopulated by invertebrates that survived the Permian extinction event Bivalves and echinoids expand into the infaunal niche	Mammals evolve from cynodonts Cynodonts become extinct Ancestral archosaur gives rise to dinosaurs Flying reptiles and marine reptiles evolve	Land flora of seedless vascular plants and gymnosperms as in Late Paleozoic	Warm-temperate to tropical Mild temperatures extend to high latitudes; polar regions may have been temperate Local areas of aridity

(Between the Geologic Period and Vertebrates columns, spanning the Cretaceous and Jurassic rows, the vertical label reads: "Greatest Diversity of Dinosaurs")

The Eocene-age Claron Formation in Bryce Canyon National Park in Utah. Bryce Canyon is actually not a canyon but rather the eroded margin of a plateau. The park is noted for its colorful spires, hoodoos, arches, and other erosional features. The formation consists of conglomerate, sandstone, limestone, and shale that were deposited in stream channels, on floodplains, and in lakes. The pink to reddish color results from small amounts of oxidized iron.

CHAPTER 23

Cenozoic Earth and Life History

James S. Monroe

OUTLINE

23.1 Introduction

23.2 Cenozoic Plate Tectonics

23.3 Cenozoic Orogenic Belts

23.4 Paleogene and Neogene Evolution of North America

GEO-INSIGHT: Cascade Range Volcanoes

23.5 The Pleistocene and Holocene Epochs

23.6 Cenozoic Mineral Resources

23.7 Paleogene and Neogene Life History

GEO-FOCUS: A Messel Pit Fossil Site in Germany

23.8 Pleistocene Faunas

GEO-RECAP

HAVE YOU EVER WONDERED?

- How plate tectonics played a role in the distribution of Cenozoic continents?

- What events led to the origin of the present-day Rocky Mountains and Appalachian Mountains?

- Why western North America has a chain of active volcanoes?

- When and how the San Andreas fault originated?

- What caused the extensive glaciers of the Ice Age?

- Why the Cenozoic Era is called the Age of Mammals?

- What caused the extinctions at the end of the Ice Age?

- What the evolutionary history of humans is (i.e., who are our ancestors, and how complete the fossil record is of our species and primates in general)?

23.1 Introduction

At 66 million years long, the Cenozoic Era is only 1.4% of all geologic time, or just 20 minutes on our hypothetical 24-hour clock (see Figure 19.1). Such a "short" interval is short only in the context of geologic time, but by any other measure it is longer than we can even imagine. Certainly the Cenozoic Era was long enough for significant evolution of Earth and its biota.

Based on recent work by the International Commission on Stratigraphy, geologists divide the Cenozoic Era into three periods of unequal duration—the *Paleogene Period* (66 to 23 million years ago), the *Neogene Period* (23 to 2.6 million years ago), and the *Quaternary Period* (2.6 million years ago to the present). Each of these periods is further divided into epochs (Figure 17.1). In your studies of geology, you may see publications with the Cenozoic Era made up of the Tertiary and Quaternary periods, because Paleogene and Neogene have only recently been proposed. In any event, the time scale in Figure 17.1 is the one now recommended for subdivisions of the Cenozoic.

Geologists know more about Cenozoic Earth and life history than about any other interval of geologic time because the rocks, being the youngest, are the most easily accessible at or near Earth's surface. In fact, Cenozoic-age sedimentary and igneous rocks are widespread in western North America, but exposures other than Ice Age deposits are limited in the east (see chapter opening photo and ▍ Figure 23.1). Furthermore, the fossil record for Cenozoic life, although not truly representative of all life of that time, is very good for some groups of organisms. For example, we know much about the evolution of many groups of mammals, a topic that we emphasize in this chapter. Among the mammals we examine are the primates—in particular the origin and evolution of humans. We are learning so much so quickly about human evolution that by the time you read this chapter, new discoveries may have changed some our interpretations.

Although we emphasize mammals here, important events took place in other groups of organisms. The angiosperms continued to dominate land-plant communities and now constitute more than 90% of all land plants. Birds evolved during the Jurassic, but the families now common appeared during the Paleogene and Neogene, reached their maximum diversity during the Quaternary, and have declined slightly since then. After the mass extinction at the end of the Mesozoic Era, marine invertebrates diversified, giving rise to the familiar fauna of the seas.

It is important that we understand Cenozoic Earth and life history because the present distribution of land and sea, oceanic and atmospheric circulation patterns, and Earth's distinctive topography and its biota evolved during this time. The Appalachian Mountains, for instance, began their overall evolution during the Proterozoic Eon, but their present topography developed largely from Cenozoic uplift and erosion. Likewise, the San Andreas fault of California, the

▍ **Figure 23.1 Cenozoic-Age Rocks** Outcroups of Cenozoic rocks are common in the western part of the country (**a, b**), but, other than Pleistocene-age deposits, are found in only a few areas in the east (**c**).

a The Paleocene Fort Union Formation in Montana.

b Outcrops of basalt from a Miocene-age volcano at Cape Foulweather, Oregon.

c Miocene and Pliocene-age sedimentary rocks in the Calvert Cliffs of Maryland.

Himalayas in Asia, and the Andes of South America owe their characteristics to Cenozoic events.

23.2 Cenozoic Plate Tectonics

The ongoing fragmentation of Pangaea accounts for the present distribution of Earth's landmasses. Moving plates also directly affect the biosphere because the geographic locations of continents profoundly influence the atmosphere and hydrosphere.

We noted in Chapter 22 that as the Americas separated from Europe and Africa, the Atlantic Ocean basin opened, first in the south and later in the north. Spreading ridges such as the Mid-Atlantic Ridge and East Pacific Rise were established, along which new oceanic crust formed and continues to form. However, the age of the oceanic crust in the Pacific is very asymmetric, because much of the crust in the eastern Pacific Ocean basin has been subducted beneath the westerly moving North and South American plates (see Figure 2.15).

Rifting and the separation of landmasses is not restricted to the Triassic. In fact, Neogene rifting began in East Africa, the Red Sea, and the Gulf of Aden (see Figure 2.18). Rifting in East Africa is in its early stages, because the continental crust has not yet stretched and thinned enough for new oceanic crust to form from below. In the Red Sea, rifting and the Late Pliocene origin of oceanic crust followed vast eruptions of basalt, and in the Gulf of Aden, Earth's crust had stretched and thinned enough by Late Miocene time for upwelling basaltic magma to form new oceanic crust. The Arabian plate is moving north, so it, too, causes some of the deformation taking place from the Mediterranean through India.

In the meantime, the North and South American plates continued their westerly movement as the Atlantic Ocean basin widened (Figure 23.2). Subduction zones bounded both continents on their western margins, but the situation changed in North America as it moved over the northerly extension of the East Pacific Rise, and it now has a transform plate boundary, a topic we discuss more fully in a later section.

23.3 Cenozoic Orogenic Belts

Cenozoic orogenic activity took place in two major zones or belts: the *Alpine–Himalayan orogenic belt* and the *circum-Pacific orogenic belt* (Figure 23.3). Both belts are made up of smaller segments known as **orogens,** each of which shows the features of an orogeny—deformation, metamorphism, emplacement of plutons, and thickening of Earth's crust.

> **CONNECTION LINK**
>
> For a review of orogenies, see pages 245–252 in Chapter 10.

The Alpine–Himalayan Orogenic Belt

The **Alpine–Himalayan orogenic belt** extends eastward from Spain through the Mediterranean region, the Middle East and India, and on into Southeast Asia (Figure 23.3). Ongoing deformation, volcanism, and seismicity remind us that it remains active. Remember that during the Mesozoic, the Tethys Sea separated much of Gondwana from Eurasia, but this sea closed during the Cenozoic when the African plate collided with the huge landmasses to the north.

During the *Alpine orogeny,* deformation occurred in a linear zone extending from Spain eastward through Greece and Turkey. Mountain building in this region yielded the Pyrenees between Spain and France and the Apennines of Italy, as well as the Alps of mainland Europe. Plate convergence also produced an almost totally isolated sea in the Mediterranean basin where Late Miocene deposition in an arid environment accounts for evaporite deposits up to 2 km thick.

The collision of the African plate with Eurasia also accounts for the origin of the Atlas Mountains of northwest Africa, and further to the east, Africa continues to force oceanic lithosphere northward beneath Greece and Turkey. Erupting volcanoes in Italy and Greece and seismic activity throughout the entire region indicate that southern Europe and the Middle East remain geologically active.

The *Himalayan orogen* also resulted from plate convergence, but in this case, two continental plates collided, giving rise to the loftiest mountains on Earth (see Chapter 10 opening photo and Figure 10.21b). During the Early Cretaceous, India broke away from Gondwana and moved north, and as it did so, oceanic lithosphere was consumed at a subduction zone along the southern margin of Asia. The descending plate partially melted, yielding magma that rose to form granitic plutons and a chain of volcanoes in what is now Tibet. The Indian plate eventually collided with and became sutured to Asia, resulting in deformation and uplift of the Himalayan orogen.

India's collision probably began about 40 to 50 million years ago when the northward movement of the Indian plate decreased from 15 to 20 cm/yr to about 5 cm/yr. This decrease probably marks the time of collision and India's resistance to subduction, but the Indian plate has been thrust about 2,000 km beneath Asia, causing crustal thickening. This underthrusting continues at a rate of about 5 cm/yr.

The Circum-Pacific Orogenic Belt

The **circum-Pacific orogenic belt** is an area of tectonism made up of several orogens along the western margins of South, Central, and North America, as well as the Aleutian Islands, the eastern margin of Asia, the islands north of Australia, and New Zealand (Figure 23.3). Subduction of oceanic lithosphere causes deformation and volcanism in the orogens in the western and northern Pacific. For example, the Pacific plate is subducted beneath the

▌**Figure 23.2** Paleogeography of the World for the Eocene and Miocene Epochs

a Eocene Epoch.

b Miocene Epoch.

Aleutian Islands, Japan, and the Philippines, accounting for continuing tectonism in these areas.

In the eastern part of the Pacific, the Cocos and Nazca plates moved west from the East Pacific Rise only to be consumed at subduction zones along the west coasts of Central and South America, respectively (see Figure 10.20). This entire region also remains tectonically active. In South America, for instance, the Andes Mountains, with about 50 peaks higher than 6,000 m, also have many active volcanoes, and it is an area of seismic activity. Another important part of the circum-Pacific orogenic belt is the *North American Cordillera,* which we discuss in the next section.

23.4 Paleogene and Neogene Evolution of North America

We mentioned in the Introduction that many of Earth's features have long histories, but they owe their present-day distinctive characteristics to events that took place during the Cenozoic Era. Here we concentrate on the events of the Paleogene and Neogene periods, although we should note that the evolution of Earth's surface continued during the Quaternary Period, which we cover later in this chapter.

▌ Figure 23.3 **Present-Day Orogenic Activity** Most of Earth's geologically recent and present-day orogenic activity is concentrated in the circum-Pacific and Alpine–Himalayan orogenic belts.

》 Critical Thinking Question
Where are orogenies taking place at convergent boundaries where oceanic lithosphere collides with continental lithosphere?

The North American Cordillera

The **North American Cordillera,** a complex mountainous region in the West, is a huge segment of the circum-Pacific orogenic belt extending from Alaska southward through Canada and the continental United States and into central Mexico. In the United States, it widens to 1,200 km, stretching east-west from the east flank of the Rocky Mountains to the Pacific Ocean (▌ Figure 23.4). Although the geologic evolution of this region began during the Proterozoic, a protracted episode of deformation known as the *Cordilleran orogeny* began during the Late Triassic as the Nevadan, Sevier, and Laramide orogenies progressively affected areas from west to east (see Figure 22.11). The first two of these orogenies were discussed in Chapter 22, so here we are concerned with the Laramide orogeny.

The Late Cretaceous to Eocene **Laramide orogeny** was the last event in the Cordilleran orogeny, but it differed from its two predecessors in important ways. First, it took place much farther inland from a convergent plate boundary, and neither volcanism nor emplacement of plutons was very common. Second, deformation was mostly vertical, fault-bounded uplifts rather than the compression-induced folding and thrust faulting typical of most orogenies.

To account for these differences, geologists have modified their model for orogenies at convergent plate boundaries. During the preceding Nevadan and Sevier orogenies, the oceanic Farallon plate was subducted

beneath North America at about a 50-degree angle, and volcanism and plutonism took place 150 to 200 km inland from the oceanic trench. Most geologists now agree that by Early Paleogene time, there was a change in the subduction angle from steep to gentle, and the Farallon plate moved nearly horizontally beneath the continent. However, they disagree on what may have caused this change in angle of subduction.

According to one hypothesis, a buoyant oceanic plateau that was part of the Farallon plate was carried beneath the continent and resulted in shallow subduction. Another hypothesis holds that North America overrode the Farallon plate beneath which was the deflected head of a mantle plume (▌ Figure 23.5a). The lithosphere above the plume was buoyed up, accounting for shallow subduction. As a result, igneous activity shifted farther inland and finally ceased because the descending plate no longer penetrated to the mantle (▌ Figure 23.5b).

The changing angle of subduction also resulted in a change in the style of deformation that involved large-scale buckling and fracturing, which yielded fault-bounded vertical uplifts. Erosion of the uplifted blocks yielded mountainous topography and erosion of the mountains supplied sediments to the intervening basins, and it also accounted for huge overthrust faults, especially in the Northern Rocky Mountains. By Middle Eocene time, Laramide deformation ceased and volcanism resumed when the mantle plume beneath the lithosphere disrupted the

Figure 23.4 The North American Cordillera and the Major Provinces in the United States and Canada
The Cordillera in western North America is a segment of the circum-Pacific orogenic belt.

overlying oceanic plate (▌ Figure 23.5c). By the Neogene, the mountains had been deeply eroded and the basins filled with sediment; however, renewed uplift and erosion during the Late Neogene accounts for the present ranges.

Cordilleran Igneous Activity

The enormous batholiths in British Columbia, Canada, Idaho, and the Sierra Nevada of California were emplaced during the Mesozoic (see Chapter 22), but intrusive activity

Figure 23.5 Laramide Orogeny The Late Cretaceous to Eocene Laramide orogeny took place as the Farallon plate was subducted beneath North America.

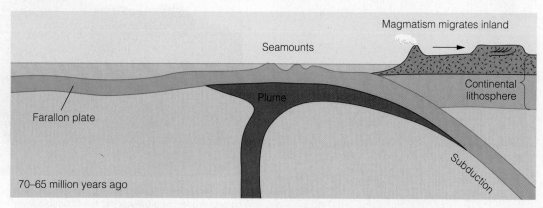

a As North America moved westward over the Farallon plate, beneath which was the deflected head of a mantle plume, the angle of subduction decreased, and the igneous activity shifted inland.

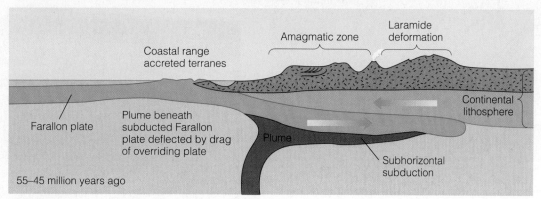

b With nearly horizontal subduction, igneous activity ceased and the continental crust was deformed, mostly by vertical uplift.

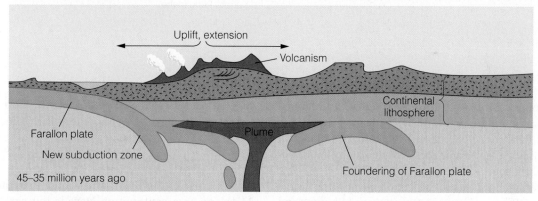

c Disruption of the oceanic plate by the mantle plume marked the onset of renewed igneous activity.

continued into the Paleogene Period. Numerous small plutons formed, including copper- and molybdenum-bearing stocks in Utah, Nevada, Arizona, and New Mexico.

Volcanism was more or less continuous in the Cordillera, but it varied in location, intensity, and eruptive style, and it ceased temporarily in the area of the Laramide orogen (Figure 23.5b). In the Pacific Northwest, the Columbia Plateau (▌ Figure 23.6a) is underlain by 200,000 km³ of Miocene lava flows called the Columbia River basalts that have an aggregate thickness of about 2,500 m. These vast lava

flows are well exposed in the walls of the canyons eroded by the Columbia and other rivers (▌ Figure 23.6b). The relationship of this huge outpouring of lava to plate tectonics remains unclear, but some geologists think that it resulted from a mantle plume beneath western North America.

The Snake River Plain (Figure 23.6a), which is mostly in Idaho, is actually a depression in the crust that was filled by Miocene and younger rhyolite, volcanic ash, and basalt (▌ Figure 23.6c). These rocks are youngest in the southwest part of the area and become older toward the northeast,

Figure 23.6 Cenozoic Volcanism

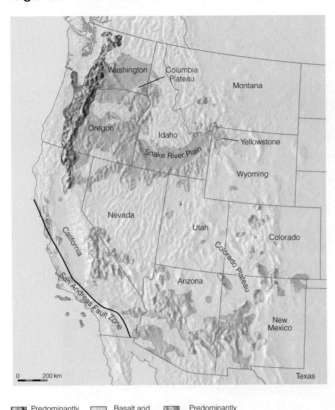

Predominantly andesitic · Basalt and andesite · Predominantly basaltic

a Distribution of Cenozoic volcanic rocks in the western United States.

>> **Critical Thinking Question**
Where on this map are the active volcanoes of the Cascade Range?

Sue Monroe

b Columbia River basalts are exposed in the walls of this canyon eroded by a tributary of the Columbia River in Oregon. Multnomah Falls plunge 189 m from a small tributary valley.

James S. Monroe

c Basalt lava flows of the Snake River Plain at Malad Gorge State Park, Idaho.

leading some to propose that North America has migrated over a mantle plume that now lies beneath Yellowstone National Park in Wyoming. Other geologists disagree, thinking that these volcanic rocks erupted along an intracontinental rift zone.

Bordering the Snake River Plain on the northeast is the Yellowstone Plateau (Figure 23.6a), an area of Pliocene and Pleistocene volcanism. Perhaps a mantle plume lies beneath the area, as just noted, that accounts for the ongoing hydrothermal activity there, but the heat may come from an intruded body of magma that has not yet completely cooled.

Some of the most majestic and highest mountains in the Cordillera are in the **Cascade Range** of northern California, Oregon, Washington, and southern British Columbia, Canada (see the Geo-Insight on pages 622 and 623). Thousands of volcanic vents are present, the most impressive of which are the dozen or so large composite volcanoes and Lassen Peak in California, the world's largest lava dome. Volcanism in this region is related to subduction of the Juan de Fuca plate beneath North America. Volcanism in the Cascade Range

goes back at least to the Oligocene, but the most recent episode began during the Late Miocene or Early Pliocene.

Basin and Range Province

Earth's crust in the **Basin and Range Province** (Figure 23.4)—an area centered on Nevada but extending into adjacent states and northern Mexico—has been stretched and thinned, yielding north-south–oriented mountain ranges with intervening valleys or basins. The ranges are bounded on one or both sides by steeply dipping normal faults that probably curve and dip less steeply with depth (Figure 23.7a).

Before faulting began, the region was deformed during the Nevadan, Sevier, and Laramide orogenies. Then, during the Paleogene, the entire area was highlands undergoing extensive erosion, but Early Miocene eruptions of rhyolitic lava flows and pyroclastic materials covered large areas. By the Late Miocene, large-scale faulting had begun, forming the basins and ranges. Sediment derived from the ranges

Figure 23.7 The Basin and Range Province

a The Basin and Range is mostly in Nevada, but it extends into adjacent states and northern Mexico. The ranges and the intervening basins trend north-south. Each range is bounded on one or both sides by normal faults.

b The Sierra Nevada at the western margin of the Basin and Range has risen along normal faults so that it is more than 3,000 m above the valley to the east.

Nicholas M. Short, Sr/HIS/NASA

Sue Monroe

was transported into the adjacent basins and accumulated as alluvial fan and playa lake deposits.

At its western margin, the Basin and Range Province is bounded by normal faults along the east face of the Sierra Nevada (Figure 23.7b). Pliocene and Pleistocene uplift tilted the Sierra Nevada toward the west, and its crest now stands 3,000 m above the basins to the east. Before this uplift took place, the Basin and Range Province had a subtropical climate, but the rising mountains created a rain shadow, making the climate increasingly arid.

Colorado Plateau and Rio Grande Rift

The vast, elevated **Colorado Plateau** in Colorado, Utah, Arizona, and New Mexico, has volcanic mountains rising above it, brilliantly colored rocks, and deep canyons. In Chapters 20 and 22, we noted that during the Permian and Triassic, the Colorado Plateau region was the site of extensive red bed deposition; many of these rocks are now exposed in the uplifts and canyons (Figure 23.8).

Cretaceous-age marine sedimentary rocks indicate that the Colorado Plateau was below sea level then, but during the Paleogene Period, Laramide deformation yielded broad anticlines and arches and basins, and a number of large normal faults. Neogene uplift elevated the region from near sea level to the 1,200 to 1,800-m elevations seen today, and as uplift proceeded, streams and rivers began eroding deep canyons.

Geologists disagree on the details of just how the deep canyons so typical of the region developed—such as the Grand Canyon. Some think that the streams were *antecedent,* meaning that they existed before the present topography developed, in which case they simply eroded downward as uplift proceeded. Others think that the streams were *superposed,* implying that younger strata covered the area on which streams were established. During uplift, the streams stripped away these younger rocks and eroded down into the underlying strata. In either case, the landscape continues to evolve as erosion of the canyons and their tributaries deepens and widens them.

The **Rio Grande rift** extends north to south about 1,000 km from central Colorado through New Mexico and into northern Mexico (Figure 23.9) where Earth's crust has been stretched and thinned. It is bounded on both sides by normal faults, volcanoes and calderas are present in the rift, and seismic activity continues. Rifting began 29 million years ago and persisted for 10 to 12 million years; after a hiatus, renewed rifting began 17 million years ago and continues even now, but only at about 2 mm per year. As rifting took place, the rift was filled with huge amounts of sediments and volcanic rocks.

GEO-INSIGHT   Cascade Range Volcanoes

Several large volcanoes and hundreds of smaller volcanic vents are present in the Cascade Range of the Pacific Northwest. Medicine Lake Volcano and Newberry Volcano are shield volcanoes that lie just east of the main trend of the Cascade Range. Volcanism began in the range during the Oligocene, but the most recent episode started about 5 million years ago, and of course, it continues to the present.

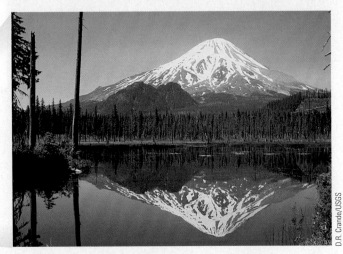

▲ **2.** Mount St. Helens in Washington as it appeared before it erupted violently in May 1980.

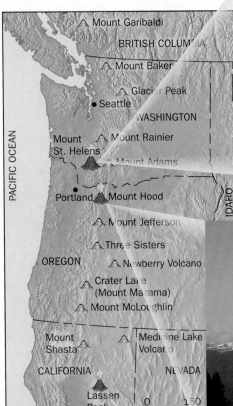

▲ **1.** The Cascade Range extends from northern California into southern British Columbia, Canada. This map shows the locations of the major volcanoes in the range. All are composite volcanoes with the exceptions of Lassen Peak (lava dome) and two shield volcanoes, Medicine Lake Volcano and Newberry Volcano.

▲ **3.** Mount Hood in Oregon last erupted in 1865, and perhaps in 1903. It lies west of Portland, Oregon, a city of about 2 million people. Although Mount Hood shows no activity at present, it is still a dangerous mountain. Several climbers have lost their lives on the mountain in recent years.

▶ **4.** Lassen Peak is a lava dome that formed about 27,000 years ago on the northeast flank of a deeply eroded composite volcano.

Courtesy of Keith Ronnholm

PhotoDisc/Getty Images

Lynn Topinka/ USGS

5. Mount St. Helens's lateral blast on May 18, 1980, took place when a bulge on the volcano's north face collapsed, reducing the pressure on gas-charged magma.

6. Mount St. Helens's lateral blast killed about 63 people and leveled 600 km^2 of forest. These trees in Smith Creek Valley are in part of the blast zone. For scale, see the two men within the red circle.

7. Shortly after the lateral blast on May 18, 1980, this 19-km-high ash and steam cloud erupted from Mount St. Helens.

Photo by Chester Mullen; courtesy National Park Service/USGS

Sue Monroe

8. Lassen Peak erupted numerous times from 1914 to 1917. This eruption took place in 1915.

9. A huge steam explosion called the Great Hot Blast leveled the area in the foreground in 1915. In the 96 years since the eruption, trees are becoming reestablished in this area, known as the Devastated Area.

Figure 23.8 The Colorado Plateau Rocks of the Colorado Plateau range from Proterozoic to Cenozoic. Cenozoic uplift has elevated the plateau to its present elevation. As uplift occurred, the area was deeply eroded, so now we see deep canyons and erosional remnants of ancient rocks such as these at Monument Valley Navajo Tribal Park in Arizona.

>> **Critical Thinking Question**
Why are the lower slopes of the mesas gentle whereas those of the upper parts are steep?

Figure 23.9 The Rio Grande Rift Location of the basins making up the Rio Grande rift. A complex of normal faults is present on both sides of the rift.

Pacific Coast

Before the Eocene, the entire Pacific Coast was a convergent plate boundary where the **Farallon plate** was consumed at a subduction zone that stretched from Mexico to Alaska. Now there are only two small remnants of the Farallon plate—the Juan de Fuca and Cocos plates (■ Figure 23.10). Continuing subduction of these small plates accounts for the present seismic activity and volcanism in the Pacific Northwest and Central America, respectively.

Another consequence of plate interactions in this region was the westward movement of the North American plate and its collision with the Pacific–Farallon ridge. Because the Pacific–Farallon ridge was at an angle to the margin of North America, the continent–ridge collision took place first during the Eocene in northern Canada

Figure 23.10 Origin of the San Andreas and Queen Charlotte Faults Three stages in the westward movement of North America and its collision with the Pacific–Farallon Ridge. As North America overrode the ridge, its margin became bounded by transform faults except in the Pacific Northwest.

▌ Figure 23.11 Cenozoic Sedimentary and Igneous Rocks in the Great Plains

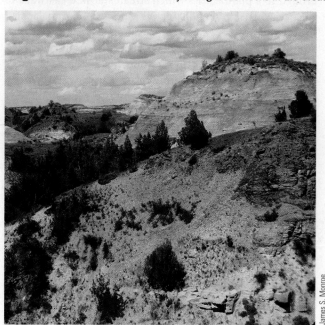

James S. Monroe

a Huge amounts of sediments shed from the Laramide highlands were deposited on the Great Plains. Paleocene sedimentary rocks seen from Scoria Point in Theodore Roosevelt National Park in North Dakota. The scoria, the reddish rock, is not the volcanic rock, but rather scoria-like material that formed when an ancient coal bed burned and baked clay and silt in the surrounding beds.

b Sierra Grande is a shield volcano in northeastern New Mexico. It stands 600 m high and is 15 km in diameter. It was active from 4.0 to 2.6 million years ago.

》 Critical Thinking Question
Why do you think the flanks of this volcano slope at only a few degrees?

and only later during the Oligocene in southern California (Figure 23.10). In southern California, two triple junctions formed, one at the intersection of the North American, Juan de Fuca, and Pacific plates, and the other at the intersection of the North American, Cocos, and Pacific plates. Continued westward movement of the North American plate over the Pacific plate caused the triple junctions to migrate, one to the north and the other to the south, giving rise to the **San Andreas transform fault** (Figure 23.10). A similar occurrence along Canada's west coast produced the *Queen Charlotte transform fault.*

Seismic activity on the San Andreas fault results from continuing movements of the Pacific and North American plates along this complex zone of shattered rocks. Indeed, where the fault cuts through coastal California, it is actually a zone as much as 2 km wide, and it has numerous branches. Movements on such complex fault systems subject blocks of rocks adjacent to and within the fault zone to extensional and compressive stresses, forming basins and elevated areas, the latter supplying sediments to the former. Many of the fault-bounded basins in the southern California area subsided below sea level and soon filled with turbidites and other deposits. A number of these basins are areas of prolific oil and gas production.

The Continental Interior

Figure 23.4 shows that much of central North America is a vast area called the **Interior Lowlands,** which in turn

are made up of the *Great Plains* and the *Central Lowlands.* During the Cretaceous, the Great Plains were covered by the **Zuni epeiric sea** (see Figure 20.5), but by Early Paleogene time, this sea had largely withdrawn except for a sizable remnant that remained in North Dakota. Sediments eroded from the Laramide highlands were transported to this sea and were deposited in transitional and marine environments (▌ Figure 23.11a).

The only local sediment source within the Great Plains was the Black Hills in South Dakota. This area has a history of marine deposition during the Cretaceous, followed by the origin of terrestrial deposits derived from the Black Hills that are now well exposed in Badlands National Park, South Dakota. Judging from the sedimentary rocks and their numerous fossil mammals and other animals, the area was initially covered by semitropical forest, but grasslands replaced the forests as the climate became more arid.

Igneous activity was not widespread in the Interior Lowlands, but was significant in some parts of the Great Plains (▌ Figure 23.11b). For instance, igneous activity in northeastern New Mexico was responsible for volcanoes and numerous lava flows, and several small plutons were emplaced in Colorado, Wyoming, Montana, South Dakota, and New Mexico. Indeed, one of the most widely recognized igneous bodies in the entire continent, Devils Tower in northeastern Wyoming, is probably an Eocene volcanic neck, although some geologists think that it is an eroded laccolith. Another prominent volcanic feature is Shiprock in New Mexico, a volcanic neck that dates from the Late Oligocene.

Pleistocene glacial deposits are present in the northern part of the Central Lowlands, as well as in the northern Great Plains; however, during most of the Cenozoic Era, nearly all of the Central Lowlands was an area of active erosion rather than deposition.

Cenozoic History of the Appalachian Mountains

> **CONNECTION LINK**

For a review of the Paleozoic and Mesozoic history of the Appalachian region, see pages 526–532 in Chapter 20 and pages 581–583 in Chapter 22.

The Appalachian Mountain region has a history of deformation during the Proterozoic and Paleozoic, and during the Mesozoic, the region experienced block-faulting. By the end of the Mesozoic, though, the mountains had been eroded to a plain across which streams flowed eastward to the ocean. The present distinctive aspect of the Appalachian Mountains developed as a result of Cenozoic uplift and erosion (▌Figure 23.12). As uplift proceeded, upturned resistant rocks formed northeast-southwest–trending ridges with intervening valleys eroding into less resistant rocks. The preexisting streams eroded downward while uplift took place, were superposed on resistant rocks, and cut large canyons across the ridges, forming *water gaps,* deep passes through which streams flow, and *wind gaps,* that no longer contain streams.

Erosion surfaces at different elevations in the Appalachians are a source of continuing debate among geologists. Some are convinced that these more or less planar surfaces show evidence of uplift followed by extensive erosion and then renewed uplift and another cycle of erosion. Others think that each surface represents a differential

Cretaceous

Late Triassic

Recent

Cenozoic

▌ **Figure 23.12** Evolution of the Present Topography of the Appalachian Mountains Although these mountains have a long history, their present topographic expression resulted mainly from Cenozoic uplift and erosion.

Figure 23.13 Cenozoic Deposition Along the Gulf and East Coasts

a Depositional provinces and the surface geology of the Gulf and Atlantic coasts. Note the carbonate shelf deposits near Florida and the Yucatán Peninsula of Mexico.

b The coastal plain and continental margin of New Jersey are covered mostly by Cenozoic sandstones and shales.

North America's Southern and Eastern Continental Margins

In a previous section, we mentioned that much of the Interior Lowlands eroded during the Cenozoic. Even in the Great Plains, where vast deposits of Cenozoic rocks are present, sediment was carried across the region and into the drainage systems that emptied into the Gulf of Mexico. Likewise, sediment eroded from the western margin of the Appalachian Mountains ended up in the Gulf, but these mountains also shed huge quantities of sediment eastward that was deposited along the Atlantic Coastal Plain. Notice in Figure 23.4 that the Atlantic Coastal Plain and the **Gulf Coastal Plain** form a continuous belt extending from the northeastern United States to Texas.

The Gulf Coastal Plain

After the withdrawal of the Cretaceous to Early Paleogene Zuni Sea, the Cenozoic **Tejas epeiric sea** made a brief appearance on the continent (see Figure 20.5). But even at its maximum extent, it was largely restricted to the Atlantic and Gulf Coastal plains and parts of coastal California. It did, however, extend up the Mississippi River Valley, where it reached as far north as southern Illinois.

The overall Gulf Coast sedimentation pattern was established during the Jurassic and persisted throughout the Cenozoic. Sediments derived from the Cordillera, western Appalachians, and the Interior Lowlands were transported toward the Gulf of Mexico, where they were deposited in terrestrial, transitional, and marine environments. The sediments form seaward-thickening wedges, grading from terrestrial facies in the north to marine facies in the south (**Figure 23.13a**).

Sedimentary facies development was controlled mostly by regression of the Tejas epeiric sea. After its maximum extent onto the continent during the Paleogene, this sea began its long withdrawal toward the Gulf of Mexico. Its regression, however, was periodically reversed by minor transgressions—eight transgressive–regressive episodes are recorded in Gulf Coastal Plain sedimentary rocks, accounting for the intertonguing among the various facies.

response to weathering and erosion. According to this view, a low—elevation erosion surface developed on softer strata that eroded more or less uniformly, whereas higher surfaces represent weathering and erosion of more resistant rocks.

Most of the Gulf Coastal Plain was dominated by detrital deposition, but in the Florida section of the region and the Gulf Coast of Mexico, significant carbonate deposition took place. Florida was a carbonate platform during the Cretaceous and continued as an area of carbonate deposition into the Early Paleogene; carbonate deposition continues even now in Florida Bay and the Florida Keys.

The Atlantic Continental Margin

The east coast of North America includes the **Atlantic Coastal Plain** and extends seaward across the continental shelf, slope, and rise (❙ Figure 23.13b). It is a classic example of a passive continental margin. When Pangaea began fragmenting during the Triassic, continental crust rifted, and a new ocean basin began to form. Remember that the North American plate moved westerly, so its eastern margin was within the plate, where a passive continental margin developed.

The Atlantic continental margin has a number of Mesozoic and Cenozoic basins, formed as a result of rifting, in which sedimentation began by Jurassic time. Even though Jurassic-age rocks have been detected in only a few deep wells, geologists assume that they underlie the entire continental margin. The distribution of Cretaceous and Cenozoic rocks is better known, because both are exposed on the Atlantic Coastal Plain, and both have been penetrated by wells on the continental shelf.

Sedimentary rocks on the broad Atlantic Coastal Plain, as well as those underlying the continental shelf, slope, and rise, were derived from the Appalachian Mountains. Numerous rivers and streams transported sediments toward the east, where they were deposited in seaward-thickening wedges (up to 14 km thick) that grade from terrestrial deposits on the west to marine deposits farther east. For instance, the Calvert Cliffs in Maryland consist of rocks deposited in marginal marine environments (Figure 23.1c).

23.5 The Pleistocene and Holocene Epochs

The *Pleistocene* and *Holocene* or *Recent* epochs constitute the most recent 2.6 million years of the Quaternary Period (see Figure 17.1). The Pleistocene, which began 2.6 million years ago and ended 10,000 years ago, is the focus of our discussion here because it was the time of extensive glaciation; hence, the Pleistocene is commonly called the *Ice Age*.

Pleistocene and Holocene Tectonism and Volcanism

The Pleistocene Epoch is best known for widespread glaciers, but it was also a time of continuing tectonic activity and volcanism. Orogeny continued in the Himalayas and the Andes, and deformation at convergent plate boundaries proceeded unabated in the Aleutian Islands, Japan, the Philippines, and elsewhere. Interactions between the North American and Pacific plates persisted along the San Andreas fault (a transform plate boundary), yielding folds, faults, and a number of basins and uplifted areas (❙ Figure 23.14a). For example, several east-west–trending mountain ranges in southern California owe their existence to stresses created along a bend in the San Andreas fault.

Ongoing subduction of remnants of the Farallon plate beneath Central America and the Pacific Northwest accounts for Pleistocene and present-day volcanism in these two regions. Although the Cascade Range began evolving far back in the Paleocene, the large composite volcanoes such as Mount Shasta and Mount Rainier, as well as the lava dome known as Lassen Peak, are mostly Pleistocene and Recent (see the Geo-Insight on pages 628 and 629). Volcanism also took place at many other locations in the western United States (❙ Figure 23.14b), and huge calderas formed in what is now Yellowstone National Park in Wyoming.

Pleistocene Glaciation

In 1837, the Swiss naturalist Louis Agassiz argued that large boulders (erratics), polished and striated bedrock, U-shaped valleys, and deposits of sand and gravel in parts of Europe resulted from huge glaciers moving over the land. Although the idea initially met with considerable resistance, scientists finally came to realize that Agassiz was correct and accepted the idea that an Ice Age had taken place in the recent geologic past.

The Distribution and Extent of Pleistocene Glaciers

We know today that the Pleistocene, or what is commonly called the Ice Age, began 2.6 million years ago and ended about 10,000 years ago. During this time, several intervals of widespread continental glaciation took place, especially on the Northern Hemisphere continents, each separated by warmer interglacial periods (❙ Figure 23.15). In addition, valley glaciers were more common at lower elevations and latitudes, and many extended much farther than their shrunken remnants do today.

As one would expect, the climatic effects responsible for Pleistocene glaciers were worldwide. Nevertheless, Earth was not as frigid as portrayed in movies and cartoons, nor was the onset of the climatic conditions leading to glaciation very rapid. Indeed, evidence from several types of

Figure 23.14 Pleistocene Deformation and Volcanism

Marli Bryant Miller

a These deformed Pliocene-age rocks are only a few hundred meters from the San Andreas fault in southern California.

Sue Monroe

b View of the Cima volcanic field in Mojave Desert National Preserve in California, which was active between 7.6 million (Late Miocene) and 10,000 years ago (Late Pleistocene). Basalt lava flows and 52 cinder cones are present here.

investigations indicates that the climate gradually cooled from the Eocene through the Pleistocene. Furthermore, evidence from oxygen isotope data (the ratio of O^{18} to O^{16}) from deep-sea cores shows that 20 major warm–cold cycles have occurred during the last 2.6 million years.

The Effects of Glaciation Glaciers moving over Earth's surface have produced distinctive landscapes in much of Canada, the northern tier of states, and the mountains of the West (see Chapter 14). Sea level has risen and fallen with the formation and melting of glaciers, and these changes in turn

have affected the margins of continents. Glaciers have also altered the world's climate, causing cooler and wetter conditions in some areas that are arid to semiarid today.

More than 70 million km^3 of snow and ice blanketed the continents during the maximum glacial coverage of the Pleistocene. The storage of ocean waters in glaciers lowered sea level 130 m and exposed large areas of the present-day continental shelves, which were soon covered by vegetation. Lowering of sea level also affected the base level of rivers and streams. When sea level dropped, streams eroded downward as they sought to adjust to a lower base level.

> **CONNECTION LINK**

See Chapter 14 for a review of how glaciers form and move, and how they modify Earth's surface by erosion, transport, and deposition.

From such glacial features as terminal moraines, erratics, and drumlins, it seems that at their greatest extent, Pleistocene glaciers covered about three times as much of Earth's surface as they do now (Figure 23.15a). That is, they covered more than 40 million km², and like the vast ice sheets now present in Greenland and Antarctica, they were probably 3 km thick. Geologists have identified four major Pleistocene glacial episodes that took place in North America—the *Nebraskan, Kansan,* *Illinoian,* and *Wisconsinan,* each named for the states in which the most southerly glacial deposits are well exposed. The three interglacial stages are named for localities of well-exposed soils and other deposits (Figure 23.15b). In Europe, six or seven major glacial advances and retreats are recognized.

Recent studies show that there were an as-yet-undetermined number of pre-Illinoian glacial advances and retreats in North America and that the glacial history of this continent is more complex than previously thought.

During the Pleistocene, stream channels in coastal areas were extended and deepened along the emergent continental shelves. When sea level rose with the melting of the glaciers, the lower ends of stream valleys along the East Coast of North America were flooded and are now important harbors (see Figure 16.22a), whereas just off the West Coast, they form impressive submarine canyons. Great amounts of sediment eroded by the glaciers were transported by streams to the sea and thus contributed to the growth of submarine fans along the base of the continental slope.

We noted in Chapter 10 that as the Pleistocene ice sheets formed and increased in size, the weight of the ice caused the crust to slowly subside deeper into the mantle. In some places, Earth's surface was depressed as much as 300 m below the preglacial elevations. As the ice sheets retreated by melting, the downwarped areas gradually rebounded to their former positions.

During the Wisconsinan glacial stage, many large lakes formed in what are now dry basins in the southwestern United States (■ Figure 23.16a). These so-called *pluvial lakes* formed far from the glaciers because of the overall greater precipitation and cooler temperatures, which lowered the evaporation rate, and they fluctuated as the glaciers elsewhere expanded contracted. The largest was Lake Bonneville, which attained a maximum size of 50,000 km² and was at least 335 m deep (Figure 23.16a). The Great Salt Lake that lies just west of Salt Lake City, Utah, is the shrunken remnant of this once-vast lake. Lake Lahontan, mostly in Nevada (■ Figure 23.16b), covered 22,000 km², and even Death Valley, California, which is now the driest place in North America, had a large lake.

In contrast to pluvial lakes, *proglacial lakes* form when meltwater accumulates along the margins of glaciers. Lake Agassiz was a large proglacial lake that covered about 250,000 km² of North Dakota, Manitoba, Saskatchewan, and Ontario. It persisted until the glacial ice along its northern margin melted and it drained northward into Hudson Bay.

■ Figure 23.15 Pleistocene Glaciers

a Centers of ice accumulation and maximum extent of Pleistocene glaciers in North America.

| Wisconsinan Glaciation |
| Sangamon Interglacial |
| Illinoian Glaciation |
| Yarmouth Interglacial |
| Kansan Glaciation |
| Aftonian Interglacial |
| Nebraskan Glaciation |
| Pre-Nebraskan |

Sangamon soil
Yarmouth soil
Aftonian soil

b Standard terminology for Pleistocene glacial and interglacial stages in North America.

Figure 23.16 Pleistocene Lakes in the Western United States

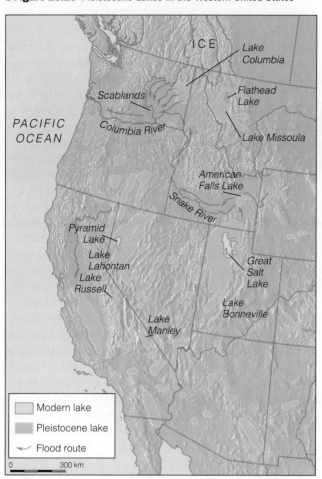

a Pyramid Lake and Great Salt Lake are shrunken remnants of much larger ancient lakes. Of all the lakes shown, only Lake Columbia and Lake Missoula were proglacial lakes. When the 600-m-high ice dam impounding Lake Missoula failed, it drained westward and scoured out the scablands of eastern Washington.

b The area in this image east of Fallon, Nevada, was covered by Lake Lahontan. In fact, the image was taken from Grimes Point Archaeological Site where Native Americans lived near the lakeshore.

c This rumpled surface of gravel ridges, or so-called giant ripple marks, formed when Lake Missoula drained across this area near Camas Prairie, Montana.

>> **Critical Thinking Question**
Why were there large lakes in places such as Death Valley, California, during the Pleistocene?

A proglacial lake in Montana called Lake Missoula (Figure 23.16a) was impounded by an ice dam in Idaho that periodically failed, and the lake drained catastrophically. Many depositional and erosion features in western Montana, Idaho, and eastern Washington formed during these gigantic floods (Figure 23.16c).

23.6 Cenozoic Mineral Resources

The United States is the third-largest producer of petroleum, accounting for about 19% of the world's total. Much of this production comes from Cenozoic reservoirs of the Gulf Coastal Plain and the adjacent continental shelf. On the Gulf Coastal Plain, most of the petroleum is in structural traps that formed adjacent to salt domes and other structures. Several basins in southern California are also important areas of petroleum production. The Green River Formation of Wyoming, Utah, and Colorado has huge reserves of oil shale and evaporites.

Diatomite is a sedimentary rock composed of the microscopic shells of diatoms, which are single-celled marine and freshwater plants that secrete skeletons of silica (SiO_2). This rock, also called diatomaceous earth, is used chiefly in gas purification and to filter liquids such as molasses, fruit juices, water, and sewage. The United States is the leader in diatomite production, mostly from Cenozoic deposits in California, Oregon, and Washington.

Huge deposits of low-grade lignite and subbituminous coal in the northern Great Plains are becoming increasingly important resources. These coal deposits are Late Cretaceous to Early Paleogene in age and are most extensive in the Williston and Powder River basins of Montana, Wyoming, and

North and South Dakota. In addition to having a low sulfur content, some of these coal beds are 30 to 60 m thick!

Gold production from the Pacific Coast, particularly California, comes mostly from Cenozoic gravels. The gold is found in placer deposits, which formed as concentrations of minerals separated from weathered debris by fluvial processes.

A variety of other mineral deposits is also important. For example, the United States must import almost all manganese used in the manufacture of steel. The largest manganese deposits are in Cenozoic rocks in Russia. One molybdenum deposit in Colorado accounts for much of the world production of this element. Cenozoic sand and gravel, as well as evaporites, building stone, and clay deposits, are quarried from areas around the world.

Sand and gravel deposits resulting from glacial activity are a valuable resource in many formerly glaciated areas. Most Pleistocene sands and gravels originated as floodplain deposits, outwash sediment, or esker deposits. The bulk of the sand and gravel in the United States and Canada is used in construction and as roadbase and fill for highway and railway construction.

The periodic evaporation of pluvial lakes in the Death Valley region of California during the Pleistocene led to the concentration of many evaporite minerals such as borax. During the 1880s, borax was transported from Death Valley by the famous 20-mule-team wagon trains.

Another Neogene resource is peat, a vast potential energy resource that has been developed in Canada and Ireland. Peatlands formed from plant assemblages as the result of particular climate conditions.

23.7 Paleogene and Neogene Life History

Earth's flora and fauna continued to evolve during the Cenozoic Era as more and more familiar plants and animals made their appearance. We emphasize the evolution of mammals in this chapter, but you should be aware of other important life events.

Marine Invertebrates and Phytoplankton

The Cenozoic marine ecosystem was populated by those plants, animals, and single-celled organisms that survived the Mesozoic extinctions. Especially prolific Cenozoic invertebrate groups were foraminifera, radiolarians, corals, bryozoans, mollusks, and echinoids. The marine invertebrate community in general became more provincial during the Cenozoic because of changing ocean currents and temperature differences with latitude.

Only a few species in each major group of phytoplankton survived into the Cenozoic. The coccolithophores,

diatoms, and dinoflagellates all recovered from their Late Cretaceous reduction in numbers. The diatoms were particularly abundant during the Miocene (❙ Figure 23.17a), probably because of increased volcanism during this time. Volcanic ash provided increased dissolved silica in seawater and was used by the diatoms to construct their skeletons.

The foraminifera were a major component of the Cenozoic marine invertebrate community. Though dominated by relatively small forms (❙ Figure 23.17b, c), it included some exceptionally large forms that lived in the warm waters of the Cenozoic Tethys Sea. Shells of these larger foraminifera accumulated to form thick limestones, some of which were used by the ancient Egyptians to construct the Pyramids of Giza.

The corals, having relinquished their reef-building role to the rudists during the mid-Cretaceous, again became the dominant reef builders during the Cenozoic. Other suspension feeders such as bryozoans and crinoids were also abundant and successful during the Cenozoic. Perhaps the least important of the Cenozoic marine invertebrates were the brachiopods, with fewer than 60 genera surviving today.

Just as during the Mesozoic, bivalves and gastropods were two of the major groups of marine invertebrates, and they had a markedly modern appearance. Following the extinction of the ammonites and belemnites at the end of the Cretaceous, the Cenozoic cephalopod fauna consisted of nautiloids and cephalopods lacking shells, such as squids and octopuses.

The echinoids continued their expansion in the infaunal habitat and were particularly prolific during the Cenozoic. New forms such as sand dollars evolved during this time from biscuit-shaped ancestors.

Cenozoic Vegetation and Climate

During the Cenozoic, angiosperms continued to diversify, although seedless vascular plants and gymnosperms were also present in large numbers. In fact, many Paleogene plants would be familiar to us today, but their geographic distribution was not what it is now, because changing climatic conditions along with shifting plant distributions were occurring.

The makeup of ancient floras and the types of leaves are good climatic indicators. Some plants today are confined to the tropics, whereas others have adapted to drier conditions, and we have every reason to think that climate was a strong control on plant distribution during the past. Furthermore, leaves with entire or smooth margins, many with pointed drip-tips, are dominant in areas with abundant rainfall and high annual temperatures. Smaller leaves with incised margins are more typical of cooler, drier areas (❙ Figure 23.18a).

In a recently discovered Paleocene flora in Colorado with about 100 species of trees, nearly 70 percent of the leaves are smooth margined, and many have drip-tips. The nature of the leaves, coupled with the diversity of plants, is much like that found in today's rain forests. In fact, the Early Oligocene fossil plants at Florissant Fossil Beds National Monument in Colorado, indicate that a warm, wet climate persisted then.

Seafloor sediments and geochemical evidence indicate that about 55 million years ago an abrupt warming trend

▌Figure 23.17 Cenozoic Diatoms and Foraminifera

Reed Wicander

Reed Wicander

a Diatoms from the Miocene-age Monterey Formation of California.

Courtesy of B. A. Masters

b *Cibicides americanus* from the Early Miocene of California is a benthic foraminifera.

Courtesy of B. A. Masters

c *Globigerinoides fistulosus* is a Pleistocene planktonic foraminifera from the south Pacific Ocean.

▌Figure 23.18 Vegetation and Climate

a Plants adapted to cool climates typically have small leaves with incised margins (top), whereas in humid, warm habitats, they have larger, smooth-margined leaves, and many have drip-tips.

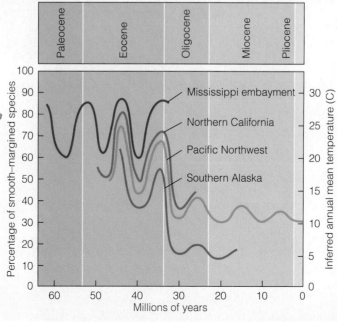

b Climatic trends for four areas in North America based on the percentages of plant species with smooth-margined leaves.

took place. During this **Paleocene–Eocene Thermal Maximum,** large-scale oceanic circulation was disrupted so that heat transfer from equatorial regions to the poles diminished or ceased. As a result, deep oceanic water became warmer, resulting in extinctions of many deepwater foraminifera. Some scientists think that this deep, warm oceanic water released methane from seafloor methane hydrates, contributing a greenhouse gas to the atmosphere and either causing, or contributing to the temperature increase at this time.

Subtropical conditions persisted into the Eocene in North America, probably the warmest of all the Cenozoic epochs. Fossil plants in the Eocene John Day Beds in Oregon include species that today live in much more humid regions, as in parts of Mexico and Central America. Yellowstone National Park in Wyoming has a temperate climate now, with warm, dry summers and cold, snowy winters, certainly not an area where you would expect avocado, magnolia, and laurel trees to grow. Yet their presence there during the Eocene indicates that the area then had a considerably warmer climate than it does now.

A major climatic change took place at the end of the Eocene, when mean annual temperatures dropped as much as 7°C in about 3 million years (Figure 23.18b). Since the Oligocene, mean annual temperatures have varied somewhat worldwide, but overall, they have not changed much in the middle latitudes except during the Pleistocene Epoch.

A general decrease in precipitation during the last 25 million years took place in the mid-continent region of North America. As the climate became drier, the vast forests of the Oligocene gave way first to *savannah* conditions (grasslands with scattered trees) and finally to *steppe* environments (short-grass prairie of the desert margin). Many herbivorous mammals quickly adapted to these new conditions by developing chewing teeth suitable for a diet of grass.

Paleogene and Neogene Birds

The first members of many of the living orders of birds, including owls, hawks, ducks, penguins, and vultures, evolved during the Early Paleogene. Beginning during the Miocene Epoch, a marked increase in the variety of songbirds took place, and by 5 to 10 million years ago, many of the existing genera of birds were present. Birds adapted to numerous habitats and continued to diversify during the Pleistocene, but since then, their diversity has decreased slightly.

Several varieties of large flightless birds existed during the Pleistocene (discussed later) and today, but some remarkable predatory birds lived during the Paleogene and Neogene. For instance, *Diatryma,* a heavily built, flightless bird that stood about 2 m high, lived during the Paleocene and Eocene in North America and Europe (Figure 23.19). About 25 species of flightless birds up to 3 m tall are known from South America, where they were the dominant predators until they were replaced by big cats and dogs that migrated from North America.

 Figure 23.19 Restoration of *Diatryma* *Diatryma* was a flightless, predatory bird that stood more than 2 m tall. It lived during the Paleocene and Eocene in North America and Europe.

James S. Monroe

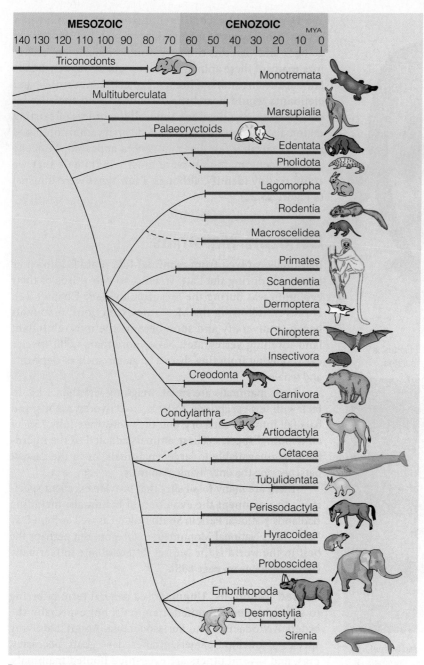

Figure 23.20 Diversification of Mammals Mammals existed during the Mesozoic, but most pacental mammals diversified during the Paleocene and Eocene epochs. Among the living orders of mammals, all are placentals except for the monotremes and marsupials. Several extinct orders are not shown. Bold lines indicate actual geologic ranges, whereas the thinner lines indicate the inferred branching of the groups.

Diversification of Mammals

Recall from Chapter 22 that reptiles, birds, and mammals are *amniotes*—that is, animals that have an amniotic egg in their reproductive cycle, although it is highly modified in some mammals. Nevertheless, the mammals known as **monotremes,** such as the platypus and spiny anteater, lay amniotic eggs and thus are called egg-laying mammals. **Marsupial mammals** and **placental mammals** give birth to live young, but in marsupials (pouched mammals) the young are born in an immature, almost embryonic, condition and then undergo further development in their mother's pouch. The amnion of the amniotic egg is fused to the wall of the uterus in placental mammals. In fact, the membranes in the amniotic egg form the placenta through which nutrients and oxygen are carried from the mother to the developing embryo, allowing the embryo to develop much more fully before birth.

Judging from the fossil record, monotremes have never been very common, and the only living ones are found in the Australian region. Marsupials have been more successful in terms of number of species and geographic distribution, but even they have been mostly restricted to South America and the Australian region. In marked contrast, the success of placental mammals is indicated by the fact that more than 90% of all mammal species, fossil and living, are placental.

Although mammals had evolved during the Triassic, a major diversification began during the Paleocene and continued throughout the Cenozoic (▌ Figure 23.20). In fact, many of the mammals of the Paleocene are called archaic, calling attention to the fact that they were holdovers from the Mesozoic or they were mammals that did not give rise to any of today's mammals. For example, the rodent-like multituberculates persisted from the Jurassic until their extinction during the Oligocene (▌ Figure 23.21a). Among the predators were creodonts that slightly resembled wolves and weasel-like "miacids," as well as the large, predatory birds we mentioned in the previous section. Among the herbivores were several groups that survived only until the Eocene.

Also among these Paleocene mammals were the first rodents, rabbits, primates, and hoofed mammals. They had not become fully differentiated from their ancestors, and differences between carnivores and herbivores were slight— even the carnivores were short-legged, flat-footed creatures that undoubtedly were not very speedy. Furthermore, most of these early mammals were small, although some fairly large ones were present by the Late Paleocene, but the first land-dwelling giant mammals did not appear until the Eocene (▌ Figure 23.21b).

CONNECTION LINK

You can review the origin and Mesozoic evolution of mammals by referring to pages 602–604 in Chapter 22.

Figure 23.21 Paleocene and Eocene Mammals

a This squirrel-sized animal is the Paleocene genus *Ptilodus,* a member of a group of mammals known as multituberculates.

b Scene from the Eocene showing the rhinoceros-sized animal known as *Unitatherium,* one of the first giants among mammals. It had three pairs of horns and saber-like upper canine teeth.

What Would You Do ❓

Diversification continued during the Eocene when several more types of mammals appeared, but if we could go back and visit this time, we would probably not recognize many of these animals. Some would be vaguely familiar, but the ancestors of horses, camels, rhinoceroses, and elephants would bear little resemblance to their living descendants. By Oligocene time, all the orders of existing mammals were present; however, diversification continued as more familiar families and genera appeared. Miocene and Pliocene mammals were mostly mammals that we could readily identify, although a few types were unusual (Figure 23.22).

Cenozoic Mammals

Mammals evolved from mammal-like reptiles known as *cynodonts* during the Late Triassic, so two-thirds of their evolution was during the Mesozoic Era (see Chapter 22). However, following the Mesozoic extinctions, mammals began to diversify and soon became the most abundant land-dwelling vertebrates. Now, more than 4,000 species exist, ranging from tiny shrews to giants such as elephants and whales.

Most mammals are small, weighing less than 1 kg. In fact, with few exceptions, rodents, insectivores, rabbits, and bats fall into this category, and they constitute fully 75% of all mammal species. These animals adapted to the microhabitats unavailable to larger mammals, or in the case of bats, became the only flying mammals.

There are many fossil sites that provide excellent specimens to document the evolution of mammals, including Badlands National Park in South Dakota as well as John Day Fossil Beds National Monument in Oregon. But perhaps the best in the world is the Messel Pit Fossil Site in Germany (see Geo-Focus on page 638).

Hoofed Mammals   **Ungulate** is a general term referring to several types of hoofed mammals, but especially the orders Artiodactyla and Perissodactyla. About 220 living species of antelope, camels, giraffes, deer, goats, peccaries, pigs, and several others are even-toed hoofed mammals, or **artiodactyls,** the most common living ungulates. In marked contrast, only 18 species of horses, rhinoceroses, and tapirs are **perissodactyls,** or odd-toed hoofed mammals. As even and odd-toed imply, artiodactyls have two or four toes, whereas perissodactyls have one or three (Figure 23.23a).

Some ungulates are small and depend on concealment to avoid predators; others, such as rhinoceroses, are so large that size alone is enough to discourage predators, at least for adults. But many of the more modest-sized ungulates are speedy runners. Adaptations for running include elongation of some of the limb bones, as well as reduction in the number of bony limb elements, especially toes. Accordingly, the limbs of speedy ungulates are long and slender (Figure 23.23b).

All ungulates are herbivores, but some are **grazers,** meaning that they feed on grasses, and others are **browsers,**

Rotated text on right edge: Chase Studio/Photo Researchers, Inc.

Figure 23.22 Miocene Mammals of Nebraska This scene depicts some of the mammals that lived in Nebraska during the Miocene Epoch, about 21 million years ago. Although a few Miocene mammals are unfamiliar, most of them look much like mammals living today. Shown in this image are relatives of today's elephants (background), a clawed, horselike animal called a chalicothere (right center), oredonts (now extinct) in the water, and small camels (left front) being chased by a bear dog. A saber-toothed cat in the tree feasts on its prey and cranes circle overhead.

feeding on the tender shoots, twigs, and leaves of trees and bushes. When grasses grow through soil, they pick up tiny pieces of sand that are quite abrasive to teeth, so the grazing ungulates developed high-crowned chewing teeth resistant to abrasion (Figure 23.23c). Browsers, on the other hand, never developed these kinds of chewing teeth.

Rabbit-sized ancestral artiodactyls of the Early Eocene differed little from their ancestors, but gave rise to numerous families, several of which are now extinct. Small, four-toed camels, for instance, appeared early in this diversification and were common in North America well into the Pleistocene. In fact, most of their evolution took place on this continent, and only during the Pliocene did they migrate to South America and Asia where they now exist.

The perissodactyls—horses, rhinoceroses, tapirs—and the extinct titanotheres and chalicotheres are united by several shared characteristics (see Figure 18.19). Furthermore, the fossil record shows that they all evolved from a common ancestor during the Eocene. Their diversity increased through the Oligocene, but since then, they have declined markedly and now constitute less than 10% of the world's hoofed mammal fauna.

Several groups of mammals such as rhinoceroses, whales, elephants, and camels have good fossil records, but certainly one of the best known is the fossil record of horses. Fossils indicate that present-day *Equus* evolved from a tiny Eocene ancestor called *Hyracotherium* (Figure 23.24). Most of horse evolution took place in North America although now they survive in the wild only in Asia and Africa. Fossils of horses reveal several trends such as

increased size, lengthening of the limbs, reduction of the toes to one, and development of high-crowned chewing teeth among others as horses became speedy grazing animals. There was, however, another branch in horse evolution that led to three-toed browsers, which became extinct during the Pleistocene (Figure 23.24a).

Other Mammals—Carnivores, Elephants, and Whales During the Paleocene, the *miacids,* small carnivorous mammals with short heavy limbs, made their appearance (Figure 23.25a). These small creatures were ancestors to all later members of the order Carnivora, which includes, among others, today's dogs, cats, hyenas, bears, weasels, and seals. All have well-developed canine teeth for slashing and tearing, and most also developed a pair of large shearing teeth (carnassials) (Figure 23.25b). Some of the better-known fossil carnivores are the saber-toothed cats, or what are more commonly called saber-toothed tigers.

Elephants (order Proboscidea), the largest land mammals, evolved from pig-sized ancestors during the Eocene. And by Oligocene time, they clearly showed the trend toward large size, a long snout (proboscis), and large tusks. Mastodons, with teeth adapted for browsing, were present by the Miocene, and during the Pliocene the present-day elephants and mammoths diverged from a common ancestor. During most of the Cenozoic, elephants were widespread on the northern continents, but now only two species exist in southern Asia and Africa.

GEO-FOCUS The Messel Pit Fossil Site in Germany

We have mentioned several outstanding fossil localities in this chapter such as John Day Fossil Beds National Monument in Oregon, but here we concentrate on the Messel Pit Fossil Site, near Frankfurt, Germany, which is one of the most remarkable fossil localities anywhere in the world. Indeed, it was designated a World Heritage Site, which is "an area or structure designated by UNESCO as being of global significance and conserved by a country that has signed a United Nations convention pledging its protection."* Of the 878 sites so designated, 679 are cultural (the Pyramids at Giza in Egypt), 174 are natural (the Messel Pit Fossil Site in Germany), and 25 are mixed (Mount Athos in Greece).

Actually, fossils have been known from the Messel Pit Fossil Site since 1875, but they have attracted the most attention only during the last few decades. The fossil site was originally a large pit excavated from 1886 to 1971 for oil shale that was used to produce crude oil. We noted in Chapter 6 that oil shale is a rock made up of mud and the organic compound kerogen from which

liquid oil and combustible gases are extracted. These organic-rich mudrocks were deposited during the Middle Eocene, about 50 million years ago, in a small, but deep lake in a fault valley. Fossils of palms, citrus fruits, laurels, beech, and other plants indicate that the climate was tropical or subtropical.

Streams carried mud and some sand into this small lake, which had oxygen-depleted deep waters where nothing but anaerobic bacteria could exist. As a consequence, the remains of organisms that sank into the deeper waters did not decompose completely, and many have been preserved so well that even stomach contents and the outlines of internal organs are visible, as well as impressions of bird feathers, bat wings, and, most remarkably, the iridescent color of beetle wing covers. Although a wide variety of organisms existed in and around the lake (most of the vertebrate fossils are of fish), it is the mammals that are the most notable. For some of these animals every bone has been preserved, including all of the toe bones and every bony segment of their tails. (Figure 1).

A complete inventory of all the animal fossils found at Messel is not practical, but we will mention that in addition to 10,000 or so fossil fish, thousands of insects have been found as well as 12 genera of birds, including a "proto-ostrich" and a large, flightless, predatory bird similar to *Diatryma* (Figure 23.19); several types of amphibians and reptiles; and a wide assortment of mammals. Indeed, the mammals include anteaters, pangolins, marsupials, bats, carnivores, rodents, and an early relative of horses. The fossil in Figure 1 is known as *Messelobunodon,* which is a 78-cm-long early artiodactyl (even-toed hoofed mammal) that was probably close to the ancestry of pigs, cattle, and camels.

The fossils at Messel are abundant and of great scientific interest, but they present a particular problem for collecting and preservation. The host oil shale contains up to 40% water, so when the rock dries it crumbles and the fossils may be lost. Accordingly, scientists developed a technique of coating the rock surface with a synthetic resin in which the fossils are preserved.

Such a remarkable association of fossils gives us a unique glimpse of what the biota was like 50 million years ago in what is now Germany. Fortunately for us, the Messel fossil locality was designated a World Heritage Site, because when the oil shale excavation ceased, it was slated to become a sanitary landfill.

*http://encarta.msn.com/dictionary
_1861714109/world_heritage_site.html

© Jonathan Blair/CORBIS

Figure 1 This fossil mammal known as *Messelobunodon* was recovered from one of the most remarkable fossil sites anywhere—the Eocene-age Messel Pit near Frankfurt, Germany. This fossil shows a small (78-cm-long) animal in a death pose with all of its bones preserved. *Messelobunodon* was one of the earliest known artiodactyls.

Figure 23.23 Evolutionary Trends in Hoofed Mammals

b In many hoofed mammals, long, slender limbs evolved as bones between the wrist and toes, and the ankle and toes, became longer.

a Perissodactyls have one or three functional toes, whereas artiodactyls have two or four. Note that in perissodactyls the weight is borne on the third toe, but artiodactyls walk on toes three and four.

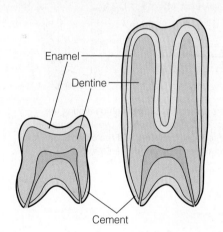

c High-crowned, cement-covered chewing teeth (right) evolved in hoofed mammals that adapted to a diet of grass. Low-crowned teeth are found in many other mammals, including primates and pigs, both of which have a varied diet.

Until recently, little was known about the transition of whales from land-dwelling ancestor to fully aquatic whales. Although a number of questions remain unanswered, the fossils now available show that whales appeared during the Early Eocene and by the Late Eocene had become diverse and widespread. Eocene whales still possessed vestigial rear limbs, their teeth resembled those of their land-dwelling ancestors, their nostrils (blowhole) were not on top of the head, and they were proportioned quite differently from living whales (▮ Figure 23.26). By Oligocene time, both groups of living whales—the toothed whales and the baleen whales—had evolved.

23.8 Pleistocene Faunas

We devote much of this section to the evolution of primates, particularly the hominids, which include present-day humans. Primates as an order evolved by Late Cretaceous time, but the ones of interest to us here date from the Pliocene and Pleistocene.

As for mammals other than primates, most of the present-day genera had evolved by Pleistocene time. Indeed, we would have little difficulty recognizing most Pleistocene mammals; only a few unusual types that persisted from earlier times are now extinct. A good example is the chalicotheres, a group of horselike mammals with claws on their forefeet. Likewise, we would recognize most Pleistocene birds, but some large ground-dwelling species are now extinct.

Mammals and Birds

One of the most remarkable aspects of the Pleistocene fauna is that so many large species of mammals and birds existed. In North America, for example, there were mastodons and mammoths, giant bison, huge ground sloths, giant camels, and beavers nearly 2 m long. Kangaroos standing 3 m tall, wombats the size of rhinoceroses, leopard-sized marsupial lions, and large platypuses characterized the Pleistocene fauna of Australia. Cave bears, elephants, and the giant deer, commonly called the Irish elk, with an antler spread of 3.35 m, lived in Europe and parts of Asia. The evolutionary

Figure 23.24 Evolution of Horses

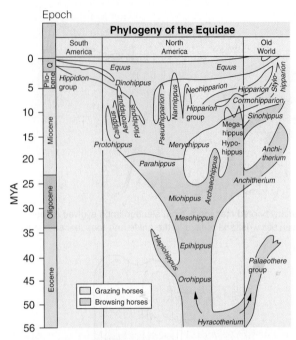

a Summary chart showing the relationships among the genera of horses. During the Oligocene two lines emerged, one leading to three-toed browsers and the other to one-toed grazers, including the present-day horse *Equus*.

Equus
(Pleistocene)

Pliohippus
(Pliocene)

Merychippus
(Miocene)

Mesohippus
(Oligocene)

Hyracotherium
(Eocene)

b Simplified diagram showing some trends in horse evolution. Trends include a size increase, a lengthening of the limbs and a reduction in the number of toes, and the development of high-crowned teeth with complex chewing surfaces.

Figure 23.25 Carnivorous Mammals

© Jim Melli/San Diego Natural History Museum

a This miacid, known as *Topocyon*, was a coyote-sized animal that lived during the Eocene. All weasels, otters, skunks, badgers, martins, wolverines, seals, sea lions, walruses, bears, raccoons, dogs, hyenas, mongooses, and cats evolved from an ancestor very much like this creature.

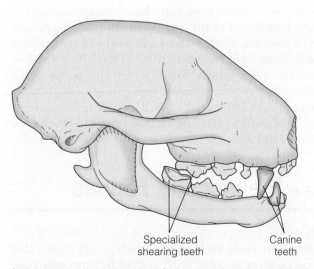

Specialized
shearing teeth

Canine
teeth

b This present-day skull and jaw of a large cat show the specialized sharp-crested shearing teeth, or carnassials, of carnivorous mammals. Note that carnivorous mammals also have well-developed canine teeth.

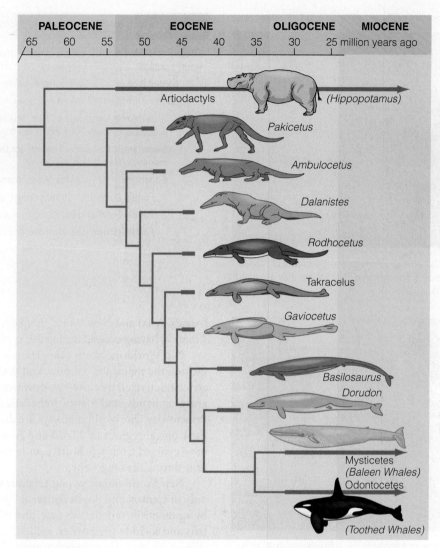

Figure 23.26 The Evolution of Whales The land-dwelling ancestors of whales were among the even-toed hoofed mammals, the artiodactyls. Note that *Pakicetus* had well-developed hind limbs, but some of the others, such as *Basilosaurus,* do not. Although *Basilosaurus* was a fully aquatic whale, it differed considerably from today's whales.

trend toward large body size was perhaps an adaptation to the cooler temperatures of the Pleistocene. Large animals have proportionately less surface area compared to their volume and thus retain heat more effectively than do smaller animals.

In addition to mammals, some other Pleistocene vertebrate animals were of impressive proportions. The giant moas of New Zealand and the elephant birds of Madagascar were very large, and Australia had giant birds standing 3 m tall and weighing nearly 500 kg, and a lizard 6.4 m long and weighing 585 kg. Huge ground sloths and armored mammals known as glyptodonts have been found in Florida (Figure 23.27a) The tar pits of Rancho La Brea in southern California contain the remains of at least 200 kinds of animals. Many of these are fossils of dire wolves, saber-toothed cats, and other mammals (Figure 23.27b), but some are the remains of birds, especially birds of prey, and a giant vulture with a wingspan of 3.6 m.

Primate Evolution

Primates are difficult to characterize as an order because they lack the strong specializations found in most other mammalian orders. We can, however, point to several trends in their evolution that help define primates and are related to their *arboreal,* or tree-dwelling, ancestry. These include changes in the skeleton and mode of locomotion; an increase in brain size; a shift toward smaller, fewer, and less-specialized teeth; and the evolution of stereoscopic vision and a grasping hand with an opposable thumb. Not all of these trends took place in every primate group, nor did they evolve at the same rate in each group. In fact, some primates have retained certain primitive features, whereas others show all or most of these trends.

The primate order is divided into two suborders, the Prosimii and Anthropoidea (Figure 23.28, Table 23.1). The *prosimians,* include the lemurs, lorises, tarsiers, and

Figure 23.27 Pleistocene Fossils from Florida and California

Erika Simons/Florida Museum of Natural History

a Among the diverse Pliocene and Pleistocene mammals of Florida were 6-m-long ground sloths and armored glyptodonts that weighted more than 2 metric tons.

Sue Monroe

b Restoration of a mammoth trapped in the sticky tar (asphalt) at a present-day oil seep at the La Brea Tar Pits in Los Angeles, California.

TABLE 23.1	Classification of the Primates

Order Primates: Lemurs, lorises, tarsiers, tree shrews, monkeys, apes, humans

    Suborder Prosimii: Lemurs, lorises, tarsiers, tree shrews

    Suborder Anthropoidea: Monkeys, apes, humans

        Superfamily Cercopithecoidea: Macaque, baboon, proboscis monkey (Old World monkeys)

        Superfamily Ceboidea: Howler, spider, and squirrel monkeys (New World monkeys)

        Superfamily Hominoidea: Apes, humans

            Family Pongidae: Chimpanzees, orangutans, gorillas

            Family Hylobatidae: Gibbons, siamangs

            Family Hominidae: Humans and their extinct ancestors

tree shrews (Figure 23.28a). They are generally small, ranging from species the size of a mouse to those as large as a house cat. They are arboreal, have five digits on each hand and foot with either claws or nails, and are typically omnivorous. They have large, forwardly directed eyes specialized for night vision—hence, most are nocturnal. As their name implies (*pro* means "before," and *simian* means "ape"), they are the oldest primate lineage, with a fossil record extending back to the Paleocene.

During the Eocene, prosimians were abundant, diversified, and widespread in North America, Europe, and Asia. As the continents moved northward during the Cenozoic and the climate changed from warm tropical to cooler midlatitude conditions, the prosimian population decreased in both abundance and diversity. Presently, prosimians are found only in the tropical regions of Asia, India, Africa, and Madagascar.

Sometime during the Late Eocene, the *anthropoids,* primates which include monkeys, apes, and humans, evolved from a prosimian lineage. By the Oligocene, the anthropoids were a well-established group with both Old World monkeys

(Africa, Asia) and New World monkeys (Central and South America) having evolved during this epoch.

Old World monkeys (superfamily Cercopithecoidea) include the macaque, baboon, and proboscis monkey and are characterized by close-set, downward-directed nostrils, grasping hands, and a non-prehensile tail (Figure 23.28b). Present-day Old World monkeys are distributed throughout the tropical regions of Africa and Asia and are thought to have evolved from a primitive anthropoid ancestor sometime during the Oligocene.

New World monkeys (superfamily Ceboidea) are found only in Central and South America. They are characterized by a prehensile tail, flattish face, and widely separated nostrils and include the howler, spider, and squirrel monkeys (Figure 23.28c). They probably evolved from African monkeys that migrated across the widening Atlantic sometime during the Early Oligocene, and they have continued evolving in isolation to this day.

The *hominoids* (superfamily Hominoidea), the group containing apes (Figure 23.28d) and humans, diverged from Old World monkeys sometime before the Miocene, but exactly when is still being debated. It is generally accepted, however, that hominoids evolved in Africa from an ancestral anthropoid group. Although there is still no agreement on the evolutionary relationships among the early hominoids, fossil evidence and molecular DNA similarities between modern hominoid families is providing a clearer picture of the evolutionary pathways and relationships among the hominoids.

Hominids

The **hominids** (family Hominidae), the primate family that includes present-day humans and their extinct ancestors (Table 23.1), have a fossil record extending back almost 7 million years. Several features distinguish them from other hominoids. Hominids are bipedal; that is, they have an upright posture, which is indicated by several modifications

❚ Figure 23.28 Primates Primates are divided into two suborders: the prosimians **(a)** and the anthropoids **(b)–(d)**, which are further subdivided into three superfamilies: Old World monkeys **(b)**, New World monkeys **(c)**, and hominoids, which include the apes **(d)** and humans.

a Tarsier.

b Baboon.

c Spider monkey.

d Chimpanzee.

in their skeleton (❚ Figure 23.29a). In addition, they show a trend toward a large and internally reorganized brain (Figure 23.29b). Other features include a reduced face and reduced canine teeth, omnivorous feeding, increased manual dexterity, and the use of sophisticated tools.

Many anthropologists think that these hominid features evolved in response to major climatic changes that began during the Miocene and continued into the Pliocene. During this time, vast savannas replaced the African tropical rain forests where the prosimians and Old World monkeys had been so abundant. As the savannas and grasslands continued to expand, the hominids made the transition from true forest dwelling to life in an environment of mixed forests and grasslands.

At present, there is no clear consensus on the evolutionary history of the hominid lineage. This lack of agreement is due, in part, to the incomplete fossil record of hominids, as well as new discoveries, and to the fact that some species are known only from partial specimens or fragments of bone.

Because of these factors, there is even disagreement on the total number of hominid species. A complete discussion of all of the proposed hominid species and the various competing schemes of hominid evolution is beyond the scope of

this chapter. In fact, new discoveries of fossil hominids, as well as new techniques for scientific analysis, are leading to new hypotheses about our ancestry. By the time you read this chapter, it is possible that new discoveries may change some of the conclusions stated here. Such is the nature of paleoanthropology—and one reason why the study of hominids is so exciting. Nevertheless, we will briefly discuss the generally accepted taxa (❚ Figure 23.30) and present some of the current theories of hominid evolution.

Remember that although the fossil record of hominid evolution is not complete, what exists is well documented. Furthermore, the interpretation of that fossil record precipitates the often vigorous and sometimes acrimonious debates concerning our evolutionary history.

Discovered in northern Chad's Djurab Desert, the nearly 7-million-year-old skull and dental remains of *Sahelanthropus tchadensis* (❚ Figure 23.31) make it the oldest known hominid yet unearthed and at or very near the time when humans diverged from our closest living relative, the chimpanzee. Currently, most paleoanthropologists accept that the human–chimpanzee stock separated from gorillas about 8 million years ago and that humans separated from chimpanzees about 5 million years ago.

The next oldest hominid is *Orrorin tugenensis,* whose fossils have been dated at 6 million years old and consist of bits of jaw, isolated teeth, and finger, arm, and partial upper leg bones. At this time, there is still debate as to exactly where *Orrorin tugenensis* fits in the hominid lineage.

Sometime between 5.8 and 5.2 million years ago, another hominid, *Ardipithecus kadabba*, was present in eastern Africa (Figure 23.30). Recently, the discovery of a second species of *Ardipithecus* was announced. *Ardipithecus ramidus* consists of the skeleton and skull of a female, 119 cm tall, weighing approximately 50 km, and nicknamed "Ardi" (❚ Figure 23.32). Ardi, who lived approximately 4.4 million years ago, displays an interesting mosaic of evolutionary characteristics that include a dexterous hand for grasping and a foot with an opposable big toe but which lacked the flexibility of an ape's foot, which is used for climbing and maneuvering in trees. Based on these and other features, scientists have concluded that Ardi was able to walk upright on the ground, while retaining the ability to climb and maneuver in trees.

❙ **Figure 23.29** Comparison of Gorilla and Human Locomotion and Hominid Brain Size

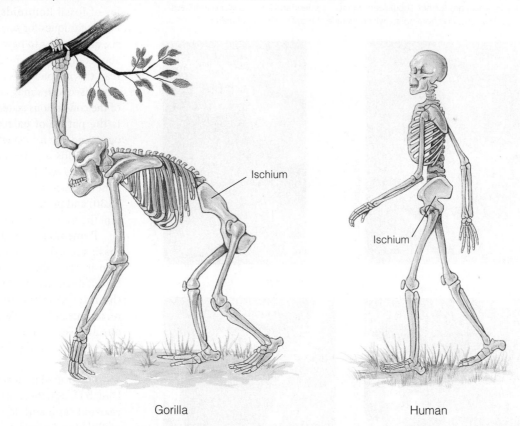

Gorilla

Human

a In gorillas, the ischium bone is long, and the entire pelvis is tilted toward the horizontal. In humans, the ischium bone is much shorter, and the pelvis is vertical.

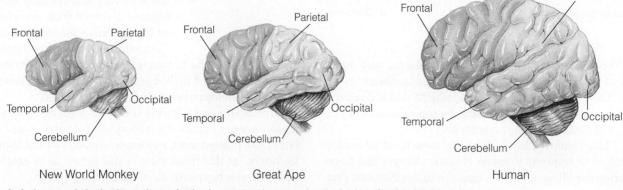

New World Monkey

Great Ape

Human

b An increase in brain size and organization is apparent in comparing the brains of a New World monkey (superfamily Ceboidea), a great ape (superfamily Hominoidea; family Pongidae), and a present-day human (superfamily Hominoidea; family Hominidae)

Australopithecines **Australopithecine** is a collective term for all members of the genus *Australopithecus*. Currently, five species are recognized: *A. anamensis, A. afarensis, A. africanus, A. robustus,* and *A. boisei*. Many paleontologists accept the evolutionary scheme in which *A. anamensis,* the oldest known australopithecine, is ancestral to *A. afarensis,* who in turn is ancestral to *A. africanus* and the genus *Homo,* as well as the side branch of australopithecines represented by *A. robustus* and *A. boisei*.

The oldest known australopithecine is *Australopithecus anamensis*. Discovered at Kanapoi, a site near Lake Turkana, Kenya, by Meave Leakey of the National Museums of Kenya and her colleagues, this 4.2-million-year-old bipedal species has many features in common with its younger relative, *Australopithecus afarensis,* yet is more primitive in other characteristics, such as its teeth and skull. *A. anamensis* is estimated to have been between 1.3 and 1.5 m tall and weighed 33 to 50 kg.

Figure 23.30 **The Stratigraphic Record of Hominids** The geologic ranges for the commonly accepted species of hominids (the branch of primates that includes present-day humans and their extinct ancestors).

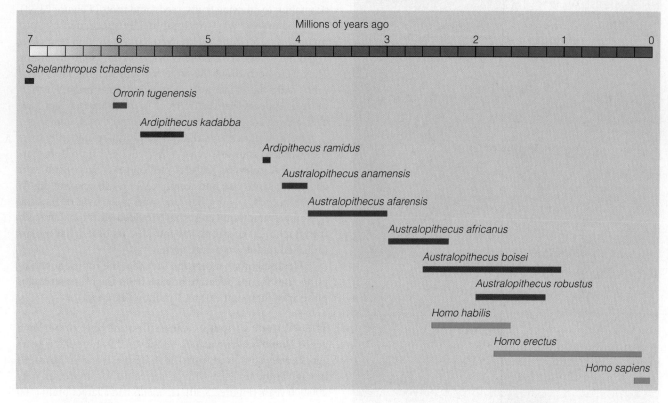

Figure 23.31 *Sahelanthropus tchadensis* Discovered in Chad in 2002, and dated at nearly 7 million years, this skull of *Sahelanthropus tchadensis* is presently the oldest known hominid.

© M.P.F.T.

Figure 23.32 *Ardipithecus ramidus* Frontal view of a female *Ardipithecus ramidus*, known as "Ardi." *A. ramidus*, who lived approximately 4.4 million years ago in what is now Ethiopia shows an interesting variety of features, such a dexterous hand for grasping, and a foot with an opposable big toe (clearly shown here) that allowed it to walk upright, yet also climb and maneuver in trees.

HO/AFP/Getty Images/Newscom

» **Critical Thinking Question**
With features that allow for climbing in trees like a gorilla, and also walking upright on the ground like a hominid, should "Ardi" be considered a "missing link?" Why do you think "missing link" is not a good term to use in evolution?

David L. Brill

▎**Figure 23.33** Skeleton of Lucy *(Australopithecus afarensis)* A reconstruction of Lucy's skeleton by Owen Lovejoy and his students at Kent State University, Ohio. Lucy, whose fossil remains were discovered by Donald Johanson, is an approximately 3.5-million-year-old *Australopithecus afarensis* individual. This reconstruction illustrates how adaptations in Lucy's hip, leg, and foot allowed for a fully bipedal means of locomotion.

A discovery of fossils of *Australopithecus anamensis* from the Middle Awash area in northeastern Ethiopia has shed light on the transition between *Ardipithecus* and *Australopithecus*. Prior to this discovery, the origin of *Australopithecus* had been hampered by a sparse fossil record. The discovery of *Ardipithecus* in the same region of Africa and at the same time as the earliest *Australopithecus* provides strong evidence that *Ardipithecus* evolved into *Australopithecus* and links these two genera in the evolutionary lineage leading to humans.

Australopithecus afarensis (▎ Figure 23.33), who lived 3.9–3.0 million years ago, was fully bipedal and exhibited great variability in size and weight. Members of this species ranged from just over 1 m to about 1.5 m tall and weighed between 29 and 45 kg. They had a brain size of 380–450 cubic centimeters (cc), larger than the 300–400 cc of a chimpanzee, but much smaller than that of present-day humans (1,350 cc average).

The skull of *A. afarensis* retained many apelike features, including massive brow ridges and a forward-jutting jaw,

but its teeth were intermediate between those of apes and humans. The heavily enameled molars were probably an adaptation to chewing fruits, seeds, and roots (▎ Figure 23.34).

A. afarensis was stratigraphically succeeded by *Australopithecus africanus,* who lived 3.0–2.3 million years ago. The differences between the two species are relatively minor. They were both about the same size and weight, but *A. africanus* had a flatter face and a somewhat larger brain. Furthermore, it appears that the limbs of *A. africanus* may not have been as well adapted for bipedalism as those of *A. afarensis.*

Both *A. afarensis* and *A. africanus* differ markedly from the so-called robust species *A. boisei* (2.6–1.0 million years ago) and *A. robustus* (2.0–1.2 million years ago). Both were about the same size and weight (1.1–1.4 m tall and 32–49 kg). *A. robustus* had a flat face, and the crown of its skull had an elevated bony crest that provided additional area for the attachment of strong jaw muscles. Its broad, flat molars indicated that it was a vegetarian.

Most scientists accept the idea that the robust australopithecines form a separate lineage from the other australopithecines that went extinct 1 million years ago.

The Human Lineage The earliest member of our own genus **Homo** is *Homo habilis,* who lived 2.5–1.6 million years ago. *H. habilis* evolved from the *A. afarensis* and *A. africanus* lineage and coexisted with *A. africanus* for approximately 200,000 years (Figure 23.30). *H. habilis* had a larger brain (700 cc average) than its australopithecine ancestors, but smaller teeth. It was about 1.2–1.3 m tall and only weighed 32–37 kg.

The evolutionary transition from *H. habilis* to *H. erectus* appears to have occurred in a short period of time, between 1.8 and 1.6 million years ago. However, based on new findings published in 2007, which suggest that *H. habilis* and *H. erectus* apparently coexisted for approximately 500,000 years, some scientists think that *H. habilis* and *H. erectus* may have evolved from a common ancestor and represent separate lineages of *Homo,* rather than the traditional linear view of *H. erectus* evolving from *H. habilis.*

In contrast to the australopithecines and *H. habilis,* which are unknown outside Africa, *Homo erectus* was a widely distributed species, having migrated from Africa during the Pleistocene. Specimens have been found not only in Africa, where it evolved 1.8 million years ago, but also in Europe (▎ Figure 23.35), India, China ("Peking Man"), and Indonesia ("Java Man").

Although *H. erectus* developed regional variations in form, the species differed from modern humans in several ways. Its brain size of 800–1,300 cc, although much larger than that of *H. habilis,* was still less than the average for *Homo sapiens* (1,350 cc). The skull of *H. erectus* was thick walled, its face was massive, it had prominent brow ridges, and its teeth were slightly larger than those of present-day humans. *H. erectus* was comparable in size to modern humans, standing between 1.6 and 1.8 m tall and weighing between 53 and 63 kg.

The archaeological record indicates that *H. erectus* was a toolmaker. Furthermore, some sites show evidence that its the "out of Africa" view. According to this view, early modern humans evolved from a single woman in Africa, whose offspring then migrated from Africa, perhaps as recently as 100,000 years ago, and populated Europe and Asia, driving the earlier hominid population to extinction.

The alternative explanation, the "multiregional" view, maintains that early modern humans did not have an isolated origin in Africa, but, rather, that they established separate populations throughout Eurasia. Occasional contact and interbreeding between these populations enabled our species to maintain its overall cohesiveness while still preserving the regional differences in people that we see today. Regardless of which theory turns out to be correct, our species, *H. sapiens,* most certainly evolved from *H. erectus.*

▌Figure 23.34 **African Pliocene Landscape** Recreation of a Pliocene landscape showing members of *Australopithecus afarensis* gathering and eating various fruits and seeds.

members used fire and lived in caves, an advantage for those living in more northerly climates (Figure 23.35).

Debate still surrounds the transition from *Homo erectus* to our own species, *Homo sapiens.* Paleoanthropologists are split into two camps. On the one side are those who support

Perhaps the most famous of all fossil humans are the **Neanderthals,** who inhabited Europe and the Near East from about 200,000 to 30,000 years ago and, according to the best estimates, never exceeded 15,000 individuals in western Europe (▌Figure 23.36). Some paleoanthropologists regard the Neanderthals as a variety or subspecies (*Homo sapiens*

▌Figure 23.35 **European Pleistocene Landscape with *Homo erectus*** Recreation of a Pleistocene setting in Europe in which members of *Homo erectus* are using fire and stone tools.

Painting by Ronald Bowen/Robert Harding Picture Library

❙ **Figure 23.36** Pleistocene Cave Setting with Neanderthals Archaeological evidence indicates that Neanderthals lived in caves and participated in ritual burials, as depicted in the painting of a burial ceremony such as occurred approximately 60,000 years ago at Shanidar Cave, Iraq.

neanderthalensis), whereas others consider them a separate species (*Homo neanderthalensis*). In any case, their name comes from the first specimens found in 1856 in the Neander Valley near Düsseldorf, Germany.

The most notable difference between Neanderthals and present-day humans is in the skull. Neanderthal skulls were long and low, with heavy brow ridges, a projecting mouth, and a weak, receding chin. Their brain was slightly larger, on average, than our own and somewhat differently shaped. The Neanderthal body was more massive and heavily muscled than ours, with a flaring rib cage and rather short lower limbs, much like those of other cold-adapted people of today. Neanderthal males averaged between 1.6 and 1.7 m tall and weighed about 83 kg.

In 2007, scientists announced that they had isolated a pigmentation gene from a segment of Neanderthal DNA that indicated that at least some Neanderthals had red hair and light skin. Furthermore, the gene is different from that of modern red-haired people, suggesting that perhaps Neanderthals and present-day humans developed the trait independently, in response to similar higher-northern-latitude environmental pressures.

Based on specimens from more than 100 sites, we now know that Neanderthals were not much different from us, only more robust. Europe's Neanderthals were the first humans to move into truly cold climates, enduring miserably long winters and short summers as they pushed north into tundra country. Their remains are found chiefly in caves and hutlike rock shelters, which also contain a variety of specialized stone tools and weapons (Figure 23.36). Furthermore, archaeological evidence indicates that Neanderthals

commonly took care of their injured and buried their dead, frequently with such grave items as tools, food, and perhaps even flowers.

As more fossil discoveries are made and increasingly sophisticated techniques of DNA extraction and analysis are carried out, our view of Neanderthals and their society as well as their place in human evolution will continue to change.

About 30,000 years ago, humans closely resembling modern Europeans moved into the region inhabited by Neanderthals and completely replaced them. **Cro-Magnons** (❙ Figure 23.37), the name given to the successors of Neanderthals in France, lived from about 35,000 to 10,000 years ago; during this period, the development of art and technology far exceeded anything the world had seen before. Using paints made from manganese and iron oxides, Cro-Magnon people painted hundreds of scenes on the ceilings and walls of caves in France and Spain, where many of them are still preserved.

Cro-Magnons were also skilled nomadic hunters, following the herds in their seasonal migrations. They used a variety of specialized tools in their hunts, including perhaps the bow and arrow. They sought refuge in caves and rock shelters and formed living groups of various sizes.

With the appearance of Cro-Magnons, human evolution has become almost entirely cultural rather than biological. Since the evolution of the Neanderthals approximately 200,000 years ago, humans have gone from a stone culture to a technology that has allowed us to visit other planets with space probes and land men on the Moon. It remains to be seen how we will use this technology in the future and

▌Figure 23.37 **Pleistocene Cro-Magnon Camp in Europe** Cro-Magnons were highly skilled hunters who formed living groups of various sizes.

whether we will continue as a species, evolve into another species, or become extinct as many groups have before us.

Pleistocene Extinctions

Extinctions are part of the evolutionary history of life, but during times of mass extinction, Earth's biotic diversity sharply declined, as at the end of the Paleozoic and Mesozoic eras. In marked contrast, the Pleistocene extinctions were rather modest, but they did have a profound effect on genera of large terrestrial mammals (those weighing more than 40 kg), and some large flightless birds. Particularly hard hit were the mammalian faunas of Australia and the Americas.

In Australia, 15 of the continent's 16 genera of large mammals died out. North America lost 33 of 45 genera of large mammals, and in South America, 46 of 58 large genera went extinct. In contrast, Europe had only 7 of 23 large genera go extinct, and in Africa south of the Sahara, only 2 of 44 died out.

These facts lead to three questions: (1) What caused Pleistocene extinctions? (2) Why did these extinctions eliminate mostly large mammals? and (3) Why were extinctions most severe in Australia and the Americas? There is currently no completely satisfactory explanation, but two competing hypotheses are currently being debated. One holds that rapid climatic changes at the end of the Pleistocene

caused extinctions, whereas another, called *prehistoric overkill,* considers that human hunters were responsible.

Rapid changes in climate and vegetation occurred over much of Earth's surface during the Late Pleistocene as glaciers began retreating. The North American and northern Eurasian open-steppe tundras were replaced by conifer and broadleaf forests as warmer and wetter conditions prevailed. The Arctic region flora changed from a productive herbaceous one that supported a variety of large mammals to a relatively barren, water-logged tundra that supported a far sparser fauna. The southwestern U.S. region also changed from a moist area with numerous lakes, where saber-toothed cats, giant ground sloths, and mammoths roamed, to a semiarid environment unable to support a diverse fauna of large mammals.

Rapid changes in climate and vegetation can certainly affect animal populations, but there are several problems with the climate hypothesis. First, why didn't the large mammals migrate to more suitable habitats as the climate and vegetation changed? After all, many other animal species did. For example, reindeer and the Arctic fox lived in southern France during the last glaciation and migrated to the Arctic when the climate became warmer. The second problem is the apparent lack of correlation between extinctions and earlier glacial advances and retreats throughout the Pleistocene Epoch. Previous changes in climate were not marked by episodes of mass extinction.

Proponents of the prehistoric overkill hypothesis argue that the mass extinctions in North and South America and Australia coincided closely with the arrival of humans. Perhaps hunters had a tremendous impact on the faunas of North and South America about 11,000 years ago because the animals had no previous experience with humans. The same thing happened much earlier in Australia soon after people arrived about 40,000 years ago. No large-scale extinctions occurred in Africa and most of Europe because animals in those regions had long been familiar with humans.

One problem with the prehistoric overkill hypothesis is that archaeological evidence indicates that early human inhabitants of North and South America, as well as Australia, probably lived in small, scattered communities, gathering food and hunting. How could a few hunters decimate so many species of large mammals? However, humans have caused major extinctions on oceanic islands. For example, in a period of about 600 years after arriving in New Zealand,

humans exterminated several species of the large, flightless birds called moas.

A second problem is that present-day hunters concentrate on smaller, abundant, and less dangerous animals. The remains of horses, reindeer, and other small animals are found in many prehistoric sites in Europe, whereas mammoth and woolly rhinoceros remains are scarce.

Finally, few human artifacts are found among the remains of extinct animals in North and South America, and there is little evidence that the animals were hunted. Countering this argument is the assertion that the impact on the previously unhunted fauna was so swift as to leave little evidence.

The reason for the extinctions of large Pleistocene mammals is still unresolved and probably will be for some time. It may turn out that the extinctions resulted from a combination of different circumstances. Populations that were already under stress from climatic changes were perhaps more vulnerable to hunting, especially if small females and young animals were the preferred targets.

GEO-RECAP

» Geo-Image

▎**Figure 1** The image below shows Mount Shasta in California, which is one of a dozen or so large volcanoes in the Cascade Range. What kind of volcano is shown in this image? If you were to sample a lava flow from Mount Shasta, what kind of volcanic rocks would you expect to find? In addition to lava flows, what other materials are found in volcanoes such as this?

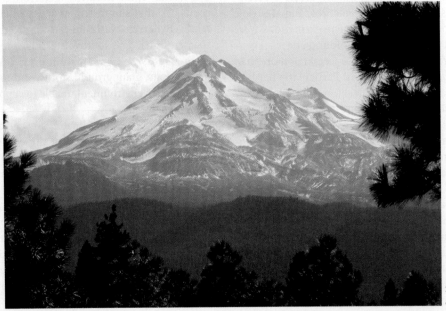

Sue Monroe

Chapter Summary

- Cenozoic tectonism was concentrated in the Alpine–Himalayan and circum-Pacific belts.

- The Cenozoic evolution of the North American Cordillera included deformation during the Laramide orogeny, extensional tectonics that yielded basin-and-range structures, extensive intrusive and extrusive igneous activity, and uplift and erosion.

- One model for the Laramide orogeny involves near-horizontal subduction of the Farallon plate beneath North America, resulting in fault-bounded uplifts in the area of the present-day Rocky Mountains.

- As the North American plate drifted westward, it collided with the Pacific–Farallon Ridge, at which time subduction ceased and the continent became bounded by large transform faults, except in the Pacific Northwest where subduction continues.

- Sediments eroded from Laramide uplifts were deposited in intermontane basins and in the Great Plains, whereas a wedge of sediments pierced by salt domes is found on the Gulf Coastal Plain.

- Cenozoic uplift and erosion were responsible for the present topography of the Appalachian Mountains. As the Appalachians eroded, much of the sediment was deposited on the Atlantic Coastal Plain.

- Vast glaciers covered about 30% of Earth's land surface during the Pleistocene. About 20 warm–cold Pleistocene climatic cycles are recognized from evidence found in deep-sea cores.

- Large parts of North America and other Northern Hemisphere continents were covered by continental glaciers several times during the Pleistocene.

- Cenozoic mineral resources include sand and gravel, placer deposits of gold, some evaporite minerals such as borax, and oil and natural gas.

- Marine invertebrate groups that survived the extinctions at the end of the Mesozoic continued to diversify, giving rise to the present-day marine fauna.

- The Paleocene mammalian fauna was composed of Mesozoic holdovers and several new orders, and by Eocene time, most living mammal orders had evolved.

- Shrewlike placental mammals that evolved during the Late Cretaceous were the ancestors for the placental mammalian orders that evolved during the Cenozoic.

- Among the hoofed mammals (artiodactyls and perissodactyls), adaptations include modifications of the teeth for grinding vegetation and changes in their limbs for speed.

- The evolutionary history of horses is particularly well documented by fossils, but scientists also know much about the evolution of other hoofed mammals, as well as elephants, whales, and some carnivores.

- Primates evolved during the Paleocene. Several trends help characterize primates and differentiate them from other mammalian orders, including a change in overall skeletal structure and mode of locomotion, an increase in brain size, stereoscopic vision, and evolution of a grasping hand with opposable thumb.

- The primates are divided into two suborders: prosimians and anthropoids. The prosimians are the oldest primate lineage and include lemurs, lorises, tarsiers, and tree shrews. The anthropoids include the New and Old World monkeys, apes, and hominids, which are humans and their extinct ancestors.

- The oldest known hominid is *Sahelanthropus tchadensis,* dated at nearly 7 million years old. It was followed by *Orrorin tugenensis* at 6 million years, and then two subspecies of *Ardipithecus,* at 5.8 and 4.4 million years old, respectively. These early hominids were succeeded by the australopithecines, a fully bipedal group that evolved in Africa 4.2 million years ago. Recent discoveries indicate *Ardipithecus* evolved into various species of *Australopithecus.*

- The human lineage began about 2.5 million years ago in Africa with the evolution of *Homo habilis,* which survived as a species until about 1.6 million years ago.

- *Homo erectus* evolved from *H. habilis* sometime between 1.8 million and 1.6 million years ago and was the first hominid to migrate out of Africa, spreading to Europe, India, China, and Indonesia, between 1.8 and 1 million years ago. The transition from *H. erectus* to *H. sapiens* is still unresolved because there is currently insufficient evidence to determine which hypothesis—the "out of Africa" or the "multiregional"—is correct. Nevertheless, *H. erectus* used fire, made tools, and lived in caves.

- Neanderthals inhabited Europe and the Near East between 200,000 and 30,000 years ago and were not much different from present-day humans. They were, however, more robust and had differently shaped skulls. In addition, they made specialized tools and weapons, apparently took care of their injured, and buried their dead.

- The Cro-Magnons were the successors of the Neanderthals and lived from about 35,000 to 10,000 years ago. They were highly skilled nomadic hunters, formed living groups of various sizes, and were also skilled cave painters.

- Modern humans succeeded the Cro-Magnons about 10,000 years ago and have spread throughout the world, as well as having set foot on the Moon.

- Although modest by comparison to the mass extinction events at the end of the Paleozoic and Mesozoic eras, the Pleistocene extinctions had a profound effect on genera of large terrestrial mammals. Two competing hypotheses have been advanced to explain these Pleistocene extinctions: climatic change and prehistoric overkill.

Important Terms

Alpine–Himalayan orogenic belt
  (p. 615)
artiodactyl (p. 636)
Atlantic Coastal Plain (p. 628)
australopithecine (p. 644)
Basin and Range Province (p. 620)
browser (p. 636)
Cascade Range (p. 620)
circum-Pacific orogenic belt (p. 615)
Colorado Plateau (p. 621)
Cro-Magnon (p. 648)

Farallon plate (p. 624)
grazer (p. 636)
Gulf Coastal Plain (p. 627)
hominid (p. 642)
Homo (p. 646)
Interior Lowlands (p. 625)
Laramide orogeny (p. 617)
marsupial mammal (p. 635)
monotreme (p. 635)
Neanderthal (p. 647)
North American Cordillera (p. 617)

orogen (p. 615)
Paleocene–Eocene Thermal Maximum
  (p. 634)
perissodactyl (p. 636)
placental mammal (p. 635)
primate (p. 641)
Rio Grande rift (p. 621)
San Andreas transform fault (p. 625)
Tejas epeiric sea (p. 627)
ungulate (p. 636)
Zuni epeiric sea (p. 625)

Review Questions

1. The Basin and Range province covers a large area in North America, but it is centered mostly on
   a. ___ Canada.
   b. ___ the Appalachian Mountains.
   c. ___ Nevada.
   d. ___ South Dakota.
   e. ___ Mexico.

2. A complex part of the circum-Pacific orogenic belt in the United States is the
   a. ___ Tejas sedimentary sequence.
   b. ___ North American Cordillera.
   c. ___ Rio Grande rift.
   d. ___ Atlantic Coastal Plain.
   e. ___ Pacific–Farallon Ridge.

3. The human lineage began with the evolution of which species?
   a. ___ *Orrorin tugenensis*.
   b. ___ *Ardipithecus ramidus*.
   c. ___ *Sahelanthropus tchadensis*.
   d. ___ *Homo habilis*.
   e. ___ *Australopithecus boisei*.

4. Which of the following statements is correct?
   a. ___ All of North America was ice-covered during the Pleistocene.
   b. ___ Continental glaciers are long, narrow tongues of ice in mountain valleys.
   c. ___ Varves are glacial deposits that form when a lake drains catastrophically.
   d. ___ The Pleistocene Epoch began 2.6 million years ago and ended 10,000 years ago.
   e. ___ Valley glaciers cover vast areas and flow out in all directions from a center of accumulation.

5. One of the defining characteristics of the carnivorous mammals is
   a. ___ carnassials teeth.
   b. ___ hooves.
   c. ___ short, massive limbs.
   d. ___ teeth adapted to grazing.
   e. ___ increase in the number of toes.

6. An area in the western United States where rifting is taking place is the
   a. ___ Gulf Coastal Plain.
   b. ___ Farallon rift system.
   c. ___ Pacific orogenic belt.
   d. ___ Tejas epeiric sea.
   e. ___ Rio Grande rift.

7. The Himalayas formed when the _____ plate collided with the _____ plate.
   a. ___ Farallon, Pacific.
   b. ___ Nazca, Cocos.
   c. ___ African, European.
   d. ___ Indian, Asian.
   e. ___ Australian, South American.

8. The only living egg-laying mammals are
   a. ___ multituberculates.
   b. ___ megadonts.
   c. ___ monotremes.
   d. ___ marsupials.
   e. ___ placentals.

9. Which of the following features distinguish hominids from other hominoids?
   a. ____ a large and internally reorganized brain.
   b. ____ a reduced face and reduced canine teeth.
   c. ____ bipedalism.
   d. ____ use of sophisticated tools.
   e. ____ all of these.

10. Lakes that existed during times of glaciation because of increased precipitation and decreased evaporation are called _____ lakes.
    a. ____ alluvial fan.
    b. ____ outwash.
    c. ____ varve.
    d. ____ pluvial.
    e. ____ cirque.

11. What were the dominant land-dwelling predators during the Early Cenozoic, and what happened to them?

12. Briefly discuss the major mammal diversification that took place during the Cenozoic Era.

13. Twenty or 30 years from now it is doubtful that you will remember many of the terms and names of events you learned in this course, unless you become a geologist or work in some related field. So, how would you describe the overall evolution of the North American Cordillera in terms that a nongeologist could readily understand?

14. How does the Cenozoic geologic history of the Colorado Plateau differ from the history of other parts of the North American Cordillera?

15. What is the Little Ice Age, when did it take place, and what impact did it have on humans?

16. Where is the Cascade Range, what kinds of volcanoes are found there, and what accounts for ongoing volcanism in this area?

17. Discuss the merits of both the *climate change* and the *prehistoric overkill* hypotheses in explaining the extinction of large mammalian faunas at the end of the Pleistocene Epoch.

18. How did the Laramide orogeny differ from other orogenies at convergent plate boundaries?

19. Now that we have covered the evolution of vertebrate animals from fish to mammals, can you summarize the major events that occurred without using technical jargon? In other words, how would you explain the overall evolution of vertebrates to an interested but uninformed audience?.

20. What factors do you think will influence the future of human evolution? Do you think it is possible that we can control the direction that evolution takes?

Urban sprawl in Mexico City, Mexico. Overpopulation is considered one of the greatest environmental problems facing the world today. Not only are we covering the land with our structures, we are contributing to the pollution of the atmosphere and possibly influencing climate change. Mexico City is also subsiding because of the removal of groundwater faster than it can be replenished. This is due to the ever increasing demand for water from its burgeoning population.

CHAPTER 24

Geology in Perspective

OUTLINE

24.1 Introduction

24.2 Geology and the Environment

24.3 A Final Word

MARK EDWARDS/PhotoLibrary

24.1 Introduction

Throughout this book, we have emphasized that Earth is a complex, dynamic planet that has changed continuously since its origin some 4.6 billion years ago. These changes, and the present-day features we observe, are the result of interactions among the various interrelated internal and external Earth systems, subsystems, and cycles. In addition, these interactions have influenced the evolution of the biosphere, as is obvious from an examination of the fossil record. Perhaps the most important lesson to be learned from your study of geology is that Earth is a dynamic and changing planet in which continual change is taking place, although not always at a rate we can readily appreciate from a human lifetime or perspective.

The rock cycle (see Figure 1.15), with its recycling of Earth materials to form the three major rock groups (see Figure 1.16), illustrates the interrelationships between Earth's internal and external processes. The hydrologic cycle (see Figure 12.2) explains the continuous recycling of water from the oceans to the atmosphere to the land and eventually back to the oceans again. Changes within this cycle can have profound effects on Earth's topography as well as its biota. For example, a rise in global temperature will cause the ice caps to melt, contributing to rising sea level, which will greatly affect coastal areas where many of the world's large population centers are presently located. We have seen the effect of changing sea level on continents in the past, which resulted in large-scale transgressions and regressions (see Chapter 20). Some of these were caused by growing and shrinking continental ice caps when land masses moved over the South Pole as a result of plate movements.

On a larger scale, the movement of plates has had a profound effect on the formation of landscapes, the distribution of mineral resources, and atmospheric and oceanic circulation patterns, as well as the evolution and diversification of life.

The launching in 1957 of *Sputnik 1*, the world's first artificial satellite, ushered in a new global consciousness in terms of how we view Earth and our place in the global ecosystem. Satellites have provided us with the ability to view not only the beauty of our planet but also the fragility of Earth's biosphere and the role humans play in shaping and modifying the environment. The pollution of the atmosphere, oceans, and many of our lakes and streams; the denudation of huge areas of tropical forests; the scars from strip mining; the depletion of the ozone layer—all are visible in the satellite images beamed back from space and attest to the impact humans have had on the global ecosystem.

Accordingly, we must understand that the changes we make to the global ecosystem can have wide-ranging effects of which we may not even be aware. For this reason, an understanding of geology, and science in general, is of paramount importance if we are to be good stewards of Earth. On the other hand, we must also remember that humans are part of the global ecosystem, and, like all other life-forms, our presence alone affects the ecosystem. We must therefore act in a responsible manner, based on sound scientific knowledge, so future generations will inherit a habitable environment.

One of the most important objectives of this book, as well as much of your secondary education, is to develop your critical thinking skills. As opposed to simple disagreement, critical thinking involves evaluating the supporting evidence for a particular point of view. Although your exposure to geology is probably still limited, having read this book, you now have the fundamental knowledge needed to appraise why geologists accept plate tectonic theory, why they think that Earth is 4.6 billion years old, and why scientists are convinced that the theory of evolution is well supported by evidence. Furthermore, your abilities as a critical thinker will probably help you more effectively evaluate the arguments about global warming, ozone depletion, groundwater contamination, and many other environmental issues.

When such environmental issues as acid rain (see Chapter 6), the depletion of the ozone layer, and the greenhouse effect and global warming (see Figure 1.4) are discussed and debated, it is important to remember that they are not isolated topics but are part of a larger system that involves the entire Earth (see Figure 1.1). In addition, it is important to remember that Earth goes through cycles of much longer duration than the human perspective of time (see Chapter 17). Although they may have disastrous effects on the human species, global warming and cooling are part of a larger cycle that has resulted in numerous glacial advances and retreats during the past 2.6 million years (see Chapter 14).

In fact, geologists can make important contributions to the debate on global warming because of their geologic perspective. Long-term trends can be studied by analyzing deep-sea sediments, ice cores, changes in sea level during the geologic past, and the distribution of plants and animals through time. As we have seen throughout this book, such studies have been done, and the results and synthesis of that information can be used to make intelligent decisions about how humans can better manage the environment and the effect we are having in altering the environment.

24.2 Geology and the Environment

Geologists who specialize in environmental geology are involved in understanding the relationship between humans and their geologic environment. Put another way, environmental geologists are concerned with studying and resolving the problems that human occupation has caused to the environment. As we discussed in Chapter 1, many scientists consider overpopulation to be the greatest environmental problem facing the world today (Figure 24.1). The world's population reached 7 billion in 2011, and it is projected to

Figure 24.1 Overpopulation Overpopulation is perhaps the greatest environmental problem facing the world today. Until the world's increasing population is brought under control, people will continue to strain Earth's limited resources. Shown here is the crush of people in front of Jama Masjid in New Delhi, India, which currently has a population of nearly 14 million people.

grow to 9 billion by 2045. Much of this population growth will be in areas that are already at risk from earthquakes, tsunami, volcanic eruptions, and floods. In addition, adequate and clean water supplies, as well as additional energy resources, must be found and properly managed, and all with a minimal impact on the environment.

Interestingly, most of the major environmental issues currently being discussed and debated, such as acid rain, ozone depletion, and global warming, seemingly have as a common denominator humans and their effect on these global problems. The role that humans play in many of these issues as well as the degree to which they play that role are being examined and debated not only by scientists, but by economists, sociologists, and politicians, to name a few. It is, therefore, more important than ever that citizens become versed in what science is, and in what it can, and cannot do, as well as becoming critical thinkers in the arguments regarding the impact of humans on the global ecosystem.

We have already discussed in previous chapters such environmental problems and concerns as the greenhouse effect (Chapter 1), soil degradation (Chapter 6), acid rain (Chapter 6), flood control (Chapter 12), groundwater contamination (Chapter 13), and climate change (Chapter 14).

For example, you learned in Chapter 6 how soils form, how they are categorized, and the factors that control soil formation and degradation. Having learned the basic science of soils and their formation, it is important to remember that our future as a species is clearly linked to the wise management of our soil resources. Like other conservation efforts, using soils sustainably depends on a combination of scientific understanding and political will. Your generation will

probably see new pressures on world soil resources and may potentially contribute new solutions to perennial soil problems.

Additionally, you have learned what acid rain is, how it forms, and why it is such a multifaceted problem that knows no national boundaries. Acid rain is just one of many environmental issues that are interconnected and are part of a dynamic and ever-changing Earth (∎ Figure 24.2).

Probably the major environmental issue currently being discussed is global warming, a topic we covered in Chapter 1 and continued to emphasize in subsequent chapters, not only because of its importance to humans, but because it is an excellent example of how Earth's systems, subsystems, and cycles are interrelated.

Recall that carbon dioxide is produced as a by-product of respiration and the burning of organic material. As such, it is a component of the global ecosystem and is constantly being recycled as part of the carbon cycle. The concern in recent years over the increase in atmospheric carbon dioxide has to do with its role in the greenhouse effect (see Figure 1.4). The issue is not whether we have a greenhouse effect, because we do, but to what degree human activity, such as the burning of fossil fuels, has contributed to global warming.

Another theme basic to geology and affecting all of us, regardless of where we live, is geologic hazards. Any geologic phenomenon that endangers property and people is a geologic hazard. Earthquakes, volcanic eruptions, landslides, and river floods and coastal flooding during hurricanes immediately come to mind, because they are spectacular geologic hazards that commonly cause property damage, injuries, and fatalities (∎ Figure 24.3). However, some of the more subtle hazards, such as soil creep and expansive soils, cause more property damage, but rarely rate headlines in newspapers or on news reports.

During 2010 and early 2011 alone, we have witnessed major earthquakes in Haiti, New Zealand, and Japan, as well as devastating tsunami. Landslides have been in the news in Brazil, and massive flooding has devastated Brisbane and environs in Australia. Even as we write this section, the Mississippi River is flooding large areas in Tennessee, and some parts of Alabama and Missouri have been devastated by tornados.

Geologic hazards account for thousands of injuries and fatalities each year, as well as billions of dollars in property damages. You may read in the more sensational press that geologic hazardous events, especially earthquakes, have been increasing. The incidence of these events has not increased, but there has been an increase in the amount of

▌Figure 24.2 Acid Rain

Yoram Lehman/Photo Researchers, Inc.

a Pollution from the smoke stacks of steel mills in Pittsburgh, Pennsylvania during the 1970s contributed to the formation of acid rain in the eastern United States and Canada.

Mary Terriberry/shutterstock.com

b The effects of acid rain can be seen in these dead trees in Grayson Highlands State Park, Virginia.

property damages and the number of injuries and fatalities because more and more people live in disaster-prone areas. Population density and poor construction practices are obvious contributors to the devastation caused by flooding and earthquakes in developing countries.

Of course, we cannot eliminate geologic hazards, but we can better understand these phenomena, enact prudent zoning and land-use regulations, and at the very least, decrease the amount of human suffering and monetary costs associated with disasters.

Unfortunately, geologic information that is readily available is often ignored. A case in point is the Turnagain Heights subdivision in Anchorage, Alaska, that was heavily damaged when the soil beneath it liquefied during the 1964 earthquake (see Figure 11.18). Not only were reports on soil stability ignored or overlooked before homes were built there, but since 1964 many new homes have been built on part of the same site!

La Conchita, California, lies at the base of steep, unstable slopes that collapsed in 1995 and again in 2005, killing 10 people in the latter event (see Figures 1 and 2 in the Chapter 11 Geo-Focus). There are many other examples of building subdivisions on known landslide masses in the Los Angeles area, and development in some coastal areas continues where erosion will surely destroy homes and other structures.

Although most of you aren't going to become licensed geological engineers, you now have the knowledge to make informed decisions on the suitability of the geology of the area where you might buy or build a house. Hopefully, with this knowledge, you will not make the same mistakes many people made in the examples given throughout this book.

24.3 A Final Word

In conclusion, the most important lesson to be learned from the study of geology is that Earth is an extremely complex and ever-changing planet. Interactions among its various systems and subsystems have resulted in changes in the atmosphere, lithosphere, and biosphere through time. By studying how Earth has evolved in the past, we can better understand how the different Earth systems and subsystems work and interact with one another, and, more importantly, how our actions affect the delicate balance between these various components.

We tend to view our planet from the perspective of a human lifetime and commonly overlook the fact that Earth has changed markedly in the context of geologic time and that it continues to do so. Geology, therefore, is not a static science, but one, like the dynamic Earth it seeks to understand, that constantly evolves as new information and methods of investigation become available. It is thus incumbent on all of us to apply the geologic lessons we have learned to make Earth a more sustainable planet for both the current and future generations.

▌**Figure 24.3** Geologic Hazards Earthquakes are just one of the numerous geologic hazards that can cause loss of human life as well as significant property damage. One example is the May 12, 2008 7.9-magnitude earthquake that devastated Sichuan province in China, killing approximately 69,000 people, and leaving millions homeless.

iStockphoto.com/pestanarui

Appendix A

	English Unit	Conversion Factor	Metric Unit	Conversion Factor	English Unit
Length	Inches (in.)	2.54	Centimeters (cm)	0.39	Inches (in.)
	Feet (ft)	0.305	Meters (m)	3.28	Feet (ft)
	Miles (mi)	1.61	Kilometers (km)	0.62	Miles (mi)
Area	Square inches (in.^{2})	6.45	Square centimeters (cm^2)	0.16	Square inches (in.^{2})
	Square feet (ft^2)	0.093	Square meters (m^2)	10.8	Square feet (ft^2)
	Square miles (mi^2)	2.59	Square kilometers (km^2)	0.39	Square miles (mi^2)
Volume	Cubic inches (in.^{3})	16.4	Cubic centimeters (cm^3)	0.061	Cubic inches (in.^{3})
	Cubic feet (ft^3)	0.028	Cubic meters (m^3)	35.3	Cubic feet (ft^3)
	Cubic miles (mi^3)	4.17	Cubic kilometers (km^3)	0.24	Cubic miles (mi^3)
Weight	Ounces (oz)	28.3	Grams (g)	0.035	Ounces (oz)
	Pounds (lb)	0.45	Kilograms (kg)	2.20	Pounds (lb)
	Short tons (st)	0.91	Metric tons (t)	1.10	Short tons (st)
Temperature	Degrees Fahrenheit (°F)	$-32° \times 0.56$	Degrees centigrade (Celsius) (°C)	$\times 1.80 + 32°$	Degrees Fahrenheit (°F)

Examples:

10 inches = 25.4 centimeters; 10 centimeters = 3.9 inches

100 square feet = 9.3 square meters; 100 square meters = 1080 square feet

50°F = 10.1°C; 50°C = 122°F

Appendix B
TOPOGRAPHIC MAPS

Nearly everyone has used a map of one kind or another and is probably aware that a map is a scaled-down version of the area depicted. For a map to be of any use, however, one must understand what is shown on a map and how to read it. A particularly useful type of map for geologists, and people in many other professions, is a *topographic map*, which shows the three-dimensional configuration of Earth's surface on a two-dimensional sheet of paper.

Maps showing relief—differences in elevation in adjacent areas—are actually models of Earth's surface. Such maps are available for some areas, but they are expensive, difficult to carry, and impossible to record data on. Thus, paper sheets that show relief by using lines of equal elevation known as *contours* are most commonly used. Topographic maps depict (1) relief, which includes hills, mountains, valleys, canyons, and plains; (2) bodies of water such as rivers, lakes, and swamps; (3) natural features such as forests, grasslands, and glaciers; and (4) various cultural features, including communities, highways, railroads, land boundaries, canals, and power transmission lines.

Topographic maps known as *quadrangles* are published by the U.S. Geological Survey (USGS). The area depicted on a topographic map is identified by referring to the map's name in the upper right and lower right corners, which is usually derived from some prominent geographic feature (Lincoln Creek Quadrangle, Idaho) or community (Mt. Pleasant Quadrangle, Michigan). In addition, most maps have a state outline map along the bottom margin, and shown within the outline is a small black rectangle indicating the part of the state represented by the map.

Contours

Contour lines, or simply contours, are lines of equal elevation used to show topography. Think of contours as the lines formed where imaginary horizontal planes intersect Earth's surface at specific elevations. On maps, contours are brown, and every fifth contour, called an *index contour*, is darker than adjacent ones and labeled with its elevation (▌ Figure B1). Elevations on most USGS topographic maps are in feet, although a few use meters; in either case, the specified elevation is above or below mean sea level. Because contours are defined as lines of equal elevation, they cannot divide or cross one another, although they will converge and appear to join in areas with vertical or overhanging cliffs. Notice in Figure B1 that where contours cross a stream they form a V that points upstream toward higher elevations.

The vertical distance between contours is the *contour interval*. If an area has considerable relief, a large contour interval is used, perhaps 80 or 100 feet, whereas a small interval such as 5, 10, or 20 feet is used in areas with little relief. The values recorded on index contours are always multiples of the map's contour interval, shown at the bottom of the map. For instance, if a map has a contour interval of 10 feet, index contour values such as 3600, 3650, and 3700 feet might be shown (Figure B1). In addition to contours, specific elevations are shown at some places on maps and may be indicated by a small X, next to which is a number. A specific elevation might also be shown adjacent to the designation BM (benchmark), a place where the elevation and location are precisely known.

Contour spacing depends on slope, so in areas with steep slopes, contours are closely spaced because there is a considerable increase in elevation in a short horizontal distance. In contrast, if slopes are gentle, contours are widely spaced (Figure B1). Furthermore, if contour spacing is uniform, the slope angle remains constant, but if spacing changes, the slope angle changes. However, one must be careful in comparing slopes on maps with different contour intervals or different scales.

Topographic features such as hills, valleys, plains, and so on, are easily shown by contours. For instance, a hill is shown by a concentric pattern of contours with the highest elevation in the central part of the pattern. All contours must close on themselves, but they may do so beyond the confines of a particular map. A concentric contour pattern also might show a closed depression, but in this case, special contours with short bars perpendicular to the contour pointing toward the central part of the depression are used (Figure B1).

Map Scales

All maps are scaled-down versions of the areas shown, so to be of any use, they must have a scale. Highway maps, for example, commonly have a scale such as "1 inch equals 10 miles," by which one can readily determine distances. Two types of scales are used on topographic maps. The first and most easily understood is a graphic scale, which is simply a bar subdivided into appropriate units of length (Figure B1). This scale appears at the bottom center of the map and may show miles, feet, kilometers, or meters. Indeed, graphic scales on USGS topographic maps generally show both English and metric distance units.

A ratio or fractional scale, which represents the degree of reduction of the area depicted, appears above the graphic scale. On a map with a ratio scale of 1:24,000, for instance, the area shown is 1/24,000th the size of the actual land area (Figure B1). Another way to express this relationship is to say that any unit of length on the map equals 24,000 of the same units on the ground. Thus, 1 inch on the map equals 24,000 inches on the ground, which is more meaningful if one converts inches to feet, making 1 inch equal to 2,000 feet. A few maps have scales of 1:63,360, which converts to 1 inch equals 5,280 feet, or 1 inch equals 1 mile.

USGS topographic maps are published in a variety of scales such as 1:50,000, 1:62,500, 1:125,000, and 1:250,000.

SCALE 1:24,000

CONTOUR INTERVAL 10 FEET
DATUM IS MEAN SEA LEVEL

▌ Figure B1 Part of the Bottomless Lakes Quadrangle, New Mexico, which has a contour interval of 10 feet; every fifth contour is darker and labeled with its elevation. Notice that contours are widely spaced where slopes are gentle and more closely spaced where they are steeper, as in the central part of the map. Hills are shown by contours that close on themselves, whereas depressions are indicated by contours with hachure marks pointing toward the center of the depression. The dashed blue lines on the map represent intermittent steams; notice that where contours cross a stream's channel they form a V that points upstream.

One should also realize that large-scale maps cover less area than small-scale maps, and the former show much more detail than the latter. For example, a large-scale map (1:24,000) shows more surface features in greater detail than does a small-scale map (1:125,000) for the same area.

Map Locations

You can locate features on topographic maps in two ways. First, the borders of maps correspond to lines of latitude and longitude. Latitude is measured north and south of the equator in degrees, minutes, and seconds, whereas the same units are used to designate longitude east and west of the prime meridian, which passes through Greenwich, England. Maps depicting all areas within the United States are noted in north latitude and west longitude. Latitude and longitude

are noted in degrees and minutes at the corners of maps, but usually only minutes and seconds are shown along the margins. Many USGS topographic maps cover 7 1/2 or 15 minutes of latitude and longitude and are thus referred to as 7 1/2- and 15-minute quadrangles.

Beginning in 1812, the General Land Office (now known as the Bureau of Land Management) developed a standardized method for accurately defining the location of property in the United States. This method, known as the General Land Office Grid System, has been used for all states except those along the eastern seaboard (except Florida), and parts of Ohio, Tennessee, Kentucky, West Virginia, and Texas.

As new land acquired by the United States was surveyed, the surveyors laid out north–south lines they called *principal meridians* and east–west lines known as *base lines*. These intersecting lines form a set of coordinates for locating specific pieces of property. The basic unit in the General Land Office Grid System is the *township,* an area measuring 6 miles on a side and thus covering 36 square miles (▌ Figure B2). Townships are numbered north and south of base lines and are designated as T.1N., T.1S., and so on. Rows of townships known as *ranges* are numbered east and west of principal meridians—R.2W and R.4E, for example. Note in Figure B2 that each township has a unique designation of township and range numbers.

Townships are subdivided into 36 1-square-mile (640-acre) *sections* numbered from 1 to 36. Because of surveying errors and the adjustments necessary to make a grid system conform to Earth's curved surface, not all sections are exactly 1 mile square. Nevertheless, each section can be further subdivided into half sections and quarter sections designated NE 1/4, NW 1/4, SE 1/4, and SW 1/4, and each quarter section can be further divided into quarter-quarter sections. To show the complete designation for an area, the smallest unit is noted first (quarter-quarter section) followed by quarter section, section number, township, and range. For example, the area shown in Figure B2 is the NW 1/4, SW 1/4, Sec. 34, T.2N., R.3W.

Because only a few principal meridians and base lines were established, they do not appear on most topographic maps. Nevertheless, township and range numbers are printed along the margins of 7 1/2- and 15-minute quadrangles, and a grid consisting of red land boundaries depicts sections. In addition, each section number is shown in red within the map. However, small-scale maps show only township and range.

Where to Obtain Topographic Maps

Many people find topographic maps useful. Land use planners, personnel in various local, state, and federal agencies, as well as engineers and real estate developers might use these maps for a variety of reasons. In addition, hikers, backpackers, and others interested in exploring undeveloped areas commonly use topographic maps because trails are shown by black dashed lines. Furthermore, map users can readily determine their location by interpreting the topographic features depicted by contours, and they can anticipate the type of terrain they will encounter during off-road excursions.

Topographic maps for local areas are available at some sporting goods stores, at National Park Visitor Centers, and from some state geologic surveys. Free index maps showing the names and locations of all quadrangles for each state are available from the USGS to anyone uncertain of which specific map is needed. Any published topographic map can be obtained from the United States Geologic Survey (USGS) at http://topomaps.usgs.gov/ or from many commercial map dealers.

▌ **Figure B2** The General Land Office Grid System. Each 36-square-mile township is designated by township and range numbers. Townships are subdivided into sections, which can be further subdivided into quarter sections and quarter-quarter sections.

Appendix C

MINERAL IDENTIFICATION

To identify most minerals, geologists use physical properties such as color, luster, crystal form, hardness, cleavage, specific gravity, and several others (see Chapter 3, Section 3.5). The mineral identification table (C1) is divided into two parts: The first part lists minerals with a metallic luster, and the second part has minerals with a nonmetallic luster. After determining luster, ascertain hardness and note that each part of the table is arranged with minerals with increasing hardness. Thus, if you have a nonmetallic mineral with a hardness of 6, it must be augite, hornblende, plagioclase, or one of the two potassium feldspars (orthoclase or microcline). If this hypothetical mineral is dark green or black, it must be augite or hornblende, and if it has 2 cleavage plane intersecting at nearly right angles it is augite.

TABLE C1	Mineral Identification Tables

Metallic Luster

Mineral	Chemical Composition	Color	Hardness / Specific Gravity	Other Features	Comments
Graphite	C	Black	1–2 / 2.09–2.33	Greasy feel; writes on paper; 1 direction of cleavage	Used for pencil "leads." Mostly in metamorphic rocks.
Galena	PbS	Lead gray	$2\frac{1}{2}$ / 7.6	Cubic crystals; 3 cleavages at right angles	The ore of lead. Mostly in hydrothermal rocks.
Chalcopyrite	$CuFeS_2$	Brassy yellow	$3\frac{1}{2}$–4 / 4.1–4.3	Usually massive; greenish black streak; iridescent tarnish	The most common copper mineral. Mostly in hydrothermal rocks.
Magnetite	Fe_3O_4	Black	$5\frac{1}{2}$–$6\frac{1}{2}$ / 5.2	Strong magnetism	An ore of iron. An accessory mineral in many rocks.
Hematite	Fe_2O_3	Red brown	6 / 4.8–5.3	Usually granular or massive; reddish brown streak	Important iron ore. An accessory mineral in many rocks.
Pyrite	FeS_2	Brassy yellow	$6\frac{1}{2}$ / 5.0	Cubic and octahedral crystals	Found in some igneous and hydrothermal rocks and in sedimentary rocks associated with coal.

Nonmetallic Luster

Mineral	Chemical Composition	Color	Hardness / Specific Gravity	Other Features	Comments
Talc	$Mg_3Si_4O_{10}(OH)_2$	White, green	1 / 2.82	1 cleavage direction; usually in compact masses; soapy feel	Formed by the alteration of magnesium silicates. Mostly in metamorphic rocks.
Clay minerals	Varies	Gray, buff, white	1–2 / 2.5–2.9	Earthy masses; particles too small to observe properties	Found in soils, mudrocks, slate, phyllite.
Chlorite	$(Mg,Fe)_3(Si,Al)_4O_{10}(OH)_8$	Green	2 / 2.6–3.4	1 cleavage; occurs in scaly masses	Common in low-grade metamorphic rocks such as slate.
Gypsum	$CaSO_4 \cdot 2H_2O$	Colorless, white	2 / 2.32	Elongate crystals; fibrous and earthy masses	The most common sulfate mineral. Found mostly in evaporite deposits.
Muscovite (Mica)	$KAl_2Si_3O_{10}(OH)_2$	Colorless	2–$2\frac{1}{2}$ / 2.7–2.9	1 direction of cleavage; cleaves into thin sheets	Common in felsic igneous rocks, metamorphic rocks, and some sedimentary rocks.
Biotite (Mica)	$K(Mg,Fe)_3AlSi_3O_{10}(OH)_2$	Black, brown	$2\frac{1}{2}$ / 2.9–3.4	1 cleavage direction; cleaves into thin sheets	Occurs in both felsic and mafic igneous rocks, in metamorphic rocks, and in some sedimentary rocks.

Calcite	$CaCO_3$	Colorless, white	3 2.71	3 cleavages at oblique angles; cleaves into rhombs; reacts with dilute HCl	The most common carbonate mineral. Main component of limestone and marble.
Anhydrite	$CaSO_4$	White, gray	$3\frac{1}{2}$ 2.9–3.0	Crystals with 2 cleavages; usually in granular masses	Found in limestones, evaporite deposits, and the cap rock of salt domes.
Halite	$NaCl$	Colorless, white	3–4 2.2	3 cleavages at right angles; cleaves into cubes; cubic crystals; salty taste	Occurs in evaporite deposits.
Dolomite	$CaMg(CO_3)_2$	White, yellow, gray, pink	$3\frac{1}{2}$–4 2.85	Cleavage as in calcite; reacts with dilute hydrochloric acid when powdered	The main constituent of dolostone. Also found associated with calcite in some limestones and marble.
Fluorite	CaF_2	Colorless, purple, green, brown	4 3.18	4 cleavage directions; cubic and octahedral crystals	Occurs mostly in hydrothermal rocks and in some limestones and dolostones.
Augite	$Ca(Mg,Fe,Al)(Al,Si)_2O_6$	Black, dark green	6 3.25–3.55	Short 8-sided crystals; 2 cleavages; cleavages nearly at right angles	The most common pyroxene mineral. Found mostly in mafic igneous rocks.
Hornblende	$NaCa_2(Mg,Fe,Al)_5$ $(Si,Al)_8O_{22}(OH)_2$	Green, black	6 3.0–3.4	Elongate, 6-sided crystals; 2 cleavages intersecting at 56° and 124°	A common rock-forming amphibole mineral in igneous and metamorphic rocks.
Plagioclase feldspars	Varies from $CaAl_2Si_2O_8$ to $NaAlSi_3O_8$	White, gray, brown	6 2.56	2 cleavages at right angles	Common in igneous rocks and a variety of metamorphic rocks. Also in some arkoses.
Potassium Feldspars — Microcline	$KAlSi_3O_8$	White, pink, green	6 2.56	2 cleavages at right angles	Common in felsic igneous rocks, some metamorphic rocks, and arkoses.
Potassium Feldspars — Orthoclase	$KAlSi_3O_8$	White, pink	6 2.56	2 cleavages at right angles	
Olivine	$(Fe,Mg)_2SiO_4$	Olive green	$6\frac{1}{2}$ 3.3–3.6	Small mineral grains in granular masses; conchoidal fracture	Common in mafic igneous rocks.
Quartz	SiO_2	Colorless, white, gray, pink, green, black	7 2.67	6-sided crystals; no cleavage; conchoidal fracture	A common rock-forming mineral in all rock groups and hydrothermal rocks. Also occurs in varieties known as chert, flint, agate, and chalcedony.
Garnet	$Fe_3Al_2(SiO_4)_3$	Dark red, green	7–$7\frac{1}{2}$ 4.32	12-sided crystals common; uneven fracture	Found mostly in gneiss and schist.
Zircon	Zr_2SiO_4	Brown, gray	$7\frac{1}{2}$ 3.9–4.7	4-sided, elongate crystals	Most common as an accessory in granitic rocks.
Topaz	$Al_2SiO_4(OH,F)$	Colorless, white, yellow, blue	8 3.5–3.6	High specific gravity; 1 cleavage direction	Found in pegmatites, granites, and hydrothermal rocks.
Corundum	Al_2O_3	Gray, blue, pink, brown	9 4.0	6-sided crystals and great hardness are distinctive	An accessory mineral in some igneous and metamorphic rocks.

Answers

Chapter 1
1. c; 2. e; 3. c; 4. c; 5. e; 6. c; 7. d; 8. d; 9. e; 10. a.

Chapter 2
1. b; 2. b; 3. b; 4. b; 5. c; 6. e; 7. c; 8. a; 9. a; 10. c.

Chapter 3
1. a; 2. d; 3. d; 4. e; 5. a; 6. c; 7. b; 8. a; 9. e; 10. d.

Chapter 4
1. b; 2. a; 3. e; 4. a; 5. d; 6. c; 7. b; 8. b; 9. c; 10. a.

Chapter 5
1. e; 2. b; 3. a; 4. c; 5. b; 6. d; 7. d; 8. a; 9. c; 10. c.

Chapter 6
1. b; 2. a; 3. e; 4. b; 5. b; 6. c; 7. c; 8. d; 9. b; 10. d.

Chapter 7
1. c; 2. d; 3. b; 4. a; 5. e; 6. b; 7. a; 8. a; 9. b; 10. b.

Chapter 8
1. b; 2. d; 3. d; 4. a; 5. c; 6. c; 7. e; 8. a; 9. b; 10. c.

Chapter 9
1. d; 2. d; 3. d; 4. c; 5. d; 6. a; 7. b; 8. c; 9. e; 10. c.

Chapter 10
1. d; 2. c; 3. b; 4. a; 5. a; 6. b; 7. e; 8. d; 9. b; 10. b.

Chapter 11
1. c; 2. e; 3. e; 4. b; 5. d; 6. b; 7. d; 8. a; 9. c; 10. e.

Chapter 12
1. a; 2. b; 3. d; 4. e; 5. a; 6. c; 7. c; 8. b; 9. d; 10. c.

Chapter 13
1. c; 2. a; 3. d; 4. d; 5. c; 6. b; 7. c; 8. e; 9. b; 10. a.

Chapter 14
1. b; 2. c; 3. e; 4. d; 5. a; 6. c; 7. c; 8. b; 9. d; 10. d.

Chapter 15
1. d; 2. c; 3. a; 4. c; 5. d; 6. b; 7. b; 8. d; 9. d; 10. a.

Chapter 16
1. d; 2. c; 3. d; 4. a; 5. c; 6. c; 7. d; 8. c; 9. a; 10. e.

Chapter 17
1. e; 2. b; 3. e; 4. c; 5. d; 6. a; 7. d; 8. e; 9. b; 10. b.

Chapter 18
1. c; 2. a; 3. d; 4. a; 5. c; 6. d; 7. e; 8. c; 9. c; 10. d.

Chapter 19
1. c; 2. d; 3. a; 4. b; 5. e; 6. b; 7. c; 8. a; 9. b; 10. a.

Chapter 20
1. c; 2. c; 3. a; 4. b; 5. e; 6. b; 7. c; 8. c; 9. a; 10. d.

Chapter 21
1. e; 2. c; 3. e; 4. c; 5. d; 6. e; 7. c; 8. d; 9. d; 10. b.

Chapter 22
1. c; 2. a; 3. a; 4. a; 5. c; 6. d; 7. d; 8. c; 9. c; 10. b.

Chapter 23
1. c; 2. b; 3. d; 4. d; 5. a; 6. e; 7. d; 8. c; 9. e; 10. d.

Answers

Chapter 1

15. The Big Bang is a model for the origin and evolution of the universe. According to the Big Bang, the universe originated about 14 billion years ago when all matter and energy were compressed into an infinitely small high-temperature and high-density state in which both time and space were set at zero. This was followed by expansion, cooling, and a less dense state.

    The evidence for the universe beginning about 14 billion years ago is the fact that the universe is expanding and it is permeated by a background radiation of 2.7 Kelvin above absolute zero. This background radiation is thought to be the fading afterglow of the Big Bang.

19. Earth is considered a dynamic planet because it has been continuously changing during its 4.6-billion-year existence. Furthermore, both internal and external forces have been at work during Earth's existence shaping and reshaping both its external and internal features.

Chapter 2

16. The distance from the most distant part of a 30-million-year-old aseismic ridge is 1,000 km from an oceanic ridge. There are 100,000 cm in a kilometer, therefore 1,000 km x 100,000 cm/km = 100,000,000 cm. Thus, 100,000,000 cm/30,000,000 years = 0.33 cm per year. This is the average rate of movement per year for this plate.

18. The present distribution of plants and animals is controlled largely by climate and geographic barriers. The movement of plates has created and destroyed ocean basins, as well as suturing together continental blocks, breaking apart continents, and moving continents around the globe. This movement has affected weather patterns and oceanic current patterns, and also created barriers for the migration of plants and animals. When continental masses are separated, newly formed ocean basins act as barriers to the migration of land animals. The formation of land bridges, such as the Isthmus of Panama, resulting from a convergent plate boundary, creates barriers for ocean species, leading to the evolution of new species.

Chapter 3

12. Some minerals are designated rock-forming minerals because they are common enough to be essential for naming and identifying rocks, but accessory minerals are found in such small quantities that they can be ignored. The most common rock-forming minerals are the silicates such as quartz, the feldspars, clay minerals, and biotite. Among the other mineral groups only the carbonate minerals calcite and dolomite are common.

18. Only about two dozen minerals are common because even though more than 90 naturally occurring elements are known, only eight of these are abundant in Earth's crust. And even of these, oxygen and silicon are by far the most common, making up more than 80% of all atoms available to make minerals.

Chapter 4

12. Mafic magma at spreading centers cools to form basalt and gabbro depending on the depth at which the magma cools. At depth, magma cools more slowly and forms a mosaic of large (seen without magnification) minerals, whereas at or near the surface, magma cools more rapidly and forms an aphanetic or fine-grained texture of small minerals.

18. Specimen 2 would be darker and denser because its predominant minerals (olivine, pyroxene, and hornblende) are ferromagnesian silicates that contain iron or magnesium or both and tend to be dark colored.

Chapter 5

14. Shield volcanoes are found mostly at or near divergent plate boundaries because the magma formed beneath these boundaries is mafic and thus has low viscosity. When erupted, mafic magma tends to form thin but extensive flows that do not built up steep slopes.

20. Geologists monitor several physical and chemical aspects of volcanoes to better anticipate eruptions. For example, they monitor gas emissions, changes in hot spring activity, temperature, and bulging as magma is injected into a volcano, as well as changes in the local magnetic and electrical fields. They are also keenly aware of volcanic tremor and a volcano's eruptive history.

Chapter 6

13. Mechanical weathering involves the breakdown of Earth materials with no change in composition, whereas chemical weathering decomposes rocks and minerals by chemical changes in parent material. The former contributes to the latter because chemical alteration takes place faster on small particles.

17. Frost action takes place when water expands as it freezes in cracks and crevasses in rocks. It is especially effective on granular rocks, such as sandstone, and is much more vigorous in areas where the temperature commonly fluctuates below and above freezing.

Chapter 7

11. Sedimentary rocks preserve evidence of how they were deposited in the form of textures, sedimentary structures, and fossils. The only requirement for geologists to interpret ancient sedimentary rocks is that the

evidence has been preserved in the geologic record. If it has, they use their knowledge of present-day processes and deposits to make inferences about ancient deposits.

16. Ancient current directions can be determined by cross bedding (cross beds dip in the current direction) and by current ripple marks, which have an asymmetric profile with the steep side in the downcurrent direction.

Chapter 8

14. Foliated metamorphic rocks are not suitable for dam foundations because their platy and elongated minerals are arranged in a parallel fashion, thus producing a foliated texture. This means that such rocks as slate, phyllite, and schist easily split into somewhat flat pieces along the cleavage, or foliation planes of the rocks.

    When such rocks are subjected to pressure, such as the water behind a dam, they can easily fail along their foliation. Water seeping along grain boundaries can also contribute to degradation of the rock and slippage along the foliation planes.

    Quartzite would make a good foundation for a dam. It is composed of quartz, which lacks cleavage and has a hardness of 7, and it is a nonfoliated metamorphic rock. Quartzite is a hard, compact rock that lacks cleavage and, unless it is fractured, is impermeable.

18. Contact metamorphism occurs wherever magma flows over the land surface as a lava flow, or is intruded into crustal rocks such as along mid-ocean ridges, or is an igneous intrusion. Contact metamorphism causes thermal alteration to the affected rock body.

Chapter 9

12. Insurance companies use the Modified Mercalli Intensity Scale because it measures actual damage done to structures and areas. Therefore, insurance companies can determine the type of damage to expect from a given earthquake in a particular area based on the amount and type of damage that resulted from previous earthquakes in that area, and thus adjust their rates accordingly.

18. The difference between an earthquake's focus and epicenter is that the focus is the location within Earth's lithosphere where fracturing begins—that is, the point at which energy is first released, whereas the epicenter is the point on Earth's surface directly above the focus. An earthquake's epicenter is the location usually reported by the news media because it is a geographic location that can be identified by longitude and latitude coordinates.

Chapter 10

13. Stress is the force applied to a given area of rock, usually expressed in kilograms per square centimeter (kg/cm²). For instance, the stress exerted by a person walking on an ice-covered pond is a function of the person's weight and the area beneath his or her feet. Strain is a measure of the amount of deformation

In our ice-covered pond example, the strain is the bending or fracturing of the ice caused by stress.

18. Synclines are elongate folds whereas basins are circular or nearly so. In both cases, however, the youngest exposed rocks will be in the center of a syncline or basin. Also, all strata will dip inward toward the axis of a syncline, but in the case of a basin, all strata dip inward from a central point.

Chapter 11

16. Slope stability can be maintained by a variety of measures, depending on the underlying geology and topography. Such measures as reducing the slope angle of a hillside, surface and subsurface water drainage, construction of retaining walls, and even rock bolts, all can help in maintaining or increasing slope stability.

18. Slumps are usually the result of erosion along the base of a slope, thus increasing the slope angle. The construction of highways and railroad beds in areas of relief, such as hillsides, causes the slope to become oversteepened near its base, resulting in increased pressure along the highway or railroad bed, and increasing the likelihood of slope failure, or slumping.

Chapter 12

11. A river with headwaters 2,000 m above sea level that flows 1,500 km to the sea has an average gradient of 1.33 m/km. Its gradient is found by dividing the vertical drop of the river by the horizontal distance that the river flows, that is 2,000/1,500 = 1.33. However, the gradient at any particular location along the river may be greater or less than this average value, because rivers typically have steeper gradients in their upper reaches and lower gradients as one proceeds downstream.

14. The deposits of meandering streams are dominated by fine-grained materials (mud) that were deposited during floods, but there are also distinctive channel deposits of sand that were deposited on point bars. In contrast, braided streams deposit mostly sheets of sand and gravel, and mud is conspicuous by its absence.

Chapter 13

11. Groundwater is one component of the hydrologic cycle and accounts for approximately 22% of the world's supply of freshwater. Although the oceans are the ultimate source for groundwater, its more immediate source is the precipitation that infiltrates the ground and seeps down through openings in the soil, sediment, and rocks. In addition to precipitation, groundwater may also come from infiltration by streams, lakes, swamps, artificial recharge ponds, water-treatment systems, and even seawater.

19. In arid regions, such as Nevada, the water table is very low, in some cases tens and hundreds of meters below the surface. Furthermore, rainfall is scant, and what little does fall and percolates into the ground usually evaporates before it gets very far. Thus, the likelihood of water

nuclear waste material where it could corrode the containers or become polluted by the radioactive material is virtually nonexistent for the foreseeable future.

However, if the climate should change and the region become humid, water from rainfall will freely percolate through the zone of aeration into the zone of saturation, and eventually cause the water table to rise, possibly to the level of the buried radioactive nuclear waste material, where the groundwater system could then become polluted by the radioactive material.

The average rate of groundwater movement during the next 5,000 years would have to be 8 cm per year to reach radioactive waste canisters buried at a depth of 400 m (400 m = 40,000 cm [100 cm/m x 400 m = 40,000 cm]; 40,000 cm/5,000 years = 8 cm/year).

Chapter 14

11. A glacier's budge is essentially a balance sheet of gains and losses. If gains exceed losses, the glacier has a positive budget, it actively grows, and its terminus advances. However, should losses exceed gains, the glacier has a negative budget, it will shrink, and its terminus will retreat toward its area of accumulation although it still actively flows. If a negative budget persists long enough, the glacier will recede and thin until it no longer flows, and it will eventually disappear.

13. The depth to which running water can erode is limited by sea level because the upstream part of a stream cannot be lower than where it discharges into the sea. Glaciers, because of their density, can stay in contact with the seafloor and erode some distance below sea level. For instance, a 500-m-thick glacier could effectively erode about 450 m below sea level before the buoyant effect of seawater would cause it to float.

Chapter 15

14. Many desert rock formations are red because of the large amount of iron in the rock formations. Iron exposed to the atmosphere becomes oxidized, which produces a red color.

18. Desert pavement is important in a desert environment because it prevents further erosion and transport of a desert's surface materials by forming a protective layer of close-fitting larger rocks.

Chapter 16

15. Upwelling involves vertical circulation of cold, nutrient-rich seawater from depth to the surface. It is important because it carries nutrients, especially phosphates and nitrates, into the photic zone where it supports large communities of organisms. Also, much of the sedimentary rock containing phosphorous was deposited in areas of upwelling.

17. No matter how long or hard the wind blows, large waves will not form on small ponds and lakes because of the fetch, the distance the wind blows over a con-

is unrestricted and huge wind-generated waves develop if the wind blows hard and for a long time.

Chapter 17

12. In the carbon-14 dating technique, one is measuring the ratio between non-radioactive carbon 12 and radioactive carbon 14. Carbon 14 is produced in the upper atmosphere when high-energy particles strike the nucleus of a nitrogen atom causing a neutron to be absorbed and a proton emitted. Thus, the ratio being measured is that of two isotopes of carbon. In the five long-lived radioactive pairs (uranium 238–lead 206; uranium 235–lead 207; thorium 232–lead 208; rubidium 87–strontium 87; potassium 40–argon 40), one is measuring the ratio between the unstable radioactive parent and its non-radioactive daughter product, which is a different, and stable element.

17. In reconstructing the geologic history of an area, it is important to differentiate between a sill and a lava flow because they formed differently in relation to the rocks found in association with them.

A sill is a tabular igneous intrusion, and thus is younger than the rocks it intrudes. A lava flow moving over the land surface will be younger than rocks that it is flowing over, and will be older than any rocks deposited on it after it cools and solidifies.

To identify a sill, one should look for baking (contact metamorphism) of the rocks immediately above and below the sill. A lava flow, however, will only show the effects of baking in the rocks below it, because the rocks above it were deposited on the cool, solidified flow.

Chapter 18

14. If the theory of evolution is correct:
    a. Fossils in ancient rocks should be of organisms very different from those existing now.
    b. The first appearance of fish in the fossil record should take place before the first amphibians, the first amphibians before the first reptiles, the first reptiles before the first mammals and birds.
    c. The fossil record should show that closely related organisms were very similar when they appeared, but that they became increasingly different as they diverged from a common ancestor.

20. Inheritance of acquired characteristics holds that organisms acquire features during their lifetimes that are then passed on to their descendants. According to natural selection, variations exist in interbreeding populations, some of which are more favorable, and that those individuals with favorable variations are more likely to survive and pass on these variations. Both hypotheses are testable but the former has failed its tests whereas the latter has not.

Chapter 19

11. The two processes that released oxygen to Earth's

photosynthesis. In photochemical dissociation, water molecules are disrupted in the atmosphere by ultraviolet radiation and split into hydrogen (H) and oxygen (O). This is a self-limiting process, though, because as the oxygen so formed reacts to form only about 2% of today's free oxygen, enough ozone forms to block most of the ultraviolet radiation and photochemical dissociation is no longer effective. Photosynthesis is a metabolic process whereby some organisms extract carbon dioxide and water from the atmosphere to manufacture organic molecules and release oxygen as a waster product.

15. Laurentia is a landmass made up mostly of North America and Greenland. Much of Laurentia assembled during the Proterozoic when smaller landmasses (minicontinents) collided along Laurentia's margins, a process called continental accretion. By the end of the Proterozoic Eon, about 75% of present-day North America had formed.

Chapter 20

11. The sequence is the basic unit of sequence stratigraphy and represents a succession of rocks of related facies bounded by unconformities. These boundaries result from a relative drop in sea level, and hence represent local, regional, and global events. Thus, sedimentary rocks can be subdivided into related units on the basis of time-stratigraphic unconformable surfaces. Therefore, global correlations can be made on the basis of a global drop in sea level, such as occurred near the end of the Ordovician Period.

15. The evidence that the Iapetus Ocean began closing during the Middle Ordovician is based on the observation that the western side of the Iapetus plate began subducting beneath Laurentia along an oceanic–continental convergent plate boundary. This marked the beginning of the Taconic orogeny and reduction of the Iapetus Ocean. The evidence for this oceanic-continental convergent plate boundary is volcanism in the rock record and downwarping of the eastern Laurentian carbonate platform as revealed by faulted and folded carbonate rocks. Further evidence for this orogeny, and consequent closing of the Iapetus Ocean, is the Queenston Delta clastic wedge. This clastic wedge is the result of erosion of the uplifted Taconic Highlands.

Chapter 21

16. Several major problems had to be overcome before organisms could completely colonize the land. One of these was the problem of desiccation, or drying out. In the terrestrial environment, organisms must evolve mechanisms to retain internal moisture or they will dry up and die. Another problem is that of gravity. In a marine or freshwater environment, pressure is exerted equally on an organism. On land, the force of gravity operates on organisms and an organism must be able to resist that force, such as by evolving a strong backbone and limbs, in the case of animals. For animals, the ability to extract oxygen from the atmosphere rather than water had to be addressed. This problem was solved by the evolution of lungs.

18. The Pennsylvanian Period was a time of repetitive transgressive-regressive cycles that resulted in widespread coal-forming swamps. These swamps provided the perfect environment for amphibians and seedless vascular plants, both of which required moist conditions as part of their reproductive cycle.

Chapter 22

12. One trend during the transition from mammal-like reptiles to mammals was the reduction in the number of bones making up the lower jaw. Reptiles typically have several bones in their jaw but mammals have only one, the dentary. The jaw-skull joint in reptiles is between the articular bone of the jaw and the quadrate bone of the skull, whereas in mammals the joint is formed between the dentary and the squamosal bone of the skull. The articular and quadrate bones of the reptile jaw-skull joint became the malleus and the incus of the mammal middle ear.

16. The Mesozoic rifting taking place on the East Coast of North America pushed North America westward, resulting in convergence occurring in the Cordilleran mobile belt. Whereas the East Coast of North America was a passive continental margin during the Mesozoic Era, the West Coast of North America became an active oceanic–continental convergent plate boundary.

Chapter 23

16. The Cascade Range extends from northern California, through Oregon, Washington,and into British Columbia, Canada. The volcanoes comprising this range are a dozen or so large composite volcanoes, and the world's largest lava dome, Lassen Peak. The ongoing volcanism is the result of subduction of the Juan de Fuca plate beneath North America.

17. According to the climate change hypothesis, rapid changes in climate at the end of the Pleistocene Epoch caused the extinction of the large mammalian faunas. Changes in climate and vegetation did occur over much of Earth's surface during the Late Pleistocene, and such changes can affect animal populations. However, one of the problems with this hypothesis is why didn't the large mammals migrate to more suitable habitats during these changes? Furthermore, there were several glacial-interglacial periods during the Pleistocene and the animal biota was not affected by these earlier climate changes.

   The prehistoric overkill hypothesis considers that human hunters caused the mass extinction of large mammal populations. Supporting this hypothesis is the fact that the mass extinctions in North and South America, as well as Australia coincided closely with the arrival of humans. However, it is argued that these early human inhabitants probably initially lived in small, scattered communities, and probably lacked the numbers needed to decimate so many species of large mammals. Furthermore, modern humans typically concentrate on smaller, abundant, and less dangerous animals, and, to date, there are few human artifacts found associated

Glossary

aa Lava flow with a surface of rough, angular blocks and fragments.

abiogenesis The origin of life from nonliving matter.

abrasion The process whereby rock is worn smooth by the impact of sediment transported by running water, glaciers, waves, or wind.

Absaroka Sequence Widespread Upper Mississippian to Lower Jurassic sedimentary rocks bounded above and below by unconformities; deposited during a transgressive–regressive cycle of the Absaroka Sea.

absolute dating Using various radioactive decay dating techniques to assign ages in years before the present to rocks. (*See also* relative dating)

abyssal plain Vast flat area on the seafloor adjacent to the continental rise of a passive continental margin.

Acadian orogeny An episode of Devonian deformation in the northern Appalachian mobile belt resulting from the collision of Baltica with Laurentia.

active continental margin A continental margin with volcanism and seismicity at the leading edge of a continental plate where oceanic lithosphere is subducted. (*See also* passive continental margin)

allele Alternative form of a single gene controlling different versions of the same trait.

allopatric speciation Model for the evolution of a new species from a small, geographically isolated part of a larger parent population.

alluvial fan A cone-shaped accumulation of mostly sand and gravel deposited where a stream flows from a mountain valley onto an adjacent lowland.

alluvium A collective term for all detrital sediment transported and deposited by running water.

Alpine-Himalayan orogenic belt An area of mountain building extending from Spain eastward to Southeast Asia.

amniote Any vertebrate animal that produces amniotic eggs in which the embryo develops in a fluid-filled amnion; includes reptiles, birds, and mammals.

amniote egg Egg in which the embryo develops in a fluid-filled cavity called the amnion. The egg also contains a yolk sac and a waste sac.

anaerobic Refers to organisms that do not need oxygen for respiration.

analogous structure Body part, such as wings of insects and birds, that serves the same function but differs in structure and development. (*See also* homologous structure)

Ancestral Rockies Late Paleozoic uplift in the southwestern part of the North American craton.

angiosperm Vascular plants that have flowers and seeds; the flowering plants.

angular unconformity An unconformity below which older rocks dip at a different angle (usually steeper) than overlying strata. (*See also* disconformity and nonconformity)

anticline A convex upward fold in which the oldest exposed rocks coincide with the fold axis and all strata dip away from the axis.

Antler orogeny Late Devonian to Mississippian deformation that affected the Cordilleran mobile belt from Nevada to Alberta, Canada.

aphanitic texture A texture in igneous rocks in which individual mineral grains are too small to be seen without magnification; results from rapid cooling of magma and generally indicates an extrusive origin.

aphotic zone The depth in the ocean below which sunlight does not penetrate. (*See also* photic zone)

Appalachian mobile belt A mobile belt along the eastern margin of North America extending from Newfoundland to Georgia.

archosaur A term referring to the ruling reptiles—dinosaurs, flying reptiles (pterosaurs), crocodiles, and birds.

arête A narrow, serrated ridge between two glacial valleys or adjacent cirques.

artesian system A confined groundwater system with high hydrostatic pressure that causes water to rise above the level of the aquifer.

artificial selection The practice of selectively breeding plants and animals for desirable traits.

artiodactyl The even-toed hoofed mammals, such as goats, cattle, antelope, and swine; members of the order Artiodactyla.

aseismic ridge A ridge or broad area rising above the seafloor that lacks seismic activity.

ash Pyroclastic materials that measure less than 2 mm.

assimilation A process whereby magma changes composition as it reacts with country rock.

asthenosphere The part of the mantle that lies below the lithosphere; it behaves plastically and flows slowly.

Atlantic Coastal Plain The broad, low-relief area of eastern North America extending from the Appalachian Mountains to the Atlantic shoreline.

atom The smallest unit of matter that retains the characteristics of an element.

atomic mass number The number of protons plus neutrons in the nucleus of an atom.

atomic number The number of protons in the nucleus of an atom.

aureole A zone surrounding a pluton in which contact metamorphism took place.

australopithecine A term referring to several extinct species of the genus *Australopithecus* that existed during the Pliocene and Pleistocene epochs.

autotrophic Describes organisms that synthesize organic molecules from inorganic raw materials, as in photosynthesis. (*See also* heterotrophic)

back-arc basin A marine basin, such as the Sea of Japan, that lies between a volcanic island arc and a continent.

Baltica One of six major Paleozoic continents; composed of Russia west of the Ural Mountains, Scandinavia, Poland, and northern Germany.

banded iron formation (BIF) Sedimentary rocks made up of thin alternating bands of silica (chert) and iron minerals (mostly hematite and magnetite).

barchan dune A crescent-shaped sand dune with its tips pointing downwind.

barrier island A long, narrow island of sand parallel to a shoreline but separated from the mainland by a lagoon.

basal slip Movement involving a glacier sliding over its underlying surface.

basalt plateau A plateau built up by horizontal or nearly horizontal overlapping lava flows that erupted from fissures.

base level The level below which a stream or river cannot erode; sea level is ultimate base level.

basin An oval to circular fold in which all strata dip inward toward a central point and the youngest exposed strata are in the center.

Basin and Range Province An area of Cenozoic block-faulting centered on Nevada, but extending into adjacent states and northern Mexico.

batholith An irregularly shaped, discordant pluton with at least 100 km² of exposed surface area.

baymouth bar A spit that has grown until it closes off a bay from the open sea or lake.

beach Any deposit of sediment extending landward from low tide to a change in topography or where permanent vegetation begins.

bed An individual layer of rock, especially sediment or sedimentary rock. (*See also* strata)

bed load That part of a stream's sediment load, mostly sand and gravel, transported along its bed.

benthos Any organism that lives on or in the bottom sediments of seas or lakes.

Big Bang A model for the evolution of the universe in which a dense, hot state was followed by expansion, cooling, and a less dense state.

biochemical sedimentary rock Any sedimentary rock produced by the chemical activities of organisms. (*See also* chemical sedimentary rock)

biostratigraphic unit An association of sedimentary rocks defined by its fossil content.

bipedal Walking on two legs as a means of locomotion, as in humans, birds, and some dinosaurs. (*See also* quadrupedal)

black smoker A type of submarine hydrothermal vent that emits a black plume of hot water colored by dissolved minerals.

body fossil The actual remains of an organism, such as bones, teeth, shells, and, rarely, the soft parts. (*See also* trace fossil)

bonding The process whereby atoms join to other atoms.

bony fish Fish with an internal skeleton of bone; the class Osteichthyes.

Bowen's reaction series A series of minerals that form in a specific sequence in cooling magma or lava; originally proposed to explain the origin of intermediate and felsic magma from mafic magma.

braided stream A stream with multiple dividing and rejoining channels.

breaker A wave that steepens as it enters shallow water until its crest plunges forward.

browser Any animal that eats tender shoots, twigs, and leaves. (*See also* grazer)

butte An isolated, steep-sided, pinnacle-like hill formed when resistant cap rock is breached, allowing erosion of less resistant underlying rocks.

caldera A large, steep-sided, oval to circular depression usually formed when a volcano's summit collapses into an underlying partially drained magma chamber.

Caledonian orogeny A Silurian-Devonian episode of deformation along the northwestern margin of Baltica as it collided with Laurentia.

Canadian shield The exposed part of the North American craton.

carbon 14 dating technique Absolute dating technique relying on the ratio of carbon 12 to carbon 14 in an organic substance; useful back to about 70,000 years ago.

carbonate mineral A mineral with the carbonate radical $(CO_3)^{-2}$, as in calcite $(CaCO_3)$ and dolomite $[CaMg(CO_3)_2]$.

carbonate rock Any rock, such as limestone and dolostone, made up mostly of carbonate minerals.

carnivore-scavenger Any animal that depends on other animals, living or dead, for nutrients.

cartilaginous fish Sharks, rays, and skates and their extinct relatives that have an internal skeleton of cartilage.

Cascade Range A mountain range with several active volcanoes in northern California, Oregon, Washington, and southern British Columbia, Canada.

Catskill Delta A Devonian clastic wedge deposited adjacent to highlands that formed during the Acadian orogeny.

cave A natural subsurface opening generally connected to the surface and large enough for a person to enter.

cementation The process whereby minerals crystallize in the pore spaces of sediment and bind the loose particles together.

chemical sedimentary rock Sedimentary rock made up of minerals that were dissolved during chemical weathering and later precipitated from seawater, more rarely lake water, or extracted from solution by organisms. (*See also* biochemical sedimentary rock)

chemical weathering The decomposition of rocks by chemical alteration of parent material.

China One of six major Paleozoic continents; composed of southeast Asia, including China, Indochina, part of Thailand, and the Malay Peninsula.

chordate Animals of the phylum Chordata; all have a notochord, dorsal hollow nerve cord, and pharyngeal pouches at some time during their life cycle.

chromosome Double-stranded, helical molecule of deoxyribonucleic acid (DNA); specific segments of chromosomes are genes.

cinder cone A small, steep-sided volcano made up of pyroclastic materials resembling cinders that accumulate around a vent.

circum-Pacific belt A zone of seismic and volcanic activity and mountain building that nearly encircles the Pacific Ocean basin.

circum-Pacific orogenic belt An area of mountain building along the west coasts of South, Central, and North America; the Aleutians Islands; Japan; the Philippines; and the islands north of Australia and New Zealand.

cirque A steep-walled, bowl-shaped depression on a mountainside at the upper end of a glacial valley.

cladistics A type of analysis of organisms in which they are grouped together on the basis of derived, as opposed to primitive, characteristics.

cladogram A diagram that shows the probable evolutionary relationships among members of a clade, a group of organisms including their most recent common ancestor.

clastic wedge An accumulation of detrital sediments eroded from and deposited adjacent to an uplifted area (e.g., the Catskill Delta).

cleavage The breaking or splitting of mineral crystals along planes of internal weakness.

Colorado Plateau An elevated area in Colorado, Utah, Arizona, and New Mexico with brilliantly colored rocks, deep canyons, and volcanic mountains raising above it.

columnar joint Columns in igneous rocks bounded by fractures that formed when lava or magma cooled and contracted.

compaction Reduction in the volume of a sedimentary deposit that results from its own weight and the weight of any additional sediment deposited on top of it.

complex movement A combination of different types of mass movements in which no single type is dominant; usually involves sliding and flowing.

composite volcano (stratovolcano) A volcano composed of lava flows and pyroclastic layers, typically of intermediate composition, and mudflows.

compound Any substance resulting from the bonding of two or more different elements (e.g., water, H_2O, and quartz, SiO_2).

compression Stress resulting when rocks are squeezed by external forces directed toward one another.

concordant pluton Intrusive igneous body whose boundaries parallel the layering in the country rock. (*See also* discordant pluton)

cone of depression A cone-shaped depression around a well where water is pumped from an aquifer faster than it can be replaced.

contact (thermal) metamorphism Metamorphism of country rock adjacent to a pluton.

continental accretion The process in which continents grow by additions of Earth materials along their margins.

continental–continental plate boundary A convergent plate boundary along which two continental lithospheric plates collide.

continental drift The theory that the continents were joined into a single landmass that broke apart with the various fragments (continents) moving with respect to one another.

continental glacier A glacier that covers a vast area (at least 50,000 km^2) and is not confined by topography; also called an ice sheet.

continental margin The area separating the part of a continent above sea level from the deep seafloor.

continental red bed Sedimentary rock, mostly sandstone and shale, with red color due to the presence of iron oxides.

continental rise The gently sloping part of the continental margin between the continental slope and the abyssal plain.

continental shelf The very gently sloping part of the continental margin between the shoreline and the continental slope.

continental slope The relatively steeply inclined part of the continental margin between the continental shelf and the continental rise or between the continental shelf and an oceanic trench.

convergent evolution The development of similar features in two or more distantly related groups of organisms resulting from adapting to a comparable lifestyle. (*See also* parallel evolution)

convergent plate boundary The boundary between two plates that move toward each other.

Cordilleran mobile belt A large region of deformation along the western margin of North America bounded on the west by the Pacific Ocean and on the east by the Great Plains.

Cordilleran orogeny An episode of deformation affecting the western margin of North America from Jurassic to Early Cenozoic time; divided into three separate phases called the Nevadan, Sevier, and Laramide orogenies.

core The interior part of Earth beginning at a depth of 2,900 km that probably consists mostly of iron and nickel.

Coriolis effect The apparent deflection of a moving object from its anticipated course because of Earth's rotation. Winds and oceanic currents are deflected clockwise in the Northern Hemisphere and counterclockwise in the Southern Hemisphere.

correlation Demonstration of the physical continuity of rock units or biostratigraphic units, or demonstration of time equivalence as in time-stratigraphic correlation.

country rock Any preexisting rock that has been intruded by a pluton or altered by metamorphism.

covalent bond A chemical bond formed by the sharing of electrons between atoms.

crater An oval to circular depression at the summit of a volcano resulting from the eruption of lava, pyroclastic materials, and gases.

craton The stable nucleus of a continent consisting of a shield of Precambrian rocks and a platform of buried ancient rocks.

cratonic sequence A widespread sequence of sedimentary rocks bounded above and below by unconformities that was deposited during a transgressive–regressive cycle of an epeiric sea.

creep A widespread type of mass wasting in which soil or rock moves slowly downslope.

Cretaceous Interior Seaway A Late Cretaceous arm of the sea that effectively divided North America into two large landmasses.

Cro-Magnon Humans that lived mostly in Europe from 35,000 to 10,000 years ago.

cross-bedding A type of bedding in which layers are deposited at an angle to the surface on which they accumulate, as in sand dunes.

crossopterygian A specific type of lobe-finned fish that had lungs; the ancestors of amphibians were among the crossopterygians.

crust Earth's outermost layer; the upper part of the lithosphere that is separated from the mantle by the Moho; divided into continental and oceanic crust.

crystal A naturally occurring solid of an element or compound with a specific internal structure that is manifested externally by planar faces, sharp corners, and straight edges.

crystal settling The physical separation and concentration of minerals in the lower part of a magma chamber or pluton by crystallization and gravitational settling.

crystalline solid A solid in which the constituent atoms are arranged in a regular, three-dimensional framework.

Curie point The temperature at which iron-bearing minerals in cooling magma or lava attain their magnetism.

cyclothem A sequence of cyclically repeating sedimentary rocks resulting from alternating periods of marine and nonmarine deposition; commonly contains a coal bed.

cynodont A type of carnivorous therapsid (advanced mammal-like reptile); ancestors of mammals are among cynodonts.

debris flow A type of mass wasting that involves a viscous mass of soil, rock fragments, and water that moves downslope; debris flows have larger particles than mudflows and contain less water.

deflation The removal of sediment and soil by wind.

deformation A general term for any change in shape or volume, or both, of rocks in response to stress; involves folding and fracturing.

delta An alluvial deposit formed where a stream or river flows into the sea or a lake.

density The mass of an object per unit volume; usually expressed in grams per cubic centimeter (g/cm³).

deoxyribonucleic acid (DNA) The chemical substance of which chromosomes are composed; the genetic material in all organisms.

depositional environment Any site such as a floodplain or beach where physical, biologic, and chemical processes yield a distinctive kind of sedimentary deposit.

desert Any area that receives less than 25 cm of rain per year and that has a high evaporation rate.

desert pavement A surface mosaic of close-fitting pebbles, cobbles, and boulders found in many dry regions; results from wind erosion of sand and smaller particles.

desertification The expansion of deserts into formerly productive lands.

detrital sedimentary rock Sedimentary rock made up of the solid particles (detritus) of preexisting rocks.

differential pressure Pressure that is not applied equally to all sides of a rock body.

differential weathering Weathering that occurs at different rates on rocks, thereby yielding an uneven surface.

dike A tabular or sheetlike discordant pluton.

dinosaur Any of the Mesozoic reptiles belonging to the orders of Saurischia and Ornithischia.

dip A measure of the maximum angular deviation of an inclined plane from horizontal.

dip-slip fault A fault on which all movement is parallel with the dip of the fault plane. (*See also* normal fault and reverse fault)

discharge The volume of water in a stream or river moving past a specific point in a given interval of time; expressed in cubic meters per second (m³/sec) or cubic feet per second (ft³/sec).

disconformity An unconformity above and below which the rock layers are parallel. (*See also* angular unconformity and nonconformity)

discontinuity A boundary across which seismic wave velocity or direction of travel changes abruptly, such as the mantle–core boundary.

discordant pluton Pluton with boundaries that cut across the layering in the country rock. (*See also* concordant pluton)

dissolved load The part of a stream's load consisting of ions in solution.

divergent evolution The diversification of a species into two or more descendant species.

divergent plate boundary The boundary between two plates that are moving apart.

divide A topographically high area that separates adjacent drainage basins.

dome A rather circular geologic structure in which all rock layers dip away from a central point and the oldest exposed rocks are at the dome's center.

downwelling The slow transfer of ocean surface water to depth. (*See also* upwelling)

drainage basin The surface area drained by a stream or river and its tributaries.

drainage pattern The regional arrangement of channels in a drainage system.

drumlin An elongate hill of till formed by the movement of a continental glacier or by floods.

dune A mound or ridge of wind-deposited sand.

dynamic metamorphism Metamorphism in fault zones where rocks are subjected to high differential pressure.

earthflow A mass wasting process involving the downslope movement of water-saturated soil.

earthquake Vibrations caused by the sudden release of energy, usually as a result of displacement of rocks along faults.

ectotherm Any of the cold-blooded vertebrate animals, such as amphibians and reptiles; an animal that depends on external heat to regulate body temperature. (*See also* endotherm)

Ediacaran faunas Name for all Late Proterozoic faunas with animal fossils similar to those of the Ediacara fauna of Australia.

elastic rebound theory An explanation for the sudden release of energy that causes earthquakes when deformed rocks fracture and rebound to their original undeformed condition.

elastic strain A type of deformation in which the material returns to its original shape when stress is relaxed.

electron A negatively charged particle of very little mass that encircles the nucleus of an atom.

electron shell Electrons orbit an atom's nucleus at specific distances in electron shells.

element A substance composed of atoms that all have the same properties; atoms of one element can change to atoms of another element by radioactive decay, but otherwise they cannot be changed by ordinary chemical means.

emergent coast A coast where the land has risen with respect to sea level.

end moraine A pile or ridge of rubble deposited at the terminus of a glacier. (*See also* terminal moraine and recessional moraine)

endosymbiosis A symbiotic relationship in which one symbiont lives permanently within another organism. Organelles in cells such as plastids and mitochondria are thought to have resulted from endosymbiosis.

endotherm Any of the warm-blooded vertebrate animals, such as birds and mammals, that maintain their body temperature within narrow limits by internal processes. (*See also* ectotherm)

epeiric sea A broad, shallow sea that covers part of a continent, such as the Sauk Sea.

epicenter The point on Earth's surface directly above the focus of an earthquake.

erosion The removal of weathered materials from their source area by running water, wind, glaciers, or waves.

esker A long, sinuous ridge of stratified drift deposited by running water in a tunnel beneath stagnant ice.

eukaryotic cell A cell with internal structures such as mitochondria and an internal membrane-bounded nucleus containing chromosomes. (*See also* prokaryotic cell)

expansive soil A soil in which the volume increases when water is present.

evaporite Any sedimentary rock, such as rock salt, formed by inorganic chemical precipitation of minerals from evaporating water.

Exclusive Economic Zone (EEZ) An area extending 371 km seaward from the coast of the United States and its possessions in which the United States claims rights to all resources.

exfoliation dome A large, rounded dome of rock resulting when concentric layers of rock are stripped from the surface of a rock mass.

Farallon plate A Late Mesozoic-Cenozoic oceanic plate that was largely subducted beneath North America; the Cocos and Juan de Fuca plates are remnants.

fault A fracture along which rocks on opposite sides of the fracture have moved parallel with the fracture surface.

fault plane A fault surface that is more or less planar.

felsic magma Magma with more than 65% silica and considerable sodium, potassium, and aluminum, but little calcium, iron, and magnesium. (*See also* intermediate magma, mafic magma, and ultramafic magma)

ferromagnesian silicate Any silicate mineral that contains iron, magnesium, or both. (*See also* nonferromagnesian silicate)

fetch The distance the wind blows over a continuous water surface.

fiord An arm of the sea extending into a glacial trough eroded below sea level.

firn Granular snow formed by partial melting and refreezing of snow; transitional material between snow and glacial ice.

fission track dating The absolute dating process in which small linear tracks (fission tracks) resulting from alpha decay are counted in mineral crystals.

fissure eruption A volcanic eruption in which lava or pyroclastic materials issue from a long, narrow fissure (crack) or group of fissures.

floodplain A low-lying, flat area adjacent to a channel that is partly or completely water-covered when a stream or river overflows its banks.

fluid activity An agent of metamorphism in which water and carbon dioxide promote metamorphism by increasing the rate of chemical reactions.

focus The site within Earth where an earthquake originates and energy is released.

fold A type of geologic structure in which planar features in rock layers such as bedding and foliation have been bent.

foliated texture A texture in metamorphic rocks in which platy and elongate minerals are aligned in a parallel fashion.

footwall block The block of rock that lies beneath a fault plane. (*See also* hanging wall block)

fossil Remains or traces of prehistoric organisms preserved in rocks. (*See also* body fossil and trace fossil)

fracture A break in rock resulting from intense applied pressure.

frost action The disaggregation of rocks by repeated freezing and thawing of water in cracks and crevasses.

gene A specific segment of a chromosome constituting the basic unit of heredity. (*See also* allele)

geologic agent Any geologic process that brings about change such as running water, glaciers, and tectonism.

geologic record The record of prehistoric physical and biological events preserved in rocks.

geologic structure Any feature in rocks that results from deformation, such as folds, joints, and faults.

geologic time scale A chart arranged so that the designation for the earliest part of geologic time appears at the bottom followed upward by progressively younger time designations.

geology The science concerned with the study of Earth materials (minerals and rocks), surface and internal processes, and Earth history.

geothermal energy Energy that comes from steam and hot water trapped within Earth's crust.

geothermal gradient Earth's temperature increase with depth; it averages 25°C/km near the surface but varies from area to area.

geyser A hot spring that periodically ejects hot water and steam.

glacial budget The balance between expansion and contraction of a glacier in response to accumulation versus wastage.

glacial drift A collective term for all sediment deposited directly by glacial ice (till) and by meltwater streams (outwash).

glacial erratic A rock fragment carried some distance from its source by a glacier and usually deposited on bedrock of a different composition.

glacial ice Water in the solid state within a glacier; forms as snow partially melts and refreezes and compacts so that it is transformed first to firn and then to glacial ice.

glacial polish A smooth, glistening rock surface formed by the movement of sediment-laden ice over bedrock.

glacial striation A straight scratch rarely more than a few millimeters deep on a rock caused by the movement of sediment-laden glacial ice.

glacial surge A time of greatly accelerated flow in a glacier. Commonly results in displacement of the glacier's terminus by several kilometers.

glaciation Refers to all aspects of glaciers, including their origin, expansion, and retreat, and their impact on Earth's surface.

glacier A mass of ice on land that moves by plastic flow and basal slip.

***Glossopteris* flora** A Late Paleozoic association of plants found only on the Southern Hemisphere continents and India; named for its best-known genus, *Glossopteris*.

Gondwana A major Paleozoic continent composed of South America, Africa, Australia, India, and Antarctica.

graded bedding Sedimentary layer that shows a decrease in grain size from bottom to top.

graded stream A stream that has an equilibrium profile in which a delicate balance exists among gradient, discharge, flow velocity, channel characteristics, and sediment load so that neither significant deposition nor erosion takes place within its channel.

gradient The slope over which a stream or river flows expressed in m/km or ft/mi.

granite-gneiss complex The most common association of rocks found in areas of Archean rocks.

gravity anomaly A departure from the expected force of gravity; anomalies may be positive or negative.

grazer An animal that eats low-growing vegetation, especially grasses. (*See also* browser)

greenstone belt A linear or podlike association of igneous and sedimentary rocks particularly common in Archean terrains.

Grenville orogeny An episode of deformation that took place in the eastern United States and Canada during the Neoproterozoic.

ground moraine The layer of sediment released from melting ice as a glacier's terminus retreats.

groundwater Underground water stored in the pore spaces of soil, sediment, and rock.

guide fossil Any easily identified fossil with an extensive geographic distribution and short geologic range useful for determining the relative ages of rocks in different areas.

Gulf Coastal Plain The broad, low-relief area along the Gulf Coast of the United States and Mexico.

guyot A flat-topped seamount of volcanic origin rising more than 1 km above the seafloor.

gymnosperm A flowerless, seed-bearing plant.

gyre A system of ocean currents rotating clockwise in the Northern Hemisphere and counterclockwise in the Southern Hemisphere.

half-life The time necessary for half of the original number of radioactive atoms of an element to decay to a stable daughter product; for example, the half-life for potassium 40 is 1.3 billion years.

hanging valley A tributary glacial valley whose floor is at a higher level than that of the main glacial valley.

hanging wall block The block of rock that overlies a fault plane. (*See also* footwall block)

hardness A term used to express the resistance of a mineral to abrasion.

headland Part of a shoreline commonly bounded by cliffs that extends out into the sea or a lake.

heat An agent of metamorphism.

herbivore An animal that depends directly on plants as a source of nutrients.

Hercynian-Alleghenian orogeny Pennsylvanian to Permian orogenic event during which the Appalachian mobile belt in eastern North America and the Hercynian mobile belt of southern Europe were deformed.

heterotrophic Any organism that depends on preformed organic molecules from its environment for nutrients. (*See also* autotrophic)

hominid Abbreviated form of Hominidae, the family of bipedal primates that includes *Australophithecus* and *Homo*.

Homo The genus of hominids consisting of *Homo sapiens* and their ancestors *Homo erectus* and *Homo habilis*.

homologous structure Body part in different organisms that has a similar structure, similar relationship to other structures, and similar development, but does not necessarily serve the same function; for example, forelimbs in whales, dogs, and bats. (*See also* analogous structure)

horn A steep-walled, pyramid-shaped peak formed by the headward erosion of at least three cirques.

hot spot Localized zone of melting below the lithosphere that probably overlies a mantle plume.

hot spring A spring in which the water temperature is warmer than the temperature of the human body (37°C).

hydraulic action The removal of loose particles by the power of moving water.

hydrologic cycle The continuous recycling of water from the oceans, through the atmosphere, to the continents, and back to the oceans, or from the oceans, through the atmosphere, and back to the oceans.

hydrolysis The chemical reaction between hydrogen (H^+) ions and hydroxyl (OH^-) ions of water and a mineral's ions.

hydrothermal A term referring to hot water as in hot springs and geysers.

hypothesis A provisional explanation for observations that is subject to continual testing. If well supported by evidence, a hypothesis may be called a theory.

Iapetus Ocean A Paleozoic ocean between North America and Europe that closed when the continents collided during the Late Paleozoic.

ice cap A dome-shaped mass of glacial ice that covers less than 50,000 km^2.

ice-scoured plain A low relief bedrock surface with glacial striations and polish eroded by a glacier.

ichthyosaur Any of the porpoise-like, Mesozoic marine reptiles.

igneous rock Any rock formed by cooling and crystallization of magma or lava or the consolidation of pyroclastic materials.

incised meander A deep, meandering canyon cut into bedrock by a stream or river.

index mineral A mineral that forms within specific temperature and pressure ranges during metamorphism.

infiltration capacity The maximum rate at which soil or sediment absorbs water.

inheritance of acquired characteristics Jean-Baptiste de Lamarck's mechanism for evolution; holds that characteristics acquired during an individual's lifetime can be inherited by descendants.

intensity The subjective measure of the kind of damage done by an earthquake as well as people's reaction to it.

Interior Lowlands An area in North America made up of the Great Plains and the Central Lowlands.

intermediate magma Magma with a silica content between 53% and 65% and an overall composition intermediate between mafic and felsic magma.

ion An electrically charged atom produced by adding or removing electrons from the outermost electron shell.

ionic bond A chemical bond resulting from the attraction between positively and negatively charged ions.

iridium anomaly The occurrence of a higher-than-usual concentration of the element iridium at the Cretaceous-Paleogene boundary.

isostasy See principle of isostasy.

isostatic rebound The phenomenon in which unloading of the crust causes it to rise until it attains equilibrium with the underlying upper mantle.

joint A fracture along which no movement has occurred or where movement is perpendicular to the fracture surface.

Jovian planet Any of the four planets (Jupiter, Saturn, Uranus, and Neptune) that resemble Jupiter. All are large and have low mean densities, indicating that they are composed mostly of lightweight gases, such as hydrogen and helium, and frozen compounds, such as ammonia and methane.

kame Conical hill of stratified drift originally deposited in a depression on a glacier's surface.

karst topography Landscape consisting of numerous caves, sinkholes, and solution valleys formed by groundwater solution of rocks such as limestone and dolostone.

Kaskaskia Sequence A widespread sequence of Devonian and Mississippian sedimentary rocks bounded above and below by unconformities that was deposited during a transgressive–regressive cycle of the Kaskaskia Sea.

Kazakhstania One of six major Paleozoic continents; a triangular-shaped continent centered on Kazakhstan.

labyrinthodont Any of the Devonian to Triassic amphibians characterized by teeth with complexly folded enamel.

laccolith A concordant pluton with a mushroom-like geometry.

lahar A mudflow composed of pyroclastic materials such as ash.

Laramide orogeny Late Cretaceous to Early Cenozoic phase of the Cordilleran orogeny; responsible for many structural features of the present-day Rocky Mountains.

lateral moraine Ridge of sediment deposited along the margin of a valley glacier.

laterite A red soil, rich in iron or aluminum, or both, resulting from intense chemical weathering in the tropics.

Laurasia A Late Paleozoic Northern Hemisphere continent made up of North America, Greenland, Europe, and Asia.

Laurentia A Proterozoic continent composed of North America, Greenland, parts of Scotland, and perhaps part of the Baltic shield of Scandinavia.

lava Magma that reaches Earth's surface.

lava dome A bulbous, steep-sided volcano formed by viscous magma moving upward through a volcanic conduit.

lava flow A stream of magma flowing over Earth's surface.

lava tube A tunnel beneath the solidified surface of a lava flow through which lava moves; also, the hollow space left when the lava within a tube drains away.

lithification The process of converting sediment into sedimentary rock by compaction and cementation.

lithosphere Earth's outer, rigid part, consisting of the upper mantle, oceanic crust, and continental crust.

lithostatic pressure Pressure exerted on rocks by the weight of overlying rocks.

lithostratigraphic unit A body of rock, such as a formation, defined solely by its physical attributes.

Little Ice Age An interval from about 1500 to the mid- to late-1800s during which glaciers expanded to their greatest historic extent.

loess Wind-blown deposit of silt and clay.

longitudinal dune A long ridge of sand generally parallel to the direction of the prevailing wind.

longshore current A current resulting from wave refraction found between the breaker zone and a beach that flows parallel to the shoreline.

Love wave (L-wave) A surface wave in which the individual particles of material move only back and forth in a horizontal plane perpendicular to the direction of wave travel.

luster The appearance of a mineral in reflected light. Luster is metallic or nonmetallic, although the latter has several subcategories.

macroevolution Evolutionary changes that account for the origin of new species, genera, orders, and so on.

mafic magma Magma with between 45% and 52% silica and proportionately more calcium, iron, and magnesium than intermediate and felsic magma.

magma Molten rock material generated within Earth.

magma chamber A reservoir of magma within Earth's upper mantle or lower crust.

magma mixing The process whereby magmas of different composition mix together to yield a modified version of the parent magmas.

magnetic anomaly Any deviation, such as a change in average strength, in Earth's magnetic field.

magnetic field The area in which magnetic substances are affected by lines of magnetic force emanating from Earth.

magnetic reversal The phenomenon involving the complete reversal of the north and south magnetic poles.

magnetism A physical phenomenon resulting from moving electricity and the spin of electrons in some solids in which magnetic substances are attracted toward one another.

magnitude The total amount of energy released by an earthquake at its source. (*See also* Richter Magnitude Scale)

mantle The thick layer between Earth's crust and core.

mantle plume A cylindrical mass of magma rising from the mantle toward the surface; recognized at the surface by a hot spot, an area such as the Hawaiian Islands where volcanism takes place.

marine regression The withdrawal of the sea from a continent or coastal area, resulting in the emergence of the land as sea level falls or the land rises with respect to sea level.

marine terrace A wave-cut platform now above sea level.

marine transgression The invasion of a coastal area or a continent by the sea, resulting from a rise in sea level or subsidence of the land.

marsupial mammal Pouched mammals, such as kangaroos and wombats, that give birth to their young in a very immature state.

mass extinction Greatly accelerated extinction rate that results in a marked decrease in biodiversity, such as the mass extinction at the end of the Cretaceous Period.

mass wasting The downslope movement of Earth materials under the influence of gravity.

meandering stream A stream that has a single, sinuous channel with broadly looping curves.

mechanical weathering Disaggregation of rocks by physical processes that yields smaller pieces that retain the composition of the parent material.

medial moraine A moraine carried on the central surface of a glacier; formed where two lateral moraines merge.

Mediterranean belt A zone of seismic and volcanic activity extending through the Mediterranean region of southern Europe and eastward to Indonesia.

meiosis Cell division that yields sex cells, sperm and eggs in animals and pollen and ovules in plants, in which the number of chromosomes is reduced by half. (*See also* mitosis)

mesa A broad, flat-topped erosional remnant bounded on all sides by steep slopes.

metamorphic facies A group of metamorphic rocks characterized by particular minerals that formed under the same broad temperature and pressure conditions.

metamorphic grade The degree to which parent rocks have undergone metamorphic change; the higher the grade the greater the change, as in low-, medium-, and high-grade.

metamorphic rock Any rock that has been changed from its original condition by heat, pressure, and the chemical activity of fluids, as in marble and slate.

metamorphic zone The region between lines of equal metamorphic intensity known as isograds.

metamorphism The phenomenon of changing rocks subjected to heat, pressure, and fluids so that they are in equilibrium with a new set of environmental conditions.

microevolution Evolutionary changes within a species.

Midcontinent rift A Late Proterozoic rift in Laurentia in which volcanic and sedimentary rocks accumulated.

Milankovitch theory An explanation for the cyclic variations in climate and the onset of ice ages as a result of irregularities in Earth's rotation and orbit.

mineral A naturally occurring, inorganic, crystalline solid that has characteristic physical properties and a narrowly defined chemical composition.

mitosis Cell division that results in two cells that have the same number of chromosomes as the parent cell; takes place in all cells except sex cells. (*See also* meiosis)

mobile belt An elongate area of deformation generally at the margins of a craton, such as the Appalachian mobile belt.

modern synthesis A combination of ideas of scientists regarding evolution that includes the chromosome theory of inheritance, mutations as a source of variation, and gradualism, but rejects the idea of inheritance of acquired characteristics.

Modified Mercalli Intensity Scale A scale with values from I to XII used to characterize earthquakes based on damage.

Mohorovičić discontinuity (Moho) The boundary between Earth's crust and mantle.

monocline A bend or flexure in otherwise horizontal or uniformly dipping rock layers.

monomer A comparatively simple organic molecule, such as an amino acid, that can link with other monomers to form more complex polymers such as proteins.

monotreme The egg-laying mammals; includes only the platypus and spiny anteater of the Australian region.

mosaic evolution The concept that not all parts of an organism evolve at the same rate, thus yielding organisms with features retained from the ancestral condition, as well as more recently evolved features.

mosasaur A term referring to a group of Mesozoic marine lizards.

mud crack A crack in clay-rich sediment that forms in response to drying and shrinkage.

mudflow A flow consisting mostly of clay- and silt-sized particles and up to 30% water that moves downslope under the influence of gravity.

multicelled organism Any organism made up of many cells, as opposed to a single cell; possesses cells specialized to perform specific functions.

mutation A change in the genes of organisms; yields some of the variation on which natural selection acts.

native element A mineral composed of a single element, such as gold.

natural levee A ridge of sandy alluvium deposited along the margins of a channel during floods.

natural selection A mechanism that accounts for differential survival and reproduction among members of a species; the mechanism proposed by Darwin and Wallace to account for evolution.

Neanderthal A type of human that lived in the Near East and Europe from 200,000 to 30,000 years ago; may be a subspecies of *Homo* (*Homo sapiens neanderthalensis*) or a separate species (*Homo neanderthalensis*).

nearshore sediment budget The balance between additions and losses of sediment in the nearshore zone.

nekton Actively swimming organisms, such as fish, whales, and squid. (*See also* plankton)

neutron An electrically neutral particle found in the nucleus of an atom.

Nevadan orogeny Late Jurassic to Cretaceous phase of the Cordilleran orogeny; most strongly affected the western part of the Cordilleran mobile belt.

nonconformity An unconformity in which stratified sedimentary rocks overlie an erosion surface cut into igneous or metamorphic rocks. (*See also* angular unconformity and disconformity)

nonferromagnesian silicate A silicate mineral that has no iron or magnesium. (*See also* ferromagnesian silicate)

nonfoliated texture A metamorphic texture in which there is no discernable preferred orientation of minerals.

normal fault A dip-slip fault on which the hanging wall block has moved downward relative to the footwall block. (*See also* reverse fault)

North American Cordillera A complex mountainous region in western North America extending from Alaska into central Mexico.

nucleus The central part of an atom consisting of protons and neutrons.

nuée ardente A fast moving, dense cloud of hot pyroclastic materials and gases ejected from a volcano.

oblique-slip fault A fault showing both dip-slip and strike-slip movement.

oceanic–continental plate boundary A convergent plate boundary along which oceanic lithosphere is subducted beneath continental lithosphere.

oceanic–oceanic plate boundary A convergent plate boundary along which two oceanic plates collide and one is subducted beneath the other.

oceanic ridge A mostly submarine mountain system composed of basalt found in all ocean basins.

oceanic trench A long, narrow feature restricted to active continental margins and along which subduction occurs.

ooze Deep-sea sediment composed mostly of shells of marine animals and plants.

ophiolite A sequence of igneous rocks representing a fragment of oceanic lithosphere; composed of peridotite overlain successively by gabbro, sheeted basalt dikes, and pillow lava.

organic evolution See theory of evolution.

organic reef A wave-resistant limestone structure with a framework of animal skeletons.

Ornithischia One of the two orders of dinosaurs, characterized by a birdlike pelvis; includes ornithopods, stegosaurs, ankylosaurs, pachycephalosaurs, and ceratopsians. (*See also* Saurischia)

orogen A linear part of Earth's crust that was, or is, being deformed during an orogeny; part of an orogenic belt.

orogeny An episode of mountain building involving deformation, usually accompanied by igneous activity, metamorphism, and crustal thickening.

ostracoderm The "bony-skinned" fish characterized by bony armor but no jaws or teeth; appeared during the Late Cambrian, making them the oldest known vertebrates.

Ouachita mobile belt An area of deformation along the southern margin of the North American craton.

Ouachita orogeny A period of mountain building that took place in the Ouachita mobile belt during the Pennsylvanian Period.

outgassing The process whereby gases released from Earth's interior by volcanism formed an atmosphere.

outwash plain The sediment deposited by meltwater discharging from a continental glacier's terminus.

oxbow lake A cutoff meander filled with water.

oxidation The reaction of oxygen with other atoms to form oxides or, if water is present, hydroxides.

pahoehoe A type of lava flow with a smooth ropy surface.

Paleocene-Eocene thermal maximum A warming trend that began abruptly about 55 million years ago.

paleomagnetism Residual magnetism in rocks, studied to determine the intensity and direction of Earth's past magnetic field.

paleontology The use of fossils to study life history and relationships among organisms.

Pangaea The name Alfred Wegener proposed for a supercontinent consisting of all of Earth's landmasses at the end of the Paleozoic Era.

Pannotia A supercontinent that existed during the Neoproterozoic.

Panthalassa Ocean A Late Paleozoic ocean that surrounded Pangaea.

parabolic dune A crescent-shaped dune with its tips pointing upwind.

parallel evolution The development of similar features in two or more closely related but separate lines of descent as a consequence of similar adaptations. (*See also* convergent evolution)

parent material The material that is chemically and mechanically weathered to yield sediment and soil.

passive continental margin A continental margin within a tectonic plate as in the East Coast of North America where little seismic activity and no volcanism occur; characterized by a broad continental shelf and a continental slope and rise.

pediment An erosion surface of low relief gently sloping away from the base of a mountain range.

pelagic clay Brown or red deep-sea sediment composed of clay-sized particles.

pelycosaur Pennsylvanian to Permian reptiles that had some mammal characteristics; many species had large fins on the back.

perissodactyl The order of odd-toed hoofed mammals that includes present-day horses, rhinoceroses, and tapirs.

permafrost Ground that remains permanently frozen.

permeability A material's capacity to transmit fluids.

phaneritic texture Igneous rock texture in which minerals are easily visible without magnification.

photic zone The sunlit layer in the oceans where plants photosynthesize. (*See also* aphotic zone)

photochemical dissociation A process by which water molecules in the upper atmosphere are disrupted by ultraviolet radiation to yield free oxygen (O_2) and hydrogen (H_2).

photosynthesis The metabolic process in which organic molecules are synthesized from water and carbon dioxide, using the radiant energy of the Sun captured by chlorophyll-containing cells.

phyletic gradualism The concept that a species evolves gradually and continuously as it gives rise to new species. (*See also* punctuated equilibrium)

pillow lava Bulbous masses of basalt, resembling pillows, formed when lava is rapidly chilled under water.

placental mammal All mammals with a placenta to nourish the developing embryo, as opposed to egg-laying mammals (monotremes) and pouched mammals (marsupials).

placoderm Late Silurian through Permian "plate-skinned" fish with jaws and bony armor, especially in the head and shoulder area.

plankton Aquatic organisms that float passively, such as phytoplankton (plants) and zooplankton (animals). (*See also* nekton)

plastic flow The flow that takes place in response to pressure and causes deformation with no fracturing.

plastic strain Permanent deformation of a solid with no failure by fracturing.

plate An individual segment of the lithosphere that moves over the asthenosphere.

plate tectonic theory The theory holding that large segments of Earth's outer part (lithospheric plates) move relative to one another.

platform The part of a craton that lies buried beneath flat-lying or mildly deformed sedimentary rocks.

playa A dry lakebed found in deserts.

plesiosaur A type of Mesozoic marine reptile; short-necked and long-necked plesiosaurs existed.

pluton An intrusive igneous body that forms when magma cools and crystallizes within the crust, such as a batholith or sill.

plutonic (intrusive igneous) rock Igneous rock that formed from magma intruded into or formed in place within the crust.

point bar The sediment body deposited on the gently sloping side of a meander loop.

polymer A comparatively complex organic molecule, such as nucleic acids and proteins, formed by monomers linking together. (*See also* monomer)

porosity The percentage of a material's total volume that is pore space.

porphyritic texture An igneous texture with minerals of markedly different sizes.

pressure release A mechanical weathering process in which rocks that formed under pressure expand on being exposed at the surface.

primary producer Organism in a food chain, such as a bacterium or green plant, that manufactures its own organic molecules, and on which other members of the food chain depend for sustenance. (*See also* autotrophic)

primate Any of the mammals that belong to the order Primates; includes prosimians (lemurs and tarsiers), monkeys, apes, and humans.

principle of cross-cutting relationships A principle holding that an igneous intrusion or fault is younger than the rocks it intrudes or cuts across.

principle of fossil succession A principle holding that fossils, and especially groups or assemblages of fossils, succeed one another through time in a regular and predictable order.

principle of inclusions A principle holding that inclusions or fragments in a rock unit are older than the rock unit itself; for example, granite inclusions in sandstone are older than the sandstone.

principle of isostasy The theoretical concept of Earth's crust "floating" on a dense underlying layer. (*See also* isostatic rebound)

principle of lateral continuity A principle holding that rock layers extend outward in all directions until they terminate.

principle of original horizontality According to this principle, sediments are deposited in horizontal or nearly horizontal layers.

principle of superposition A principle holding that in a vertical sequence of undeformed sedimentary rocks, the relative ages of the rocks can be determined by their position in the sequence—oldest at the bottom followed by successively younger layers.

principle of uniformitarianism A principle holding that past events can be interpreted by understanding present-day processes; based on the idea that natural processes have always operated in the same way.

prokaryotic cell A cell that lacks a nucleus and organelles such as mitochondria and plastids; the cells of bacteria and cyanobacteria. (*See also* eukaryotic cell)

proton A positively charged particle found in the nucleus of an atom.

protorothyrid A loosely grouped category of small lizard-like reptiles that include the earliest reptiles.

pterosaur Any of the Mesozoic flying reptiles.

punctuated equilibrium A concept holding that a new species evolves rapidly, in perhaps a few thousand years, and then remains much the same during its several million years of existence. (*See also* phyletic gradualism)

P-wave A compressional, or push-pull, wave; the fastest seismic wave and one that can travel through solids, liquids, and gases; also called a primary wave.

P-wave shadow zone An area between 103 and 143 degrees from an earthquake focus where little P-wave energy is recorded by seismographs.

pyroclastic (fragmental) texture A fragmental texture characteristic of igneous rocks composed of pyroclastic materials.

pyroclastic materials Fragmental substances, such as ash, explosively ejected from a volcano.

pyroclastic sheet deposit Vast, sheetlike deposit of felsic pyroclastic materials erupted from fissures.

quadrupedal A term referring to locomotion on all four limbs, as in dogs and horses. (*See also* bipedal)

Queenston Delta A clastic wedge resulting from erosion of the highlands formed during the Taconic orogeny.

quick clay A clay deposit that spontaneously liquefies and flows when disturbed.

radioactive decay The spontaneous change of an atom to an atom of a different element by emission of a particle from its nucleus (alpha and beta decay) or by electron capture.

radon A colorless, odorless naturally occurring gas that results from the decay of uranium 238 to lead 206.

rain-shadow desert A desert found on the lee side of a mountain range because precipitation falls mostly on the windward side of the range.

rapid mass movement Any kind of mass wasting that involves a visible downslope displacement of material.

Rayleigh wave (R-wave) A surface wave in which individual particles of material move in an elliptical path within a vertical plane oriented in the direction of wave movement.

recessional moraine An end moraine that forms when a glacier's terminus retreats, then stabilizes, and a ridge or mound of till is deposited. (*See also* end moraine and terminal moraine)

reef A mound-like, wave-resistant structure composed of the skeletons of organisms.

reflection The return to the surface of some of a seismic wave's energy when it encounters a boundary separating materials of different density or elasticity.

refraction The change in direction and velocity of a seismic wave when it travels from one material into another of different density or elasticity.

regional metamorphism Metamorphism that occurs over a large area, resulting from high temperatures, tremendous pressure, and the chemical activity of fluids within the crust.

regolith The layer of unconsolidated rock and mineral fragments and soil that covers most of the land surface.

relative dating The process of determining the age of an event as compared to other events; involves placing geologic events in their correct chronological order, but does not involve consideration of when the events occurred in number of years ago. (*See also* absolute dating)

reserve The part of the resource base that can be extracted economically.

resource A concentration of naturally occurring solid, liquid, or gaseous material, in or on Earth's crust in such form and amount that economic extraction of a commodity from the concentration is currently or potentially feasible.

reverse fault A dip-slip fault on which the hanging wall block has moved upward relative to the footwall block. (*See also* normal fault)

Richter Magnitude Scale An open-ended scale that measures the amount of energy released during an earthquake.

Rio Grande rift A linear depression made up of interconnected basins extending from Colorado into Mexico.

rip current A narrow surface current that flows out to sea through the breaker zone.

ripple mark Wavelike (undulating) structure produced in granular sediment, especially sand, by unidirectional wind and water currents or by oscillating wave currents.

rock A solid aggregate of one or more minerals, as in limestone and granite, or a consolidated aggregate of rock fragments, as in conglomerate, or masses of rocklike materials, such as coal and obsidian.

rock cycle A group of processes through which Earth materials may pass as they are transformed from one major rock type to another.

rockfall A type of extremely fast mass wasting in which rocks fall through the air.

rock-forming mineral Any mineral common in rocks that is important in their identification and classification.

rock slide Rapid mass wasting in which rocks move downslope along a more or less planar surface.

Rodinia The name of a Neoproterozoic supercontinent.

rounding The process by which the sharp corners and edges of sedimentary particles are abraded and smoothed during transport.

runoff The surface flow in streams and rivers.

salinity A measure of the dissolved solids in seawater, commonly expressed in parts per thousand.

salt crystal growth A mechanical weathering process in which salt crystals growing in cracks and pores disaggregate rocks.

San Andreas transform fault A transform fault extending from the Gulf of California in Mexico to its termination in the Pacific Ocean off the north coast of California.

sandstone-carbonate-shale assemblage An association of sedimentary rocks typically found on passive continental margins during the Proterozoic Eon.

Sauk Sequence A widespread association of sedimentary rocks bounded above and below by unconformities that was deposited during the latest Proterozoic to Early Ordovician transgressive–regressive cycle of the Sauk Sea.

Saurischia An order of dinosaurs characterized by a lizard-like pelvis; includes theropods, prosauropods, and sauropods. (*See also* Ornithischia)

scientific method A logical, orderly approach that involves gathering data, formulating and testing hypotheses, and proposing theories.

seafloor spreading The theory that the seafloor moves away from spreading ridges and is eventually consumed at subduction zones.

seamount A submarine volcanic mountain rising at least 1 km above the seafloor.

sediment Loose aggregate of solids derived by weathering from preexisting rocks, or solids precipitated from solution by inorganic chemical processes or extracted from solution by organisms.

sediment-deposit feeder Animal that ingests sediment and extracts nutrients from it.

sedimentary facies Any aspect of a sedimentary rock unit that makes it recognizably different from adjacent sedimentary rocks of the same or approximately the same age.

sedimentary rock Any rock composed of sediment, such as limestone and sandstone.

sedimentary structure Any feature in sedimentary rock that formed at or shortly after the time of deposition, such as cross-bedding, animal burrows, and mud cracks.

seedless vascular plant Plant with specialized tissues for transporting fluids and nutrients that reproduces by spores rather than seeds, as in ferns and horsetail rushes.

seismic profiling A method in which strong waves generated at an energy source penetrate the layers beneath the seafloor. Some of the energy is reflected back from various layers to the surface, making it possible to determine the nature of the layers.

seismograph An instrument that detects, records, and measures the various waves produced by earthquakes.

seismology The study of earthquakes.

sequence stratigraphy The study of rock relationships within a time stratigraphic framework of related facies bounded by widespread unconformities.

Sevier orogeny Cretaceous phase of the Cordilleran orogeny that affected the continental shelf and slope areas of the Cordilleran mobile belt.

shear strength The resisting forces that help maintain a slope's stability.

shear stress The result of forces acting parallel to one another but in opposite directions; results in deformation by displacement of adjacent layers along closely spaced planes.

sheet joint A large fracture more or less parallel to a rock surface resulting from pressure release.

shield A vast area of exposed ancient rocks on a continent; the exposed part of a craton.

shield volcano A dome-shaped volcano with a low, rounded profile built up mostly by overlapping basalt lava flows.

shoreline The area between mean low tide and the highest level on land affected by storm waves.

Siberia One of six major Paleozoic continents; composed of Russia east of the Ural Mountains, and Asia north of Kazakhstan and south of Mongolia.

silica A compound of silicon and oxygen.

silica tetrahedron The basic building block of all silicate minerals; consists of one silicon atom and four oxygen atoms.

silicate A mineral that contains silica, such as quartz (SiO_2).

sill A tabular or sheetlike concordant pluton.

sinkhole A depression in the ground that forms by the solution of the underlying carbonate rocks or by the collapse of a cave roof.

slide Mass wasting involving movement of material along one or more surfaces of failure.

slow mass movement Mass movement that advances at an imperceptible rate and is usually detectable only by the effects of its movement.

slump Mass wasting that takes place along a curved surface of failure and results in the backward rotation of the slump mass.

soil Regolith consisting of weathered materials, water, air, and humus that can support vegetation.

soil degradation Any process leading to a loss of soil productivity; may involve erosion, chemical pollution, or compaction.

soil horizon A distinct soil layer that differs from other soil layers in texture, structure, composition, and color.

solar nebula theory A theory for the evolution of the solar system from a rotating cloud of gas.

solifluction Mass wasting involving the slow downslope movement of water-saturated surface materials; especially the flow at high elevations or high latitudes where the flow is underlain by frozen soil.

solution A reaction in which the ions of a substance become dissociated in a liquid and the solid substance dissolves.

Sonoma orogeny A Permian-Triassic orogeny caused by the collision of an island arc with the southwestern margin of North America.

sorting A term referring to the degree to which all particles in a sedimentary deposit or sedimentary rock are about the same size.

species A population of similar individuals that in nature can interbreed and produce fertile offspring.

specific gravity The ratio of a substance's weight, especially a mineral, to an equal volume of water at 4°C.

spheroidal weathering A type of chemical weathering in which corners and sharp edges of rocks weather more rapidly than flat surfaces, thus yielding spherical shapes.

spit A fingerlike projection of a beach into a body of water such as a bay.

spring A place where groundwater flows or seeps out of the ground.

stock An irregularly shaped discordant pluton with a surface area smaller than 100 km².

stoping A process in which rising magma detaches and engulfs pieces of the country rock.

storm surge The surge of water onto a shoreline as a result of a bulge in the ocean's surface beneath the eye of a hurricane and wind-driven waves.

strain Deformation caused by stress.

strata (singular, stratum) Refers to layering in sedimentary rocks. (*See also* bed)

stratified drift Glacial deposits that show both stratification and sorting; deposited by streams that discharge from glaciers.

stream terrace An erosional remnant of a floodplain that formed when a stream was flowing at a higher level.

stress The force per unit area applied to a material such as rock.

strike The direction of a line formed by the intersection of an inclined plane and a horizontal plane.

strike-slip fault A fault involving horizontal movement of blocks of rock on opposite sides of a fault plane.

stromatolite A biogenic sedimentary structure, especially in limestone, produced by the entrapment of sediment on sticky mats of photosynthesizing bacteria.

submarine canyon A steep-walled canyon best developed on the continental slope, but some extend well up onto the continental shelf.

submarine fan A cone-shaped sedimentary deposit that accumulates on the continental slope and rise.

submarine hydrothermal vent A crack or fissure in the seafloor through which superheated water issues. (*See also* black smoker)

submergent coast A coast along which sea level rises with respect to the land or the land subsides.

Sundance Sea A wide seaway that existed in western North America during the Middle Jurassic Period.

superposed stream A stream that once flowed on a higher surface and eroded downward into resistant rocks while maintaining its course.

suspended load The smallest particles (silt and clay) carried by running water that are kept suspended by fluid turbulence.

suspension feeder Animal that consumes microscopic plants and animals or dissolved nutrients from water.

S-wave A shear wave that moves material perpendicular to the direction of travel, thereby producing shear stresses in the material it moves through; also known as a secondary wave; S-waves travel only through solids.

S-wave shadow zone Those areas more than 103 degrees from an earthquake focus where no S-waves are recorded.

syncline A down-arched fold in which the youngest exposed rocks coincide with the fold axis and all strata dip toward the axis.

system A combination of related parts that interact in an organized fashion; Earth systems include the atmosphere, hydrosphere, biosphere, and solid Earth.

Taconic orogeny An Ordovician episode of mountain building that resulted in the deformation of the Appalachian mobile belt.

talus Accumulation of coarse, angular rock fragments at the base of a slope.

Tejas epeiric sea A Cenozoic sea largely restricted to the Gulf and Atlantic Coastal Plains, coastal California, and the Mississippi Valley.

tension A type of stress in which forces act in opposite directions but along the same line, thus tending to stretch an object.

terminal moraine An end moraine consisting of a ridge or mound of rubble marking the farthest extent of a glacier. (*See also* end moraine and recessional moraine)

terrane A small lithospheric block with characteristics quite different from those of surrounding rocks. Terranes probably consist of seamounts, oceanic rises, and other seafloor features accreted to continents during orogenies.

terrestrial planet Any of the four innermost planets (Mercury, Venus, Earth, and Mars). They are all small and have high mean densities, indicating that they are composed of rock and metallic elements.

theory An explanation for some natural phenomenon that has a large body of supporting evidence. To be scientific, a theory must be testable — for example, plate tectonic theory.

theory of evolution The theory holding that all organisms are related and that they descended with modification from organisms that lived during the past.

therapsid Permian to Triassic mammal-like reptiles; the ancestors of mammals are among one group of therapsids known as cynodonts.

thermal convection cell A type of circulation of material in the asthenosphere during which hot material rises, moves laterally, cools and sinks, and is reheated and continues the cycle.

thermal expansion and contraction A type of mechanical weathering in which the volume of rocks changes in response to heating and cooling.

thrust fault A type of reverse fault in which a fault plane dips less than 45 degrees.

tide The regular fluctuation of the sea's surface in response to the gravitational attraction of the Moon and Sun.

till All sediment deposited directly by glacial ice.

time-stratigraphic unit A body of strata that was deposited during a specific interval of geologic time; for example, the Devonian System (a time-stratigraphic unit) was deposited during the Devonian Period (a time unit).

time unit Any of the units such as eon, era, period, epoch, and age used to refer to specific intervals of geologic time.

Tippecanoe Sequence A widespread body of sedimentary rocks bounded above and below by unconformities that was deposited during an Ordovician to Early Devonian transgressive-regressive cycle of the Tippecanoe Sea.

tombolo A type of spit that extends out from the shoreline and connects the mainland with an island.

trace fossil Any indication of prehistoric organic activity such as tracks, trails, and nests preserved in rocks. (*See also* body fossil)

Transcontinental Arch Area extending from Minnesota to New Mexico that stood above sea level as several large islands during the Cambrian transgression of the Sauk Sea.

transform fault A fault along which one type of motion is transformed into another; commonly displaces oceanic ridges; on land, recognized as a strike-slip fault, such as the San Andreas fault.

transform plate boundary Plate boundary along which plates slide past one another and crust is neither produced nor destroyed.

transverse dune A ridge of sand with its long axis perpendicular to the wind direction.

tree-ring dating The process of determining the age of a tree or wood in a structure by counting the number of annual growth rings.

tsunami A large sea wave that is usually produced by an earthquake, but can also result from submarine landslides and volcanic eruptions.

turbidity current A sediment-water mixture, denser than normal seawater, that flows downslope to the deep seafloor.

ultramafic magma Magma with less than 45% silica. (*See also* mafic, intermediate, and felsic magma)

unconformity A break in the geologic record represented by an erosional surface separating younger strata from older rocks. (*See also* nonconformity, disconformity, *and* angular unconformity)

ungulate An informal term referring to the hoofed mammals, especially the orders Artiodactyla and Perissodactyla.

upwelling The slow circulation of ocean water from depth to the surface. (*See also* downwelling)

U-shaped glacial trough A valley with steep or vertical walls and a broad, rather flat floor formed by the movement of a glacier through a stream valley.

valley A linear depression bounded by higher areas such as ridges or mountains.

valley glacier A glacier confined to a mountain valley or an interconnected system of mountain valleys.

valley train A long, narrow deposit of stratified drift confined within a glacial valley.

vascular Land plants that have specialized tissues for transporting nutrients and fluids.

velocity A measure of distance traveled per unit of time, as in the flow velocity in a stream or river.

ventifact A stone with a surface polished, pitted, grooved, or faceted by wind abrasion.

vertebrate Any animal that has a segmented vertebral column, as in fish, amphibians, reptiles, birds, and mammals.

vesicle A small hole or cavity formed by gas trapped in cooling lava.

vestigial structure Structure in an organism that no longer serves any, or a limited function, such as dewclaws in dogs and wisdom teeth in humans, or a structure that serves a completely different function, such as two of the middle ear bones in mammals.

viscosity A fluid's resistance to flow.

volcanic ash Pyroclastic materials that measure less than 2 mm.

volcanic explosivity index (VEI) A semi-quantitative scale for the size of a volcanic eruption based on evaluation of criteria such as volume of material explosively erupted and height of eruption cloud.

volcanic (extrusive igneous) rock An igneous rock formed when magma is extruded onto Earth's surface where it cools and crystallizes, or when pyroclastic materials become consolidated.

volcanic neck An erosional remnant of the material that solidified in a volcanic pipe.

volcanic pipe The conduit connecting the crater of a volcano with an underlying magma chamber.

volcanic tremor Ground motion lasting from minutes to hours, resulting from magma moving beneath the surface, as opposed to the sudden jolts produced by most earthquakes.

volcanism The processes whereby magma and its associated gases rise through the crust and are extruded onto the surface or into the atmosphere.

volcano A hill or mountain formed around a vent as a result of the eruption of lava and pyroclastic materials.

water table The surface that separates the zone of aeration from the underlying zone of saturation.

water well A well made by digging or drilling into the zone of saturation.

wave An undulation on the surface of a body of water, resulting in the water surface rising and falling.

wave base The depth corresponding to about one-half wavelength, below which water in unaffected by surface waves.

wave-cut platform A beveled surface that slopes gently seaward; formed by the erosion and retreat of a sea cliff.

wave refraction The bending of waves so that they move nearly parallel to the shoreline.

weathering The physical breakdown and chemical alteration of rocks and minerals at or near Earth's surface.

Wilson cycle The opening and then closing of an ocean basin as a result of plate movements.

zone of accumulation The part of a glacier where additions exceed losses and the glacier's surface is perennially covered with snow. Also refers to horizon B in soil where soluble material leached from horizon A accumulates as irregular masses.

zone of aeration The zone above the water table that contains both air and water within the pore spaces of soil, sediment, or rock.

zone of saturation The area below the water table in which all pore spaces are filled with water.

zone of wastage The part of a glacier where losses from melting, sublimation, and calving of icebergs exceed the rate of accumulation.

Zuni epeiric sea A widespread sea present in North America mostly during the Cretaceous; however, it persisted into the Paleogene.

Index

Note: *Italicized* page numbers indicate illustrations.

Aa lava flow, 113, *114*
Abiogenesis, 491
Abrasion
    defined, 159, 292, 351, 367
    glaciers and, 350, 351, *351*, 354
    wave erosion and, 398, *398*
    wind, *367*, 367–368
Absaroka Sequence, *511*, 523–526, 581
Absolute dating, 416, 418, *439*
Absolute dating methods, 432–437
Abyssal plains, 33, 34
Acadia National Park, Maine, *419*
Acadian orogeny, 509, *511*, 530, 532
Acanthodians, *555, 556, 557*, 558
Acanthostega, 560
Acasta Gneiss, Canada, *182, 476, 477,*
    480, 481
Accessory minerals, 75
Accretionary wedge, 248, *249*, 251
Accumulation, zone of, 144, *146*, 346, 347,
    347, 350
Acid fog, 144
Acid rain, 144–145, *145*, 656, 657, *658*
Acid snow, 144
Acidity, defined, 144
Acids, weathering and, 140
Acquired characteristics, 453
Acritarchs, *484, 495, 495*
Actinolite, 480
Active continental margins, 46
Adenosine triphosphate (ATP), 494
Adirondack Highlands, New York, 188, *188*
Adirondack Mountains, 245, 479
African plate, 39, *39, 49, 125*, 615
Aftershocks, 200
Agassiz, Louis, 628, 630
Agate, 175
Age, defined, 440
Age of Reptiles. *See* Mesozoic Era
Agnostid trilobites, 548, *549*
Agriculture, 314, 326, 327, 366
Aine Corona, *42*
Air-pressure belts, 372, 374
Airy, George, 252
Alaska earthquake, *200*, 211
Alaska Volcanoes Observatory, 126
Alcohol, fermentation and, 494
Aletsch Glacier (Switzerland), *338, 343*
Algae
    in Burgess Shale, *540*
    calcareous, 552, 553
    green, 560, 565
    in Permian reef, *553*
    plankton and, 544

in Silurian reef, *551*
    trilobites and, 548
Alkali soils, 147, *148*
Allantois, 562
Alleghenian orogeny, 509, *511, 532*, 582
Alleles, defined, 456
Allopatric speciation, 459, *459*
Allosaurus, 169
Alluvial fans, 298, *298, 379, 380,*
    381, 487
Alluvium, 293, 298
Alpha decay, 432, *432*, 433
Alpine glacier, 340. *See also* Valley glaciers
Alpine orogeny, 615
Alpine-Himalayan orogenic belt,
    615, *617*
Alps, Europe, 234, *338*, 341, *417*, 615
Alps, New Zealand, 341
Aluminum
    Bowen's reaction series and, 90
    in Earth's crust, 87
    importance of, 60
    magma and, 87, *87*
    as resource, 79
    in rock cycle, 17
    weathering and, 153
Aluminum ore, 60, 79, 80
Alvin, 389
Amazon River, South America, 290
Amber, 77
American Falls Lake, *631*
Amino acids, 492, 493
Ammonia, 12, 490, 492
Ammonites, 592
Ammonoid cephalopod, *552, 558*
Ammonoidea, 592
Ammonoids, 551
Amnion, 562, 635
Amniote egg, defined, 562, *562*
Amniotes, defined, 593, *595*
Amorphous, defined, 64
Amphibians
    evolution and, 467, 468, *470*
    fish to, 560
    geologic time and, *417*
    in Messel Pit Fossil Site, 638
    in Paleozoic history, 542, 558, *559,*
        560–562, *562*
    plants and, 565
    reptiles and, 562–563
Amphibole, 69, *69*, 74, *75*, 89, *89*, 96
Amphibolite, *190*, 191
Amphibolite facies, 194
Anaerobic cells, 494
Analogous structures, 465, *466*
Anatolepis, 555
Ancestral Rockies, 526, *527*

Ancient Denvers: Scenes from the Past 300
    Million Years of the Colorado Front
    Range (Johnson), 442
Andalusite, 189, 195
Andes Mountains (South America), *45, 46,*
    96, 234, 249, *250*, 251, 341, 509
Andesite, 95, *95*, 96, *96*, 97, *478*
Andrews, Sarah, 6
Angiosperms, 592, 593, *594*, 614, 632
Angle of repose, *260*, 261
Angular unconformity, 420, *425*
Anhydrite, 75, 516, 517, *517*, 534, *665*
Animals, distribution of, 656
Ankylosaur, 596, 597
Ankylosaurus, 599
Anning, Mary, 605, *605*
Anomalocaris, 547
Anorthosite, *487*
Antarctic plate, 39, *39, 125*
Antecedent streams, 621
Anthracite coal, 165, 173, *190*, 193, *534,*
    535, 590
Anthropoids, 642, *643*
Anticlines, 237–238, *239*, 240, *240*, 241
Antler Highlands, 532
Antler orogeny, 509, 532, *532*, 584
Antrim Shale, 521, *522. See also*
    Black shales
Apatite, *75*
Apatosaurus, 442, 596
Apes, 642, *644*
Aphanitic texture, 93, *95*, 96, *96*
Aphelion, 360
Aphotic zone, 390
Aphrodite of Melos, 182
Appalachian mobile belt
    black shales and, 521
    in Cambrian paleogeography, *512,*
        513, *513*
    defined, 505
    as major mobile belt, *505*
    in Ordovician paleogeography, *514*
    as Paleozoic mobile belt, 526, 528–532,
        530
    in Silurian paleogeography, *516*
    Taconic Highlands and, 515
    Tippecanoe Sequence and, 520
Appalachian Mountains
    Cenozoic history and, 614, *626,* 626–
        627, 628
    deformation and, 234, 245, 252
    geologic time and, *417*
    mass wasting and, 265
    Mesozoic history and, *582, 583,* 590
    plate tectonics and, 29–30
    Precambrian history and, 480, 487
Appennines, Italy, 615

Aquicludes, 315, 317, *317*, 319, *319*
Aquifer, 315, 316, 318, 319, *319*, 320, 323–326, 331
Arabian plate, *39*, *125*, 252, 615
Aragonite, 69, 78
Arboreal ancestry, 641
Archaeocyathids, 515, 546, *547*, 550
Archaeopteryx, 601, *601*, 602
Archean cratons, 484, *485*, 486
Archean crust, 480
Archean Earh history, 480–482
Archean Eon, 234, *417*, 476, *476*, 480, *481*, 491
Archean fossils, *493*, 494
Archean geologic/biologic events, *481*
Archean organisms, 493–494
Archean pegmatites, 499
Archean plate tectonics, 480, 482
Archean resources, 497, 499
Archean rocks, 478, 479, *479*, 480, 482, 488, 497
Archean sedimentary deposits, 490
Archean volcanoes, 490
Arches, 137, *150*, 399, *399*, 520, 526, *612*
Arches National Park, Utah, *150–151*
Archimedes, 253
Archosaurs, 594–595, *595*
Arctomys Cave, Canada, 322
Ardipithecus kadabba, 643
Ardipithecus ramidus, 643, *645*
Aresenic, *328*, 328–329, *329*
Arêtes, 353–354, *354*
Argentite, 70, *73*
Argillite, 480
Argo, 389
Argon, 62, 434, *434*, 435
Arid regions, 374, *375*, 382
Aristotle, 450
Arkose, 161
Army Corps of Engineers, 407
Arsenic, *328*, 328–329, *329*
Art, 6, *6*
Artesian systems, 317, *319*, 319–320
Artesian-pressure surface, 319
Arthropods, 548–549, 551, 560
Artificial recharge ponds, 314
Artificial selection, 452, *452*, *454*, 454–455
Artiodactyls, 636, 637, 638, *638*, *640*
Asbestos, 182, 183, *183*, 195
Aseismic ridges, 34, 48
Ash fall, defined, 113
Ash flow, defined, 113
Assimilation, defined, 92, *93*, 95
Asteroid belt, *11*, 12
Asteroids, in solar system, 11, 12
Asthenosphere
    defined, 13, *13*
    earth's mantle and, 225
    magma and, 91
    plate tectonics and, 17, *17*, 39, 44, 46, *49*, *50*, *51*
Astrobleme, 498

Astronomy, 21
Aswan High Dam, 286
Athabaska tar sands, Canada, 176
Atlantic Coastal Plain, 628
Atlantic Ocean basin, 46, 615
Atlantic-Pacific Ridge, 33
Atlantis, 388, *389*
Atlas Mountains, Africa, 615
Atmosphere
    circulation pattern of Earth's, *374*
    as Earth system, 4
    evolution of, 489–491, *490*
    formation of, 12, 476
    glaciers and, 344, 348, 360, 489
    oceans and, 389, 407, 409
    origin of, 489–491, *490*, 492
    plate tectonics/rock cycle and, 20, 615
    running water and, 287, 290
    volcanism and, 110
    weathering and, 141
Atmospheric ammonia, 490
Atmospheric pollution, 144, *654*
Atolls, 392, *392*
Atomic mass number, defined, 62, 432
Atomic number, defined, 62, 432
Atoms
    daughter, 433, 434, *434*, 435, *435*
    defined, 61, 432
    elements and, 61–64
    geologic time and, 432
    lava and, 93
    magma and, 93
    parent, 433, 434, *434*, 435, *435*
    shell models for common, *61*
ATP. *See* Adenosine triphosphate (ATP)
Atrypa, 429
Augite, *68*, 74, *74*, *75*, 664, *665*
Augustine volcano, Alaska, *121*
Aureole, 185, *186*, 187
Australian plate, *125*
Australopithecines, 644, 646
Autotrophic cells, 494
Autotrophs, 545
Avalanches, 264, 277, *278*
Avalonia, 34, 505, 506, *507*, 533
Axial tilt, 360
Ayers Rock, Australia. *See* Uluru, Australia

Baboon, 642, *643*
Baby Arch, *151*
Back-arc basin, 45–46, 248, *249*, 480, *483*
Background extinction, 463
Backshore, 400, *400*
Bacon, Francis, 28
Baculites, 440
Badgers, *640*
Badlands National Park, South Dakota, 625, 636
Bajadas, 379, 381
Baking zone, 420, *421*
Balanced Rock, Utah, *151*
Balanced rocks, *150*

Balkan Mountains, Serbia, *234*
Baltica, 505, 506, *506*, *507*, 509, 530, *530*, 531, *531*, 532
Baltis Vallis, *42*
Banded iron formation (BIF), 176, *176*, 480, 486, 489, *489*, 499
Bangiomorph eukaryote, 495
Barchan dunes, 369, 370, *371*
Barnum and Bailey Circus, 121
Barrier islands, 402–403, *403*, 404, 407, *408*
Barrier reefs, 392, *392*, 517
Barrow, George, 193
Baryonyx, 598
Basal slip, 345, *345*, 347
Basalt
    on beaches, 401, *409*
    on continents, 478
    depicted, *19*
    in deserts, 382
    in Earth's crust, 13, 252
    geologic time and, *422*
    glaciers and, *352*
    igneous rocks and, *19*, 95, *95*, 96, *96*, 97, 100
    metamorphism and, 191
    mineral formation and, 78
    oceanic crust and, 227
    oceanic ridges and, 33
    seafloor spreading and, *38*
    volcanism and, 110, 116
Basalt lava flows, 96, 113, *115*, *117*, 629
Basalt plateaus, defined, 122–124, *124*
Basaltic dikes, 46
Base level, *304*, 304–305, 306
Base lines, 662, *663*
Basement rocks, 182, 504, 526
Basilosaurus, *467*, 640
Basin
    Atlantic Ocean, 44, 615
    back-arc, 45–46, 248, *249*, 480, *483*
    defined, 241, *241*
    Delaware, 534
    drainage, 302, *303*, 305, 306
    Elk Point, 534
    Illinois, 523, 525, 535
    Mesoproterozoic, *487*, 488
    Michigan, 516, 517, *517*, 520, 535
    Neoproterozoic, *487*, 488
    ocean (*See* Ocean basins)
    Pacific Ocean, *26*, 46, 615
    Powder River, 631
    Raton, *606*
    Sudbury, 498
    Triassic fault, 44
    Williston, *505*, *521*, *523*, 535
Basin and Range Province, 246, *248*, 620–621, *621*
Bates, L. W., Jr., *407*
Bathhouses, 331, *333*, 334
Batholiths
    Cordilleran igneous activity and, 618
    defined, 102

deformation and, 247, 248
emplacement of, *104*
geologic time and, 420, *422*
granite and, 100, 102
Mesozoic history and, 576, 585, *587*
origin of, 103
orogens and, 484
Sierra Nevada, *574*
stocks and, 102
Bauxite, *79*, 153
Baymouth bar, 401–402, *402*
Beach face, 400, *400*, 401
Beaches
    artificially maintaining, 408, 409
    coasts and, 404
    defined, *400*, 400–401
    nearshore sediment budget and, 403
    pocket, 400
    rising sea level and, 407
    summer, 401
    winter, 401
Bearpaw Formation, Canada, *440*
Bears, 639, *640*
Beavers, 639
Bed load, 292–293, 366–367
Bedded chert, *164,* 165
Bedrock
    deserts and, 381, *381*
    glaciers and, 351, *351,* 355, *355*
    mass wasting and, 261, 263, 265, *265,*
        277, 279, *281*
    Proterozoic glacial deposits and, 488
    running water and, 306, 307
Beds, defined, *167*, *167,* 439, *439*
Belemnoids, 592, *592*
Bell Telephone Laboratories, 11
Bellini, Giovanni, 6
Benching, 279, *280*
Benioff zone, 44, 203, *204*
Benioff-Wadati zone, 203, *204*
Benthos, 545, *545,* 546
Berms, 400, *400,* 401, 408, *409*
Beryl, *76*
Beryllium, 499
Beta decay, 432, *432,* 433
BIF. *See* Banded iron formation (BIF)
Big Bang theory, 10–11
Bighorn Mountains, Wyoming, 237, *238,*
    242, 479
Bingham Cooper Mine, *53*
Biochemical sedimentary rocks, *162,*
    162–165
Biochemistry, 464, 465
Biology, molecular, 464, 468
Biosphere
    atmosphere/hydrosphere and, 489
    as Earth system, 4
    evolution of, 656
    organisms in, 491
    plate tectonics/rock cycle and, 20, 615
    volcanism and, 110
Biostratigraphic unit, 439

Biostratigraphy, *440*
Biotic provinces, defined, 53
Biotite
    igneous rocks and, *95, 96*
    magma and, 89, *89*
    metamorphism and, *186, 188, 189,* 193
    in mineral identification table, *664*
    minerals and, 69, *69,* 74, *74, 75*
    weathering and, 137
Biotite mica, 89, *89, 97*
Biozone, 439, *439,* 440
Bipedal, 594, 642
Birds
    in Cenozoic history, 614, 634, *634,*
        639–641
    elephant, 641
    evolution and, *451, 456, 459, 460, 461,*
        462, 462, 465, 467, 468, *470*
    in Mesozoic history, 576, 594, *595, 601,*
        601–602
    at Messel Pit Fossil Site, 638
Birmann, Samuel, *342*
Bishop Tuff, California, 125
Bison, 639
Bitterroot Mountains, 245
Bituminous coal, 165, *165,* 173, *534,* 535,
    590
Bivalves, *545,* 551, 592, 593, 632
Black Hills, South Dakota, *139,* 625
Black shales, 520, 521–522, *522,* 528. *See
    also* Antrim Shale, Chattanooga Shale,
    and New Albany Shale
Black smoker, 33, *34,* 388, 390, 493, *493,* 497
Blastoids, 552
Block slide. *See* Rock slides
Block-faulting, 245, *248,* 626
Blocks, defined, 113
Blowout preventer, 411
Blowouts, 368
Blue Haze Famine, 112, 122
Blueschist facies, 194, 248
Bodega Bay, California, *398*
Body fossil, 467, *468*
Body waves, 205, 206, *206, 208*
Bogs, 287
Bombs, defined, 113
Bonding, 62–64
Bony fishes, *470,* 557–558
Bootlegger Cove Clay, 275, *275*
Bornite, 195, *195*
Bothriolepis, 556, 557
Bottomset beds, 296
Bowen, N. L., 89
Bowen's reaction series, *89,* 89–90, 92, 142
BP, 411
Brachiopods, 514, *529,* 546, *550,* 551, *551,*
    552, *553*
Brachiosaurus, 596, 606
Braided streams, 293–294, *294,* 355
Breaker zone, 396, 397, *397,* 400, *401*
Breakers, *395,* 396
Bridalveil Falls, California, 352, *354*

Bright Angel Shale, *166*
Brittle rocks, 236
Brook, 290
Browsers, 636, 637
Bryce Canyon National Park, Utah, *419,*
    430, 612
Bryophytes, 565
Bryozoans, 515, *529,* 550, *550,* 551, *551,*
    553, 632
Buchan Cave, Australia, *325*
Bucheria, 566
Buckland, William, 605
Buffon, Georges Louis de, 416
Building codes, 278
Bullard, Edward, 29
Bulldozing, 350
Bureau of Land Management, 662
Burgess Shale biota, 547, *547,* 550
Burgess Shale, Canada, *540,* 542, *547*
Burgess Shale fossils, 547, *547,* 550
Bushveld complex, South Africa, 499
Buttes, 245, 382, *382*
Button Mold Knob Quarry, Kentucky, *522*
Byron Glacier, Alaska, *346*

Calamites, 568
Calcareous algae, 552, 553
Calcareous ooze, *391*
Calcareous tubes, 542, 543
Calcareous tufa, 334
Calcite
    lava and, *87*
    in mineral identification table, *665*
    minerals and, *68,* 69, *70,* 74, *74,* 75, *75,*
        77, 78
    nonfoliated metamorphic rocks
        and, 192
    sedimentary rocks and, *162,* 163,
        174, 175
    stalactites and, 323
    weathering and, 140, 141
Calcium
    Bowen's reaction series and, 90
    Cambrian explosion and, 543
    in Earth's crust, 87
    in groundwater, 331
    magma and, 87, *87*
    oceans and, 390
    in seawater, 162
Calcium carbonate
    Cambrian explosion and, 543
    glaciers/glaciation and, 440, *441*
    groundwater and, 334
    minerals and, 78
    oceans and, 391, 401
    sedimentary rocks and, 160, *163,*
        174, *174*
Calcium phosphate, 543
Calcium plagioclase, 96
Caldera
    Crater Lake as, *116*
    at Deception Island, 125

defined, 115
igneous activity and, 486
Long Valley, 125, 129, 130
Mount Mazama as, 118
pyroclastic flows and, 125
Yellowstone, 125
Caledonian Highlands, *507, 521, 523,* 530, *530,* 531, *531*
Caledonian orogeny, 506, 530, 532
Caliche, 145, *148*
Calico Mountains, California, *238*
California gold rush, 173, 576, 591
Calvert Cliffs, Maryland, *614,* 628
Calving, 344, 346, *347*
Calymmian Period, *477*
Cambrian explosion, 542–543, 547, 550
Cambrian facies, *513*
Cambrian fossils, 542, 543, 550
Cambrian marine community, 546, *546*
Cambrian paleogeography, *512*
Cambrian Period
　Burgess Shale fossils and, 550
　evolution and, *470*
　fish during, 555, *555*
　geologic time and, *430,* 435, 440
　Grand Canyon and, *519*
　marine community in, 546
　in Paleozoic geologic system, 504
　Paleozoic paleogeography and, 506, *506*
　polar wandering in, *36*
　Sauk Sequence and, 511, *512*
　trilobites in, 548–549
Cambrian rocks, 493
Cambrian System, 440
Camels, 636, 637, 639
Cameron International, 411
Canadian Minerals Yearbook, 80
Canadian shield, *479,* 479–480, 482, *505,* 512
Cancer, 183, *183*
Cap rock, defined, 173
Cape Cod, Massachusetts, 398
Cape Foulweather, Oregon, *399, 614*
Capillary fringe, 316
Carat, defined, 77
Carbon
　in atmosphere, 492
　in biochemical sedimentary rocks, *162*
　coal and, 165
　geologic time and, 436
　minerals and, 62, 63, 64, *64,* 66, 75, 77
　in seawater, 493
Carbon 12, 62, *62,* 436
Carbon 13, 62, *62*
Carbon 14, 62, *62,* 436
Carbon dioxide
　atmosphere and, 489, 490, *490,* 492
　caves and, 322
　fermentation and, 494
　glacial ice and, 489
　in global ecosystem, 657
　greenhouse effect/global warming and, 8, 348

magma/lava and, 88
metamorphism and, 185
minerals and, 62, 75
photosynthesis and, 491
volcanism and, 110, 112, *129,* 130
weathering and, 140, 141
Carbon monoxide, 110, 490
Carbon-14 dating technique/method, 222, 436, *437*
Carbonate deposition, 513, 514, 522–523, 528, 628
Carbonate facies, 165
Carbonate minerals
　carbonitite and, *87*
　caves and, 322, *324*
　chemical sedimentary rocks and, 163
　defined, 69–70
　in greenstone belts, 480
　karst landscape and, *323*
　Kaskaskia rocks as, 520
　in Michigan Basin, 516, *517*
　as mineral group, *67, 75,* 77
　most common, *70*
　in seawater, 162
　sedimentation and, 487
　Wopmay orogen and, 485
Carbonate radical, 67, *68*
Carbonic acid, 140
Carboniferous coal swamp, 562, 566
Carboniferous marine community, 552–553
Carboniferous Period
　amphibians during, 561, 562
　continental drift and, *30*
　fish during, 555
　geologic time and, *417*
　Mississippian Period and (*See* Mississippian Period)
　in Paleozoic geologic system, 504
　Paleozoic paleogeography and, *508,* 509, *510*
　Pennsylvanian Period and (*See* Pennsylvanian Period)
　plants during, 564, 566, 568, *568*
Carbonitite, *87*
Caribbean plate, *39,* 125
Carlsbad Caverns, New Mexico, 322
Carnivores, 562, 635
Carnivore-scavengers, 545, *545*
Carnotite, 176, 591
Cartilaginous fishes, *470, 555,* 557, *557,* 558
Cascade Range
　in Cenozoic history, *618,* 620, 622–623, 628
　defined, 126, *127,* 620
　deformation and, 246
　drainage patterns in, 303
　forecasting eruptions and, 129
　glaciers in, 349, *349,* 353
　Mount Shasta in, 349
　volcano belts and, 125
Cascade Range volcanoes, 622–623

Cascades Volcano Observatory, 126
Cassiterite, *195*
Castle Crags State Park, California, *103*
CAT scans. *See* Computerized axial tomography (CAT) scans
Cathedral Group, *474*
Cats, *640,* 641
Catskill Delta, 530, 531, *531*
Caudipteryx, 598
Cave deposits, *312,* 322–323, *325*
Cave formation, *324*
Cavern, 322, *324*
Caves, 321, 322–323, *324,* 399, *399*
Cedaria minor, 548
Celerity, 395
Cell cleavage, 554, *554*
Cells
　anaerobic, 494
　autotrophic, 494
　convection, 13, 50, *50, 51*
　earliest, 494
　eukaryotic, *494,* 494–496, *496*
　heterotrophic, 494
　mantle convection, *50*
　multicelled organisms and, 496
　plants and, 565
　prokaryotic, 494, *494,* 495
　protobionts and, 492
　sex, 457, *457,* 458
　thermal convection, 37, 50, *50*
Cellulose, 565
Cementation, 160, *161*
Cenozoic climate, 629, 632, *633,* 634
Cenozoic deposition, *627*
Cenozoic Earth history
　of Appalachian Mountains, 614, *626,* 626–627, 628
　Cascade Range and, *618,* 620, 622–623, 628
　glaciation in, 628–631
　Holocene Epoch in, 628–631
　introduction to, 614–615
　mineral resources in, 631–632
　Neogene Period in, 614, 616–621, 624–628
　orogenic belts in (*See* Cenozoic orogenic belts)
　Paleogene Period in, 614, 616–621, 624–628
　plate tectonics in, 615
　Pleistocene Epoch in, 628–631, 639–650
Cenozoic Era
　distribution of life and, 54, *54*
　evolution and, 461, *470*
　fish during, 555, 558
　geologic time and, *417,* 429, *476*
　Paleozoic history and, 505
　Pangaea and, 578
　petroleum and, 52
Cenozoic life history
　birds in, 614, 634, *634,* 639–641
　climate in, 629, 632, *633,* 634

hominids in, 639, 642–649, *645*
introduction to, 614–615
mammals in, 614, *635*, 635–637, *637*, 639–641, *640*
marine invertebrates in, 632
Messel Pit Fossil Site and, 638
Neogene Period in, 614, 632–637, 639
Paleogene Period in, 614, 632–637, 639
phytoplankton in, 632
Pleistocene faunas in, 639–650
primate evolution in, 641–642
vegetation in, 632, *633*, 634
Cenozoic mineral resources, 631–632
Cenozoic orogenic belts
Alpine-Himalayan orogenic belt as, 615, *617*
circum-Pacific orogenic belt as, 615–616, 617, *618*
Cenozoic plate tectonics, 615
Cenozoic vegetation, 632, *633*, 634
Central Lowlands, 625, 626
Cephalopods, *545*, *550*, 551, *551*, *553*, 592, 593, 632
Ceratites, 592
Ceratopsians, 597
CFCs. *See* Chloroflurocarbons (CFCs)
Chalcedony, 175
Chalcopyrite, 195, *195*, *664*
Chalet Enzian, Switzerland, *190*
Chalicotheres, 637, 639
Channel flow, 290
Chattanooga Shale, 176, 521, *522*. See also Black shales
Checkerboard Mesa, Utah, *241*
Cheirolepis, 556, *557*
Chemical deterioration, 152
Chemical sediment, 158, 160
Chemical sedimentary rocks, *159*, 162, 162–165, *163*, *164*
Chemical waste, 330
Chemical weathering
acid rain and, 144
for chemical/biochemical sedimentary rocks, 162
as decomposition, 139–142
defined, 137, 139
deserts and, 376
at Grimes Point Archaeological Site, *139*
laterite and, *149*, 153
mass wasting and, 265, 270
oceans and, 390
sediment sources and, *159*
spheroidal weathering and, *143*
tropics and, 147
Chert, *162*, *164*, 164–165, 480, *489*
Chimpanzee, *643*
China, in Paleozoic history, 505, 506, *506*, *507*, *508*, 510
Chinle Formation, 586
Chloride ions, in seawater, 390
Chlorine
minerals and, 62, 63, *63*, 66

in seawater, 162, 163, 409
volcanism and, 110, 122, *123*
Chlorite, 187, *188*, 189, 193, 480
Chlorite schist, 191
Chloroflurocarbons (CFCs), 122–123, *123*
Chordate, 554, *554*, 555
Chromium, *79*, 80, 499
Chromosomal mutation, 458
Chromosomes, defined, 457
Chrysotile asbestos, 183, *183*
Cibicideds americanus, *633*
Cincinnati Arch, 526
Cinder cones
in Cascade Range, 126
composite volcanoes *vs.*, 119
defined, 118, *118*
Eldfell as, 118, *118*
Helgafel as, *118*
in Mojave Desert, *629*
Paricutín as, *118*
Wizard Island as, *116*
Circulation
deep-ocean, 390, 391
downwelling as, 391
geologic time and, 442
horizontal, 391
oceanic, 390–392, 576
upwelling as, 390, 391
vertical, 391
Circulation patterns, 578–580, *579*, 656
Circum-Pacific orogenic belt
in Cenozoic history, 615–616, 617, *618*
defined, 125, *125*
earthquakes in, 203, *204*, 214, *218*
Cirques, 353–354, *354*
Clack, Jennifer, 560
Clade, defined, 462
Cladistics, 461–462, 465
Cladograms, 461–462, *462*, *465*, 470
Cladoselache, 557, *557*
Clams, 467, 592
Clarity, for diamond, 77
Claron Formation, Utah, *134*, *612*
Classic texture, 161
Classification
of Burgess Shale fossils, 547
evolutionary theory and, 463, 464–465
of fish, *566*
of igneous rocks, 93–98, *95*, *98*
texture, *98*
Clastic wedge, 529, 531
Clasts, 161
Clay deposits, 137, 172
Clays
geologic time and, 434
graywacke and, 480
mass wasting and, 263, 264, *264*, *266*, *274*, 274–275, *275*
metamorphism and, 185, 187
minerals and, 69, *75*, 78, 79, *664*
pelagic, 391, *391*

permeability and, 315
polymerization and, 493
quick, *266*, *274*, 274–275, *275*
running water and, 296
sediment and, 158, *158*, *159*, 161, *161*
in soil, 143
weathering and, 139, 141, 153
wind and, 368
Claystone, 161
Cleavage
in augite, *74*
cell, 554, *554*
chert and, 164
defined, 64, 74
in diamond, 77
in hornblende, *74*
as physical property, 664
radial, 554, *554*
slaty, 189
spiral, 554, *554*
types of, *74*
Climatic events, 360
Climate/climatic change
in Cenozoic history, 629, 632, *633*, 634
continental drift and, *32*
deserts and, *364*, 366
as environmental issue, 8–9, 657
geologic time and, 440–444, *441*, *444*
glaciers and, 340, 344, 346, 348, 359, 360, *360*
global, 442
in Grand Canyon, *518*
mass wasting and, 261, 263
Mesozoic, 576, 604, 606
Pangaea and, 578–580
running water and, 306
soil formation and, 145, *146–147*
in United States, 9
weathering and, 137, 141
wind/ocean currents and, 53
Clinton Formation, *517*, 535
Closed system, 435
Clouds, 287
Coachella Valley, California, *378*
Coal
anthracite, 165, 173, *190*, 193, *534*, 535, 590
bituminous, 165, *165*, 173, *534*, 535, 590
global warming and, 348
mass wasting and, 270
in Mesozoic history, 590
ocean resources and, 409
origin of, *165*
Paleozoic paleogeography and, 509
plants and, 566
running water and, *288*
sedimentary rocks and, 158, *162*, 165, *165*, 173
subbituminous, *534*, 631
Coal deposits, *534*, 535, 579
Coal mine, *502*
Coal swamp, 535, *538*, 562, 566, *567*, 568

Coast Mountains, Canada, *357, 358*
Coastal communities, 408
Coastal development, 408
Coastal flooding, *300,* 389, 405–406, 407, 657
Coasts
    depositional, 404
    drowned, 404 (*See also* Submergent
      coasts)
    emergent, 404–405, *405*
    erosional, 404
    shorelines *vs.,* 392, *393*
    submergent, 404–405, *405*
    types of, 403–405
Cobalt, 79, *79,* 409
Coccolithophores, 593, *593,* 632
Cocos plate, *39,* 615, 624, 625
Coelacanths, 558
Coke, defined, 173, 535
Cole, Thomas, 6
Coleoids, 592
Colonial corals, *550*
Colonial organisms, 496
Color
    for diamond, 77
    for marble, 182
    for minerals, 70, 71
    as physical property, 664
    of seawater, 390, *390*
    of sedimentary rocks, 167
Colorado Plateau, 429, *430,* 518, 586, 591,
    621, *624*
Columbia River, Washington, 99, 122, *124,*
    619, *620, 631*
Columbus, Christopher, 389, 450
Columnar joints
    defined, 113, *115*
    deformation and, 242
Columns, in caves, *312,* 314, 323, *324, 325*
Comets, 11, 12, 476, 491, 498
Compaction, 160, *161,* 417
Comparative anatomy, 463
Complex movement, 277, *277*
Composite volcanoes, 118–119, *119*
Composition
    of coasts, 403
    of continents, 478
    of core, 224–225
    of Earth, 223, *223*
    of igneous rocks, 95, 96, 97, *98*
    of mantle, 225–226
    of metamorphic rocks, 182
    of seawater, 390
    of sedimentary rocks, 165, 167
    of soils, *146*
Compounds, 62–64, 70
Compression, defined, 235, *236*
Compsognathus, 596
Computerized axial tomography (CAT)
    scans, 227
Comstock Lode, *73*
Concordant ages, 435
Concordant pluton, 99, *101*

Concretions, *174,* 174–175, *175*
Concurrent range zones, 429, *431, 439*
Condensation, running water and, 287
Condie, K. C., *483*
Cone of ascension, 325
Cone of depression, 318, *318*
Cones, 566, *567*
Configuration, of coasts, 403
Conglomerate
    deformation and, *237*
    depicted, *19*
    in greenstone belts, 480
    groundwater and, 315
Coniferous forest soil, *147*
Conifers, 593
Constancy of interfacial angles, defined,
    65, *65*
Contact metamorphism, 185, 187, *187,*
    192, 195, *195,* 420
Continental accretion, 234, 252, 478, 482
Continental crust
    basalt in, 96
    composition of, *223*
    on continents, 478
    defined, 13
    density of, *223*
    in Earth's interior, 227
    gabbro in, 96
    Hadean and, 476, 477, 478, *478*
    igneous activity and, 486
    igneous rocks in, 182
    Laurasia and, *531*
    as loading/unloading, 254
    magma and, 91
    metamorphic rocks in, 182
    to oceanic crust, 32, 39, 47, *50,* 252
    Ouachita mobile belt and, *533*
    rifting and, 44
    sedimentary crust in, 182
Continental depositional setting, 159,
    160, 293
Continental divide, 302
Continental drift
    continental fit and, 29, *30*
    data supporting, 39
    early ideas about, 28–29
    evidence for, 29–31
    fossil evidence for, 31, *32*
    glacial evidence for, 30–31, *31*
    mountain ranges and, 29–30
    rock sequences and, 29–30, *30*
Continental drift hypothesis, 29, 50
Continental fit, 30, *31*
Continental glaciers/glaciation
    defined, 340, 341, *341,* 344
    erosion/sediment transport and, 351,
      354–355, *355*
    glacial budget and, 346, 347, 350
    glacial deposits and, 355, *356,* 357, 358
    movement of, 345, 347
    Paleozoic history and, 504
    stratified drift and, 358

Continental interior, 581, 625–626
Continental lithosphere, 46, 92, 251
Continental margin
    active, 46
    Cenozoic history and, 627, *627,* 628
    defined, 32, 388
    features of, *33*
    hydrocarbons on, 410
    oceans and, 391
    rifting and, 485, 488
    shelf space on, 552
Continental movement, 50
Continental red beds, 486, 489, 491, 526
Continental rise, 32–33, *33,* 388, 391, 628
Continental shelf, 32–33, *33,* 46, 388, 391,
    394, 409, 628
Continental slope, 32–33, *33,* 388, 391
Continental-continental convergent plate
    boundary, *17, 45,* 46, *251,* 251–252, 530
Continental-oceanic convergent plate
    boundary, *17*
Continent-continent collision, 485
Continents
    defined, 488
    foundations of, 478–480
    mobile belts and, 505
    Paleozoic history and, 505, 542
Continuous branch, 89, *89*
Contour interval, 661
Contour plowing, 153
Contour spacing, 661
Contours, 661, *662*
Convection cells, 13, 50, *50, 51*
Convergent evolution, 459–461, *461*
Convergent plate boundary
    Archean rocks and, 482
    continental-continental, *17, 45,* 46, *251,*
      251–252, 530
    continental-oceanic, *17*
    defined, 44–47
    deformation and, 238, 245, 247
    earthquakes and, 203, *204*
    igneous rocks and, 86, 89, 91
    metamorphic rocks and, 194, *195*
    oceanic-oceanic, *17,* 44–46, *45,*
      248, *249*
    ophiolites and, 488
    volcanoes near, *125,* 126, 128
Cook, James, 389
Cooksonia, 565, *565*
Copper
    bonding/compounds and, 63
    importance of, 60
    metamorphism and, 187, *195*
    as mineral resource, 534
    ocean resources and, 409
    percentage imported, *79*
    plate tectonics and, 52, 53, *53*
    plutons and, 86
    porphyry, 591
    as resource, 79
    submarine hydrothermal vents and, 33

at Sudbury Basin, 498, 499
weathering and, 153
Copper Ridge, Utah, *469*
Coprolite, *170*
Coquina, 163, *163*
Coral reefs. *See* Reefs
Corals
Cenozoic history and, 632
colonial, *550*
evolution and, 467, 468
Paleozoic history and, 514, 515, *529, 545*
Precambrian history and, *497*
rugose, 550, 551, 553
tabulate, 550, 551, 553
Cordaites, 568, *568*
Cordilleran igneous activity, 618–620
Cordilleran mobile belt
in Cambrian paleogeography, *512*
defined, 505
in Devonian paleogeography, *521*
as major mobile belt, *505*
in Mesozoic history, *580, 581, 582, 583,
585, 587*
in Mississippian paleogeography, *523*
in Ordovician paleogeography, *514*
as Paleozoic mobile belt, 532, *532*
in Pennsylvanian paleogeography, *524*
in Permian paleogeography, *528*
in Silurian paleogeogrphay, *516*
Cordilleran orogeny, 581, 584–585, *587,* 617
Core
composition of, 224–225
defined, 13, *13*
density of, 224–225
as Earth system, 4
Hadean and, 476, 478
Coriolis effect, 374, 390
Correlating rock units, 428–432, *429*
Correlation, 428–429, *430, 431,* 431–432
Corundum, 75, 195, *665*
Corythosaurus, 599
Cosmology, 10
Country rock
batholiths and, 103, *104*
defined, 92, *92, 93*
metamorphism and, *186,* 187
plutons and, 99, *101*
Covalent bonding, 62, 63, *64*
Crater, 114–115, 498
Crater Glacier, Iceland, *343,* 349
Crater Lake, Oregon, 115, *116,* 125
Cratonic sequence, 509, 511, *511,* 581
Cratonic uplift, 526
Cratons
Archean, 484, *485,* 486
defined, 504, 526
in Mesozoic Era, 580, 581
in Paleozoic history, 504–505, *505,*
509, 511–514, *513,* 520–524, 526,
531, 532, 546
in Precambrian history, 478–480, *479,*
482, 484, 486, 488

Creek, 290
Creep, *266, 276,* 276–277, 657
Crest, 395, *395,* 396
Cretaceous Interior Seaway, 589
Cretaceous Period
concretions and, *175*
dinosaurs and, 576
fish during, *555, 558*
geologic time and, *417, 430*
Gulf Coastal region in, 583
marine communities and, 551
meteorite in, 553
paleogeography during, *577, 578, 582*
Pangaea and, 576, 578, *579,* 580
petroleum and, 52
plants during, *564*
polar wandering in, *36*
rifting in, 583
Crevice Cave, Missouri, 444
Crinoidal limestones, 522
Crinoids, *545, 550, 551, 551, 552, 552,* 632
Critical thinking, 656
Crocidolite, 183
Crocodiles, 576, 600
Crocodilians, *595*
Cro-Magnons, 648, *649*
Crop rotation, 153
Cross-bedding, 167, *167, 172,* 174, *371,*
442, 480, 485
Cross-dating, 436
Crossopterygians, 558, *559,* 560
Crude oil, 80
Crust
Archean, 480
in Basin and Range Province, 620
batholiths and, *104*
Cenozoic orogenic belts and, 615
classifying igneous rocks and, 95
continental (*See* Continental crust)
defined, 13, *13,* 223, *223*
deformation and, 237, 245, 247,
252–254
elastic strain and, 235
elements in, 66, *67,* 68
geologic time and, 416
Hadean and, 478
isostatic rebound and, 254, *254*
lithostatic pressure and, 184, *184*
magma in, 86, 87, 88
minerals in, 76, 87
oceanic (*See* Oceanic crust)
Precambrian, 480
principle of isostasy and, 252–253, *253*
sedimentary rocks and, 158
temperature of, 227
weathering and, 136
Crustal evolution, 482
Crustal movement, 526
Cryogenian Period, *477*
Crystal form, 71, 664
Crystal settling, 92, *92,* 95
Crystalline rock, 158

Crystalline solid, 60, 64, *162*
Crystalline texture, 163
Crystals, 64, 71, *71*
Cubic crystals, 71, *71*
Curie, Marie, 432
Curie, Pierre, 432
Curie point, 36
Current ripple marks, 168, *168*
Currents
longshore, 396, 400, *401,* 402, 409
nearshore, 393, 396–397, 401
ocean/oceanic, 28, *388*
ooids and, 172
rip, 396, 397, *397,* 400
turbidity, 167, *168,* 480
Curtains, at Mammoth Cave, *312,* 314
Custer, George Armstrong, *73*
Custer State Park, South Dakota, *84*
Cut bank, 294
Cut-and-fill method, 279, *280*
Cutin, 565
"Cutting," 77
Cuvier, Georges, 450
Cyanobacteria, 493, 494
Cycads, 593
Cycles, of Earth, 656, 657
Cyclothems, *502,* 524–526, *525*
Cynodonts, 602–604, *603,* 636
Cynognathus, 32

da Vinci, Leonardo, 6, 467
Dacite, 95
Daggett Rock, Maine, 355
Dams, 288–289, 298, *302,* 304, 403, 409
Darwin, Charles
evolution and, 17, 389, 418, 450–454,
451, 452, 456, 457, 459, 463
*On the Origin of Species by Means of
Natural Selection* by, 21, 451, 453,
465, 542
Daughter atoms, 433, 434, *434,* 435, *435*
Daughter element, 433
Davis, Jefferson, 100
Dawn (cruise ship), 396
Dawsonites, 566
de la Beche, Henry Thomas, 605
Dead Horse State Park, Utah, *308*
Deadwood Formation, Wyoming, 555
Death Valley, California, *162,* 298, *300,
368, 369, 370, 379, 380,* 630, 632
Death Valley National Park, California, *380*
Debris avalanche, 277, *278*
Debris flow, *266,* 271–272, *273,* 277
Deception Island, 125
Deciduous forest soil, *147*
Dedham Granodiorite, 100
Deep-focus earthquakes, 202, 203, *204*
Deep-ocean circulation, 390, 391
Deep-Sea Drilling Project, 39
Deep-Sea sediment, 44, 46, *46,* 656
Deepwater Horizon oil spill, 410–411, *411*
Deer, 639

Deflation, 368
Deflation hollows, 368, *369*
Deformation. *See also* Mountain building
    Archean and, 482
    Cenozoic history and, 615, 617, 628, *629*
    continents and, 478
    defined, 234, *235*
    dip and, 236–237, *237, 238, 239, 240, 242,* 243
    Earth's crust and, 237, 245, 247, 252–254
    effects of, *234*
    engineering/geology and, 246–247
    folded rock layers and (*See* Folded rock layers)
    geologic structures and (*See* Geologic structures)
    joints and, *241,* 241–242
    in Mesozoic history, 584, 585, 589
    orientation of, 236–237, *237*
    origin of mountains and, 245–252
    orogeny and, 487
    Pleistocene, *629*
    stress/strain and, *235,* 235–236, *236*
    strike and, 236–237, *237*
Deinonychus, 596, 598
Delaware Basin, Texas, 534
Delicate Arch, *150*
Deltas, 290, 296, *297, 402,* 404
Denali National Park, Alaska, *294*
Dendritic drainage, 302, *303*
Density
    body waves and, 206
    of core, 224–225
    defined, 74–75
    of Earth, 223, *223*
    of mantle, 225–226
Denver Museum of Nature and Science, 442
Deoxyribonucleic acid (DNA), 457, 492
Department of Commerce, 324
Deposition
    carbonate, 513, 514, 522–523, 528, 628
    Cenozoic, *627*
    defined, 158
    deformation and, *254*
    geologic time and, *423, 424, 425, 426, 427,* 431
    glaciers and, 340, 341
    groundwater, *320,* 320–323
    by running water, 293–298, *304,* 305, 306
    of Sauk Sequence, 513
    sediment/sedimentary rocks and, 158–159, 160, 165, 166, *167,* 172, 174
    along shorelines, 398–403
    sites of, *159*
    of Tippecanoe Sequence, 514
Depositional coasts, 404
Depositional environment
    defined, 159, *160,* 293
    determining, 172–173
    fossils and, 169
    Proterozoic rocks and, 488
    of sandstone, 161

of sedimentary rocks, 159, *160,* 165, 167, 169, 172–173
Deranged drainage, 303, *303*
Descartes, René, 11
Desert landforms, 378–382
Desert pavement, 368, *369*
Desert soil, *147,* 376–377
Desertification, 366
Deserts
    characteristics of, 376–377
    defined, 374
    distribution of, 374–375
    dunes in (*See* Dunes)
    expansion of, 366
    groundwater and, 366, 377, 381
    loess in, 368, 372
    mass wasting and, 377
    precipitation of, *375,* 376, 378
    rain-shadow, 375, *375*
    soils and, *147,* 376–377
    streams and, 377
    temperature of, 376, 378
    vegetation of, 366, 376, *376,* 377
    weathering and, 376–377, 382
    wind and (*See* Wind)
    windmills/wind power and, *378,* 378–379
Detrital sediment, 158, 160
Detrital sedimentary rocks, *159,* 161, *161, 162*
Devastated Area, *623*
Devil's Postpile National Monument, California, *115, 295, 352*
Devil's Slide, California, *247*
Devil's Tower, Wyoming, *102*
Devonian marine community, *551,* 551–552, *557*
Devonian paleogeography, *521*
Devonian Period
    amphibians during, 560, *560,* 561
    black shales in, 521
    continental drift and, *30*
    evolution and, *470*
    fish during, 555, *555, 556, 557,* 558
    geologic time and, *417, 430*
    Paleozoic history and, 504, *508,* 509
    plants during, 564, *564,* 565–566, *566, 566*
    polar wandering in, *36*
    reef development in, 520
    Tippecanoe Sequence and, 516, 520
Devonian Reef complex, Canada, *522*
Diamonds
    hardness of, *75*
    minerals and, *65,* 66, 70, 74, *74, 75,* 76, *77, 77*
    sedimentary rocks and, 173
    in South Africa, 591
    weathering and, 137, 153
Diatomite, 631
Diatomite layers, 263
Diatoms, 544, 593, *593,* 631, 632, *633*

Diatryma, 634, 634, 638
Dickens, Charles, 605
Dicynodon, 564
Diseases, climate change and, 9
Differential erosion, 137, *150,* 245, 368
Differential pressure, 185, *185, 188*
Differential weathering, 137, *150,* 245, *381*
Dikes
    basaltic, 46
    defined, 99
    example of, *101*
    in Mesozoic history, 582
    rifting and, 44
    sheeted, 46, *46*
Dimetrodon, 562, 563
Dinoflagellates, 544, 593, *593,* 632
Dinosaur bones, 582, *589*
Dinosaur Museum, New Mexico, *169*
Dinosaur National Monument, Utah, *377, 589*
Dinosaurs
    archosaurs and, 594–595
    cladogram of, *596*
    defined, 594
    evolution and, 462, 463, 467, 468, *468, 469*
    geologic time and, *417,* 443
    in Mesozoic history, 594–599, *596*
    origin of, 594–595
    ornithischian, 596–597
    in Precambrian history, 498
    relationships among, *596*
    as reptiles, 594
    Saurischian, 595–596
    sediment/sedimentary rocks and, 158, *170,* 174
    warm-blooded, 597
Diorite, 95, *95,* 96, *96,* 97, 100, 498
Diorite pegmatites, 97
Dip, 236–237, *237, 238, 239, 240, 242,* 243
Dip needle, defined, *35*
Diplodocus, 596
Dipolar magnetic field, 35
Dip-slip faults, *243,* 243–244, *244*
Disappearing streams, 321, *322*
Discharge, running water and, 290–292, *291*
Discordant ages, 435
Disconformity, 420, *424*
Discontinuity, 223, 225, *226*
Discontinuous branch, 89, *89*
Discordant pluton, 99, *101,* 102
Dissolved load, defined, 292, *293*
Distributary channels, defined, 296
Divergent evolution, 459–461, *460*
Divergent plate boundaries. *See also* Spreading ridges
    defined, 40, 44
    earthquakes and, 203
    history of, *41,* 44
    igneous rocks and, 86, 91
    metamorphism and, 194
    resources from, 53, *53*
    volcanoes near, *125,* 127–128

Diversification
  in Cambrian explosion, 543
  in Cenozoic history, 635, 636, 637
  of fusulinids, 552–553
  of jawed fish, 558
  of life, 656
  of mammals, *635,* 635–636
  of plants, 565
  of reptiles, 593–597, 600–601
Diversity
  Phanerozoic, *553*
  plants and, 565
Divide, defined, 302
DNA. *See* Deoxyribonucleic acid (DNA)
Dogs, *640*
Dolichosoma, 562
Dolomite
  in mineral identification table, *665*
  minerals and, 70, *74, 75, 75,* 77
  nonfoliated metamorphic rocks and, 192
  sedimentary rocks and, *162,* 163
Dolostone
  carbonate rocks and, 485
  in Grand Canyon, 488
  marble from, 182, 192, *193*
  metamorphic facies and, 194
  as mineral resource, 534
  minerals and, 70
  permeability of, 315
  porosity of, 314, 315
  sedimentary rocks and, 153, 163
  Tippecanoe Sequence and, 515
Domes, 138, *138,* 241, *241,* 520, 526, 576, 583
Doppler effect, 10, *10*
Dormant volcanoes, 110
Double Arch, *151*
Double refraction, defined, 75
Downcutting, 305
Downwelling, 391
Dragline, *502*
Drainage, 278–279, 302, *303,* 355
Drainage basin, 302, *303,* 305, 306
Drainage patterns, 302, *303*
Drainage systems, 302–305
Drinking water, arsenic in, 329
Drip curtain, 323, *324*
Dripstone, 322
Dropstones, 359, *359*
Droughts, 366
Drowned coasts, 404. *See also*
    Submergent coasts
Drumlin fields, 357
Drumlins, 355, 357–358, 630
du Toit, Alexander, 29, 31
Ductile rocks, 236
Dune migration, 369, *370*
Dune types, 369–372
Dunedoo Formation, Australia, *28*
Dunes
  barchan, 369, 370, *371*
  defined, 368, 369
  formation of, 369

geologic time and, 442
  longitudinal, 369, 370–371, *372*
  migration of, 369, *370*
  oceans and, *400*
  Pangaea and, 579
  parabolic, 369, 371, *373*
  star, 371, *373*
  transverse, 369, 371, *372*
Dunkleosteus, 557, 557
Durand, Asher Brown, *6*
Dust Bowl, 152, *152*
Dwarf planet, 11, *11, 14*
Dynamic equilibrium, 261
Dynamic metamorphism, 185, 187–188, *188*

Earth
  age of, 416–417, 491
  appearance of, *477*
  Archean history, 480–482
  atmospheric circulation of, *374*
  climatic changes on, 340
  composition of, *223*
  core of, 224–225
  cross section of, *13*
  density of, 223, *223*
  as dynamic/evolving planet, *11,* 12–13,
    16–17
  highest peak on, *233*
  hydrosphere and, 388
  interior of, *223,* 223–227
  internal heat of, 227
  magnetic field of, 35, *35,* 37, *38*
  origin of, 4, *5,* 477, 656
  origin of universe and, 9–11
  ozone depletion and, 122
  running water and, 286, *286,* 287, 290
  satellite-based image of, *2*
  snowball, 488–489
  solar system and, 9–12, *14,* 14–15, *15*
  subsystems of, 4, *5,* 20, 22
  system concept of, 4
  tectonics of, *42, 43*
  theme regarding, 4
  tides and, 393, 394, *394*
Earthflow, *266,* 272, *274,* 277, *277*
Earthquake control, 219–220, *221*
Earthquake prediction programs, 219
Earthquake prediction/precursors, 217,
    219, *219*
Earthquakes
  in Alaska, *200,* 211
  in Armenia, *200*
  in California, *200,* 201, *201,* 210, 211,
    213, *214*
  in Chile, *200,* 211
  in China, *200,* 217, *659*
  in circum-Pacific belt, 203, *204,* 214, *218*
  danger of, 406
  deep-focus, 202, 203, *204*
  defined, 200
  distance from, *208*
  distribution of, 203, 205

Earth's core and, 224–225
  Earth's interior and, *223,* 223–227
  Earth's mantle and, 225–226
  effects of, 212–213, 216–217
  elastic rebound theory and, 201, *201*
  epicenter of (*See* Epicenter)
  fire in, 213
  focus of, 202–203, *203*
  as geologic hazard, 657, *659*
  Good Friday, 275
  ground failure in, 217, *218*
  ground shaking in, 212–213, *213*
  in Haiti, *198,* 200, *200, 203,* 657
  in India, *200*
  in Indonesia, *200,* 212
  intensity of, 208, *210,* 211
  intermediate-focus, 202, 203, *204*
  in Iran, *200*
  in Japan, *26,* 28, 200, *200,* 203, *205,* 212,
    213, 657
  locations of, 203, 205, 207
  Loma Prieta, 203, 211, 213, *219*
  magnitude of, 208, *211,* 211–212
  mass wasting and, 264
  in Mexico, *200*
  in New Zealand, *200,* 657
  Northridge, 205, 246
  overpopulation and, 657
  in Pakistan, 205
  paleoseismology and, 222
  in Peru, 217, 277, *278*
  population and, 658
  in Portugal, *200,* 217
  preparing for, *220*
  San Andreas fault and, *48*
  San Fransisco, *200,* 201, *201,* 210, 211,
    211, 213, *214*
  seismic tomography and, 226–227
  seismic waves and, 202, 205–206, *206, 207*
  seismology and, 202–203
  shallow-focus, 202, 203, *204*
  significant, *200*
  strength of, 208, 210–211
  Sumatra, 212
  tsunami and, 200, 213, *216,* 216–217
  in Turkey, *200,* 213
East African Rift system, 128
East African Rift Valley, 40, *41, 44*
East Pacific Rise, 33, *493,* 615
Ebb tide, 393, 409
Echinoderms, 555
Echinoids, 592, 632
Echo sounder, 389
Ecosystems, 366, 544–546, *545,* 554, 558,
    561, 632, 656, 657
Ectasian Period, *477*
Ectotherms, 597
Edaphosaurus, 562, 563
Ediacaran fauna, *484, 496,* 496–497,
    542–543
Ediacaran Period, *477*
EEZ. *See* Exclusive Economic Zone (EEZ)

Einstein, Albert, 10
Elastic rebound theory, 201, *201*
Elastic strain, 235, 236
Eldfell, as cinder cone, 118, *118*
Eldgja fissure, Iceland, 99
Electricity, 6, 145, 286, 288, *288, 289, 302,*
    335, *335*
  ocean resources and, 409
  windmills and, 379
Electromagnetic force, Big Bang and, 11
Electron capture, 432, *432*
Electron shells, defined, 62
Electrons, defined, 62, 432
Elements
  atoms and, 61–64
  daughter, 433
  defined, 61
  in Earth's crust, 66, *67*
  geologic time and, 432, 433
  native, *67, 70, 73*
  parent, 433
  periodic table of, *61*
Elephant birds, 641
Elephants, 636, 637, 639
Elevations, 661
Elizabeth II, Queen, 77
Elk Point Basin, Canada, 534
Ellesmere orogeny, 509
Eluviation, 144
Embryology, 463, 464
Emerald, *76, 77*
Emergent coasts, 404–405, *405*
Emperor Seamount-Hawaiian Island chain,
    48, *49, 91*
Enchanted Rock State Natural Area,
    Texas, 100
End moraines, 355, 357, *357*
Endoskeleton, 543
Endosymbiosis, 495–496, *496*
Endothermic, defined, 563
Endotherms, 597
Energy
  earthquakes and, 200, 201, 202, 205, 211
  fermentation and, 494
  geologic time and, 432
  geothermal, 334–335, *335*
  kinetic, 292
  for life, 492
  nuclear, *288*
  potential, 292
  running water and, 286, 289, 292, 305
  solar, 360, 374
  wave, 388, 389, 401, 403
Energy resources, 6, 7, 60, 78, 80,
    534, 657
Engineering, geology and, 246, *247*
English-metric conversion chart, *660*
Environment
  depositional (*See* Depositional
    environment)
  deserts and, 366
  humans and, 656

of metamorphism, *186*
  steppe, 634
Environmental geology, 7–9, 656–659
Environmental Protection Agency (EPA),
    183, 329
Environmental refugees, 366
Eoarchean Era, *477, 481,* 488, 491
Eocene Epoch, *417,* 443, 463, *616*
Eomaia, 604
Eon, defined, 440
Eoraptor, 595
EPA. *See* Environmental Protection
    Agency (EPA)
Epeiric seas, 504, 509, 513, 522, 523, *531,*
    542, 578, 625, 627
Epicenter
  defined, 202–203, *203*
  locating earthquakes and, *204,* 207, *209,*
    211, *216*
  plate boundaries and, *204*
Epidote, 187, 480
Epifauna, 545, *545*
Epiflora, 545, *545*
Epoch, defined, 440
Equisetum, 568, *568*
Equus, 637
Era, defined, 440
Erosion
  in Arches National Park, *150–151*
  in Canadian shield, *479,* 480
  Cenozoic history and, 617, 626, *626,* 627
  cratons and, 523
  cyclothem and, 525
  defined, 136–137
  deformation and, 245, 254, *254*
  deserts and, 366, 382
  differential, 137, *150,* 245, 368
  Dust Bowl and, 152, *152*
  geologic time and, 417, 420, *424,* 425,
    425, 426, 427
  glaciers and, 158, 340, 341, 350–355
  groundwater, *320,* 320–323
  Hadean and, 476
  headward, 305, *307*
  lateral, 305, *307*
  mass wasting and, *261, 263,* 266, 277
  in Mesozoic Era, 582
  Paleozoic paleogeography and, 509
  in Palo Duro Canyon State Park, *156*
  rill, 152, *153*
  running water and, *156,* 290, 292–293,
    304, 304–306, *307*
  Sauk Sequence and, 512
  sediment and, 158
  sheet, 152, 290
  along shorelines, *398,* 398–403, *399,*
    408–409
  soil formation and, 152
  water, 153
  wind, 367–368
Erosional coasts, 404
Erratics, 630

Erta Ale, Ethiopia, 125, 128
Eryops, 562
Esker deposits, 632
Eskers, 358, *358*
Estes Park, Colorado, *273*
Estuaries, 405, 409
Eukaryote, *484, 495, 495*
Eukaryotic cells, *494,* 494–496, *496*
Eurasia, *577, 579,* 615
Eurasian plate, 39, *39, 49, 125, 251*
Eurypterids, 551, *551,* 558
Eusthenopteron, 558, *559,* 560
Evans, David M., 219
Evaporation
  deserts and, 377
  geologic time and, 442
  glaciers and, *347*
  running water and, 287, 292
  seawater and, 409, 517
Evaporite deposits, 526, 576, 579
Evaporite mineral resources, 534
Evaporite sedimentation, *520*
Evaporites
  Cenozoic history and, 631, 632
  defined, 164
  evolution and, 442
  Mesozoic history and, *578,* 583
  Paleozoic history and, 505, 509,
    515–517, *517,* 520, 526, 534
  Pangaea and, *578*
  sedimentary rocks and, *162,* 164,
    172–173
  Silurian, 534
  Zeichstein, 534
Evolution
  artificial selection in, 452, *452, 454,*
    454–455
  of atmosphere, 489–491, *490*
  biological evidence supporting,
    465–467
  of biosphere, 656
  cladistics and, 461–462, 465
  cladograms and, 461–462, *462, 465, 470*
  convergent, 459–461, *461*
  of corn, *454*
  crustal, 482
  Darwin on (*See* Darwin, Charles)
  divergent, 459–461, *460*
  evidence supporting, 465–467,
    468–469
  extinctions and, 463
  fossil record and, *470*
  fossils and, 454–455, *455,* 467–468,
    468, 469
  genetics and, 456–457
  hominid, 643
  of horses, 463, *463*
  human, 614
  of hydrosphere, 489, 491
  Lamarck on, *448,* 450–451, 453, 458
  of Laurentia, *485*
  Lysenko on, 453

of mammals, 443, 450, 456, 459, *460*, 461, 463–468, *466, 470*, 602–604
marine invertebrates and, 542
Mendel on, 456–457
mosaic, 462–463, 565, 602, 604
natural selection in (*See* Natural selection)
of North America, 509, 511
organic (*See* Organic evolution)
Paleozoic, 509, 511, 542
parallel, 459–461, *461*
plant, *564*, 564–568, *566*
in Precambrian history, 476
primate, 641–642
rate of, 458–459
of reptiles, 461, 465, *466, 467*, 468, *470*, 542, 562–563
of solar system, 11–12
speciation and, 458–459, *460*
theory of, 450, 458, 463–470, *464*, 542
variation in, 458, 463, 467
vertebrate, 554–555
Wallace on, 451–453, 454, 456, 457
why study, 450
Evolutionary novelties, 461, 462
Evolutionary relationships, *451, 455, 563*
Evolutionary theory. *See* Theory of evolution
Evolutionary trends, 462–463, *463*
Excavations, 261, *262*
Exclusive Economic Zone (EEZ), 409
Exfoliation domes, 138, *138*
Exit Glacier, Alaska, *356, 357*
Exoskeleton, 543
Expansive soils, 148–149, 152–153, 657
Explosions, 264
Extinct volcanoes, 110
Extinctions
  background, 463
  of Ediacaran fauna, 543
  mass, 463, 542, 549–554, 563, 591, 592, 606–607
  Mesozoic, *606*
  Pleistocene, 649–650
  of pterosaurs, 600
Extrusive igneous rocks, 17, *18, 19*, 20, *20*, 87, 93
Eyjafjallajökull, Iceland, *92, 108*, 110, 128, *343*

Falls, *266*, 271
Famines, 340
Farallon plate, 583, 584, 585, 617, *619*, 624, 628
Fault breccia, 242, *242*
Fault plane, 242, *242, 243, 243*
Fault scarp, 242, *242, 243*
Faulting, *173*, 200, *236*, 526, 580, 620. *See also* Folding
Faults. *See also* San Andreas fault, California
  defined, 242, *242*
  dip-slip, *243*, 243–244, *244*

joints and, 242
left-lateral strike-slip, 244
living along, 406
normal, 243, *243*, 244, *244*, 245, *248*
oblique-slip, *243*, 244–245, *245*
reverse, *243*, 244, *244*
rifting and, 44
right-lateral strike-slip, 244
strike-slip, 243, *243*, 244, *245*
thrust, *243*, 244, 248, 480, *588*
transform, 47, *47*, 244
types of, *243*
Feather duster, *497*
Feldspars
  defined, 69
  in Earth's crust, 87
  graywacke and, 480
  metamorphism and, 189, 191
  plagioclase (*See* Plagioclase feldspars)
  potassium (*See* Potassium feldspars)
  as resource, 79
  weathering and, *136*, 141, 153
Felsic igneous rocks, 97
Felsic magma
  Bowen's reaction series and, 89, *89*
  change and, 92, *93*
  classifying igneous rocks and, 95, *96*, 97
  composition and, 95
  defined, 87, *87*, 88, *88*
  granite and, 103
  metamorphism and, 185
  spreading ridges and, *90*
  subduction zones and, 91, *91*
  volcanism and, 111, 119, 127
Fermentation, 494
Ferromagnesian silicates
  Bowen's reaction series and, 89, *89*, 90
  classifying igneous rocks and, 95, *96, 96*
  in Earth's crust, 87
  magma and, 87, 90, 91, 92, *92*
  mantle and, 226
  mineral formation and, 78
  minerals and, 69, 70, 74, 75, *75*, 78
  weathering and, *142*
Fertilization, 566, *567*
Fertilizers, 70, 409, 520
Fetch, 396
Fiords, 352, *353*
Fire, 213
Firn, 345, *347*
Firn limit, 346, *347*
Fish
  to amphibians, 560
  bony, *470*, 557–558
  cartilaginous, *470, 555, 557, 557*, 558
  classification of, *556*
  evolution and, 459, 465, *470*
  jawed, *470, 555, 557*, 558
  jawless, *470, 555, 556*
  lobe-finned, *555, 558, 558, 559, 561*
  in Mesozoic history, *592, 592*
  at Messel Pit Fossil Site, 638

in Paleozoic history, 542, *545, 555*, 555–559
  ray-finned, *555, 556, 558, 558*
Fishapod, 560, *561*
Fission-track dating, 436, *436*
Fissure eruptions, defined, 122–124, *124*
Fixity of species, 450
Flash floods, *300*
Flathead Lake, *631*
Flint, 164
Flood control, *288*, 288–289, 298–302
Flood tide, 393, 409
Floodplain deposits, 296, *296*, 632
Floodplains
  defined, 296, *296*
  meandering river and, *294*
  stream terraces and, 306, *308*
Floods/flooding
  causes of, 658
  coastal, *300*, 389, 405–406, 407, 657
  controlling, 298–302
  as environmental issue, 657
  examples of, *286*, 289, 294, 296, 300–301, 631
  flash, *300*
  geologic time and, 418
  predicting, 298–302
Floodwalls, 299, *302*, 406
Floodwaters, *284*
Floodways, 299, *302*
Florissant Fossil Beds National Monument, Colorado, *171, 455*, 632
Flows
  ash, 113
  channel, 290
  debris, *266*, 272–272, *273*, 277
  fluid, 290
  glaciers and, 341
  laminar, 290
  lava (*See* Lava flows)
  mass wasting and, *266*, 271–277
  plastic, 345, *345, 347*
  sheet, 290, 292
  turbulent, 290, *290*
Flue-gas desulfurization, 144
Fluid activity, defined, 185
Fluid flow, 290
Fluoride, 331
Fluorite, *71, 74, 75*, 665
Flying reptiles, 576, 600, *600*
Focus
  defined, 202–203, *203*
  locating earthquakes and, 207, *208*
Folded rock layers
  anticlines as, 237–238, *239*, 240, *240*, 241
  basins as, 241, *241*
  domes as, 241, *241*
  monoclines as, 237
  plunging folds as, 240, *240*
  synclines as, 237–238, *238, 239*, 240, *240*, 241
Folding, *173*, 235, 236, *236*, 237, 242, 410. *See also* Fracturing

Folds
defined, 237
inclined, 238, *239*
joints and, 242
nonplunging, 240
overturned, 238, *239*
plunging, 240, *240*
recumbent, 238, *239*
types of, 237–238, *238, 239, 240,*
240–241, 241
upright, 238
Foliated metamorphic rocks, 18, *19,*
189–191, *190*
Foliated texture, 189, *189*
Foliation, 18, 191
Footwall block, *242,* 243, *243,* 244, *244*
Foraminifera, 545, 552, 593, 632, *633,* 634
Forceful injection, 103
Foreset beds, 296
Foreshore, 400, *400*
Formation, defined, 438, *439*
Fort Union Formation, Montana, *614*
Fossil assemblages, 420
Fossil content, 439
Fossil evidence
for continental drift, 31, *32,* 36
for evolution, 450
for limbs, 560
Fossil fuels
acid rain and, 144
burning of, 8
global warming and, 348, 657
windmills and, 379
Fossil lineages, 462
Fossil record
amphibians and, 562
of angiosperms, 593
biosphere and, 656
of birds, 602
for Cenozoic history, 614, 635, 637,
642, 643
continental drift and, 31
Darwin and, 21
evolutionary theory and, 464, 465,
467, *470*
extinctions and, 463
macroevolution and, 461
Mesozoic history and, 593, 596,
597, 602
missing links and, 450
multicelled organisms and, 496
Paleozoic history and, 542, 543, 544
photosynthesis and, 491
plants and, 565, 566
punctuated equilibrium and, 459
reptiles and, 562
Fossiliferous limestone, *163*
Fossilization, *170–171*
Fossils
amphibians and, 560
Archean, *493,* 494
of Bearpaw Formation, *440*

body, 467, *468*
Burgess Shale, 547, *547,* 550
Cambrian, 542, 543, 550
in Cenozoic history, 637, *642,* 643
defined, 21, 169, 467
in England, 504
evolution and, 454–455, *455,* 463,
467–468, *468, 469*
geologic time and, 418, 420, 429
glaciation and, 30
guide, 429, *431,* 592
of horses, 637
index, 429
in Messel Pit Fossil Site, 638
Neoproterozoic, *497*
paleontology and, 450
plant, 28, *28*
in Precambrian history, 476
Proterozoic, *495, 497*
of reptiles, 562
seafloor spreading and, 39
sedimentary rocks and, 169, *169, 172*
trace, 467, *468, 469,* 546
transitional, 468
wormlike, *484*
Fractional scale, 661
Fracture, defined, 74, 236
Fracturing, *234, 235,* 236, 241, *315,* 410, 617
Fragmental texture, 94, *94, 98*
Franciscan Complex, 584, 585, *586*
Frankfort Shale, New York, *548*
Franklin mobile belt, *505*
Freshwater, plants and, 564, 565
Fringing reefs, 392, *392*
Frogs, 561
Front Range, 442
Frost action, 137, 270
Frost Fair, *342*
Frost wedging, 137, *137*
Fukushima Prefecture, Japan, *26*
Fumeroles, 331
Fungi, 565
Fusulinids, 552–553

Gabbro
on continents, 478
in Earth's crust, 13, 32, 46, *46,* 252
geologic time and, *419*
igneous rocks and, 95, *95,* 96, *96,* 97,
100
oceanic crust and, 227
oceanic ridges and, 33
at Sudbury Basin, 498
Galaxies, Big Bang and, 10
Galena, 65, 70, *70,* 71, *74,* 75, 195, *195, 664.*
See also Lead
Galilean moons, *43*
Galileo, 43
Gamete-bearing plant, 566
Gametophyte, 566, *567*
Ganges-Brahmaputra delta, *297*
Garden of the Gods, Colorado, 526, *527*

Garnet-mica schist, 191, *191*
Garnets, 71, 77, 182, *185, 188,* 189, 193,
195, *665*
Gases
in atmosphere, 489, 490, *490,* 492, 493
ATP and, 494
defined, 60
greenhouse, 348, 379, 407, 634
noble, 62, 63
running water and, 290
volcanic, 110–112, 122–123
weathering and, 140
Gastropods, 551, *551,* 592, 632
Gemstones, 76–77. *See also* Specific
gemstones
General Land Office, 662
General Land Office Grid System, 662,
663, *663*
Genes, 451, 456, 543
Genetic drift, 458
Genetic research, 453
Genetics, birth of, 456–457
Geode, *58,* 174–175, *175*
Geographic distribution, 463
Geologic hazards, 246, 657, 658, *659*
Geologic map, 504
Geologic resources, 6
Geologic structures
defined, 234, 237
faults as, *242,* 242–245, *244, 245*
folded rock layers as (*See* Folded
rock layers)
joints as, *241,* 241–242
Geologic time
absolute dating methods and, 432–437
Archean Eon and, 480
climate change and, 440–444, *441, 444*
correlating rock units and, 428–432
early concepts of, 416–417
Earth's age and, 416–417
frame of reference for, 476
Hutton and, 417–418
Ice Age and, 340
importance of, 416
measuring, 416
Precambrian history and, 476
recognition of, 417–418
relative dating methods and, 418–428
represented by 24-hour clock, *476*
soil degradation and, 149
stratigraphy/stratigraphic terminology
and, 437–440
uniformitarianism and, 21–22
Geologic time scale, 416, *417,* 418, 437, *477*
Geologic time scale, defined, 9, 21, *22*
Geological Survey of India, 224
Geology
affecting everyday lives, 6–7
art and, *6*
benefits of, 22, 659
defined, 4
earthquake intensity and, *210*

engineering and, 246, *247*
of England, 504
environmental, 7–9, 656–659
father of, 417
global issues regarding, 7–9
historical, 4
human experience and, 6
physical, 4
slope stability and, 264
theories and, 4–6
Geometry, 167
Georgia Warm Springs Foundation, 331
Geothermal energy, 334–335, *335*
Geothermal gradient, 90, 227, 331
Geothermal heat, 347
Geothermal power plants, 288, *335*
Germination, *567*
Gettysburg National Military Park,
    Pennsylvania, 144
Geyserite, 334
Geysers, 331, 332, *333,* 334, *335*
Geysers, The, California, 335, *335*
Gibson Desert, Australia, *372*
Gigantosaurus, 598
Gigjökull, Iceland, *343*
Gills, 560
Ginkgos, 593
Glacial budget, 345–347, *347,* 348, 350
Glacial deposits, 36, 355–359, *359,* 479,
    488–489, 626
Glacial drift, 355
Glacial erratics, 355, *356*
Glacial evidence, 30–31, *31*
Glacial ice, 344, *344,* 345, 346, 355, 357, 489
Glacial lakes, 359, *359*
Glacial movement, *31*
Glacial polish, 351, *352*
Glacial striations, *31,* 351, *352*
Glacial surges, 347, 350
Glacial troughs, 351–352, *353, 353,* 354, *354*
Glacier National Park, Montana, *169,* 346,
    348, 349, *487*
Glacier Peak, Washington, 349, *349*
Glaciers/glaciation. *See also* Ice
    Aletsch, *338, 343*
    alpine, 340
    Byron, *346*
    in Cenozoic history, 628–631
    coasts and, 404
    continental (*See* Continental glaciers/
        glaciation)
    Crater, *343,* 349
    defined, 340, 344
    distribution of, 345
    effects of, 629–631
    erosion by, 158, 340, 341, 350–355
    Exit, *356, 357*
    global warming and, 340, 346, 348–349
    Grinnel, *348*
    hydrologic cycle and, 344
    ice ages and, *338,* 340, 349, 359–360
    Illinoian, 630, *630*

Kansan, 630, *630*
in Karakoram Range, *232, 233*
kinds of, 340–341, 344
Kutiah, 350
Little Ice Age and, 340, *342,* 342–343,
    343, 348, 348–350, 359, 360
Lowell, 350, *350*
Mendenhall, *352, 356*
mountain, 340
movement of, 344–345, *345,* 347, 349
Nebraskan, 630, *630*
oceans and, *408*
origin of, 344–345
outlet, 347
Paleozoic history and, 504, 509, 551, 553
Pleistocene, 628–631
Precambrian history and, *484,* 486,
    488, *489*
quick clays and, 274
running water and, 290, 303
Salmon, *357, 358*
sediment transport by, 350–355
stagnant, 346
tidewater, 341, *341*
valley (*See* Valley glaciers)
Whitechuck, 349, *349*
Wisconsinan, 630, *630*
Glass Mountains, Texas, *553*
Glassy texture, *98*
Glauconite, 434, 435
Global Seismic Hazard Assessment
    Program, 217, *218*
Global warming
    during Cambrian, 543
    critical thinking and, 656
    deserts and, 366
    as environmental issue, *8,* 8–9,
        656, 657
    geologic time and, 21, 440
    glaciers and, 340, 346, 348–349
    industrialization and, 144, 145
    sea level and, 407
Global weather patterns, 28
Global wind patterns, 372, 374
Globigerinoides fistulosus, 633
Glomar Challenger, 39
Glossopteris flora, 28, *28, 30,* 31, *32,* 568
Gneiss, *19, 188,* 189, *190,* 191, *191, 474,*
    479, 482
Gold
    bonding/compounds and, 62, 63, 66
    Cenozoic history and, 632
    discovery of, *72–73*
    granite and, 102
    metamorphism and, 187
    as mineral resource, 534
    percentage imported, *79*
    plate tectonics and, 52
    in Precambrian rocks, 497
    as resource, 79
    sedimentary rocks and, 173
    at Sudbury Basin, 498

at Sutter's Mill, 576
weathering and, 137, 153
Gold rush, 173, 576, 591
Gondwana
    Cenozoic orogenic belts and, 615
    continental drift and, 28–31, *30, 31, 32*
    cratonic uplift and, 526
    cyclothems and, 526
    defined, 28, *28,* 505, 506
    Hercynian-Alleghenian orogeny and, 532
    Ordovician marine community and, 551
    Ouachita mobile belt and, 533, *533*
    plants and, 568
Goniatites, 592
Good Friday earthquake, 275
Gorges, 305
Gorilla, 644
Gossan, 153
Grabens, 246, *248*
Graded bedding, 167, *168,* 480
Graded streams, *304,* 305
Gradient, 290–292, *291,* 305, 306, 318
Grand Canyon, Arizona
    Archean/Proterozoic rocks in, 480
    inside walls of, *429*
    metamorphic rocks in, *426*
    microfossils in, 495, *495*
    in Paleozoic history, 513, 518–519
    Powell and, *414,* 416
    running water in, 290, 305
    sandstone in, *487*
    sedimentary rocks in, *166,* 172
    springs in, *318*
Grand Canyon National Park, Arizona,
    430, 518, *518,* 519
Grand Canyon Suite (Grofé), 6
Grand Ronde River, Washington, *124*
Grand Teton, Wyoming, 354
Granite
    batholiths and, 103
    at Castle Crags State Park, California, *103*
    classifying igneous rocks and, 95, *95,*
        96–97, *97*
    on continents, 478, *479*
    in Custer State Park, *84*
    defined, 100
    depicted, *19*
    in Earth's crust, 252
    geologic time and, *422*
    gneiss and, 191
    magma and, 86, *86,* 92
    minerals and, 60, *75,* 78
    pink, 100
    plutons and, 100
    in Sierra Nevada, *86*
    as source of gold, 102
    in Teton Range, *474*
    weathering of, *136,* 137, 138, *138,*
        143, *143*
Granite porphyry, 97
Granite-gneiss complexes, 480, 482,
    482, 488

Granodiorite, 97
Granophyre, 498
Granulite facies, 194
Graphic scale, 661
Graphite, 64, 66, 70, 74, 77, 79, 189, 195, 664
Graptolites, 551
Grassland soil, 147
Gravel
    Cenozoic history and, 632
    glaciers and, 340, 355, 358, 359
    gold and, 576
    as mineral resource, 534
    as ocean resource, 409
    running water and, 292, 296, 298
    sedimentary rocks and, 158, 158, 159,
        160, 161, 161, 172
    wave erosion and, 398
Gravimeter, 252
Gravity
    Big Bang and, 11
    mass wasting and, 260, 260, 261, 264
    Newton and, 252
    running water and, 292, 308
    specific, 74–75, 664
    wind erosion and, 368
Gravity anomalies, 252, 253
Graywacke, 480
Grazers, 636
Great Barrier Reef, Australia, 392
Great Dark Spot, 15
Great Flood of 1993, 298, 299, 301
Great Glen fault, Scotland, 244
Great Hot Blast, 623
Great Plains, 625, 627, 631
Great Red Spot, 15
Great Salt Lake, Utah, 630, 631
Great Sand Dunes National Monument,
    Colorado, 372
Great Smokey Mountains, 245
Great Valley Group, 584, 585
Green algae, 560, 565
Green Mountains, Vermont, 245
Green River Formation, 631
Greenhouse effect, 8, 8–9, 286, 656, 657
Greenhouse gases, 348, 379, 407, 634
Greenschist facies, 194
Greenstone, 190, 192
Greenstone belts, 480, 481, 482, 483, 488
Grenville orogeny, 487
Grenville rocks, 487
Grimes Point Archaeological Site, Nevada,
    139, 631
Grinnel Glacier, 348
Grofé, Ferde, 6
Groins, 401, 409
Ground failure, 217, 218
Ground moraine, 355, 357
Ground shaking, 212–213, 213
Groundmass, 93
Groundwater
    arsenic and, 328, 328–329, 329
    artesian systems and, 317, 319, 319–320

defined, 314
demand for, 314
deserts and, 366, 377, 381
hydrologic cycle and, 314
hydrothermal activity and (See
    Hydrothermal activity)
percentage of, 323
permeability of, 314–315
porosity of, 314–315, 315
removal of, 654
rising sea level and, 407
sources of, 4
springs and, 317, 318, 321, 322, 331,
    332, 334
uses of, 314
water table and, 316, 316, 317, 317, 318,
    322–326, 324, 330, 333
water wells and, 317–319
Groundwater arsenic, 328, 328–329, 329
Groundwater contamination, 326–327,
    330, 330, 656
Groundwater deposition, 320, 320–323
Groundwater erosion, 320, 320–323
Groundwater movement, 315, 316–317, 317
Groundwater quality, 331
Groundwater system
    glaciers and, 344
    lowering water table and, 323–325
    modification of, 323–327, 330–331
    saltwater incursion and, 325, 326
    subsidence and, 325–326, 327
Groundwater withdrawal, 324, 326, 327
Groups, defined, 439, 439
Guadalupe Mountains, Texas, 526, 529
Guide fossils, 429, 431, 592
Gulf Coastal Plain, 627–628, 631
Gulf Stream, 391
Gullies, 152, 305, 306
Gunflint Iron Formation, Canada, 493
Guyots, 34, 34, 388
Gymnosperms, 566, 567, 593, 632
Gypsum
    metamorphism and, 185
    mineral formation and, 78
    minerals and, 70, 70, 71, 75, 77, 78
    rock, 162, 164, 164, 173, 390, 534
    sedimentary rocks and, 162, 162, 164,
        164, 173
Gyres, 391

Hadean Eon, 476, 476–478, 481
Half-lives, 432–433, 434, 436
Haliburton Company, 411
Halide minerals, 67, 70, 70, 75, 77
Halite
    in Michigan Basin, 517, 517
    mineral formation and, 78
    in mineral identification table, 665
    minerals and, 62, 63, 66, 71, 71, 74, 75,
        75, 77
    sedimentary rocks and, 162, 164
    solution of, 140, 140–141

Hallet's Cove, Australia, 31
Hallucigenia, 547
Hanging valleys, 352, 354
Hanging wall block, 242, 243, 243, 244
Hansen, Em, 6
Hard water, 331
Hardness, 71, 74, 75, 664
Harmonic tremor, 129
Harney Peak Granite, 84
Hawaiian Volcano Observatory, 112
Hawaiian volcanoes, 117
Hawaiian-type eruptions, 116. See also
    Shield volcanoes
Headlands, 399, 399
Headward erosion, 305, 307
Heat
    contact metamorphism and, 420
    Curie and, 432
    defined, 184
    in Earth, 227, 234, 476, 477, 480, 486,
        490, 491
    geothermal, 347
    Kelvin and, 418
    metamorphic rocks and, 435, 436
    radiogenic, 480, 486, 490
    residual, 490, 491
    seawater and, 493
Helgafel, 118
Helium, 11, 12, 62
Hematite, 70, 75, 141, 160, 195, 195, 489, 664
Hemicyclaspis, 555, 556
Henry Mountains, Utah, 99
Herbivores, 545, 545, 546, 562, 635
Hercynian mobile belt, 509, 533
Hercynian orogeny, 509, 532
Hercynian-Alleghenian orogeny, 532
Hess, Harry, 36, 37
Heterotrophic cells, 494
Hiatus, 420, 423
High Plains aquifer, 323–324, 325
High tide, 393, 394, 400
Highlands, 526
Highway maps, 661
Himalayan mountain system, 232
Himalayan orogen, 615
Himalayas, Asia, 45, 234, 251, 251, 252,
    253, 341, 417, 615
Historical geology, 4
HMS Beagle, 389, 450, 452
Hole-in-the-Wall, 151
Holmes, Arthur, 37
Holocene Epoch, 417, 628–631
Holocene Maximum, 340
Holocene volcanism, 628
Holochroal eyes, 548, 549
Holy Cross Mountains, Poland, 561
Home, Everard, 605
Hominid evolution, 643
Hominids, 639, 642–649, 645
Hominoids, 642, 643
Homo (genus), 646
Homologous structures, 465, 466

Hoodoos, 137, *612*
Hook, defined, 401
Hooker Chemical Compnay, 330
Hoover Dam, 288, *288*
Horizontal circulation, 391
Hornblende, *68,* 74, *74, 75, 95,* 191, *664, 665*
Hornblende porphyry, *96*
Hornblende mica schist, *191*
Hornfels, *190,* 192–193
Horns, 353–354, *355*
Hornworts, 565
Horses, *470,* 636, 637, *640,* 650
Horsts, 246, *248*
Hot spots, 34, *39,* 48, *49,* 91, *91*
Hot springs, 331, *332,* 334, *334*
Hox genes, 543
Hubble, Edwin, 10
Human evolution, 614
Human life. *See* Life
Human lineage, 646–649
Humans, *417,* 639, *643, 644,* 649–650,
      656, 657
Humus, 143, *146, 147,* 148
Hurricane Isabel, 406
Hurricane Katrina, *300,* 406, *407*
Hurricane Rita, 406
Hurricanes, 389, 405–406
Hutton, James, 417–418, 424, *425,* 432
Hydraulic action, 292, *292,* 398, *398*
Hydrocarbon production, 240
Hydrocarbons, 173, 410, 535, 591
Hydroelectric plants, 286, *288,* 288–289
Hydrofluoric acid, 436
Hydrogen
      in atmosphere, 490, *490,* 492
      Big Bang and, 11
      in seawater, 493
      solar system and, 12
      volcanism and, 110
Hydrogen chloride, 122
Hydrogen sulfide, 78, 110, 111
Hydrologic cycle, 287, *287,* 314, 340, 344,
      349, 656
Hydrolysis, defined, 141–142
Hydrophones, *389*
Hydrosphere
      as Earth system, 4
      evolution of, 489, 491
      glaciers and, 344
      oceans and, 388, 389
      origin of, 489, 491
      plate tectonics/rock cycle and, 20, 615
      volcanism and, 110
Hydrostatic pressure, 184, 319, 320
Hydrothermal activity
      defined, 331
      geothermal energy as, 334–335, *335*
      geysers as, 331, 332, *333,* 334, *335*
      hot springs as, 331, *332,* 334, *334*
Hydrothermal alteration, 187
Hydrothermal vents, 388, 390, 489, 493, *493*
Hydroxide minerals, *67,* 70

Hyenas, *640*
Hylonomus, 562
Hypocenter, 202. *See also* Focus
Hypotheses, defined, 5
Hyracotherium, 637

Iapetus Ocean, 528, 530, *530*
Iberia-Armorica, 533
Ice, 60, 344, *344,* 345, 346, 355, 357, 489.
      See also Glaciers/glaciation
Ice Age, 304, *338,* 340, 349, 359–360, *417,*
      628. *See also* Pleistocene Epoch
Ice caps, *341,* 344, 346, 656
Ice cores, 656
Ice sheets, *341,* 346, 347, 509, 630. *See also*
      Continental glaciers/glaciation
Ice shelves, 341
Ice-scoured plains, 355, *355*
Ichthyosaurs, 576, *595,* 600, *601,* 605
Ichthyostega, 560, *560,* 561
Igneous activity
      Cenozoic history and, 618–620, 625
      Cordilleran, 618–620
      in Paleoproterozoic/Mesoproterozoic, 486
Igneous rock textures, 93–94
Igneous rocks
      batholiths and, 100, 102, 103, *104*
      Cenozoic history and, *625*
      characteristics/classification of, 93–98,
            95, 98
      common, *19*
      crust and, 13
      crystalline rock and, 158
      defined, 17, 86
      examples of, *99*
      extrusive, 17, *18, 19,* 20, *20,* 87, 93
      felsic, *97*
      geodes and, 175
      geologic time and, 420, 424, 425, 427,
            427, 428, *439*
      granite as, 100
      hydrothermal activity and, 331, 332
      intermediate, *96*
      intrusive, 17, *18, 19,* 20, *20,* 86, 87, 93,
            115 (See also Plutonic rocks)
      lava and, 86–88, 113
      mafic, *96*
      magma and, 86–92, *90, 93*
      mantle and, 13
      metamorphic rocks and, 182, 183,
            189, 191
      minerals and, *75,* 78
      permeability of, 315
      plate boundaries and, 46
      plutons as (*See* Plutons)
      porosity of, 314, *315*
      Precambrian, 504
      in rock cycle, 17, *18*
      at Sudbury Basin, 498
      weathering and, 136
Iguanadon, 597
Illinoian glaciation, 630, *630*

Illinois Basin, 523, 525, 535
Imatra stones, 174, *174*
Inactive volcanoes, 110
Incised meanders, 306–307, *308*
Inclined folds, 238, *239*
Inclusions
      batholiths and, 103, *104*
      defined, 92, *93*
      geologic time and, 420, *421, 422,* 425, 427
      principle of, 420, *422,* 425
Index contour, 661
Index fossils, 429
Index maps, 663
Index minerals, 188–189
Indian plate, *125,* 615
Indian Ridge, 33, 128
Indian-Australian plate, 39, *39*
Industrial Revolution, 504
Industrialization, 9, 144–145, *145,* 378
Industry, 326, 331
Infauna, 545, *545*
Infiltration, 292, 377
Infiltration capacity, 290
Inheritance, 453, 456, 464
Insecticides, 152
Insects, 560, 600, 638
Inselbergs, 381
Institute of Genetics, 453
Intensity, 208, *210,* 211, 401
Intensity maps, 211
Intensity value, 211
Intensity zones, 211
Interglacial stage, 359
Interior Lowlands, 625, 627
Intermediate igneous rocks, *96*
Intermediate magma
      Bowen's reaction series and, 89, *89*
      classifying igneous rocks and, *95*
      composition and, 95
      defined, 87, *87*
      subduction zones and, 91
      volcanism and, 119, 127
Intermediate-focus earthquakes, 202, 203, *204*
Internal drainage, 377
International Astronomical Union, *11*
International Commission on Stratigraphy,
      22, 476, *477,* 614
International Stratigraphic Chart, *22*
Intrusive igneous bodies, 98. *See also* Plutons
Intrusive igneous rocks, 17, *18, 19,* 20, *20,*
      86, 87, 93, *115. See also* Plutonic rocks
Invertebrate groups, *544,* 547
Invertebrate marine life, 544–547, 550–554.
      See also Marine invertebrates
Invertebrates, 467, 543, 544, 550, 551, 552,
      554, 592–593, 606, 614, 632
Ionic bonding, 62–63, *63*
Ions
      defined, 66–67
      mineral groups and, *67*
      sedimentary rocks and, *159*
      sizes/charges of, *66*

Iridium, 606
Iridium anomaly, 606
Irish elk, 639
Iron
    in Earth's core, 225
    in Earth's crust, 87
    in groundwater, 331
    importance of, 60
    magma and, 87, *87*, 92, *92*
    metamorphism and, *195*
    minerals and, 66, 75, 79
    in Nagaunee Iron Formation rocks, *489*
    Paleoproterozoic history and, 486
    as resource, 79
    in solar system, 12
    submarine hydrothermal vents and, 33
    weathering and, 153
Iron ore, *79*, 172, *176*, 187, 499
Iron oxide, 160, 164, 189, 195, 226, 409, 489
Irrigation, 320, 323, 324, 325, *325*, 326
Island arc, 248, *249*, 478, *478*, 480, 505,
    532, 533, 580, 584, *584*
Isograd, 193
Isostatic rebound, 254, *254*
Isotopes
    carbon, 62, *62*
    defined, 62, 432
    geologic time and, 432, *434*
    radioactive, 62, 435–436
Isthmus of Panama, 54, *54*

Jackson, Andrew, 100
James, Jesse, 322
Jasper, 164
Jasper Knob, Michigan, *489*
Jawed fishes, *470*, 555, 557, 558
Jawless fishes, *470*, 555, 556
Jefferson, Thomas, *84*, 100
Jeffreys, Harold, 224
Jellyfish, 545
Jet Propulsion Laboratory, California, 110
Jetties, 402, *408*
Joggins Formation, Canada, 562
John Day Fossil Beds National Monument,
    Oregon, 634, 636, 638
Johnson, Kirk R., 442
Joints, *241*, 241–242
Joly, John, 417
Jordan Sandstone, 534
Joshua Tree National Park, California, *143*
Journey to the Center of the Earth (Verne), 6
Jovian planets, 12, 14–15
Juan de Fuca plate, *125*, *127*, 624, 625
Juneau Icefield, 344
Jungfrau Firn, Switzerland, *338*
Jupiter, 11, *11*, 12, 14, *14*, *15*, 43, 110
Jurassic Period
    Absaroka Sequence and, 523
    birds in, 576
    continental drift and, *30*
    dinosaurs in, 582
    evolution and, *470*

fish during, *555*
geologic time and, *417*, *424*, *430*
paleogeography during, *577*, *581*
Pangaea in, 576, 580
plants during, *564*, *568*
rifting in, 583

K2, *232*
Kames, 358, *358*
Kangaroos, 639
Kansan glaciation, 630, *630*
Kaolinite, 141, 153
Karakoram Range, *232*, *251*
Karst topography, *320*, 320–322, *322*, *323*
Kaskaskia Sequence, *511*, 520–523, 524
Kazakhstania, 505, 506, *506*, *507*, *508*,
    509, *510*
Kelvin, Lord, 418
Kenai Fjords National Park, Alaska, *114*,
    346, *356*, *357*
Keoua, Chief, 116
Kerogen, 176
Kesseiring, Joseph, 328
Kettle Moraine State Park, Wisconsin, 358
Kettles, 358, *358*
Key beds, 429
Kharga Depression, Egypt, *368*
Kilamanjara volcano, Tanzania, 125
Kilauea Volcano, Hawaii, *112*, 116–117, 128
Kimberella, 497, *497*
Kindred Spirits (Durand), 6, *6*
Kinetic energy, 292
Kircher, Anthanasium, *389*
Kluane National Park, Canada, *275*, 350, *350*
Komatites, 95, 480
Krakatau eruption, 110, 216
Kutiah Glacier, Pakistan, 350
Kyanite, *188*, 189, 193, 195

La Brea Tar Pits, California, *171*, 641, *642*
La Conchita, California, 268, *268*, *269*, 659
Labyrinthodonts, *559*, 561–562
Laccoliths, 99
Lacy bryozoans, 552, *552*
Lagoon, 392, 402, *402*, *403*, *408*
Laguna Beach, California, 268, 269, 270, *271*
Lahar, 118, *119*
Lake Bonneville, Utah, 630, *631*
Lake Columbia, Washington, *631*
Lake Lahontan, Nevada, 630, *631*
Lake Manley, California, *631*
Lake Missoula, Montana, 631, *631*
Lake Nyos, Cameroon, 112
Lake Russell, California, *631*
Lakes
    base level and, 304, 305
    Cenozoic history and, 630, 631, *631*, 632
    Crater, 115, *116*, 125
    deltas and, 296
    drainage systems and, 303
    Flathead, *631*
    glacial, 359, *359*

groundwater and, 314, 316, 326
oxbow, 294, *294*, *295*, 296
playa, 378–379, *380*
pluvial, 630, 632
proglacial, 630, 631, *631*
saline, 388
seawalls and, 408
shorelines and, 392–393, 396, 401
spits and, 401
subglacial, 489
Laki fissure, Iceland, 99
Lamarck, Jean-Baptiste de, *448*, 450–451,
    453, 458
Laminar flow, 290
Landfills, 314, 330, *330*, 331
Landslides. *See also* Slides
    causes of, 260, 263, 264
    defined, 260
    along highway, *262*
    near Rio de Janeiro, *258*
    preventing, 277, 279
    reducing, *281*
    South California, *268*, 268–269, *269*
Lange, Dorothea, *152*
Lapilli, defined, 113
Laramide orogeny, 443, 585, 617, *619*, 620
Laramide-Alleghenian orogeny, 532
Large-scale maps, 662
Lassen Peak, California, 88, 95, *120*, 126,
    127, 620, *622*, *623*, 628
Lassen Volcanic National Park, California,
    112, *120*, 332
Lateral erosion, 305, *307*
Lateral gene transfer, 458
Lateral moraines, 357, *357*
Laterite, 147, 148, *149*
Latitude, 662
Laurasia
    Mesozoic history and, 576, *577*, 606
    Paleozoic history and, 506, *508*, 509,
        510, 526, *531*, 532, 533, 561
    understanding Earth and, 29, 31
Laurentia
    Paleoproterozoic history of, 484–486
    Paleozoic history and, *505*, 506, *506*,
        507, 509, 528–533, *530*, *531*
    Precambrian history and, 484, *484*,
        484–486, *485*, 487
    Proterozoic evolution of, *485*
Lava
    defined, 78, 86
    deformation and, 242
    metamorphism and, 184, 187
    pillow, 44, 46, *46*, 113, *114*, 127, 480, *482*
    properties/behavior of, 86–88
    sheet, 46, *46*
    temperature and, 87, *87*, 88, *88*, 93
    viscosity of, 88, *88*
Lava domes, 87, 119–121, *120*, *349*, 620, *622*
Lava flows
    aa, 113, *114*
    basalt, 96, 113, *115*, 117, 629

cooling of, 93
defined, 86, 112–113
geologic time and, 420, *421*, 437, *439*
in Hawaii, *87*
from Kilauea, 117
mafic, 116, 127
magma chambers and, 89
metamorphism and, 187, *187*
pahoehoe, 113, *114*
Permian, 36
plate tectonics and, 36, *42, 43*
rifting and, 44
Roza, 123
silica tetrahedra networks and, 88
Silurian, 36
soils and, 148
submarine, 127
in Zaire, 110
Lava tube, 112–113, *113*
Law of universal gravitation, 252
Layering, 167
Leaching, 153
Lead. *See also* Galena
asbestos and, 183
bonding/compounds and, 64
geologic time and, 433, *434,* 435
metamorphic rocks and, 183, 195, *195*
as mineral resource, 534, 535
minerals and, 70, *70, 79*
percentage imported, *79*
plate tectonics and, 52
weathering and, 153
Leaning Tower of Pisa, Italy, 326, *327*
Leaves, 565
Lee, Robert E., 100
Left-lateral strike-slip fault, 244
Lehman Cave, Nevada, 322
Lehmann, Inge, 224
Lemurs, 641
Lepidodendron, 566
Levees, 298, *301, 302, 327,* 406
Lewis and Clark Caverns, Montana, 322
Lewis overthrust, Montana, 244
Liberty Cap, *334*
Lienhard, John H., *605*
Life
chemical evidence for, *481*
distribution of, 53–54, *54*
extinctions of (*See* Extinctions)
history of, 21, 491–497
during Mesozoic Era, 591–597, 600–604
mineral usage for, 7
origin of, 450, 491–497
plate tectonics and, 53–54, *54*
in Precambrian history, 491–497
Lightning, 492
Lignin, 565
Lignite, 165, *165,* 173, *534,* 590, 631
Limbs
amphibians and, 560
of folded rocks, 237–238, *238, 239*

Limestone
Absaroka Sequence and, 525, 526
acid rain and, 144
crinoidal, 522
cyclothem and, 525
depicted, *19*
deserts and, 382
fossiliferous, *163*
groundwater and, 314, 315, 320, *320,* 334
marble from, 182, 192, *193*
metamorphic facies and, 194
as mineral resource, 534
minerals and, 69
oceans and, 391
oolitic, 163, *163,* 522
origin of, 163
permeability of, 315
porosity of, 314, 315
sedimentary rocks and, 160, *162,* 163, *163, 164,* 172, 174
Tippecanoe Sequence and, 514–515
types of, 163
weathering and, 140, 153
Limestone facies, 165, *166*
Limonite, 141, 160
Lincoln, Abraham, *84,* 100
Lingula, 429, *431*
Linnaeus, Carolus, 464–465, *465*
Lions, 639
Liquefaction, 212
Liquids
defined, 60
running water and, 290
S-waves and, 206, 223, 224
viscosity and, 88
Lithificaiton, 160, *161*
Lithium, 499
Lithosphere
continental, 46, 92, 251
defined, 13, *13*
as Earth system, 4
earthquakes and, 202, 226, 227
Earth's crust and, 227
oceanic, 33, 40, 44, 46, *51,* 92, 248, 249
plate tectonics and, 17, *17, 34, 38,* 39, 44–47, 50, *50,* 51
Lithosphere-atmosphere-hydrosphere system, 17
Lithostatic pressure, 184, *184*
Lithostratigraphic boundaries, *440*
Lithostratigraphic units, 438, *439*
Little Ice Age, 340, *342,* 342–343, *343, 348,* 348–350, 359, 360
Liverworts, 565
Lizards, 576, *595*
Llao, Native American god, 114
Lo, defined, *43*
Lobe-finned fish, *555,* 558, *558, 559, 561*
Local base level, 304
Lodgepole Formation, Montana, *439*
Loess, 368, 372, *373*
Loihi Seamount, 48, *49,* 128

Loma Prieta earthquake, 203, 211, 213, *219*
Long Valley, California, 121, 125, 129, 130
Longitude, 662
Longitudinal dunes, 369, 370–371, *372*
Longitudinal profile, 305
Longshore currents, 396, 400, *401,* 402, 409
Longshore drift, *401*
Lorises, 641
Louann Salt, 591
Love, A. E. H., 206
Love Canal, New York, 330
Love wave, 206, *207, 208*
Lovejoy, Owen, *646*
Low tide, *393, 394,* 400, *400*
Lowell Glacier, Canada, 350, *350*
Lower mantle, 13, *13*
Low-velocity zone, 225
Lung cancer, 183, *183*
Lungfish, 558
Lungs, 560
Luster, 70, 71, 664
Luxor Temple, Egypt, *100*
L-wave. *See* Love wave
Lycopsids, 566, 568
Lyell, Charles, 417, 432
Lysenko, Trofim Denisovich, 453
Lystrosaurus, 32

Macaque, 642
Mackinac Breccia, 246, *247*
Mackinac Bridge, Michigan, 246, *247*
Macroevolution, 461
Madison Boulder, New Hampshire, 355
Madison Group, Montana, *439*
Mafic igneous rocks, *96*
Mafic lava flows, 116, 127
Mafic magma
Bowen's reaction series and, 89, *89*
change and, 92
classifying igneous rocks and, *95, 96*
composition and, 95
defined, 87, *87,* 88, *88*
metamorphism and, 185
spreading ridges and, 90, *90*
subduction zones and, 91
volcanism and, 111, 126–127
Magellan spacecraft, *42*
Magma
batholiths and, 103, *104*
Bowen's reaction series and, 89, *89*–90, 92
at Bushveld complex, 499
change and, 88–92, *92*
composition of, 87
defined, 13, 78, 86
deformation and, 242, *248*
earthquakes and, 225
felsic (*See* Felsic magma)
geologic time and, *426, 434*
geothermal energy and, 334
granite and, 100
hot spots and, 91, *91*
hydrothermal activity and, 331

intermediate (*See* Intermediate magma)
mafic (*See* Mafic magma)
metamorphism and, 184, 185, 187, 194
origin of, 88–92, *90*
orogens and, 484
plate boundaries and, 44, 46, 52, *53*
properties/behavior of, 86–88
rifting and, 44
spreading ridges and, 89, *90*, 90–91
subduction zones and, 91
temperature and, 87, *87*, 88, *88*, 90, 93
types of, *87*
ultramafic, 87, *87*, 90, 91, 95, *95*, 480
viscosity of, 88, *88*, 97
viscous, *120*
Magma chambers, 89, 102
Magma mixing, 92, *93*, 95
Magnesium
    in Earth's crust, 87
    magma and, 87, *87*, 92, *92*
    minerals and, 66, 75
    sedimentary rocks and, 163
    Tippecanoe Sequence and, 515
Magnesium oxide, 12, 226
Magnetic anomalies, 37, *38*, 49, *49*
Magnetic field, 35, *35*, 37, *38*, 40, 490
Magnetic reversals, 36–39, *37*
Magnetism, 35
Magnetite, 70, 75, 195, *195*, *664*
Magnetosphere, 490. *See also*
    Magnetic field
Magnitude, 208, *211*, 211–212
Maiasaura, 597, *599*
Malaria, 9
Malthus, Thomas, 452, 453
Mammals
    in Cenozoic history, 614, *635*, 635–637,
        637, 639–641, *640*
    cynodonts and, 602–604
    diversification of, *635*, 635–636
    evolution of, 443, 450, 456, 459, *460*,
        461, 463–468, *466*, *470*, 576,
        602–604
    geologic time and, *417*
    hoofed, 635, 636–637, 638, *639*
    marsupial, 604, *604*, 635, *635*
    in Mesozoic history, 576, 594, *595*, *602*,
        602–604, *603*, *604*
    in Messel Pit Fossil Site, 638
    origin of, 602–604
    placental, 604, *604*, 635
    reptiles to, *602*
Mammoth Cave, Kentucky, *312*, 314,
    317, 322
Mammoth Hot Springs, 334
Mammoth Mountain volcano, California,
    129, 130
Mammoths, 639, *642*
Manganese, 60, 79, *79*, 80, 153, 409, 632
Manganese nodules, 409
Mantell, Gideon, 597
Mantell, Mary Ann, 597

Mantle
    composition of, *223*, 225–226
    defined, 13, *13*, 223, *223*
    density of, *223*, 225–226
    as Earth system, 4
    Hadean and, 476, 478
    igneous activity and, 486
    lower, 13, *13*
    ophiolites of, 488
    structure of, 225–226
    upper (*See* Upper mantle)
Mantle convection cell, *50*
Mantle plumes, 34, 48, *49*, 91, *91*, 121
Map locations, 662–663
Map scales, 661–662
Marathon Mountains, Texas, 532, 533
Marble, 140, 144, *180*, *182*, *190*, 192, *193*, 194
Marianas Trench, 33
Marine community
    Cambrian, 546, *546*
    Carboniferous, 552–553
    Devonian, *551*, 551–552, *557*
    Mississippian, *532*
    Ordovician, *550*, 550–551
    Permian, 552–554, *553*
    Silurian, *551*, 551–552
Marine depositional setting, 159, *160*, 293
Marine ecosystem, 544–546, *545*, 632
Marine food web, *546*
Marine invertebrates, 592–593, 606, 614,
    632. *See also* Invertebrate marine life
Marine regression, 166, *166*, 167
Marine reptiles, 576, 600–601, *601*
Marine terraces, 398, 405
Marine transgression, 165, *166*, 167
Mariner 10, *42*
Marjum Formation, Utah, *549*
Marlekor, 174, *174*
Mars, 11, *11*, 12, 14, *14*, *15*, *42*, 110, 286, 476
Marshall, James, *72*, 576
Marsupial mammals, 604, *604*, 635, *635*
Martins, *640*
Mass extinctions, 463, 542, 549–554,
    563, 591, 592, 606–607. *See also*
    Extinctions
Mass movement, 260, 265, *266*, 271, 277,
    278. *See also* Mass wasting
Mass spectrometer, 433
Mass wasting. *See also* Mass movement
    cause of, 261
    defined, 260
    deserts and, 377
    effects of, 277–281
    factors influencing, 260–264
    glaciers and, 354
    types of, 265–277
    valleys and, 305
Mass-wasting processes, *156*
Mastodons, 639
Matter, 60–64
Matterhorn, Switzerland, 354, *355*
Mauna Loa Volcano, Hawaii, *43*, 117, *117*, 119

Mayon volcano, Philippines, 112, 119, *119*
Mazatzal-Pecos orogen, *484*, 486
McCloud River, California, *293*
Meander scars, 296
Meandering river, *294*
Meandering streams, 294, *295*
Meanders
    defined, 294, *295*
    incised, 306–307
Mechanical weathering, 137–139, *159*, 376
Medial moraine, 357, *357*
Medicine Lake Volcano, California, 126, *622*
Medieval Warm Period, 340
Mediterranean belt, 125
Mediterranean-Asiatic belt, *204*, 205,
    214, *218*
Megafossil, 495, *495*, 565
Meiosis, 457, *457*
Mélange, 46
Melting, *188*, 317, 344, 346, 355, *483*, 630
Meltwater, 358
Members, defined, 439, *439*
Mendel, Gregor, *456*, 456–457
Mendenhall Glacier, Alaska, *352*, *356*
Mer de Glace, *342*
Meramec Caverns, Missouri, 322
Mercury
    running water and, 286
    in solar system, *11*
    tectonics of, *42*
    as terrestrial planet, 12, 14, *14*, *15*
    volcanism and, 110
Mesa, 245, 382, *382*
Mesoarchean Era, *477*, *481*
Mesoproterozoic basins, 488
Mesoproterozoic Era, *477*, *484*, 495
Mesoproterozoic igneous activity, 486
Mesoproterozoic orogeny, 486–487
Mesoproterozoic rifting, 486–487
Mesoproterozoic rocks, *495*
Mesoproterozoic sedimentary basin, *487*
Mesoproterozoic sedimentation, 487–488
Mesosaurus, 31, *32*
Mesothelioma cancer, *183*
Mesozoic Earth history
    climate in, 576, 604, 606
    mass extinctions in, 606–607
    minerals in, 590–591
    of North America, 580–590
    paleogeography in, *577–578*, *580*, *581*,
        604, 605, 606
    Pangaea and, 576, 578–580, 582, 591,
        592, 604, 606
    terranes and, 589–590
Mesozoic Era
    continental drift and, 29, 31
    dinosaur extinction in, 463
    evolution and, *470*
    fish during, *555*, 558
    geologic time and, *417*, *429*, 476
    Paleozoic history and, 505
    petroleum and, 52

Mesozoic life history
  birds in, 576, 594, *595, 601,* 601–602
  dinosaurs in, 594–599, *596*
  mammals in, 576, 594, *595, 602,* 602–
    604, *603, 604*
  marine invertebrates in, 592–593
  mass extinctions in, 606–607
  phytoplankton in, 592–593
  plants in, 592, 593, *594*
  reptiles in, 576, 582, 591, 593–597, *600,*
    600–601, *601*
Mesozoic mineral resources, 590–591
Mesozoic paleogeography, *577–578,* 604,
    605, 606
Mesozoic sedimentation, 580, 582,
    585–586, 589
Messel Pit Fossil Site, Germany, 636, 638, *638*
Messelobunodon, 638, *638*
Metallic bonding, 62, 63–64
Metallic luster, 70, 664
Metallic mineral deposits, 534
Metallic ore deposits, 52
Metallic resources, 78
Metamorphic facies, 193–194, *194*
Metamorphic grade, *188,* 188–189
Metamorphic rocks
  asbestos as, 182, 183, *183,* 195
  classification of, 189–193, *190*
  crystalline rock and, 158
  defined, 182
  deformation and, 254
  foliated, 18, *19,* 189–191, *190*
  geologic time and, 424, 435
  glaciers and, 344
  igneous rocks and, 86, 100, 189
  minerals and, *75*
  nonfoliated, 18, *19, 190,* 192–193
  permeability of, 315
  porosity of, 314, *315*
  Precambrian, 504
  understanding Earth and, 17, 18, *18,*
    19, 20, *20*
  weathering and, 136
Metamorphic zones, 193–194, *194*
Metamorphism
  agents of, 184–185
  asbestos and, 183
  Cenozoic orogenic belts and, 615
  contact, 185, 187, *187,* 192, 195, *195,* 420
  continents and, 478
  defined, 182
  deformation and, 234, 246, 247, 248,
    249, 254
  dynamic, 185, 187–188, *188*
  environments of, *186*
  fluid activity in, 185
  greenstone belt and, 480
  heat in, 184
  index minerals and, 188–189
  marble and, *193*
  metamorphic grade and, *188,* 188–189
  mineral formation and, 78

natural resources and, 195
plate tectonics and, 46, 194
pressure in, *184,* 184–185, *185*
of Proterozoic rocks, 488
quartzite and, *193*
on radiometric dating, 435, *435*
regional, 185, 188, 189, 190, 192, 193
sandstone and, *487*
types of, 185–189
understanding Earth and, 18, *18,* 20
Meteorites, 476, *481,* 491, 498, 553, *606,*
    606–607
Methane
  atmosphere and, 490, 492
  Cenozoic history and, 634
  global warming and, 348
  ocean resources and, 409
  solar system and, 12
Methane hydrate, 409
Miacids, 637
Miami Beach, Florida, 407
Mica Dam, Canada, 288
Micas, *68,* 69, *74,* 189, 191
Michigan Basin, 516, 517, *517,* 520, 535
Microcline, 69, *75,* 664, *665*
Microcontinents, 34
Microevolution, 461
Microfossils, 172, 495, *495*
Microplates, 533–534
Microspheres, 492, *492*
Mid-Atlantic Ridge, 33, 40, *41,* 49, *49,*
    249, 615
Midcontinent rift, *484,* 487, 499
Migmatite, *188, 190,* 191, *192*
Milankovitch, Milutin, 359
Milankovitch theory, 359–360, *360*
Military tactics, 6
Milky Way galaxy, 11
Miller, Stanley, 492
Millipedes, 560
Mineral commodities, *79*
Mineral Commodity Summaries, 80
Mineral crystals, 64–65
Mineral deposits, 52–53, 534
Mineral grains, 93
Mineral groups, 66–70, *70*
Mineral identification, 70–71, 74–75,
    664–665
Mineral identification tables, 664, *664–665*
Mineral production, *78*
Mineral properties, 75
Mineral resources, 6, 60, 80, 86, 656. *See
    also* Resources
  Cenozoic, 631–632
  evaporite, 534
  Mesozoic, 590–591
  nonfuel, *78,* 79
  Paleozoic, 534–535
Mineral usage, *7*
Minerals
  accessory, 75
  alteration of, 137

in atmosphere, 490
Bowen's reaction series and, 90
carbonate (*See* Carbonate minerals)
chemical composition of, 66
chemical sediment and, 158
defined, 17, 60, 64
exploring world of, 64–66
formation of, 78
gemstones as, 76–77 (*See also* Specific
    gemstones)
geologic structures and, 245
geologic time and, 435
in groundwater, 331
halide, *67,* 70, *70, 75,* 77
hydroxide, *67,* 70
igneous rocks and, 86, 89, 93, *95*
importance of, 60
index, 188–189
matter for, 60–64
natural resources and, 78–80
oxide, *67,* 70
phosphate, *67,* 80, 153, 494
physical properties of, 66
precious metals and, 72–73
reserves and, 78–80
rock-forming, *75,* 75–77
rock-forming silicate, 69
silicate, *67,* 67–69, *68, 69,* 76, *142,*
    183, 184
sulfate, *67,* 70, *75,* 77
sulfide, *67,* 70, *70, 73,* 153
Minerva Terrace, *334*
Miocene Epoch, *417,* 616
Missing links, 450, 468
Mission Canyon Formation, Montana, *439*
Mississippi River delta, 290, 296, *297*
Mississippian marine community, *552*
Mississippian paleogeography, *523*
Mississippian Period
  carbonate deposition and, 522
  continental collision during, *533*
  continental drift and, 30
  evolution and, *470*
  Gondwana/Laurasia and, 532
  Grand Canyon and, *519*
  organic reefs during, 523
  reptiles during, 542
Mitosis, 457, *457*
Mixed rocks, 191
Moas, 641, 650
Mobile belts
  Appalachian (*See* Appalachian
    mobile belt)
  Cordilleran (*See* Cordilleran
    mobile belt)
  Franklin, *505*
  Hercynian, 509, 533
  Ouachita, 505, *505,* 509, 513, 524, 526,
    532–533, *533*
  Paleozoic, 526, 528–533
  Uralian, 509
Mobile benthos, 545

Modern synthesis, 458
Modified Mercalli Intensity Scale, 208, *209*
Moho, 225
Mohorovičić, Andrija, 225, *226*
Mohorovičić discontinuity, 225
Mohs, Friedrich, 74
Moine Thrust Zone, Scotland, 187–188
Mojave Desert, California, *380, 629*
Mojave orogen, *484*
Molecular biology, 464, 468
Mollusks, 592, 632
Molybdenum deposit, 632
Mongooses, *640*
Monkeys, 642, *643*
Monoclines, 237
Monomers, 492, 493
Monotremes, 604, 635, *635*
Monsoon rains, 286, *286*
Montreal Protocol, 122, 123
Monument Valley Navajo Tribal Park,
    Arizona, *382, 624*
Moon, *14, 15, 43,* 110, 389, 393, *394,* 476, 478
Moon rocks, *481*
Moraines
    defined, *357*
    end, 355, 357, *357*
    ground, 355, 357
    lateral, 357, *357*
    medial, 357, *357*
    recessional, *356,* 357
    stratified drift and, 358
    terminal, *356,* 357, *630*
Morning Glory hot spring, *332*
Morphology, 560
Morrizon Formation, Utah, *589, 589*
Mosaic evolution, 462–463, 565, 602, 604
Mosasaurs, 576, 601, *601*
Moschops, 564
Mosquitoes, 9
Mosses, 565
Mother lode, 591
Mount Adams, *127*
Mount Airy Granite, North Carolina, *138*
Mount Assiniboine, 354
Mount Baker, Washington, 345
Mount Erbus, Antarctica, 125
Mount Etna, Sicily, 117, 125, 128
Mount Everest, 117
Mount Hekla, Iceland, 113
Mount Hood, Oregon, 126, *127, 622*
Mount Kenya, Africa, 348
Mount Kilimanjaro, Africa, 348
Mount Loa, Hawaii, *88,* 128
Mount Mazama, 118
Mount Merapi, 28
Mount Pelée, Martinique, 120–121, *121,* 124
Mount Pinatubo, Philippines, 110, 117,
    119, *119,* 122, 128, 129, *129*
Mount Rainier, Washington, 126, 628
Mount Rushmore National Memorial,
    South Dakota, *84,* 100
Mount Shasta, California, 119, 349, 628

Mount Sinabung, Indonesia, 110
Mount Spurr, Alaska, 126
Mount St. Helens, Washington, 87, 119, *119,*
    126, *127,* 128, 129, 349, *349, 622, 623*
Mount Vesuvius, Italy, 110, *111,* 125
Mount Whitney, California, *86, 574*
Mountain building, 234, 245–252, 476,
    478, 505, 509, 530, 534, 615. *See also*
    Deformation
Mountain glacier, 340. *See also* Valley
    glaciers
Mountain ranges
    cratonic uplift and, 526
    defined, 245
    in Mesozoic history, 581
    mountain building theories and, 247
    origin of Earth's, 234
    Paleozoic history and, 542
    similarity of rock sequences and, 29–30
Mountain system, defined, 245
Mountains
    defined, 245
    mid-ocean ridges as, 246
    origin of, 245–252
    volcanic, 246
Mt. Carmel Junction, Utah, *244*
Muav Limestone, *166*
Mud, 158, *158,* 160, 292, 296, 359
Mud cracks, defined, 168, *169*
Mud facies, 165
Mud pot, 331, *332*
Mudflats, *393, 565*
Mudflows, 263, *266,* 271–277, *273*
Mudrock, 161
Mudslides, 260, 270
Mudstone, 161
Mudstone facies, 165
Multicelled organisms, 496, 550
Mundus Subterraneus (Kircher), *389*
Muscovite, 60, *68, 69, 69, 74, 75,* 79, *188*
Muscovite mica, *89*
Music, 6
Mutagens, 458
Mutations, 457, 458, 463, 467
Myanmar, 28
Mylonites, 187, *188*
Myrtle Beach, South Carolina, *400*

Namib-Naukluft Park, Namibia, *373*
Napoleon Bonaparte, 6
Nashville Dome, 526
National Park Visitor Centers, 663
National Resources Conservation Service
    (NRCS), 144–145
Native elements, *67, 70, 73*
Natural Bridges National Monument,
    Utah, *167*
Natural disasters, 6, 286, 406
Natural gas
    in China/Siberia, 499
    deformation and, 234, 240, 245
    geologic structures and, 245

    in Mesozoic history, 591
    Mesozoic history and, 576
    as mineral resource, 535
    minerals and, 80
    oceans and, 409, 410
    running water and, *288*
    sedimentary rocks and, 158, 173, *173,* 176
Natural hazards, 246
Natural History Museum, Austria, *169*
Natural levees, 296
Natural resources
    groundwater as, 314, 323
    metamorphism and, 195
    minerals and, 78–80
    plate tectonics and, 28, 52–53
    understanding Earth and, 6
    weathering and, 137
Natural selection
    defined, 21, 453
    Lamarck and, 458
    misconceptions about, 456
    significance of, 453–456, *454*
    two-step process of, 463
    variation and, 458
Nautiloids, 592, 632
Navajo Sandstone, United States, 172,
    586, *588*
Nazca plate, *39,* 46, *125,* 615
Neander Valley, Germany, 648
Neanderthals, 647, 648, *648*
Neap tides, 394, *394*
Nearshore currents, 393, 396–397, 401
Nearshore facies, 165
Nearshore sediment budget, 403, *404*
Nearshore zone, 396, 397
Nebraskan glaciation, 630, *630*
Needles, The, *84*
Negative budget, 403, *404*
Negative gravity anomalies, 252, *253*
Negaunee Iron Formation, Michigan, *484,*
    489, 495
Nekton, 544, *545*
Neoarchean Era, *477, 481, 484*
Neoarchean rocks, 488
Neogene Period, *417, 555, 564,* 614, 616–
    621, 624–628, 632–637, 639
Neogene rifting, 615
Neon, 62
Neoproterozoic basins, 488
Neoproterozoic Era, *477, 484,* 495, 504,
    511, 543, 547
Neoproterozoic fossils, *497*
Neoproterozoic glacial deposits, 488
Neoproterozoic rocks, *495, 497*
Neoproterozoic sedimentary basin, *487*
Neoproterozoic sedimentation, 487–488
Neoproterozoic supercontinent, *488,* 504
Neptune, *11,* 12, 14, *14, 15,* 110
Neptunism, 100
Neutrons, 432
Nevadan orogeny, 585, 617, 620
Nevado del Ruiz, Colombia, 118, 128

New Albany Shale, 521, *522. See also*
     Black shales
New World monkeys, *643, 644*
Newark Group, 582
Newark rocks, 582
Newberry Volcano, Oregon, 126, *622*
Newton, Isaac, 4, 252
Newts, 561
Niagara Falls Board of Education, 330
Niagara Formation, *517*
Nickel, *79*, 153, 409, 498, 499
Nile River delta, *297*
Niobium, *79*
Nitrogen, 110, 141, *490, 492, 493*
Nitrous oxide, 348
Noble gases, 62, 63
Nonconformity, 424–425, *426, 427, 439*
Nonferromagnesian silicates
     igneous rocks and, 89, 90, 95, 97, *97*
     minerals and, 69, 70, *75*, 78
     weathering and, *142*
Nonfoliated metamorphic rocks, 18, *19,*
     190, 192–193
Nonfoliated texture, 192
Nonfuel mineral resources, *78, 79*
Nonmetallic luster, 70, 664
Nonmetallic mineral deposits, 534
Nonmetallic resources, 78
Nonplunging folds, 240
Nonrenewable energy resources, 6
Nonrenewable resources, 4, 7, 79
Nonvascular plants, 565
Norite, 498
Normal faults, 243, *243*, 244, *244*, 245, *248*
North America
     active volcanoes in, 126
     continental interior of, 581
     Eastern Coastal region of, 581–583
     growth of, 589–590
     Gulf Coastal region of, 583–589
     Mesozoic history of, 580–590, *590*
     Neogene evolution of, 616–621
     Paleogene evolution of, 616–621
     Paleozoic evolution of, 509, 511
     satellite-based image of, *2*
     Western region of, 583–589
North American Cordillera, 615,
     617–618, *618*
North American plate, 39, *39*, 47, *48, 49,*
     125, 127, 615, 625, 628
North Carolina Museum of Natural
     Sciences, 597
Northridge earthquake, 205, 246
No-till planting, 153
Novarupta, Alaska, *88*, 126, 128
NRCS. *See* National Resources
     Conservation Service (NRCS)
Nuclear energy, *288*
Nuclear waste, 366
Nucleus, 62, 432
Nuée ardente, 120, *121*
Nyiragongo volcano, Zaire, 125

Obama administration, 411
Oblique-slip faults, *243*, 244–245, *245*
Obsidian, 75, 86, 93, *94*, 98, *98, 99*, 110
Occipital condyle, 603
Ocean basins. *See also* Basins
     age of, *38, 39*
     Paleozoic history and, 542
     Paleozoic paleogeography and, 506
     plate tectonics and, 488
Oceanic circulation, 390–392, 576
Oceanic circulation patterns, 578–580, *579*
Oceanic crust
     composition of, *223*
     continental crust to, 32, 39, 47, *50*, 252
     on continents, 478
     defined, 13
     density of, *223*
     in Earth's interior, 227
     gabbro in, 96
     Hadean and, 478
     igneous rocks and, 86
     magma and, 91, *91*, 92
     oceans/seas and, 388
     ophiolites and, *46*, 488
     Paleozoic history and, 505, *533*
     plate boundaries and, 46, 47, 615
     seafloor spreading and, 36–37, *38*, 39
     sedimentary rocks and, 158
     subduction and, 50, 91, *91*
     volcanism and, 110, 127
Oceanic lithosphere, 33, 40, 44, 46, *51*, 92,
     248, 249
Oceanic ridges
     defined, 33, *33*
     expansion of, 578
     plate boundaries and, 40
     plate tectonic theory and, 39–40, 47, 49
     as seafloor aspect, 388
     seafloor spreading and, 37, 38, *38*, 39
Oceanic trenches
     defined, 33, *33*
     plate tectonic theory and, 40, 44, 46,
     47, *50*
Oceanic-continental plate boundaries, *45,*
     46, 249, 251, *530*
Oceanic-oceanic convergent plate
     boundary, *17*, 44–46, *45*, 248, *249*
Ocean/oceanic currents, 28, *388*
Oceanographic research, 389
Oceans
     Cambrian explosion and, 543
     coasts of (*See* Coasts)
     exploring, 389
     fish and, 555
     formation of, 12
     geologic time and, 417
     glaciers and, 344
     groundwater and, 314, 316, 325, 326
     hydrologic cycle and, 287, 290
     hydrosphere and, 491
     introduction to, 388–389
     map of, *388*

oil spills in, 410–411
     resources from, 409
     running water and, 293, 296
     shorelines and (*See* Shorelines)
     study of, 389
Octopus, 592, 632
Offshore facies, 165
Oil, 173, *173*, 245, 409, 499, 576
Oil production, *529*
Oil rigs, 410, *410*
Oil shales, 176
Oil spills, 410–411, *411*
Oil stains, 431
Okefenokee Swamp, Georgia, *525*
Old Faithful Geyser, Wyoming,
     332, *333*
Old Man of the Mountain, New
     Hampshire, 100
Old Red Sandstone, 530, *531*, 531–532
Old World monkeys, 642, *643*
Oldham, R. D., 224
Oligocene Epoch, *417*
Olivine
     Bowen's reaction series and, 89, *89*
     classifying igneous rocks and, 95, *95*
     magma and, 92
     metamorphism and, 185
     in mineral identification table, *665*
     minerals and, 66, 68, *68*, 69, *69*, 70,
     75, 77
     understanding Earth and, 46
Olivine phenocrysts, 96
Olympic National Park, Washington, *399*
Olympus Mons (Mars), *43*, 117, *117*
On the Origin of Species by Means of
     Natural Selection (Darwin), 21, 451,
     453, 465, 542
Ooids, 163, *163*, 172
Oolitic limestone, 163, *163*, 522
Oolitic textures, 513
Ooze, 391, *391*
Opal, 76, 175
Ophiolites, 46, *46*, 482, 488
Ophir Creek, Nevada, *273*
Orbital eccentricity, 359
Ordovician marine community, *550,*
     550–551
Ordovician paleogeography, *514*
Ordovician Period
     evolution and, *470*
     fish during, 555, *555*
     geologic time and, 442
     Paleozoic history and, 504, 506, *506,*
     507, 513, 516, 546
     Pangaea and, 578
     plants during, *564*, 565
     trilobites and, 549
Ordovician Taconic orogeny, 487
Ore deposits, 195, *195*
Organ Pipe National Monument,
     Arizona, *376*
Organelles, 496

Organic evolution. *See also* Evolution
    defined, 21
    meaning of, 450–453, 456
    modern view of, 458–463
    Wallace and, 451–453, 454, 456, 457
Organic reefs, defined, 515, *515, 523*
Organisms
    amino acids in, 492
    Archean, 493–494
    in biosphere, 491
    colonial, 496
    distribution of, 576
    first, 476
    Hadean and, 477
    multicelled, 496, 550
    pelagic, 544
    photosynthesis and, 491
    in reef communities, 515
    single-celled, 542
    soils and, 148
    weathering and, 139, *139,* 140, 141
Origin of Continents and Oceans, The
    (Wegener), 29
Ornithischia, 595, *596*
Ornithischian dinosaurs, 596–597
Ornithopod, *596,* 597, *599*
Orogenic belts
    Alpine-Himalayan, 615, *617*
    Cenozoic (*See* Cenozoic orogenic belts)
    circum-Pacific (*See* Circum-Pacific
        orogenic belt)
Orogens, 484, *484,* 485, *486,* 615
Orogeny
    Acadian, 509, *511,* 530, 532
    Alleghenian, 509, *511,* 532, 582
    Alpine, 615
    Antler, 509, 532, *532,* 584
    Caledonian, 506, 530, 532
    Cordilleran, 581, 584–585, *587,* 617
    defined, 246–247
    Ellesmere, 509
    Grenville, 487
    Hercynian, 509, 532
    Hercynian-Alleghenian, 532
    Laramide, 443, 585, 617, *619,* 620
    Laramide-Alleghenian, 532
    Mesoproterozoic, 486–487
    Nevadan, 585, 617, 620
    Ordovician Taconic, 487
    Ouachita, 509, *511,* 532
    Pan-African, 488
    Sevier, 585, *588,* 617, 620
    Sonoma, 580, *584*
    Taconic, 506, *511,* 528–530, 532
Orosinian Period, *477*
Oroville Dam, California, 288, *302*
Orthoclase, 67, 69, *75,* 90, 141, 664, *665*
Ostracoderms, 555, *555, 556,* 557, 558
Otters, *640*
Ouachita mobile belt, 505, *505,* 509, 513,
    524, 526, 532–533, *533*
Ouachita Mountains, 532, 533

Ouachita orogeny, 509, *511,* 532
Our Wandering Continents (du Toit), 29
Outgassing, 490, *490,* 491
Outlet glaciers, 347
Outwash plain, 358
Outwash sediment, 632
Overloading, 261, 263–264, 279
Overpopulation, 7–8, *654,* 656–657, *657*
Overthrust, 244
Overturned folds, 238, *239*
Owen, Richard, 594
Oxbow lakes, 294, *294, 295,* 296
Oxidation, 141, 153, 330
Oxide minerals, *67,* 70
Oxygen
    amphibians and, 560
    atmosphere and, 489, 490, *490,* 491, 492
    Bowen's reaction series and, 90
    in Earth's crust, 66, 87
    geologic time and, *441*
    in rock cycle, 17
    in seawater, 493
    in solar system, 12
    weathering and, 140, 141
Oysters, 592
Ozark Dome, 526
Ozone, 490, *490*
Ozone depletion, 110, 122–123, 554, 656, 657
Ozone hole, 122, 123
Ozone layer, 122, 123, 144, 489, 490
"Ozymandias" (Shelley), 6

Pachycephalosaur, 596, 597, *599*
Pacific Ocean basin, *26,* 46, 615
Pacific Palisades, California, 268
Pacific plate, 39, *39,* 47, 48, *48, 49,* 54, *125,*
    127, 615, 625, 628
Pahoehoe lava flows, 113, *114*
Pakicetus, 640
Paleoanthropology, 643
Paleoarchean Era, *477, 481*
Paleocene Epoch, *417*
Paleocene-Eocene Thermal Maximum, 634
Paleogene Period, *417, 430, 555, 564,* 614,
    616–621, 624–628, 632–637, 639
Paleogeography
    Anning and, 605, *605*
    Cambrian, *512*
    Cretaceous, *582*
    Devonian, *521*
    Jurassic, *581*
    Mesozoic, *577–578,* 604, 605, 606
    Mississippian, *523*
    Ordovician, *514*
    Paleozoic, 505–509, *506, 507, 508, 510*
    Pennsylvanian, *524*
    Permian, *528*
    Silurian, *516*
    Triassic, *580*
    of world for Eocene Epoch, *616*
    of world for Miocene Epoch, *616*
Paleomagnetism, 36

Paleontology, *448,* 450, 468
Paleoproterozoic assemblages, 488
Paleoproterozoic Era, 480, *484, 485,* 486,
    488, 498, 499
Paleoproterozoic history of Laurentia,
    484–486
Paleoproterozoic igneous activity, 486
Paleoproterozoic rocks, 488
Paleoproterozoic sedimentary rocks, *486*
Paleoseismology, 222
Paleozoic arthropods, 548–549, 551, 560
Paleozoic Earth history
    Absaroka Sequence in, *511,* 523–526
    continental architecture in, 504–505
    cratons in, 504–505, *505,* 509, 511–514,
        513, 520–524, 526, 531, 532
    Grand Canyon and, 513, 518–519
    introduction to, 504
    Kaskaskia Sequence in, *511,* 520–523, 524
    microplates and, 533–534
    mobile belts in (*See* Mobile belts)
    Paleozoic evolution of North America
        in, 509, 511
    Paleozoic paleogeography in, 505–509,
        506, 507, 508, 510
    Pangaea and, 505, 509, *510,* 532,
        533–534
    Sauk Sequence in, *511,* 511–513, 514
    terranes and, 533–534
    Tippecanoe Sequence in, *511,*
        514–517, 520
Paleozoic Era
    amphibians during, 561
    atmosphere and, 491
    Cambrian explosion and, 542
    continental drift and, 30, *32*
    evolution and, 463, *470*
    extinction in, 463, 606
    fish during, *555,* 558
    fossils during, *169*
    geologic time and, *417, 426, 429, 476*
    global mountain building in, 530
    mobile belts and, 526
    Pangaea and, 488
    reefs in, 583
    reptiles during, 563, *563*
    seafloor features and, 34
Paleozoic evolution of North America, 509,
    511, 542
Paleozoic geologic systems, 504
Paleozoic invertebrate marine life,
    544–547, 550–554
Paleozoic life history
    amphibians and, 542, 558, *559,*
        560–562, *562*
    arthropods in, 548–549
    Cambrian explosion and, 542–543
    introduction to, 542
    invertebrate marine life in, 544–547,
        550–554, 592
    plant evolution and, *564,* 564–568, *566*
    reptiles and, 542, 562–563

shelly fauna in, *543*, 543–544, 550
trilobites and, 546, *548*, 548–549, *550, 551*
vertebrate evolution in, 554–555
vertebrates and, 554–555, 560–562
Paleozoic mineral resources, 534–535
Paleozoic mobile belts, 526, 528–533
Paleozoic paleogeography, 505–509, *506, 507, 508, 510*
Palladium, 498
Palo Duro Canyon State Park, Texas, *156, 158*
Pan-African orogeny, 488
Pan-American Highway, Mexico, *262*
Pangaea
  breakup of, 576, 578–580, 582, 591, 592, 604, 606, 615
  circulation patterns and, 578–580, *579*
  continental drift and, 29, 31, *32*
  deformation and, 249
  evaporites and, *578*
  formation of, 17, 533–534
  geologic time and, *417, 442*
  in Paleozoic history, 505, 509, *510*, 532, 533–534, 553
   plate movement and, 50
  as supercontinent, 488
Pannotia, *484*, 488, 504
Panthalassa, 509, 579, *579*
Parabolic dunes, 369, 371, *373*
Paradoxides, 429, *431*
Parallel evolution, 459–461, *461*
Parasaurolophus, *599*
Parent atoms, 433, 434, *434*, 435, *435*
Parent element, 433
Parent material, 136, 139, 140, 141, *146*
Parexus, *556*
Paricutín, *118*
Particle size, 141, 142, *142*
Parvanconrina, *496*
Passive continental margins, 40
Patch reefs, 552, *553*
Peace Monument, Washington, DC, 182
Pearls, 77
Peat, 165, *165*
Peatlands, 632
Pediments, 381, *381*
Pegmatite, 97, *98*, 153, 499
Pegmatite texture, 97, *98*
Pelagic clay, 391, *391*
Pelagic communities, 551
Pelagic organisms, 544
Pele, Hawaiian goddess, 114
Peleoproterozoic history of Laurentia, 484–486
Pelycosaurs, 562, 563, *563*, 594
Pennsylvanian paleogeography, *524*
Pennsylvanian Period
  Absaroka Sequence and, 523
  continental collision during, *533*
  continental drift and, 30, 31, *32*
  cratonic uplift and, 526
  evolution and, *470*
  geologic time and, *417, 430, 442*

Gondwana/Laurasia and, 532
  Grand Canyon and, *519*
  ice ages and, 359
  Okefenokee Swamp and, *525*
  plants during, 566, 568
  reptiles during, *563*
  sedimentary rocks and, 173
Penny Ice Cap, Canada, *341, 344*
Penzias, Arno, 11
Perched water table, 317
Peridotite
  asbestos and, 183
  defined, 13, 46
  in granite-gneiss complexes, 480
  mantle and, 226
Perihelion, 360
Period, defined, 440
Periodotite, 95, *95*
Perissodactyls, 636, 637, *639*
Permafrost, *275*, 275–276, *276*
Permafrost soils, 409
Permeability, 314–315, 431
Permian lava flows, 36
Permian marine community, 552–554, *553*
Permian paleogeography, *528*
Permian Period
  Absaroka Sequence and, 526
  amphibians during, 561, 562
  continental collision during, *533*
  continental drift and, *30, 31, 32*
  evolution and, *470*
  fish during, 555
  geologic time and, *417, 430, 442*
  Gondwana/Laurasia and, 532
  Grand Canyon and, *519*
  ice ages and, 359
  mass extinction during, 553–554
  Paleozoic paleogeography and, 509, *510*
  Pangaea and, 532, 578
  plants during, *564*, 566, 568
  polar wandering in, *36*
  reptiles during, 562, 563, *563, 564*
  sedimentary rocks and, 158
Peru earthquake, 217, 277, *278*
Peru-Chile Trench, 46, 249
Perunica, 533
Pesticides, 152
Petrified Forest National Park, Arizona, 586, *588*
Petroglyphs, 377, *377*
Petroleum
  Cenozoic history and, 631
  deformation and, 234, 240
  global warming and, 348
  groundwater and, *327*
  in Mesozoic history, 590–591
  as mineral resource, 535
  minerals and, 79, 80
  oceans and, 410
  oil spills and, 410
  reef development and, 520
  running water and, *288*

sedimentary rocks and, 158, 173, 176, 223
pH scale, 144
Phaneritic texture, 93, 95, *95*, 96, *96*
Phanerozoic age, 489
Phanerozoic Eon, *417, 476*, 487, 488, 497, 504, 516
Phanerozoic marine diversity, *553*
Phanerozoic sediments, 504
Phenocrysts, 93
Phoenix, River, *151*
Phorphorite, 409
Phosphate minerals, 67, 80, 153, 494
Phosphate radical, 70
Phosphorus, 70
Photic zone, 390, 391, *546*
Photochemical dissociation, 490, *490*
Photosynthesis, 489, *490*, 491, 494, 565
Phyletic gradualism, 459
Phyllite, *188*, 190, *190, 191*
Phylogenetic tree, 461, *462*
Phylogeny, 463, 543
Physical geology, 4
Phytoplankton, 544, 545, *545, 546*, 592–593, 632
Phytosaurs, 158
Pillow lavas, 44, 46, *46*, 113, *114*, 127, 480, *482*
Pink granite, 100
Pinnacle reefs, 516
Pinnacles, *150*
Pitted outwash plains, 358
Place movement/motion, 48–50
Placental mammals, 604, *604*, 635, *635*
Placer deposits, 173, 576, 591
Placoderms, *555*, 556, 557, *557, 558*
Plagioclase, 96, 97, 191, 664
Plagioclase feldspars
  in anorthosite, *487*
  Bowen's reaction series and, 89, *89*, 90
  classifying igneous rocks and, *95*
  granite and, 86
  in mineral identification table, *665*
  minerals and, 68, 69, *69, 74, 75, 76*
  weathering and, 137, 139
Planetesimals, 11, 12, 476
Plankton, 544, *545*
Plant evolution, *564*, 564–568, *566*
Plant fossils, 28, *28*
Plants
  in Cenozoic history, 614
  distribution of, 656
  gamete-bearing, 566
  in Mesozoic history, 592, 593, *594*
  nonvascular, 565
  seedless vascular, 565–566, *567*, 568, 593, 632
  vascular, 565
Plasma, defined, 60
Plastic flow, 345, *345*, 347
Plastic strain, 236, 237

Plate boundaries
  convergent (*See* Convergent plate boundary)
  divergent (*See* Divergent plate boundaries)
  earthquake epicenters and, *204*
  hot spots/mantle plumes and, 48
  mineral deposits and, 52
  oceanic-continental, *45, 46,* 249, 251, *530*
  plate tectonics and, 16, 17, *17*
  transform, *17,* 47, *47, 48,* 203
  types of, 40–48
Plate movement/motion
  continents/ocean basins and, *2,* 656
  distribution of life and, 53
  Hadean and, 476
  heat and, 476
  natural resources and, 52
  Paleozoic history and, 504
  in plate tectonic theory, 48–50, *50, 51*
  plate tectonics and, *2, 5, 20,* 21
  seafloor spreading and, 38
  visualizing, 40
Plate positions, *49, 50*
Plate tectonic theory/plate tectonics
  Archean, 480, 482
  Cenozoic, 615
  continental drift and (*See* Continental drift)
  defined, 16–17, 39–40
  distribution of life and, 53–54, *54*
  driving mechanism of, 50–52
  Earth's magnetic field and, 35, *35,* 37, *38*
  hot spots and, 34, *39,* 48, *49*
  magnetic anomalies and, 37, *38,* 49, *49*
  magnetic reversals and, 36–39, *37*
  mantle plumes and, 34, 48, *49*
  metamorphism and, 194
  mountain building and, 246–252
  natural resources and, 52–53
  ocean basins and, 488
  paleomagnetism and, 36
  Paleozoic history and, 505, 542
  plate boundaries and (*See* Plate boundaries)
  plate movement/motion and (*See* Plate movement/motion)
  polar wandering and, 36, *36*
  rock cycle and, 20, *20*
  seafloor features and, 32–34
  seafloor spreading and, 36–39, *38*
  of terrestrial planets, *42–43*
  understanding Earth and, 6, 16–17, 20–22
  volcanoes and, 126–128
Plates, defined, 13
Platforms, 478–480, *479, 480,* 504, *505*
Platinum, 62, 66, 70, *79,* 498, 499
Plato, 388, *389*
Platypuses, 639
Playa lakes, 378–379, *380*
Pleistocene deformation, *629*

Pleistocene Epoch. *See also* Ice Age
  in Cenozoic history, 628–631, 639–650
  deserts and, 377, 404, 405, *405*
  geologic time and, *417,* 436, 442
  glaciers/glaciation and, 340, 349, 352, 357, 359, 360
  groundwater and, 324
  running water and, 304
Pleistocene extinctions, 649–650
Pleistocene faunas, 639–650
Pleistocene glacial deposits, 479
Pleistocene glaciation, 628–631
Pleistocene tectonism, 628
Pleistocene volcanism, 628, *629*
Plesiosaurs, 576, *592, 595,* 601, *601,* 605
Pliny the Elder, *111*
Pliny the Younger, *111*
Pliocene Epoch, *417*
Plucking, 350, 351, *351,* 354
Plug domes, 119. *See also* Lava domes
Plunging breaker, *395,* 396
Plunging folds, 240, *240*
Pluto, 11, *11,* 14, *14*
Plutonic rocks, 87, 92, 93, 95, 479, 480, *487.* *See also* Plutons
Plutons. *See also* Igneous rocks
  Cenozoic history and, 615, 619
  concordant, 99, *101*
  defined, 98
  deformation and, 234, 242, 247, 248, 251, 254
  diagram of, *101*
  discordant, 99, *101,* 102
  granite and, 100
  greenstone belts and, 480
  igneous activity and, 486
  as intrusive igneous bodies, 98–99, 101–102
  metamorphism and, 185, *186*
  orogens and, 484
  volcanoes and, 99, 126–128
Pluvial lakes, 630, 632
Plymouth Rock, Massachusetts, 100, *100*
Pocket beaches, 400
Point bar, 294, *295*
Point Fermin, California, 263, *263,* 268, 270, *271,* 279
Point mutation, 458
Polar wandering, 36, *36*
Pollutants, in groundwater, 326–327, 330, *330*
Pollution
  acid rain and, *658*
  atmospheric, 144, *654*
  as chemical deterioration, 152
  groundwater and, 314
  hydroelectric power and, 289
  impact of, 656
  water, 4
  water supplies and, 6, 7
Polymerization, 492, *493*
Polymers, 492
Polyploidy, 459

Pompeii, Italy, 110, *111,* 125
Ponds, 314, 317, 325, 396
Popping, 138
Population
  deserts and, 366
  as environmental issue, 7–8
  groundwater and, 326
  of Mexico City, *654*
  Pangaea and, 576
Population density, 658
Porcelain, 187, 195
Pore spaces, defined, 160
Porosity
  geologic time and, 431
  of groundwater, 314–315, *315*
Porphyritic texture, 93, *94, 96*
Porphyry copper, 591
Portage Lake, Alaska, *344*
Port-au-Prince, Haiti, *198,* 200, *203*
Portland Museum of Science and Industry, *455*
Positive budget, 403, *404*
Positive gravity anomalies, 252, *253*
Postolonnec Formation, France, *419*
Potassium
  Bowen's reaction series and, 90
  in Earth's crust, 87
  geologic time and, 434, 435
  magma and, 87, *87,* 92, *92*
  in seawater, 162
Potassium feldspars
  Bowen's reaction series and, *89,* 90
  classifying igneous rocks and, *95, 96,* 97
  granite and, 86
  magma and, 91
  mineral identification and, 664, *665*
  minerals and, *68,* 69, *69, 74, 75,* 76
  weathering and, 137, 139
Potassium-argon method, 435
Potential energy, 292
Potholes, 292, *293*
Powder River basin, 631
Powell, John Wesley, *414,* 416, 519, *519*
Powell Memorial, 416
Power plants, 144, 246, 288–289, *302, 335,* 409
Pratt, J. H., 252
Precambrian Earth history
  Archean Earth history during, 480–482
  atmosphere and, 489–491, *490,* 492
  foundations of continents during, 478–480
  Hadean and, *476,* 476–478, *481*
  hydrosphere and, 489, 491
  introduction to, 476
  Proterozoic Earth history during, 482, 484–489
  Sudbury Basin and, 498
Precambrian Eon
  continental architecture and, 504
  defined, *476, 476*
  events during, 476

geologic time and, *417, 426*, 429, *430*, 442
  Grand Canyon and, *519*
Precambrian geologic time scale, *477*
Precambrian life history, 491–497
Precambrian rocks, *479, 480*, 497–499
Precession of the equinoxes, 360
Precious gemstones, 76
Precious metals, 72–73
Precipitation
  Cenozoic history and, 634
  deserts and, *375, 376*, 378
  glaciers and, 348
  groundwater and, 314, 316, *319*, 323
  running water and, 287
  from seawater, *520*
Prediction, defined, 464, *464*
Pressure
  differential, 185, *185, 188*
  Earth and, 223
  folding and, 237
  geothermal energy and, 335
  geysers and, *333*
  hydrostatic, 184, 319, 320
  hydrothermal activity and, 334
  lithostatic, 184, *184*
  in metamorphism, *184*, 184–185, *185, 189*
  rock layers and, 234, 236, 242
Pressure release, defined, 138
Primary producers, 545
Primary sedimentary structures, defined, 174
Primary waves, 206, *206*, 207, *208, 211*, 223–227
Primate evolution, 641–642
Primates, 614, 635, 639, *639*, 641, 642, *642, 643*
Principal meridians, 662
Principle of buoyancy, 253
Principle of cross-cutting relationships, 418, *419*
Principle of faunal and floral succession, 420
Principle of fossil succession, 420, *423*
Principle of inclusions, 420, *422*, 425
Principle of isostasy, 252–253, *253*
Principle of lateral continuity, 418, *419*, 428
Principle of original horizontality, 236, 418, *419*, 427
Principle of superposition, 418, *419*, 420, 427
Principle of uniformitarianism.
  See Uniformitarianism
Principles of Geology (Lyell), 417
Prism, *65*
Proboscis monkey, 642
Productid, *553*
Proglacial lakes, 630, 631, *631*
Progradation, 296, 526
Prokaryotes, *484, 493*, 494, 496
Prokaryotic cells, 494, *494*, 495
Prosimians, 641–642, *643*
Protein synthesis, 496
Proterozoic Earth history, 482, 484–489

Proterozoic Eon
  Archean Earth history and, 480
  atmosphere and, 491
  Cambrian explosion and, 542, 543
  continental architecture and, 504
  geologic time and, *417, 476*
  ice ages and, 359
  life of, 494–497
  plate tectonics since, 40, 482
  Precambrian time and, 476
Proterozoic evolution of Laurentia, *485*
Proterozoic fossils, *495, 497*
Proterozoic geologic/biologic events, *484*
Proterozoic glacial deposits, 488–489
Proterozoic glaciation, *489*
Proterozoic resources, 499
Proterozoic rocks, 478, 479, 480, 482, *487*, 488–489, 493, 495, 497, 499
Proterozoic supercontinents, 486, 488
Protoavis, 602
Protobionts, 492
Protocontinents, 478
Protolepidodendron, 566
Protons, defined, 62, 432
Protorothyrids, 562, *563*, 593
P-S time interval, 207, *208, 209*
Pteranodon, 600, *600*
Pterodactylus, 600
Pterosaurs, 576, 595, 600, *600*, 605
Pumice, 98, *98, 99*
Punctuated equilibrium, 459
P-wave shadow zone, 224, *225*
P-waves. *See* Primary waves
Pyramid Lake, *631*
Pyramids of Giza, 632
Pyrenees, 615
Pyrite, *65*, 70, 71, *71*, 153, 195, 490, *664*
Pyroclastic materials
  in Cascade Range, 126
  cinder cones and, 118, *118*
  composite volcanoes and, 118, *119*
  crystalline rock and, 158
  defined, 87, 89
  forecasting eruptions and, *130*
  igneous activity and, 127, 128
  lava domes and, 120
  overview of, 113
  pyroclastic sheet deposits and, 125
  shield volcanoes and, 116, *117*
  soil and, 142
  supervolcano eruptions and, 120
Pyroclastic sheet deposits, 122, 124–125
Pyroclastic texture, 94, *98*
Pyroxene
  Bowen's reaction series and, 89, *89*
  classifying igneous rocks and, 95, *95*
  minerals and, *68*, 69, *69*, 74, *75*

Quadrangles, 661, 663
Quadrupedal, 594
Quarrying, 350, 351

Quartz
  in beach sand, 401
  Bowen's reaction series and, *89*, 90
  classifying igneous rocks and, 96, 97, *97*
  in Earth's crust, 87
  geodes and, 175
  granite and, 86
  graywacke and, 480
  hardness of, *75*
  magma and, 91
  metamorphism and, 189, 191
  in mineral identification table, *665*
  minerals and, 60, *60*, 64, *65*, 66, *68*, 69, *69*, 70, *73, 75*
  as resource, 79
  weathering and, *136*, 137
Quartz sandstone, 161, *193*, 194
Quartzite, *19*, 142, *190*, 192, *193*, 194
Quaternary Period, *417*, 614, 615, 628
Queen Charlotte Fault, Canada, *624*, 625
Queenston Delta, 529, 530, *530, 531*
Quick clays, *266, 274*, 274–275, *275*

Rabbits, 635
Raccoons, *640*
Radial cleavage, 554, *554*
Radial drainage, 303, *303*
Radiation
  Big Bang and, 10–11
  exoskeleton and, 543
  marine invertebrates and, 542
  of multicelled organisms, 550
  solar, 489
  ultraviolet, 122, 489, 490, 492
Radicals
  defined, 67, *68*
  mineral groups and, *67*
  phosphate, 70
Radioactive decay, 432, *432*, 432–433, *433*, 435, 477, 491
Radioactive isotopes, 62, 435–436
Radioactivity, 416, 418
Radiocarbon dating methods, 436
Radiogenic heat, 480, 486, 490
Radiolarians, 545, 632
Radiometric dates, *440*
Radiometric dating, 416, 418, *434, 435*
Rae craton, 484
Rain
  acid, 144–145, *145*, 656, 657, *658*
  monsoon, 286, *286*
  oceans and, 390
  wind erosion and, 368
Rainfall, 317, 376, 377, 405
Rainfall patterns, 9
Rain-shadow deserts, 375, *375*
Raleigh waves, 206, 207
Rancho La Brea, California, 641
Ranges, 663
Rapid mass movements, 265
Ratio scale, 661
Raton Basin, New Mexico, *606*

Ray-finned fish, *555, 556,* 558, *558*
Rayleigh, Lord, 206
Rayleigh waves, 206, *207*
Recent Epoch. *See* Holocene Epoch
Recessional moraines, *356,* 357
Recharge, 317, *319,* 323, 324, 325
Recharge ponds, 317, 325
Recovery, of groundwater, 315
Recrystallization, 192
Rectangular drainage, 302, *303*
Recumbent folds, 238, *239*
Recurved spit, 401
Red beds, 486, 489, 491, 509, 526, 579
Red Canyon, *134*
Red Planet, *43. See also* Mars
Red Rocks Amphitheatre, Colorado, 526
Red Sea, as divergent plate boundary, *44*
Redoubt volcano, Alaska, 113
Reef communities, 551, 552
Reef development, 520
Reefs
   atolls as, 392, *392*
   barrier, 392, *392, 517*
   defined, 392
   Devonian, 551
   evaporite deposits and, 526
   fringing, 392, *392*
   in Gulf Coastal region, 583
   Kaskaskia Sequence and, 520
   in Mesozoic history, 592
   oil production and, *529*
   Ordovician, 550
   organic, 515, *515,* 523
   Paleozoic history and, 515–517,
      520, 526
   patch, 552, *553*
   pinnacle, 516
   Silurian, 551
Reforestation, 299
Refraction, 75, 223, *224,* 396, *397,* 400
Regional metamorphism, 185, 188, 189,
   190, 192, 193
Regolith, 142
Reid, H. F., 201
Reindeer, 650
Relative dating, 416
Relative dating methods, 418–428
Relative dating principles, *428*
Relative geologic time scale, 416
Relative movement, of blocks, 243
Relief, defined, 148, 661
Repenomamus giganticus, 604
Reproduction, sexual, 494
Reptiles
   cynodonts and, 603
   diversification of, 593–597, 600–601
   evolution and, 461, 465, *466,* 467, 468,
      470, 542, 562–563
   flying, 576, 600, *600*
   geologic time and, *417*
   to mammals, *602*
   marine, 576, 600–601, *601*

   in Mesozoic history, 576, 582, 591,
      593–597, *600,* 600–601, *601*
   in Messel Pit Fossil Site, 638
   in Paleozoic life history, 542, 562–563
   stem, 594, *595*
Reserves, 78–80
Reservoir rock, defined, 173
Reservoirs, 288–289, 290, *302,* 305, 314,
   317, 326, *327,* 344, 349, 355, 523
Residual bond, 64. *See also* van der Waals
   bonds
Residual concentrations, 153
Residual heat, 490, 491
Residual soils, 143
Resources
   Archean, 497, 499
   defined, 78
   energy, 6, 7, 60, 78, 80, 534, 657
   geologic, 6
   metallic, 78
   mineral (*See* Mineral resources)
   natural (*See* Natural resources)
   nonmetallic, 78
   nonrenewable, 4, 7, 79
   nonrenewable energy, 6
   from oceans, 409
   in Precambrian rocks, 497, 499
   Proterozoic, 499
   in sedimentary rocks, 173–174, 176
   weathering and, 153
Retaining walls, 279, *281*
Reverse fault, *243,* 244, *244*
Rhinoceroses, *470,* 636, 637, 650
Rhipidistian crossopterygian, *559,* 560
Rhipidistians, 558
Rhizome, 565
Rhyacian Period, *477*
Rhyolite, 95, *95,* 96–97, *97,* 175, 486
Rhyolite tuff, 98
Richter, Charles F., 211
Richter Magnitude Scale, 211, *211,* 212
Ridgeley Formation, 534
Ridge-push, 51, *51,* 52
Rifting, 40, 44, 52, *483,* 485, 486–487, 576,
   582–583, *583,* 615
Right-lateral strike-slip fault, 244
Rill erosion, 152, *153*
Rills, 152, 290, 305
Ring of Fire, *26, 125*
Rio Grande Rift, 621, *624*
Rio Reventado, Costa Rica, *153*
Rip currents, 396, 397, *397,* 400
Ripple marks, 168, *168,* 174
Riprap, 407, *409*
Rivers
   Cenozoic history and, 621
   deltas and, 296
   flooding, 299
   floodplain deposits and, 296
   longshore currents and, 400
   meandering, *294*
   as running water (*See* Running water)

   shorelines and, *402*
   valleys and, 305
RNA, 492
Roche moutonnée, 351, *351*
Rock bolts, 279, *281*
Rock bursts, 138
Rock cycle, 17–20, *18, 20,* 93, 136, 656
Rock flour, 351, *352*
Rock gypsum, *162, 164, 164,* 173, 390, 534
Rock layers
   anticlines as, 237–238, *239*
   folded (*See* Folded rock layers)
   geologic time and, 416, 420
   in Grand Canyon, *414*
Rock salt, *162,* 163, 164, *164,* 390, 442, 534
Rock sequences, 29–30, *30,* 36, 319, 416
Rock slides, *266,* 268, 269, *269, 270,*
   270–271, *271, 272*
Rock type
   geologic time and, 431, 438
   soils and, 148
   weathering and, 137
Rock varnish, 376, *377*
Rockfalls, 265, *265, 266*
Rock-forming minerals, 75, 75–77
Rock-forming silicate minerals, *69*
Rocks
   alteration of, 137
   Archean, 478, 479, *479,* 480, 482, 488, 497
   balanced, *150*
   basement, 182, 504, 526
   brittle, 236
   Cambrian, 493
   cap, 173
   in Cenozoic Era, 614, *614*
   country (*See* Country rock)
   crystalline, 158
   defined, 17, 75, 86
   ductile, 236
   in England, 504
   in Franciscan Complex, 584
   geodes in, *58*
   of Grand Canyon, *519*
   Grenville, 487
   heat conduction of, 87
   igneous (*See* Igneous rocks)
   Mesoproterozoic, *495*
   metamorphic (*See* Metamorphic rocks)
   mixed, 191
   Moon, *481*
   Neoarchean, 488
   Neoproterozoic, *495,* 497
   Newark, 582
   Paleoproterozoic, 488
   plutonic, 87, 92, 93, 95, 479, 480, *487*
      (*See also* Plutons)
   Precambrian, *479,* 480, 497–499
   Proterozoic, 478, 479, 480, 482, *487,*
      488–489, 493, 495, 497, 499
   reservoir, 173
   sedimentary (*See* Sedimentary rocks)
   Silurian, 557

source, 173, 176
transform faults and, 47
ultramafic, 95
volcanic (*See* Volcanic rocks)
Rocky Mountain Arsenal, 219–220, *221*
Rocky Mountain National Park, Colorado, *192*, *273*
Rocky Mountain system, 245
Rocky Mountains, 265, 345, 442, 443, *479*, *480*, 576, 581, 585
Rodents, 635
Rodeo Beach, California, *402*
Rodinia, *484*, 488, *488*, 504
Rogue waves, 396
Roosevelt, T., *84*, 100, 518
Roots, 565
Ross Ice Shelf, 341
Rotational slides, 266. *See also* Slumps
Rounding, 159, *159*, 167, 172
Roza lava flow, 123
Rubidium, 499
Rubidum-strontium pair, 435
Ruby, 76, 77
Rudists, 592
Rugose corals, 550, 551, 553
Running water
channel flow and, 290
dams and, 288–289
defined, 290
deposition by, 293–298, *304*, 305, 306
deserts and, 377, 378
discharge and, 290–292, *291*
drainage systems for, 302–305
on Earth, 286, *286*, 287, 290
erosion and, *156*, 290, 292–293, *304*, 304–306, *307*
floods/flooding and (*See* Floods/flooding)
fluid flow and, 290
glaciers and, 344, 352
gradient and, 290–292, *291*
hydroelectric power plants and, 288–289
introduction to, 286
reservoirs and, 288–289, 290
sediment transport and, 292–293
sheet flow and, 290
valleys and, 305–309, *306*, *309*
velocity and, 290–292, *291*
wind and, 366
Runoff, 287, 290, 299, 305, 316, 344, *347*
Russian River, California, *402*
Russo-Japanese War, 6
R/V Chikyu ("Earth"), 389
R-waves. *See* Rayleigh waves

Safe Drinking Water Act, 328
Sahara Desert, Africa, *364*, 366, 442, *442*, 506
Sahel area, Africa, 366
Sahelanthropus tchadensis, 643, *645*
Salamanders, 561
Saline lakes, 388

Salinity, 390, 393, 417, 491
Salinization, 152
Salmon Glacier, Canada, *357*, *358*
Salt
geologic time and, 417
Louann, 591
in Michigan Basin, 516
minerals and, 62
rock, 162, 163, 164, *164*, 390, 442, 534
seawater and, 409
table, 173
Salt crystal growth, 139
Salt deposits, 576
Salt domes, 576, 583
Salt pan, 379. *See also* Playa lakes
Saltation, 293, 366–367, *367*, 369, *370*
Saltpeter, 314
Saltwater
hydrosphere and, 388
plants and, 564, 565
quick clays and, 274
Saltwater contamination, 314
Saltwater incursion, 325, *326*, 407
Samuel H. Boardman State Park, Oregon, *399*
San Andreas fault, California
in Cenozoic history, 614, *624*, *625*, 628, *629*
deformation and, 244, *245*, 246, *247*
earthquakes and, 201, *214*, 214–215, *215*, 217, 219, 222
metamorphic rocks and, 188
plate tectonics and, 47, *48*
San Francisco earthquake, *200*, 201, *201*, *210*, 211, *211*, 213, 214
San Joaquin Valley, California, 326, *327*
Sand
Cenozoic history and, 632
glaciers and, 340, 355, *358*
gold and, 576
as mineral resource, 534
as ocean resource, 409
replacing, 409
running water and, 292, 296, *298*
sedimentary rocks and, 158, *158*, 160, 161, *161*, 172
silica, 534
wave erosion and, 398
Sand dollars, 632
Sand facies, 165
Sand seas, 371
Sandblasting, 367
Sandstone
continental red beds and, 489
deformation and, *237*
deserts and, 382
geologic time and, 420, *422*
in Glacier National Park, *487*
in Grand Canyon, *487*, 488
graywacke as, 480
groundwater and, 315, 317, *318*, 319
minerals and, 60

oceans and, 391
Paleoproterozoic history and, 486
quartz, 161, *193*, 194
rifting and, 487
sedimentary rocks and, 161, *162*, 172, 174, 176
sedimentation and, 487
at Shore Acres State Park, *386*
Sandstone facies, 165, *166*
Sandstone-carbonate-shale assemblages, 485, 488
Sangre de Cristo Mountains, 245
Santorini volcano, Greece, 125
Sapphire, 76, 77
Saturn, *11*, 12, 14, *14*, 15
Sauk Sequence, *511*, 511–513, 514
Saurischia, 595, *596*
Saurischian dinosaurs, 595–596
Sauropod, 596, *596*, 597, 598
Savannah conditions, 634
Scharth, A., *121*
Scheelite, *195*
Schist, *188*, 190, *190*, 191, *191*, 474
Schistose foliation, 191
Schistosity, 191
Schizochroal eyes, 549, *549*
Scientific method, 5, 12
Scleractinians, 592
Scoria, 98, *98*, 99
Scorpions, 560
Sea arch, 399, *399*
Sea caves, 399, *399*
Sea cliffs, 398, 399, *399*, 404
Sea level
atmosphere at, *286*
base level and, 304, *304*
changes in, 656
coastal management and, 407–409, *408*
continents and, 478
geologic time and, 443
landscapes and, 305
Paleozoic history and, 504, 511, 512, 516
rising, 349, 352, 407–409, *408*, 576, 580, 629, 656
stream terraces and, 306
Sea lions, *640*
Sea of Ice, *342*
Sea of Japan, 46, 480
Sea stacks, 399, *399*, 404
Seafloor features, 32–34, *391*, 391–392
Seafloor spreading, 36–39, *38*
Seafloor spreading theory, 37, 39
Seals, *640*
Seamounts, 34, *34*
Seas, defined, 396
Seawall construction, *406*
Seawalls, 406, 407–409
Seawater
color of, 390, *390*
composition of, 390, 409
elements for life in, 493
oceanic circulation and, *390*, 390–392

Seaweed, *545*

Secondary sedimentary structures, defined, 174

Secondary waves, 206–208, *211,* 223–225, 253

Sections, 663

Sediment
  becoming sedimentary rock, 158, 160, *161*
  chemical, 158, 160
  deep-sea, 44, 46, *46,* 656
  defined, 158
  deformation and, 248
  detrital, 158, 160
  geologic time and, 418, 434
  outwash, 632
  Phanerozoic, 504
  rounding in, 159, *159*
  running water and, 293, 294, 296, 298
  Sauk Sequence and, 513
  sorting in, 159, *159*

Sediment sources, 158–159

Sediment transport
  glaciers and, 340, 350–355
  running water and, 292–293, *293*
  sedimentary rocks and, 158–159
  on shorelines, 400
  by wind, 366–367

Sedimentary breccia, 161, *162*

Sedimentary facies, 165–166

Sedimentary rocks
  BIFs as (*See* Banded iron formation (BIF))
  biochemical, *162,* 162–165
  Cenozoic history and, *625, 628,* 631
  chemical, *159, 162,* 162–165, *163, 164*
  in Claron Formation, *134*
  concretions and, *174,* 174–175, *175*
  continental crust and, 182
  continents and, 479
  defined, 158
  deformation and, 236, *237,* 247, 251
  deserts and, 382
  detrital, *159,* 161, *161, 162*
  Earth's interior and, 223
  fossilization/fossils and, 169, *169, 170–171,* 172
  geodes and, 174–175, *175*
  geologic time and, 416, 418, *419,* 420, 424–425, *427,* 429, 434, 435, 437, *439,* 480
  in granite-gneiss complexes, 480
  in greenstone belt, 480
  igneous rocks and, 86
  magma and, 91
  mass wasting and, 263
  metamorphism and, 185, 187, 190, 192, 193
  minerals and, 70, *75,* 77
  natural gas and, 158, 173, *173,* 176
  oceans and, *391,* 391–392
  orogens and, 484
  Paleoproterozoic, *486*

Paleozoic history and, 511, 565
  at Palo Duro Canyon State Park, *156*
  petroleum and, 158, 173, 176
  plate boundaries and, 46
  plate tectonics and, 30
  plutons and, 99
  porosity of, 315, *315*
  resources in, 172–173, 176
  rounding and, 159, 167, 172
  sediment becoming, 158, 160, *161*
  sorting and, 159, 167, 172
  story in, 167–169, 172
  subsidence and, 325
  at Sudbury Basin, 498
  types of, 161–165
  understanding Earth and, 17, 18, *18, 19,* 20, *20*
  uranium and, 158, 176
  weathering and, 136
  Wopmay orogen and, 485

Sedimentary structures, 167–169, 172, 174

Sedimentation
  cyclothems and, 526
  Devonian Period and, 521
  evaporite, *520*
  Mesoproterozoic/Neoproterozoic, 487–488
  Mesozoic Era, 580, 582, 585–586, 589

Sediment-deposit feeders, 545, *545,* 546

Seedless vascular plants, 565–566, *567,* 568, 593, 632

Seismic gaps, 217, *219*

Seismic profiling, 389, *389*

Seismic rays, 223

Seismic risk maps, 217

Seismic sea wave, 216. *See also* Tsunami

Seismic tomography, 226–227

Seismic waves, 202, 205–207, *206, 207,* 211, 212, 223–227

Seismic zones, 203

Seismic-moment magnitude scale, 212

Seismogram, 202, *202*

Seismograph stations, 207, *208, 209*

Seismographs, 202, *202, 208,* 224

Seismology, 202–203

Self-exciting dynamo, 35

Semiarid regions, 374, *375,* 382

Semiprecious gemstones, 76, 77

Septaria, 174

Septarian nodules, 174, *174*

Septic systems, 331

Sequence boundaries, defined, 511

Sequence, defined, 511

Sequence stratigraphy, 511

Series, defined, 440

Serpentine, 185

Serpentine asbestos, 183, *183*

Serpentinite, 46

Sessile benthos, 545

Sevier orogeny, 585, *588,* 617, 620

Sex cells, 457, *457,* 458

Sexual reproduction, 494

Shale
  Antrim, 521, *522* (*See also* Black shales)
  black, 520, 521–522, *522,* 528 (*See also* Antrim Shale, Chattanooga Shale, and New Albany Shale)
  Bright Angel, *166*
  Cenozoic history and, 631
  Chattanooga, 176, 521, *522* (*See also* Black shales)
  continental red beds and, 489
  cyclothem and, 525
  defined, 161
  deformation and, *237*
  in Grand Canyon, 488
  groundwater and, 317, *318, 319*
  metamorphic rocks and, 187, *187,* 187–188, 189
  New Albany, 521, *522* (*See also* Black shales)
  oceans and, 391
  Paleoproterozoic history and, 486
  rifting and, 487
  sedimentary rocks and, 161, *162,* 165, *173,* 174, 176
  sedimentation and, 487
  Tippecanoe Sequence and, 515

Shale facies, *166*

Shallow-focus earthquakes, 202, 203, *204*

Shark Bay, Australia, *493*

Sharks, 557, *557*

Shear strength, 261

Shear stress, 235, *236,* 244

Shear waves, 206

Shearing, 241, 487

Sheep Mountain anticline, Wyoming, *240*

Sheep Rock, *151*

Sheet erosion, 152, 290

Sheet flow, 290, 292

Sheet joints/jointing, 138, *138,* 242

Sheet lava, 46, *46*

Sheet silicates, 139, 141

Shelf-slope break, 33

Shell Creek, Wyoming, *479*

Shell Oil Company, 410, *410*

Shelley, Percy B., 6

Shelly fauna, *543,* 543–544, 550

Shermer, Michael, 464

Shield volcanoes, 116–117, *117,* 119, 126, 128, *625*

Shields, 182, 478–480, *479,* 504

Shore Acres State Park, Oregon, *386*

Shoreline processes, 392–397

Shorelines
  coasts *vs.,* 392, *393*
  defined, 392
  deposition along, 398–403
  erosion along, *398,* 398–403, *399,* 408–409
  perils of living along, 405–409
  shoreline processes and, 392–397
  study of, 389

Shrews, 642

Shrimp, *551*
Siberia, 506, *506, 507, 508,* 509, *510*
Siberian Traps, 554
Siccar Point, Scotland, *425*
Siderian Period, *477*
Sierra Grande, New Mexico, *625*
Sierra Nevada, California, 97, 265, 345, *417, 574, 576,* 581, *585,* 591, 621, *621*
Sigillaria, 568
Silica
    Cenozoic history and, 631
    chert and, 164
    classifying igneous rocks and, 95, 96
    defined, 67
    groundwater and, 334
    Hadean and, 478
    magma and, 87, 88, 90, 91, 92, *92,* 95, 97
    oceans and, 390, 391
    in seawater, 162
Silica radicals, *68*
Silica sand, 534
Silica tetrahedron, 68, *68,* 69, 74, 88
Silicate minerals, *67, 67*–69, *68, 69,* 76, *142,* 183, 184
Silicates
    defined, 67
    in Earth's crust, 87
    ferromagnesian (*See* Ferromagnesian silicates)
    nonferromagnesian (*See* Nonferromagnesian silicates)
    sheet, 139, 141
    weathering and, 139, 141
Siliceous ooze, 391, *391*
Siliceous sinter, 334
Silicon
    Bowen's reaction series and, 90
    in Earth's crust, 66, 87
    in rock cycle, 17
    in solar system, 12
Silicon dioxide, 78, 160, 175, 226
Sillimanite, 187, *188,* 189, 193, 195
Sills
    defined, 92, 99
    example of, *101*
    geologic time and, 420, *421*
    in Mesozoic history, 582
    rifting and, 44
Silt, 158, *158, 159,* 161, *161,* 296, 315, 368
Siltstone, 161
Silurian evaporites, 534
Silurian lava flows, 36
Silurian marine community, *551,* 551–552
Silurian paleogeography, *516*
Silurian Period
    evolution and, *470*
    fish during, 555, *555,* 557
    geologic time and, *417*
    Michigan Basin and, *517,* 520
    in Paleozoic history, 504, 506, *506, 507*
    plants during, 564, *564,* 565–566
    polar wandering in, *36*

Silurian rocks, 557
Silver, 63, 66, 70, *73, 79,* 498, 534
Silver sulfide, 70
Sinkholes, 320–322, *321, 322, 324*
Sinodelphys, 604
Skel, Native American god, 114
Skunks, *640*
Skylight, defined, 113
Slab-pull, 51, *51,* 52
Slate, 182, *186, 188,* 189, 190, *190, 195*
Slaty cleavage, 189
Slave craton, 482, 484
Slide-flow landslides, 277
Slides. *See also* Landslides
    defined, 265–267
    quick-clay, *274*
    rock, *266, 268, 268, 269, 270,* 270–271, *271, 272*
Slope
    Atlantic Coastal Plain and, 628
    continental, 32–33, *33,* 388, 391
    contours and, *661*
    defined, 148
    of dunes, 369
    glaciers and, 341
    running water and, 290
Slope angle, 148, 261, *261*
Slope direction, 148
Slope equilibrium, 264
Slope failure, 260, *261,* 263, 277, 278. *See also* Mass wasting
Slope instability, 264
Slope stability, 260, *260,* 263, 264, *264*
Slope-stability maps, 277, *278*
Sloss, Laurence, 509
Sloths, 639, 641
Slow mass movements, 265
Slumps, *266, 266,* 267, *267,* 271
Smerkar, Suzanne, 110
Smith, Robert Angus, 144
Smith, William, 420, 504
Smog, 144
Snails, 560
Snake River Plain, Idaho, 123, *124,* 619, *620, 620, 631*
Snakes, 576, *595*
Snow, 144, 344–345, *347*
Snowball Earth, 488–489
Snowfields, 290
Sodium
    Bowen's reaction series and, 90
    in Earth's crust, 87
    hard water and, 331
    magma and, 87, *87,* 92, *92*
    minerals and, 62, 63, *63,* 66
    in seawater, 162, 163, 409
Sodium chloride, 62
Sodium ions, in seawater, 390
Soil conservation practices, *153*
Soil degradation, 148, 149, 152–153, *153,* 657
Soil formation, 145, *146–147,* 147, 148, 152, 657

Soil horizons, 143–144, *146–147*
Soil orders, 145
Soil profile, 143–145
Soild bacteria metabolism, 144
Soils
    alkali, 147, *148*
    in arid regions, *148*
    composition of, *146*
    coniferous forest, *147*
    deciduous forest, *147*
    defined, 142
    desert, *147,* 376–377
    deterioration of, 152
    expansive, 148–149, 152–153, 657
    glaciers and, 350, 351
    grassland, *147*
    organisms and, 148
    origin of, 136–137, 142–148
    permafrost, 409
    residual, 143
    running water and, 290, 292
    sedimentary rocks and, 158
    transported, 143
    tropical rain forest, *147*
    weathering and, 136–137
Solar energy, 360, 374
Solar nebula theory, 11, 12, *12*
Solar radiation, 489
Solar system
    diagrammatic representation of, *11*
    evolution of, 11–12
    origin of, 9, 11–12
Solar wind, 490
Solids
    defined, 60
    glaciers and, 350
    running water and, 290
Solifluction, *266,* 275, *275*
Solnhofen Limestone, Germany, 601, *601*
Solution, defined, *140,* 140–141
Solution valleys, 321, *322, 323*
Somalia, 28
Sonoma orogeny, 580, *584*
Sonoran Desert, North America, *376, 454*
Sorting, 159, *159,* 167, 172
Sound waves, 389
Source rock, 173, 176
South American plate, 39, *39,* 46, *125,* 615
Southern Superior craton, 482, *483*
Soviet Academy of Agricultural Sciences, 453
Speciation, 458–459, *460*
Species, defined, 458
Specific gravity, 74–75, 664
Sphalerite, 74, *74,* 195, *195. See also* Zinc
Sphenopsids, 566, 568, *568*
Spheroidal weathering, 142, *143*
Spider monkey, *643*
Spiders, 560
Spilling breaker, 395, 396
Spindles, *497*
Spiral cleavage, 554, *554*
Spires, 137, *150, 612*

Spits, 401–402, *402*

Sponges, *529, 540, 553, 553*

Sporangia, *565*

Spore-generating plant, 566

Sporophyte, 566, *567*

Spreading ridges, 40, 50, *50*, 89, *90*, 90–91, *127*, 158, *204*, 225, 227, *530*, 615. See *also* Divergent plate boundaries

Sprigg, R. C., 496

Spriggina, 496, *496*

Spring tides, 394, *394*, 408

Springs, 317, *318*, 321, *322*, 331, 332, 334, *334*

Sputnik 1, 656

Squid, 592, 632

Sri Lanka, 28

St. Elias Mountains, Canada, *354*

St. Francis Dam, California, *300*

St. Peter Sandstone, 534

Stacks, 399

Stage, defined, 440

Stagnant glacier, 346

Stalactites, *312, 323, 324, 325*

Stalagmites, *312,* 323, *324, 325,* 440, *441,* 444, *444*

Star dune, 371, *373*

Starfish, *545*

Stars, 11

Staurolite, *188,* 193

Steam, *333, 334, 335*

Steam power, 378

Steel, 172

Stegoceras, 599

Stegosaur, *596,* 597, 606

Stegosaurus, 597, *599*

Stem reptiles, 594, *595*

Stenian Period, *477*

Steno, Nicholas, 65, 236, 418

Steppe environments, 634

Stillwater Complex, Montana, 497

Stillwater Range, Nevada, *248*

Stocks, defined, 102

Stone Forest, China, *323*

Stone Mountain, Georgia, 100, *138*

Stoping, defined, 103, *104*

Storm surge, 406

Strain. See *also* Deformation

    defined, *235,* 235–236, *236,* 345

    elastic, 235, 236

    plastic, 236, 237

Strata

    basin and, 241

    defined, 167, *167*

    dome and, 241

    geologic time and, 420, *421, 423*

Stratherian Period, *477*

Stratified drift, 355, 358

Stratigraphic section, *440*

Stratigraphic terminology, 437–440

Stratigraphic traps, 176, 240

Stratigraphic units, *439*

Stratigraphy, 437–440, 511

Stratosphere, ozone depletion and, 122

Stratovolcanoes, 118, *119.* See *also* Composite volcanoes

Stream piracy, 305, *307*

Stream terraces, 306, *308*

Stream-dominated deltas, 296, *297*

Streams

    amphibians in, 561

    antecedent, 621

    base level of, 304, 305

    braided, 293–294, *294,* 355

    Cenozoic history and, 621, 626, 628, 629, 630

    coasts and, 404

    deltas and, 296

    deserts and, 377

    development of, *307*

    disappearing, 321, *322*

    drainage systems and, 302, 303

    flooding, 299

    floodplain deposits and, 296

    geologic time and, 417

    glaciers and, 347, 351, 352

    graded, *304,* 305

    groundwater and, 314, 316, 321, 326

    incised meanders and, 306–307, *308*

    meandering, 294, *295*

    as running water (See Running water)

    superposed, 308, *309,* 621

    valleys and, 305–306

Stress, 235, *235, 236,* 244, 345

Strike, 236–237, *237, 238, 239, 240, 242,* 243

Strike-slip faults, 243, *243,* 244, *245*

Stromatolites, *481, 484,* 485, *486,* 488, *493,* 493–494, 513, *517,* 528

Stromatoporoid, 515, 516, *517,* 550, 551, 552

Stromboli, Italy, 128

Strong nuclear force, 11

Strontium, *434*

Structural traps, 176

Subbituminous coal, *534,* 631

Subduction zones

    blueschist facies rocks and, 194

    distribution of life and, 53, 54

    Earth's plates and, *39*

    magma at, *90,* 91

    in Mesozoic history, 584

    natural resources and, 52, *53*

    plate boundaries and, 44, 45, *45,* 46, 47, *48,* 615

    plate movement and, 50, *50*

    rifting and, 485

    seafloor spreading and, 33, 39

    volcanism and, *127*

Subglacial lakes, 489

Sublimation, 344, 346, *347*

Submarine hydrothermal vents, 33, *34*

Submarine lava flows, 127

Submergent coasts, 404–405, *405*

Subsidence, 325–326, *327,* 392, 407

Subsoil, 144, *146*

Subsurface correlation, 431–432

Subsystems, of Earth, 4, *5,* 8, 20, 22, *364,* 656, 657

Sudbury Basin, 498

Sudbury Crater, Canada, *498*

Sudbury meteorite, 498

Sudbury mining district, Canada, 499

Suess, Edward, 28

Sulfate, in seawater, 162

Sulfate minerals, *67, 70, 75,* 77

Sulfide minerals, *67,* 70, *70, 73,* 153

Sulfur, 78, 110, 112

Sulfur dioxide, 78, 110, 112, 144, 490

Sulfuric acid, 144, 153

Sumatra earthquake, 212

Summer beaches, 401

Sun

    atmosphere and, 490, *490*

    Earth's orbital period around, 478

    oceans and, 389, 393

    ozone depletion and, 122

    running water and, 286

    in solar system, *11,* 11–12, *14, 15*

Sundance Formation, 589

Sundance Sea, 589

Supercontinents

    defined, 488

    Neoproterozoic, *488,* 504

    Proterozoic, 486, 488

Supergroups, defined, 439, *439*

Superior craton, 482, *483,* 484

Superposed streams, 308, *309,* 621

Surf zone, 396

Surface waters, 476, 478, 491

Surface waves, 205, 206, *208*

Suspended load, 292, *293,* 366, 367

Suspended water, 316

Suspension feeders, 545, *545,* 550, 592, 632

Sutter Buttes, California, 110

Sutter, John, 576

Sutter's Mill, California, 576

Swamps

    coal, 535, *538, 562,* 566, *567,* 568

    geologic time and, *417*

    groundwater and, 314, 316, 317

    running water and, 287, 303

S-wave shadow zone, 224, *225*

S-waves. See Secondary waves

Swells, 396

Sylvania Sandstone, 534

Symbiosis, 495

Synclines, 237–238, *238, 239,* 240, *240,* 241

System

    defined, 4, 439, 440

    of Earth, *364,* 656, 657

Table salt, 173

Tabulate corals, 550, 551, 553

Taconic Highlands, *507, 514,* 515, *516, 529, 530, 531*

Taconic orogeny, 506, *511,* 528–530, 532

Taj Mahal, India, *180,* 182

Talc, 75, *75*, 182, 187, 195, *664*
Talus, 137, *137*, 265
Tambora eruption, 110, 112, 128, 360
Tapeats Sandstone, *166*
Tapirs, *470*, 637
Tar sands, 176
Tarns, 354
Tarsiers, 641, *643*
Taupo, New Zealand, 121
Tectonic plate, *50*
Tejas epeiric sea, 627
Temperature
    Cenozoic history and, 634
    of crust, 227
    of deserts, 376, 378
    Earth and, 223, 227, 656
    folding and, 237
    geologic time and, 418
    of geysers, *333, 334*
    glaciers and, 342, 345, 347, 348, 349, 360
    global warming and, 348
    of hot springs, 331, *332*
    of mantle, 486
    oceans and, 393
    orogeny and, 248
    Pangaea and, 579–580
    of reptiles, 563
    rock layers and, 234, 236
Temporary base level, 304
Tension
    defined, 235, *236*
    deformation and, 235, *236*, 241
    glaciers and, 353
    groundwater and, 316, *316*
    oceans and, *408*
Terminal moraine, *356, 357*, 630
Terraces, 153, 323, *324, 334*, 398, 405
Terranes, 252, 533–534, 589–590, *590*
Terrestrial planets, 12, 14–15, *42–43*, 476
Tertiary Period, 614
Teton Dam, Idaho, 298
Teton Range, Wyoming, 245, *474*
Tetrapods, 558, 560, 561, *561*
Texture
    aphanitic, 93, *95, 96, 96*
    classic, 161
    crystalline, 163
    defined, 93
    foliated, 189, *189*
    fragmental, 94, *94, 98*
    glassy, *98*
    of igneous rocks, 93–94, *94, 95, 99*
    for marble, 182
    metamorphism and, 187, 189, *189*, 192
    nonfoliated, 192
    oolitic, 513
    pegmatite, 97, *98*
    phaneritic, 93, *95, 95, 96, 96*
    porphyritic, 93, *94, 96*
    pyroclastic, 94, *98*
    of sedimentary rocks, 165, 167, 172
    vesicular, *94, 98*

Texture classification, *98*
Thecodontians, *563*
Theodore Roosevelt National Park, North
    Dakota, *625*
Theory
    Big Bang, 10–11
    defined, 4–6
    elastic rebound, 201, *201*
    evolutionary (*See* Theory of evolution)
    Milankovitch, 359–360, *360*
    plate tectonic (*See* Plate tectonic theory/
        plate tectonics)
    seafloor spreading, 37, 39
    solar nebula, 11, 12, *12*
Theory of evolution, 450, 458, 463–470,
    464, 542
Theory of inheritance of acquired
    characteristics, 450–451, 453, 458
Theory of Relativity, 10
Therapsids, 563, *563, 564*, 594
Therizinosaurs, 596
Thermal convection cells, 37, 50, *50*
Thermal expansion and contraction, 139
Thermal proteins, 492, *492*
Thermal spring, 331. *See also* Hot springs
Theropods, 595, 596, *596*, 597, 601–602
Three Gorges Dam, China, 289
Thrust fault, *243*, 244, 248, 480, *588*
Thunder, 264
Thunder eggs, 175, *175*
Thunder River Spring, Arizona, *318*
Tidal bulges, *394*
Tidal ranges, 394
Tidal wave, 216
Tide-dominated deltas, 296
Tides
    defined, *393, 393–395, 394*
    ebb, 393, 409
    flood, 393, 409
    high, *393, 394, 400*
    low, *393, 394*, 400, *400*
    neap, *394, 394*
    spring, 394, *394*, 408
Tidewater glacier, 341, *341*
Tiktaalik roseae, 560–561, *561*
Till, 355, 357–358
Tillites, 442
Tilting, *425*
Time
    defined, 476
    geologic (*See* Geologic time)
Time units, 440
Time-distance graphs, 207, *208, 209*
Time-stratigraphic units, 439, 440
Tin
    importance of, 60
    metamorphism and, *195*
    as mineral resource, 534
    percentage imported, *79*
    plate tectonics and, 52
    sedimentary rocks and, 173
    weathering and, 137, 153

Tippecanoe Sequence, *511*, 514–517, 520, 550
Titanic, 389
Titanotheres, 463, *463*, 637
Toads, 561
Toba caldera, Sumatra, 115
Toba, Indonesia, 121
Tombolos, 401–402, *402*
Tonian Period, 477
Topaz, *75*, 77, *665*
Topocyon, 640
Topographic maps, 661–663
Topography, 6, *320*, 320–322, *322, 323*, 341
Topset beds, 296
Topsoil, 143, *146*
Tornadoes, 657
Tourmaline, 77
Township, 663
Toxic chemicals, 330
Toxic waste sites, 314, 330, 331
Trace fossil, 467, *468, 469*, 546
Transcontinental Arch, 512, 513
Transform faults, 47, *47*, 244
Transform plate boundaries, *17*, 47, *47,
    48*, 203
Trans-Hudson orogen, 484
Transition zone, 226
Transitional depositional setting, 159,
    160, 293
Transitional fossils, 468
Transocean, 411
Transpiration, 290
Transported soils, 143
Transverse dunes, 369, 371, *372*
Travertine, 334
Travertine terraces, 323, *324, 334*
Tree-ring dating methods, 436, *438*
Trellis drainage, 302, *303*
Triassic fault basins, United States, 44
Triassic Peak, Texas, *156*
Triassic Period
    amphibians during, 562
    continental drift and, *30, 32*
    dinosaurs and, 576
    evolution and, *470*
    fish during, *555*
    geologic time and, *417, 430, 442*
    Gulf Coastal region in, 583
    mammals during, 576
    paleogeography during, *577*
    Pangaea and, 576, 579, *579*, 582
    plants during, *564*, 568
    polar wandering in, *36*
    reptiles during, 563
    sedimentary rocks and, 158
Tribachidium, 496, 496
Tributaries, 302–303, *303*, 305, 340, 347,
    357, 358
Triceratops, 597, *598*
Triconodonts, 604
Trieste, 389
Trilobites, *169*, 467, 468, 546, *548*, 548–549,
    550, 551

Triton, 110
Trophic levels, 545
Tropical rain forest soil, *147*
Troughs, 351–352, 353, *353*, 354, *354*, 395, *395*
Truncated spurs, 352
Tsunami, *26*, 28, 200, 213, *216*, 216–217, 657
Tuataras, *595*
Tuff, 98, *98*, *99*, 110, 175
Tungsten, *79*, *195*
Tunguska Event, Siberia, 498
Turbidity current, 167, *168*, 480
Turbulent flow, 290, *290*
Turkey earthquake, *200*, 213
Turnagain Heights subdivision, Alaska, 246, 275, *275*, 659
Turquoise, 77
Turtle Mountain, 270, 271, *272*
Turtles, 576, *595*, 600
Tuscarora Sandstone, 534
2007 Fourth Intergovernmental Panel on Climate Change, 9
Tyrannosaurs, 596, 597
Tyrannosaurus rex, 598

Ultimate base level, 304
Ultramafic magma, 87, *87*, 90, 91, 95, *95*, 480
Ultramafic rocks, 95
Ultraviolet radiation, 122, 489, 490, *490*, 492
Uluru, Australia, 381, *381*
Unconformities, 420, *423*, 424–425, *425*, 428, 511, *511*, 520
Undercutting, 261, *261*, *262*, 264, 265, *270*
Ungulate, 636
Uniform Building Code, 278
Uniformitarianism, 21–22, 172, 417–418, 450
United States Department of Agriculture, 145
United States National Museum of Natural History, 542
Universe, origin of, 9–11
Unterer Grindelwald (Birmann), *342*
Unzen volcano, Japan, 119
Uplift. *See also* Subsidence
    in Canadian shield, *479*, 480
    Cenozoic history and, 615, 617, 621, *624*, 626, *626*
    cratonic, 527
    geologic time and, *424*, *425*, *426*, 427
    shorelines and, 405, 407
Upper mantle
    classifying igneous rocks and, 95
    defined, 13, *13*
    Earth's crust and, 227
    magma and, 88
    ophiolites and, 46, *46*, 47
Upright folds, 238
Upwelling, 390, 391, 582
Uralian mobile belt, 509
Urals, Russia, 252
Uraninite, 176, 490
Uranium, 62, 80, 158, 176, 435, 436, 591

Uranium-lead series, 435
Uranus, *11*, 12, 14, *14*, *15*
Ursa, 410, *410*
U.S. Army Corps of Engineers, *301*
U.S. Geological Survey (USGS), 78, 80, 121, 129, 219, 299, 328, 661, 662, 663
U.S. National Park Service, *332*
U-shaped glacial trough, 351–352, *353*, 354
Ussher, James, 416

Valley glaciers
    defined, *340*, 340–341, 344
    erosion/sediment transport and, 351–354, *352*, *353*, *354*
    glacial budget and, 346, 347, 350, *350*
    glacial deposits and, 355, 357
    movement of, 345, 347
    stratified drift and, 358
Valley of Ten Thousand Smokes, Alaska, 126
Valley of The Kings, Egypt, *164*
Valley trains, 358, *358*
Valleys
    coasts and, 404
    defined, 305
    evolution of, 305–309, *306*, *309*
    hanging, 352, *354*
    solution, 321, *322*, *323*
van der Waals bonds, 62, 63–64, *64*
Vanadium, *79*
Variation, 458, 463, 467
Varves, 359
Vascular plants, 565
Vegetation
    in Cenozoic history, 632, *633*, 634
    deserts and, 366, 376, *376*, 377
    dunes and, 370, *371*, *372*
    in Grand Canyon, *518*
    groundwater and, *318*
    mass wasting and, 261, 263, 277, 279, *281*
    Pangaea and, 579
    running water and, 298, 305
    Sauk Sequence and, 512
    soil formation and, *146–147*, 147
VEI. *See* Volcanic explosivity index (VEI)
Velociraptor, 596
Velocity
    of earthquakes, 206, 223, 224, *224*, 225, *226*, 227
    of glaciers, 347, *350*
    of groundwater, 316
    of running water, 290–292, *291*, *293*, *294*, *295*, *296*, 298
    of tides, 395
    of waves, 395, 396, 397, 400
Ventifacts, 367, *368*
Venus
    running water and, 286
    tectonics of, *42*, *43*
    understanding Earth and, *11*, 12, 14, *14*, 234
    volcanism and, 110

Venus de Milo, 182
Verne, Jules, 6
Vertebrate evolution, 554–555
Vertebrate jaw, *557*
Vertebrates, 456, 461, *462*, 463, *465*, 467, 554–555, *557*, 560–562
Vertical circulation, 391
Vesicles, 94, 98
Vesicular texture, *94*, *98*
Vestigial structures, 466, *467*
Virgin and Child with Saint Anne (da Vinci), 6
Virgin of the Rocks (da Vinci), 6
Viruses, 492
Viscosity
    defined, 88
    of lava, 88, *88*
    of magma, 88, *88*, 97
Viscous magma, *120*
Vog, 112
Volcanic arc, 46
Volcanic ash
    defined, 113
    glaciers and, *343*
    mass wasting and, 263, 271
Volcanic breccia, 98, *98*
Volcanic domes, 119. *See also* Lava domes
Volcanic eruptions
    atmosphere and, 490
    duration of, 128
    earthquakes and, 200, 214
    forecasting, 128–130
    as geologic hazard, 657
    glaciers and, 360
    mass wasting and, 264
    notable, 120
    overpopulation and, 657
    size of, 128
Volcanic explosivity index (VEI), *130*
    defined, 128
Volcanic gases, 110–112, 122–123
Volcanic glass, 98
Volcanic hazards, 128–130, *129*
Volcanic island arc, 248, *249*
Volcanic landforms, 122–125
Volcanic mountains, 246
Volcanic necks, 102, *102*
Volcanic pipe, 102
Volcanic rocks. *See also* Extrusive igneous rocks
    defined, 87
    deformation and, 246, 248, 252
    in greenstone belt, 480
    igneous rocks and, 93, 98, 99
    Precambrian history and, 479
Volcanic smog, 112
Volcanic tremor, 129
Volcanism. *See also* Volcanoes
    acid rain and, 144
    Cenozoic history and, 615, 619, 620, *620*, 628, *629*
    defined, 110

deformation and, 234
glaciers and, 347
greenstone belts and, 480
Hadean and, 476, 477
Holocene, 628
ice ages and, 360
magma and, 86, 91
on Mars, 43
Pleistocene, 628, 629
plutons and, 98
on Venus, 42
volcanoes and, 110–114
Volcano belts, 125, 125
Volcano monitoring, 128–130, 130
Volcano Sapas Mons, 43
Volcanoes. See also Volcanism
active, 110, 125, 126
of Andes Mountains, 96
Appalachian mobile belt and, 529
Archean, 490
Augustine, 121
Cascade Range, 622–623
composite, 118–119, 119
defined, 110
dormant, 110
extinct, 110
forecasting erupting, 128–130
frequency of eruptions of, 117
geodes and, 175
guyots as, 34
Hawaiian, 117
hot spots and, 48, 91, 91
inactive, 110
Kilimanjaro, 125
Kilauea, 112, 116–117, 128
lava temperatures at, 87
living along, 406
magma and, 86, 91, 91
Mammoth Mountain, 129
Mauna Loa, 43, 117, 117, 119
Mayon, 112, 119, 119
Medicine Lake, 126, 622
Newberry, 126, 622
North America's active, 126
Nyiragongo, 125
oceans and, 390, 406
ozone depletion and, 122
plate tectonics and, 126–128
plutons and, 99, 126–128
Redoubt, 113
Santorini, 125
shield, 116–117, 117, 119, 126,
128, 625
types of, 114–121
Unzen, 119
volcanism and, 110–114
of world, 125
Voyager, 43
Vulcan, Roman deity, 114
Vulcan's Fury: Man Against the Volcano
(Scharth), 121
Walcott, Charles, 493, 542, 547

Wallace, Alfred Russel, 451–453, 454,
456, 457
Walruses, 640
Wanapiei Crater, Canada, 498
Waptia, 547
Warm spring, 331. See also Hot springs
Warm Springs, Georgia, 331
Warm-blooded dinosaurs, 597
Wasatch Mountains, Utah, 245
Washington, George, 84, 100, 343
Waste disposal, 6
Water
deserts and, 366
drinking, 329
on Earth's surface, 286, 489, 491
erosion and, 158
geologic time and, 441
hard, 331
magma and, 90
mass wasting and, 261, 263, 264, 268,
270, 279
metamorphism and, 185
photosynthesis and, 491
plants and, 564–565
running (See Running water)
surface, 476, 478, 491
suspended, 316
weathering and, 140
Water depth, 400
Water erosion, 153
Water gaps, 308, 626
Water molecule, 64
Water pollution, 4
Water supplies, 7, 9
Water table, 316, 316, 317, 317, 318, 323–
326, 324, 330, 333, 358
Water treatment systems, 314
Water vapor
atmosphere and, 490, 490, 492
on early Earth, 489
glaciers and, 344
hydrologic cycle and, 287
hydrosphere and, 491
oceans and, 390
volcanism and, 110
Water wells, 317–319
Wave advance, 395, 396
Wave base, 395, 395, 396
Wave energy, 388, 389, 401, 403
Wave generation, 396
Wave height, 395, 397
Wave intensity, 401
Wave period, 395
Wave rays, 223
Wave refraction, 396, 397, 400
Wave size, 396
Wave-built platform, 398, 399
Wave-cut platforms, 398, 399, 399,
404, 405
Wave-dominated deltas, 296, 297
Wave-formed ripple marks, 168, 168
Wavelength, 395, 395, 396

Waves
body, 205, 206, 206, 208
defined, 393, 395, 395–396
love, 206, 207, 208
primary, 206, 206, 207, 208, 211,
223–227
Raleigh, 206, 207
rogue, 396
secondary, 206–208, 211, 223–225
seismic, 202, 205–207, 206, 207, 211,
212, 223–227
shear, 206
sound, 389
surface, 205, 206, 208
tidal, 216
Weak nuclear force, 11
Weasels, 640
Weather patterns, 28
Weathering
acid rain and, 144–145, 145
alteration of minerals/rocks
and, 137
Arches National Park and,
150–151
caves and, 322
Cenozoic history and, 627
chemical (See Chemical weathering)
cratons and, 523
defined, 136–137
deserts and, 376–377, 382
detrital sediment and, 158
differential, 137, 150, 245, 381
Dust Bowl and, 152
geologic time and, 425
of granite, 136, 137, 143, 143
Hadean and, 476, 478
mass wasting and, 260, 261, 263,
265, 277
mechanical, 137–139, 159, 376
in Mesozoic Era, 582
oceans and, 390, 400
organisms and, 139, 139, 140, 141
resources and, 153
Sauk Sequence and, 512
sediment and, 158
soil and, 142
spheroidal, 142, 143
Weeks Formation, Utah, 548
Wegener, Alfred, 29, 29, 31, 36, 53
Welded tuff, 98, 98, 125
Well cuttings, 172, 431
Well logs, 431, 431
Wellesley Glacier, Alaska, 341
Werner, Abraham Gottlob, 100
Westlothiana, 562
Whales, 637, 639, 640
Whitechuck Glacier, 349, 349
Why Darwin Matters (Shermer), 464
Williston Basin, 505, 521, 523, 535
Wilson cycle, 484, 487
Wilson, J. Tuzo, 91, 485
Wilson, Robert, 11

Wind
  cross-bedding and, 167, *167*
  deserts and (*See* Wind)
  erosion and, 158
  sediment transport by, 366–367, 403
  shorelines and, 405
  solar, 490
  waves and, 396
Wind abrasion, *367*, 367–368
Wind deposits, 368–372
Wind erosion, 367–368
Wind farms, 379
Wind gaps, 308, 626
Wind patterns, 372, 374
Wind power, 317, 378–379, *379*
Wind turbines, 379
Windmills, *378*, 378–379
Winter beaches, 401
Winter Park, Florida, *321*
Wisconsinan glaciation, 630, *630*
Wiwaxia, *547*
Wizard Island, Oregon, *116*, 118

Wolframite, *195*
Wolverines, *640*
Wolves, 641
Wombats, 639
Woodward, John, 467
Wopmay orogen, 484, 485
Worms, *545*, *551*

Yardangs, 367–368, *368*
Yavapai orogen, *484*, 486
Yellowstone caldera, Wyoming, 125
Yellowstone National Park, Wyoming, 121,
  295, 332, *333*, 334, *334*, *421*, 628, 634
Yellowstone Tuff, *124*
Yosemite National Park, California, *265*,
  352, *354*
Yucca Mountain, Nevada, 366
Yunnanozoon lividum, *554*

Zagros Mountains, Iran, 252
Zebroid, 459
Zeichstein evaporites, 534

Zinc. *See also* Sphalerite
  metamorphism and, 187, 195, *195*
  as mineral resource, 534, 535
  minerals and, 74, *79*
  plate tectonics and, 52
  submarine hydrothermal vents and, 33
  weathering and, 153
Zion National Park, Utah, *172*, *241*, *371*,
  430, *588*
Zircons, 477, *481*, 665
Zone of accumulation, 144, *146*, 346, 347,
  347, 350
Zone of aeration, 316, *316*, 317, *317*, 322
Zone of collision, 530
Zone of saturation, 316, *316*, 317, *317*, 318,
  324, 330, *330*
Zone of wastage, 346, *347*, 350
Zoology, *448*
Zooplankton, 544–545, *545*, *546*
Zuni epeiric sea, 625
Zuni Sequence, 581

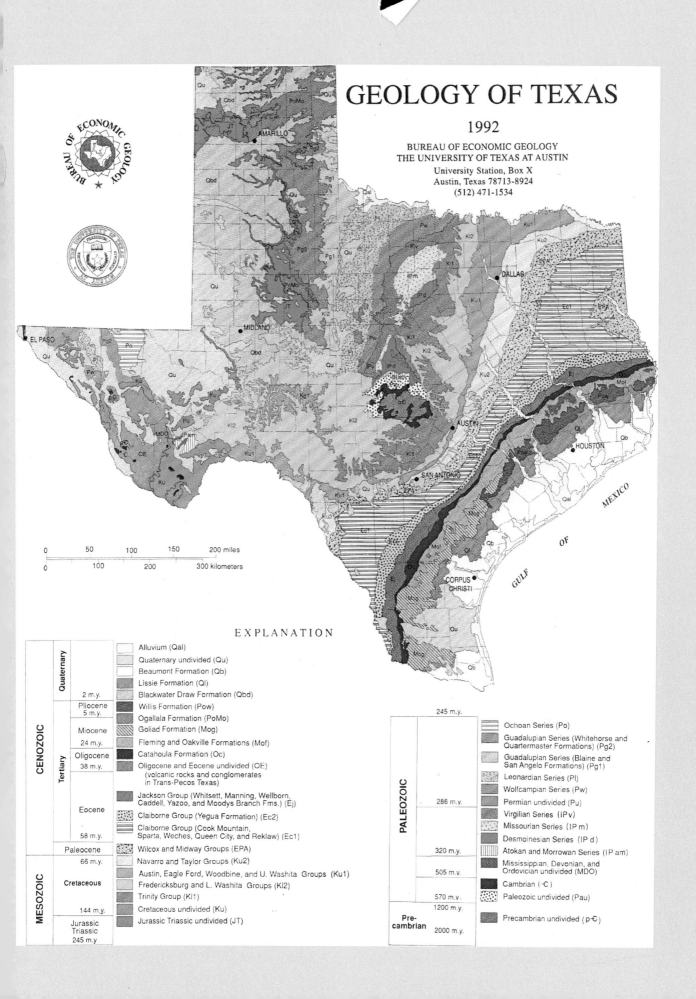

GEOLOGY OF TEXAS

1992

BUREAU OF ECONOMIC GEOLOGY
THE UNIVERSITY OF TEXAS AT AUSTIN

University Station, Box X
Austin, Texas 78713-8924
(512) 471-1534

Scale:

| 0 | 50 | 100 | 150 | 200 miles |
| 0 | 100 | 200 | | 300 kilometers |

EXPLANATION

CENOZOIC	Quaternary	2 m.y.	Alluvium (Qal)
			Quaternary undivided (Qu)
			Beaumont Formation (Qb)
			Lissie Formation (Ql)
			Blackwater Draw Formation (Qbd)
	Tertiary	Pliocene 5 m.y.	Willis Formation (Pow)
			Ogallala Formation (PoMo)
		Miocene 24 m.y.	Goliad Formation (Mog)
			Fleming and Oakville Formations (Mof)
		Oligocene 38 m.y.	Catahoula Formation (Oc)
			Oligocene and Eocene undivided (OE) (volcanic rocks and conglomerates in Trans-Pecos Texas)
		Eocene	Jackson Group (Whitsett, Manning, Wellborn, Caddell, Yazoo, and Moodys Branch Fms.) (Ej)
			Claiborne Group (Yegua Formation) (Ec2)
		58 m.y.	Claiborne Group (Cook Mountain, Sparta, Weches, Queen City, and Reklaw) (Ec1)
		Paleocene 66 m.y.	Wilcox and Midway Groups (EPA)
MESOZOIC	Cretaceous		Navarro and Taylor Groups (Ku2)
			Austin, Eagle Ford, Woodbine, and U. Washita Groups (Ku1)
			Fredericksburg and L. Washita Groups (Kl2)
			Trinity Group (Kl1)
		144 m.y.	Cretaceous undivided (Ku)
	Jurassic Triassic 245 m.y.		Jurassic Triassic undivided (JT)

PALEOZOIC	245 m.y.		Ochoan Series (Po)
			Guadalupian Series (Whitehorse and Quartermaster Formations) (Pg2)
			Guadalupian Series (Blaine and San Angelo Formations) (Pg1)
			Leonardian Series (Pl)
	286 m.y.		Wolfcampian Series (Pw)
			Permian undivided (Pu)
			Virgilian Series (IPv)
			Missourian Series (IP m)
	320 m.y.		Desmoinesian Series (IP d)
			Atokan and Morrowan Series (IP am)
	505 m.v.		Mississippian, Devonian, and Ordovician undivided (MDO)
	570 m.y.		Cambrian (-C)
	1200 m.y.		Paleozoic undivided (Pau)
Pre-cambrian	2000 m.y.		Precambrian undivided (p-C)

ESSENTIAL QUESTIONS TO ASK

Texas.1 Introduction

- Why is the geology of Texas important to students of physical geology and to all the inhabitants of the state today?

Texas.2 Precambrian Texas

- What tectonic events shaped the Proterozoic geologic record of Texas?
- When and how were the metamorphic and igneous rocks of the Llano region accreted to the ancient continent Laurentia?

Texas.3 The Paleozoic Tectonic Cycle

- What kinds of sediment were deposited during the Lower Paleozoic?
- What is the economic significance of these rocks today?

Texas.4 The Ouachita Orogeny and Pangaea

- What were the geologic results in Texas of the collision of Gondwana with Laurasia?
- How do the sedimentary rocks and structural styles of the Marathon area illustrate the collision?

Texas.5 The Fort Worth Basin in the Late Paleozoic Era

- What sedimentary environments characterized the Fort Worth Basin during and after the Ouachita orogeny?
- Why is this basin rich in oil and gas resources?

Texas.6 Permian Red Beds and Evaporites

- Why are red beds and evaporites characteristic of the Permian basins of central and west Texas?
- How does the geology of that region affect present-day land-use patterns?

Texas.7 Breakup of Pangaea and Cretaceous Transgressions

- What happened in Texas during and after the breakup of Pangaea?
- What were the results of the Cretaceous transgressions?

Texas.8 Basin and Range Texas

- What types of tectonic activity are indicated by the volcanic rocks in trans-Pecos Texas?
- How did the basin and range structural features form in that region?

Texas.9 Llano Estacado

- Which geologic processes led to the formation of the Llano Estacado?
- What is the Ogallala aquifer, and how is it recharged?

Texas.10 Edwards Plateau and Balcones Fault Zone

- When and how did the Edwards Plateau and Balcones Fault Zone form?
- How were these geologic features important to the initial settlement of Texas, and how are they important to Texans today?

Texas.11 The Gulf Coastal Plain

- How have transgressions and regressions shaped the geologic evolution of the Gulf Coastal Plain?
- What are the economic and environmental implications of using lignite to generate electricity?
- How did salt domes form, and why are economic resources concentrated around them?

Texas.12 The Texas Gulf Coast

- What are the principal features of the Texas Gulf Coast, and how did they form?
- How do hurricanes affect the coastal areas of Texas?

Texas.1
Introduction

The decision to add this chapter on the geology of Texas to your textbook was based on the fact that you are taking this course in Texas and on the expectation that this discussion will increase the value you derive from learning about physical geology. There are practical reasons for illustrating physical geology concepts with examples from Texas. Perhaps most important is that these examples have been chosen to show the ways in which the geologic processes described in this textbook fit together in a story of how the Texas landscape developed over a vast period of time. These examples provide an excellent means of illustrating the theory of plate tectonics, translating it from a theoretical perspective into a regional one that is relevant and useful for a better understanding of the geologic history of the Texas region.

A substantial part of this chapter is devoted to illustrating why Texas is endowed with such vast energy resources, which define much of its economic structure and employ thousands of Texans. Although Texas is energy rich, it is water poor; therefore, we explore the distribution and character of the vital water resources in Texas. The factors of surface geology, climate, soils, and availability of groundwater are discussed in terms of the historical development of the state's agriculture and ranching industries, as well as patterns of settlement and urban growth.

Texas is part of the passive margin of the North American continent; thus, it is spared from geologic hazards such as large earthquakes and volcanic eruptions. But we will see that hurricanes and floods are frequent and often devastating events. Environmental protection in Texas entails considerable contributions from geologists, particularly with respect to minimizing the adverse environmental impacts realized through land-use activities such as lignite mining in the Gulf Coastal Plain. Along the way, you can pause to see examples of how the surface geology of our state adds to its diverse scenic beauty, which enriches our lives and contributes to the recreational and tourism industries. Indeed, one of the goals of this chapter is to nudge you to go see these places and enjoy their geologic stories firsthand.

The objective of your course in physical geology is to gain an understanding of the geologic processes that are characteristic of our planet and how the geology of our planet is important to our lives and livelihood. In terms of introducing geology to students, our curricula still follow the plan Charles Lyell first used in the 1830s. Lyell's *Principles of Geology* became the course you are taking now. It explores the nature and results of interactions among the geosphere, biosphere, and atmosphere.

A fundamental premise of Lyell's approach was that the geologic processes that operate today have not changed over time; therefore, we must study and understand the geologic present to interpret the geologic past. This approach is called uniformitarianism, and it is a cornerstone of the earth sciences. This chapter compares several modern environments with ancient settings in the rock record of Texas. These

comparisons are supported by discussions in your text, including specific examples of how earth scientists formally apply this principle.

Lyell's companion volume, *Elements of Geology*, is taught today as historical geology, which examines the earth and life through time. Lyell eloquently labeled the sequences of ancient rocks and their included fossils as the "earth's autobiography." This chapter treats the geology of Texas chronologically, mainly because it enables us to take advantage of plate tectonics theory as we examine the autobiography of Texas.

The Geologic Map of Texas

The front piece for this chapter is the geologic map of Texas. You should refer to this map frequently as you read the following sections, and especially when you consider the photographs of surface geologic scenes. The geology map and the physiographic map (Figure Texas.1) show much of the geologic history and modern environmental diversity of the state. Notice the strong correspondence between the features of these two maps. They demonstrate that the rocks and sediments of Texas are not just parts of a chronologic sequence, but in many areas are also associated with distinct landforms and environments. While examining the photographs in subsequent sections, you will see different patterns of erosion and markedly different vegetation communities, defined by the dramatic decrease in rainfall from east to west across this huge land area.

Geologic maps are marvelous tools for exploring the geologic history and character of the areas they portray. The ages of the rock units are indicated by their color so that our eyes can search for patterns in their spatial arrangements that make both physical and chronologic sense. For example, notice that the Paleozoic rocks in north central Texas (see shades of blue in the front piece illustration) are progressively younger from east to west. In contrast, the rocks of the coastal plain are progressively younger from west to east. The boundary between these two regions is best marked by the Balcones Fault Zone (see red in Figure Texas.1). Can you explain these patterns by identifying the processes that produced them?

Another interesting feature is the nature of the contact between Cretaceous rocks (see green in Figure Texas.1) and adjacent rocks in the central part of the state. In this region, they directly overlie older rocks of the Precambrian and Paleozoic eras, and the Triassic Period. What could account for this? How much time is "missing" between the Cretaceous and older rocks? You will find the answers to these questions, and hopefully some of your own questions, in the following discussions and accompanying figures.

Section Texas.1 Summary

● Learning about the geology of Texas provides local examples of the main themes of physical geology, especially the theory of plate tectonics.

● Texas has abundant geologic resources, including petroleum, coal, minerals, groundwater, and fertile soils.

▶ **Figure Texas.1** This map shows the principal physiographic regions of Texas. *Stars* indicate the locations of localities discussed in this chapter. Note the close correspondence between these physiographic regions and the surface geology of the state.

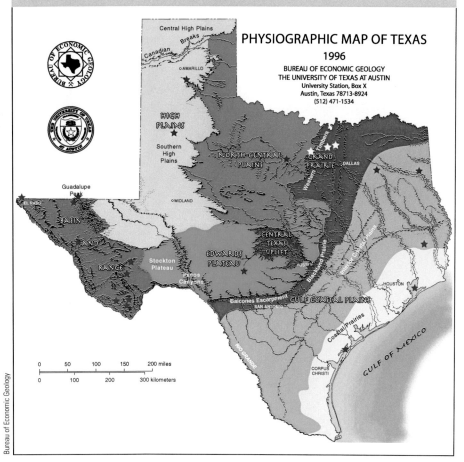

Bureau of Economic Geology

Texas.2 Precambrian Texas

The tectonic cycle that begins in the late Proterozoic is illustrated by rocks exposed in the Central Texas (Llano) Basin, the Van Horn area, and the Franklin Mountains near El Paso (see Figure Texas.1). The first two areas show evidence of an Andean-type orogeny, whereas the rocks in the Franklin Mountains indicate a passive continental margin. In the Llano Basin, sedimentary rocks were subjected to regional metamorphism, resulting in the formation of gneisses, schists, and a few zones with marble, a contact metamorphic rock. The igneous rocks associated with this mountain building phase are not known and presumably were removed during later rifting.

The Packsaddle Schist is an example of these rocks in the Llano area (Figure Texas.2). It was formed from sediments that derived from weathering and erosion of mafic igneous rocks, possibly associated with an island arc. Metamorphism of those sediments occurred about 1.3 billion years ago (1.3 Ga). About 1.1 Ga, the

▶ **Figure Texas.2**

Reid Ferring

ⓐ The Packsaddle Schist, exposed here south of Llano, is one of several metamorphic rocks that formed about 1.3 billion years ago (1.3 Ga). A transgression that began in the middle Cambrian Period is registered by the sandstones in the background of this photo (see Figure Texas.7).

ⓑ The fabric of the Packsaddle Schist is not defined by sedimentary layers, but rather by schistosity, which developed through segregation of dark and light minerals that crystallized during metamorphism.

▶ **Figure Texas.3** Located just north of Fredericksburg, the prominent granite dome in this cluster is called Enchanted Rock. Granitic magmas intruded metamorphic rocks of the Llano area, including the Packsaddle Schist shown in Figure Texas.2a.

▶ **Figure Texas.5** This pegmatite near Llano, Texas, has three minerals that also dominate the Town Mountain granite: microcline (a potassium feldspar), biotite (a mica), and quartz. Notice that the quartz crystallized last, enveloping the mica and feldspar crystals, which formed at higher temperatures.

metamorphic rocks at Llano were intruded by granitic magmas. All of these rocks are part of the Grenville Province, which was accreted to Laurentia in the late Proterozoic.

The intrusive rocks are expressed as small dikes and veins (see Figure Texas.2), up to quite large laccolithic bodies, such as Enchanted Rock (Figures Texas.3 and Texas.4). These are classic examples of exfoliation domes, illustrating physical weathering. A pegmatite exposed in a quarry near Llano nicely illustrates the sequence of mineral crystallization as predicted by Bowen's reaction series described in your textbook (Figure Texas.5). Note the faces of the large hexagonal crystal of microcline, surrounded by quartz. How can you tell which of these minerals formed last?

The state capital in Austin is made almost entirely from granite quarried near Marble Falls, and this stone can be seen in many of the late Victorian Period courthouses around the

▶ **Figure Texas.4** Looking west from the summit of Enchanted Rock, one can see this classic example of an exfoliation sheet that has broken up into slabs, which are sliding down the dome. Limestone on the horizon formed during Cretaceous transgressions, and the flat surface is the Edwards Plateau. Faulting during the Tertiary Period contributed to the breakup and removal of the limestone that once covered the Precambrian granite.

state. Gold is found in low quantities in the Llano area, but the semiprecious mineral topaz is quite common.

Section Texas.2 Summary

- During the Proterozoic eon, metamorphic and igneous rocks were accreted to Laurentia as part of the Grenville Terrane.

- Rocks of Proterozoic age near Llano, Van Horn, and the Franklin Mountains demonstrate the nature of metamorphic and intrusive igneous processes.

Texas.3
The Paleozoic Tectonic Cycle

All of the Proterozoic rocks in Texas were subjected to weathering and erosion, which lasted until the Cambrian Period, when rifting in this part of Laurentia took place. Rocks associated with this rifting are exposed near Van Horn, and notably in Oklahoma's Arbuckle aulacogen, which was a failed arm of the Cambrian rift system.

A postrifting transgression began in the Cambrian Period and continued throughout almost all of the Ordovician Period, when a shallow sea covered all of Texas. Shallow seas that cover part of the craton are called epeiric seas; sometimes they are called subsiding seas, because by the process of

▶ **Figure Texas.6**

Ⓐ These rocks, exposed in a road cut overlooking Lake LBJ in the Llano area, were deposited near the shoreline at the beginning of the Cambrian transgression. The Lion Mountain sandstone derives its green color from the iron-rich mineral glauconite, and it also has thin beds and pods of limestone with numerous fossils of trilobites and brachiopods.

Ⓑ These symmetric wave ripples in the Welge Sandstone were formed along a Cambrian beach. The lower part of this bed shows extensive bioturbation through burrowing by organisms, whereas the upper part provides an excellent example of cross-bedded sedimentary structures.

subsidence, thick sequences of sedimentary rock can be preserved, even with little change in the sedimentary environment. The early stages of the Cambrian transgression in Texas are marked by beach and near-shore sandstones, including those dominated by quartz grains, which were derived from weathering of Precambrian granites. Cambrian deposits exposed near Kingston, Texas, include beds dominated by green, glauconite sands and contain abundant trilobite and brachiopod fossils (Figure Texas.6). As the transgression progressed, a shift from clastic to biogenic sedimentation was completed, and thick limestones formed across virtually all of Texas (and surrounding parts of Laurentia) during the

Ordovician. Most of these rocks are deeply buried, but surface exposures can be seen in the central basin, the Beach Mountains near Van Horn (Figure Texas.7), and the Franklin Mountains (Figures Texas.8 and Texas.9).

The ancient sedimentary environments of the Ordovician epeiric sea are comparable with the "carbonate factories" of the present-day Bahamas, where calcareous algae thrive in warm, clear ocean water. When these flourishing algae die, they break down into tremendous volumes of calcareous mud, which is then buried and lithified into limestone. The weight of these new rocks causes subsidence, and by this mechanism, tremendously thick accumulation of

▶ **Figure Texas.7** Located in the Beach Mountains near Van Horn, the limestone and dolomite of the El Paso Formation formed in the shallow epeiric sea that covered Texas during the Ordovician Period. These rocks overlie sandstones of latest Precambrian to Cambrian age.

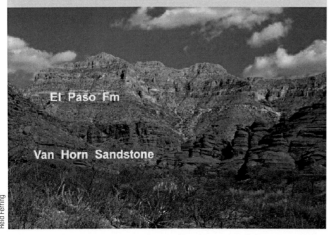

▶ **Figure Texas.8** The Bliss Sandstone (Cambrian) and the overlying Ordovician dolomites are exposed in the Franklin Mountains near El Paso. These sedimentary rocks formed during the early Paleozoic transgression and contain fossils of corals, mollusks, and trilobites. This east-facing slope formed along a Miocene Epoch normal fault that separates this uplifted block (a horst) from the broad Hueco Basin (a graben).

▶ **Figure Texas.9**   This is the west-facing slope of the Franklin Mountains. The rocks at the summit are Devonian-Mississippian in age, and those below belong to the Pennsylvanian Magdalena Formation. All of these rocks formed in shallow, clear marine environments. Because the major source of these carbonates was calcareous algae, these are good examples of biogenic rocks.

limestone occurs with no change in water depth. It is estimated that the depth of Texas's Ordovician sea was always on the order of less than 10 meters.

After deposition, many of the Ordovician limestones were changed diagenetically into dolomite. In this process, the limestone was altered to dolomite by substitution of about half of the calcium ions in calcite with the smaller magnesium ion. This process increases the porosity of the rocks by about 15%, which makes them ideal reservoir rocks for hydrocarbons. Thus, the Ordovician rocks have abundant algae-derived organic matter as a source for hydrocarbons and have abundant, highly porous reservoirs. These rocks are among the most important sources of oil and gas in this region. Of special importance are natural gas reservoirs in west Texas that are up to 6,000 meters deep. Royalties on much of those reserves are owned by the State of Texas and by law are used to subsidize (your) higher education!

Section Texas.3 Summary

• During the Lower Paleozoic, a transgression resulted in creation of an epeiric sea that covered all of Texas.

• Thick sequences of Ordovician dolomites contain large reserves of oil and natural gas.

Texas.4
The Ouachita
Orogeny and Pangaea

The Ouachita orogeny in Texas brought a dramatic end to the Paleozoic tectonic cycle. Over a very long period, perhaps as long as 50 million years, the ocean between Texas (the southern end of

Laurentia) and South America (the northern end of Gondwana) was closed, and finally, those continents collided and became part of the supercontinent Pangaea. When the collision actually took place, rocks that formed during the collision, as well as older Paleozoic rocks, were pushed up onto the continent as part of the Ouachita Mountains. The ancient Ouachita Mountains extended from Marathon to northeast Texas and into Oklahoma. At the same time, Africa and Europe also were colliding with Laurentia, resulting in the Appalachian orogeny in the eastern United States.

Between Marathon and Oklahoma, the traces of these mountains are deeply buried, and they can be reached only by drilling. But the former axis of the Ouachita Mountains follows almost exactly the path of the Balcones Fault Zone (see Figure Texas.1). The Ouachita orogeny had both immediate and long-lasting effects, and as explained in the following discussions, the geologic history of the Texas interior west of the Ouachita Mountains is very different from the history of the Gulf of Mexico Basin, which was formed after the rifting of South America away from Texas.

The effects of the Ouachita orogeny in Texas are, in part, structural, as considerable faulting took place. But the main effects, and the most important ones for people today, were in the patterns of sedimentation that accompanied and followed this collision, for those late Paleozoic sediments contain important geologic resources that are vital to the economy of Texas.

Rocks of the Mississippian and Pennsylvanian periods in Texas reveal that the second tectonic cycle was beginning. In dramatic exposures in the hills near Marathon, there are rhythmically arranged thin beds of sandstone and siltstone called flysch (Figure Texas.10). These flysch deposits have a total thickness of more than 3 kilometers (km) near

▶ **Figure Texas.10**   Two major events in the geologic history of Texas are recorded in these rocks. The flysch sediments of the Pennsylvanian Haymond Formation were deposited as turbidites, then folded and faulted during the Ouachita orogeny. After an enormous interval of erosion, the ocean transgressed over Texas during the Cretaceous Period. Rock creep caused the distinctive bending of the rocks near the surface.

CRETACEOUS

HAYMOND Fm

▶ **Figure Texas.11** These hogbacks near Marathon were formed in rocks that were intensely folded and faulted during the Ouachita orogeny, when South America collided with North America. At the end of that collision, Texas became part of the supercontinent Pangaea.

CAMBRO-ORDOVICIAN STRATA   CABALLOS NOVACULITE

Reid Ferring

▶ **Figure Texas.12** In the past few years, drilling for natural gas in the Mississippian Barnett Shale west of Dallas-Fort Worth has been intense. These wells are 5 km west of Denton. The one in the foreground has been completed.

Reid Ferring

Marathon. They were deposited by turbidite flows, which were supplied by great volumes of sediment being shed off of mountains that were being created as South America approached Laurentia from the south. The flysch deposits at Marathon were overlain by coarse-textured sediments called molasse. These sandstones and conglomerates indicate a shift from marine to terrestrial environments at the completion of the continental collision. Discussions in your textbook covering plate tectonics have excellent illustrations of this kind of convergent plate boundary.

The structural features of the Marathon area include numerous anticlines and synclines interrupted by high-angle reverse faults, all of which are underlain by a long thrust fault, indicating that these rocks were pushed up onto the continent. These have been eroded into distinctive landforms evident today, such as the hogbacks formed by the Silurian Caballos Novaculite (Figure Texas.11). The entire package of Paleozoic rocks near Marathon was pushed up onto the continent along a long, low-angle thrust fault.

Section Texas.4 Summary

● During the late Paleozoic, South America collided with Texas, creating the Ouachita Mountains and adding these continents to the supercontinent Pangaea.

● Rocks exposed in the Marathon area reveal the nature and patterns of sedimentation and structural deformation that accompanied the collision.

Texas.5
The Fort Worth Basin in the Late Paleozoic Era

In north central Texas, sediments were shed from the west front of the Ouachita Mountains into the Fort Worth Basin during and after the Ouachita orogeny, providing another example of tectonic controls on the character and rate of sedimentation. Although small deposits of coal have been mined in these deposits, the Fort Worth Basin is much more important as a major producer of oil and gas.

At the base of this basin, overlying early Paleozoic rocks, is the Mississippian Period Barnett Shale, which is buried about 2,000 meters below the surface. This organic-rich shale has abundant natural gas reserves, which are now being developed extensively with technologies including horizontal drilling and slick water fracking (Figure Texas.12). Horizontal drilling greatly increases the area of gas extraction, and injection of water (without the sand usually used in fracking) into the shale enhances permeability, increasing the flow of gas to the well.

The overlying Pennsylvanian Period deposits in the Fort Worth Basin were delivered from both the Ouachita and Arbuckle mountains, which underwent episodic uplift and erosion during the very long Ouachita orogeny (Figure Texas.13a). This resulted in the progradation of numerous deltas on to the shelf, with sandy deposits near the delta, and muds settling on the shelf and in the interdelta bays. Delta progradation resulted in a vertical sequence of sedimentary facies of organic-rich prodelta muds, overlain by increasingly coarser delta front and channel deposits (see Figure Texas.13b). Quite rapid subsidence and lateral shifting of delta positions through time resulted in emplacement of an impermeable shale or limestone "cap" over these deltaic deposits, making the channel sands ideal petroleum reservoirs.

Carbonate banks and larger platforms developed away from the deltas, where clearer water supported algae, which were the base of the food chain (see Figure Texas.13c). Reefs were colonized by filter-feeding animals that also required clear water; these include corals, crinoids, bryozoans, and brachiopods, as well as numerous bottom-dwelling and burrowing organisms, such as urchins and mollusks, including numerous gastropods that "grazed" on surface coatings of algae and organic debris. Many of these buried reefs are targets for oil exploration as well.

Thick Pennsylvanian limestone northwest of Fort Worth has been actively quarried because the regional supplies of stream gravel were largely exhausted. Rock crushers break the hard limestone into aggregate, which is used to make concrete. The Dallas-Fort Worth region is one of the fastest-growing areas of the country, creating huge

▶ **Figure Texas.13**

(a) During and after the Ouachita orogeny, rivers flowed west into the Fort Worth Basin in north central Texas. Deltas prograded into the shallow shelf, which also supported carbonate banks and platforms where the water was clearer.

(b) These rocks near Mineral Wells, Texas, were formed from sediments deposited by a Pennsylvanian delta that prograded westward (right to left) into the Fort Worth Basin (a). Similar rocks, buried thousands of feet deep, are important reservoirs of oil and gas.

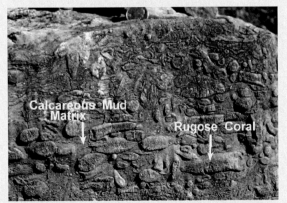

(c) This Pennsylvanian coral reef is a good example of the "muddy" reefs of the late Paleozoic Era. The calcareous mud matrix was contributed by algae, which formed the base of the food chain in these reefs and supported colonies of filter-feeding organisms, such as the solitary corals shown here.

demands for concrete by the residential, public, and commercial sectors.

Section Texas.5 Summary

• The Fort Worth Basin has thick Mississippian and Pennsylvanian deposits that formed in response to the Ouachita orogeny.

• There are substantial deposits of oil and natural gas in the Fort Worth Basin.

Texas.6
Permian Red Beds and Evaporites

In the early and middle Permian Period, thick limestone deposits accumulated in central and west Texas, in the sea that was bounded on the east and south by the Ouachita Mountains. In west Texas, deposition during the Permian Period was concentrated in basins that were created by faulting perpendicular to the trend of the Ouachita collision. The greatest amount of deposition centered on the Midland Basin, where thousands of meters of Permian limestone overlay the Lower Paleozoic rocks. By the late Permian Period, however, cyclical sea level changes, driven by the expansion and retreat of glaciers in Gondwana, resulted in periodic isolation of separate closed basins. In the Delaware Basin, the one closest to the open sea, the Capitan Reef flourished around the rim of the basin when sea levels were high (Figure Texas.14). When sea level fell, thick beds of evaporitic calcite and gypsum filled

▶ **Figure Texas.14**   El Capitan Peak, a prominent feature of the Guadalupe Mountains in west Texas, is part of the Permian reef that encircled the Delaware Basin. Just across the New Mexico state line, Carlsbad Caverns formed in this limestone.

▶ **Figure Texas.15** These red beds and gypsum layers exposed in Palo Duro State Park are part of the Caprock Escarpment, which is the eastern edge of the Llano Estacado (see Figure Texas.26). The red beds were formed during the arid conditions in the Permian Period. The highest light beds are part of the Triassic Dockum Group.

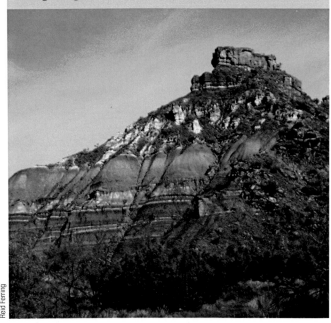

Reid Ferring

▶ **Figure Texas.16** Gypsum is mined in this quarry near Sweetwater, in the Red Bed Plains. Formed in evaporitic "ponds" during the arid Permian Period, this gypsum is used to manufacture Sheetrock wallboard and Portland cement.

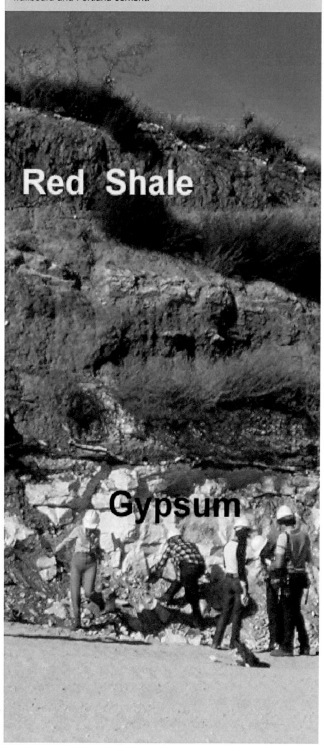

Reid Ferring

the closed basin. The Capitan Reef rocks are now part of the Guadalupe Mountains. Notably, the famous Carlsbad Caverns formed in these limestones, just across the border in New Mexico.

In the late Permian Period, Texas was near the equator and tropical climates contributed to deep weathering of sediments exposed around the interior basins during times of low sea level. Lasting evidence of this weathering is preserved as "red beds," which derive their color from oxidized forms of iron and aluminum. Excellent examples of these are exposed in the walls of Palo Duro Canyon, just below the Caprock Escarpment of the Llano Estacado (Figure Texas.15). This section of sediments includes multicolored beds of sandstone and shale, locally called Spanish skirts. The reddest beds are those that were most intensely weathered.

Water trapped in these interior basins evaporated, resulting in the precipitation of gypsum and halite. Shallowly buried deposits of gypsum are mined in open pits near Sweetwater (Figure Texas.16). The gypsum is used for manufacturing Portland cement and also wallboard, which is made by creating a muddy slurry from the gypsum, which is then extruded between sheets of paper, where it forms a hard layer of interlocking gypsum crystals (hence the name Sheetrock).

In the North Central Plains (see Figure Texas.1), also referred to as the Red Bed Plains, Permian salts are dissolved easily by percolating water. As a result, much of the groundwater and surface water in this region is too salty for drinking or irrigation. Even the water in Lake Texoma, which is on the Red River downstream from the Red Bed Plains, is too salty for municipal use.

At Palo Duro Canyon, the light-colored Triassic sandstone at the top of the section conformably overlays the

Permian beds. This section thus preserves the Permo-Triassic contact, which signals the end of the Paleozoic and the beginning of the Mesozoic. Erosion dominated the rest of the Triassic Period and continued into the Jurassic, when rifting created the Gulf of Mexico Basin. Therefore, no further deposition occurred over most of Texas west of the Gulf Basin until the postrifting transgression of the early Cretaceous Period. The next section addresses this subject.

Section Texas.6 Summary

- During the Permian Period, thick deposits of limestone, red beds, and evaporites accumulated in deep basins in central and west Texas.

- Dissolution of Permian salts in north central Texas greatly reduces freshwater supplies.

Texas.7
The Breakup of Pangaea and Cretaceous Transgressions

During the Jurassic Period, the continents of North and South America rifted apart, as part of the breakup of Pangaea. In Texas, no evidence of the rifting is preserved at the surface, but parts of Jurassic rift basins have been reached by drilling on the Gulf Coastal Plain. Most notably, the early Jurassic Gulf Basin was closed, and thick deposits of halite, called the Louann Salt, formed there by evaporation. We revisit the Louann Salt in Section Texas.11.

After erosion of the continent for much of the Triassic and Jurassic periods, and after the Gulf was connected to the Caribbean-Atlantic, there were two Cretaceous transgressive cycles, which at their highest stands not only covered all of Texas but also continued northward to the Arctic. Everywhere in Texas, except deeply buried under the Gulf Coastal Plain, Cretaceous rocks overlie much older, eroded rocks. We have seen examples of this near Enchanted Rock (see Figure Texas.4) and at Marathon (see Figure Texas.10). Since the end of the Cretaceous, the Texas part of the Gulf of Mexico Basin has been filling in quite rapidly with sediments. This phase of Texas geologic history is called the Gulfian Cycle. As the Gulf Coastal Plain was constructed, Tertiary tectonics caused dramatic changes in the landscapes of central and west Texas.

Lower Cretaceous rocks over most of the Edwards Plateau are limestone and marl, as can be seen in a section of the Glen Rose Formation near Kerrville (Figure Texas.17a). As mentioned earlier, these thick sequences of quite similar sediments developed because they formed in environments where carbonate sediments were rapidly generated and where subsidence maintained those environments, mainly by keeping water depths relatively constant. Mangrove swamps and shallow lagoons, far from any rivers in Yucatan, are good modern analogs for many of the Lower Cretaceous sedimentary environments in Texas (see Figure Texas.17b). Some of the erosional surfaces in these sections are the result of hurricanes, which were apparently common in Texas during the Cretaceous Period, because global climates then were significantly warmer than today. This is illustrated by the fact that

Ⓐ These limestone and marl beds, exposed near Kerrville, belong to the Cretaceous Glen Rose Formation. These rocks form part of the Edwards Plateau, which is the recharge zone for the Edwards Aquifer.

Ⓑ This mangrove swamp in western Yucatan, Mexico, is a modern analog for the sedimentary environments in which the Glen Rose limestone was formed. Like the ancient environments, these swamps and shallow lagoons are inhabited by clams, oysters, urchins, and crustaceans.

during the Cretaceous Period, coral reefs formed as far north as 70° North latitude, well above the Arctic Circle. Today, reefs cannot form much above 35°, because the oceans are too cold for corals. Globally warm conditions during the Cretaceous Period also supported phenomenal populations of rapidly evolving phytoplankton, which was the base of the food chain for the Cretaceous oceans. In fact, the term *Cretaceous* refers to the common chalk deposits of that period, which are made of the calcareous skeletons of those phytoplankton.

Lower Cretaceous rocks in north central Texas and in the trans-Pecos region (especially Big Bend) include terrestrial deposits because they were closer to coastlines. These are the rocks that contain Texas' dinosaur fossils and track ways such as near Glen Rose and Grapevine, Texas. Cretaceous marine sediments, exposed on the western edge of the Gulf Coastal Plain, preserve many fossils of large marine reptiles, including plesiosaurs and mosasaurs.

Cretaceous limestone is quarried extensively, especially along the eastern edge of the Edwards Plateau, closest to rapidly growing cities (Figure Texas.18). The hard, white varieties of limestone and chalk, usually called Austin stone, were widely used for public buildings and big homes in

▶ **Figure Texas.18** This quarry in Cretaceous rocks near Austin is one of many that are used to produce crushed stone aggregate, lime for cement, and dimension stone for construction and landscaping.

the nineteenth century. This stone is still popular among architects, giving even modern-looking structures a connection to Texas' geologic and historic past.

The K/T Boundary and the Age of Mammals in Texas

It is well known that at the end of the Cretaceous Period, there was a profound mass extinction, which ended the Age of Reptiles (the Mesozoic Era) and ushered in the Age of Mammals (the Cenozoic Era). This is often called the K/T boundary, based on the geologic map symbols for Cretaceous and Tertiary, respectively. On the geologic map of Texas, the K/T boundary can be seen on the western edge of the Gulf Coastal Plain as a line separating the last Cretaceous rocks (Ku2, green) with the first Tertiary rocks (PWA, stippled orange).

During the Cenozoic Era, mammals diversified rapidly on all the continents and in the oceans as well. Texas has an excellent fossil record of mammalian evolution, but it is contained mainly in sediments in the western part of the state. There the Tertiary landscape evolved in quite dramatic ways, in response to a sequence of compressional and then extensional intraplate tectonics. These tectonic changes appear to be related to an eastward extension of the zone of subduction of the Pacific Plate under the western edge of North America, which resulted in compressional uplift and volcanism. A further lowering of the subduction angle may have contributed to the subsequent extensional regimen, when basin and range features developed not only in trans-Pecos Texas but also over most of the western United States. The regional manifestations of this phase of intraplate tectonism are discussed in the following sections.

Section Texas.7 Summary

• In the Jurassic Period, South America rifted away from Texas, forming the Gulf of Mexico Basin.

• In the Cretaceous Period, marine transgressions covered Texas, leaving thick limestone deposits.

Texas.8 Basin and Range Texas

In the Tertiary Period, the compressional phase of intraplate tectonism mentioned earlier transformed Texas west of the Pecos River (the trans-Pecos) into a dynamic landscape studded with scores of volcanoes. The peak of this volcanic activity was in the Oligocene Epoch, when all of the older rocks in this region were intruded by magmas, leaving some small plutons, but many more volcanic features, including a number of calderas.

Most of the trans-Pecos volcanoes preserved today are eroded remnants, much smaller than when they were active. Resistant volcanic necks contribute to the beautiful desert landscape, especially in Big Bend National Park (Figure Texas.19). Most of the volcanoes in this region were fed by felsic magmas, producing common rhyolite and rhyolite porphyry (Figure Texas.20). But there are also many mafic rocks, especially basalt, indicating a pattern called bimodal volcanism; that is, the chemical composition of the magmas that fed different volcanoes ranges from felsic to mafic. Bimodal volcanism is characteristic of intraplate tectonism, where deep magmatic sources had mafic, mantle-like compositions, whereas nearby volcanoes were fed by shallower, continental sources with felsic composition. Formation of the Tertiary Rio Grande Rift in nearby New Mexico was accompanied by bimodal volcanism, as can be seen in the juxtaposition of andesitic and basaltic rocks near Santa Fe.

The high silica content of the felsic magmas caused extremely violent eruptions that produced huge volumes of pyroclastic ash. Winds carried those ashes to the east, where they became buried within the Ogallala Formation under the present-day High Plains (see Section Texas.9). Ash beds are also preserved in a number of the Tertiary formations of

▶ **Figure Texas.19** The Chisos Mountains, in the heart of Big Bend National Park, are a series of necks that have been exposed by erosion of the softer lavas and ashes that formed these once tall volcanoes.

▶ **Figure Texas.20** Willow Mountain, near Terlingua, Texas, is the neck of a rhyolitic volcano. Note the well-expressed columnar jointing of these rocks and the talus slopes formed by erosion.

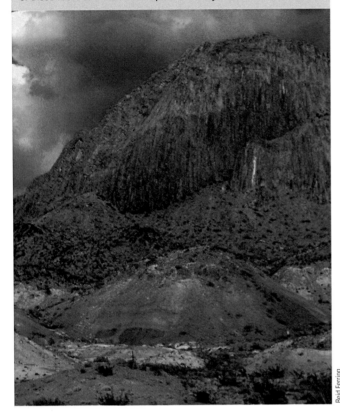

Reid Ferring

▶ **Figure Texas.21** Big Bend National Park is within the Chihuahuan Desert. This desert pavement, studded with cacti and the spindly Ocatillo, was formed by wind erosion (deflation) of the surface sediments, leaving behind an "armored" surface of pebbles and cobbles.

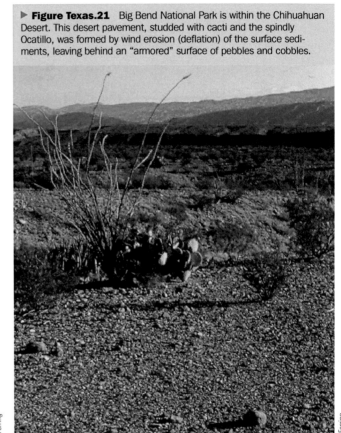

Reid Ferring

the Gulf Coastal Plain. Ash weathers quickly into fertile soils, and the extensive grasslands of that period nourished the rapidly evolving communities of grazing animals, including horses, which populated the ancestral Great Plains environments.

Today, much of trans-Pecos Texas is within the Chihuahuan Desert, which is just one of the low-latitude deserts of the Northern Hemisphere. Visitors to Big Bend are treated to a scenic landscape that has formed in the desert climate. The student of physical geology will quickly recognize the results of weathering and erosion typical of deserts, including extensive sheets of sediment eroded from the volcanic peaks. Those surfaces are capped by desert pavements, composed of pebbles and cobble "armor" that have been left behind after long periods of wind erosion called **deflation** (Figure Texas.21).

After the major period of volcanic activity in the trans-Pecos, extensional tectonics resulted in formation of the Texas portion of the vast basin and range province of the western United States. The result was a series of northerly trending uplifted blocks, called ranges or horsts. Many older rocks are exposed in these uplifts, including the Precambrian through

▶ **Figure Texas.22** These salt flats, on the floor of a graben-type valley, are separated from the Guadalupe Mountains in the background by a series of normal faults. The salt crusts are deflated and blown to the edges of the flat, forming dunes in the distance.

Reid Ferring

Paleozoic sequences of the Franklin and Beach Mountains (see Figures Texas.7 and Texas.8) and the Permian rocks of the Guadalupe Mountains (see Figure Texas.14).

The basins between these uplifted blocks are graben-type valleys formed between pairs of normal faults. The Hueco Basin (also called by the Spanish name *Bolson*), lies between the Franklin and Hueco uplifts near El Paso. The Salt Basin formed below El Capitan Peak west of the Guadalupe Mountains (Figure Texas.22). Up to several kilometers of sediment fill these basins. The deposits include stream deposits, windblown sand and silt, and also evaporites, such as the gypsum and halite seen on the surface of the Salt Basin.

One of the Tertiary faults flanking a basin near the Rio Grande River was reactivated in 1931, producing the Valentine earthquake. This earthquake, the strongest recorded in Texas, registered 6.0 on the Richter scale. Buildings were damaged near the epicenter, and the earthquake was felt over much of Texas. Since then, Texas has experienced a number of smaller earthquakes, rarely above Richter 4.0.

Section Texas.8 Summary

- Intraplate tectonics resulted in volcanism and creation of basin and range features in the trans-Pecos region of Texas.

Texas.9
The Llano Estacado

In the Tertiary Period, uplift of the southern Rocky Mountains increased rates of erosion. As a result, a ramp of sediment called the High Plains was deposited on the eastern front of the Rockies, from Texas to Canada. The part of the High Plains in Texas is called the Llano Estacado. The Ogallala Formation includes most of the sediments that formed this feature. The lower part of the Ogallala is composed of fluvial sand and gravel. The upper part is composed of eolian sand and silt, as well as thick zones of

▶ **Figure Texas.23** This map shows the eastern edge of the Llano Estacado (Southern High Plains), with the Caprock Escarpment surrounding Palo Duro Canyon. The extremely flat surface of the High Plains is dotted by more than 20,000 playa lakes. Lunette dunes are usually on the east sides of playa depressions because of prevailing westerly winds.

Reid Ferring

▶ **Figure Texas.24** This playa lake has been exposed in a borrow pit on the north side of Lubbock. The Randall Clay has filled the basin over the last 10,000 years. Mammoth bones and 11,000 year-old Clovis Paleoindian spear points are found in some of these playa lake deposits.

caliche (locally called the Caprock). Caliche is hardened calcium carbonate formed by soil processes characteristic of arid and semiarid landscapes. The lower part of the Ogallala Formation is the Ogallala aquifer, which provides almost all of the water needs for this region, where irrigation is necessary for crop farming.

The extremely flat surface of the Llano Estacado formed as the Pecos River cut northward, intercepting the Rocky Mountain drainages and isolating the Llano Estacado. In geomorphic terms, the streams that formerly crossed the Llano Estacado were beheaded by the Pecos River. After that, only wind could bring sediments to the Llano Estacado, and a flat blanket of silt and sand accumulated (Figure Texas.23). On the western edge of the Llano Estacado, there are number of periodically active sand dune fields, and at its southern end is Monahans State Park, which contains a large dune complex that began accumulating in the late Pleistocene.

Today, more than 20,000 playa lakes are present on the Llano Estacado (Figure Texas.24), and many older lake deposits are buried. These ephemeral lake basins formed by a combination of deflation, as indicated by dunes around their margins, and dissolution of underlying carbonates (a karstic process). In the late Pleistocene, extinct mammals such as mammoth, horses, camels, saber-toothed cats, and large, fast bears populated this region. Playa lakes were frequented by these animals, and 11,000-year-old Clovis spear points have been found in association with mammoth bones at the Miami Site, a late Pleistocene playa north of Amarillo. These playas collect seasonal rainfall, some of which percolates down, recharging the Ogallala aquifer. But many parts of the Ogallala Aquifer have experienced progressive lowering of the water table, making water resources management critical for sustaining this region's farming and cattle industries.

Section Texas.9 Summary

- In the late Tertiary Period, the Texas High Plains was constructed by deposition of fluvial and eolian sediments.

- The Ogallala aquifer is lowered by periodic droughts, reducing supplies of water for irrigation.

Texas.10
The Edwards Plateau and Balcones Fault Zone

During the Miocene Epoch, while the basin and range features were formed by extensional tectonics in the trans-Pecos, the central part of the state was subjected to epeirogenic uplift, creating the Edwards Plateau (see Figure Texas.1). This uplift was accommodated by up to 500 meters of displacement along the Balcones Fault Zone, which is a series of normal faults, with downward movement towards the coastal plain. The Balcones Escarpment defines the eastern margin of the Edwards Plateau, extending from west of San Antonio to near Dallas, where it bends into east Texas.

During epeirogenic uplift, larger rivers will erode deeply into resistant bedrock, forming what are called entrenched valleys. Perhaps the most famous example of this process is the Grand Canyon, which formed as a result of epeirogenic uplift of the Colorado Plateau. The same kinds of forces caused the Brazos River to form a long series of entrenched meanders across the Paleozoic and Cretaceous rocks of northern Texas (Figure Texas.25).

From near Austin to San Antonio, streams also adjusted to the uplift by eroding deep valleys just upstream from the Balcones Fault Zone. In a local twist in perspective, that area of deep valleys is called the Hill Country, which is one of the most scenic parts of the state, popular with tourists, campers, and hunters (Figure Texas.26).

As Texas was colonized, first by Spaniards and then by Euro-Americans, major settlements were established and persisted in a repeated pattern, where larger rivers crossed the Balcones Escarpment. The nineteenth-century settlements of San Antonio, Austin, Waco, and Dallas-Fort Worth were ideally situated for growth, because they were located at the convergence of waterways, springs, and fords, where

▶ **Figure Texas.25** While Tertiary volcanoes were erupting in west Texas, the central part of the state was experiencing broad regional uplift. This caused larger rivers, such as the Brazos River, to become entrenched deeply into bedrock. Here, a large sandy point bar of the Brazos River stands opposite the cut bank in an entrenched meander.

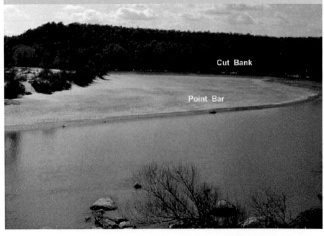

▶ **Figure Texas.26** Popular with tourists, the scenic "Hill Country" of central Texas is actually an area of deeply eroded valleys. These rivers cut down into Cretaceous bedrock, as the Edwards Plateau was uplifted along the Balcones Fault Zone.

wagons could cross the rivers that flowed over bedrock. This string of cities also grew along the boundary between the agriculturally rich coastal plain and the ranch country to the west. Thus, they were ideally situated to become centers of trade, transportation, culture, and government.

The western margins of the Edwards Plateau merge with the arid Basin and Range Province and the semiarid High Plains. Rain falling on the central part of the plateau percolates down into the basal Cretaceous rocks, forming the Edwards Aquifer, one of the most important sources of water in the state. Groundwater from the Edwards Aquifer is released to the surface along the Balcones Fault Zone in a series of springs that supply water to cities, including San Antonio and Austin. As recent decades have shown, the level of the Edwards Aquifer drops quite rapidly during droughts, and conservation of groundwater resources is a major concern for all of the communities of that region.

The geology of the Edwards Plateau has influenced land-use patterns in other ways. Relatively low rainfall and hard limestone combine to form thin, poor soils, which are characteristic of the Edwards Plateau. Historically, that area is noted for sheep and goat herding, which have enhanced soil erosion through overgrazing. Thin soils also promote extremely high rates of surface runoff during storms, especially hurricanes that frequently penetrate this part of Texas after crossing the coast. Over the years, many lives have been lost, especially at low water crossings, where people attempted to escape approaching flash floods. As a result, arrays of flood detectors have been installed that warn unsuspecting people in downstream areas of an advancing flood.

Section Texas.10 Summary

- In the Miocene Epoch, the epeirogenic uplift created the Edwards Plateau and the Balcones Fault Zone.

- The Edwards aquifer is the major supply of water for cities in central Texas.

Texas.11
The Gulf
Coastal Plain

The Cenozoic Gulf Coastal Plain Province covers about one fourth of Texas (see Figure Texas.1). The geology of this region is especially important because it supports agriculture and forestry industries, holds significant groundwater resources, and contains enormous energy reserves of petroleum and lignite.

Examination of the geologic map of Texas demonstrates that the coastal plain began forming at the end of the Cretaceous Period. Since then, the coastline has shifted far into the Gulf of Mexico Basin. In east Texas, the Cenozoic coastal plain is about 425 km (265 miles) wide. Between Dallas and Galveston, the Trinity River valley has a gradient of just 0.4 meters/km (1.8 ft/mile)! This extremely gentle relief extends far out into the Gulf, following Texas's broad continental shelf. The Cenozoic package of sediments underlying the coastal plain and the shelf is extremely thick. Just offshore from Corpus Christi, those deposits are approximately 12 km (40,000 ft) thick.

Construction of the Cenozoic coastal plain was clearly dominated by progradation of sediments into the Gulf Basin. Stated in terms of relative sea level, this was a long period of marine regression. But within this overall pattern, there were also transgressions, when sediment supply decreased and the position of the shoreline either remained stationary or shifted inland. Those conditions were similar to the present Texas coast, which has a system of salt marshes, bays, and muddy lagoons separated from the continental shelf environments by sandy barrier islands (Figure Texas.27). Shelf deposits composed of organic-rich clay form the most fertile soils of the coastal plain. These black, clayey soils follow the coastal prairie zone from Corpus Christi to the Louisiana border, supporting cotton and dry rice agriculture (see Figure Texas.1).

In contrast to periods of marine transgression, large deltaic systems prograded into the Gulf during recessional

▶ **Figure Texas.27** Many of the Tertiary deposits of the Gulf Coastal Plain formed in environments such as these near the Aransas National Wildlife Refuge. A salt marsh is in the foreground, and San Antonio Bay is in the distance.

▶ **Figure Texas.28** This exposure of Tertiary Wilcox sediments in east Texas illustrates short-term changes in sedimentary environments caused by processes such as shifting positions of river channels and flood plains.

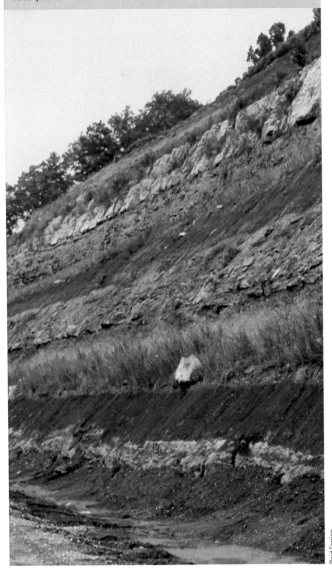

Reid Ferring

▶ **Figure Texas.29** This cypress swamp near the Red River is a good analog for the hardwood swamp environments in which lignite formed on the Tertiary coastal plain of Texas.

Reid Ferring

stages. This is well illustrated by the diverse sequences of sedimentary rocks of the Wilcox Group that were deposited during the Paleocene and Eocene epochs. The sedimentary facies of the Wilcox Group rocks reflect the complex mosaic of sedimentary environments of the coastal plain and adjacent shelf (Figure Texas.28). Especially common in fluvial environments were tropical hardwood swamps, in which thick accumulations of plant debris were buried and transformed into lignite. A contemporary cypress swamp near the Red River is an excellent analog for the sedimentary environments in which the organic matter for lignite accumulated (Figure Texas.29). Similar environments can be seen today at Caddo Lake, located near Marshall. Remarkably, Caddo Lake is the only natural lake in Texas.

Texas is the greatest consumer of coal in the United States, and most of that coal is lignite, which is mined in Texas and

used to generate electricity. Shallow lignite seams, covered by sediments or soft rocks, are mined using the dragline side-casting method (Figure Texas.30a). This method involves removal of the overburden with enormous draglines, exposing a long strip of lignite. The lignite is removed, then hauled to the nearby generating plant for immediate use (see Figure Texas.30b). Overburden from the next strip is used to backfill the section just mined, and the process continues until the leased area has been completed. The mine area is reclaimed by resurfacing with soil that had been stacked separately during overburden removal, followed by planting of vegetation according to the needs of the landowners who leased their property to the mining companies.

Both the mining and burning of lignite pose risks to the environment. In addition to surface reclamation, shallow groundwater must be protected from contamination during and after mining. Lignite has higher sulfur content than bituminous coal or anthracite, thus increasing the risk for acid rain that can contaminate nearby bodies of water. This problem has been reduced, but not eliminated, by use of "scrubbers" that use dissolved lime to help neutralize smokestack emissions (see Figure Texas.30b). Burning lignite entails significant emissions of CO_2, which contribute to the greenhouse effect and global warming; needless to say, this issue is one of increasing international concern and a focus for lively debates at home.

It is certain that lignite mining will become even more common in the future. At current rates of consumption, the shallow lignite reserves are estimated to last about 400 years. Deeper lignite reserves are even larger, but exploitation of those deposits by any method can threaten the extremely important aquifers of the Wilcox Group. No doubt, lignite mining and the associated efforts to preserve the quality of Texas' environment will be familiar issues to the next generation of Texans.

► **Figure Texas.30**

(a) The South Hallsville Mine near Tyler provides lignite for the nearby electrical generating plant. These lignite beds are exposed and removed by colossal-sized machinery such as the dragline at the far end of the strip that is about to be mined.

(b) Protection of the environment is a major challenge for the lignite industry, yet significant strides have been made to maintain air and water quality near mines and generating plants.

The Gulf Coastal Plain has enormous petroleum reserves. Although these reserves are widely distributed, active oil fields are concentrated around two principal geologic features. The first is a series of fault zones that roughly follow contours around the Gulf Basin. These differ from the tectonically induced faults discussed earlier in this chapter. They are called growth faults, and they formed along trends where high rates of sedimentation induced rapid subsidence. The lower parts of growth faults commonly bend toward the Gulf Basin, and in some cases, huge blocks of sedimentary rocks slide along nearly horizontal fault planes. All of this faulting results in formation of numerous traps for the hydrocarbons, not only on the coastal plain, but also out on the continental shelf where numerous offshore drilling platforms have been constructed.

The other geologic features that concentrate petroleum reserves are salt domes. These domes are common throughout east Texas, forming the western part of their more general distribution about the lower Mississippi valley. As mentioned in Section Texas.7, thick deposits of halite accumulated in the early Gulf Basin that was opened by rifting during the Jurassic Period. Burial under thousands of meters of younger sedimentary rocks mobilized the salt, forcing it upward as discrete plumes. The rising salt intrusions warped and faulted the surrounding rocks, creating ideal traps for oil and gas (Figure Texas.31).

► **Figure Texas.31** As thick Cretaceous-Tertiary sediments accumulated rapidly in the Gulf of Mexico basin, the deeply buried Jurassic Louann Salt was compressed, forcing numerous plumes of salt up through the overlying rocks, creating salt domes and giant oil fields.

Oil wells
Calcite
Gypsum
Salt Intrusion
Sandstone
Oil Reservoir

Salt domes have been targets for oil exploration ever since the Spindletop discovery well was brought in near Beaumont in 1901. In 9 days, this first Texas "gusher" blew an estimated 1 million barrels of oil into the air, forever changing Texas and the oil industry. In addition to oil and natural gas, salt domes concentrate significant sulfur deposits, which are dissolved and brought to the surface. Because the salt has almost no permeability and cannot be dissolved by oil, huge caverns created within large salt bodies are used to store large portions of our nation's strategic oil reserves. Last, it should be mentioned that these salt bodies are mined for their salt, mainly for industrial purposes and for deicing roads.

Section Texas.11 Summary

- The Gulf Coastal Plain was constructed during prolonged Tertiary regression of the Gulf of Mexico.

- This region contains vital resources including petroleum, lignite, and groundwater.

Texas.12
The Texas Gulf Coast

The last topic of this chapter is the present coast of Texas, which is an excellent example of a submergent coast. Its most obvious feature is the nearly continuous barrier island complex, which is separated from the mainland by salt marshes and shallow lagoons (see Figure Texas.27). These are excellent habitats for sport fishing and also serve as feeding areas for migratory ducks and geese in winter. The Intracoastal Canal, which extends from Brownsville to Florida behind the barrier islands, is a vital corridor for barge traffic. Galveston and Corpus Christi bays are among the busiest seaports in the United States, serving the agricultural, petroleum, and manufacturing sectors of Texas' economy. All of the canals and ship channels in Texas must be maintained by dredging because of infilling sediment that is delivered by rivers and as washover during storms. At Port Aransas, long jetties extend out into the Gulf, along the ship channel leading to Corpus Christi. The jetties, made of huge blocks of Precambrian granite quarried near Marble Falls, were constructed in the 1930s to keep longshore drift from filling in the channel.

Rivers with high sediment loads, especially the Rio Grande and Brazos rivers, have deltas, but the dominant forces of waves and longshore currents prevent the deltas from prograding beyond the barrier island chain (see Figure Texas.1). The bays of smaller rivers, including the Nueces, Colorado, and Trinity rivers, are good examples of drowned estuaries, created as sea levels rose during the last deglaciation (Figure Texas.32). These bays are separated from the Gulf of Mexico by barrier islands. The barriers have "passes" that are created and maintained by washover

▶ **Figure Texas.32** Barrier islands fringe almost the entire Texas coast, which is dominated by waves, currents, and storms. Galveston Bay is a drowned estuary, as are all the major bays farther south. In 1900, before construction of a protective seawall, more than 6,000 inhabitants of Galveston died during a hurricane. And still today, these low coastal areas are threatened by potentially devastating storms, such as hurricanes Katrina and Rita in 2005.

during hurricanes and by the seaward return of storm surge waters after the hurricanes move inland.

All too often, hurricanes make their landfall along the Texas coast, and over the past century, the loss of lives and property has been devastating. The tragedies caused by hurricanes Katrina and Rita persist today for thousands of displaced persons, and the damage done to offshore drilling rigs and submarine pipeline networks is still being repaired. Texas is especially vulnerable to hurricanes because of the extensive low-lying areas that can be flooded easily by storm surges. Barrier islands retard the return flow of storm surge waters, lengthening the period of flooding and hampering recovery and rescue efforts. Larger storms are able to overtop the seawalls constructed in Galveston and Corpus Christi, and the populations in these areas are clearly at risk any time that a storm approaches.

Section Texas.12 Summary

- Texas has a submergent coast, with bays, marshes, and lagoons behind an extensive barrier island system.

- Frequent hurricanes threaten the coastal environments and the populations who live along the coast.

Review Workbook

SUMMARY

Texas.1 Introduction

■ *Why is the geology of Texas important to students of physical geology and to all the inhabitants of the state today?*
Learning about the geology of their state helps Texas students better understand the principles of physical geology and makes them more aware of how the state's geologic past is related to the ways in which its residents interact with its landscape and exploit its resources today.

Texas.2 Precambrian Texas

■ *What tectonic events shaped the Proterozoic geologic record of Texas?*
The genesis of Proterozoic rocks in Texas is indicated by process sequences such as erosion, sedimentation, regional metamorphism, and postorogenic igneous intrusion.

■ *When and how were the metamorphic and igneous rocks of the Llano region accreted to the ancient continent Laurentia?*
The metamorphic suite of Proterozoic rocks in Texas was accreted to Laurentia and became part of the much larger accretionary Grenville Terrane because of the highly compressional tectonic regimen about 1.3 billion years ago (Ga). The intrusive suite was accreted about 0.2 billion years later.

Texas.3 The Paleozoic Tectonic Cycle

■ *What kinds of sediment were deposited during the Lower Paleozoic?*
Sandstone overlain by thicker deposits of limestone, much dolomitized, were deposited in the Cambrian and Ordovician periods.

■ *What is the economic significance of these rocks today?*
These rocks hold tremendous petroleum reserves.

Texas.4 The Ouachita Orogeny and Pangaea

■ *What were the geologic results in Texas of the collision of Gondwana with Laurasia?*
The collision caused prolonged orogeny and proportionate structural and sedimentary responses.

■ *How do the sedimentary rocks and structural styles of the Marathon area illustrate the collision?*
The rocks at Marathon show a sequence from rapid preorogenic marine sedimentation to intense thrust faulting and lastly to terrestrial sedimentation.

Texas.5 The Fort Worth Basin in the Late Paleozoic Era

■ *What sedimentary environments characterized the Fort Worth Basin during and after the Ouachita orogeny?*
The Fort Worth Basin was flanked on the east by the Ouachita Mountains, from which active deltas prograded onto a shelf that had carbonate banks and platforms.

■ *Why is this basin rich in oil and gas resources?*
Those environments produced abundant hydrocarbons, as well as ideal reservoirs and traps.

Texas.6 Permian Red Beds and Evaporites

■ *Why are red beds and evaporites characteristic of the Permian basins of central and west Texas?*
Periodic isolation of evaporitic basins surrounded by weathered and eroded landforms were the depositional environments for those rocks.

■ *How does the geology of that region affect present-day land-use patterns?*
Today, those salts make their way into the surface and ground-water systems, dramatically reducing water quality and water use potentials.

Texas.7 Breakup of Pangaea and Cretaceous Transgressions

■ *What happened in Texas during and after the breakup of Pangaea?*
Rifting in the Jurassic Period created the Gulf of Mexico Basin, which became an open ocean just before Cretaceous transgressions.

■ *What were the results of the Cretaceous transgressions?*
The transgressions left thick packages of sedimentary rocks over most of the state and deeply buried below the Gulf Coastal Plain.

Texas.8 Basin and Range Texas

■ *What types of tectonic activity are indicated by the volcanic rocks in trans-Pecos Texas?*
Extensive bimodal volcanism is indicated by mainly Oligocene Epoch lava, and ash resulting from a compressional phase of intraplate tectonics.

■ *How did the basin and range structural features form in that region?*
The basin and range features primarily formed during the Miocene Epoch when tensional conditions promoted normal faulting.

Texas.9 The Llano Estacado

■ *Which geologic processes led to the formation of the Llano Estacado?*
The Llano Estacado was constructed by accumulating fluvial sediments from the Rocky Mountains, followed by finer sediments that were deposited by wind.

■ *What is the Ogallala aquifer, and how is it recharged?*
The fluvial deposits in the lower part of the Ogallala Formation are the Ogallala aquifer. Abundant playa lake basins promote some recharge of the aquifer, which still falls during droughts.

Texas.10 The Edwards Plateau and Balcones Fault Zone

■ *When and how did the Edwards Plateau and Balcones Fault Zone form?*
The Edwards Plateau formed in the Miocene Epoch by epeirogenic uplift, which was accommodated along the Balcones Fault Zone, separating the plateau from the Gulf Coastal Plain.

■ *How were these geologic features important to the initial settlement of Texas, and how are they important to Texans today?*
Locations where springs fed by the Edwards aquifer were near rivers crossing the fault zone were chosen by early settlers of cities from San Antonio to Dallas. The Edwards aquifer is a major source of water today.

Texas.11 The Gulf Coastal Plain

■ *How have transgressions and regressions shaped the geologic evolution of the Gulf Coastal Plain?*

Those different sedimentary cycles promoted varied accumulations of organic matter that were sources for rich accumulations of petroleum and lignite.

■ *What are the economic and environmental implications of using lignite to generate electricity?*

Lignite can be used to fire generating plants almost indefinitely, and at a reasonable cost; but both the mining and burning of lignite are, without any remediation, harmful to the environment. This threat is the source of considerable efforts to minimize those problems.

■ *How did salt domes form, and why are economic resources concentrated around them?*

Salt domes formed when halite was compressed under thick sediments in the Gulf Basin, forcing it to rise through the overlying deposits. As the salt column rose, it created numerous traps by faulting.

Texas.12 The Texas Gulf Coast

■ *What are the principal features of the Texas Gulf Coast, and how did they form?*

The Gulf Coast of Texas is marked by its low-gradient and high-wave energy. It exhibits bays, salt marshes, and lagoons behind an extensive series of barrier islands.

■ *How do hurricanes affect the coastal areas of Texas?*

Hurricanes attack the fragile landforms along the coast, delivering large volumes of both sediment and saltwater, together with massive erosion potential. Their threats to the lives and homes of people who live in low coastal areas are realized too often, as shown by the devastation of hurricanes Katrina and Rita.

ESSENTIAL TERMS TO KNOW

Aulacogen The failed arm of a triple-junction rift, left behind on the edge of the continent after the new ocean opens.

Balcones Fault Zone The zone of normal faults that defines the eastern edge of the Edwards Plateau and the Coastal Plain of Texas.

Basin and range Low mountain forms that consist of upraised blocks bounded by normal faults and separated by graben-type valleys.

Bimodal volcanism Volcanism in a region usually subjected to extensional tectonics, characterized by magmatic compositions ranging from felsic to mafic.

Calcareous algae Algae that produce large amounts of carbonate mud composed of minute calcite crystals that help support the algal organism during life.

Caldera A large volcano with a central crater that formed when collapse was induced by evacuation of the magma chamber below.

Caliche A hard, thick zone of calcium carbonate that formed by soil activity, usually in regions with arid to semiarid climates.

Chalk A rock composed of the calcareous skeletons (tests) of marine phytoplankton.

Deflation Removal of fine-grained sediments by wind.

Drowned estuary The mouth of a river valley that has been inundated by ocean waters during the sea level increase accompanying deglaciation.

Entrenched valley A valley that has been incised into bedrock in response to regional uplift.

Eolian Adjective applied to sedimentary processes and deposits associated with wind.

Epeirogenic uplift More or less uniform uplift of large areas in the absence of compressional mountain building.

Ephemeral Adjective applied to streams or bodies of standing water that carry or contain water only after rains or rainy seasons.

Flysch Thick accumulations of thinly bedded sandstones, shales, and marls that were rapidly deposited in basins in advance of a continental convergence.

Fracking Fracturing of rocks that surround a gas or oil well, induced by injecting water and sand into the rocks under high pressure to increase the porosity of the reservoir and the flow of hydrocarbons to the well.

Glauconite A green silicate mineral that commonly occurs as granular clasts in marine sandstones.

Gondwana The continent that formerly included Africa, South America, Antarctica, India, Australia, and Madagascar.

Grenville Province Huge masses of rocks that were added to the eastern and Texan edges of Laurentia in the middle Proterozoic Era.

Hogback A sharp, steep ridge created by erosion of steeply dipping rocks.

Intraplate tectonics Tectonic activity that occurs usually within continents, well away from plate margins.

Laurentia Name given to the North American continent as it appeared until its collision with Baltica in the Silurian Period.

Lignite A low-grade coal formed of plant remains that were concentrated, buried, and altered more than peat but less than higher grades of coal.

Marl A calcareous shale.

Microcline A low-temperature potassium feldspar, commonly found as pink to red crystals in granites and especially in pegmatites.

Molasse Terrigenous sediments deposited as a continental collision neared completion.

Orogeny A mountain building event.

Physiographic map A map divided into provinces that reflect a similar geologic history of deposition and erosion; it usually has distinctive soils, vegetation, and climate.